Advances in
Hillslope Processes

British Geomorphological Research Group
Symposia Series

Geomorphology in Environmental Planning
Edited by **J. M. Hooke**

Floods
Hydrological, Sedimentological and Geomorphological Implications
Edited by **Keith Beven** and **Paul Carling**

Soil Erosion on Agricultural Land
Edited by **J. Boardman, J. A. Dearing** and **I. D. L. Foster**

Vegetation and Erosion
Processes and Environments
Edited by **J. B. Thornes**

Lowland Floodplain Rivers
Geomorphological Perspectives
Edited by **P. A. Carling** and **G. E. Petts**

Geomorphology and Sedimentology of Lakes and Reservoirs
Edited by **J. McManus** and **R. W. Duck**

Landscape Sensitivity
Edited by **D. S. G. Thomas** and **R. J. Allison**

Process Models and Theoretical Geomorphology
Edited by **M. J. Kirby**

Environmental Change in Drylands
Biogeographical and Geomorphological Perspectives
Edited by **A. C. Millington** and **K. Pye**

Rock Weathering and Land Form Evolution
Edited by **D. A. Robinson** and **R. B. G. Williams**

Geomorphology and Land Management in a Changing Environment
Edited by **D. F. M. McGregor** and **D. A. Thompson**

Advances in Hillslope Processes Volumes 1 and 2
Edited by **M. G. Anderson** and **S. M. Brooks**

Advances in Hillslope Processes

VOLUME 1

Edited by

Malcolm G. Anderson and Susan M. Brooks
University of Bristol, UK

JOHN WILEY & SONS

Chichester · New York · Brisbane · Toronto · Singapore

Other Wiley Editorial Offices

John Wiley & Sons, Inc., 605 Third Avenue,
New York, NY 10158-0012, USA

Jacaranda Wiley Ltd, 33 Park Road, Milton,
Queensland 4064, Australia

John Wiley & Sons (Canada) Ltd, 22 Worcester Road,
Rexdale, Ontario M9W IL1, Canada

John Wiley & Sons (Asia) Pte Ltd, 2 Clementi Loop #02-01,
Jin Xing Distripark, Singapore 129809

British Library Cataloguing in Publication Data

A catalogue record for this book is available from the British Library

ISBN 0-471-96774-2

Typeset in 10/12pt times by Dobbie Typesetting Ltd, Tavistock, Devon
Printed and bound in Great Britain by Bookcraft (Bath) Ltd, Midsomer Norton, Somerset
This book is printed on acid-free paper responsibly manufactured from sustainable forestation, for which at least two trees are planted for each one used for paper production

In 1995 Professor Dick Chorley became Emeritus Professor at the Department of Geography, University of Cambridge.

We would like to acknowledge the outstanding worldwide impact that he had, and continues to have, on the development of research in geomorphology. His inspired teaching, both at Sidney Sussex College and in the Cambridge Department, was, of course, much involved with hillslope processes.

It is particularly appropriate therefore that these two volumes on this theme represent one of the many ways of noting Dick's contribution to the subject.

Contents

VOLUME 2

x   Contents

List of Contributors

A. D. Abrahams Department of Geography, State University of New York at Buffalo, Buffalo, New York 14261, USA

C. T. Agnew Department of Geography, University College London, 26 Bedford Way, London WC1H 0AP, UK

H. J. Åkerman Department of Physical Geography, Solvegatan 13, S-22362 Lund, Sweden

C. D. Allen National Biological Service, Jemez Mountain Field Station at Bandelier National Monument, Los Alamos, New Mexico 87544, USA

M. G. Anderson Department of Geography, University of Bristol, University Road, Bristol BS8 1SS, UK

J. R. M. Arah Institute of Terrestrial Ecology, Bush Estate, Penicuik, Midlothian EH26 0QB, UK

A. C. Armstrong ADAS Land Research Centre, Gleadthorpe, Meden Vale, Mansfield, Notts NG20 9PF, UK

J. Ball Department of Geography, University of Sheffield, Weston Bank, Sheffield S10 2TN, UK

C. K. Ballantyne Department of Geography, University of St Andrews, St Andrews, Fife KY16 9ST, UK

D. I. Benn Department of Geography, University of Aberdeen, Aberdeen AB9 2UF, UK

J. Boardman Environmental Change Unit and School of Geography, University of Oxford, 5 South Parks Road, Oxford OX1 3UB, UK

D. D. Brammer Water Resources Research Group, College of Environmental Science and Forestry, State University of New York, Marshan, Syracuse, NY 1321, USA

P. C. Brookes Soil Science Department, IACR-Rothamsted, Harpenden, Herts AL5 2JQ, UK

S. M. Brooks Department of Geography, University of Bristol, University Road, Bristol BS8 1SS, UK

L. A. Bruijnzeel Vrije Universiteit Amsterdam, Faculteit der Aardwetenschappen, De Boelelaan 1085, 1081 HV Amsterdam, Netherlands

D. Brunsden Department of Geography, University of London, King's College London, The Strand, London WC2R 2LS, UK

R. Bryan Soil Erosion Laboratory, University of Toronto, 1265 Military Trail, Scarborough, Ontario M1C 1A4, Canada

P. A. Bull School of Geography, University of Oxford, Mansfield Road, Oxford OX1 3TB, UK

T. P. Burt Department of Geography, Science Laboratories, South Road, Durham DH1 3LE, UK

K. Castle Soils Department, SAC, West Mains Road, Edinburgh EH9 3JG, UK

J. H. Chandler Department of Civil Engineering, Loughborough University of Technology, Loughborough, Leicestershire LE11 3TU, UK

A. Chappell Department of Geography, University College London, 26 Bedford Way, London WC1H 0AP, UK

M. Charlton Department of Physics and Astronomy, University College London, Gower Street, London WC1E 6BT, UK

J. E. Clark Forest Advisor, ODA, 94 Victoria Street, London SW1E 5JL, UK

A. J. C. Collison Department of Geography, King's College London, The Strand, London WC2R 2LS, UK

W. R. S. Critchley Vrije Universiteit Amsterdam, Faculteit der Aardwetenschappen, De Boelelaan 1085, 1081 HV Amsterdam, Netherlands

M. C. Da Silva Júnior Universidade de Brasília, Departamento de Engenharia Florestal, Cx.P. 4357, 70910–000 Brasília, Brazil

D. W. Davenport Environmental Science Group, MS J495, Los Alamos National Laboratory, Los Alamos, New Mexico 87545, USA

T. J. A. Davie Department of Geography, Queen Mary & Westfield College, University of London, Mile End Road, London E1 4NS, UK

M. C. R. Davies School of Engineering, University of Wales Cardiff, PO Box 917, Cardiff CF2 1XH, UK

P. G. B. De Louw Institute of Geography, Utrecht University, Heidelberg Laan 2, 3508 TC, Utrecht, The Netherlands

A. P. J. De Roo Department of Physical Geography, Utrecht University, PO Box 80.115, 3508 TC Utrecht, The Netherlands

R. M. Dils Department of Geography, University of Sheffield, Winter Street, Sheffield S10 2TN, UK

A. P. Dykes Department of Geographical and Environmental Sciences, The University of Huddersfield, Queensgate, Huddersfield HD1 3DH, UK

L. Ebbs Macaulay Land Use Research Institute, Craigiebuckler, Aberdeen AB9 2QJ, UK

A. C. Edwards Macaulay Land Use Research Institute, Craigiebuckler, Aberdeen AB9 2QJ, UK

H. Elsenbeer GIUB/Hydro-Biogeochemistry, Hallerstrasse 12, 30112 Berne, Switzerland

R. Evans Division of Geography, Anglia Polytechnic University, East Road, Cambridge CB1 1PT, UK

H. Faulkner The University of North London, School of Geography and Environmental Studies, 62–66 Highbury Grove, London N5 2AD, UK

D. T. Favis-Mortlock Environmental Change Unit, University of Oxford, 5 South Parks Road, Oxford OX1 3UB, UK

P. A. Furley Department of Geography, University of Edinburgh, Drummond Street, Edinburgh EH8 9XP, UK

W. J. Gburek Pasture Systems and Watershed Management Research Laboratory USDA-ARS, University Park, Pennsylvania, USA

K. W. T. Goulding Soil Science Department, IACR-Rothamsted, Harpenden, Herts AL5 2JQ, UK

G. Govers Laboratory for Experimental Geomorphology, Catholic University of Leuven, 16 Redingenstraat, Leuven, Belgium

J. Gunn Limestone Research Group, Department of Geographical and Environmental Sciences, University of Huddersfield, Queensgate, Huddersfield HD1 3DH, UK

P. Hardwick Limestone Research Group, Department of Geographical and Environmental Sciences, University of Huddersfield, Queensgate, Huddersfield HD1 3DH, UK

C. Harris Department of Earth Sciences, University of Wales Cardiff, PO Box 914, Cardiff CF1 3YE, UK

G. L. Harris ADAS Land Research Centre, Anstey Hall, Maris Lane, Trumpington, Cambridge CB2 2LF, UK

S. Harrison Division of Geography, Coventry University, Coventry CV1 5FB, UK

J. Hartshorne Department of Geography, University of Bristol, University Road, Bristol BS8 1SS, UK

A. M. Harvey Department of Geography, University of Liverpool, PO Box 147, Liverpool L69 3BX, UK

P. M. Haygarth Institute of Grassland and Environmental Research, North Wyke, Okehampton, Devon EX20 2SB, UK

A. L. Heathwaite Department of Geography, University of Sheffield, Winter Street, Sheffield S10 2TN, UK

G. Heckrath Soil Science Department, IACR-Rothamsted, Harpenden, Herts AL5 2JQ, UK

R. A. A. Hetterschijt Institute of Geography, Utrecht University, Heidelberg Laan 2, 3508 TC, Utrecht, The Netherlands, UK

R. A. Hodgkinson ADAS Land Research Centre, Anstey Hall, Maris Lane, Trumpington, Cambridge CB2 2LF, UK

S. C. Jarvis Institute of Grassland and Environmental Research, North Wyke, Okehampton, Devon EX20 2SB, UK

A. Lack GIUB/Hydro-Biogeochemistry, Hallerstrasse 12, 3012 Berne, Switzerland

C. J. Lawrance Overseas Centre, Transport Research Laboratory, Crowthorne, Berkshire, UK

D. M. Lloyd Department of Geography, University of Bristol, University Road, Bristol BS8 1SS, UK

A. Malmer Department of Forest Ecology, Swedish University of Agricultural Sciences, S-901 83 Umea, Sweden

L. S. Matchett School of Geography, Oxford University, Mansfield Road, Oxford OX1 3TB, UK

J. J. McDonnell Water Resources Research Group, College of Environmental Science and Forestry, State University of New York, Marshan, Syracuse, NY 1321, USA

P. Meitz Martin-Luther-Universität Halle-Wittenberg, Institut für Geographie, Domstr. 5, 06108 Halle, Germany

R. P. C. Morgan Silsoe College, Cranfield University, Silsoe, Bedford MK45 4DT, UK

M. Mulligan Department of Geography, King's College London, The Strand, London WC2R 2LR, UK

M. A. Oliver Department of Soil Science, University of Reading, Whiteknights, PO Box 233, Reading RG6 6DW, UK

A. Park Department of Geography, University of Bristol, University Road, Bristol BS8 1SS, UK

S. J. Park School of Geography, University of Oxford, Mansfield Road, Oxford OX1 3TB, UK

A. J. Parsons Department of Geography, University of Leicester, University Road, Leicester LE1 7RH, UK

N. E. Peters Water Resources Division, United States Geological Survey, Atlanta GA 30360, USA

J. Pethick Cambridge Coastal Research Unit, Department of Geography, University of Cambridge, 62 Sidney Street, Cambridge CB2 3JW, UK

D. Petley Department of Geology, University of Portsmouth, Burnaby Building, Portsmouth PO1 2UP, UK

H. B. Pionke Pasture Systems and Watershed Management Research Laboratory USDA-ARS, University Park, Pennsylvania, USA

J. Pitlick Department of Geography, University of Colorado, Boulder, Colorado 80309, USA

E. M. Porter School of Geosciences, The Queen's University of Belfast, Belfast BT7 1NN, UK

S. Porter Department of Plant and Soil Science, Aberdeen University, Aberdeen AB9 2UF, UK

I. P. Prosser Cooperative Research Centre for Catchment Hydrology, Division of Water Resources, CSIRO, GPO Box 1666, Canberra City, Australia

J. Puigdefábregas Estación Experimental de Zonas Áridas (CSIC), General Segura, 1.04001 Almeria, Spain

T. A. Quine Department of Geography, University of Exeter, Amory Building, Rennes Drive, Exeter EX4 4RJ, UK

E. B. Ratcliffe Department of Geography, University of Bristol, University Road, Bristol BS8 1SS, UK

J. A. Ratter Royal Botanic Garden, Inverleith Row, Edinburgh EH3 5LR, UK

B. Rawlins Department of Geography, University of Sheffield, Weston Bank, Sheffield S10 2TN, UK

B. R. Rea School of Geosciences, The Queen's University of Belfast, Belfast BT7 1NN, UK

R. M. Rees Soils Department, SAC, West Mains Road, Edinburgh EH9 3JG, UK

R. J. Rickson Department of Rural Land Use, Silsoe College, Cranfield University, Silsoe, Bedford MK45 4DT, UK

M. D. Ron Vaz Instttuto De Recursos Naturales y Agrobiologia De Sevilla, Sevilla, Spain

G. Sánchez Estación Experimental de Zonas Áridas (CSIC), General Segura, 1.04001 Almeria, Spain

M. R. Savabi US Department of Agriculture–Agricultural Research Service, National Soil Erosion Research Laboratory, 1196 SOIL Building, West Lafayette, IN 47907-1196, USA

K. H. Schmidt Martin-Luther-Universität Halle-Wittenberg, Institut für Geographie, Domstr. 5, 06108 Halle, Germany

A. N. Sharpley Pasture Systems and Watershed Management Research Laboratory USDA-ARS, University Park, Pennsylvania, USA

M. C. Slattery Department of Geography, East Carolina University, Greenville, NC 27858, USA

P. M. Stone Department of Geography, University of Exeter, Amory Building, Rennes Drive, Exeter, Devon EX4 4RJ, UK

M. T. J. Terlien International Institute for Aerospace Survey and Earth Sciences (ITC), Kanaalweg 3, 2628 EB, Delft, Netherlands

J. B. Thornes Department of Geography, King's College London, The Strand, London WC2R 2LS, UK

M. Tranter Department of Geography, University of Bristol, University Road, Bristol BS8 1SS, UK

S. Trudgill Department of Geography, University of Cambridge, Downing Place, Cambridge CB2 3EN, UK

Th. W. J. Van Asch Institute of Geography, Utrecht University, Heidelberg Laan 2, 3508 TC, Utrecht, Netherlands

J. Wainwright Department of Geography, King's College London, The Strand, London WC2R 2LS, UK

D. E. Walling Department of Geography, University of Exeter, Amory Building, Rennes Drive, Exeter EX4 4RJ, UK

A. Warren Department of Geography, University College London, 26 Bedford Way, London WC1H OAP, UK

C. P. Webster Soil Science Department, IACR-Rothamsted, Harpenden, Herts AL5 2JQ, UK

W. B. Whalley School of Geosciences, The Queen's University of Belfast, Belfast BT7 1NN, UK

B. P. Wilcox Environmental Science Group, MS J495, Los Alamos National Laboratory, Los Alamos, New Mexico 87545, USA

K. Wu Department of Geography, University of Liverpool, PO Box 147, Liverpool L69 3BX, UK

Foreword

More than 30 years ago, in discussing the nodal position of slope studies in geomorphology (*Geog. Jour.* 1964, pp. 70–73), I ventured the opinion that research to do with slopes had long provided something of a touchstone regarding the methodological state of the discipline as a whole. Assuming that this opinion still contains some elements of truth, it is instructive to view the contents of the present volumes as in some way indicative of British geomorphological work at the close of the 20th century.

An overriding impression is that British geomorphologists are numerous, active and very well organized. Even at the time of publication of the pivotal work by Carson and Kirkby, some quarter of a century ago, it would have been difficult to envisage the future production of such a massive research effort devoted to slope studies. These studies are very much alive, and continuing to occupy a nodal position in our discipline.

In 1964 I identified "the assumed minor importance of studies of present processes" as a major impediment to slope research. A major message of the current volumes is that the study of reasonably contemporary processes lies close to the heart of modern concern. What studies of slope form exist stem from studies of slope processes, rather than forming a focus in their own right. Most research is concerned with the effects of fluvial processes and, indeed, there is a marked merging of hillslope studies with those concerned with the analysis of whole river basins (as in Gregory and Walling 1973). This, of course, has been the natural, if delayed, consequence of the work of Horton half a century ago, significantly stimulated by the research of Strahler and his group in the 1950s. This also explains in part the great overlap evidenced in the present volume of geomorphic slope studies and those of hillslope hydrology which was given an immense boost by Kirkby's edited work of almost 20 years ago. It also explains the increased emphasis on chemical aspects of slope studies.

The emphasis on the operation and effects of present processes, as distinct from the postulated influences of long-past historical events permeates the present work – some would say indicating too great a swing of the pendulum in the past 30 years. Indeed, historical depth is contributed by emphases on recurrence intervals, the Holocene and the remainder of the Quaternary, rather than with events of greater antiquity. Even the section devoted to periglacial hillslope processes has a relentlessly contemporary focus. I do not complain about this in that I believe it to be a very true reflection of modern slope studies, but it is interesting to note that, after 30 years, attempts at a balanced evaluation on the geomorphic roles of 'present' versus 'past'

Advances in Hillslope Processes, Volume 1. Edited by M. G. Anderson and S. M. Brooks.

events still elude us. Indeed, this dichotomy has largely been replaced by a concern with matters to do with slope stability versus instability, as evidenced by the recent volumes edited by Anderson and Richards (1987) and Brunsden and Prior (1984).

A further feature of contemporary geomorphology exemplified by the present volumes is its overridingly empirical, practical and utilitarian outlook. In this it contrasts most strikingly with the work of 30 years ago. Today theory enters into slope studies mainly through a concern with mathematical modelling, noted by me in 1964 as having "progressed little in the past 100 years". Since that time, the earlier work of Scheidegger has blossomed with theory being securely anchored in the empirical evidence used as the basis for modelling.

In 1949 and 1958 Russell was at pains to separate "geographical geomorphology" from "geological geomorphology". It is clear that in the past 30 years the geological basis of slope studies has declined due to concerns with process rather than form, and with contemporary and recent happenings rather than those in the more remote geological past. It is noteworthy, for example, that tectonic influences on slope development, which would have dominated any publication prior to 1950, find an echo in only one chapter of these volumes. Although geographical concerns are apparent in the climatic foci of the volume, as well as in the stress upon the practical significance of contemporary processes, it is clear that, together with studies of river geomorphology, slope studies have moved into a field which may be most appropriately designated as "engineering geomorphology".

R. J. Chorley

REFERENCES

Anderson, M.G. and Richards, K.S. 1987. *Slope Stability – Geotechnical Engineering and Geomorphology*. Wiley, Chichester.

Brunsden, D. and Prior, D.B. 1984. *Slope Instability*. Wiley, Chichester.

Carson, M.A. and Kirkby, M.J. 1972. *Hillslope Form and Process*. Cambridge University Press, Cambridge.

Gregory, K.J. and Walling, D.E. (eds) 1973. *Drainage Basin Form and Process*. Edward Arnold, London.

Kirkby, M.J. 1978. *Hillslope Hydrology*. Wiley, Chichester.

Russell, R.J. 1949. Geographical Geomorphology. *Ann. Assoc. Am. Geog.*, **39**, 1–11.

Russell, R.J. 1958. Geological Geomorphology. *Bull. Geol. Soc. Am.*, **69**, 1–22.

Section 1

INTRODUCTION

1   Hillslope Processes: Research Prospects

M. G. ANDERSON and S. M. BROOKS
Department of Geography, University of Bristol, UK

1.1 HISTORICAL PERSPECTIVES ON ADVANCES IN GEOMORPHOLOGY

We should perhaps be cautious when using the term "advances" in respect of research, particularly in the environmental field. This text seeks to represent *current* research themes, within the field of hillslope process; however, in the only major history yet written on the study of landforms (Chorley *et al*. 1964), important warnings for current research directions can be seen in the *historical* context of earlier scientific "advances". In reading volume 1 of *The History of the Study of Landforms* we learn, in Chapter 8, of "distractions" that, unrecognized at the time, led to the slowing of "true advances" in respect of mainstream scientific enquiry:

> Yet whereas most sciences originally adopted calculation and measurement as a natural form of discipline, their rigid application does not appear to have been generally introduced into geomorphology until comparatively modern times. The reasons for the belated growth of scientific facts on physical processes was partly due to distractions: first the vertical superimposition of strata proved of more interest to geologists: and second, the horizontal distribution of rocks and landscapes also attracted much attention (p. 86).

This then resulted in an emphasis of study that placed the study of landscape first and of physical processes second. And yet during the 17th and 18th centuries there was such an understanding of river flow mechanics that

> this could well have led to the full acceptance of Huttonian fluvialism but this store of hydrological knowledge was either utterly ignored or quite overlooked by contemporary geologists. Hydraulic engineers might then have founded a tradition of truly scientific investigation into landscape development had not geologists been so determined that rock structure and rock composition were to assume the dominant role in their studies (p. 87).

The notion of advances in geomorphological investigations is further commented upon in historical perspectives of more recent origin. Illustrated lecture notes from W. M. Davis which he used at Texas A & M in 1927 have been compiled and edited by King and Schumm (1980). In the editorial to this volume, King remarks:

Advances in Hillslope Processes, Volume 1. Edited by M. G. Anderson and S. M. Brooks.
© 1996 John Wiley & Sons Ltd.

> During the nearly half-century gap between the original note taking and the present, Davis's concepts have gone into gradual eclipse among geomorphologists, and many of them, it is true, now seem quaint, old fashioned and oversimplified. But many of the great principles on which they were based though overlooked today, are still valid – especially the concept of a continual evolution of landforms from the past, through the present, into the future (p. xvii).

In the foreword to the present volume, Chorley echoes this theme: "It is interesting to note that attempts at a balanced evolution on the geomorphic roles of present versus past events still elude us". A more direct challenge to the foci of theoretical development and change in the final quarter of the 20th century was given by Chorley (1978):

> A recent questionnaire from the British Geomorphological Research Group on the teaching of geomorphology in University Geography Departments in the United Kingdom requested comments regarding the "methodological component" of each course – in itself an interesting phrase. The responses were unenthusiastic. Two of the most common attitudes were either that there is no need to distinguish methodology from techniques, or that the scientific method is obvious and therefore needs no discussion. I do not agree with these views and, to adapt an aphorism, believe that the only true prisoners of theory are those who are unaware of it. It is indeed disturbing that attitudes to theory are so negative at a time when the bases of our theories are changing so rapidly and drastically with, as I believe, profound implications for the future of geomorphology.

In this paper Chorley, writing almost 20 years ago, makes some points by way of conclusion that are critical for many current research "programmes" that now characterize geomorphology:

> It is now generally recognised, even by the most ardent logical positivists, that generation of hypothesis precedes accumulation of data, at least in part. If data are not generated in a theoretical vacuum then it is clear that the growth of theory must derive in some measure for the existing intellectual climate. Therefore, not only the types of scientific work in which it is thought proper to engage, but also their theoretical underpinnings, are vulnerable to the pressures of convention. Those of you who have attended this paper prepared to discuss hydrodynamics, the physics of distributed shear, or other empirical relationships under the guise of geomorphological theory (Scheidegger, 1961) may find this suggestion rather fanciful, but perhaps I can elaborate my meaning. The quest for utility not only conditions the objects of our work, but also the manner in which theory may emerge. Utilitarian approaches to geomorphology will either result in large-scale work of which the intellectually-sterile taxonomic morphological mapping is the most depressing precursor, or in a piecemeal concentration on small-scale realist systems. The latter will encourage treatment in a manner which attempts to broaden spatial or temporal scales by processes of aggradation and which will commonly produce, as have attempts in ecology, highly-interlocked complex models whose empirical linkages predestine a catastrophic behaviour outside arbitrary equilibrium limits. The construction of such models in geomorphology would, of course, be very convenient for the syllabus integration of the Geography Departments which produce the majority of our geomorphologists, but it may not be a secure base for geomorphological theory. Fortunately, geomorphology in the United States (not least by virtue of the lead given to the United States Geological Survey by Leopold), the land classification of the

CSIRO in Australia and, perhaps, ironically even, the development of landscape science in Russia (predating even Engels' *Dialectics of Nature*, 1872–82) have preserved in my view the proper distinction between nature and society which is necessary to set apart objective management from subjective symbiosis. Conventional social attitudes to the environment affect theory in science, not only indirectly by means of their influence on preferred subject matter, but by much more direct and insidious controls over basic attitudes. Perhaps my fears are groundless and our colleagues are right in believing in a simplistic role for theory in geomorphology. On the other hand, we must not ignore the possibility that the wheel of theory may be coming full circle from one teleology to another, from an old religious, to a new social orthodoxy (p. 11).

The distractions that we spoke of in relation to the 17th and 18th centuries that impacted upon advances have their modern day equivalents of course. As we write this chapter a correspondent in *The Times* letter page writes

What has gone in the mad dash to amass large research contracts with successful professors being traded like football stars, is the pursuit of academic excellence. Harassed lecturers just publish as many research papers as possible, often of pedestrian quality (Fells 1996).

The real concern of course is that distractions so engendered are *additional* to those of the, some would say, more natural distortions of discipline dominance which we have already illustrated. At one level we need to assess and develop new methodologies for examining processes in the landscape. At another level we need to assess whether social attitudes to the environment are affecting the theory of our science with the possible consequences that Chorley (1978), quoted above, cautioned could occur. This discussion has thus sought to remind us of important perspectives that have clear relevance to what we consider advances in the discipline.

1.2 THEMES OF RECENT HILLSLOPE PROCESSES RESEARCH

Whilst as geomorphologists we focus on a *temporal* context, that is the "continual evolution of landforms from the past, through the present to the future", there is a generally recognized *spatial* ordering classification which accommodates the land-form features of concern to the geomorphologist. Table 1.1 shows the hierarchical spatial ordering of landforms after Chorley *et al.* (1984). They observe that geomorphologists are more generally concerned with third-to ninth-order features, but of perhaps greater significance, "this classification according to scale is useful because it enables us to see how differing conceptual bases for geomorphological theory-building rely on differing scale emphases" (p. 15).

In this section we identify five themes that emerge from the research reported in these two volumes. The themes principally cover *spatial* orders 4 to 9 of Table 1.1 and in each case certain chapters relate to landform evolution – the geomorphological *temporal* context. We may therefore consider such research, in encompassing the full spatial–temporal context of geomorphology, as sound at least in scope. Continuing critical examination of the developing methodologies is warranted however in the context of potential distractions (see subsection 1.1).

Table 1.1 Hierarchical spatial ordering of landforms (after Chorley *et al.* 1984)

Order	Examples
1st	Continents, oceans, plates, convergence zones
2nd	Physiographic provinces, mountain ranges, massifs, plateaus, lowlands, accumulative plains, tectonic depressions (termed "morphostructural" units by the Russians)
3rd	Medium-scale geological units, such as folded sequences, fault blocks, domes and volcanoes
4th	Large-scale erosional/depositional units, such as large valleys, deltas and long continuous beaches
5th	Medium-scale erosional/depositional units, such as smaller valleys, floodplains, alluvial fans, cirques, and moraines
6th	Small-scale erosional/depositional units, such as small valleys, offshore bars, and sand dunes
7th	Hillslopes, stretches of stream channel
8th	Slope and flat facets, pools, riffles
9th	Stream bed and aeolian sand ripples, slope terracettes
10th	Microroughness represented by the diameter of individual pebbles or sand grains

The first significant area of progress relates to possibilities for adopting increasingly *long temporal and large spatial scales* of analysis, and for investigating process operation in circumstances where direct observation is not possible, for example in relation to long-term change in hillslope systems. Several chapters emphasize this theme, including those by Park *et al.* (Chapter 17), Brooks and Collison (Chapter 21), Brunsden and Chandler (Chapter 40), Harvey (Chapter 33), as well as Quine *et al.* (Chapter 25). In each, different temporal scales are emphasized, ranging from millennia to decades, but we see several pertinent issues emerging from this research. Park *et al.*, for example, stress the significance of relict features in the soil profile in influencing contemporary soil hydrology, while Chandler and Brunsden provide ample evidence of the way increasingly detailed field monitoring can drastically alter previous conceptual models concerning the long-term behaviour of hillslope systems. In emphasizing long-term issues we are increasingly required to consider the way hillslopes might respond to changing external conditions. Mulligan (Chapter 50) presents an innovative modelling strategy for soil–vegetation responses to climatic change, while Harvey discusses evidence for long-term change in gully response to hydrological and vegetation changes. Finally, where emphasis on large spatial scales is concerned, the chapter by Anderson *et al.* (chapter 36) considers the potential for large spatial coverage through an integrated GIS-modelling approach.

Links between soil profiles and hillslope hydrological and erosional processes from a second major research area. Although such links have been discussed in earlier research, it is only recently that this area has gained wider consideration. Brammer and McDonnell (Chapter 2), for example, consider macropore–matrix interactions within the soil profile, as well as the role of microtopographic variations in the soil–

bedrock interface, in relation to water delivery to the stream, while Brooks and Collison (Chapter 21) emphasize the significance of soil profile differentiation to shallow mass movement, thereby stressing potential feedback between soil profile development, hillslope hydrology and mass movement. Such linkages do have the broader context of evolution explanation at their heart – evolution as offering landscape development from the past, through the present to the future; as we have seen in subsection 1.1 this is a vital aspect of geomorphological theory to develop and advance further.

Linking the chemical and physical elements of hillslope hydrological and erosional processes is an important issue, since soil hydrological and chemical processes operate at different scales. There is growing interest in phosphorus loss in particular, with modelling beginning to provide important new insights. The chapter by Gburek *et al.* (Chapter 12) indicates the extent to which it is possible to elucidate the chemical–physical links using a combination of runoff and erosion modelling within a GIS framework to ascertain the main areas of phosphorus delivery within a small catchment. Further attempts to use combined chemical and physical field data to investigate the behaviour of hillslope systems are considered in the chapter by Elsenbeer and Lack (Chapter 43) for small subcatchments in the Amazon Basin. From chemical signals in stream water, information concerning pathways of delivery can be ascertained and linked to soil profile characteristics.

The fourth major area is the necessity to examine the *extent to which available physically based models accurately represent process mechanisms* at an appropriate scale. In modelling hillslope hydrology this issue is frequently exposed, as most models are based on Darcian Flow. Both the chapter by Ratcliffe *et al.* (Chapter 6) as well as that by Brammer and McDonnell (Chapter 2) focus on the significance of macropore flow processes and macropore–matrix interactions. The chapter by Brammer and McDonnell also shows how hydrological monitoring over decades has revealed increasing complexity in the way we believe hillslope hydrological systems operate. Previously, field measurement supported the use of Darcian Flow models, but continuous field monitoring reveals that this is clearly no longer adequate. We now require three-dimensional macropore–matrix models to elucidate further the hillslope hydrological system, which can take into account spatial variability in both parameter values and process mechanisms.

The fifth issue involves the need to provide *appropriate model parametrization and validation data*. Field data are scale-dependent and this has important implications for compatibility with the scale at which models need to be parametrized. This is especially significant in the light of increasing availability of commercial software in which little attention has been paid to parametrization scales, discussed by Anderson *et al.* (Chapter 36). Compatibility is also required for validation exercises, and issues related to whether possibilities exist for model validation on the basis of field data are discussed by DeRoo (Chapter 30) in respect of soil erosion.

Although these five major issues run through the chapters presented in these volumes, their arrangement is systematic, involving a number of additional themes pertinent to hillslope processes. We start with hillslope hydrology as the fundamental control on other aspects of the hillslope system. This leads into a series of chapters on hillslope solute processes. Finally, in Volume 1, the balance between soil development, and related soil profile properties, and soil removal is considered in two sections. The soil erosion theme is expanded in Volume 2, through sections on gully development and slope stability. Finally, we consider how issues related to hillslope processes in different regions are being addressed in three sections related to tropical, semi-arid and periglacial regions.

1.3   MONITORING FLOW PATHWAYS ON HILLSLOPES AS A BASIS FOR MODEL FORMULATION

The issue of process representation in hillslope hydrological models is raised in the opening section on hillslope hydrology. Brammer and McDonnell (Chapter 2), for example, show how ideas concerning flow pathways have altered over the course of a 25-year period, following the development of increasingly detailed means of field monitoring. Of particular interest is the way in which early research in their instrumented catchment considered small-scale, isolated sections of the hillslope (single pits), producing results which emphasized the role of subsurface preferential flow pathways, while later instrumentation permitted a larger-scale monitoring framework in which several interrelated elements of an integrated hillslope system were included. With increasingly sophisticated monitoring, generating new ideas concerning the operation of hillslope hydrology, the need for models capable of representing an increasingly complex array of processes is emphasized. In Figure 1.1 this development is apparent, with the early work (a) suggesting the need for a simple model which represents the hillslope system wholly as pipeflow channelling of new water past the soil matrix to the channel. Displacement of old water in the matrix represents a further development (b), while later monitoring suggests yet greater complexity via a response in the groundwater system initiated through vertical (as opposed to lateral) bypassing of the matrix (c). Finally, the most recent monitoring provides evidence for the possible existence of a significant role for bedrock microtopography in storing water in isolated pockets for extended periods between events (d). Increases in water-table elevation during "high-magnitude" events connects this isolated system, enabling flushing of the stored water from the hillslope system. The implications of this ongoing research for providing suggestions for formulation of mathematical models to represent adequately the spatial and temporal dimensions in this system are considerable, involving increasing complexity and a concomitant need for model discretization into smaller spatial and temporal units.

In terms of process representation, the significance of soil physics is also stressed in the chapters by Brammer and McDonnell as well as that by Ratcliffe *et al.* (Chapter 6). The former chapter shows how it is only through detailed tensiometric measurement that it is possible to evaluate internal system behaviour. While early work at the Maimai catchment involved sampling stream and subsurface water for chemical/isotopic analysis, conclusive evidence concerning source areas and flow paths was only possible through detailed tensiometric measurements. As well as highlighting the need for better process representation in mathematical models, some of these findings also suggest an urgent need for more detailed consideration of issues

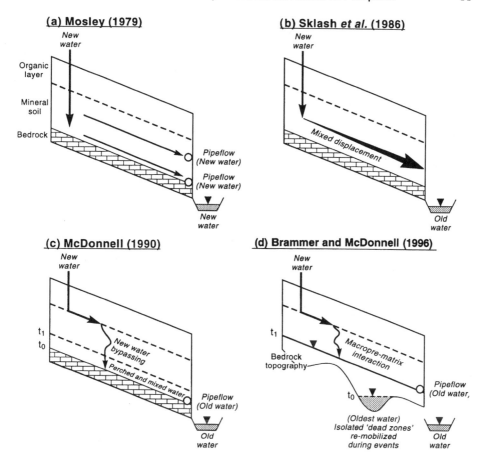

Figure 1.1   Conceptual models of hillslope flow (see Chapter 2)

related to model parametrization and validation. For example, the distinction made in the work of McDonnell (1990) between k_{sat} (soil hydraulic conductivity) and k^* (matric hydraulic conductivity) has serious implications for the way in which parameter values are selected for modelling exercises, as well as for selection of field data appropriate for model validation. Traditionally, k_{sat} is used to parametrize hydrological models. However, this may only be appropriate for considering large-scale system behaviour. By examining k^* it is possible to envisage localized ponding at rainfall intensities considerably lower than those required to produce widespread ponding if k_{sat} applies. Thus system response, as modelled at the large-scale, might be both qualitatively and quantitatively different from that found from a smaller-scale formulation. The issue of selection of effective parameter values appropriate to the scale of analysis is one deserving greater attention.

Focusing on small-scale detailed measurement to enable appropriate process representation and parametrization represents an important first step for developing larger-scale models, although this will only be successful provided that detailed sub-hillslope scale units can be connected accurately. Ratcliffe *et al.* provide an integration of tensiometric and thermal monitoring to examine the detailed response to rainfall inputs for a soil profile in the Panola Mountain Research Watershed, Georgia. Figure 1.2 shows schematically how the two sets of measurements interrelate, suggesting a combination of both bypassing flow in the upper soil horizons and groundwater rise at deeper levels in the profile is responsible for the delivery of "old" water to the stream channel. However, the seasonal effects raise important concerns about the care needed in assessing hydrological mechanisms from field data. The hydrological response depends on prevailing thermal contrasts in "old" (groundwater) and "new" (macropore) water. In summer, groundwater is cooler than rainwater, promoting cooling in the soil as the water-table rises. However, preceding this cooling is a notable rise in temperature, suggesting rapid delivery of warmer rainwater from the surface. Finally lowering of the water-table following the cessation of rain raises temperatures to their pre-storm levels. This pattern reverses in winter. In both cases, groundwater rise is preceded by a marked temperature change indicating the role of macropores in rapid delivery of new water, but interpreting thermograph data requires close attention to seasonal and possible diurnal effects. Chapters 2 and 6 adopt different scales and methods of field monitoring to determining the processes controlling hydrological response. In each case it is clear how ideas about process operation, which provides the basis for model formulation, are dependent on the nature and scale of field monitoring. We are also witnessing a period in which model requirements for parametrization and validation drive the field monitoring programmes, so care is needed in avoiding circularity of argument in which modelling assumptions and field monitoring become mutually reinforcing.

1.4 LINKS BETWEEN SOLUTE PROCESSES, HYDROLOGICAL PATHWAYS AND EROSION MECHANISMS

Hydrological pathways determine whether solutes are delivered rapidly in bulk to stream channels, or persist in the hillslope system affecting both surface and subsurface regions. This is of particular significance in the light of additions of considerable amounts of fertilizers to many agricultural systems, combined with a growing awareness of the role of different tillage practices in affecting delivery routes and fluxes. Use of models to elucidate the integration of chemical with physical hillslope processes is particularly challenging, given the different scales which apply to each. The problem is therefore one of identifying mechanisms that connect plot-scale behaviour with large-scale response. The chapter by Gburek *et al.* (Chapter 12) provides a GIS-physically-based modelling framework which offers potential for considering the mechanisms whereby phosphorus is delivered to stream channels. In this study, phosphorus export, linked to problems of downstream eutrophication, is assessed initially via development of an index describing sites of potential vulnerability. However, the index is inappropriate in that it gives little idea of actual loss. The development of the GIS-model basis produces a series of results for a small catchment having mixed land use, highlighting considerable spatial variability

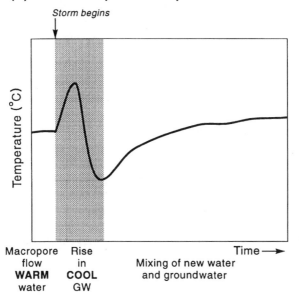

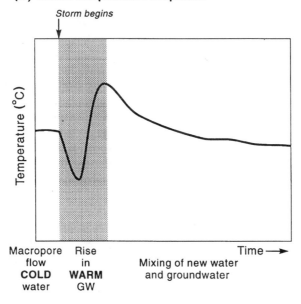

Figure 1.2   Integration of tensiometric and thermal monitoring to identify sources of hillslope waters (see Chapter 6)

in phosphorus loss (Figure 1.3(a)). The chapter emphasizes the existence of a two-phase system in which phosphorus loss is linked to both runoff generation (dissolved P in Figure 1.3(b)) and to sediment erosion (sediment bioavailable P in Figure 1.3(c)). Where land is grassed or untilled, dissolved P accounts for 60% of the phosphorus loss, while under tillage practices sediment-bound phosphorus accounts for the majority of the phosphorus loss (around 90%). Thus different mechanisms of phosphorus loss relate to different parts of the catchment. Where runoff is most readily generated (near the stream channel and on hydrologically active breaks in slope) dissolved P dominates, while areas undergoing active erosion have more sediment-bound P. In either case it is apparent that only a small proportion of the catchment is participating in phosphorus delivery to the stream channel (around 6–10% supplies 98% of the P). In terms of modelling solute processes on hillslopes, it is clear that small-scale studies are again necessary in order to understand the mechanisms responsible, and that a lumped catchment approach is of limited value in elucidating the mechanisms of solute loss.

1.5 SOILS AND SLOPES: CONTEMPORARY LINKS AND LONG-TERM CHANGES

Heterogeneity in soil profile hydrology is difficult to incorporate into physically based hillslope- or catchment-scale models due to the difficulty of discretizing such small variations over such large areas. However, it is increasingly recognized that soil profile hydrology is fundamental to hillslope behaviour on both long and short timescales. In recognizing preferential flow pathways within the soil cover, Mosley (1979), for example, stressed the significance of the base of the B horizon as a location for widespread lateral throughflow, determining short-term response to rainstorms. It is evident from research over the past 20–30 years that, via the links between hydrology and erosion, long-term change in hillslope form is dependent on the balance between erosion and deposition, linked to soil profile development. The integration of soil development and removal with changing hillslope form over the longer term, and the development of models which can elucidate these long-term changes is an exciting challenge for future research. The main issue highlighted by such model development is that of model validation. It is not possible to validate models in applications to past events, and there is growing emphasis on the view that validation exercises in contemporary applications are highly problematic. These points are raised by a series of chapters.

Park et al. (Chapter 17) examine the connections between past and present processes by examining the changing dominance of current and relict features prevailing in soil profiles in different topographic locations. Around a hollow in Somerset the current humid climate and permeable parent material favour the development of podsols. However, there is considerable variation in the degree to which podsolization dominates, related to the presence of relict features. Two main scenarios are evident (Figure 1.4). Firstly, on the interfluve and upper convexities there is considerable evidence for gleying and an absence of the clearly defined horizons commonly associated with podsolization. Impeded drainage resulting from

15

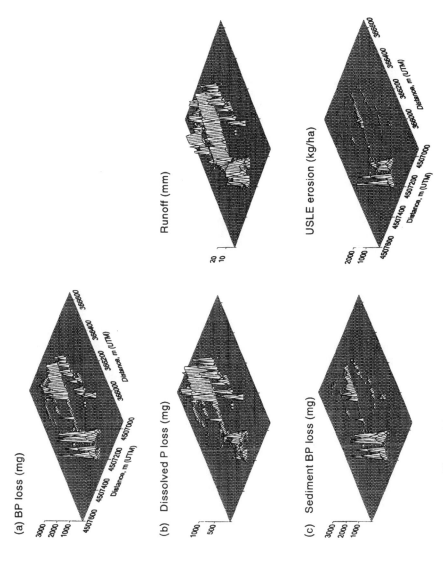

(a) BP loss (mg)

(b) Dissolved P loss (mg)

(c) Sediment BP loss (mg)

Runoff (mm)

USLE erosion (kg/ha)

Figure 1.3 Phosphorous export and links to runoff and erosion (see Chapter 12)

16

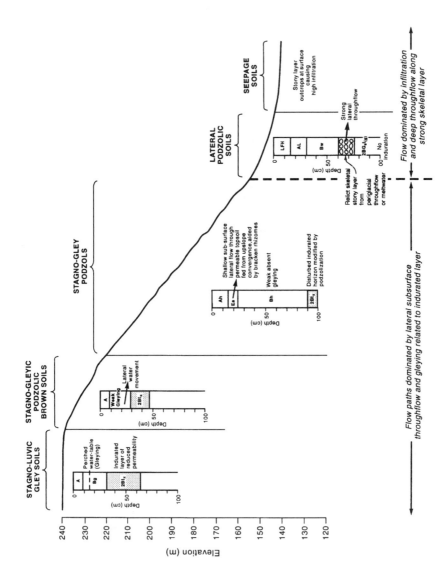

Figure 1.4 Relationships between past and present hillslope soil and hydrological processes (see Chapter 17)

the presence of a relict indurated horizon formed during past periglacial conditions has resulted in stagnoluvic gleys and stagnogleyic podsolic brown soils in this region of the hillslope. The surface of the indurated horizon is irregular, giving rise to a varying balance between vertical and lateral water movement. This is clearly significant to the nature of the resulting soil profile, but also to the hydrological response further downslope. On progressing over the convexity onto steeper slopes the indurated horizon becomes more disturbed and broken, resulting in higher vertical percolation and the opportunity for (i) stagnogley podsols and (ii) non-hydromorphic podsols to develop. Both the intensity of gleying and the extent of lateral eluviation is determined by the continuity of the indurated horizon. Towards the middle hollow and into the upper part of the basal concavity, soils termed "lateral podsolic" are found. These show evidence of strong lateral eluviation related to water movement via a preferential pathway in a loosely fabricated, stony layer at about 60 cm below the ground surface. Again this layer is thought to be of periglacial origin, but clearly is influencing both soil development and subsurface flow. Thus a modelling scheme for long-term evolution of hillslopes must include the influence of variations in the soil profile on hydrological response, as well as the persistence of various relict features in different parts of the hillslope system via the way they determine both water flow pathways as well as soil profile characteristics.

Over the long-term it is also necessary to consider the changing balance between soil accumulation and removal to develop an integrated scheme which connects soil evolution, hydrological process operation and soil removal. Brooks and Collison (Chapter 21) provide a very preliminary exploration of this issue through use of a physically based model for shallow planar slope failures. Here the role of soil profile differentiation is emphasized through the way increasing textural differentiation influences pore water pressure (in this case negative) changes under different rainfall

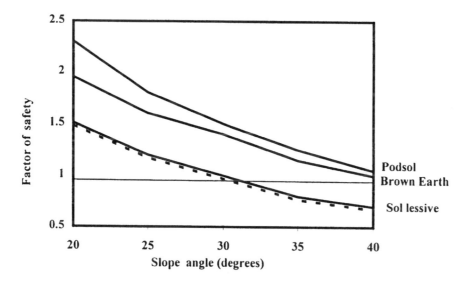

Figure 1.5 Effect of soil profile differentiation on factor of safety (see Chapter 21)

events. The influence of these distributions on shallow planar slope failure is then considered. The "worst case" pore-water pressures are most conducive to failure in soils having a high degree of textural differentiation, and the consequences of this for angles of limiting stability for different soil profiles are shown in Figure 1.5. Such results emphasize the connections between soil profile development and long-term slope stability, but also highlight the need for models which can provide better integration of the development of the whole hillslope system over long time periods. Such use of models also raises the question of the extent to which the results are testable in the field, particularly when used to examine behaviour of hillslopes under past conditions, possibly involving a changing balance between relict and contemporary features, as well as incorporating hillslope behaviour under substantially different climatic regimes.

1.6   SOIL EROSION: USE OF FIELD DATA FOR MODEL VALIDATION

The extent to which it is possible to validate models is increasingly dominating modelling research in both geomorphology and hydrology. Traditionally, "good fits" have been sought between field data and model outputs. In this context we are increasingly forced to address issues related to the accuracy and compatibility of means of data acquisition. Such issues related to validation are raised in the chapters by DeRoo (Chapter 30) and Quine et al. (Chapter 25); emphasis is placed on poor correlations which exist between model simulations of erosion compared with field estimates based on ^{137}Cs concentrations in the soil profile (Figure 1.6). De Roo suggests that model validation might be impossible in the light of data measurement uncertainty for input parameters (particularly spatial and temporal variability in rainfall intensity and infiltration), calibration difficulties, deficiencies in theoretical understanding of hydrological and erosional processes (e.g. failure to consider operation of biological agencies), as well as errors associated with implementation of mathematical solutions. However, it is equally probable that the field data against which models are being tested is also in error. Soil erosion rates estimated from ^{137}Cs measurements fail to take into account significant uncertainties arising from the operation of tilage practices; for example, producing estimates of soil erosion which do not reflect actual rates. This latter point is addressed by Quine et al. (Chapter 25), where a model of ^{137}Cs redistribution (Figure 1.7) enables prediction of its likely spatial variability in relation to soil erosion. Interestingly, this invites debate concerning the extent to which it is possible/acceptable to be comparing predictions from one model against those of another model applied to the same problem, and whether this constitutes an acceptable validation methodology.

1.7   EPISODIC EROSION ON HILLSLOPES

In considering long-term change in hillslopes it is necessary to have a framework for including changing boundary conditions in respect of inputs, such as climatic conditions as well as outputs at the slope base. Basal conditions are inextricably

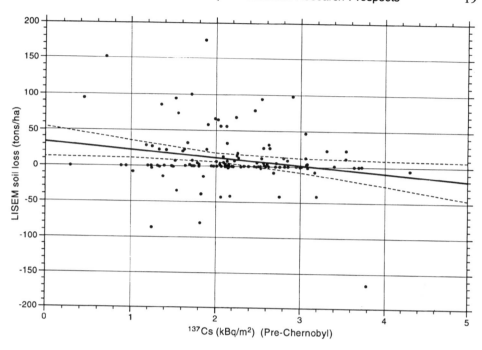

Figure 1.6 Comparison of soil loss based on ^{137}Cs concentrated with that calculated using the LISEM model (see Chapter 30)

linked to the hillslope processes and this is particularly apparent in recent research related to long-term gully behaviour. Gully systems in different environments provide indications of extrinsic environmental change, but also undergo alterations linked to intrinsic factors. Gully systems are sensitive to change in both slope and channel conditions, providing a record of past conditions prevailing in both. Harvey (Chapter 33) identifies a two-fold gully system including stabilized large midslope gullies which fed debris cones in previous periods, and smaller basal gullies which are currently feeding sediment to the stream channel. These latter gullies represent a situation of intrinsic coupling between the slope and stream. The larger mid slope gullies, on the other hand, have only been active in past periods related to the crossing of an extrinsic threshold under changed environmental conditions. The locations and characteristics of these two sets of gullies have received little attention, although their morphology and hydrological setting can provide important clues as to whether gully activation in past conditions is likely to have been linked to climatic (hence hydrologic) or vegetational (hence sedimentological) changes. Figure 1.8 shows the way in which these systems have been classified, according to gully size, pattern and drainage density for gullies at different locations in the drainage basin (valley-head or valley-side). By plotting the relationship between slope length or upslope drainage area (which determine the amount of runoff being delivered) and slope gradient (which determines runoff power) derived from topographic maps, thresholds required for the different forms of gullies to be initiated are ascertained.

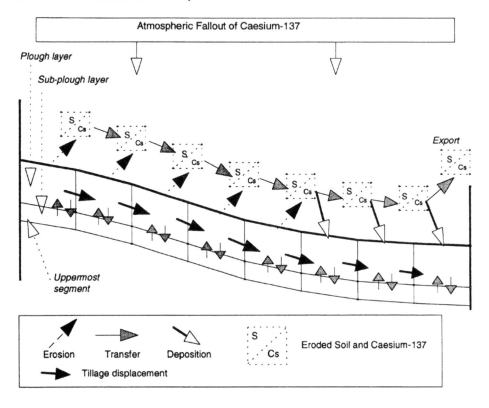

Figure 1.7   Model of ^{137}Cs redistribution (see Chapter 25)

For hillslope gullies, gradient has a strong influence, indicated by the association between a declining threshold for gully formation and an increase in slope length; hence these gullies seem to be related more to changes in hillslopes than in channels. Whatever extrinsic control is operational, climatic or vegetational (in this case the latter), it is clear that the links between the hillslope hydrological system and gully formation require closer investigation and inclusion in larger-scale erosion models. Given that gullies can be a significant source of sediment delivery, erosion models based on plot-scale experiments cannot accurately represent erosion rates for slopes with gully systems.

The role of gullies in providing a hydrological and sedimentological link between hillslopes and channels is further stressed by Faulkner (Chapter 32). Field evidence supports a model of the way in which gully sediments affect sediment patterns in the main channel of ephemeral streams as well as on the surrounding terraces. In this system sediment inputs to the main channel from integrated gully systems are high, particularly when the type 1 (Hortonian) and type 2 (seepage) gullies become connected. Sedimentation patterns are influenced both by the material supplied from the gullies and by channel constrictions. Around such constrictions flow becomes "superelevated", causing dumping of coarse-grained sediment and scouring of existing sediments downstream of the constriction (figure 1.9). These findings, set alongside arguments concerning additional downstream transmission losses as width depth ratios increase, challenge previous models of downstream sediment coarsening. In fact, a general trend for downstream fining with superimposed complexities surrouding channel constrictions and oversteepened bed sections is deemed to provide a better model for understanding the sedimentation patterns in ephemeral channels. The role of sediment delivery from valley-side slopes via the gully network is though to be a highly significant influence on channel sediments.

1.8 TECHNOLOGICAL ADVANCES IN FIELD MONITORING AND MODEL DEVELOPMENT

The issue of episodic erosion is one which is continued in the section on slope stability, although this section also highlights advances in field monitoring and model development. Brunsden and Chandler (Chapter 40) present a new model for episodic erosion on coastal cliffs based on DEMs derived from digital photogrammetry. This technique enables far greater detail in data capture than previously possible, with results suggesting a revision of traditional ideas about the evolution of the Black Ven system. Previous ideas have been based on the proportion of the slope system occupying different slope angle groups. The original model, based on episodic evolution of the Black Ven complex, involved an initial phase of basal erosion leading to an increase in mean slope angle, followed by a reduction in mean slope angle as the system stabilizes. Through time, little change in the overall slope angle distribution was observed despite loss of a significant amount of sediment from the system. Several difficulties were identified with the techniques used to derive the data on which the model was based. First, the DEM grid only delivered rectangular shapes, involving distortion of the boundaries and consequent inaccuracy in slope angle histograms. Secondly, it was not possible to separate individual mudslide complexes within the DEM, and to study their morphology separately. Thirdly, the model involved mass-movement triggers based only on slope steepening through basal erosion, although other controls might be important. New developments in photogrammetric monitoring and data capture enabled assessment of small subunits of the whole complex (Figure 1.10), as well as observation of a high-magnitude event, necessitating refinement of the original model. In the new model the frequency of wet and dry years is shown to play a significant role in determining changes in the landslide complex between major erosional events. It is only in years with a moisture balance below the average that basal erosion is able to oversteepen the mudslide complex. In wet years such a situation cannot arise since, as soon as slope angles begin to build up, failure occurs. Thus a two-phase model is required for inter-event periods in which the frequency of wet and dry years is instrumental.

The application of increasingly refined computer software to enhance our understanding of slope instability processes is highlighted in the chapters by Wu and Thornes (Chapter 35) as well as Anderson et al. (Chapter 36). As in the previous example, the role of short-term hydrological fluctuations is stressed as a major control on slope stability. Aided by increasingly sophisticated computer hardware

(a)

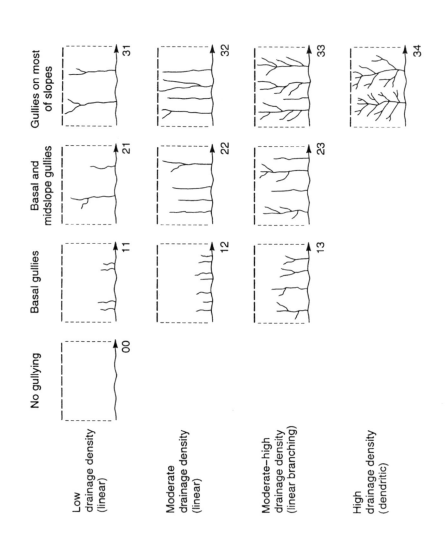

(b)

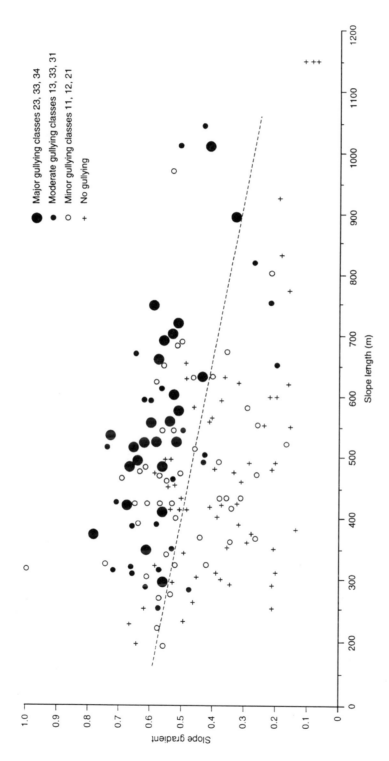

Figure 1.8 (a) Classification of gully systems. (b) Relationship between slope length and gradient with different gully classes indicated (see Chapter 33)

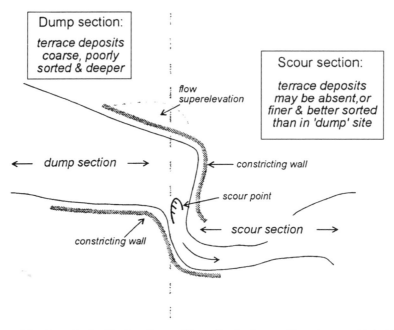

Figure 1.9 Hypothetical hydraulic conditions (in plan) prevailing at a constricted bend in an ephemeral rambla (see Chapter 32)

and software, it is becoming possible to assess the net effect of opposing influences on slope stability, as well as to identify short-term changes in pore-water pressures which might otherwise be overlooked in models requiring a high degree of spatial and temporal lumping. The grid resolution used by Wu and Thornes, for example, provides a high level of detail in trying to establish the likely effect of different terrace constructions on slope stability. Parallelling this trend towards increasingly sophisticated models is the growing availability of an ever-increasing number of "off the shelf" models designed to be run in a relatively user-friendly environment. The chapter by Anderson *et al.* discusses some of the issues related to the increasing circumstances in which the researcher has little or no control over either model coding or parametrization requirements. This raises considerable difficulties associated with applications to field sites where parametrization data are limited and processes operating are different from those specified in model subroutines. Issues of model validation are also highlighted in the light of recent trends towards internal validation. In situations in which commercial software provides one main output, internal validation and the use of such models for establishing process dominance are not possible, restricting their use to design (engineering) applications rather than their more geomorphologically relevant role of elucidating complex hillslope process dominance. Whilst future development of "in house" models is crucial for providing process model and internal mechanics representation, the potential seduction of researchers by commercial software is a real issue of concern in this context.

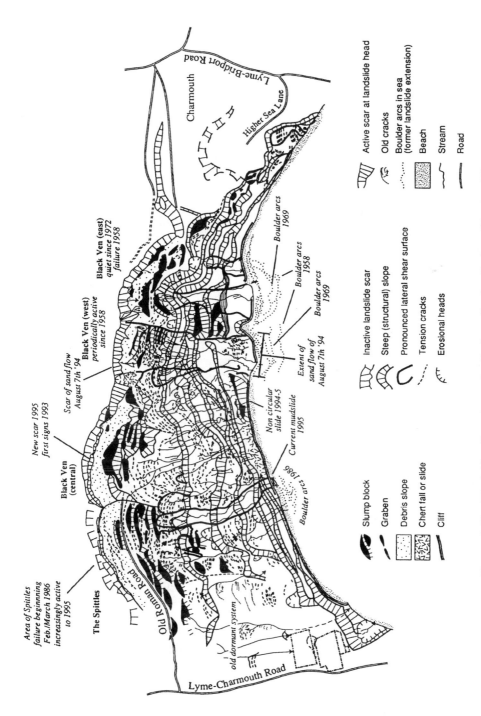

Figure 1.10 Black Ven Complex – detail of small subunits (see Chapter 40)

1.9 SPATIAL VARIABILITY IN HILLSLOPE SYSTEMS

As well as temporal variability in erosion rates, frequently expressed as episodic behaviour, hillslope systems exhibit a high level of spatial variability. This is emphasized by Elsenbeer and Lack (Chapter 43) who discuss variability within the Amazon Basin, a region traditionally viewed as being largely spatially homogeneous. The consequences for the hydrology of the River Amazon are evaluated in some detail, particularly given the fact that many hydrological point measurements are taken as representative of larger areas. Connections between soil profile character- istics and hydrological response are again stressed, raising issues related to the scale of model discretization appropriate for modelling the behaviour of large catchments. Large drainage basins cannot simply be assumed to consist of a series of homogeneous subunits. Considerable variability exists at the hillslope scale. For example, in the Amazon there has been a traditional emphasis on Oxisol hydrology, involving predominance of high saturated hydraulic conductivity in the soils producing considerable infiltration and throughflow volumes, applicable to the whole drainage basin. However, substantial parts of the Amazon Basin have Ultisols with saturated hydraulic conductivity values in the subsoil of up to five orders of magnitude lower. In such cases considerable volumes of overland flow are generated.

It is further suggested that the two soil types represent opposite ends of a spectrum and that significant variations in stream water chemistry might be expected from the contrasts in flow paths operating along this spectrum. Using the Ultisol dominated region of the catchment, Figure 11.1 indicates the dilution effect of rapid delivery of silica-poor overland flow to the stream. As discharge rises (due to overland flow) silica concentration reduces. With potassium the situation is more complex, involving a significant role for antecedent moisture conditions. In the dry season the potassium-rich overland flow causes a rise in concentrations with discharge, but during the rainy season when antecedent moisture levels are high, the contribution from the soil water depleted in potassium is sufficient to offset this effect. Thus potassium levels display little variation with discharge. Further complications are suggested in relation to the existence of catenary sequences on hillslopes in the drainage basin in which both flow pathways and chemical dilution effects are likely to vary over sub-hillslope scales. Thus there is considerable scope for research into "soilscape" variations, the relationship between these variations and water flow pathways, as well as the links to stream hydrochemistry, and the implications for large-scale catchment models.

1.10 SENSITIVITY ANALYSIS OF THE VEGETATIONAL AND SOIL HYDROLOGICAL RESPONSES TO CLIMATIC CHANGE

Semi-arid hillslopes present research opportunities and challenges for monitoring and modelling the impact of climatic and land-use changes. Traditionally, when modelling the impact of climatic change, it has been assumed that all parts of the hillslope system are affected. Scope for inclusion of differential (or zero) response

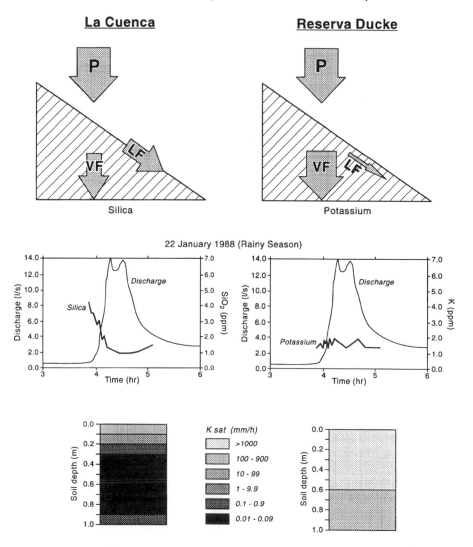

Figure 1.11 Relationships between hydrological properties, waterflow pathways and Si/K concentrations for two contrasting tropical rain forest locations (see Chapter 40)

between the different components has been limited. However, the chapter by Mulligan (Chapter 50) suggests that such a simple approach fails to capture the true impact of climatic change as exacerbated or modified by non-linear responses in soil and vegetation, including feedbacks between different parts of the hillslope system. In this chapter a model (PATTERN) is developed and applied to the problem of climatic change for semi-arid hillslopes to assess the significance of differential

28

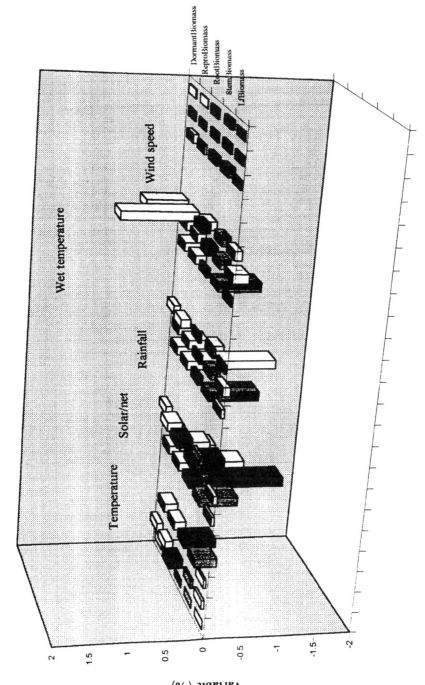

(a) Change in output variable per unit change in input variable (%)

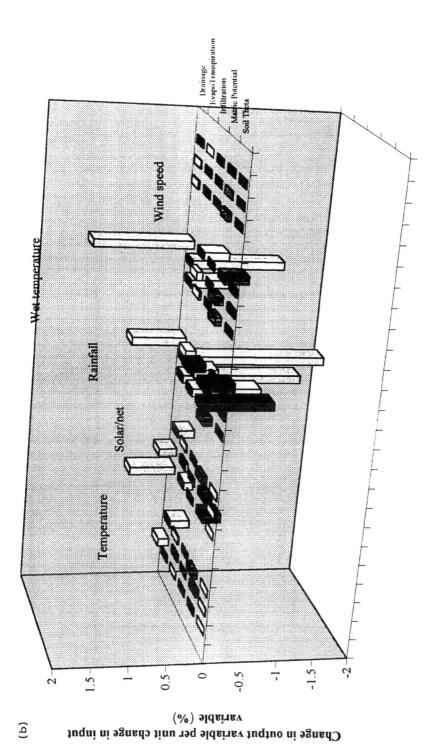

Figure 1.12 Using the PATTERN model. Hydrological responses (see right-hand side) to changes in temperature, radiation, rainfall and windspeed. (b) Biomass sensitivity to the same meteorological factors as in (a) (see Chapter 50)

changes in hydrological and vegetational responses to the generation of significant spatial variability in the system.

Detailed sensitivity analysis, involving both soil and vegetation responses, reveals considerable variation in the parameters which undergo the most extensive changes. Figure 1.12 indicates that soil response is clearly related far more to variations in rainfall than to either windspeed or temperature, while vegetation appears to vary with solar radiation to a far greater extent. This is a useful exercise in that it clearly demonstrates the value of having modelling capabilities which enable the simultaneous net effect of several variables to be evaluated. However, a further point raised is that it is not simply sensitivity to change in a particular parameter which is important, but that it is also necessary to consider the propensity for that parameter to change. Such modelling exercises have a valuable role in generating priority research questions, as well as in providing a structure around which field experiments can be designed.

1.11   USE OF HARDWARE MODELLING IN EVALUATING HILLSLOPE BEHAVIOUR

The emphasis of recent research has clearly been on utilizing increasingly sophisticated technology to address increasingly complex research questions via the interrelationship between monitoring and modelling. In the case of modelling it is very evident that mathematical models are currently at the forefront of research into hillslope systems. The final section of the volumes raises some highly pertinent issues related to the use of hardware models. Many of the issues raised in connection with mathematical modelling are applicable to hardware modelling, in particular the extent to which the model outputs adequately reflect field behaviour. In this, scale issues again dominate the discussion.

The chapter by Harris and Davies (Chapter 51) provides one of the few examples of the use of hardware modelling in assessing hillslope processes, and highlights the value of adopting such a research methodology. In their example, four cycles of freezing and thawing are simulated in the laboratory for a 12° slope constructed from different soils (fine sand and fine sandy silt). Four main properties were monitored at 30-minute intervals. These were pore-water pressures, downslope soil displacement, soil temperatures and frost heave/thaw settlement. The data thereby obtained enabled conclusions to be drawn regarding the effective processes controlling downslope movement of the surface layer. Previous ideas, related to failure of a saturated layer of material with reduced frictional strength, were discounted, since pore-water pressures remained negative and were consequently insufficient to cause such downslope movement. However, localized slope failure occurred near the surface and displacement measurements suggested that downslope movement was predominantly a slow creep process. Such results, impossible to obtain under field conditions, indicate the value of hardware modelling, although how closely laboratory conditions relate to those operating under field conditions requires careful consideration.

1.12   DISCUSSION

The review of the contents of these two volumes in the above subsections (1.3–1.11) has sought to draw attention to key research developments in hillslope processes. Critical in this context is the need to be absolutely clear on the hypotheses being developed and tested. As Chorley *et al.* (1973) observe: "In an exact science *laws* provide stepping stones for progress, whereas in geomorphology, an inexact science, *hypotheses* provide themes for future investigations" (p. 753). It is important that neither suites of available software nor technological advances cause hypotheses (self-generated from such sources) to be uncritically accepted as laws, since this may lead to a lessening of scientific investigation. Neither should the pressure to provide "solutions" to environmental questions be allowed to become a distortion to scientific inquiry for the reasons we outlined in subsection 1.1. Again, a commentary by Chorley *et al.* (1984) provides a postscript to this debate;

> One of the reasons that the understanding of the development and evolution of hillslopes has been slow is due to the complexity of the phenomena investigated. The effects of climate, lithology and vegetative cover and the differences between erosional and depositional slopes produce forms that cannot be generally related one to the other and, therefore, generalisations about hillslopes on a global basis are impossible to make. And yet, from a practical point of view, it is essential that we understand how slopes erode and what form they will take over long periods of time. In the past practical applications of hillslope erosion were related primarily to agricultural activities. However, more recently the disposal of tailings from oil shale and uranium mines has led to a need for slope stability of the order of 1000 years (p. 275).

Current "applied" questions can themselves demand long-term process understanding. Such an awareness is not without its consequences: it will help to maintain scientific investigative standards and to generate hypotheses appropriate to the "inexact science" to which Chorley has referred.

REFERENCES

Brammer, D.D. and McDonnell, J.J. 1996. *An evolving perceptual model of hillslope flow at the Maimai Catchment.*

Chorley R.J. 1978. Bases for theory in geomorphology. In Embleton, C. Brunsden, D. and Jones, D.K.C. (eds), *Geomorphology Present Problems and Future Prospects*, Oxford University Press, 1–13.

Chorley, R.J., Dunn, A.J. and Beckinsale, R.P. 1964. *The History of the Study of Landforms. Volume 1, Geomorphology before Davis*. Methuen, London.

Chorley, R.J., Beckinsale, R.P. and Dunn, A.J. 1973. *The History of the Study of Landforms. Volume 2, The Life and Work of William Morris Davis*. Methuen, London.

Chorley, R.J., Schumm, S.A. and Sugden, D.E. 1984. *Geomorphology*. Methuen, London.

Embleton, C., Brunsden, D. and Jones, D.K.C. (eds) 1978. *Geomorphology: Present Problems and Future Prospects*. Oxford University Press.

Fells, I. 1996. Universities in an age of efficiency. Letter to *The Times*, 4 March 1996.

King, P.B. and Schumm, S.A. (eds) 1980 *The Physical Geography (Geomorphology) of William Morris Davis*. Geobooks, Norwich.

McDonnell, J.J. 1990. A rationale of old water discharge through macropores in a steep humid catchment. *Water Resour. Res.*, **26**, 2821–2832.

Mosley, M.P. 1979. Streamflow generation in a forested watershed, New Zealand. *Water Resour. Res.*, **15**, 795–806.

Sklash, M.G., Stewart, M.K. and Pearce, A.J. 1986. Storm runoff generation in humid headwater catchments 2. A case study of hillslope and low order stream response. *Water Resour. Res.*, **22**, 1273–1282.

Section 2

HILLSLOPE HYDROLOGICAL PROCESSES

2 An Evolving Perceptual Model of Hillslope Flow at the Maimai Catchment

DEAN D. BRAMMER and JEFFREY J. MCDONNELL
Water Resources Research Group, College of Environmental Science and Forestry, State University of New York

2.1 INTRODUCTION

Hillslope hydrological investigations have been conducted in a variety of geographical, hydrogeological and climatic settings and have been reviewed recently by Bonell (1993) and Buttle (1994). While extensive in location and scope, most catchment-based hillslope investigations have not continued much beyond a typical two- to three-year funding cycle. This frequently limits the opportunity for formal testing of hypotheses proposed by previous researchers at a site and limits the cumulative understanding of a single hillslope. Therefore, hillslope hydrological observations or *perceptual models* (Beven 1991) are rarely "challenged" by other research scientists carrying on increasingly intensive work on the same hillslope or catchment. One notable exception is the Maimai research catchment in New Zealand. Maimai has been the site of ongoing hillslope research by several research teams since the late 1970s. These studies have facilitated the evolution and development of a very detailed yet qualitative perceptual model of hillslope hydrology at Maimai. This perceptual model has now grown in complexity to defy analytical description; none the less it provides a very useful case study of hillslope hydrological processes and encapsulates all that the field hydrologists have come to recognize as the dominant hillslope runoff processes in steep, humid catchments.

The goal of this chapter is to synthesize the development of ideas relating to an evolving perceptual model of subsurface flow at the Maimai catchment. We hope that this chapter will (1) provide a comprehensive overview of studies on subsurface flow within a well-characterized humid, temperate-forested catchment and (2) chronicle how the evolution of the Maimai perceptual model has been affected by the methods used, the magnitude and frequency of events studied and the scale of inquiry of specific studies. We hope to show the value of working in a cumulative fashion at a single research site. We feel that this approach has yielded, in the case of the Maimai catchment, a rich understanding of hydrological processes.

Advances in Hillslope Processes, Volume 1. Edited by M. G. Anderson and S. M. Brooks.
© 1996 John Wiley & Sons Ltd.

2.2 MAIMAI: THE QUINTESSENTIAL STEEP HUMID HILLSLOPE

2.2.1 History of the Maimai Catchment

In 1974, the Forest Research Institute initiated a multi-catchment study at Maimai in the Tawhai State Forest, near Reefton, North Westland, on the South Island of New Zealand (Rowe *et al.* 1994; Rowe and Pearce 1994). This was done in response to concerns regarding the possible adverse effects on the quantity and quality of water supplies and aquatic habitats (large-scale forestry management operations). Data were gathered on streamflow characteristics, stream sediment yield and slope stability of mixed beech forest catchments. Though the conversion of native forests to exotics and the harvesting of mixed forest stands proved to be economically inviable, the long-term studies of the Maimai catchments continued. The data collected provides a useful comprehensive and long-term study of hydrological processes in forested catchments.

2.2.2 Physical Characteristics of the Maimai Catchment

The Maimai study area consists of eight small catchments (1.63–8.26 ha) located to the east of the Paparoa Mountain Range situated on south-facing slopes draining into Powerline Creek (informal name) (Figure 2.1). The catchments lie parallel to each other and share similar topographic characteristics. Slopes are short (<300 m) and steep (average 34°C) with local relief of 100–150 m. Stream channels are deeply incised and lower portions of the slope profiles are strongly convex. Areas that could contribute to storm response by saturation overland flow are small and limited to 4–7% of catchment areas (Mosley 1979; Pearce *et al.* 1986). The M8 catchment has been the most intensely studied and is typical of the physical characteristics of the other Maimai watersheds (Figure 2.2).

Passage of frontal systems from the Tasman Sea across the Paparoa Range (from westerly and northerly directions) occurs regularly and creates a climatic regime with frequent and occasionally prolonged periods of rainfall. Mean annual precipitation is approximately 2600 mm, producing an estimated 1550 mm of runoff. The summer months are the driest; monthly rainfall from December to February averages 165 mm and for the rest of the year between 190 and 270 mm. On average, there are 156 rain days per year with little temperature extreme and only about two snow days per year (Rowe *et al.* 1994).

In addition to being wet, the catchments are highly responsive to storm rainfall. Quickflow (QF as defined by Hewlett and Hibbert 1967) comprises 65% of the mean annual runoff and 39% of annual total rainfall (P) (Pearce *et al.* 1986). The quickflow response ratio (QF/P) is roughly double that of the most responsive basins documented in eastern United States (Hewlett and Hibbert 1967). Pearce *et al.* (1986) note that the R index for QF/P averaged for runoff events from rainfalls of greater than 25 mm is 46% compared with 3–35% for 11 basins distributed between Georgia and New Hampshire that Hewlett *et al.* (1977) studied.

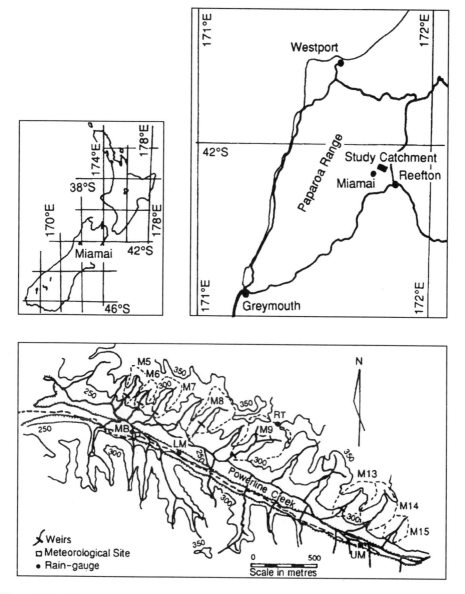

Figure 2.1 Maimai study catchment location, South Island, New Zealand (from Rowe *et al.* 1994)

2.2.3 Vegetation and Soils of the Maimai Catchment

The vegetation is a mixed evergreen beech forest (*Nothofagus* spp.), podocarps, and broadleafed hardwoods. It is multi-stored, with a canopy of 20–36 m high, a dense fern and shrub understory and a fern and moss ground cover. Annual interception

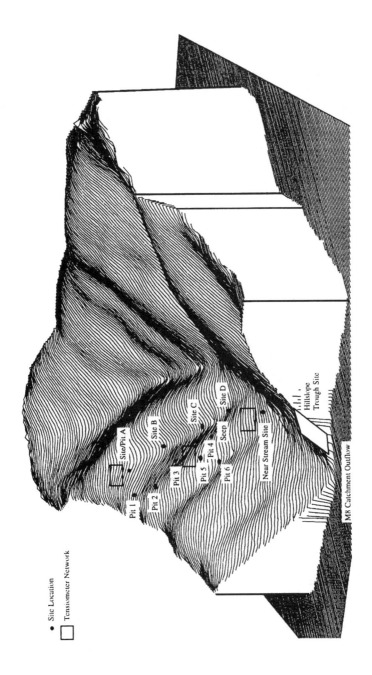

Figure 2.2 M8 catchment study site locations. Sites A–D and Pits 1–6 are original study locations from Mosley (1979). Pearce *et al.* (1986) and Sklash *et al.* (1986) reactivated seven of Mosley's measurement sites and added suction lysimeters and max-rise piezometers at the near-stream site (SL and Well B) and near the M8 stream weir (SL and Well A). McDonnell (1990a) extended the instrumentation with tensiometer networks installed at the near-stream site, and upslope of Pit 5 and Pit A

losses are estimated to be 26% for undisturbed mixed evergreen forest (Rowe 1979). Mean evaporation rates for the M8 catchment are 0.46 and 0.28 mm/h for summer and winter, respectively (Pearce and Rowe 1981).

The study area is underlain by a firmly compacted, moderately weathered, early Pleistocene conglomerate (Old Man Gravels). The conglomerate comprises clasts of sandstone, granite and schist in a clay–sand matrix and is effectively impermeable with seepage losses to deep groundwater estimated at 100 mm year (O'Loughlin *et al.* 1978; Pearce and Rowe 1979). Overlying soils are classified as Blackball hill soils. The typical soil horizon is characterized by a thick, well-developed organic horizon (average 17 cm), thin, slightly stony, dark greyish brown A horizons and a moderately thick, very friable mineral layer of podsolized, stony, yellow-brown earth subsoils (average 60 cm). Silt loam textures predominate. Study profiles examined by Webster (1977) showed that the organic humus layer averaged a total porosity and macroporosity of 86% and 39% by volume, respectively, with an infiltration rate of 6100 mm/h. The mineral soils are very permeable and promote rapid translocation of materials in suspension or solution (Rowe *et al.* 1994). The total porosity averaged 70% by volume, with average bulk densities of $0.80 \, t/m^3$ and saturated hydraulic conductivities of 250 mm/h. The wet and humid climatic environment, in conjunction with topographic and soil characteristics, result in the soils normally remaining within 10% of saturation (Mosley 1979). As a result, the soils are strongly weathered and leached, with low natural fertility.

The thin nature of the soils promotes the lateral development of root networks and channels. Soil profiles reveal extensive macropores and preferential flow pathways at vertical pit faces in the Maimai M8 catchment which form along cracks and holes in the soil and along live and dead root channels (Mosley 1979, 1982). Lateral root channel networks are evident in the numerous tree throws that exist throughout the catchments. Preferential flow also occurs along soil horizon planes and the soil–bedrock (Old Man Gravels) interface.

2.3 EARLY STUDIES OF MACROPORE FLOW

2.3.1 Hydrometric Observations at the Catchment and Hillslope Scale

The first comprehensive hydrologic study of the Maimai catchments was conducted by Mosley (1979). This study included a series of hydrometric and dye-tracing experiments in the Maimai M8 catchment to monitor streamflow and subsurface flow through the soil mantle at a variety of topographic positions. Streamflow was measured at three sites along the stream channel. Seven 2–3 m long pits were dug into the underlying gravels, and troughs were installed at the base of the pits to intercept and measure subsurface flow (Figure 2.2).

Hydrometric observations were made during 12 storm events of varying characteristics. At the three stream channel sites (Sites B–D) and one upslope hollow site (Site A), stream and subsurface flow hydrographs were closely aligned in time and increased in a downslope direction (Figure 2.3). There was a close coincidence in the time of the peaks and an increase in discharge and total flow volumes downslope. This indicated that water was moving considerable distances

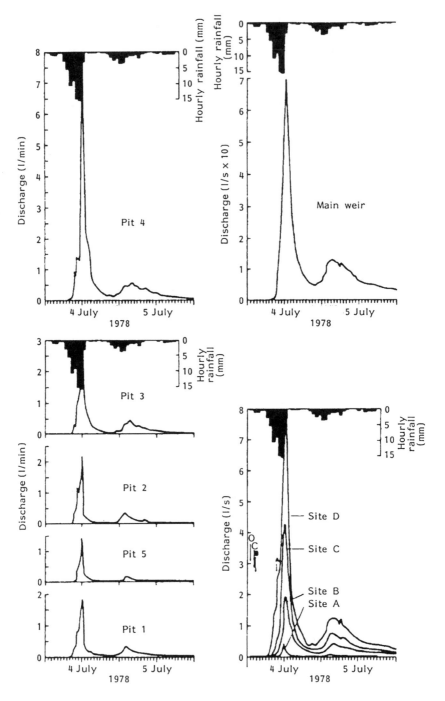

Figure 2.3   Comparison of hydrograph response from the throughflow pits (1–6) zero-order stream sites (A–D) and at the M8 main weir as shown in Figure 2.2, 4–5 July 1978 (from Mosley 1979)

through the soil during the rising limb of the hydrograph. Mosley (1979) noted that if subsurface flow at each site was derived from the immediate surrounding contributing area only, peak discharge and total flow volume would be independent of distance from the catchment divide.

2.3.2 Hydrometric Observations at the Pit Face Scale

In addition to flow measurements, Mosley (1979) made a number of important visual observations of hillslope flow. He noted that the pit faces displayed points of significant seepage during storm events, usually at the base of the B horizon, at which high rates of outflow were observed. At one site, Mosley (1979) observed that water gushed out of two pipes discovered at the base of the B horizon, at a rate approaching 20 l/s. The significance of preferential/macropore flow mechanism was evaluated via eight dye-trace trials, conducted at various pit faces (Figure 2.2). The first four trials were conducted at the end of a low intensity storm event. Dye (sodium dichromate, rhodamine B and lissamine green) was applied to the soil surface 1–4 m upslope of the pits using a cylinder inserted into the humus layer. The first traces of dye emerged through root holes in the pit faces at arrival times of 1.5–23 min following application. Dye appeared later along the base of the B horizon. Dye velocities were calculated as 0.17–0.81 cm/s. Lateral spread of the dye of up to 1 m was also noted. The fourth trial was conducted at a newly excavated pit with additions of water (79 l total) applied 4 m upslope prior to dye application. Under these conditions, dye appeared 30 s after application indicating a travel velocity of 1.1 cm/s.

Further dye-tracer experiments were conducted at the end of a higher intensity storm event with rhodamine B dye simply sprinkled onto the soil surface 1 m upslope from the pits. Dye first appeared at the base of the humus layer and within seeps in the B horizon 60–82 s after application. Travel velocities ranged from 1.2–2.1 cm/s. In general, Mosley (1979) found that maximum dye travel velocities were up to 300 times greater than the measured saturated hydraulic conductivity (252 mm/h) for the mineral soil. This indicated that there was some downslope movement of water occurring over distances of up to several tens of metres during the rising limb of the storm hydrograph. Mosley reasoned that this could not be accomplished by saturated flow through the mineral soil alone. Where there was a lack of root density and channels in the A and B horizons, water appeared downslope at the base of the A horizon. At sites with a greater density of roots and channels in the mineral soil, water reappeared from seeps and at the soil–bedrock interface; losses to the matrix were minimal. Mosley concluded that this flow was of "new" water and that subsurface flow through macropores and along long discontinuities in the soil profile was capable of contributing to storm period streamflow.

2.3.3 Subsurface Flow Velocity Estimations

Mosley (1982) continued the subsurface flow trial by sprinkling sites under different harvesting conditions; undisturbed, logged and logged/burned/planted. Water was applied as a line source 1 m upslope from pits, 2 m long and 0.5–1 m wide. Once

42

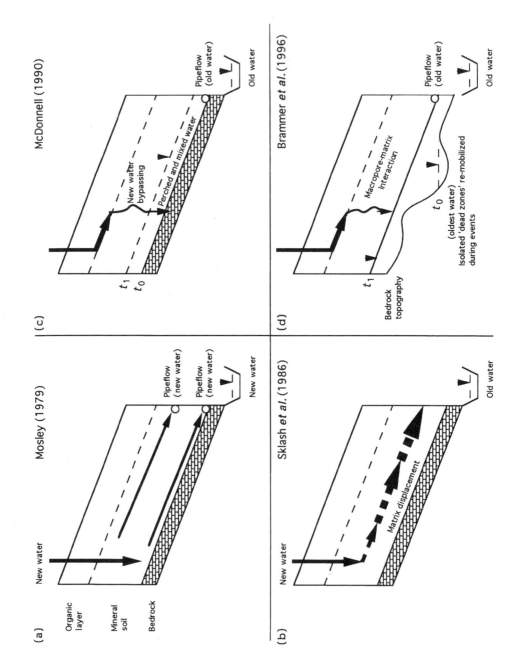

(a) Mosley (1979)

New water

Organic layer

Mineral soil

Bedrock

Pipeflow (new water)
Pipeflow (new water)

New water

(b) Sklash et al. (1986)

New water

Matrix displacement

Old water

(c) McDonnell (1990)

New water bypassing

Perched and mixed water

t_1
t_0

Pipeflow (old water)

Old water

(d) Brammer et al. (1996)

Macropore-matrix interaction

t_0
(oldest water)
Isolated 'dead zones' re-mobilized during events

t_1

Bedrock topography

Pipeflow (old water)

Old water

again, the pits were excavated to bedrock with intercepting troughs installed at the base. A maximum of 30 l water was applied, but subsurface flow was often observed before more than 15–20 l had been applied. Rates of application were within the range of subsurface flow discharges observed by Mosley (1979).

A comparison of the values for mean velocity, maximum velocity and ratio of output volume to input volume were conducted. High mean subsurface velocities were measured under each site type; 0.45 cm/s for undisturbed, 0.69 cm/s for slash-covered, and 0.64 for burned sites. The hydrographs of all sites were shown to be very similar in shape and if plotted on a common time base, similar in peak times. The rapid flow velocities and visual observations showed that there were preferential flow pathways, along cracks and holes in the soil and roots, both live and dead. Profile wetness increased downslope and vertically; saturated wedge thinning was observed in an upslope direction.

2.3.4 A Proposed Perceptual Model of Hydrologic Response

From these preliminary studies, a schematic/perceptual model of catchment hydrologic response (subsurface flow paths) was proposed (Figure 2.4(a)). It was recognized that all of the illustrated flow pathways were probably active at each site, but the relative importance varied significantly. "Variability in flow velocity and the proportion of the input appearing as rapid outflow is a function of antecedent moisture conditions and of the relative importance of the various pathways at a given site, which in turn is a function of soil characteristics, macropore network and parent material at base of soil" (Mosley 1982, p. 65). The model considered macropore flow to be a totally "short-circuiting" process by which "new" water could rapidly appear in the stream hydrograph.

With hindsight, we know that Mosley's perceptual model was limited by the lack of fluorescence intensity measurements (future investigations would show considerable dilution of the applied dye, indicating much more mixing). Mosley's model relied on the assumption of a continuous, well-conducted macropore flow system present within the soil. Also, the trace application intensity of simulated "events" was that of very long recurrence interval "storms"; hence application rates may have actually *induced* macropore flow.

2.4 ISOTOPE TRACING AND OLD WATER DISPLACEMENT

2.4.1 Application of the Isotope Hydrograph Separation Technique (Long-Term Catchment Sampling)

Pearce *et al.* (1986) sampled weekly M8 rainfall, soil water and streamflow for electrical conductivity (EC), chloride (Cl^-), Deuterium (δD) and oxygen-18 ($\delta^{18}O$) composition, from 1977 to 1980. Streamflow was sampled from the M8 catchment,

---

Figure 2.4 (*opposite*) Maimai conceptual models: (a) Mosley (1979); (b) Sklash *et al.* (1986); (c) McDonnell (1990a); (d) Brammer *et al.* (1996)

the undisturbed control catchment, M6 and Powerline Creek. Rainfall was sampled at two sites within the study area. Seven measurement sites (Pits 1–3, 5, Sites A and D and the seep) used by Mosley (1979) in M8 were reactivated and instrumented with suction lysimeters and piezometers. Two additional sites within M8 (near stream site and at the catchment outflow) were similarly instrumented and sampled (Figure 2.2).

Pearce et al. (1986) found that the $\delta^{18}O$ values for the weekly rain samples ranged from -3 to -12 per mil (‰) and displayed some seasonality; isotopically heavier values occurred during the summer and lighter values were observed in the winter months. Both the stream and groundwater samples followed rainfall trends but with smaller seasonal variations. The groundwater samples followed M8 $\delta^{18}O$ stream values indicating similar water. The decreased temporal variations in stream and groundwater suggested to Pearce et al. (1986) that (1) most of the mixing of old and new waters occurred on the hillslope and (2) subsurface water discharge to the stream was an isotopically uniform mixture of stored water. Samples collected at higher than normal flow rates showed no deviation from the seasonal isotopic trends or from other times despite large associated fluctuations in rainfall δD or $\delta^{18}O$, indicating that only small contributions of new water occurred at high flow rates. This observation directly refuted Mosley's determination that rapid transmission of new water formed the majority of stream runoff.

2.4.2 Sampling at the Hillslope and Catchment Scale

Pearce et al. (1986) sampled two small storms immediately after logging of the M8 catchment in April 1979. The $\delta^{18}O$ values of the storm runoff fluctuated only slightly from baseflow $\delta^{18}O$; new water inputs were only 3% of storm runoff. The electrical conductivity and chloride data confirmed the low contribution of rain (new water) to streamflow. Stream EC rose to 80–$100\,\mu S/cm$ from an initial value of $47\,\mu S/cm$ during the first event. EC rose to over $140\,\mu S/cm$ during the second event the following day, lagging the hydrograph peak. The changes in EC indicated increases in total solutes rather than dilution, which would be expected if water from the storm rainfall had dominated the runoff. Cl^- concentrations in streamflow remained constant throughout the events. The increases in EC and the consistency of Cl^- concentrations indicate that the storm runoff response was predominantly water which had a substantial period of contact with the soil. The contribution of 3% of new water could be accounted for by direct precipitation onto the stream channel. The combination of isotope compositions and solute concentrations, provided strong evidence that at least in small events, and under moderately wet antecedent conditions, rapid throughflow of infiltrated rainwater was not the mechanism that produces storm runoff.

Sklash et al. (1986) extended the Pearce et al. (1986) hydrograph separations into two first- and one second-order stream, and six throughflow pits, for several storm events in September 1983. Events sampled had return periods of between four weeks and three months (i.e. high-frequency events). Isotope hydrograph separations of the M6 and M8 streams indicated that old water dominated runoff from all of the events. New water was approximately 15–25% of the stormflow. New water contributions to quickflow in M8 could be accounted for by flow from less than 10% of the catchment area; not much larger than the area capable of generating saturation overland flow by Mosley (1979). The EC and Cl^- data again supported isotopic results by indicating no significant dilution of the stream by new water. Flushing of water with high EC and Cl^- values into the stream was pronounced in M8. This was characterized by marked increases in EC and Cl^- on the rising limb of the hydrograph even though concentrations in the rain were considerably lower than those in the stored water (Figure 2.5). The effect was much decreased in the second of closely spaced storm events. Stored water appeared to dominate outflow as shown by variations in EC and Cl^- concentrations in subsequent storms, reflecting the differing contact times between stored water and the soil matrix.

2.4.3 Sampling at the Pit Face Scale

The collected hillslope water displayed large spatial variability in $\delta^{18}O$ composition. The pit locations (used originally by Mosley 1979) represented a cross-section of hillslope positions and topographic hydrologic conditions. The values of δD during both low-flow periods and in response to storm events showed that pit throughflow was dominated by old water (Table 2.1). Deep suction lysimeter δD values (Table 2.1; location shown in Figure 2.2) were much lighter isotopically than shallow lysimeter and throughflow δD samples. This suggested that much of the throughflow

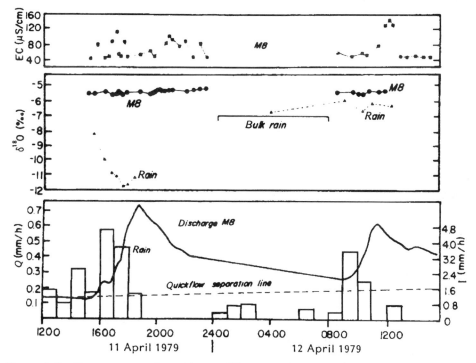

Figure 2.5 Storm response in catchment M8, 11–12 April 1979 (from Pearce *et al.* 1986)

Table 2.1 Catchment M8 old and new water contributions to throughflow 21 September 1983 (from Sklash *et al.* 1986)

Site	δD_O(%)	δD_N(%)	δD_P(%)	Time of sample, h	% Old water	% New water
Pit 1	−43.1[a]	−12.2[c]	−33.7	14:10–17:52	70	30
Pit 2	−43.1[a]	−12.2[c]	−31.4	14:10–17:52	62	38
Pit 3	−43:1[a]	−12.2[c]	−29.2	14:10–17:52	55	45
Site A	−42.3	−12.2[c]	−32.9	14:02	69	31
Pit 5	−45.5	−12.2[c]	−43.5	14:06	94	6
Seep	−44.0	−12.2[c]	−41.3	14:01	92	8
Site D	−39.1[b]	−12.2[c]	−32.4	14:18	75	25
M9	−41.3	−12.2[c]	−32.5	14:58	70	30

δD_O, old water δD; δD_N, new water δD; δD_P, δD of throughflow at peak discharge.
[a]Based on SL5S value.
[b]Low flow value on 23 September.
[c]Weighted average rain.

arrived laterally from thinner soil profiles (Figure 2.6). At the same sites, the EC and Cl⁻ values showed noticeable differences in the relative influence of solute flushing at the hillslope pit sites.

2.4.4 A Proposed Perceptual Model of Hydrologic and Isotopic Response

Sklash *et al.* (1986) measured large water-table rises in the mid-slope and near stream max-rise piezometers during storms (Figure 2.6; locations shown in Figure 2.2). The piezometers located near the valley bottom had the highest response, close to achieving surface saturation. Visual observations confirmed that overland flow occurred only in valley-bottom areas. Sklash *et al.* (1986) hypothesized that two mechanisms could possibly account for the large water-table rises: (1) conversion of capillary fringe into phreatic water (as had been observed elsewhere by Sklash and Farvolden 1979; Gillham 1984; Abdul and Gillham 1984), or (2) rapid lateral inflow of displaced old water into areas of deep soil from areas of shallower soil. Both mechanisms appear to be triggered by new water infiltration but old water from the saturated zone still dominated storm runoff (Pearce *et al.* 1986; Sklash *et al.* 1986). The response of the maximum-rise piezometers in the M8 catchment was consistent with the concept of groundwater ridging. Saturated wedges on the lower slopes and groundwater ridges in the valley bottoms were thought by Sklash *et al.* (1986) to develop quickly as infiltrating rain converted the tension-saturated zone into phreatic water. This perceptual model negated the need to invoke rapid transmission of new water downslope via macropores in order to explain the streamflow response, since stored water was the main component discharged into the stream channel during events (Figure 2.4(b)).

With hindsight, we know that the Pearce *et al.* (1986) and Sklash *et al.* (1986) perceptual model was limited by the lack of any soil physics data to confirm that M8 soils indeed had a tension saturated zone. Only small events (return periods of <6 months) were studied and no direct evidence of groundwater ridging was presented

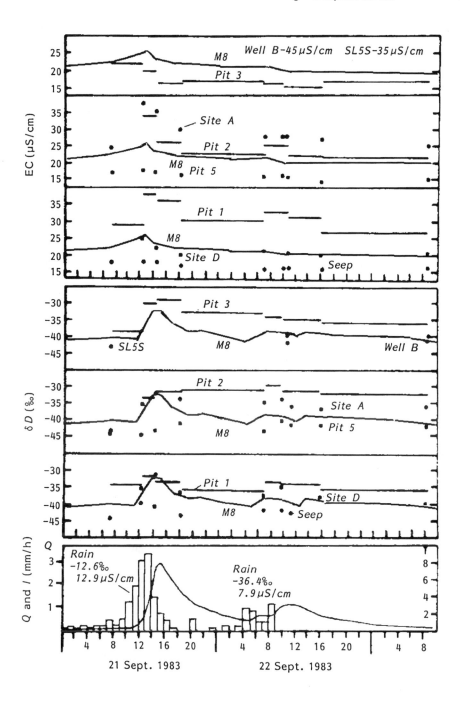

Figure 2.6 Comparison of stream (M8) and hillslope δD and EC response, 21 September 1983, storm (from Sklash *et al.* 1986)

apart from point observations from maximum-rise wells. Pearce *et al.* (1986) and Sklash *et al.* (1986) discounted pipeflow measurements of Mosley entirely, possibly because significant amounts of macropore flow did not necessarily occur during the small-magnitude events monitored.

2.5   A COMBINED HILLSLOPE SOIL PHYSICS, ISOTOPIC AND CHEMICAL APPROACH

2.5.1   Soil Potential Response in Hydrologically Active Hollows

McDonnell (1990a) and McDonnell *et al.* (1991a,b) combined isotope and chemical tracing with detailed tensiometric recording in near-stream, mid-hollow and upslope hollow positions in the M8 catchment (Figure 2.2) in an effort to explain the discrepancies between the perceptual models offered by Mosley (1982), Pearce *et al.* (1986) and Sklash *et al.* (1986). In the hydrologically active mid-slope hollows, (e.g. Pit 5 as shown in Figure 2.2), McDonnell (1990b) found that soil potential response was highly variable for different storm magnitudes, intensities and pre-storm matric-potential conditions. Tensiometric measurements revealed an erratic infiltration–potential relationship (McDonnell 1989). During a low-magnitude (25 mm) rainfall even on 23–24 October 1987 (Figure 2.7(a)), tensiometric data showing a semi-constant wetting front propagate through the profile with strong soil potential response lags with depth. Although some bypass flow seemed to occur in the upper soil horizon (< 50 cm), as evidenced by the response of tensiometer T5 (see Figure 2.7 caption for explanation), rainfall depth and soil-moisture content were low enough so that the lower soil depths did not receive appreciable moisture from above, until streamflow response had subsided (Figure 2.7(a)). Therefore, a slope water-table did not develop.

During a larger magnitude (58 mm) rainfall event on 29 October, pressure potential in the lower soil horizons (> 75 cm) responded almost instantaneously to infiltrating rain (Figure 2.7(b)). This response was a function of a disequilibrium in soil pressure potentials during wetting, caused by the presence of soil macropores (McDonnell 1991). Furthermore, much of the matrix exhibited unrequited storage during this type of wetting, indicative of a two-component flow system of rapid macropore flow and slow matrix flow. For the largest event monitored (103 mm of rainfall) on 13 October, soil-pressure potential remained, relatively constant throughout the profile, during the limited period of tensiometer coverage (Figure 2.7(c)). McDonnell (1990b) observed that most of the soil profile remained saturated during this episode.

2.5.2   Groundwater Development and Longevity

Generally, when intensities were low, but pre-storm soil water content was high, McDonnell (1989) found that additional rainfall rapidly filled the available soil-moisture storage, and perched water-table conditions quickly developed at the soil–bedrock interface. On the other hand, if short-term rainfall intensitites were high, rainfall bypassed the upper soil horizons and moved to the profile base via vertical cracks, so that tensiometers in the lower half of the soil profile responded ahead, or independent of, the upper tensiometers. Water-table longevity was very short and

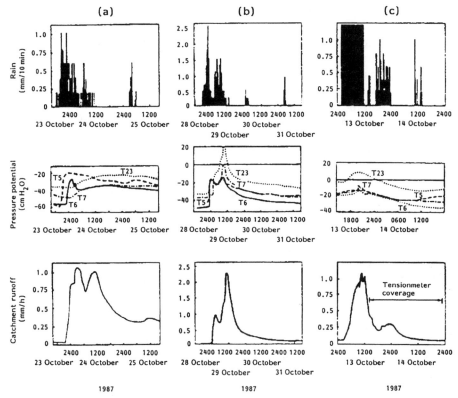

Figure 2.7 Pressure potential response for tensiometers located in instrumented hollow, showing relationship between matric potential (ψ) and rainfall-catchment runoff condition. Three storms are shown, having rainfall totals of (a) 25 mm, (b) 58 mm and (c) 103 mm. Soil depth 1–1.5 m, slope angle 35–40°. Tensiometers T5, T6, T7 and T23 inserted at 170, 410, 820 and 1080 mm below soil surface respectively (from McDonnell 1990b)

showed a close correspondence with hillslope throughflow rate, as measured by McDonnell (1989, 1990a). Downslope drainage of perched water was extremely efficient and showed no lag with recorded throughflow for select storms. McDonnell *et al.* (1991a) noted that this indicated that lateral saturated flow was rapid and moved through pipes (corroborated visually using dye tracers) formed at the soil–bedrock interface. The rapidity of tensiometric recession in the lower half of the soil profile in events with perched water-table conditions, supported the idea of rapid downslope drainage through pipes (McDonnell 1989). The interconnectedness of pipes in those zones was assumed by McDonnell (1990a) to be high enough to account for the rapidity of water-table decline.

2.5.3 Mechanics of Preferential Flow

McDonnell (1990a) reasoned that in the steeply sloping hollow zones (where much of the M8 stream runoff originated), bypass flow leads to the soil truly "releasing" water long before wetting along a measured wetting–drying curve would predict. This was due to the pressure potential disequilibrium within the soil. McDonnell (1990a) noted that it was important to distinguish between the conductivity of the matrix (K^*) and that of the soil with macropores (K_{sat}). If the flux density of the rain (V_o) is greater than K^*, local ponding would eventually occur leading to vertical bypassing, whether or not V_o is greater than K_{sat}. Therefore, McDonnell (1990a) found that it was not unrealistic for 5–10 mm/h rainstorms to create localized ponding on a soil purported by Mosley (1979) to have a K_{sat} of 100–200 mm/h. It simply meant that K^*, the appropriate matrix property, was less than 5–10 mm/h. McDonnell (1990a) argued that local bypassing required only that $V_o > K^*$.

2.5.4 A Perceptual Model of Macropore Flow and Old Water Displacement

McDonnell (1990b) noted that as invading new water moved to depth, free water perched at the soil–bedrock interface, as water "backed-up" into the matrix, where it mixed with a much larger volume of stored old matrix soil water (Figure 2.4(c)). Pressure potential evidence from the responsive mid-slope hollow (Pit 5; Figure 2.2) showed that this water-table was dissipated by the moderately well-connected system of pipes at the mineral soil–bedrock interface. The relationship between crack infiltration and lateral pipeflow was not linear, because there was a significant time delay between water-table perching and subsequent distribution of positive pore pressure in the soil. McDonnell (1990a) reasoned that this delay was the critical process necessary to shift the new water signatures to that of old water at the hillslope scale.

Isotopic data from Pit 5 throughflow (Sklash et al. 1986; McDonnell et al. 1991b) showed that old water dominated subsurface flow at these mid-slope hollow sites by up to 85%. McDonnell (1990a) reasoned that the pipes distributed this mixture of newly bypassed rainfall and mixed stored water downslope to the first-order channel bank. The shift from new to old water was expected to occur on the slope, as indicated in Figure 2.8. Stewart and McDonnell (1991) showed that between-storm matrix water varied in age from approximately one week at the catchment divide (near Site A; Figure 2.2) to over 100 days at the main M8 channel margin, supporting the notion of very short hillslope water residence time. (A subsequent hillslope excavation project in 1992 by an EarthWatch research team (McDonnell, unpublished data) revealed that soil pipes at the soil–bedrock interface are not continuous beyond about 25 cm, thus affecting the applicability and acceptance of the above-stated perceptual model.)

2.6 WHOLE HILLSLOPE TRENCHING AND FLOW COLLECTION

2.6.1 The Trench Excavation

Woods and Rowe (1996) established a subsurface collection system along the base of a hillslope hollow on the left bank of the stream, draining the M8 catchment (Figure 2.2). A vertical face 60 m long and 1.5 m high was cut across the toe of the hillslope. Thirty subsurface flow collection troughs were installed end-to-end, across the base

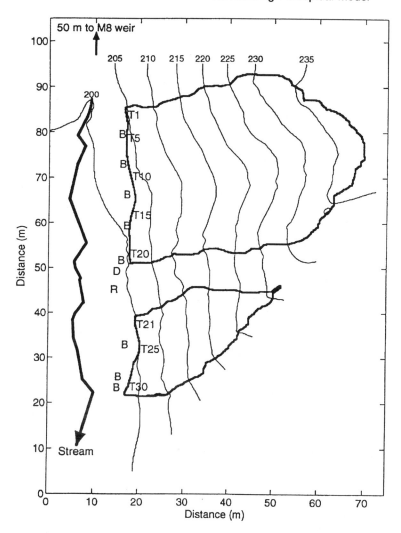

Figure 2.8 M8 hillslope trench study; topography and location of selected troughs and measurement equipment: T*n*=trough *n*, B=tipping bucket stand, D=datalogger, R=rain-gauge (from Woods and Rowe 1996)

of the excavated face at the soil–bedrock interface (Figure 2.8). The troughs were sealed to the cut face and covered. Flow collected in the troughs was routed to tipping bucket flowmeters and recorded from November 1992 to 1993.

2.6.2 Subsurface Flow Volumes

Subsurface flow from the hillslope was highly variable in both magnitude and timing. Neighbouring collection troughs showed unexpected differences in flow

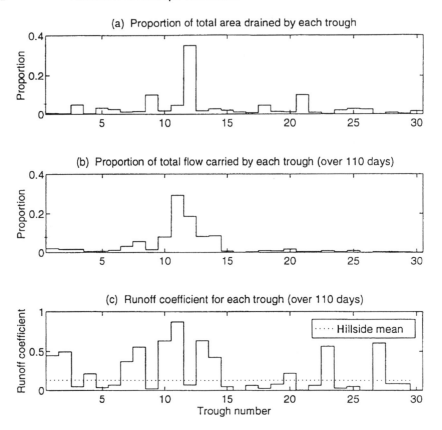

Figure 2.9 Spatial distribution of trough characteristics: (a) subcatchment area; (b) total flow over entire study period; (c) runoff coefficients (from Woods and Rowe 1996)

rates. Figure 2.9 shows the spatial distribution of the subcatchment areas and proportion of total flow among the troughs. Woods and Rowe (1996) noted that the troughs draining the central hollow displayed higher than expected volumes of flow. Consequently, the amount of subsurface flow drained by each trough did not always correspond with the estimated upslope subcatchment (trough catchment) area (Figure 2.9). In particular, Woods and Rowe (1996) estimated that trough 12 (Figure 2.9) would drain 35% of the hillslope, yet did not dominate hillslope flow. Trough 11 drained only 5% of the total hillslope area but yielded the highest flow volumes during events (Woods and Rowe 1996). This suggested that surface topography alone did not determine the trough subcatchment divide or the subsurface contributing area. Trough runoff coefficients were also variable across the hillslope (Figure 2.9(c)). When the troughs were grouped by topographic feature (convergence and divergence), variability in flow across the hillslope cut face was still observed (Woods and Rowe 1996).

2.6.3   Isotopic Sampling along Trench Face

Stewart and Rowe (1994) applied a lumped-parameter approach to model the residence time of subsurface water in the trough subcatchments. Rainfall $\delta^{18}O$ was compared to soil water $\delta^{18}O$ collected from suction lysimeters on the slope. Shallow-depth suction lysimeters were the most responsive isotopically to rainfall inputs. Residence times increased while isotopic response to rainfall input decreased with soil depth.

Results of the Stewart and Rowe (1994) model applied to the M8 stream for a sampled storm during January 1994, was discussed in Unnikrishna *et al.* (1995) (Figure 2.10). When this model was applied to the trough system, isotopic results displayed a large variation across the trough face, similar to the flow variations (Stewart, pers. comm., 1994). Low-flow trough (end troughs away from the central hollow) $\delta^{18}O$ values showed that 60% of the flow was event water, but this water had a relatively long residence time. The distribution had a peak at 4–5 hours, a mean of 28 hours; 50% of the water discharged had a residence time of less than 12 hours. Flow commenced about four hours after the onset of rain when sufficient water had infiltrated the soil. The results from the high-flow troughs were very different. The water was initially more depleted isotopically ($\delta^{18}O$ about -6.2‰) than the stream, indicating a source from groundwater with a much longer residence time. After 8 hours of rainfall, the $\delta^{18}O$ changed to -5.0‰ and remained constant for 16 hours, before gradually shifting back toward the groundwater composition after 36 hours. Stewart and Rowe (1994) assumed that groundwater supplied a near-constant flow to the stream that gradually had soil water added to it after rain has been falling for some hours. Groundwater again dominated the storm hydrograph as the soil-water flow decreased.

2.6.4   A Perceptual Model of Hillslope Flow: A Single Pit does not a Hillslope make

The spatial variability across the trough face, even when grouped by topographic features, revealed that the flow data of small troughs are difficult to extrapolate to larger scales. This study makes apparent the fact that a model of hillslope response must adequately describe the spatial variability, but also take into account the effect of scale and the physical controls on the production of runoff. For subsurface flow, Woods and Rowe (1996) reasoned that it was not an "effective" contributing area which determined the runoff (as is the case with infiltration excess or saturation excess flow), but rather the size of the saturated soil moisture storage.

Although very useful, the Woods and Rowe (1996) study treated the hillslope as a "black box" and did not collect data on flow paths and mechanisms of flow to the troughs. Stewart and Rowe (1994) restricted their isotope sampling to the stream and specific high-flow and low-flow troughs; they did not investigate the spatial variability across the trough face or soil-water isotopic compositions on the slope.

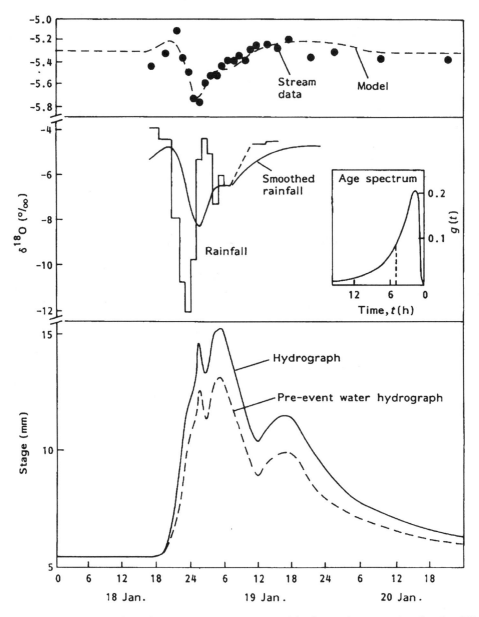

Figure 2.10 Event based water age spectrum and hydrograph separation for the M8 Maimai catchment (from Unnikrishna *et al.* 1995)

2.7 A HILLSLOPE-SCALE BROMIDE TRACER INJECTION

2.7.1 The Br⁻ Line Source Injection

During March to May 1995, Brammer (1996) and Brammer *et al.* (1996) performed a hillslope-scale Br⁻ tracer experiment on the trenched hillslope described by Woods and Rowe (1996). A 5 m grid of suction lysimeters and co-located maximum-rise wells were installed in a 30×30 m plot 5 m upslope from the trench face (Figure 2.11). A line source of 3 kg LiBr was applied 35 m upslope from the trough face. Antecedent moisture conditions were relatively high; 17 mm of rainfall occurred in the previous two days. The first rainfall event occurred 12 hours after the tracer application.

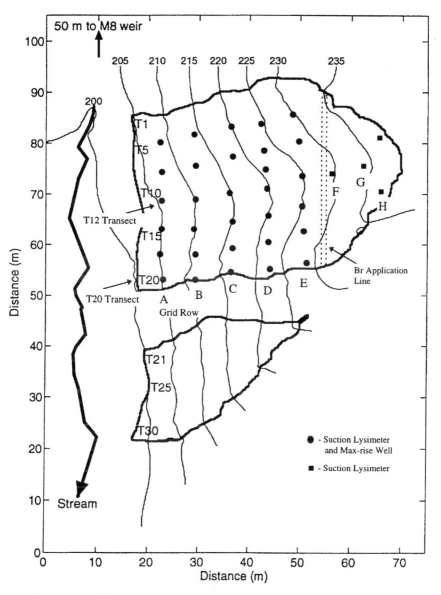

Figure 2.11   M8 hillslope plot (based on figure from Woods and Rowe 1996)

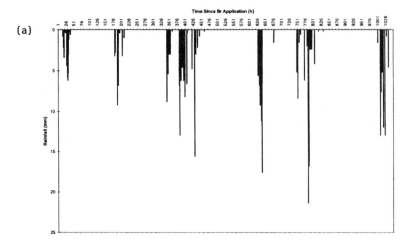

(a)

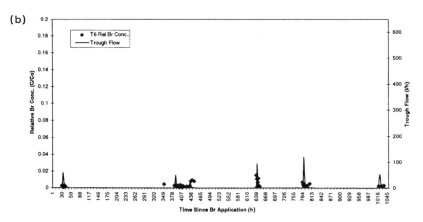

(b)

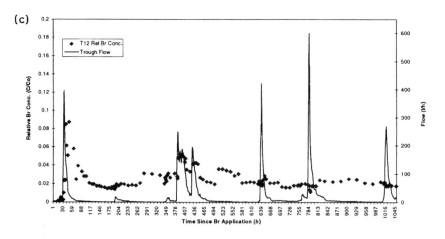

(c)

2.7.2 Surface versus subsurface controls on Br⁻ export off the plot

Surface topography alone did not fully explain the subsurface flow response (as Woods and Rowe (1996) found) nor Br⁻ concentrations across the trench face. Furthermore, spatial patterns of water-table and Br⁻ concentration on the hillslope grid showed poor correlations with topographic position. Depths to bedrock were measured at a 2.5 m grid-scale over the hillslope. Soil depths within the grid became shallower downslope near the trough face forming an impeding "lip". A midslope bedrock depression appeared to act as a subsurface storage reservoir. The mapped spatial water-table response and groundwater persistence in this zone reflected this drainage pattern. Maximum-rise wells indicated a rapid rise and fall in water-table levels along the edges of the grid. The highest levels were recorded in the hollow with a persistent water-table (of up to four days), detected up the central hollow transect, 15 m upslope from troughs 11 and 12. This may explain the long flow recession times for these troughs, as observed by Woods and Rowe (1996).

Within the grid, the tracer plume reached the first row of the suction lysimeters six days after application, consistent with the average soil hydraulic conductivity on the slope of 30 mm/h. High Br⁻ concentrations were observed at the edges of the grid, apparently due to bedrock topographic control of subsurface flow. In general, Br⁻ concentration did not follow expected spatial patterns based on surface topography. Increased tracer concentrations were not observed in the hollow wells or suction lysimeters, despite their location within the estimated subcatchment areas receiving significant drainage from upslope.

2.7.3 Br⁻ Breakthroughs: Effect of Diffusion into Matrix depending on Antecedent Wetness Conditions

Flow volume variability across the trough face was similar to observations by Woods and Rowe (1996). Similarly, Br⁻ tracer arrival and concentrations displayed spatial and temporal variability related to flow. Br⁻ was detected in the outflow from the troughs draining the central hollow (troughs 11–14) 18 hours after the start of rainfall, indicating rapid movement downslope. During the first event, Br⁻ concentrations arriving at the high-flow troughs were the highest measured during the study period, while samples from low-flow troughs (troughs 4–8 and 16–20) did not show detectable Br⁻ concentrations until the third storm event, 14 days after application (Figure 2.12). Subsequent events did show increases in Br⁻ concentration with time, indicating remobilization of Br⁻ tracer from the soil matrix; remobilization appeared sensitive to antecedent soil-moisture conditions. It is important to note, however, that tracer breakthrough at the trough face preceded appearance of Br⁻ in the upslope suction lysimeters by 2 days and maximum-rise wells by 7 days, suggesting rapid downslope preferential flow of 10's of metres.

Figure 2.12 (*opposite*) Comparison of Br⁻ concentration breakthrough curves across the hillslope: (a) study period rainfall; (b) low flow (trough 6); (c) high flow (trough 11) (from Brammer *et al.* 1996)

2.7.4 A flowpath-based perceptual model of 3D hillslope flow across catchment

The maximum-rise well response and subsurface flow Br⁻ concentrations support the notion of bedrock microtopographic depressions at the soil–Old Man Gravel interface. The less mobile water is mobilized during large events, but effectively isolated from downslope saturated–unsaturated flow between events. The interaction of these mobile and immobile regions of both soil and lateral subsurface flow system is illustrated in Figure 2.4(d)). The maximum-rise wells located in deep soil slope positions were the most responsive to rainfall, but did not contain significant tracer concentrations. The Br⁻ concentration response of the grid wells and suction lysimeters at the edge of the plot suggests opportunity for inter-microcatchment transfer of water and flow.

2.8 CONCLUSIONS

This chapter synthesizes the progression of ideas in the development and modification of a perceptual model of hillslope flow at the Maimai catchment. Each data set reviewed in the chapter reveals not only a cumulative understanding of catchment behaviour, but alternative interpretations of the hillslope subsurface flow system. The initial single-technique approaches of Mosley (1979) show the limitations and often misleading interpretations from dye tracers alone. Subsequent isotopic studies by Pearce *et al.* (1986) and Sklash *et al.* (1986) showed clearly that stored water comprised the majority of channel stormflow; notwithstanding, their isotope-oriented approach did not enable them to develop a mechanistic under-standing of the processes. Studies that followed (McDonnell 1990a,b; McDonnell *et al.* 1991a,b) demonstrated that an integration and combination of techniques was required for comprehensive hillslope characterization and reconciliation of different process interpretations. Although single throughflow pits continued to be the indicator of subsurface flow timing and magnitude, further study by Woods and Rowe (1996) showed that flow varied widely across a slope section – making the single pit observations of the previous studies highly suspect. Most recent observations by Brammer *et al.* (1996) reveal much microtopographical control on subsurface flow timing and tracer breakthrough. Whilst previous works have treated the soil–bedrock interface as a sharp boundary, Brammer *et al.* (1995) demonstrated that small depressions in the bedrock surface exert a large control on water mobility and mixing; thereby producing localized "dead zones" of longer residence time waters.

REFERENCES

Abdul, A.S. and Gillham, R.W. 1984. Laboratory studies of the effects of the capillary fringe on stream flow generation. *Water Resour. Res.*, **90**, 691–698.

Beven, K.J. 1991. Spatially distributed modeling: conceptual approach to runoff prediction. In D.S. Bowles and P.E. O'Connell (eds), *Recent Advances in the Modeling of Hydrologic Systems*. Kluwer Academic Publishers, pp. 373–387.

Bonnell, M. 1993. Progress in the understanding of runoff generation dynamics in the forests. *J. Hydrol.*, **150**, 217–275.

Brammer, D.D. 1996. Hillslope hydrology in a small forested catchment, Maimai, New Zealand. *MS thesis*, State University of New York, College of Environmental Science and Forestry, Syracuse, NY, in prep.

Brammer, D.D., McDonnell, J.J., Kendall, C. and Rowe, L.K. 1995. Controls on the downslope evolution of water, solutes and isotopes in a steep forested hillslope. *American Geophysical Union, EOS*, **76**(46), 268.

Brammer, D.D., McDonnell, J.J., Kendall, C., and Rowe, L.K. 1996. Physical controls on the downslope movement of water and tracer in a forested hillslope, Maimai, New Zealand, *Water Resour. Res.*, in prep.

Buttle, J. 1994. Isotope hydrograph separations and rapid delivery of pre-event water from drainage basins. *Prog. In Phys. Geog.*, **18**, 16–41.

Gillham, R.W. 1984. The effect of the capillary fringe on water table response. *J. Hydrol.*, **67**, 307–324.

Hewlett, J.D. and Hibbert, A.R. 1967. Factors affecting the response of small watersheds to precipitation in humid areas. *Proceedings of 1st International Symposium on Forest Hydrology*, pp. 275–290.

Hewlett, J.D., Cunningham, G.B. and Troendle, A.C. 1977. Predicting stormflow and peakflow from small basins in humid areas by the R-index method. *Water Resour. Bull.*, **13**, 231–253.

McDonnell, J.J. 1989. The age, origin and pathway of subsurface stormflow in a steep, humid headwater catchment. *PhD thesis*, Univ. of Canterbury, Christchurch, New Zealand, 270pp.

McDonnell, J.J. 1990a. A rationale of old water discharge through macropores in a steep, humid catchment. *Water Resour. Res.*, **26**, 2821–2832.

McDonnell, J.J. 1990b. The influence of macropores on debris flow initiation. *Q. J. Eng. Geol.*, **23**, 325–331.

McDonnell, J.J. 1991. Preferential flow as a control of stormflow response and water chemistry in a small forested watershed. In *Preferential Flow*, Proceedings of the National Symposium, ASAE, Chicago, Illinois, 16–17 December 1991, pp. 50–58.

McDonnell, J.J., Owens, I.F. and Stewart, M.K. 1991a. A case study of shallow flow paths in a steep zero-order basin: a physical–chemical–isotopic analysis. *Water Resour. Bull.*, **27**, 679–685.

McDonnell, J.J., Stewart, M.K. and Owens, I.F. 1991b. Effects of catchment-scale subsurface watershed mixing on stream isotopic response. *Water Resour. Res.*, **26**, 3065–3073.

Mosley, M.P. 1979. Streamflow generation in a forested watershed, New Zealand. *Water Resour. Res.*, **15**, 795–806.

Mosley, M.P. 1982. Subsurface flow velocities through selected forest soils, South Island, New Zealand. *J. Hydrol.*, **55**, 65–92.

O'Loughlin, C.L., Rowe, L.K. and Pearce, A.J. 1978. Sediment yields from small forested catchments, north Westland-Nelson, New Zealand. *J. Hydrol. (N.Z.)*, **17**, 1–15.

Pearce, A.J. and Rowe, L.K. 1979. Forest management effects on interception, evaporation and water yield. *J. Hydrol. (N.Z.)*, **18**, 73–87.

Pearce, A.J. and Rowe, L.K. 1981 Rainfall interception in a multi-storied evergreen mixed forest: estimates using Gash's analytical model. *J. Hydrol.*, **48**, 341–353.

Pearce, A.J., Stewart, M.K. and Sklash, M.G. (1986). Storm runoff generation in humid head-water catchments 1. Where does the water come from? *Water Resour. Res.*, **22**, 1263–1272.

Rowe, L.K. 1979 Rainfall interception by a beech–podocarp–hardwood forest near Reefton, north Westland, New Zealand. *J. Hydrol. (N.Z.)*, **18**, 63–72.

Rowe, L.K. and Pearce, A.J. 1994. Hydrology and related changes after harvesting native forest catchments and establishing *Pinus radiata* plantations. Part 2. The native forest water balance and changes in streamflow after harvesting. *Hydrol. Process.*, **8**, 281–297.

Rowe, L.K., Pearce, A.J. and O'Loughlin, C.L. 1994. Hydrology and related changes after harvesting native forest catchments and establishing *Pinus radiata* plantations. Part 1. Introduction to study. *Hydrol. Process.*, **8**, 263–279.

Sklash, M.G. and Farvolden, R.N. 1979. The role of groundwater in storm runoff. *J. Hydrol.*, **43**, 45–65.

Sklash, M.G., Stewart, M.K. and Pearce, A.J. 1986. Storm runoff generation in humid headwater catchments 2. A case study of hillslope and low-order stream response. *Water Resour. Res.*, **22**, 1273–1282.

Stewart, M.K. and McDonnell, J.J. 1991. Modelling baseflow soil water residence times from deuterium concentrations. *Water Resour. Res.*, **27**, 2681–2694.

Stewart, M.K. and Rowe, L.K. 1994. Water component analysis of runoff and soil water flows in small catchments. In *Tracer Modeling*, Proceedings of the Western Pacific Geophysics Meeting, p. 37.

Unnikrishna, P.V., McDonnell, J.J. and Stewart, M.K. 1995. Soil water isotopic residence time modelling. In S.T. Trudgill (ed). *Solute Modelling in Catchment Systems*. John Wiley, New York, pp. 237–260.

Webster, J. 1977. The hydrologic properties of the forest floor under beech/podocarp/hardwood forest, North Westland. *MAgSci thesis*, University of Canterbury, New Zealand.

Woods, R. and Rowe, L.K. 1996. Consistent temporal changes in spatial variability of subsurface flow across a hillside. *J. Hydrol. (NZ)*, in press.

3 Runoff and Erosion from a Rapidly Eroding Pinyon–Juniper Hillslope

BRADFORD P. WILCOX,[1] JOHN PITLICK,[2] CRAIG D. ALLEN[3] and DAVID W. DAVENPORT[1]

[1] *Environmental Science Group, Los Alamos National Laboratory, USA*
[2] *Department of Geography, University of Colorado, USA*
[3] *National Biological Service, Jemez Mountain Field Station at Bandelier National Monument, Los Alamos, USA*

3.1 INTRODUCTION

Pinyon–juniper woodlands are extensive in the western United States, covering around 24 million ha. Although the range of these woodlands has fluctuated considerably over the last 12 000 years, largely in response to climate change, their expansion and increase in density during the last 100 years has been unprecedented (Miller and Wigand 1994). A number of explanations for this expansion have been put forward, including overgrazing, fire control, climate change, and higher concentrations of carbon dioxide.

An important ramification of the spread and the increase in density of pinyon–juniper woodlands is a decline in understorey vegetation. These trees – juniper in particular – have widely spread lateral root systems and are able to out-compete many of the herbaceous species for the limited water and nutrient resources (Johnsen 1962; Gottfried *et al.* 1995). A diminished understorey often results in a dramatic acceleration of soil erosion, especially in the more xeric locations, such as south-facing slopes and areas where soils are shallow (Miller and Wigand 1994). Accelerated erosion in pinyon–juniper woodlands represents a threat to the long-term stability and productivity of these regions. But despite widespread recognition of this threat, there have been few, if any, sustained efforts to study the phenomenon and the runoff dynamics associated with it. Schmidt (1987) noted that "other than the water yield studies at Corduroy Creek and Beaver Creek in Arizona, most of the remaining work has been conducted as small plot studies in Utah and Nevada." Recognizing that an adequate understanding of runoff and erosion processes cannot be obtained without more long-term, catchment-scale studies, he makes a forceful plea for such work.

Because our ability to find effective solutions to many environmental problems depends on understanding how erosion processes behave and the factors that control those processes, we have initiated a number of hillslope- and catchment-scale hydrologic studies on the Pajarito Plateau in northern New Mexico (Wilcox and

Advances in Hillslope Processes, Volume 1. Edited by M. G. Anderson and S. M. Brooks.
© 1996 John Wiley & Sons Ltd.

Breshears 1995), one of which focuses on a rapidly eroding pinyon–juniper catchment (1 ha) within Bandelier National Monument. The vegetation in this area has changed markedly over the past century, from an open forest dominated by ponderosa pine (*Pinus ponderosa*) to a woodland entirely composed of pinyon (*Pinus edulis*) and one-seed juniper (*Juniperus monosperma*). Since July of 1993, we have been gathering data on the hydrology, geomorphology, ecology and soils of the site. These data indicate greatly accelerated erosion of the intercanopy soils. We hypothesize that this accelerated erosion was triggered by a reduction in ground cover resulting from the combined and interrelated effects of overgrazing, increase in the density of pinyon and juniper, and a severe drought in the 1950s.

The long-term objectives of the study are to (1) estimate water and sediment budgets for rapidly eroding semi-arid woodlands; (2) determine the effect of 20th century vegetation change on erosion processes; (3) develop a conceptual and quantitative understanding of the relationships between runoff and erosion; and (4) determine what effect scale has on runoff and erosion in rapidly eroding semi-arid woodlands. Given the current scarcity of data, this information will prove valuable for evaluating runoff and erosion models for these ecosystems. Although we recognize that these long-term objectives have not been fully attained less than three years into the study, our results to date have proved most interesting and have added significantly to our knowledge of runoff and erosion dynamics in these important woodlands.

3.2 STUDY AREA

The subject catchment is located in northern New Mexico, within Bandelier National Monument (Figure 3.1). The elevation of the area ranges from 1969 to 1990 m; average annual precipitation for this elevation on the Pajarito Plateau is around 360 mm (Bowen 1990). Soils on the site developed on tuff residua (although there have probably been small inputs of material from aeolian sources) and pumice. An open ponderosa pine forest dominated the site until the late 1800s, when livestock grazing and an associated reduction in fire frequency (from both fire suppression and reduced ground fuel) allowed pinyon and juniper to markedly increase in density (Allen 1997). The growing pinyon–juniper population probably caused a further decline in herbaceous cover, which reached a critical point with the severe drought of the 1950s; this drought also killed all the ponderosa pine. A large feral burro population, present in the area from 1940 to 1980, further stressed the herbaceous cover. These events apparently triggered the current episode of accelerated erosion. Today, fallen ponderosa pine logs are scattered across the area, but there are no live ponderosa trees. Clumps of pinyon and juniper trees provide canopy cover over about half the catchment. The slope of the catchment is around 5%. Ground surfaces in intercanopy patches are mostly bare soil or rock, while those beneath the canopy are mostly needle litter. In addition to the extensive patches of bare soil and the scant vegetation coverage, evidence of accelerated erosion includes numerous hillslope channels, soil pedestals, and exposed subsoils. Although channelling is extensive, the proximity of bedrock to the soil surface prevents the formation of

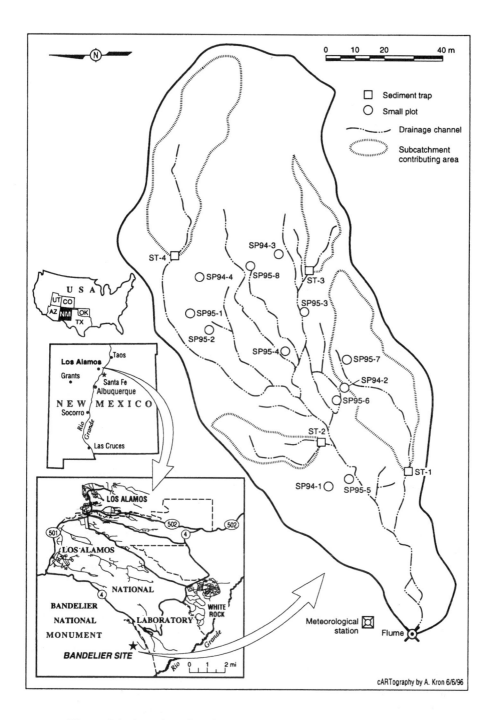

Figure 3.1   Location of study area, including details of Bandelier site

gullies or deeply incised channels. Most of these hillslope channels are less than 30 cm wide and 10 cm deep.

3.3 METHODS

3.3.1 Characterization of Vegetation and Soils

Vegetation and surface soil conditions were characterized by establishing three line-intercept transects (Mueller-Dombois and Ellenberg 1974) along the contour and across the width of the catchment, at high-, mid- and low-elevation positions. The transects, which measure 50 m, 65 m and 40 m in length, respectively, are measured by stretching a fibreglass tape along the ground between permanently marked end-points. The nature of the surface cover and of overstorey layers is recorded at 1-cm intervals along one edge of the tape. These data are electronically recorded in the field, using such categories as plant species, bare soil, rock, (plant) litter, wood, cryptogamic crust, etc.

Subsurface soil conditions were characterized from 19 soil pedons, exposed in pits and auger holes. For each horizon, we recorded colour, texture, structure (in pits), dry consistence, root density, percentage and type of coarse fragments, and boundary characteristics.

3.3.2 Hydrometric and Erosion Measurements

We have established a network for monitoring runoff, erosion and weather conditions (the types and amount of equipment that can be used for measurement are somewhat limited, because the catchment lies within a designated wilderness area and is rather remote).

Weather, microtopography, land survey, and runoff data have been collected on the site since July 1993. A solar-powered weather station continuously measures solar radiation, ambient air temperature, relative humidity, wind speed and direction, and precipitation (the latter by means of a heated tipping-bucket rain gauge).

Changes in microtopography caused by erosion are monitored at 10 permanent sites, located to represent the range of intercanopy cover conditions within the catchment. Within each site are two 2-m-long transects, about 1 m apart; the end-points of each are marked with vertically installed rebar. A carpenter's level, modified at each end to fit over the rebar and drilled through at 8-cm intervals along its length, is placed on top of the rebar. A thin aluminium rod is then slid through each of the 20 drilled holes, and the distance from the top of the rod to the level is measured (Shakesby 1993). By taking these measurements periodically (after the spring thaw, before the summer rains, occasionally during the summer rainy season, and in the late autumn), changes in surface elevations can be mapped.

The catchment has been surveyed in detail to map the major drainages and other topographic features.

Individual runoff events from the catchment are measured by means of a flume installed in a bedrock-floored segment of the main channel, above the point at which the channel drops into a canyon. Following the design of Replogle *et al.* (1990), the flume is constructed from a 4-m-long piece of 38-cm PVC pipe; the floor of the pipe has a flat concrete sill that forces the flow to critical depth (Froude number = 1.0) as it exits the pipe. Water height in the flume is measured by a pressure transducer located in an adjacent stilling well. Because of the high sediment and debris load from this catchment, however, there have been some problems in measuring runoff, especially for the larger events. The pipe connecting the flume and the stilling well becomes clogged with sediment when runoff is receding, with the result that for the last one-half to one-third of the hydrograph recession limb, the flow must be estimated. In addition, for the largest runoff events, debris may accumulate in front of the flume, forcing water around it (this has happened on at least one occasion).

Concurrent with installation of the flume, a pit was excavated immediately upstream to capture sediment being transported in the channel. However, with its 0.4-m^3 capacity, the pit was found to be too small to trap all the sediment leaving the catchment when discharge was moderate or high. Because it was not practical to enlarge the pit, which was already dug into bedrock, in 1995 we installed four additional sediment traps, each at the base of a tributary channel within the catchment (Figure 3.1). The traps are lined with wood and have a storage capacity of 1 m^3. The four contributing areas (subcatchments) range from 300 to 1100 m^2.

In 1995 we also completed a network of 12 small (1 m^2) runoff plots; each is equipped, along its downstream end, with a gutter that catches the runoff and channels it into a bucket set into the ground. After each runoff event, the volume of runoff is recorded; two litres of the water (if available) is then reserved for measurement of sediment concentration. Ten of the plots were established in the intercanopy areas having tuff residua soils, one was established under a tree canopy, and one was established in an intercanopy area having high-pumice soil.

3.4 RESULTS

3.4.1 Vegetation

Our vegetation characterizations indicate that about 45% of the catchment has a canopy of small pinyon and juniper trees. Canopy understoreys are dominated (93%) by litter (needles, dead wood), whereas intercanopy areas are mostly bare ground (66%), with about 30% litter and only 2.2% basal cover by herbaceous plants. The most important herbaceous species is blue grama (*Bouteloua gracilis*), which makes up about 1% of basal cover. Cryptogams account for less than 1% of basal cover – another striking piece of evidence of accelerated erosion (in nearby stable woodlands, we have measured cryptogamic cover as high as 50% in intercanopy areas (Wilcox 1994)).

3.4.2 Soils

Most of the soils in the catchment (dominantly Lithic and Typic Haplustalfs) are shallow, having an average depth of 35 cm. Some soils, consisting of a buried B horizon under a 50- to 75-cm layer of locally reworked pumice, are considerably deeper. The buried portions of these soils, which average 69 cm in thickness, are presumably the largely uneroded lateral equivalent of surface soils in the non-pumice areas. Assuming that to be the case, we can infer that the surface soils in the non-pumice areas have lost roughly 34 cm to erosion (in addition to any overlying pumice). The timing of the loss cannot be determined on the basis of the pumice alone, which was initially deposited some 50 000 to 60 000 years ago (Reneau *et al.* 1996) and may have been locally reworked more recently. There is growing evidence, however, for multiple periods of erosion on the Pajarito Plateau over the past 10 000 years (Longmire *et al.* 1995).

At the same time, we do find clues to recent erosion rates in the morphology of soils upslope and downslope of fallen ponderosa pine trees (the trees, which according to sequential aerial photographs, fell in the 1960s, have acted as natural sediment traps). Four of five pedons immediately upslope of these trees include buried A and B horizons, overlain by an average of 12 cm of recent sediment. For the 30-year period since the trees fell, this represents an average sediment deposition rate of about 4 mm per year. Downslope of the fallen trees, slopes are steeper and the reddish B horizons are commonly exposed in broad patches. In contrast to their upslope (buried) counterparts, these soils have been stripped of much of their A horizon (an average of 12 cm), probably in the same 30-year period that saw deposition behind the fallen trees. Thus, we have a rough estimate based on soil morphology of about 4 mm per year of erosion.

3.4.3 Runoff

3.4.3.1 Catchment-scale runoff

A summary of catchment-scale runoff and the precipitation that generated it is presented in Table 3.1. Since July 1993, when the first measurements were taken, 19 runoff events have been recorded – all but one generated by intense summer thunderstorms. The highest rainfall intensity recorded at the site was on 29 June 1995: 2.7 mm/min (such intensities are maintained only for a few minutes at a time). The debris transported by this event blocked the flume, diverting runoff around it (for this reason, the runoff volume shown in Table 3.2 for this event was estimated from the small plot data). Runoff at this site is typical of that of many semi-arid landscapes, in that it is of short duration and peak flow occurs within minutes of the onset of runoff (Figure 3.2(a)).

In terms of volume, the largest runoff event was actually produced by a fall frontal storm, which dropped 55 mm of precipitation in two days. Runoff began after the first 24 hours, during which about 15 mm of rain fell, and continued unabated for a 6-h period (Figure 3.2(b)); another 30 mm of rain fell during that time, but with rainfall intensity never exceeding 0.25 mm/min.

Because of below-freezing temperatures, the flume is not operational in the winter and spring. However, we did observe evidence (pools of water in the channel, traces of sediment in the flume) of small amounts of spring runoff on one occasion, probably from snowmelt while the soils were still frozen.

Water budget calculations are given in Table 3.2. We have assumed that at our site, as in other semi-arid areas, groundwater recharge makes up an insignificant

Table 3.1 Catchment-scale runoff (July 1993–September 1995)

Date	Precipitation (mm)	Peak precipitation (mm/min)	Catchment runoff (mm)	Peak flow (l/s)
1993				
28 July 93	36	1.8	7	3660
6 Aug. 93	9	0.8	1	960
20 Aug. 93	5	0.5	<1	480
26 Aug. 93	23	0.5	4	780
27 Aug. 93	24	2.0	12	5160
28 Aug. 93	11	0.8	3	600
6 Sept. 93	19	2.0	6	3480
13 Sept. 93	4	2.0	<1	570
1994				
2 Aug. 94	6	1.0	2	1800
21 Aug. 94	4	1.5	1	2220
5 Sept. 94	10	0.8	5	2340
15 Oct. 94	54	0.5	20	1560
1995				
29 May 95	14	1.0	4	4070
29 June 95	26	2.7	>10[a]	
17 July 95	14	0.5	1	720
18 July 95	5	0.2	<1	375
13 Aug. 95	10	1.0	2	1540
18 Sept. 95	9	1.0	2	1340
29 Sept. 95	3	0.8	1	1470

[a]Flume clogged with debris; runoff volume estimated from small-plot data.

portion of the water budget and that any water not accounted for as surface runoff is lost to evapotranspiration. Precipitation during the period of observation has been higher than the average projected for this elevation on the basis of long-term measurements in the area (Bowen 1990). Annual precipitation was especially high in water year 1995 (October 1994–September 1995), mainly as a result of heavy rainfall during October and November 1994. As a fraction of the water budget for the two complete years of observation, the contribution of runoff has been small (2% in WY 94 and 7% in WY 95). It does, however, contribute a much higher percentage of the budget for the summer rainy season (Table 3.2).

3.4.3.2 Small plot runoff

For the third summer rainy season (summer of 1995), we have data from the small plots as well as from the catchment (Table 3.3). Runoff from these plots was recorded for a total of 11 events (those on 17 and 18 July were treated as a single event), including five for which small amounts of runoff were measured from at least some of the plots, but none was measured for the catchment. There was considerable variability in runoff amounts from plot to plot; the intercanopy tuff residua plots produced the most, and the amounts from these plots were also highly variable,

Table 3.2 Annual and summer precipitation and runoff for the period of observation

Water year	Total precipitation (mm)	Summer precipitation (mm)	Yearly runoff (mm)	Summer runoff (mm)	Total runoff (%)	Summer runoff (%)
1993		195[a]	35	35		18
1994	384	131[b]	8	8	2	6
1995	549	123[b]	41	21	7	17

[a]26 July–31 Sept.
[b]1 July–30 Sept.

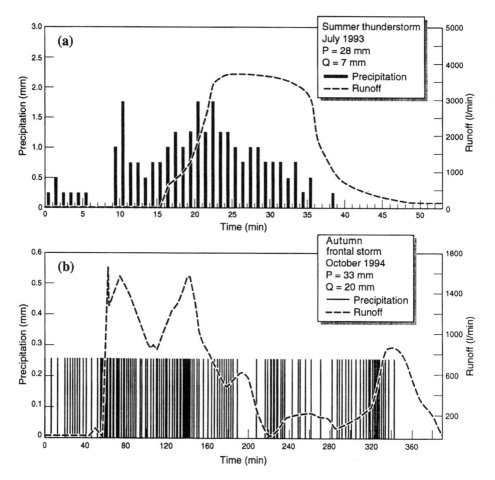

Figure 3.2 Rainfall and runoff for (a) a summer thunderstorm and (b) an autumn frontal storm

Table 3.3 Summary of small plot runoff in the summer of 1995

Plot no.	Plot description	Event day											Total
		29 May	26 June	29 June	30 June	16 July	17,18 July	5 Aug.	13 Aug.	10 Sept.	17 Sept.	29 Sept.	
		Total precipitation (mm)											
		42	6	35	4	10	31	8	14	17.5	8.5	12.5	
		Runoff (mm)											
95-5	Intercanopy – tuff residua	2.0	0.0	20.0	1.3	0.8	12.5	0.1	3.0	3.0	4.5	3.0	50.2
95-3	Intercanopy – tuff residua	4.5	0.0	20.0	0.7	0.0	10.0	0.6	4.0	1.0	4.0	4.0	48.8
95-7	Intercanopy – tuff residua	3.5	0.0	25.0	0.8	0.0	7.3	0.1	4.0	1.0	4.0	2.0	47.7
95-2	Intercanopy – tuff residua	8.0	0.2	22.0	0.0	0.0	4.5	0.1	3.5	0.1	3.0	2.5	43.9
95-8	Intercanopy – tuff residua	10.0	0.0	24.0	0.0	0.0	1.5	0.0	2.5	0.1	2.8	2.0	42.9
95-4	Intercanopy – tuff residua	5.0	0.2	20.0	0.0	0.0	3.8	0.0	2.0	0.0	1.1	2.0	34.1
94-4	Intercanopy – tuff residua	6.5	0.0	20.0	0.1	0.0	2.0	0.0	2.0	0.0	1.2	1.2	33.0
95-1	Intercanopy – tuff residua	2.0	0.1	15.0	0.0	0.0	1.0	0.0	1.0	0.0	2.5	2.0	23.6
94-3	Intercanopy – tuff residua	a	0.0	19.0	0.0	0.0	0.6	0.0	0.3	0.0	0.5	1.0	21.4
94-2	Intercanopy – tuff residua	2.0	0.0	12.0	0.0	0.0	0.0	0.0	0.1	0.0	0.1	0.1	14.3
95-6	Canopy	0.5	0.0	5.0	0.0	0.0	0.1	0.0	0.0	0.0	0.0	0.0	5.6
94-1	Intercanopy – pumice soil	0.0	0.0	3.5	0.0	0.0	0.0	0.0	0.1	0.0	0.1	0.0	3.7
		Weighted average (mm)											
		2.18	0.02	10.73	0.12	0.03	1.78	0.04	0.91	0.21	0.96	0.79	17.8

[a] Plot malfunction.

ranging from 14 to 50 mm. The tuff residua plot that produced only 14 mm, No. 94–2, differs from the others in that cryptogamic cover is still partially intact. Very little runoff came from either the canopy plot or the pumice-soil plot; the little that was measured came mostly from the large storm of 29 June.

Using the small plot data, we computed a "weighted average" for runoff from the catchment. Each plot was assigned to one of three categories on the basis of cover type (canopy, intercanopy with pumice soil, and intercanopy with tuff residua soil) and weighted according to the percentage of the catchment each cover type was estimated to represent. The canopy plot was weighted 50%, the pumice-soil plot 10%, and the tuff residua plots (using an average of the ten) 40%. The results of this computation for each runoff event were then compared with runoff measured at the flume (Figure 3.3). Although the match is quite good, catchment-scale runoff is generally slightly higher, suggesting that we are not sufficiently weighting the high-runoff-producing areas. The most important implication, however, is that little runoff is being stored on the catchment (otherwise plot-scale runoff would be greater than catchment-scale runoff).

3.4.4 Erosion

3.4.4.1 Microtopography Measurements

Microtopographic measurements have been made regularly since July 1993; since that time, surface elevations have changed in a consistent and predictable pattern, in response to (1) soil erosion during the summer rains, (2) frost heaving in the early spring, and (3) raindrop compaction in the spring and summer. We have found that over a period of one year, surface changes due to frost heaving and those due to raindrop compaction are roughly equal. Between July 1993 and September 1995, the average change in elevation of the 400 measured points at the 20 sites is −6.7 mm, most or all of which we attribute to soil erosion. Three representative micro-

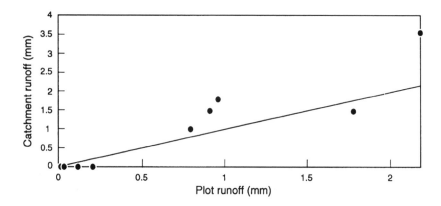

Figure 3.3 Runoff amounts calculated from weighted averages of small-plot data compared with those measured at the catchment outlet. The line represents a perfect fit

topographic profiles are shown in Figure 3.4: a rapidly eroding section from which as much as 7 cm of soil has been lost in a 26-month period (Figure 3.4(a)); a deposition zone in which up to 8 cm of sediment has accumulated, underscoring the dynamic changes occurring on the hillslope (Figure 3.4(b)); and a location at which there is still some cryptogamic cover and at which erosion has been comparatively small, showing the stabilizing effect of this cover (Figure 3.4(c)).

An estimate of erosion in t/ha is given by the following expression:

$$E = (B_d)(H)(10),   \tag{3.1}$$

where E is erosion (t/ha), B_d is soil bulk density (g/cm^3), and H is net change in surface elevation.

Assuming an average loss in surface elevation of 6.7 mm for half of the catchment (the intercanopy areas) and a bulk density of 1.4 g/cm^3, we estimate total erosion from the catchment at about 47 t/ha or 15 600 kg/ha/yr.

3.4.4.2 Small Plot Erosion

Erosion data collected from the small plot network in 1995 are shown in Table 3.4. By far, most of the erosion was produced by the largest storm, that of 29 June. The plots show dramatic differences in extent of erosion. For example, erosion was two orders of magnitude greater from many of the intercanopy plots on tuff residua than from either the canopy plot or the pumice-soil plot.

3.4.4.3 Subcatchment and Catchment Erosion

Estimates of erosion during 1995 for the catchment as a whole and for the four subcatchments are given in Table 3.5. We note trends similar to those observed from the small plots, namely, that most of the erosion was produced by the single largest event; however, erosion at the subcatchment scale is an order of magnitude greater than that at the plot scale, reflecting the influence of channel erosion. As already noted, there are numerous small channels across the catchment. In fact, in many locations, the hillslope has the appearance of a braided stream channel. In addition, there was considerable variability in erosion among the subcatchments; for example, erosion from subcatchment 2 was four times greater than that from subcatchment 1 (large portions of which consist of pumice soil).

As previously noted, because the sediment trap at the catchment outlet was too small for larger amounts of sediment, no measurements are available for the first two events of the year, both of which overtopped the trap. For the four smaller storms, erosion was greater at the subcatchment scale than at the catchment scale, indicating storage of sediment on the hillslope (at least for the smaller events); and indeed, we observe accumulations of sediment in many areas in the catchment hillslopes and channels. We hypothesize that during larger events, these sediment storage sites are evacuated – and, thus, that over a period of a year or two, the net quantities of sediment leaving the catchment are probably roughly the same as those leaving the subcatchments.

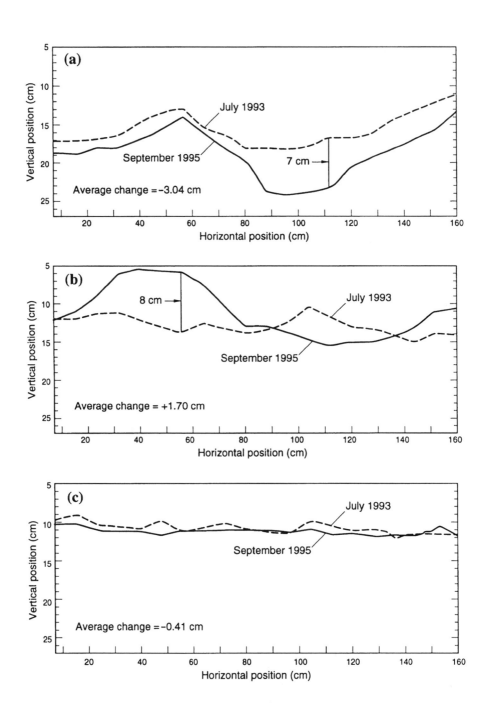

Figure 3.4 Microtopographic changes at three locations within the catchment representative of (a) an eroding surface; (b) a depositional surface; and (c) a relatively stable surface

Table 3.4 Summary of small plot erosion in the summer of 1995

Plot no.	Plot description	Event date											Total
		Erosion (g/m²)											
		29 May	26 June	29 June	30 June	16 July	18 July	5 Aug.	13 Aug.	10 Sept.	17 Sept.	29 Sept.	
95-5	Intercanopy tuff residua	13	0	179	1	2	8	0	15	3	12	10	244
95-3	Intercanopy tuff residua	32	0	393	7	0	17	15	65	2	37	167	736
95-7	Intercanopy tuff residua	12	0	224	2	0	3	0	46	3	26	15	330
95-2	Intercanopy tuff residua	34	1	167	0	0	2	0	14	0	102	23	344
95-8	Intercanopy tuff residua	88	0	250	0	0	3	0	57	0	22	68	488
95-4	Intercanopy tuff residua	24	0	93	0	0	2	0	4	0	1	12	136
94-4	Intercanopy tuff residua	43	0	257	0	0	2	0	54	0	4	38	398
95-1	Intercanopy tuff residua	6	0	84	0	0	0	0	4	0	5	15	114
94-3	Intercanopy tuff residua	[a]	0	173	0	0	0	0	3	0	1	6	185
94-2	Intercanopy tuff residua	2	0	32	0	0	0	0	0	0	0	0	34
95-6	Canopy	1	0	5	0	0	0	0	0	0	0	0	6
94-1	Intercanopy pumice soil	0	0	3	0	0	0	0	0	0	0	0	3
Weighted average		12	0	77	0	0	1	1	10	0	8	14	124

[a]Plot malfunction.

Table 3.5 Subcatchment- and catchment-scale sediment yields

	Area (m²)	29 May	29 June	18 July	13 Aug.	17 Sept.	29 Sept.	Total[b]
		Event date						
		Catchment runoff (mm)						
		3.6	>10	1.5	1.5	1.8	1	
		Erosion (g/m²)						
Subcatchment 1	1046	65	239	10	15	0	14.93	279
Subcatchment 2	308	[a]	921	15	77	0	101.44	1114
Subcatchment 3	308	294	468	152	25	16.7	25.36	687
Subcatchment 4	1107	[a]	493	56	28	31.75	14.11	623
Subcatchment average		180	530	59	36	12	39	855
Catchment	10 000	>40	>40	19.5	22.6	17.18	14.06	

[a]Trap not yet installed.
[b]Totalled for last three events only.

3.5 DISCUSSION AND CONCLUSIONS

The expansion of pinyon–juniper woodlands during the last century is unprecedented and incontrovertible (Miller and Wigand 1994). Although it is widely believed that this expansion has led to increased soil erosion, this belief remains controversial, being based mostly on anecdotal information (Gottfried *et al.* 1995). The effect of pinyon–juniper expansion on the soil resource has not been studied systematically, and the work that has been done consists mainly of small plot studies using artificial rainfall. While this work has added to our knowledge, an adequate understanding of erosion processes in pinyon–juniper woodlands will come only through larger-scale studies (in terms of both time and space). The small catchment study described here is a step in that direction, and although longer-term observations will be required to fully address our objectives, the insights gained from these first 26 months are important. In regard to the project objectives, we can say the following.

3.5.1 Objective 1: Estimate Water and Sediment Budgets for Rapidly Eroding Semi-arid Woodlands

In semi-arid environments, measurements must be taken over many years to develop representative "averages" for water budget components. The estimates we have made so far are based on the assumption that contribution to groundwater is zero (any water that does not run off will be evapotranspired). Our results to date indicate that runoff accounts for less than 10% – and perhaps closer to 5% – of the water budget of the catchment, which is consistent with data from other semi-arid environments. In the summer rainy season, however, runoff makes up perhaps 15–20% of the water budget. For many of the intercanopy areas the runoff percentage during this time may be as high as 40%.

To estimate erosion from the catchment, we are directly measuring erosion at a number of scales: plot ($1\,m^2$), subcatchment (300–$1100\,m^2$), and catchment ($10\,000\,m^2$). In addition, by monitoring the surface microtopography and documenting soil morphological changes, we have some indirect measures of erosion. The estimates based on soil morphology are probably the crudest, but they represent the longest time periods and are remarkably close to those given by the microtopographic measurements: the soil morphology data indicate an average soil loss in the intercanopy zones of about $4\,mm/yr$ over the last 30 years, and microtopographic measurements show a loss of about $2\,mm/yr$ for the past 3 years. Assuming 50% canopy area, these data suggest an average erosion rate for the catchment of $10\,000$–$20\,000\,kg/ha$. Our direct measurement of erosion at the subcatchment scale (which is comparable to the scale represented by the indirect measurements) in 1995 was about $9000\,kg/ha$.

At these rates – 50 to 100 times higher than in more stable pinyon–juniper woodlands (Wilcox 1994) – the soil resource cannot be sustained. We estimate very roughly that the intercanopy soil will be mostly stripped from this catchment in about a century.

3.5.2   Objective 2: Determine the Effect of 20th Century Vegetation Change on Erosion Processes

Vegetation in our study catchment has changed markedly over the past century. An open stand of ponderosa pine, with only a scattering of pinyon and juniper trees, characterized this site until the onset of domestic livestock grazing in the late 1800s. The grazing apparently suppressed the surface fires, which had previously burned through such low-elevation pine stands at roughly 15-yr intervals (Allen 1989), by reducing the herbaceous ground cover through which the fires propagated (Gottfried *et al.* 1995). Livestock grazing continued until 1932, after which feral burros similarly affected the system until about 1980. With fire eliminated and herbaceous competition reduced, pinyon and especially juniper tree establishment increased markedly. As the density of pinyon and juniper trees inceased, herbaceous ground cover was likely further reduced. When the severe drought occurred in the 1950s, there was inadequate water available to sustain all the woody vegetation that had developed since the fires ceased. As a result, every ponderosa pine in this catchment died, leaving the more drought-tolerant pinyon and juniper. It seems likely that the drought also caused additional declines in herbaceous ground cover. Reduced cover of vegetation and litter led to increased surface runoff. We hypothesize that these vegetation changes, caused by this history of human land-use practices and climatic variability, have led to the current episode of accelerated soil erosion.

The information gathered to date at our site, particularly the soil characterization information, supports our hypothesis. We know that the surface was stable at one time, because we find moderately well-developed argillic horizons, which in semi-arid environments require time periods on the order of 10 000 years to develop (Birkeland

1984). That the current period of instability began around the time of the ponderosa pine die-off is suggested by the presence of mostly intact soils upslope of fallen ponderosa pine logs (which act as sediment traps), whereas the soils downslope of the logs are highly eroded. Physical rather than biological processes now dominate the barren, little-vegetated intercanopy areas.

3.5.3 Objective 3: Develop a Conceptual and Quantitative Understanding of the Relationships between Runoff and Erosion

Although many more years of observation will be required to fully achieve this objective, our results to date are quite enlightening and generally strengthen those from studies in other semi-arid environments. For example, we have clear evidence that in these regions intense thunderstorms produce the most erosion per unit of runoff; gentle rains, even though prolonged and able to generate large amounts of runoff, result in comparatively little erosion. Our results to date are the beginning of a quantitative database for evaluating runoff/erosion models in semi-arid landscapes.

3.5.4 Objective 4: Determine what Effect Scale has on Runoff and Erosion in Rapidly Eroding Semi-arid Woodlands

Our results with respect to this issue are contrary to those obtained for more stable pinyon–juniper woodlands, where runoff and erosion clearly diminish as scale increases from small plot to hillslope (Wilcox 1994). Most interesting is that at this site, not only does runoff not diminish with scale (i.e., there is little if any storage of water on the hillslope), but erosion actually increases from the plot to the subcatchment scale. We hypothesize that the hillslope channels themselves – or areas very near the channels – are sources of sediment.

In conclusion, understanding the many factors that play a role in accelerated erosion in semi-arid ecosystems – and related feedback mechanisms – is obviously key to the ability to manage and sustain these ecosystems. Only through long-term, detailed gathering of relevant data, from a multitude of sites, can such an understanding be gained. This study is contributing to the development of such an understanding.

ACKNOWLEDGEMENTS

This work was supported by the Environmental Restoration Project, Los Alamos National Laboratory. We thank Marvin Gard, Kevin Reid, Shannon Smith, Randy Johnson, Nicole Gotti, Kay Beeley and John Hogan for assistance in the field. This work contributes to the Global Change & Terrestrial Ecosystem (GCTE) Core Research Program, which is part of the International Geosphere–Biosphere Program (IGBP).

REFERENCES

Allen, C.D. 1989. Changes in the landscape of the Jemez Mountains, NM. *PhD dissertation*, University of California, Berkeley.

Allen, C.D. 1997. Ecological patterns and environmental change in the Bandelier landscape. In T.A. Kohler (ed.), *Village Formation on the Pajarito Plateau, NM: The Archeology of Bandelier National Monument*. University of New Mexico Press, Albuquerque, in press.

Birkeland, P.W. 1984. *Soils and Geomorphology*. Oxford University Press, New York.

Bowen, B.M. 1990. *Los Alamos Climatology*. Los Alamos National Laboratory Report LA-11735-MS, Los Alamos, NM.

Gottfried, G.J., Swetnam, T.W., Allen, C.D., Betancourt, J.L. and Chung-MacCoutbrey, A. 1995. Pinyon–juniper woodlands of the Middle Rio Grande Basin, New Mexico. In D.M. Finch and J.A. Tainter (eds), *Ecology, Diversity, and Sustainability of the Middle Rio Grande Basin*. USDA Forest Service General Technical Report RM-268, Fort Collins, Colorado.

Johnsen, T.N. 1962. One-seed juniper invasion of northern Arizona grasslands. *Ecological Monographs*, **32**, 187–207.

Longmire, P., Reneau, S., Watt, O., McFadden, L., Gardner, J., Duffy, C. and Ryti, R. 1995. *Natural Background Geochemistry, Geomorphology, and Pedogenesis of Selected Soil Profiles and Bandelier Tuff, Los Alamos, New Mexico*. Los Alamos National Laboratory Report LA-12913-MS, Los Alamos, New Mexico.

Miller, R.F. and Wigand, P.E. 1994. Holocene changes in semiarid pinyon–juniper woodlands. *BioScience*, **447**(7), 465–474.

Mueller-Dombois, D. and Ellenberg, H. 1974. *Aims and Methods of Vegetation Ecology*. John Wiley & Sons, New York.

Reneau, S.L., Gardner, J.N. and Forman, S.L. 1996. New evidence for the age of the youngest eruptions in the Valles caldera, New Mexico. *Geology*, **24**(1), 7–10.

Replogle, J.A., Clemmens, A.J. and Bos, M.G. 1990. Measuring irrigation water. In G.J. Hoffman, T.A. Howell and K.H. Solomon (eds), *Management of Farm Irrigation Systems*. ASAE Monograph, American Society of Agricultural Engineers, St Joseph, MI, pp. 345–351.

Schmidt, L.J. 1987. Present and future themes in pinyon–juniper hydrology. Proceedings of the Pinyon–Juniper Conference, USDA-Forest Service, Intermountain Research Station. General Technical Report INT-215.

Shakesby, R.A. 1993. The soil erosion bridge: a device for microprofiling soil surfaces. *Earth Surface Processes and Landforms*, **18**(9), 823–827.

Wilcox, B.P. 1994. Runoff and erosion in intercanopy zones of pinyon–juniper woodlands. *Journal of Range Management*, **47**(4), 285–295.

Wilcox, B.P. and Breshears, D.D. 1995. Hydrology and ecology of pinyon–juniper woodlands: conceptual framework and field studies. *Desired Future Conditions for Pinyon–Juniper Ecosystems*, USDA-Forest Service, Intermountain Research Station. General Technical Report RM-258.

4 Time-Dependent Changes in Soil Properties and Surface Runoff Generation

T. P. BURT
Department of Geography, University of Durham, UK

and M. C. SLATTERY
Department of Geography, East Carolina University, USA

4.1 INTRODUCTION

It is now widely recognized that the hydrological and sedimentological response of a catchment is controlled to a large extent by the surface condition of the soil. Especially important in this regard are both the temporal changes in soil erodibility caused by changes in soil moisture content and the evolution of surface seals and crusts in response to rainsplash activity and related processes. Antecedent soil moisture conditions have been shown in many studies to be a critical control on the timing and magnitude of storm runoff and on sediment transport during these events, yet there are very few studies in which any systematic attempt has been made to monitor this regularly, particularly in the field (Bryan 1991). A major problem in relation to changing soil physical properties is that many of the available measures of antecedent soil conditions are obtained in the laboratory and their field application is not always clear (Bryan 1991). Of necessity, most laboratory studies have used artificial cohesionless sediment mixtures, or natural soils in which the field structure has been largely or completely obliterated.

The situation is further complicated in agricultural areas where soils are disturbed repeatedly by soil management practices; the resulting alteration of soil physical properties greatly affects the likelihood of surface runoff and erosion. Of particular interest is the effect on infiltration of a thin surface zone of reduced permeability. This zone develops after tillage during one or more rainstorms as a consequence of soil structure breakdown by mechanical or physico-chemical processes. This effect is often called a surface seal or, when dry, a surface crust. In either case, the effect of this layer of reduced permeability is to decrease infiltration and to increase the risk of surface runoff (Römkens *et al.* 1990). Imeson and Kwaad (1990; Figure 4.1) have identified three periods in the evolution of the tilled layer; a freshly ploughed soil is taken as the starting point.

Advances in Hillslope Processes, Volume 1. Edited by M. G. Anderson and S. M. Brooks.
© 1996 John Wiley & Sons Ltd.

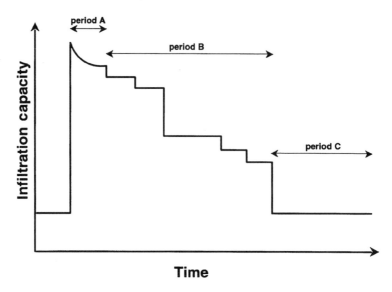

Figure 4.1   The evolution of soil surface condition from tillage to harvest (based on Imeson and Kwaad 1990; Boardman 1992)

- *Period A*. Ploughing usually results in relatively coarse clods which are loosely packed; there is a decrease in bulk density and an increase in infiltration capacity. The upper part of the ploughed layer is often subjected to secondary tillage operations (harrowing) to reduce the size of the clods before drilling. This may be followed by rolling to achieve a certain compaction of the seedbed, often to improve the efficacy of pesticides. The clods of a freshly tilled soil are usually loosely packed with large voids between them; macroporosity is high, encouraging infiltration and percolation. Once tillage operations are completed, the soil begins to settle immediately under its own weight, and aggregates may begin to break up by fissuring processes (wetting and drying, freezing and thawing). This first period does not usually last very long, only until the first significant rainfall event.
- *Period B*. Once the tilled soil is subjected to the impact of falling raindrops, the process of compaction and subsidence of the surface layer is greatly accelerated. A variety of processes contribute to this effect (Römkens *et al.* 1990). Mechanically based changes relate to the destructive impact of raindrops or to the disintegration of soil clods and aggregates when wetted. Physico-chemical changes are brought about by the swelling of clay material upon contact with water. Raindrop impact is the principal force governing the physical breakdown, rearrangement and compaction of aggregates at the surface. The effectiveness of rainfall in causing surface degradation is related in large part to the kinetic energy of droplets (Monnier and Boiffin 1986). Heavy rainfall causes the rapid degradation of the surface structure of a tilled soil, decreasing porosity and greatly lowering surface permeability. Boiffin (1984) distinguishes two phases of crust formation: structural crusts formed by the settling of splashed material and slaking of aggregates; and,

after ponding and runoff have occurred, depositional crusts formed by inwashing of soil into surface depressions and macropores. According to Imeson and Kwaad's model, degradation of the soil surface takes place in a largely discontinuous manner, phases of gradual change separated by more rapid alteration during rainfall events. The transformation in the state of the soil surface during period B greatly affects a number of soil physical properties including infiltration capacity, surface roughness and shear strength. The decrease in porosity and change in pore-size distribution affect bulk density, penetration resistance, hydraulic conductivity and water retention (Imeson and Kwaad 1990). There may be some recovery of roughness, porosity and infiltration capacity during the growing due to biological activity in the soil (e.g. by earthworms).

• *Period C*. Imeson and Kwaad consider that no further changes in the physical state of the tilled layer take place during this period. This condition is exemplified by the stubble field which remains after the harvest until the next tillage operations are undertaken.

Although the Imeson and Kwaad model was defined specifically for the loess soils of north west Europe, it provides a useful framework for interpreting the response of other soils. Monnier and Boiffin (1986) point out that the processes of superficial structural degradation by rain depend on the climatic conditions to which the soil is subjected and on the intrinsic mechanical and structural properties of the soil itself, especially its mineral composition and organic status. The precise chronology of soil surface changes, even for a single field, will change from season to season, depending on the succession of crops and management of the crop and inter-crop intervals. More importantly perhaps, interannual variability in both erodibility and erosivity depend on climatic factors (Boardman 1990).

It is clear from the preceding discussion that the dynamic character of soil physical properties must be given fuller consideration in order to develop a more complete understanding of runoff generation and soil loss from agricultural land. There is, in particular, a dearth of field observations on how soil surface conditions change over time in response to rainfall and management practices. Here, we use Imeson and Kwaad's (1990) model as a basis for analysing a year of field observations of soil properties and surface runoff generation in a small catchment in the Cotswold Hills, England.

4.2 RESEARCH DESIGN

The study catchment, the headwaters of the River Stour (which flows into the Warwickshire Avon) is located in the Cotswold Hills in south central England (Figure 4.2). It has an area of 6.2 km^2; its altitude ranges from 202 m on the northern divide to 126 m at the basin outlet (UK grid reference: SP 356 362). Slopes are gentle on the interfluves (around 1°) but with steeper slopes in the central part of the basin (maximum 13°). The catchment occupies a graben involving rocks of the Middle Jurassic. In the central portion of the basin, Northamptonshire Sandstone is overlain by *brown calcareous earths* of the Aberford Series; these are moderately stony, well-

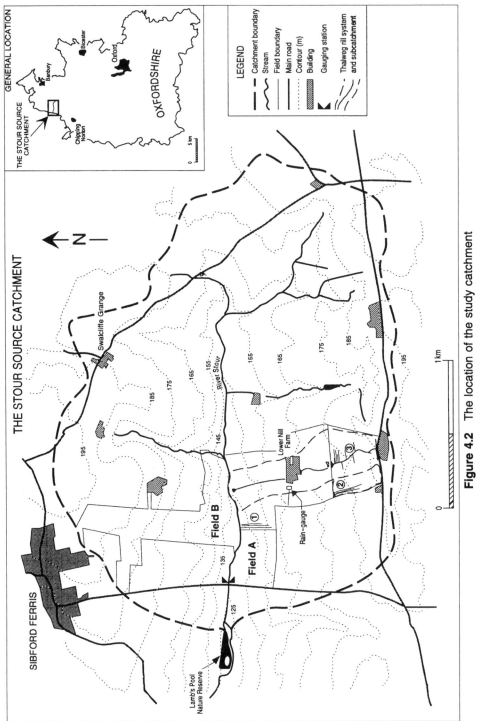

Figure 4.2 The location of the study catchment

drained fine loam soils. On the southern and northern peripheries, Great Oolite Limestone is overlain by *ferritic brown earths* of the Banbury Series; these are stony, well-drained fine loamy soils. In neither case dose horizonation significantly affect the soil hydrology.

A 90° V-notch weir was installed at the basin outlet in July 1992. A Campbell data logger was used to provide a continuous record of stage with an Ott stage recorder as back-up. Two rain-gauges were installed: a tipping bucket gauge was positioned close to the weir and attached to the logger, measuring each 0.2 mm rainfall: a Casella autographic natural siphon gauge was installed at Lower Nill Farm (Figure 4.2).

Two fields in the catchment were studied between August 1992 and July 1993:

- *Field A*. Oil seed rape was harvested in early August 1992. The field was ploughed on 18/19 August and then subject to two tillage treatments on 11 and 18 September using a CTYNE cultivator. The surface was ploughed again on 17 October and roteared in preparation for planting of winter wheat. Heavy and persistent rains throughout October and November prevented any access of heavy machinery into the field and it was subsequently left bare throughout the winter. The field was recultivated on 25 February after a prolonged dry spell, drilled with spring wheat on 27 February and rolled on 7 March.
- *Field B*. Winter wheat was harvested between 15 and 17 September 1992. It was ploughed 29/30 September and given one roterra pass and drilled with winter wheat on 10 October. It was then rolled on 15 October. The soil surface was, as a result, quite different to that of field A over the winter period as rolling resulted in a smoother surface with distinct soil aggregates of varying sizes rather than large clods.

Five soil properties were measured at each site, either directly in the field or later in the laboratory. On each occasion, three undisturbed cores (7.5 cm diameter × 7.5 cm length) were taken from the soil surface of both fields; usually, standard deviations were small so that one can have much confidence in any trends or changes identified. The properties chosen relate either to those which have an influence on infiltration and surface runoff generation (soil bulk density, water content, infiltration capacity) or to those which determine erodibility (soil shear strength, aggregate stability). Discussion here is limited to the former group of variables; further details of the erodibility study are to be found in Slattery (1994). The methodology employed was briefly as follows:

- *Soil moisture content* surface samples (c5–10 g) were collected and water content determined gravimetrically using the standard technique of oven-drying at 105°C.
- *Soil bulk density* was obtained by carefully extracting undisturbed cores in thin-walled tubes of 7.5 cm diameter and 7.5 cm length; these were dried and weighed to establish the mass of soil in each core.
- *Infiltration capacity* was measured using a ring infiltrometer designed by Burt (1987). A ring of 15 cm diameter was driven into the soil to a depth of 15 cm. Care was taken to minimize disturbance of the soil surface, although it was difficult to prevent cracking near the border of the ring, especially when the soil surface had

become dry and crusted. Any cracks which did develop were infilled and compacted down. The ring was flooded to a depth of 4 cm and the test was then run until the infiltrometer had emptied or for 60 minutes, whichever came first. Despite its availability, it was decided *not* to use a rainfall simulator (Bowyer-Bower and Burt 1989): the simulator takes a minimum of two hours to set up and operate at each site and, given the need to complete a range of other measurements on each visit to the catchment, it was decided that the ring infiltrometer would provide acceptable results despite being a less good analogue of the infiltration process.

4.3   CHANGES IN SOIL SURFACE CONDITION DURING THE STUDY PERIOD

4.3.1   Bulk Density

4.3.1.1   Field A

After ploughing on 17 October, bulk density quickly increased following a large storm on 20 October (23.1 mm rain in 9 hours) which produced a fairly uniform soil surface seal (Figure 4.3(a)). Slattery and Bryan (1992), among others, have shown that near-surface soil bulk density generally increases during sealing and accompanying compaction. The kinetic energy of rainfall during that storm would have been more than sufficient to cause disruption and compaction of surface aggregates. A series of four storms at the end of November caused further compaction of aggregates and breakdown of clods at the soil surface as a result of raindrop impact and the continued incorporation of material into the surface seal. This was the wettest time during the study period and it seems likely that there was also collapse of aggregates at depth at this time as the wet soil settled under its own weight. No measurements were possible between 18 December and 10 January owing to a heavy and almost continuous ground frost. A slight decrease in bulk density thereafter was probably related to needle ice activity. After a dry spell, the field was ploughed again on 25 February, causing bulk density to drop considerably; subsequent cultivation, drilling and rolling resulted in a significantly smoother, more compacted soil surface of higher bulk density, which increased still further after rainfall in late March and early April. The period following this was characterized by a relatively stable soil surface with bulk densities varying only marginally.

4.3.1.2   Field B

Somewhat surprisingly, drilling (10 October) and rolling (15 October) produced a mean decrease in bulk density, though the relatively large standard deviations at this time point to much spatial variability immediately after cultivation (Figure 4.3(b)). As on field A, raindrop impact during the large storm of 20 October caused much compaction and the development of a surface seal. Thereafter bulk densities remained relatively constant during the study period, rising slightly after heavy rainfall and falling during dry periods (notably in February) or after frost. As on field A, it is suggested that the increase in bulk density after rainfall is the result of

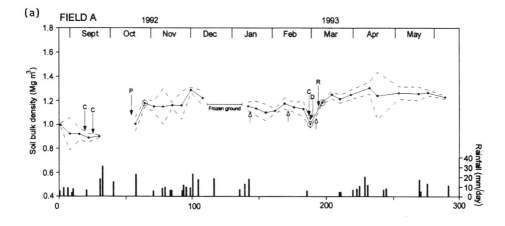

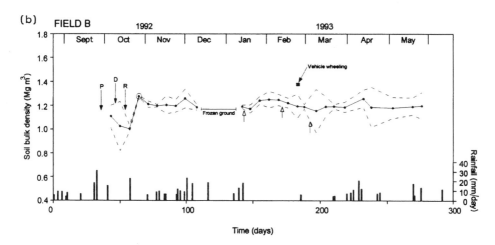

Figure 4.3 Variation in surface bulk density for two fields in the Stour Source catchment, August 1992 to June 1993. Dashed lines indicate one standard deviation either side of the mean

compaction at the soil surface and collapse of soil structure at depth. Note that Figure 4.3b includes one bulk density measurement for a tractor wheeling; this showed a considerably higher value than that for the cultivated surface. Wheelings are important components of the sediment delivery system as they tend to generate runoff more easily than adjacent surfaces.

4.3.2 Soil Moisture Content

In field A, soil moisture content increased steadily from the middle of October, reaching a peak in early December (Figure 4.4). No measurements were possible during the time when the ground was frozen. During January, moisture contents remained

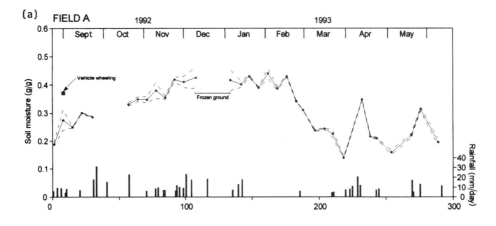

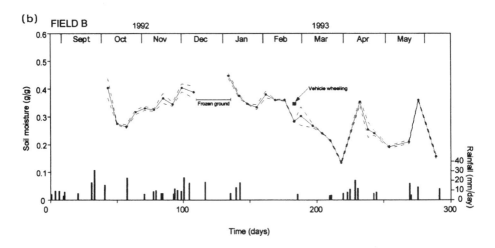

Figure 4.4 Variation in surface soil moisture for two fields in the Stour Source catchment, August 1992 to June 1993. Dashed lines indicate one standard deviation either side of the mean

high. From February onwards, the soil gradually dried out, with moisture contents increasing only temporarily after rainfall. In field B, a very similar pattern was seen, except that soil moisture content was rather higher in early January than it had been before the soil froze in December. It is not clear why the two soils behaved differently at this time. It may relate to slight differences in the drainage characteristics of the seals or to the effect of soil freezing in drawing water towards the soil surface.

4.3.3 Infiltration Capacity

The results from field B are examined first in this section as they are more readily explicable in terms of the changes in infiltration seen over time. The values plotted

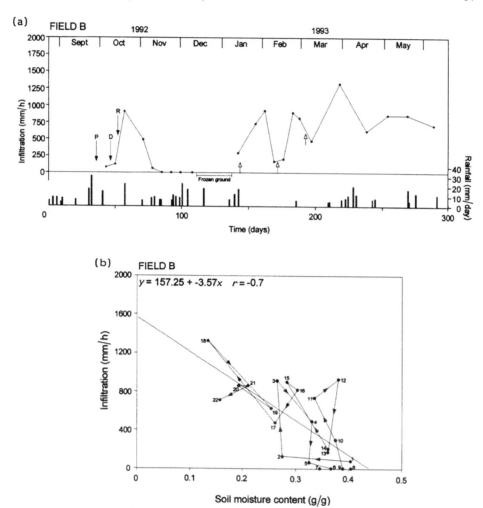

Figure 4.5 (a) Variation in infiltration capacity for field B, August 1992 to June 1993. (b) The relationship between soil moisture content and infiltration capacity for field B during the study period

are the infiltration rates measured at the end of each experiment; given that infiltration rate normally declines over time, this value usually represents the minimum rate observed during the run. It should be emphasized that these are observed minima and no attempt is made here to use estimates of final infiltration capacity (at $t=$ infinity) provided by curve fitting (Philip 1957).

4.3.3.1 Field B

Figure 4.5(a) shows the measured changes in infiltration capacity on field B during the study period. Figure 4.5(b) shows the relationship between soil moisture content and infiltration capacity; the individual data points have been joined in their temporal sequence to show what is essentially a hysteresis-type plot between the two variables. In essence, the observations may be divided in two: those periods when infiltration was high, much higher than any rainfall intensities likely to be observed in this part of the world; and the period from mid-November to mid-December when infiltration capacity fell to very low levels (below 2.5 mm/h). It is clear that, during this specific period, there was an effective surface seal, but once the seal became a crust during the drying phase, infiltration rates increased, most likely as a result of the development of surface cracks, which would expedite the penetration of water into the soil matrix. Cracking of the soil surface was most likely aided by needle ice activity which easily disrupts the delicate crust. The formation of the seal can be clearly identified on Figure 4.5(b): there is no change in moisture content between points 4 and 5 but nevertheless the infiltration capacity falls greatly between these two times. Apart from the period when the surface seal was effective, changes of infiltration capacity on field B were largely controlled by changes in surface soil moisture content, as Figure 4.5(b) shows.

4.3.3.2 Field A

Figure 4.6 shows the infiltration data for field A during the study period. There is apparently no clear control of soil moisture on infiltration on field A; it seems that changes in the structure of the soil surface are relatively more important in this field. However, if the regression is re-run, excluding those data points affected by structural changes, the correlation between soil moisture and infiltration becomes strongly negative for field A too ($r^2 = 0.52$). (A similar analysis for field B yields an increase in the r^2 value from 0.49 to 0.74.) Apart from this one difference, the pattern of change in infiltration for both fields is broadly similar, with low values for both fields in late November and early December and much higher levels in the autumn, late winter and spring. Infiltration capacities are generally lower on field A but this is of little consequence since, apart from the period when there was an effective surface seal, the infiltration capacity remains far too high for infiltration-excess overland flow to occur under British rainfall conditions.

4.3.3.3 General discussion

1. Many of the individual infiltration values seem unrealistically high. It has been suggested (Dunne and Leopold 1978) that ring infiltrometers may give values that are 2–10 times too large. This may be a particular problem when soils are dry and cracked; it can be very difficult to insert the ring without causing disturbance at such times. However, Dunne and Leopold argue that ring infiltrometers at least provide *relative* rates of infiltration which allow the importance of different controlling factors to be compared.
2. Ring infiltrometers seem to work best when the soil is wet. This is shown by the low readings recorded during the time when the surface seal was developed. Had insertion of the ring caused significant disturbance to the seal, measured infiltration rates would have been much higher. We are confident, therefore,

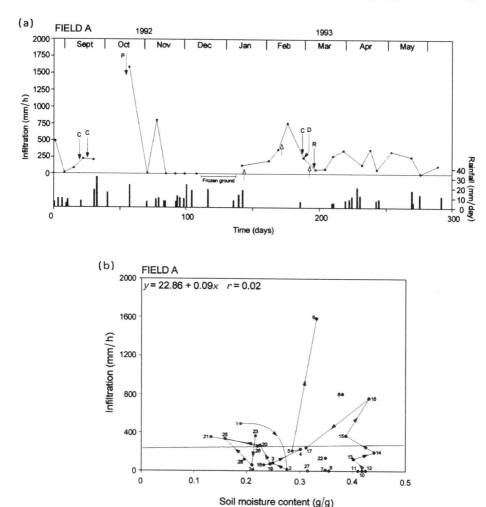

Figure 4.6   (a) Variation in infiltration capacity for field A, August 1992 to June 1993. (b) The relationship between soil moisture content and infiltration capacity for field A during the study period

that the minimum infiltration capacities recorded in November and December are legitimate values, even if the values recorded during drier periods are possibly overestimates. Confirmation of the sound performance of ring infiltrometers in wet soils was provided by later tests where an infiltration capacity of 1.6 mm/h was measured using a ring infiltrometer and a value of 2.3 mm/h was obtained using a rainfall simulator.

3. As noted above, infiltration rates in field B were higher than in field A, especially after the period of ground frost. The reason for this seems to be that the winter wheat had become well established on field B during the early part of 1993,

whereas there was no crop on field A until late March. It was estimated that crop cover had reached $c.\ 40\%$ on field B by the beginning of February. From studies on untilled soils, it is known that macropores formed by crop roots contribute to the relatively high infiltration capacity and hydraulic conductivity of such soils.

4. With hindsight, a clearer understanding of the role of soil moisture in controlling infiltration capacity might have been achieved by measuring the negative pore water pressure (suction) of the upper soil layer. It is much easier to measure the water content of the soil surface but this may be less meaningful in relation to infiltration than pore water pressure, especially since the suction–moisture relation will vary as a result of changes in bulk density and porosity.

4.4 INFILTRATION CAPACITY AND STORM RUNOFF GENERATION

Figure 4.7 shows the stream hydrograph for the study period together with the daily rainfall record. It is now recognized that all techniques of hydrograph separation are arbitrary and have little or nothing to do with the processes by which stormflow is generated, but that if one method is employed consistently then usable results are obtained (Burt 1992). After some experimentation with a number of different techniques, it was found that drawing a line from the initial point of discharge rise to a point on the recession limb one and a half times the "time to rise" after peak discharge produced a result that was sensible and worked well in most cases. This method of separation was used to calculate storm runoff volumes and the ratio of runoff to rainfall (runoff per cent or ROP). The fact that storms with similar rainfall produced very different amounts of runoff suggested that several factors were affecting the response of the basin to rainfall. Lack of positive correlations between short-term rainfall intensity (periods lasting up to one hour) and peak discharge or runoff volume shows that rainfall intensity is not an important variable in controlling stormflow response in the study catchment (cf. Hewlett et al. 1977; Taylor and Pearce 1982). A positive correlation between time to rise and runoff volume implies that the dominant mechanism generating storm runoff is subsurface stormflow with small amounts of rainfall (even if intense) yielding small hydrographs, and large responses taking some time to develop. If the dominant source of storm runoff were infiltration-excess overland flow, one would expect intense rain to produce flashy hydrographs with short times of rise and large volumes of runoff.

Early autumn storms were characterized by ROPs ranging from 0.09 to 0.89; all these events seem to have been generated solely by subsurface stormflow as there was no evidence of any surface runoff on the catchment slopes. Indeed, most slopes had been recently ploughed and infiltration capacities remained high throughout the autumn period. Of course, as in all storms, there was some runoff from roads and tracks in such storms. The storm of 25 September was unique, not only in terms of the mode of runoff generation, but also in terms of its magnitude. Persistent rain generated 31.2 mm in 19.5 hours, with the last 1.5 hours yielding 16.9 mm. There was widespread flooding and the weir was overtopped, hence the lack of a recorded peak

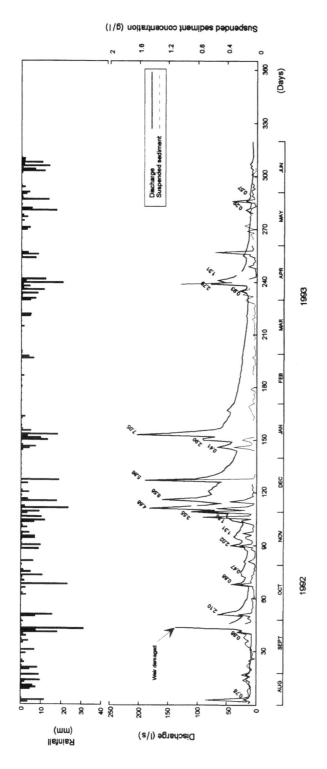

Figure 4.7 The stream hydrograph for the study period. Runoff per cent (ROP) for each storm hydrograph is indicated by the figure immediately above the discharge peak

in Figure 4.7. It is clear from the incomplete record of discharge that the rising limb was very steep; such a rapid response suggests overland flow. However, there was no evidence of this except on one field in the northeast of the basin, which had been recently harvested but not yet ploughed, where there were clear signs of stubble having been washed downslope along pseudo-channels. Infiltration capacity in this field was 3.2 mm/h at this time. There were no signs of surface runoff in fields A and B where infiltration capacities remained well above rainfall intensity. It seems likely that there was a good deal of runoff from roads and a significant contribution from subsurface flow as well during this event. A similar combination of runoff sources was observed during the storm of 3 October; additionally, flow was *observed* along vehicle wheelings for the first time. Infiltration capacity on the wheelings was generally found to be less than 5 mm/h so that it is likely that these will always provide an important source of surface runoff in any rainfall event of at least moderate intensity.

Frequent rainfall in November and December resulted in a series of storm hydrographs; ROPs ranged from 1.9 to 6.5 during these events. In addition to the runoff sources noted above, piezometer data and the results of a chemical mixing model suggest that saturation-excess overland flow was a significant source of runoff at this time (Velterop 1993). During these storms, infiltration-excess overland flow was observed in fields A and B, especially along wheelings but also more generally from within the fields. This was not surprising given the low infiltration capacity of their soils at the time. However, very few rills were eroded by this surface runoff. At point 1 (Figure 4.2) a series of rills along wheelings discharged water and sediment directly into the stream channel. The most extensive zone of infiltration-excess overland flow developed in fields on the south side of the catchment. At point 2, runoff along wheelings eroded a series of rills which developed into an extensive braided network. By far the most important zone of surface runoff was at point 3 where there was widespread runoff from wheelings; this flow became concentrated along the thalweg of a dry valley where a deep rill developed. Use of the ring infiltrometer showed that the infiltration capacity in the wheelings was below 5 mm/h compared to 64 mm/h on the cultivated slopes. During the two major events of 18 December (ROP = 5.9; maximum hourly rainfall = 30 mm) and 13 January (ROP = 7.1; maximum hourly rainfall = 5.2 mm), the thalweg rill produced considerable runoff. During the second storm, hydraulic measurements were made along the thalweg rill (see Slattery *et al.* 1994 for further details) where a maximum discharge of 31 l/s was observed. By the end of this event, the thalweg rill had extended nearly to the stream channel itself, never quite reaching it because of infiltration losses en route downslope. Extensive runoff along roads and paths was again observed during these large runoff events, underlining the importance of impermeable linear features in storm runoff production, most notably their ability to conduct water from distant parts of the basin to the stream channel (Burt 1989; De Ploey 1989). After a dry February, the major storms that occurred during the remainder of the study period all resulted in storm hydrographs generated mainly by subsurface stormflow. Surface runoff was rarely observed on catchment hillslopes at this time since soils were too dry and too well protected by an extensive crop cover.

Although not a widespread occurrence in the Stour catchment, our results confirm that infiltration-excess overland flow is not as rare in temperate basins as proponents of subsurface runoff have argued, a point also made by Church and Woo (1990). Soil surface condition rather than climatic regime determines whether this type of storm runoff happens or not. The Partial Area model of Betson (1964) remains the best guide to the location of source areas for infiltration-excess overland flow although, as Burt (1992) notes, it is still difficult to predict which partial areas distant from the stream can introduce runoff and sediment to the channel. The importance of linear features in the landscape as links between the channel and distant source areas must be emphasized. Storm runoff production was found to be a complex phenomenon with both the Variable Source Area and Partial Source Areas models being applicable; no single storm runoff mechanism dominates to the exclusion of all others.

4.5   TOWARD A CONCEPTUAL MODEL OF THE EVOLUTION OF A TILLED SOIL

It is clear that soil physical properties which have a significant influence on runoff and erosion mechanisms are far from constant on tilled land during the year: weather, tillage system, crop type and crop management are all influential factors in this regard. Most of the changes in soil surface condition identified by Imeson and Kwaad (1990) during period A (i.e. immediately following tillage but before any rainfall) were observed in this study. Through complete disruption of large aggregates, tillage produces a highly erodible soil, but infiltration capacity remains very high so that runoff and erosion are most unlikely. According to Imeson and Kwaad, the degradation of the soil in period B takes place in a stepwise manner, each period of breakdown of the soil structure being triggered by rainfall. The main problem here is that the nature of the changes are so complex that general statements are difficult to make. We concur that the process of surface sealing is most important in period B, leading to an increase in soil bulk density, a decrease in porosity, and a decrease in infiltration capacity. However, climatic conditions may reverse the direction of change in a particular soil property, for example, if soils dry out or freeze. Infiltration is then controlled by the level of soil saturation as well as by the intactness of the seal. Slattery (1994) shows that soil strength and aggregate stability also vary during this period.

The main discrepancy between the model proposed by Imeson and Kwaad (1990) and the results obtained in this study relate to period C. According to the model, no further changes of the tilled layer and its physical properties take place: crusting of the soil surface and compaction and subsidence of the tilled layer are at a maximum. In terms of soil bulk density, period B would have ended during November. However, the soil surface suffered considerable disruption during the winter as a result of wetting and drying, freezing and thawing, and the impact of soil fauna and microbial activity. Infiltration capacity was highly variable during this period and, more importantly, cracking of the soil surface raised infiltration capacity well above likely rainfall intensities, making infiltration-excess overland flow most unlikely, except on roads, tracks and wheelings.

While the model relates specifically to the changes in soil properties brought about by rainfall, cultivation practices also result in significant changes to soil surface condition, not only in the immediate effects of tillage, but also more permanent changes in soil structure and aggregate size which persist throughout the year irrespective of the total amount of rain received. All this suggests that the forecasting of soil condition, runoff generation and erosion remains a complex and uncertain task even for an individual field, depending on the exact pattern of cultivation and the vagaries of climate. The Imeson and Kwaad (1990) model provides a reasonable but general guide to the evolution of soil surface condition of tilled soils, but more field studies of this type are required before an improved model can be formulated.

ACKNOWLEDGEMENT

Both authors were members of the School of Geography, Oxford University, UK, when this research was conducted.

REFERENCES

Betson, R.P. 1964. What is watershed runoff? *Journal of Geophysical Research*, **68**, 1541–1552.

Boardman, J. 1990. Soil erosion on the South Downs: a review. In J. Boardman, I.D.L. Foster and J.A. Dearing (eds), *Soil Erosion on Agricultural Land*. John Wiley and Sons, Chichester, pp. 87–105.

Boardman, J. 1992. Agriculture and erosion in Britain. *Geography Review*, **6**(1), 15–19.

Boiffin, J. 1984. La dègradation structurale des couches superficielles des sols sous l'action des pluies. *Thése Docteur Ingenieur*, Paris. INA-PG.

Bowyer-Bower, T.A.S. and Burt, T.P. 1989. Rainfall simulators for investigating soil response to rainfall. *Soil Technology*, **2**, 1–16.

Bryan, R.B. 1991. Surface Wash. In O. Slaymaker (ed.), *Field Experiments and Measurement Programs in Geomorphology*. A.A. Balkema, Rotterdam, pp. 107–167.

Burt, T.P. 1987. Measuring infiltration capacity. *Geography Review*, **1**(5), 37–39.

Burt, T.P. 1989. Storm runoff generation in small catchments in relation to the flood response of large basins. In K. Beven and P.A. Carling (eds), *Floods: Hydrological, Sedimentological and Geomorphological Implications*. John Wiley and Sons, Chichester, pp. 11–35.

Burt, T.P. 1992. The hydrology of headwater catchments. In P. Calow and G.E. Petts (eds), *The Rivers Handbook, Volume I*. Blackwell Scientific Publications, Oxford, pp. 3–28.

Church, M. and Woo, M.-K. 1990. Geography of surface runoff: some lessons for research. In M.G. Anderson and T.P. Burt (eds), *Process Studies in Hillslope Hydrology*. John Wiley and Sons, Chichester, pp. 299–325.

De Ploey, J. 1989. Erosional systems and perspectives for erosion control in European loess areas. *Soil Technology*, **1**, 93–102.

Dunne, T. and Leopold, L.B. 1978. *Water in Environmental Planning*. W.H. Freeman and Company, New York.

Hewlett, J.D., Cunningham, G.B. and Troendle, C.A. 1977. Predicting stormflow and peakflow from small basins in humid areas by the R-index method. *Water Resources Bulletin*, **13**, 231–253.

Imeson, A.C. and Kwaad, F.J.P.M. 1990. The response of tilled soils to wetting by rainfall and the dynamic character of soil erodibility. In J. Boardman, I.D.L. Foster and J.A. Dearing (eds), *Soil Erosion on Agricultural Land*. John Wiley and Sons, Chichester, pp. 3–14.

Monnier, G. and Boiffin, J. 1986. The effect of the agricultural land use of soils on water erosion: the case of cropping systems in western Europe. In G. Chisci and R.P.C. Morgan (eds), *Soil Erosion in the European Community*, Balkema, pp. 17–32.

Philip, J.R. 1957. The theory of infiltration. 1. The infiltration equation and its solution. *Soil Science*, **83**, 345–357.

Römkens, M.J.M., Prasad, S.N. and Whisler, F.D. 1990. Surface sealing and infiltration. In M.G. Anderson and T.P. Burt (eds), *Process Studies in Hillslope Hydrology*. John Wiley and Sons, Chichester, pp. 127–172.

Slattery, M.C. 1994. Contemporary sediment dynamics and sediment delivery in a small agricultural catchment, north Oxfordshire, UK. *DPhil thesis*, University of Oxford, UK.

Slattery, M.C. and Bryan, R.B. 1992. Laboratory experiments on surface seal development and its effect on interrill erosion processes. *Journal of Soil Science*, **43**, 517–529.

Slattery, M.C., Burt T.P. and Boardman, J. 1994. Rill erosion along the thalweg of a hillslope hollow: a case study from the Cotswold Hills, central England. *Earth Surface Processes and Landforms*, **19**, 377–385.

Taylor, C.H. and Pearce, A.J. 1982. Storm runoff processes and subcatchment characteristics in a New Zealand hill country catchment. *Earth Surface Processes and Landforms*, **7**, 439–447.

Velterop, M.J.H. 1993. The Behaviour and Transport of Nitrate in a Small Agricultural Catchment, North Oxfordshire, UK. *Thesis*, Ingenieur in Milieukunde, HLO Delft, The Netherlands.

5 Hydrological Impact of a High-Magnitude Rainfall Event

R. EVANS

Division of Geography, Anglia Polytechnic University, Cambridge, UK

5.1 INTRODUCTION

Wycoller (National Grid Reference SD 932393) is a scenic hamlet within a small country park in the central Pennines; within and upstream of it are four ancient bridges for which the hamlet is famous (Bentley 1975). On 19 May 1989 a severe storm affected the headwaters of the Wycoller Beck. The resulting flood carried trees and branches which had fallen into the stream and swept away a bridge comprising one slab of gritstone about 5 m long, 0.7 m wide and 0.3 m thick, considered to be about 1000 years old, and did much damage to the hamlet itself. Although an adjacent catchment to the south, Trawden Brook, was also affected by the storm (*Colne Times*, 1989), damage was less severe.

On a visit to the area in June 1989 it was apparent that, compared to earlier in the year, there had been much damage to the stream network, but it appeared localized within the catchment. The obvious freshness of the eroded stream banks and of the boulder- to sand-size deposits in the channels and on adjacent banks, suggested that it should be possible to map and measure erosion and deposition within the whole of the 9.2 km² catchment. Also, other than undercutting of banks by streams, no mass movements had occurred on slopes so that sources of sediment external to the stream channels could be ignored; the complexities of sediment delivery (Walling 1983, 1988) did not have to be unravelled therefore. It seemed possible then that for this small catchment a sediment budget could be estimated for this one storm.

The impacts of individual large storms on upland catchments in the UK have been described in detail in many papers (Hudleston 1930; Learmonth 1950; Scott 1950; Bleasdale and Douglas 1952; Gifford 1953; Kidson 1953; Common 1954; G. W. Green 1955; Baird and Lewis 1957; F. W. H. Green 1958; 1971; Bowes 1960; Crisp *et al.* 1964; Beven *et al.* 1978; Newson 1980; Acreman 1983, 1989a, 1991; Harvey 1986; Carling 1986; Addison 1987; Jenkins *et al.* 1988; McEwan and Werritty 1988; Coxon *et al.* 1989; Collinge *et al.* 1992). Rarely were stream channels just or mainly affected by the storm (Gilligan 1908; Miller 1951; Bleasdale 1957; Duckworth 1969). Occasionally the storm affected one catchment more than adjacent ones (Werritty 1984; Reid *et al.* 1989), but usually the storm affected a considerable area of ground.

In the UK there are a few quantitative assessments of volumes of material eroded and deposited in or adjacent to channels during large storms (Carling 1986; Harvey

Advances in Hillslope Processes, Volume 1. Edited by M. G. Anderson and S. M. Brooks.

1986; Acreman 1983). In the USA, cross-sections were taken across valley floors to show depths of material eroded or deposited in a large storm, or depths of deposition across a valley floor were mapped and quantified (Wolman and Eiler 1958). But, often because of the large area affected by the storm and because both slopes and channels have suffered erosion and deposition to add to the complexity, as far as I know there has been no study done to estimate a sediment budget for an individual storm.

Sediment budgets are usually derived by making an inventory of material eroded and deposited within a catchment for a period of time (Dietrich and Dunne 1978; Trimble 1983; Roberts and Church 1986; Phillips 1987; Slaymaker 1993); they do not relate to one storm. But the amount of sediment eroded and deposited in a storm, the storm sediment budget, for a whole catchment rather than just a channel reach (Newson 1989), could give a more detailed insight into our understanding of sediment sources and sinks and lead to a better understanding of what is happening in headwaters (Lewin 1989).

Sediment transported in channels is usually measured in the suspended state or as bedload (Hadley and Walling 1984; Walling 1988), which is not appropriate for estimating a sediment budget, and rarely as a volume of material (a sediment "wedge") (Roberts and Church 1986). In this study, volumes of material eroded from stream channel banks and floors were estimated as the inputs into the sediment budget, and the volumes of sediment deposited, were estimated as the output.

5.2   STORM CHARACTERISTICS

A "wall of water several feet high" (*Colne Times*, 1989) surged through Wycoller just before 1700 hours (BST) on Friday 19 May 1989. The rainfall recorded for that day in the standard gauges nearest the Wycoller Beck catchment (Figure 5.1) were 48.4, 31.0 and 10.6 mm. Although these amounts probably fell within one short storm, they do not seem particularly high in the light of the damage done to the stream network by the flood. However, to the east and south of the catchment, on the Yorkshire side of the Pennines, a spectacular thunderstorm caused massive flooding in a series of valleys draining south and east to the River Calder. This storm and some of its impacts have been described by Acreman (1989a,b).

At The Lodge by Walshaw Dean Middle Reservoir, just over 3 km to the southeast of the watershed separating Wycoller Beck and Walshaw Dean, 193.2 mm of rain which fell in about 2 hours, much of it between 1600 hours and 1700 hours was recorded in a standard gauge. The amount was queried because it did not relate well to the estimates made using radar from a nearby station (Nicholls 1990; Collier 1991) but Collinge *et al.* (1990) considered that although the gauge did overestimate the rainfall, it was "not by much". Collinge *et al.* (1992) accepted the figure of 193 mm rainfall for the storm, and put the Walshaw Dean storm as top of their list of maximum observed rainfalls in the UK for durations of 1.5–2 h. A further 3 km to the southwest of Walshaw Dean Lodge, the rainfall was as low as 4.1 and 7.0 mm. As Acreman noted, not only was this clearly an extreme event (1989a,b), but it was very localized for despite the intensity of the rain at the centre of the storm, the average

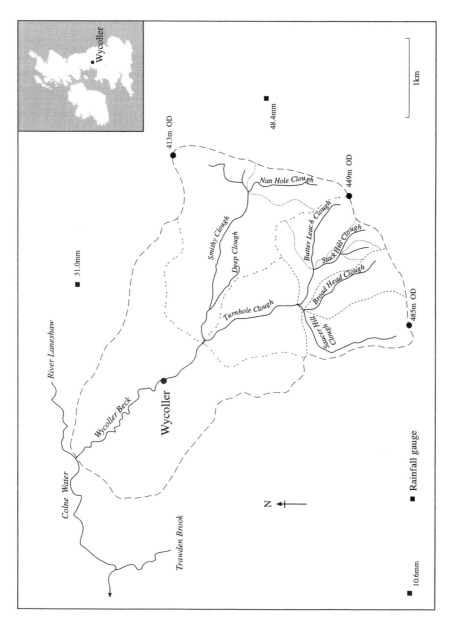

Figure 5.1 Sketch map of the Wycoller catchment

rainfall was not so extreme at the reservoir catchment scale for the runoff to pose a threat to any of the reservoir dams (Acreman 1990).

Data from rainfall gauges sited around the Wycoller catchment show that the first ten days in May were dry, about 17 mm of rain fell on 12 and 13 May, 1–2 mm on 15 May, and 0.4–1.4 mm on the two days prior to the storm. After the storm, up to and during the time fieldwork was carried out in August and September, there were no floods; indeed the summer of 1989 was notably dry, July and August having about half their normal rainfall.

5.3 WYCOLLER BECK CATCHMENT

The streams heading on the northeast flanks of Boulsworth Hill drain north and west to form Wycoller Beck (Figure 5.1). The catchment is less than 10 km^2 in extent, and three of four headwater catchments are about 1 km^2 in area or smaller (Table 5.1). Many first-order channels have not been mapped by the Ordnance Survey, but they are clearly seen on aerial photographs and on the ground; many of these flow intermittently. First-order tributaries on peat occasionally flow underground in pipes or tunnels before reappearing. Saucer Hill and Broad Head Clough are third order basins, as are the upper parts of Stack Hill Clough and Butter Leach Clough. Below the confluence of lower Stack Hill Clough, Turnhole Clough is a fifth-order stream. Smithy Clough, below Nan Hole Clough, a fourth-order stream, joins the fifth-order Turnhole Clough. Drainage ditches dug into peat drain into the headwaters of Smithy Clough and Broad Head Clough, and in the latter catchment the ditches appeared very recently excavated. Channel width varies from as little as 0.3 m near channel heads to 7–12 m along Colne Water.

Table 5.1 Catchment/section characteristics

Catchment/section	Area (ha)	Length channel (m)	Relief (m)	Mean gradient
Saucer Hill Clough	80.5	1735	177	0.10202
Broad Head Clough	55.0	1575	174	0.11048
Stack Hill Clough	102.5			
– upper west	7.5	436	70	0.16055
– upper east	12.5	470	61	0.12979
– middle	–	405	61	0.15062
– lower	–	500	37	0.07500
Butter Leach Clough	47.0	1065	117	0.10958
Smithy/Nan Hole Clough	290.5			
– Nan Hole Clough	43.0	1210	126	0.10413
– Smithy Clough	–	2305	105	0.04555
Turnhole Clough	322.5	1435	72	0.04997
Wycoller Beck	920.0	2771	42	0.01508
Colne Water	–	1914	20	0.01045

On Boulsworth Hill the watershed lies between 449 and 485 m OD, and on the west-facing side the ground rises steeply from about 300 m OD to the ridge crest (Figure 5.2). Above Wycoller the valley floor is narrow and often flanked by the steep slopes of the last phase of incision. Below Wycoller the valley floor widens out, but the sinuous stream channel is often confined by terraces. Around Wycoller and below Laneshawbridge the channel is often confined within walled banks. The confluence of Wycoller Beck and the River Laneshaw to form Colne Water is at 170 m OD, and where the latter meets Trawden Brook (Figure 5.1) the confluence is at about 153 m OD.

The Wycoller catchment is underlain by much-faulted Westphalian (Lower Coal Measures) and Namurian (Millstone Grit Series) aged rocks (Earp *et al.* 1975) of sandstone and shale lithology. However, till and head cover most of the lower and upper slopes respectively, except where sandstone and gritstone outcrop. Peat, of the Winter Hill soil association (Allison *et al.* 1983; Jarvis *et al.* 1984), occurs on the

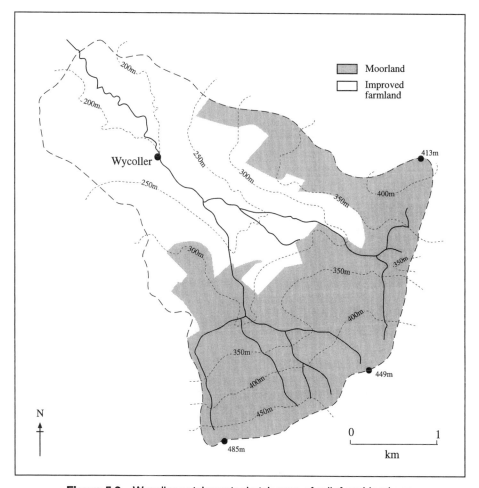

Figure 5.2   Wycoller catchment: sketch map of relief and land use

gently sloping interfluves above 300 m OD. On steep slopes soils are of the Belmont association of very acid sandy loams over sandstone with a wet peaty surface horizon and thin discontinuous ironpan, with some thin peats and bare rock and scree locally. This association is found on the steep slopes of Boulsworth and on the upper slopes of the north side of the Wycoller valley. Below the Belmont and Winter Hill associations is the Wilcocks 1 association on gentle slopes and crests. These are wet for much of the year and have a peaty surface horizon; soil textures are mostly clay loams or sandy clay loams but are occasionally coarser textured. On the lower ground, soils similar in texture occur, but without the peaty top: the Brickfield association. Silty alluvium lies on the floodplain and terraces below Wycoller.

The flanks of Boulsworth and the northeast of the catchment are dominantly unenclosed moorland (Figure 5.2), mostly of cotton-grass (*Eriophorum* spp.) with some mosses, with fine-leaved acidic grassland species (mostly *Nardus stricta*, *Deschampsia flexuosa*) and bracken (*Pteridium aquilinum*) on steeper slopes, the latter where soils are deeper; and heather (*Calluna vulgaris*) and bilberry (*Vaccinium myrtillus*) occur in places. The lower slopes of the catchment have walled fields of improved grasses except where rushes (*Juncus* spp.) have invaded, and on steep slopes, where acidic grasses, bracken and woodland can occur. From just below Wycoller, trees flank the stream up to about 270 m OD on Turnhole Clough and 230 m OD on Smithy Clough. The land use is pastoral, with dairy and beef cattle and sheep in the fields below about 325 m OD and sheep on the moors above.

5.4 ASSESSING THE IMPACTS OF THE STORM

Preliminary reconnaissance in June 1989 indicated that below Wycoller erosion and deposition had occurred at widely spaced intervals, not just to the confluence with the River Laneshaw but for another 1.914 km to the confluence with Trawden Brook. Erosion and deposition were often intermittently spaced too on Smithy Clough. When the full survey was carried out in August and September 1989, on those sections of channel the volumes eroded and deposited were estimated and mapped as they were found.

However, along Turnhole Clough and its tributaries, erosion and deposition were frequent. It was not possible to measure all the variations in depth of channel incision or bank undercutting and collapse, nor allow for the frequently varying shape, width and depth of depositional features. To overcome the problem of recording this variability, measures were taken along transects across the stream and valley floor at 60 pace intervals.

Pacing is a rapid means of estimating distance. Also, once started, measurements are taken at set intervals, not at chosen locations. Sixty of my paces on level or gently sloping ground cover about 50 m. However, this was not so where the terrain was rough, as occasionally in the bed of the stream, or when slopes were steep. Accordingly, when walking along the stream bed it was noted, for example, where bridges crossed the channel, walls abutted or ran alongside the stream, and where marked bends occurred, and then the estimated distances compared with those measured on a 1:10 000 scale Ordnance Survey map; the paced distance was then adjusted accordingly. The shortest distance measured on the map which was covered by my 60 paces was 38 m.

At each transect measurements of widths and depths of erosion were made. It was rarely difficult to decide if a bare soil scar was freshly formed or not. Generally, scars exposed at an earlier date were covered in mosses or lichens and often, they had a marked "undercut" on them showing the flood height of the May storm.

The depth and width of deposits at each transect, if there were any, were noted. The type of deposit was also noted, i.e. silt, sand, shale, gravel, pebbles, cobbles or boulders. Unfortunately, although there are size connotations here, the approximate length of the longest axis being 0.1 mm for silt, 0.1–2 mm for sand, 2–50 mm for shale fragments, 2–50 mm for gravels, 50–100 mm for stones, 100–200 mm for cobbles, and > 200 mm for boulders, these size classes do not relate precisely to those of other classifications. For example, silt and sand may be < 0.06 mm and 0.06–2 mm respectively (Hodgson 1976), and gravel, cobbles and boulders 2–64 mm, 64–256 mm and 256–4100 or 4096 mm respectively (Ward 1984; Church *et al.* 1987). Most boulders were scratched and chipped, and most stones, cobbles and boulders were unweathered on at least their uppermost face. Axial dimensions of the larger cobbles and boulders and of blocks of peat or turf were measured.

Wherever a trash-line was found, or debris of plants or twigs was lodged in trees, walls or bridges, this was noted and its height above the channel floor noted.

Photographs were taken at most transects except where there had been no erosion or deposition. Photographs were often taken of erosional and depositional phenomena which occurred between transects, as well as notes and measurements where appropriate. The photographs were used to help estimate the lengths of erosional and depositional features. Volumes of material eroded and deposited were estimated for each sample site or nominal 60-pace length on Turnhole Clough and its tributaries.

Some first- and second-order gullies in peat debouched onto slopes, and were not linked directly to other downslope channels. The gravel and boulder fans at the ends of these gullies were easily seen from a distance. The positions of these fans were noted but the amounts eroded and deposited were not estimated.

In all, 15.821 km of channel was surveyed. Deep Clough, a second-order tributary to Smithy Clough (Figure 5.1), was not examined in detail, although from a footpath a small amount of deposition was seen in the valley floor. This stream appeared less affected by the storm than Smithy Clough. Many first-order tributaries in peat or on the lower slopes were not eroded. It is considered that most of the material eroded and deposited in the catchment was accounted for.

Many of the larger depositional features noted on the ground in 1989, such as spreads of gravel and stones and boulder fans, were still visible a year later on 1:10 000 scale colour aerial photographs taken on 2 May 1990 for Lancashire County Council. The width of the valley floor flooded during the storm was also clearly visible on the photographs because of the more vigorous and greener vegetation growth where the surface had been "fertilized" by fine deposits.

The volumes of material eroded and deposited were plotted on a logarithmic scale onto the stream channel thalwegs taken from 1:10 000 scale maps (Figures 5.3–5.6). The trash-line was also plotted onto the thalwegs.

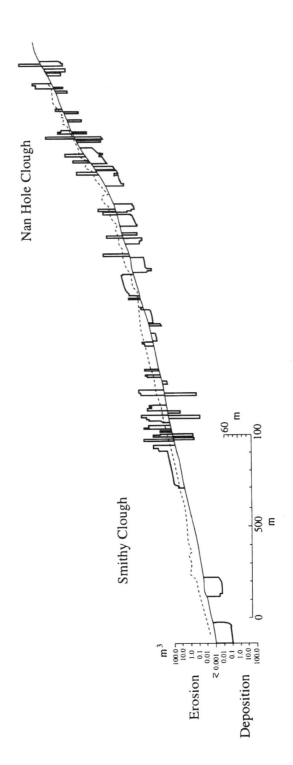

Figure 5.3 Volumes eroded and deposited, and depth of streamflow as indicated by the trash-line: (- - -) Butter Leach and Stack Hill Cloughs

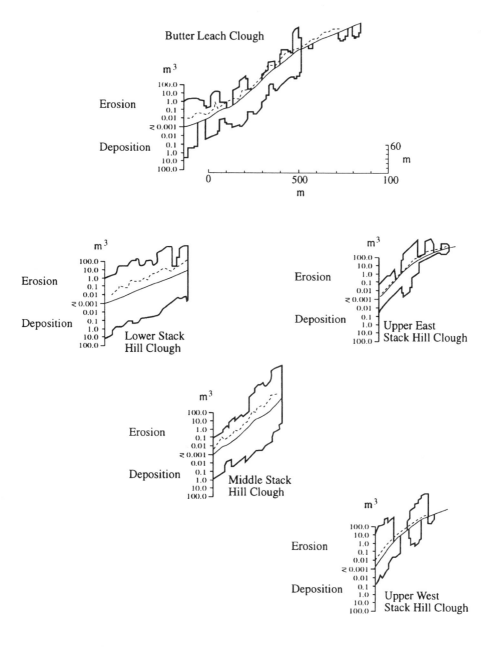

Figure 5.4 Volumes eroded and deposited, and depth of streamflow as indicated by the trash-line (- - - -): Butter Leach and Stack Hill Cloughs

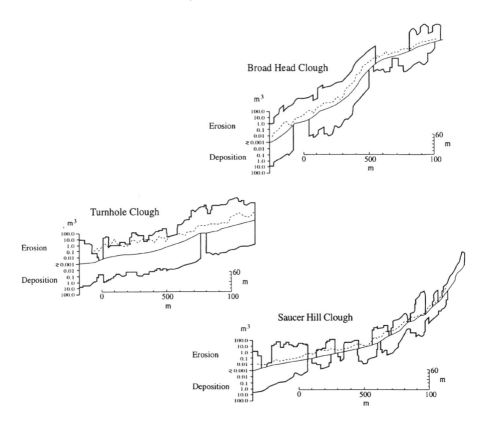

Figure 5.5 Volumes eroded and deposited, and depth of streamflow as indicated by the trash-line (- - - -): Broad Head and Saucer Hill Cloughs with Turnhole Clough below their confluence

5.5 IMPACTS OF THE STORM

5.5.1 Flood Levels

In the depression heading on the watershed ridge above the easternmost tributary of Stack Hill Clough, blades of dead grass and small fragments of peat had been deposited on the tops of clumps of bilberry, indicating a depth of flow of 250 mm between the tussocks of cotton-grass and bilberry. Elsewhere, no other traces of flow were seen in hollows above channels. Debris indicating flood levels was found further down channel in the outer streams Nan Hole Clough and Saucer Hole Clough, than it was in the other headwaters (Figures 5.3–5.5).

The height of the debris- or trash-line varied depending where the measurement was taken; for example, it was higher on the outside of a bend. The height also varied with the type of channel cross-section, i.e. whether broad and wide or shallow and deep; with channel smoothness; where dams of trees and/or boulders were present; and as the volume of water increased downstream.

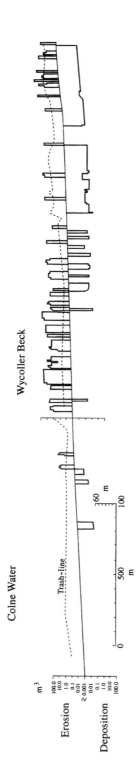

Figure 5.6 Volumes eroded and deposited, and depth of streamflow as indicated by the trash-line: Wycoller Beck and Colne Water

Generally, flood levels as shown by the trash-lines were higher in the headwaters of the two centre basins, especially Broad Head Clough (Figure 5.5) where the mean of 55 measures was 0.97 m above the channel floor. The mean heights of the trash-lines on the two tributaries of Stack Hill Clough (Figure 5.4) to the east of Broad Head Clough, were 0.40 m ($n=8$) and 0.38 m ($n=11$), becoming 0.86 m ($n=25$) below their confluence. On Butter Leach Clough (Figure 5.4), still further north and east, the mean flood level was 0.57 m ($n=53$), and below the confluence with Stack Hill Clough 1.13 m ($n=19$). The mean height of the trash-line in the westernmost basin, Saucer Hill Clough (Figure 5.5), was 0.59 m ($n=51$) and in the northern and easternmost tributary, Nan Hole Clough flowing into Smithy Clough (Figure 5.3), 0.54 m ($n=69$). Along Turnhole Clough, below the confluence of Broad Head and Stack Hill Cloughs, the mean flood level was 1.60 m ($n=55$), rising to 1.68 m ($n=26$), and was highest (1.92 m, $n=20$) above the confluence with Smithy Clough (Figure 5.5), and 1.90 m ($n=20$) to just below Wycoller (Figure 5.6). Along Colne Water (Figure 5.6) the mean height of the trash-line fell to 1.44 m ($n=8$).

5.5.2 Channel Bank Erosion

Bank erosion took place mostly where the channel was cut into head or till (Figure 5.7), but shale (Figure 5.8), peat (and its subsoil, Figure 5.9), flood loams and gravels (Figure 5.10) also eroded. Blocks of flaggy sandstone in the stream bed were moved only rarely. Shale outcropped mostly in the lower reaches of Saucer Hill Clough (Figure 5.5) and in places in the lower stretches of Nan Hole Clough (Figure 5.3). Channels in peat became undercut about 50 m below the heads of the channels,

Figure 5.7 Bank erosion in head, Turnhole Clough

Figure 5.8 Bank erosion in shales, Stack Hill Clough

except where smooth-sided drainage ditches had been dug into peat in Broad Head Clough where undercutting did not take place for about 180 m. On the steep headward section of Saucer Hill Clough, bank erosion occurred up to the point of channel incision (Figure 5.5). Where banks were undercut the turf mat had not always collapsed into the stream and occasionally had been "peeled" back from the underlying soil by the strongly flowing water (Figure 5.11).

Banks eroded rarely in Saucer Hill (Figure 5.12) and Nan Hole/Smithy Cloughs, the outermost of the four catchments, and in Butter Leach Clough and the easternmost tributary of Stack Hill Clough. This was in contrast to the much more severely eroded western headwater of Stack Hill Clough, the central and lower portions of Stack Hill Clough and of Broad Head Clough. Turnhole Clough too suffered massive erosion, but bank erosion rapidly decreased in severity in Wycoller Beck below the confluence of Turnhole and Smithy Cloughs.

Figure 5.9   Collapse of peat and subsoil in walls of gully, near head of Broad Head Clough

Figure 5.10   Bank erosion of flood loams over gravel, Wycoller Beck

Figure 5.11   Turf mat "peeled back" by flood

Mean widening of the stream channel was greatest along Turnhole and Broad Head Cloughs (Figure 5.13), and much less in the outermost basins, Butter Leach Clough and the easternmost tributary of Stack Hill Clough. The extent of widening decreased rapidly downstream of Turnhole Clough. The mean width of bank widening where erosion actually took place was 407 mm in Turnhole Clough and 370 mm in Broad Head Clough.

A new channel 40 m long, 1.0 m wide and 0.3 m deep was cut into head in Broad Head Clough, just above its confluence with Saucer Hill Clough.

Incision occurred dominantly where the stream bed was in head (Figure 5.14) or, occasionally, in peat; and once in flaggy sandstone where a block was removed above a waterfall. No erosion of the stream bed was seen in Colne Water nor Wycoller Beck, nor in Smithy and Nan Hole Cloughs (Figure 5.15). Incision was rare in Saucer Hole and Butter Leach Cloughs, and the eastern tributary of Stack Hill Clough and, in contrast to bank erosion, in Turnhole Clough. Incision took place over more than one-fifth of the channel length in the western tributary and the lower section of Stack Hill Clough, and in Broad Head Clough. Mean depths of incision reached 83 mm along Broad Head Clough and 70 mm along the western tributary of Stack Hill Clough; elsewhere it was less than 38 mm (Figure 5.16). Incision provided large proportions of the volumes of material eroded in Stack Hill Clough, especially its headwaters, and in Broad Hill Clough (Figure 5.17).

The heads (knick points) of intermittent gullied channels in peat had not often retreated. The head of the last section of gully in peat in Nan Hole Clough had

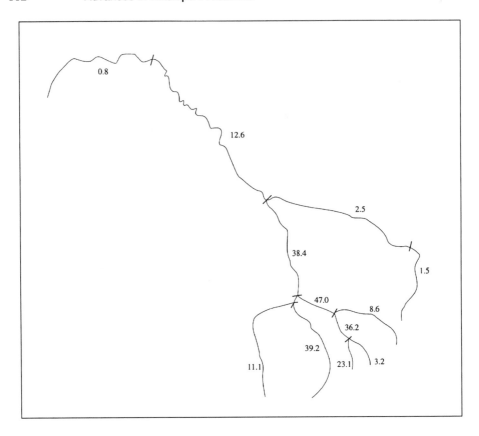

Figure 5.12   Bank erosion: per cent length of channel (both sides stream) eroded

retreated 0.2 m, and the heads of four lengths of gullies in Broad Head Clough had retreated on average 0.37 m.

Most erosion then, taking into account both channel widening and incision, occurred in Turnhole Clough, Broad Head Clough and Wycoller Beck (Figure 5.18). Mean erosion along a unit length of stream was almost 2000 times greater along Turnhole Clough than it was along Colne Water (Figure 5.19).

5.5.3   Deposition

In the upper reaches (400–500 m) channel and bank deposits were generally of gravel size or less; below this, often as gradients steepened, deposits became coarser, becoming again mostly gravelly in the lower third of Turnhole Clough as gradient lessened, and dominantly sandy below Turnhole Clough. Bouldery deposits were especially notable in Broad Head Clough (Figure 5.20), and the lower section of Stack Hill Clough, and the upper section of Turnhole Clough. Below Wycoller, sandy deposits were held back behind trees which had fallen across the channel.

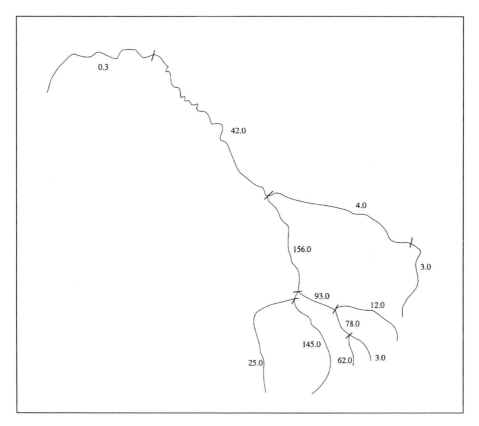

Figure 5.13   Channel erosion: mean widening (mm) channel

Most deposits were within stream channels but overbank deposits, especially of sands and shales, were common throughout the stream network.

In upper Broad Head Clough peat blocks up to $1.5 \times 1.0 \times 0.5$ m, which had collapsed from the banks into the stream, were transported 30–50 m; and in the middle section of Stack Hill Clough peat blocks $0.5 \times 0.4 \times 0.2$ m were moved at least 400 m.

In Broad Head Clough boulders up to $0.8 \times 0.55 \times 0.55$ m were piled high in boulder jams and one boulder of $0.9 \times 0.7 \times 0.55$ m was found on the grassy bank. However, many of these large boulders probably had not travelled far; they had come from undercut banks which had collapsed into the stream and their momentum as well as the power of the stream carried them only a short distance. In Turnhole Clough boulders up to $0.5 \times 0.5 \times 0.3$ m were rolled along the stream bed; on sand bars in the stream bed marked boulders up to $0.5 \times 0.3 \times 0.2$ m were deposited; and in overbank deposits on grass were boulders $0.2 \times 0.2 \times 0.2$ m. In places banks had collapsed but the blocks of soil and turf had not been transported downstream.

In Saucer Hill Clough gravel-sized newly eroded rock fragments were carried at least 250 m from their source and in the same Clough fresh shale fragments were

Figure 5.14   Stream bed incision, Broad Head Clough

transported at least 150 m beyond the shale outcrop into Turnhole Clough. In Nan Hole Clough fresh shale fragments were transported at least 750 m from the outcrop into upper Smithy Clough. Sand particles must often have been carried longer distances than these as they were most frequently found as deposits below Turnhole Clough. Rarely were finer deposits found.

Deposition was greatest in Turnhole Clough, Broad Head Clough and Wycoller Beck (Figure 5.21). The proportions of the total volumes eroded and deposited were similar in the different parts of the stream network (Figures 5.18 and 5.21).

5.5.4   The Sediment Budget

There were inputs into the stream network from outside the channels monitored. In the upper reaches of west Stack Hill Clough, Butter Leach Clough and Nan Hole Clough, deposition of peat blocks or gravel from unmonitored tributaries was recorded upstream of erosion, and totalled about 1.26 m^3. In lower Stack Hill Clough, an undercut depression on the steep right bank valley side into which a channel had incised contributed at least 15.0 m^3.

Amounts eroded and deposited along a stretch of channel were often not greatly discrepant. In Nan Hole Clough more deposition was recorded than erosion (Table 5.2), but taking into account the at least 0.94 m^3 contributed by unmonitored channels, the excess is small. The discrepancy between erosion and deposition is greater in lower Stack Hill Clough, although at least 15.0 m^3 of deposits can be accounted for from the incised right bank tributary, and possibly more, as another

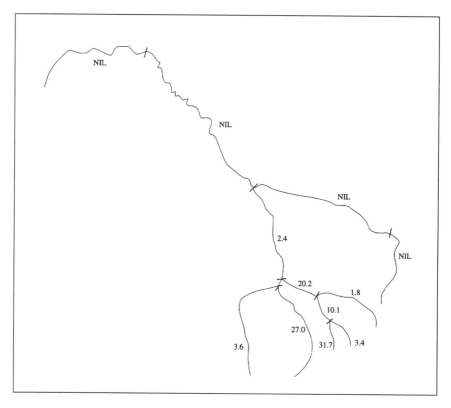

Figure 5.15   Channel incision: per cent length incised

right bank tributary above the incised gully was noted in the field to have eroded slightly. It is likely that the "excess" deposits in the upper reaches of east Stack Hill Clough and Butter Leach Clough were also contributed by unmonitored channels. However, these discrepancies in total are small when compared with the total amounts moved in these catchments, i.e. 3.7–4.4% of the total volumes eroded and deposited. That deposition exceeded erosion along the gentler gradients of Colne Water is not surprising, although the amounts recorded are small.

Much more erosion than deposition was recorded in upper West Stack Hill Clough, Broad Head Clough, from where about 51.0 m³ was laid down as overbank deposits flanking the lowermost reach of Saucer Hill Clough, and Wycoller Beck. The estimated volume eroded was about 16% (352 m³) larger than the amount estimated to have been deposited.

5.6   DISCUSSION

The pattern of erosion in the headwaters of Wycoller Beck cannot be related to channel gradient, differential erodibility of stream banks and beds, nor to land use. Slope morphology may have been a contributory factor, as the uppermost part of

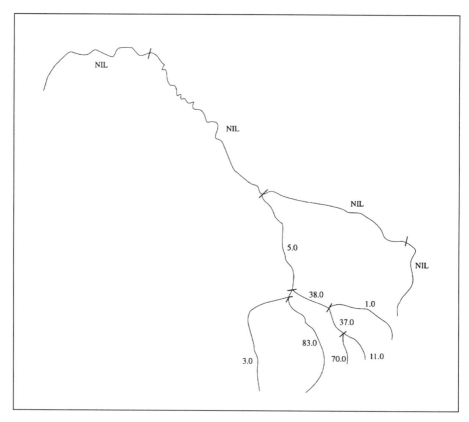

Figure 5.16   Channel incision: mean depth (mm) incised

Broad Head Clough is a gently sloping depression larger than those in which the other streams head, and within this depression a network of drainage ditches had been dug to enable water to drain away more quickly.

The pattern of erosion was related to the distribution and intensity of the rainfall during the storm. The centre of the storm must have been located very close to the headwaters of Broad Head Clough and the western tributary to Stack Hill Clough (at about Grid Reference SD 958365), with a very marked declining rainfall gradient away from this point to the west, north and south (see below). The rapid transit of rainfall through the stream networks of Broad Head and Stack Hill Cloughs led to severe erosion below their convergence in Turnhole Clough.

Generally, bank erosion occurred just downstream of the apex of bends where (1) the mean stream gradient was low, as in Wycoller Beck and Colne Water (Table 5.1); or (2) in steeper headwater catchments where flood levels were lower than those in other headstreams, in Smithy and Nan Hole Cloughs. Where gradients were steeper and/or flood levels higher, erosion took place not just on bends but also on straighter sections. In these sections, an increase in gradient over short (i.e. a few metres) rather than long (i.e. tens or hundreds of metres) lengths of channels, or a supply of eroding material, triggered erosion. Once bank erosion was initiated in head, which provided

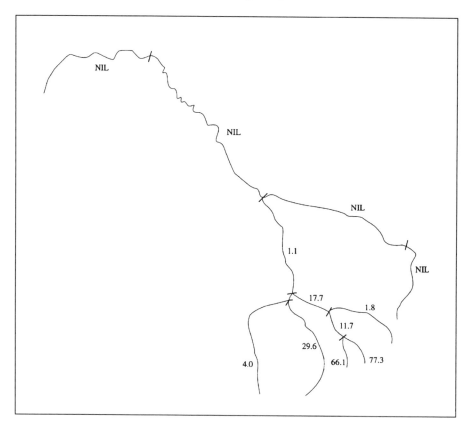

Figure 5.17 Volumes incised as a percentage of the total volume eroded

coarse material to impact on channel floors and walls, erosion occurred over a considerable distance downstream.

In the middle section of Stack Hill Clough, from below a waterfall over an outcrop of sandstone, where the stream took a right-angled bend, to the confluence with Butter Leach Clough, erosion diminished markedly, by almost two orders of magnitude (Figure 5.4). Above the waterfall, a channel which contoured along the slope had been dug, possibly around the turn of the 19th century to divert water to farms enclosed from the moorland waste at that time, and along this channel during the May storm, water flowed some distance. The ponding back of water along this section of easier gradient caused the deposition of much material in, and along the banks of, the channel above the waterfall. Hence, below the waterfall there was little coarse sediment in the water to impinge on the channel walls and initiate erosion.

Channel erosion was rare where banks were lined with trees as their roots bound and protected the soil. But neither walls nor gabions afforded much protection to stream banks as they were easily undercut by the powerfully flowing water.

Most of the sandy deposits surveyed were laid down within the channel above and below Wycoller. Above the hamlet, branches of trees jammed against a series of

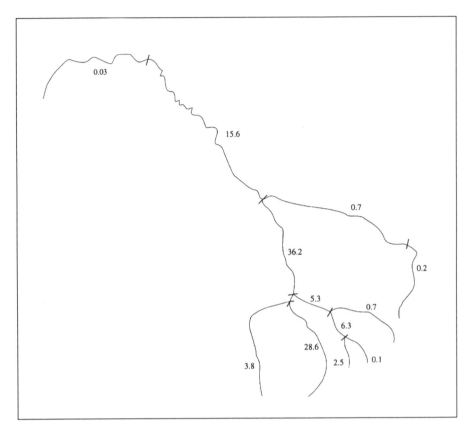

Figure 5.18 Erosion of channel network as a percentage of the total erosion

bridges so damming the flow and causing deposition. Through Wycoller the stream gradient is similar to that up- and downstream, but much of the channel is confined within walled banks. There was no deposition of sediment within the confined channel where water levels were lower (Figure 5.6) and stream velocities higher. Below Wycoller there was a marked fall in water-level of the flood-flow of about 1.2 m over a distance of about 50 m, and this fall was not accompanied by a change in channel gradient (Figure 5.6). At this point the stream flows out of a small tree-lined gorge. Within the gorge fallen trees had impounded the flood-flow and the ponded back water was probably the major cause for the flooding of Wycoller.

Although there was much bank erosion, the undercutting of steep slopes flanking the streams was rare. A number of old bankslips were undercut by the flood, but movement of these was not renewed. There are many old, largely revegetated bank slips throughout the Wycoller catchment, and they must have formed during storms of lesser intensity. They probably formed when floods coincided with saturated ground conditions. Before the May storm soils were unlikely to be saturated as the weather beforehand had been mainly dry; during the storm it is probable the rain fell

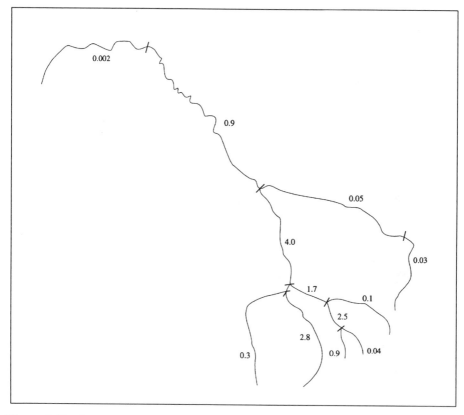

Figure 5.19 Ratio of total erosion along a stream section to total length of that section

at such an intensity that the soil infiltration capacity was exceeded and the rain ran rapidly off the land.

Although this was a major storm and its impacts on the stream network impressive, the headward retreat of gullies on peat was slight. Retreat of about 0.3 m per rare large storm could not account for the formation of these gullies over the last 250 years or so, a time period suggested from other evidence (Evans 1992). The present-day cotton-grass moor is resistant to incision, therefore. However, formerly these moors probably had a greater cover of mosses, especially in depressions, and the death of the mosses because of industrial pollution probably triggered gully erosion (Evans 1992).

The extrinsic erosional threshold (Schumm 1981) of the stream channel was crossed in all parts of the Wycoller catchment, only to a small extent in the outermost and lowest parts of the basin, but to a large extent in the central and middle sections. Indeed, in these last locations the geomorphic threshold (Schumm 1981) has been passed and bank undercutting is continuing to occur, in places causing slips, and sediment is being moved through the network at lower flood flows than previously because more fine material (<2 mm) is now available.

Figure 5.20 Boulder jam, Broad Head Clough (note the small rucksack to the left of the boulder in the foreground and just below the large upstream boulder)

The volume of soil eroded from the stream network was estimated at about 2178 m³. The volume of deposits within the stream channels and on the adjacent banks was found to be about 1826 m³. A net loss out of the catchment, as happened, would have been expected because clay and silt soil particles and aggregates eroded from the stream banks would have been transported downstream (Lewin *et al.* 1974).

About 27 m³ of peat and turf were estimated to have eroded; the same amount was found as deposits. However, some of these organic deposits originated outside the surveyed channels. Even so, the volumes of peat and turf eroded were insignificant with regard to the total volume eroded. The peat and turf blocks appear to have floated or been carried in suspension rather than being rolled on the channel bed and broken down into finer fragments and particles.

The head or till into which the streams had eroded was often stony, and an estimate made from 13 photographs of bank sections indicated a stone content of the soil of about 30% (range 20–60%). This would yield about 645 m³ of stones eroded from a volume of 2151 m³. Deposits of >2 mm diameter were estimated at 1494 m³, that is 1799 m³ of total mineral deposits minus 305 m³ of dominantly sandy as well as finer-textured deposits. Descriptions of similar soil profiles to those eroded (Hall and Folland 1970; Jarvis *et al.* 1984) indicate that of the fine earth (<2 mm) fraction, about 60% by weight is sand and about half of the volume of the soil comprises air and water. It is estimated then that the volume of sand eroded in the storm was 645 m³.

The fraction deposited as stones, therefore, appears too great, and not enough sand seems to have been deposited. However, the stony deposits often contained

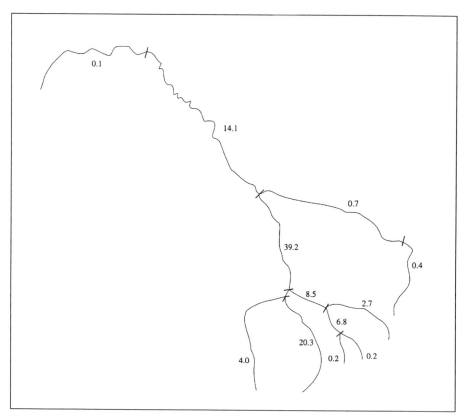

Figure 5.21   Volumes deposited in the stream network as a percentage of the total volume deposited

finer material; Carling and Reader (1982) noted that in four north Pennine streams, deposits could contain between 10 and 27% material finer than 2 mm. Also, finer-textured deposits were carried out of the catchment.

Another source of error is that the estimate of the volume of coarser material deposited in the stream channel was probably the least accurate of the estimates made, with a tendency for overestimation, because the evidence for it is less clear-cut than it is for overbank and sandy deposits.

These calculations are necessarily crude, and do not take into account variations in particle size, bulk density and porosity of the eroded soils or of the stream deposits. But they do suggest that the estimates for volumes eroded and deposited are not widely discrepant, although volumes of different size fractions appear to vary by a factor ranging from one-half to two.

It is estimated (from Lewin *et al.* 1974) that about 5–7.5% of the length of the stream channel banks in the small (0.54 km²) Maesnant catchment in mid-Wales were eroding. This is not too dissimilar from the 1.5–8.6% of the banks of the three channels in the Wycoller Beck catchment least affected by the storm (Figure 5.12). In

Table 5.2 Sediment budget

Catchment/section	Volume eroded (m³)	Volume deposited (m³)	Erosion minus deposition	% transported downstream
Saucer Hill Clough				
(to Broad Head Clough)	82.99	73.52	9.47	11.4
(to Stack Hill Clough)	82.99	124.52	(51.00)[a]	–
Broad Head Clough	622.20	370.21	251.99	40.5
Upper Stack Hill Clough				
– west	54.85	3.14	51.71	94.3
– east	2.77	4.34	(1.57)[b]	–
Middle Stack Hill Clough	137.86	124.13	13.73	10.0
Lower Stack Hill Clough	114.52	154.78	(40.26)[c]	–
Butter Leach Clough	14.33	48.68	(34.35)[b]	–
Nan Hole Clough	4.96	7.27	(2.31)[b]	–
Smithy Clough	15.22	13.02	2.20	14.4
Turnhole Clough	787.17	716.16	71.01	9.0
Wycoller Beck	340.27	257.65	82.62	24.3
Colne Water	0.67	2.47	(1.80)[d]	–
Total	2177.81	1826.37	351.44	16.1

[a]Material transported into section from Broad Head Clough.
[b]Material transported into section from unmonitored tributary.
[c]Material transported into section from upper west and middle Stack Hill Clough plus c. 15.00 m³ from unmonitored tributary.
[d]Material transported into section from Wycoller Beck.

those parts of the catchment most affected by the storm the extent of bank erosion probably increased by about an order of magnitude or somewhat less, therefore.

Acreman (1983) noted that when the Ardessie Burn catchment, northwest Scotland, was subject to severe flooding in September 1981, much of the stream channel erosion took place low down in the 13.3 km² catchment, and about 910 m³ of material was removed. This was much less than the volumes eroded in the Wycoller catchment. However, as he noted, the availability of material to be eroded is important, and there was little material available in the thin soils on the higher slopes of the Ardessie Burn catchment into which the stream was cut, unlike the deep head or till deposits in the Wycoller catchment.

In the Ardessie Burn flood the amount of suspended sediment transported was equivalent to about 14 years' normal average annual sediment output (Acreman 1983). The long-term sediment yield in the UK uplands is probably of the order of 50–60 tonnes per km², but can be as large as 204 tonnes per km² in the southern Pennines where the peat moors are more severely gullied (Labadz et al. 1990) than they are on Boulsworth Hill. Assuming a bulk density of the material eroded from the Wycoller catchment stream channels of 1.2 (Labadz et al. 1990) or 1.9–2.0 (Acreman 1983; Lewin et al. 1974), between 2613 and 4354 tonnes of material was transported in the storm. At the more likely prevailing lower rates of sediment yield quoted by Labadz *et al.* (1990), the amount transported in one storm was equivalent to between 6 and 9 years' average sediment transport.

Storms which triggered mass movements on slopes in the north Pennines (Carling 1986) and Howgill Fells (Harvey 1986) caused much more debris to be delivered to the stream network than did the Wycoller storm. Carling noted that about 5500 m^3 of gravel was deposited in a 1.86 km^2 drainage basin in the northern Pennines in contrast to 420 m^3 in the 0.55 km^2 Broad Head Clough – a ratio of almost 4:1. In the Wycoller catchment the sources of the material which mass movements of some kind could deliver to streams are restricted. There are no extensive areas of peat bog to "burst" or slide, whereas in the adjacent Walshaw Dean catchment there was a massive peat burst in the headwaters of the catchment during the May storm (Acreman 1989a,b; Evans 1992). In the north Pennines storm peat slides carried material to the streams (Carling 1986). Nor, in the Wycoller catchment are the steep slopes underlain by shattered rock very vulnerable either to fragipan (Brooks *et al.* 1995) or ironpan formation (Evans 1992) upon which mass movements can take place during storms, as happened in the Howgill Fells (Harvey 1986).

Putting such erosional events into perspective is not easy. They look spectacular but within the landscape only very small areas are affected, even when mass movements are triggered on slopes (Evans 1992). In this study, in the most affected catchment, Broad Head Clough, channel incision took place over 27% of its length, and bank erosion over 39%; impressive figures when compared with the outermost catchments and lower parts of the Wycoller basin. But channel widening appears slight when compared with the width of the channel; for example, the mean of 26 estimates of the width of Turnhole Clough after the storm was 5.033 m and there had been a mean widening of 156 mm, an increase in width of only 3.2%. Similar figures for Broad Head Clough are 1.959 m (mean of 22 estimates) and 145 mm, an increase in width of 8.0%.

The depths of incision of some of the channels were also impressive; for example, 83 mm for Broad Head Clough. But the surface lowering of the catchment itself was only 1.1 mm, i.e. volume eroded (622.2 m^3; Table 5.2) divided by catchment area (55 ha; Table 5.1). For the 9.20 km^2 Wycoller catchment, surface lowering was only 0.2 mm. These figures compare with the surface lowering caused by rilling of fields in a Sussex chalk downland catchment (Boardman and Evans 1991) of 1.7 mm for the most eroded field, about 47 ha in extent, and 0.3 mm for the whole of the 10.1 km^2 catchment. Erosion in the chalkland catchment took place twice in the autumn and winter of 1990/91 when 60.6% of the basin was largely bare of crop. The erosional events in Sussex and the central Pennines appear of similar magnitude, therefore, but their frequencies of occurrence are very different. In Sussex, a storm similar in size to that which caused the extensive rilling is likely to occur every 3–5 years, whereas the storm in the central Pennines was "unprecedented" (Acreman, 1989b).

Mapping of the damage done to the landscape and stream channel network both in the Wycoller catchment and in the adjacent Walshaw Dean catchment to the south (Evans, unpublished) suggests that the rainfall amount and intensity of the storm over Broad Head and Stack Hill Cloughs was similar to that around Walshaw Dean Lodge rain-gauge. Damage to the landscape and channels was centred around the watershed and Walshaw Dean Lodge localities, with a less damaged area between. The Wycoller catchment was on the fringe of this very localized storm, therefore, which was fortunate for the inhabitants of Wycoller, Winewall and Colne, the downstream settlements.

In the British uplands, therefore, even large storms have an impact on only a very small part of the landscape. This is largely because these slopes are vegetated; an obvious point, but one which needs to be borne in mind in view of the continuing pressure to intensify land use in the uplands, for example, ploughing ditches for afforestation or creating bare soil by overgrazing of sheep. Land-use change to arable or more intensive grazing accompanying global warming (Boardman *et al.* 1990) could have severe implications for stream channel erosion in the uplands and their margins.

5.7 CONCLUSIONS

Mapping of stream channel erosion in the Wycoller catchment showed that it was very localized, and was related to the variability within the catchment of rainfall amount and intensity. Within the storm was a small cell of extremely intense rainfall little more than 500 m in width, but of probably an amount and intensity similar to that recorded at Walshaw Dean Lodge rain-gauge, i.e. about 193 mm in 2 hours. This provided the runoff which spectacularly eroded Broad Head and Stack Hill Cloughs and, where these merged, Turnhole Clough.

The extrinsic threshold triggering channel erosion in the outermost catchments of the Wycoller Basin was a rainfall of more than 50 mm falling over 2 hours. The rainfall intensity of the storm cell caused the geomorphic threshold to be exceeded, so that erosion is continuing in parts of the basin. Elsewhere in the British uplands severe channel erosion has been initiated when the approximate average intensity for 14 storms was 49 mm per hour (Newson 1980).

The sediment budget for different parts of the Wycoller catchment broadly balanced and about 16% of the material eroded was transported out of the catchment. However, much coarse material travelled only metres to tens of metres, whereas gravels and shale fragments probably moved up to a few hundreds of metres, and sands often more than this.

Although erosion in parts of the catchment was spectacular, fortunately the frequency of occurrence of the event was probably very low (once only?), especially when compared, for example, with a similar-sized erosional event of arable land in lowland southern England.

The extrinsic threshold controlling headward retreat of channels in peat is largely a function of vegetation type, in this instance cotton-grass moor, and the threshold could well have been lower in the past when mosses were more widespread (Evans 1992).

As Newson (1989) has noted in a slightly different context, it is important to study the impacts of storms throughout a drainage basin, not just within stream sections. Only then can the storm and its impacts be put into perspective, and our understanding of mechanisms and impacts, especially those exacerbated by the land user, be improved.

The land user made worse the impacts of the May storm in at least three ways. Firstly, by digging drainage ditches in a formerly unchannelled hollow so that rainfall travelled more rapidly into and through Broad Head Clough. Secondly, in response to former storms and consequent bank erosion, sections of the stream channel have been protected by walls or gabions, but these did not protect the banks as well as did the roots of trees in those places where these still survive. One of the responses to the May storm has been to build more walls and gabions. Over the longer term, trees are likely to be better protectors. Thirdly, because trees which had fallen across the stream below Wycoller had not been removed, the flood flow ponded back, so exacerbating flooding in the hamlet. More active management of the riparian zone is needed to make sure that flood flows are not impeded.

ACKNOWLEDGEMENTS

Dr Tom Spencer made helpful comments on an earlier draft of the chapter. I am most grateful to the British Geomorphological Research Group for giving a grant to help me buy aerial photographs of the Wycoller and adjacent catchments.

REFERENCES

Acreman, M.C. 1983. The significance of the flood of September, 1981 on the Ardessie Burn, Wester Ross. *Scottish Geographical Magazine*, **99**, 150–160.
Acreman, M.C. 1989a. *The Rainfall and Flooding on 19 May 1989 in Calderdale, West Yorkshire*. Institute of Hydrology, Wallingford.
Acreman, M. 1989b. Extreme rainfall in Calderdale, 19 May 1989. *Weather*, **44**, 438–446.
Acreman, M. 1990. Extreme rainfall in Calderdale 19 May 1989 – letter to the editor. *Weather*, **45**, 156–157.
Acreman, M. 1991. The flood of July 25th 1983 on the Hermitage Water, Roxburghshire. *Scottish Geographical Magazine*, **107**, 170–178.
Addison, K. 1987. Debris flow during intense rainfall in Snowdonia, North Wales: a preliminary survey. *Earth Surface Processes and Landforms*, **12**, 561–566.
Allison, J.W., Bendelow, V.C., Bradley, R.I., Carroll, D.M., Furness, R.R., Kilgour, I.N.L., King, S.J. and Matthews, B. 1983. Soils of Northern England. Sheet 1. *Soil Map of England and Wales. Scale 1: 250 000*. Soil Survey, Harpenden.
Baird, P.D. and Lewis, W.W. 1957. The Cairngorm floods, 1956: summer solifluction and distributary formation. *Scottish Geographical Magazine*, **73**, 91–100.
Bentley, J. 1975. *Portrait of Wycoller*. Nelson Local History Society, Nelson.
Beven, K., Lawson, A. and McDonald, A. 1978. A landslip/debris flow in Bilsdale, North York Moors, September 1976. *Earth Surface Processes*, **3**, 407–419.
Bleasdale, A. 1957. Rainfall at Camelford, Cornwall on 8th June 1957. *Meteorological Magazine*, **86**, 339–343.
Bleasdale, A. and Douglas, C.K.M. 1952. Storm over Exmoor on August 15, 1952. *Meteorological Magazine*, **81**, 353–367.
Boardman, J. and Evans, R. 1991. *Flooding in the Steepdown Catchment*. Unpublished Report to Adur District Council.
Boardman, J., Evans. R., Favis-Mortlock, D.T. and Harris, T.M. 1990. Climate change and soil erosion on agricultural land in England and Wales. *Land Degradation and Rehabilitation*, **2**, 95–106.
Bowes, D.R. 1960. A bog-burst in the Isle of Lewis. *Scottish Geographical Magazine*, **76**, 21–23.

Brooks, S.M., Anderson, M.G. and Crabtree, K. 1995. The significance of fragipans to early Holocene slope failure: application of physically based modelling. *The Holocene*, **5**, 293–303.

Carling, P.A. 1986. The Noon Hill flash floods; July 17th 1983. Hydrological and geomorphological aspects of a major formative event in an upland landscape. *Transactions of the Institute of British Geographers, New Series*, **11**, 105–118.

Carling, P.A. and Reader, N.G. 1982. Structure, composition and bulk properties of upland stream gravels. *Earth Surface Processes and Landforms*, **7**, 349–365.

Church, M.A., McLean, D.G. and Wolcott, J.F. 1987. River bed gravels: sampling and analysis. In Thorne, C.R., Bathurst, J.C. and Hey, R.D. (eds), *Sediment Transport in Gravel-bed Rivers*. Wiley, Chichester, pp. 43–79.

Collier, C.G. 1991. Problems of estimating the extreme rainfall from radar and raingauge data illustrated by the Halifax storm, 19 May 1989. *Weather*, **46**, 200–208.

Collinge, V.K., Archibald, E.J., Brown, K.R. and Lord, M.E. 1990. Radar observations of the Halifax storm, 19 May 1989. *Weather*, **45**, 354–364.

Collinge, V.K., Thielen, J. and McIlveen, J.F.R. 1992. Extreme rainfall at Hewenden Reservoir, 11 June 1956. *Meteorological Magazine*, **121**, 166–171.

Colne Times 1989. Door handle saved Eddie's life! *Colne Times*, No. 6723, 26 May, p. 1.

Common, R., 1954. A report on the Lochaber, Appin and Benderloch floods, May 1953. *Scottish Geographical Magazine*, **70**, 6–20.

Coxon, P., Coxon, C.E. and Thorn, R.H. 1989. The Yellow River (County Leitrim Ireland) flash flood of June 1986. In K. Beven and P. Carling (eds), *Floods: Hydrological, Sedimentological and Geomorphological Implications*. Wiley, Chichester, pp. 199–217.

Crisp, D.T., Rawes, M. and Welch, D. 1964. A Pennine peat slide. *Geographical Journal*, **130**, 519–524.

Dietrich, W.E. and Dunne, T. 1978. Sediment budget for a small catchment in mountainous terrain. *Zeitschrift für Geomorphologie*, Supplementband **29**, 191–206.

Duckworth, J.A. 1969. Bowland Forest and Pendle floods. *Association of River Authorities Yearbook*, pp. 81–90.

Earp, J.R., Magraw, D., Poole, E.G., Land, D.H. and Whiteman, A.J. 1975. *Clitheroe. Sheet 68*. 1 : 50 000 Series. Drift Edition, Geological Survey of Great Britain (England and Wales). Institute of Geological Sciences/Ordnance Survey, Southampton.

Evans, R. 1992. Sensitivity of the British landscape to erosion. In D.S.G. Thomas and R.J. Allison (eds), *Landscape Sensitivity*. Wiley, Chichester, pp. 189–210.

Gifford, J. 1953. Landslides on Exmoor caused by the storm of 15th August, 1952. *Geography*, **38**, 9–17.

Gilligan, A. 1908. Some effects of the storm of June 3rd, 1908, on Barden Fell. *Proceedings of the Yorkshire Geological Society*, **16**, 383–390.

Green, F.W.H. 1958. The Moray floods of July and August, 1956. *Scottish Geographical Magazine*, **74**, 48–50.

Green, F.W.H. 1971. History repeats itself – flooding in Moray in August 1970. *Scottish Geographical Magazine*, **87**, 150–152.

Green, G.W. 1955. North Exmoor floods, August 1952. *Bulletin of the Geological Survey of Great Britain*, **7**, 68–84.

Hadley, R.F. and Walling, D.E. (eds) 1984. *Erosion and Sediment Yield. Some Methods of Measurement and Modelling*. Geo Books, Norwich.

Hall, B.R. and Folland, C.J. 1970. *Soils of Lancashire*. Bulletin No. 5, Agricultural Research Council, Soil Survey of Great Britain, England and Wales, Harpenden.

Harvey, A.M. 1986. Geomorphic effects of a 100 year storm in the Howgill Fells, Northwest England. *Zeitschrift für Geomorphologie*, **30**, 71–91.

Hodgson, M. 1976. *Soil Survey Field Handbook*. Technical Monograph No. 5, Soil Survey, Harpenden.

Hudleston, F. 1930. The cloudburst on Stainmore, Westmorland, June 18, 1930. *British Rainfall 1930*, pp. 287–292.

Jarvis, R.J., Bendelow, V.C., Bradley, R.I., Carroll, D.M., Furness, R.R., Kilgour, I.N.L. and

King, S.J. 1984. *Soils and Their Use in Northern England.* Bulletin No. 10, Soil Survey, Harpenden.

Jenkins, A., Ashworth, P.J., Ferguson, R.I., Grieve, I.C., Rowling, P. and Stott, A.T. 1988. Slope failures in the Ochil Hills, Scotland, November 1984. *Earth Surface Processes and Landforms,* **13,** 69–76.

Kidson, C. 1953. The Exmoor storm and the Lynmouth floods. *Geography,* **38,** 1–9.

Labadz, J.C., Burt, T.P. and Potter, A.W.R. 1990. Sediment yield and delivery in the blanket peat moorlands of the southern Pennines. *Earth Surface Processes and Landforms,* **16,** 255–271.

Learmonth, A.T.A. 1950. The floods of 12th August, 1948, in south-east Scotland. *Scottish Geographical Magazine,* **66,** 147–153.

Lewin, J. 1989. Floods in fluvial geomorphology. In K. Beven and P. Carling (eds), *Floods: Hydrological, Sedimentological and Geomorphological Implications.* Wiley, Chichester, pp. 265–284.

Lewin, J., Cryer, R. and Harrison, D. I. 1974. Sources for sediments and solutes in mid-Wales. In K.J. Gregory and D.E. Walling (eds), *Fluvial Processes in Instrumented Watersheds.* Special Publication No. 6, Institute of British Geographers, London, pp. 723–785.

McEwen, L.J. and Werritty, A. 1988. The hydrology and long-term geomorphic significance of a flash flood in the Cairngorm Mountains, Scotland. *Catena,* **15,** 361–377.

Miller, A.A. 1951. Cause and effect in a Welsh cloudburst. *Weather,* **6,** 172–179.

Newson, M.D. 1980. The geomorphological effectiveness of floods – a contribution stimulated by two recent events in mid-Wales. *Earth Surface Processes,* **5,** 1–16.

Newson, M.D. 1989. Flood effectiveness in river basins: progress in Britain in a decade of drought. In K. Beven and P. Carling (eds). *Floods: Hydrological, Sedimentological and Geomorphological Implications.* Wiley, Chichester, pp. 151–169.

Nicholls, J.M. 1990. Extreme rainfall in Calderdale 19 May 1989 – letter to the editor. *Weather,* **45,** 156.

Phillips, J.D. 1987. Sediment budget stability in the Tar River basin, North Carolina. *American Journal of Science,* **287,** 780–794.

Reid, I., Best, J.L. and Frostick, L.E. 1989. Floods and flood sediments at river confluences. In K. Beven and P. Carling (eds), *Floods: Hydrological, Sedimentological and Geomorphological Implications.* Wiley, Chichester, pp. 135–150.

Roberts, R.G. and Church, M. 1986. The sediment budget in severely disturbed watersheds, Queen Charlotte Ranges, British Columbia. *Canadian Journal of Forest Research,* **16,** 1092–1106.

Schumm, S.A. 1981. Geomorphic thresholds and complex response of drainage systems. In M. Morisawa (ed.), *Fluvial Geomorphology.* Allen and Unwin, London, pp. 299–310.

Scott, A. 1950. Fighting the deluge – the border floods of August 1948. *Scottish Agriculture,* **29,** 127–132.

Slaymaker, O. 1993. The sediment budget of the Lillooet River basin, British Columbia. *Physical Geography,* **14,** 305–320.

Trimble, S.W. 1983. A sediment budget for Coon Creek basin in the Driftless Area, Wisconsin, 1953–1977. *American Journal of Science,* **283,** 454–474.

Walling, D.E. 1983. The sediment delivery problem. *Journal of Hydrology,* **69,** 209–237.

Walling, D.E. 1988. Measuring sediment yields from river basins. In R. Lal (ed.), *Soil Erosion Research Methods.* Soil and Water Conservation Society, Ankeny, pp. 39–73.

Ward, P.R.B. 1984. Measurement of sediment yield. In R.F. Hadley and D.E. Walling (eds), *Erosion and Sediment Yield. Some Methods of Measurement and Modelling.* Geo Books, Norwich, pp. 37–71.

Werritty, A. 1984. Stream response to flash floods in upland Scotland. In T.P. Burt and D.E. Walling (eds), *Catchment Experiments in Fluvial Geomorphology.* Geo Books, Norwich pp. 537–560.

Wolman, M.G. and Eiler, J.P. 1958. Reconnaissance study of erosion and deposition produced by the flood of August 1995, in Connecticut. *Transactions of the American Geophysical Union,* **39,** 1–14.

6 Short-term Hydrological Response of Soil Water and Groundwater to Rainstorms in a Deciduous Forest Hillslope, Georgia, USA

ELIZABETH B. RATCLIFFE,[1] **NORMAN E. PETERS**[2] **and MARTYN TRANTER**[1]
[1]*Department of Geography, University of Bristol, UK*
[2]*Water Resources Division, United States Geological Survey, Atlanta, USA*

6.1 INTRODUCTION

Streamwater comprises a mixture of components from different hydrologic pathways or flowpaths (Cosby *et al.* 1985; Woolhiser *et al.* 1985; Kennedy *et al.* 1986; Eshleman 1988). Changes in the composition of streamwater are determined by changes in flowpaths and the concomitant changes in component water chemistry (Christophersen 1990; Hooper *et al.* 1990; Hooper and Christophersen 1992). Traditionally, modelling of streamwater chemistry during hydrological events has concentrated on two major components only, "old" and "new" water (Dincer *et al.* 1970; Martinec *et al.* 1974; Fritz *et al.* 1976). "Old" water is that resident in the watershed since the last rainstorm at least (e.g. as soil water or groundwater), whereas "new" water is the current rainfall. All major flowpaths within a system need to be identified and intensively sampled to accurately predict streamwater hydrochemical trends (Jenkins *et al.* 1994). Tracer investigations can be combined with these measurements to assess whether the water along these flowpaths is predominantly "old" or "new" (Shanley and Peters 1988; McDonnell 1990).

The detailed hydrology of many watersheds is poorly understood and much conflict still exists amongst hydrologists about which flowpaths contribute most to streamflow and how their contributions vary during rainstorms (Hooper *et al.* 1990; Hooper and Christophersen 1992; Jenkins *et al.* 1994). Within the unsaturated and saturated zones, many flowpaths have been identified as important routes for water transport (Horton 1933; Hewlett and Hibbert 1967; Whipkey 1967; Beven and Germann 1982; Kennedy *et al.* 1986). Thus, intensive sampling of flow in the unsaturated and saturated zones during rainstorms is necessary to determine the flowpaths followed by "old" and "new" waters.

The aim of this investigation is to determine the major flowpaths that carry "old" and "new" water vertically in a forested hillslope during rainstorms. Tipping-bucket gauges were used to measure flow at two depths in the soil (15 and 50 cm) and groundwater stage monitoring was used to measure the response of the water-table.

Advances in Hillslope Processes, Volume 1. Edited by M. G. Anderson and S. M. Brooks.
© 1996 John Wiley & Sons Ltd.

Finally, temperature was used to trace water movement, and to separate contributions of "old" and "new" waters. Prior to presenting results from the study, we outline current ideas of the flowpaths that transmit water to stream channels during rainstorms and how temperature can be used for tracing contributions of "old" and "new" waters.

6.2 FLOWPATHS DURING RAINSTORMS

6.2.1 Overland Flow

Early investigations of streamflow generation suggest that rapid routing of "new" water to the stream channel occurs as overland flow (Horton 1933; Hewlett and Hibbert 1967). Overland flow occurs when "new" water exceeds the infiltration capacity of the soil surface, or, if water infiltrates the soil surface and moves laterally, saturated overland flow occurs when the lateral flow capacity is exceeded and the water table intersects the soil surface (Hewlett and Hibbert, 1967). Consequently, water flows over the ground surface, either as quasi-laminar sheet flow, or more usually following rills or channels on the ground surface (Horton 1933). Overland flow is rarely observed because most vegetation-covered surfaces generally have a high infiltration capacity. It may occur on the lower zones of some hillslopes (Pearce *et al.* 1986; Thomas and Beasley 1986).

6.2.2 Interflow (or Subsurface Lateral Flow)

Interflow occurs in the unsaturated zone and moves water laterally downslope through the upper soil horizons, either as unsaturated flow or as perched saturated flow in the unsaturated zone (Ward 1967). Interflow occurs where the lateral hydraulic conductivity in the upper soil horizons is substantially greater than the overall vertical hydraulic conductivity through the soil profile and the vertical hydraulic conductivity is exceeded by infiltration. There may be several levels of interflow corresponding to textural changes between horizons and to the junction between the weathered mantle and bedrock (Whipkey 1967).

Interflow transports both "old" and "new" water (Dunne and Black 1970). New water may infiltrate the soil and then flow laterally. However, the movement of "new" water into the soil may lead to the lateral movement of "old" water that has been stored in the profile since the previous rainstorm (Luxmoore 1981; McDonnell 1990).

6.2.3 Macropore and Matrix Flow

Important flowpaths have been identified as instrumentation of field sites has become more intensive and more sophisticated (Mulholland *et al.* 1990; Jenkins *et al.* 1994; Peters 1994). Two major flowpaths for soil water have been identified routinely in the unsaturated zone: macropore flow and micropore (or matrix) flow. Many researchers claim that macropore flow is a major mechanism by which "new" water is introduced to the stream channel (Mosley 1974, 1982; Beven and Germann 1982;

Flury *et al.* 1994). Clothier and White (1981) consider macropores to be voids and channels $> 750 \, \mu m$ in diameter, with the remaining soil volume constituting the soil matrix. Water in the soil matrix, which is most of the pore volume, moves more slowly than water in macropores.

Mosley (1974) identified preferential flowpaths in the Maimai Catchment, New Zealand, by several dye injection tracer techniques. Preferred pathways include cracks and holes in the soil, and voids adjacent to roots and root channels. Sklash *et al.* (1986) suggested that rapid, downward transmission of "new" water via macropore or other rapid-flux routes is not consistent with $\delta^{18}O$ variations in streamwaters of the Maimai catchments, since "old" water dominated stormflow (Pearce *et al.* 1986; Sklash *et al.* 1986). Based on computer simulations (Sklash and Farvolden 1979) and physical principles (Gillham 1984), it was suggested that a rapid matrix-flow displacement mechanism occurred through saturated wedges on the lower slopes and groundwater ridges on the valley bottoms. The process involved a capillary fringe response (Ragan 1968) that increased local hydraulic gradients and promoted increased gravity drainage of "old" water to the stream channel (McDonnell 1990).

6.2.4 Monitoring Water Movement using Temperature

Temperature is a relatively conservative property of groundwater (Ward 1967). Ground temperatures vary both diurnally and seasonally and are controlled by inputs of incident radiation. Temperature variations decrease with depth, and deeper groundwaters tend to have relatively constant temperatures (Ward 1967).

The usefulness of temperature for tracing flowpaths relates to the temporal and spatial variability, and the magnitude of the difference in temperature of end-member components. Groundwater temperatures tend to be cooler than air temperatures in summer and warmer in winter. Groundwater temperatures at 10–20 m depth normally exceed mean annual air temperature by 1–2°C (Ward 1967). The lower atmosphere usually warms rainfall as it falls. Hence, rainfall is typically warmer than groundwaters in summer (Shanley and Peters 1988). Temperature is a potential tracer of new water (rainfall) inputs to the groundwater table. Macropore flow should allow the rapid transit of "new", warm water to depth in the summer (and "new", cool water in the winter).

6.3 FIELD SITE

A $20 \, m \times 20 \, m$ forested hillslope plot was chosen for study at Panola Mountain Research Watershed (PMRW). PMRW (84° 10′ W, 33°37′ N) is a 41-ha catchment in Panola State Conservation Park, Stockbridge, Georgia, about 25 km southeast of Atlanta (Figure 6.1). Elevation of the centre of the plot is 231 m above sea level. The hillslope is located on the drainage boundary of a gauged 10-ha subcatchment, which contains a 3-ha bedrock (granodiorite) outcrop. The US Geological Survey initiated field investigations in the watershed in 1985. Annual rainfall at PMRW averages 119 mm and annual temperature averages 15°C. Winds predominate from the

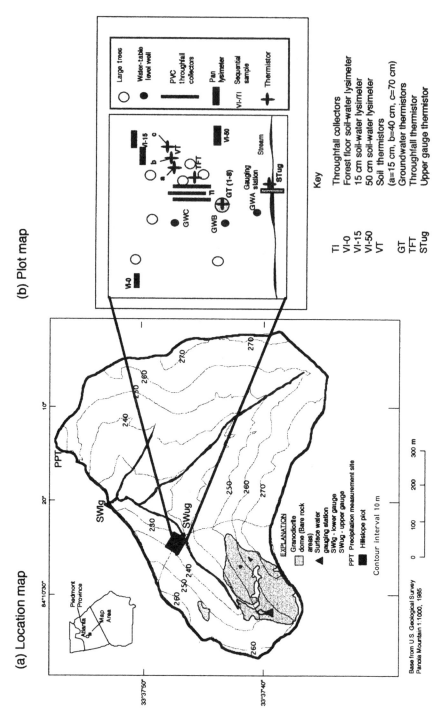

Figure 6.1 Location and relief map of the Panola Mountain Research Watershed (PMRW) and hillslope site showing instrumentation and dominant site characteristics

northwest. Summer rainstorms are convective, whereas most winter rainstorms are caused by frontal systems (Shanley and Peters 1993).

The hillslope plot is vegetated by deciduous species, the dominant of which are *Quercus falcata* (Southern red oak) and *Quercus rubra* (Northern red oak). Other species within the plot include *Carya glabra* (pignut hickory), *Carya tomenttosa* (mockernut hickory) and *Cornus florida* (dogwood). Soils are generally 60–100 cm in depth and are typically Ultisols, reddish brown in colour and consist of the Gwinnett and Pacolet series. The inclination of the hillslope averaged 15°.

6.4 METHODOLOGY

Hydrometric measurements were made at several points, henceforth referred to as nodes, along a vertical profile through a forested hillslope to determine flow rates and flow mechanisms. Measurements included water flow from zero-tension lysimeters, groundwater stage, streamflow and temperature. Results from three rainstorms occurring on 11 July 1994, 16 August 1994 and 10 February 1995 are presented herein (Table 6.1).

6.4.1 Tipping-Bucket Gauges

Tipping-bucket gauges were used to measure the timing and rate of transport of rainfall, throughfall and soil waters at various depths below land surface. Throughfall was intercepted by three 7.62×100 cm PVC troughs (Figure 6.1). Soil water was intercepted at 15-cm and 50-cm depths using zero-tension lysimeters: two 0.23-m^2 stainless-steel zero-tension lysimeters were used to measure flow at 15 cm (VI-15); and one 1-m^2 lysimeter was used to measure flow at 50 cm (VI-50). At each node, flow from the collectors was routed via a combination of PVC pipe and silastic tubing to a plexiglass box containing a Rainwise tipping bucket. Water flux was calculated from the number of tips and the area of the collectors. The tipping buckets were monitored with Campbell Scientific Model CR21X and CR10 dataloggers (Peters 1994).

6.4.2 Groundwater Stage Measurement

Groundwater stage was monitored in three wells (Figure 6.1) one at the base of the slope (GWA), another 5 m upslope (GWB) and another 10 m upslope (GWC).

Table 6.1 Duration and magnitude of select rainstorms at the PMRW, Stockbridge, Georgia, USA

Storm date	Precipitation duration	Total precipitation (mm)
11 July 1994	10 h 45 min	78
16 August 1994	14 h 39 min	59
10 February 1995	26 h 27 min	85

Groundwater stage was monitored continuously (5-min output by a Campbell Scientific Model CR21X datalogger), using potentiometers attached to a float and counterweight system (Peters 1994).

6.4.3 Temperature Measurement

Thermistors were installed to trace temperature variations in the hillslope and stream. In addition to the continuous monitoring (5-min output) of air and throughfall temperatures (TF), thermistors were installed in the soil at depths of 15, 40 and 70 cm (VTa,b and c). Thermistors also were installed in a 4.6-m deep well at depths of 2.44 (GT3), 2.59 (GT4), 2.74 (GT5), 3.35 (GT6), 3.96 (GT7) and 4.57 (GT8) m below land surface. The throughfall temperatures were measured in a 125 ml polyethene bottle that was connected to a 27.5 cm diameter funnel installed under the canopy adjacent to the PVC throughfall troughs. Streamwater temperatures were monitored using thermistors at the basin outlet (41 ha drainage; STlg) and at a gauge adjacent to the hillslope plot (10 ha drainage; STug) (Figure 6.1).

6.5 RESULTS

Results from three rainstorms are presented herein (Table 6.1). Two of these rainstorms, July and August, occurred during the growing season, and the February rainstorm occurred during the dormant season. All were relatively high-magnitude events; 78 mm for the July rainstorm, 59 mm for the August rainstorm and 85 mm for the February rainstorm. However, the antecedent moisture conditions differed markedly. The July rainstorm was preceded by extremely wet conditions caused by tropical rainstorm Alberto (163 mm rainfall in the previous week), but the stream baseflow prior to the storm was the lowest of the three storms at $3.9 \, L \, s^{-1}$. The August rainstorm was preceded by dry conditions (12 mm rainfall in the previous week), but baseflow prior to the storm was intermediate between that of July and February at $4.0 \, L \, s^{-1}$. The February rainstorm also was preceded by extremely dry conditions (1 mm rainfall in the previous week), but baseflow prior to the storm was the highest of the three at $5.4 \, L \, s^{-1}$.

6.5.1 Evidence for Macropore Flow

Soil water response: the timing of response at each node provides evidence for the occurrence of macropore flow. A response at some depth in the soil immediately after rainfall began suggests rapid infiltration of rainwater. In the unsaturated zone, this response could only occur along preferred hydrologic pathways, i.e. macropores, particularly when the antecedent soil-moisture content is very low. Also, if there is a similar temporal response at depth in the unsaturated zone (such as at 50 cm) and in the saturated zone, (i.e. a rise in water-table), then this again would suggest macropore flow through the unsaturated zone.

Figures 6.2(a) and (b) show the tipping-bucket responses to rainfall, throughfall and 15 cm and 50 cm soil waters for the August rainstorm (intermediate baseflow

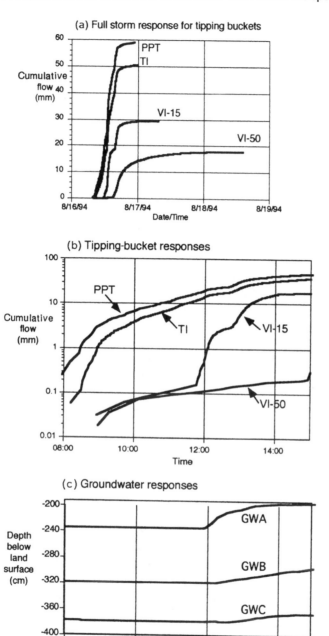

Figure 6.2 Hydrological response for a summer storm on 16 August 1994. (a) tipping-bucket response (flow) during the entire storm for precipitation (PPT), throughfall (TI), and soil-water depths of 15 (VI-15) and 50 cm (VI-50); (b) tipping-bucket response during the first 7 h; (c) response of groundwater levels during the first 7 h

and lowest pre-event rainfall). Rainfall totalled 59 mm in 15 h; however, periods of rainfall were very intense, reaching intensities in excess of 40 mm/h. Soil water flow begins within 1 h of the onset of rainfall, and the onset of flow at 15 and 50 cm depths occurred at the same time (Figure 6.2(b)), suggesting that water was in rapid transit to depth in the soil. After the initial 4 h, flow for the 15 cm depth lysimeter increased rapidly and was almost twice as much as at 50 cm (Figure 6.2(b)). Rainfall intensity at 12:40 hours was high, leading to a major response in throughfall, and at 15 cm depth. Thus, this is also indicative of macropore flow, showing that once the system becomes wetter, a brief period of intense rainfall causes water flow at depth almost instantaneously and with relatively high yields.

The majority of flow at 50 cm depth might be attributed in part to matrix flow, which occurred later in the storm. However, this flow was also concurrent with another intense 10 mm downpour at 15:00 hours, which also produced a rapid response throughout the profile, consistent with additional macropore flow.

Flow results are presented for the July rainstorm in Figure 6.3 and the February rainstorm in Figure 6.4. The July rainstorm was preceded by extremely high rainfall, but relatively low streamflow, so the infiltration capacity of the soil was low. Flow was measured at the onset of this rainstorm at both 15 and 50 cm depths (Figure 6.3(b)), occurring within 10 min of the onset of rainfall at 15 cm depth. Cumulative flow was low at 15 cm depth, totalling only 7 mm (Figure 6.3(a)), suggesting that "new" water did not travel to depth in the soil. Flow at 50 cm depth was difficult to assess during this rainstorm because water was still draining from the previous rainstorm (Figure 6.3(a)).

The February (winter) rainstorm (Figure 6.4) was preceded by low rainfall, but relatively high streamflow. However, the rapid responses that had been noted in the summer rainstorms at the onset of the event were not evident for this rainstorm. The response at 15 cm depth occurred some 9 h after rainfall began. A possible reason for this may be that rainfall was not continuously intense throughout its duration. Rather, rainfall only became intense after the first 5 h (Figure 6.4(a)). The response of the soil water correlates well to this period of intense rainfall, and significant response at 15 cm depth is recorded within an hour of the onset of intense rainfall. Flow at 50 cm depth occurs rapidly after the onset of the rainstorm, but only contributes low water volumes. Most of the response at this depth occurs 17 h after the onset of the rainstorm, which again corresponds to another period of intense rainfall.

Thus, from the responses of tipping buckets in the unsaturated zone, it would appear that macropore flow can be an important mechanism for the rapid transport of water to depth. It would also appear that the occurrence of macropore flow is controlled by antecedent moisture conditions and by rainfall intensity. In general, the wetter the system, the more rapid and substantial the flow at depth, particularly in response to intense rainfall.

Groundwater response: a comparison can be made between the timing of the response of groundwater levels and soil-water flow in the unsaturated zone. If macropore flow channels water rapidly to the saturated zone, then a rise in groundwater levels would also be expected.

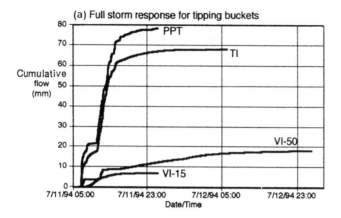

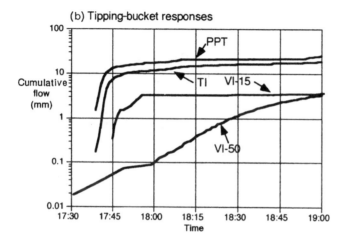

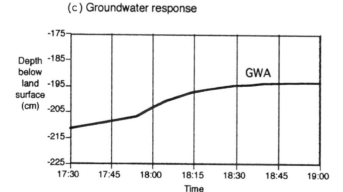

Figure 6.3 Hydrological response for a summer storm on 11 July 1994. (a) tipping-bucket response (flow) during the entire storm for precipitation (PPT), throughfall (TI), and soil-water depths of 15 (VI-15) and 50 cm (VI-50); (b) tipping-bucket response during the first 1.5 h; (c) response of groundwater levels during the first 1.5 h

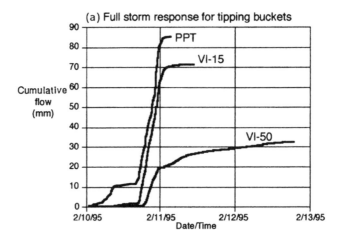

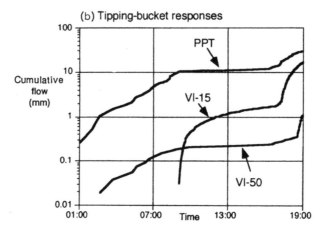

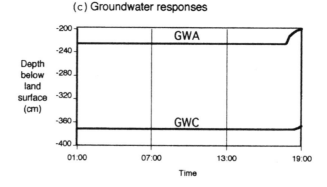

Figure 6.4 Hydrological response for a summer storm on 11 July 1994. (a) tipping-bucket response (flow) during the entire storm for precipitation (PPT), and soil water depths of 15 (VI-15) and 50 cm (VI-50); (b) tipping-bucket response during the first 17 h; (c) response of groundwater levels during the first 17 h

Figure 6.2(c) illustrates the response of groundwaters at three wells (GWA, B and C) along with flow in the unsaturated zone for the August rainstorm (Figure 6.2(b)). The largest change in groundwater levels occurred in well GWA. A water-table rise occurred in GWA at 11:50, some 4 h after the onset of the rainstorm. The responses of the other wells are not as great, but the water levels rose earlier in both cases (Figure 6.2). Thus, the responses of the wells higher on the hillslope actually preceded the passage of the wetting front at 50 cm depth by as much as 45 min. The response for the well at the base of the slope (GWA) coincided with the response at 15 cm depth. For all wells, a response occurred at least 2 h in advance of the arrival of the wetting front at 50-cm depth in the unsaturated zone, as shown by comparison of Figures 6.2(b) and 6.2(c). Thus, results suggest that macropore water moves rapidly to depth, causing rapid responses in groundwater levels.

The rising limb of the stream hydrograph at the upper gauge (SWug) coincides with the groundwater-level rise at the most downslope well (GWA), suggesting that the rise may have been caused by lateral flow from the stream. The earlier groundwater-level rises in GWB and GWC must be caused by movement of water through the hillslope in response to the rainfall because they are upslope from GWA.

For the July rainstorm, a similar lag in the responses of the saturated and unsaturated zones occurred (Figure 6.3(b)) and groundwater responded 1 h before the arrival of the wetting front (matrix flow) at 15 cm depth. For the February rainstorm, groundwater responded after the rapid response at 15 cm depth, but at least 30 min before major flow occurred at 50 cm depth (Figure 6.4(b)). Thus, for all rainstorms, groundwater responded up to 2 h before major flow occurred at 50 cm depth in the unsaturated zone, and in some cases, before major flow occurred at 15 cm depth. A possible explanation for these observations is that "new" water passes through the unsaturated zone rapidly, causing groundwater levels to rise. The tipping-bucket data suggests that the rapid transport is caused by macropores. Macropore flow can operate both vertically and laterally. Although vertical flow in the unsaturated zone was targeted for collection in this study, some flow may have been derived from horizontal flow downslope. This lateral flow may account for the apparent rise in groundwater level before the recorded arrival of the wetting front at 50 cm depth in the unsaturated zone. Alternatively, the lysimeters may not have been representative of the vertical flow across the hillslope, and some areas may transport water more rapidly than observed in these lysimeters. Also, for the well adjacent to the stream, lateral transport from the stream channel into the surficial aquifer can cause a rise in groundwater level.

6.5.2 Temperature Variations and Macropore Flow

Groundwater temperatures stratify vertically. Deeper groundwaters for this shallow groundwater system are cooler than shallower groundwaters in summer months (Ward 1967). The opposite is true for the winter period; rainfall and air temperatures are cold, causing groundwater temperatures to be cooler near the surface than at depth.

Flow data above suggest macropores may rapidly change groundwater levels. We show that the temperature difference between "old" and "new" waters is large

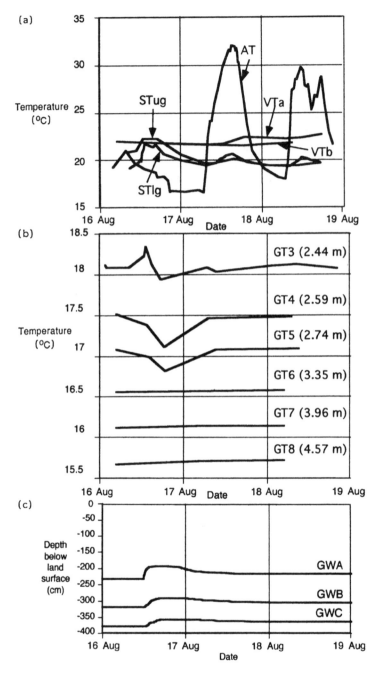

Figure 6.5 Temperature and groundwater level responses for a summer storm on 16 August 1994. Temperatures are presented for air (AT), throughfall (TFT), soil water at 15 (VTa) and 40 cm depth (VTb), streamwater at the upper (STug) and lower gagues (STlg), and groundwater at depths of 2.44 m (GT3), 2.59 m (GT4), 2.74 m (GT5), 3.35 m, (GT6), 3.96 m (GT7) and 4.57 m (GT8). For site locations see Figure 6.1

enough for the two water types to be distinguished from one another for each of the rainstorms.

Summer rainstorm: groundwater was thermally stratified during August (Figure 6.5). Groundwaters 2.44 m below land surface average 18.2°C, whereas, the deepest groundwater measured, GT8 (4.57 m), remains relatively constant at 15.7°C. Temperature variations occurred during the August rainstorm in all water bodies (Figure 6.5). The diurnal variation in air temperatures generally are dampened by the onset of the rainstorm. Streamwater temperatures at the lower and upper gauges (Figure 6.1) increased during the rainstorm from 19 to 22°C. Temperature increased at the onset of the rainstorm at GT3 (2.44 m) from 18.1 to 18.4°C, followed by a marked decrease in temperature of 0.7°C. Thereafter, the temperature slowly rose to a level similar to that measured before the onset of the rainstorm. Temperature decreased markedly from 17.5 to 17.1°C at 2.59 m depth (GT4) and from 17.1 to 16.8°C at 2.74 m depth (GT5). Temperatures remained relatively constant at depths below 3.35 m (GT6–GT8).

Groundwater responses are shown in Figure 6.2(c). The rising limb of all well hydrographs is concurrent with a temperature decrease in shallow groundwaters, i.e. from 2.44 to 2.74 m below land surface (GT3, 4 and 5). The groundwater-level rise in GWB adjacent to the well containing the thermistors was 30 cm, which is consistent with the depth range of the responding thermistors (Figures 6.2(c) and 6.5). After groundwater levels reach a maximum, they decrease slowly and temperatures in shallower groundwaters increase toward prestorm levels.

Winter rainstorm: rainfall for the February rainstorm was colder than groundwater (Figure 6.6). Streamwater temperatures at the lower gauge decreased slightly at the onset of the rainstorm, perhaps due to the contribution of "new" water falling directly on the channel (Shanley and Peters 1988). The temperature decrease was followed by an increase, which was higher than that of throughfall and is attributed to contributions of "old" water. The thermal stratification of groundwater is the reverse of that in summer. In February, groundwater temperatures at 4.6 m depth (GT8) remained relatively constant at 14.9°C, and at 2.44 m depth (GT3) varied around 14.3°C. At the onset of the rainstorm, groundwater temperatures at well GT3 decreased rapidly from 14.5 to 14.0°C and then rose gradually to 14.34°C and remained near this value. A similar pattern occurred for temperature at 2.59 and 2.74 m depth (wells GT4 and GT5, respectively). Temperatures remain relatively constant in the deepest groundwater, similar to the pattern in the summer. In addition, the timing of groundwater-level rise is concurrent with the increase in temperature observed in shallow groundwaters (wells GT3–5).

In summary, the temperature responses of shallow groundwater during storms in summer are the reverse of those in winter and are consistent with inputs of "new" water, or at least water derived from higher in the soil profile.

6.6 DISCUSSION

Soil-water flow at various depths in the unsaturated zone occurred at the onset of rainstorms (Figures 6.2–6.4), suggesting that macropore flow operates during

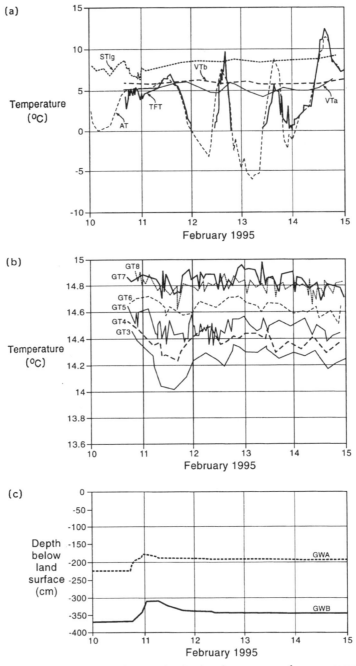

Figure 6.6 Temperature and groundwater level responses for a summer storm on 10 February 1995. Temperatures are presented for air (AT), soil water at 15 (VTa) and 40 cm depth (VTb), streamwater at the lower gauge (STlg), and groundwater at depths of 2.44 m (GT3) 2.59 m (GT4), 2.74 m (GT5), 3.35 m (GT6), 3.96 m (GT7) and 4.57 m (GT8). For site locations see Figure 6.1

rainstorms at PMRW. Periods of intense rainfall during individual rainstorms correlate well with periods of rapid flow in the unsaturated zone (Figures 6.2(a), 6.3(a) and 6.4(a)), suggesting that rainfall intensity is an important control on macropore flow. The lag time between the onset of the rainstorm and the onset of soil-water flow differed for all rainstorms (Figures 6.2(a), 6.3(a) and 6.4(a)), and the

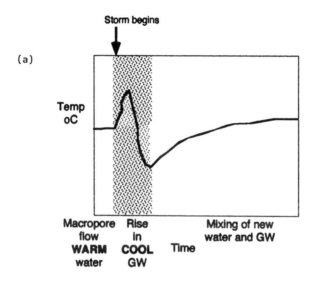

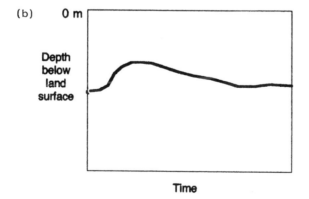

Figure 6.7 Hypothetical groundwater level and temperature trends for a summer rainstorm. (a) Temperature response at 2.4 m below land; (b) groundwater level response. For summer storms, rainfall is warm and groundwater is cool

variability in lag time is attributed to varying antecedent soil-moisture conditions (Kennedy *et al.* 1986; Poinke and De Walle 1992).

Groundwater response was more rapid than the largest increase in flow in the unsaturated zone. The rapid response is consistent with the prevalence of macropore flow. "Old" water cannot be distinguished from "new" water considering hydrometric measurements alone, but "old" water, in part, can be distinguished from "new" water by using temperature.

Macropore flow may promote groundwater displacement. If this is the case, temperature responses should be found at depth (Figure 6.7). In summer, warm "new" waters transported to depth along macropores cause shallow groundwater temperatures to increase and groundwater levels to rise. Shallow groundwater is warmer in summer than deeper groundwater, and the converse is true during the winter period. Infiltration of "new" water to depth via macropores causes the saturated zone to extend upward into the soil matrix (McDonnell 1990). The groundwater-level rise, caused by displacement of groundwater further upslope, would cause the temperature of water surrounding a thermistor to be similar to a temperature of groundwater previously at some depth below the thermistor. In this case, as the water-table rises, a thermistor's temperature changes abruptly, decreasing in the summer (since it is now surrounded by cooler groundwater) and increasing in the winter (since it is now surrounded by warmer groundwater). These temperature changes are therefore caused by the change in the water level. In addition, as groundwater drains after the rainstorm and the water-table decreases below the thermistor, temperature will return to approximately prestorm values. An alternative cause for the temperature changes is the mixing of "old" and "new" waters.

We suggest that shallow groundwater is recharged rapidly during the onset of a rainstorm with "new" water which causes an increase in temperature in summer and a decrease in winter. Figure 6.7 shows a hypothetical response for a thermistor at a depth of 2.4 m at the monitoring site. The rapid input of "new" water to depth causes a concurrent groundwater-level rise and increase in temperature. After the initial transport of "new" water and as groundwater levels continue to rise and then fall, lateral movement of groundwater from upslope will displace and mix with the "new" water causing temperatures to decrease below pre-storm values, and then to trend back to the pre-storm levels.

Temperature data supports this hypothesis (Figure 6.5). Groundwater temperatures decrease with depth, and rainfall typically is much warmer. Thus, when macropore flow introduces "new" water to the saturated zone, an increase in temperature occurs (Figure 6.5). Addition of "new" water to the saturated zone promotes lateral groundwater flow accounting for the decrease in groundwater temperatures as the water level rises. Shallow groundwater temperatures return to levels, similar to those observed before the onset of the rainstorm, either as "new" and "old" water mix, or when the water-table level declines. The reverse temperature trends were observed during the winter rainstorm (Figure 6.6) when macropore flow introduces "new" and colder water to the saturated zone in winter. A decrease in temperature occurs in the shallowest groundwater. As groundwater levels fall the temperature increases to pre-storm levels.

6.7 CONCLUSIONS

Macropore flow was found to be an important flowpath in the unsaturated zone. Evidence for its occurrence has been provided from consideration of the response of soil-water flow during rainstorms at Panola Mountain Research Watershed. This macropore flow causes lateral displacement of groundwater and its movement into the soil matrix, causing a rise in groundwater level. This has been observed from groundwater stage data, and also from temperature responses in the saturated zone. The temperature difference between "old" and "new" waters was large enough for the two water types to be distinguished from one another for each rainstorm under investigation. Thus, temperature has been used successfully as a tracer for water movement in the saturated and unsaturated zones of a forested hillslope plot at Panola Mountain Research Watershed.

ACKNOWLEDGEMENTS

This study was conducted in co-operation with the Georgia Department of Natural Resources. The authors are grateful for support provided by the staff of Panola Mountain State Park. We also are grateful to E. H. Drake, T. K. Pojunas and D. L. Booker of the US Geological Survey, Atlanta, Ga. for providing laboratory support. The use of brand names in this report is for identification purposes only and does not indicate endorsement by the US Geological Survey.

REFERENCES

Beven, K. J. and Germann, P. 1981. Water flow in soil macropores: II: A combined flow model. *Journal of Soil Science*, **32**, 15–29.

Beven, K.J. and Germann, P. 1982. Macropores and water flow in soils. *Water Resources Research*, **18**(5), 1311–1325.

Christophersen, N. 1990. Controlling mechanisms for stream water chemistry at the pristine Ingakken site in Mid Norway. *Water Resources Research*, **26**(1), 59–67.

Clothier, B.E. and White, I. 1981. Measurement of sorptivity and soil dissusivity in the field. *Soil Science Society of America Journal*, **45**, 241–245.

Cosby, B.J., Wright, R.F., Hornberger, G.M. and Galloway, J.N. 1985. Modeling the effects of acidic deposition: assessment of a lumped parameter model of soil water and streamwater chemistry. *Water Resources Research*, **21**(1), 51–63.

Dincer, T., Payne, B.R., Florkowski, T., Martinec, J. and Tongiorgi, E. 1970. Snowmelt runoff from measurements of tritium and O-18. *Water Resources Research*, **6**(1), 110–124.

Dunne, T. and Black, R.D. 1970. Partial area contribution from rainstorm runoff in a small New England watershed. *Water Resources Research*, **6**, 1296–1311.

Eshleman, K.N. 1988. Predicting regional episodic acidification of surface waters using empirical models. *Water Resources Research*, **24**(7), 1118–1126.

Feddes, R.A., Kabat, P., van Bakel, P.J.T., Bronswijk, J.J.B. and Halbertsma, J. 1988. Modelling soil water dynamics in the unsaturated zone-state of the art. *Journal of Hydrology*, **100**, 69–111.

Flury, M., Fluhler, H., Jury, W.A. and Leuenberger, J. 1994. Susceptibility of soils to preferential flow of water: A field study. *Water Resources Research*, **30**(7), 1945–1954.

Fritz, P., Cherry, J.A., Weyer, K.U. and Sharpe, W.E. 1976. Rainstorm runoff analyses using environmental isotopes and major ions. In *Isotope Techniques in Groundwater Investigations*. IAEA, Vienna, pp. 111–131.

Gillham, R.W. 1984. The capillary-fringe and its effects on water-table response. *Journal of Hydrology*, **67**, 307–324.

Hewlett, J.D. and Hibbert, A.R. 1967. Factors affecting the response of small watersheds to rainfall in humid areas. In W.E. Sopper and H.W. Hull (eds), *Forest Hydrology*, Pennsylvania State University, University Park, Penn., pp. 275–290.

Hooper, R.P. and Christophersen, N. 1992. Predicting episodic stream acidification in the southeastern United-States – combining a long-term acidification model and the end-member mixing concept. *Water Resources Research*, **28**(7), 1983–1990.

Hooper, R., Christopherson, N. and Peters, N.E. 1990. Modeling streamwater chemistry as a mixture of soil water end members – an application to the Panola Mountain Catchment, Georgia, USA. *Journal of Hydrology*, **116**, 321–343.

Horton, R.E. 1933. The role of infiltration in the hydrological cycle. *American Geophysical Union Transactions*, **14**, 446–460.

Jenkins, A., Ferrier, R.C., Harriman, R. and Ogunkoya, Y.O. 1994. A case study of catchment hydrochemistry; conflicting interpretations from hydrological and chemical observations. *Hydrological Processes*, **8**, 335–349.

Kennedy, V.C., Kendall, C., Zellweger, G.W., Wyerman, T.A. and Avanzino, R.J. 1986. Determination of the components of rainstormflow using water chemistry and environmental isotopes, Mattole River Basin, California. *Journal of Hydrology*, **84**, 107–140.

Luxmoore, R.J. 1981. Micro-, meso- and macroporosity of soil. *Soil Science Society of America Journal*, **45**, 671.

Martinec, J., Siegenthaler, H., Oescheger, H. and Tongiorgi, E. 1974. New insight into the runoff mechanism by environmental isotopes. In *Isotope Techniques in Groundwater Hydrology*. IAEA, Vienna, pp. 129–143.

McDonnell, J.J. 1990. A rationale for "old" water discharge through macropores in a steep, humid catchment. *Water Resources Research* **26**(11), 2821–2832.

Mosley, M.P. 1974. Experimental study of rill erosion. *Trans. ASCE*, **17**. 909–913.

Mosley, M.P. 1982. Subsurface flow velocities through selected forest soils, South Island, New Zealand. *Journal of Hydrology*, **55**, 65–92.

Mullholland, P.J., Wilson, G.V. and Jardine, P.M. 1990. Hydrogeochemical response of a forested watershed to rainstorms: effects of preferential flow along shallow and deep pathways. *Water Resources Research*, **26**(12), 3021–3036.

Pearce, A.J., Stewart, M.K. and Sklash, M.G. 1986. Rainstorm runoff generation in humid headwater catchments: 1: Where does the water come from? *Water Resources Research*, **22**(8), 1263–1272.

Peters, N.E. 1994. Water-quality variations in a forested Piedmont catchment, Georgia, USA. *Journal of Hydrology*, **156**, 73–90.

Poinke, H.B. and DeWalle, D.R. 1992. Intra- and inter-storm O18 trends in selected rainstorms in Pennsylvania. *Journal of Hydrology*, **138**, 131–143.

Ragan, R.M. 1968. An experimental investigation of partial area contribution. *IAHS Publication*, **76**, 241–249.

Shanley, J.B. 1992. Sulphur retention and release in soils at Panola Mountain, Georgia. *Soil Science*, **53**, 361–382.

Shanley, J.B. and Peters, N.E. 1988. Preliminary observations of streamflow generation during rainstorms in a forested Piedmont Watershed using temperature as a tracer. *Journal of Contaminant Hydrology*, **3**, 349–365.

Shanley, J.B. and Peters, N.E. 1993. Variations in aqueous sulphate concentrations at Panola Mountain, Georgia. *Journal of Hydrology*, **146**, 361–382.

Sklash, M.G. and Farvolden, R.N. 1979. The role of groundwater in rainstorm runoff. *Journal of Hydrology*, **43**, 45–65.

Sklash, M.G., Stewart, M.K. and Pearce, A.J. 1986. Rainstorm runoff generation in humid headwater catchments: 2. A case study of hillslope and low-order stream responses. *Water Resources Research*, **22**(8), 1273–1282.

Thomas, D.L. and Beasley, D.B. 1986. A physically-based forest hydrology model I: development and sensitivity of components. *Trans. ASAE*, **29**(4), 962–972.

Ward, R.C. 1967. *Principles of Hydrology*, 1st edn. McGraw-Hill, London.

Whipkey, R.Z. 1967. Theory and mechanisms of subsurface stormflow. In W.E. Sopper and H.W. Hull (eds), *Forest Hydrology*. Pennsylvania State University, University Park, Penn., pp. 255–280.

Woolhiser, D.A., Emmerich, W.E. and Shirley, E.D. 1985. Identification of water sources using normalised chemical ion balances: a laboratory test. *Journal of Hydrology*, **76**, 205–231.

7 Modelling the Influence of Afforestation on Hillslope Storm Runoff

T. J. A. DAVIE

Department of Geography, Queen Mary & Westfield College, University of London

7.1 INTRODUCTION

A forest canopy provides a layer for water to pass through within the hydrological cycle. Conceptually it can be seen as a store in the same manner as the soil mantle or snow cover. Two of the major land-use changes seen within catchments are the removal of forest cover through deforestation and the addition of a forest cover through afforestation. There have been many studies on the affects of deforestation on catchment runoff through variations on the paired catchment study. Less well studied are the affects of afforestation where it might be expected that the change is slow and gradual. In their review of catchment experiments investigating the effects on water yield of vegetation change, Bosch and Hewlett (1982) report only 12 (out of 94) studies that were concerned with afforestation or reforestation. In the more recent review of 62 experiments investigating the effects of vegetation change on stormflow, Son (1991) found only five that were concerned with afforestation or reforestation.

Long-term studies such as those at Plynlimon, UK, have revealed that afforestation has significantly altered the annual streamflow. This is largely due to the increased interception of rainfall, and subsequent evaporative loss, provided by the greater intercepting capacity of a forest canopy (Kirby *et al.* 1991). The canopy interception properties can be expected to change as a forest grows, but this is a slow and gradual process. This makes it difficult to isolate the individual effect of afforestation in a streamflow record at a particular point in time from other noise in the data. For this reason numerical modelling provides an attractive approach to investigating the affects of afforestation on hillslope and catchment runoff processes. Although numerical modelling can never hope to fully represent the extreme complexity of catchment hydrology it does provide a powerful means of isolating particular processes for detailed investigation. By varying input parameters within a set of values likely to be found in the field a range of different scenarios can be modelled that may take many years of hydrological record in many different locations to provide a similar field database. Burt and Butcher (1985) used this approach to investigate the generation of delayed peaks in storm discharge. They

Advances in Hillslope Processes, Volume 1. Edited by M. G. Anderson and S. M. Brooks.
© 1996 John Wiley & Sons Ltd.

estimate that it would have required "at least thirty years' data to provide a similar sample of events" (Burt and Butcher 1985, p. 364) to the several hundred model runs. A similar approach is used in this chapter, where individual processes important to hillslope hydrology are isolated within the model structure, and through a series of different scenarios the relative importance of each is assessed. In particular the role of canopy growth is investigated to assess whether it has a slow and gradual effect on storm runoff events.

7.2 MODELLING SCHEME

The modelling scheme used here is an amalgamation of several different numerical models. The basic hillslope and catchment hydrology is modelled through the Variable Source Area Simulator (VSAS), developed by Troendle (1979), following on from earlier work by Hewlett and Nutter (1970) and Hewlett and Troendle (1975). The name VSAS derives from its original function: to mathematically represent the variable source area concept of Hewlett and Hibbert (1967). The model has been developed further by Bernier (1982), and used by Bernier (1985), Whitelaw (1988), Prevost et al. (1990), and Fawcett et al. (1995) amongst others. The model is physically based, and the processes are simulated in a distributed manner.

In order to investigate the effects of vegetation change a more detailed canopy hydrology routine has been incorporated into the VSAS structure. This is a three-dimensional rainfall partitioning routine described in detail by Davie and Durocher (In prep a). The rainfall partitioning is based upon the Rutter model (Rutter et al. 1971), with the canopy being subdivided into a series of representative areas for throughfall, stemflow and interception loss. The use of three-dimensional canopy is important so that the physical structure can be manipulated to simulate the growth of a changing canopy. Whilst it may be possible to manipulate the parameters of a single canopy store, it would be difficult to maintain any physical meaningfulness.

The simulation of the growth of a three-dimensional canopy is carried out through the use of a forest growth model. The model is based partly upon work by Leps and Kindlmann (1987) and Ford and Diggle (1981). It is a distant dependant, individual-tree based model using circular zones of influence to simulate inter-tree competition.

The various components of the overall modelling scheme to investigate the affects of afforestation, can be thought of as:

1. a pre-processing forest growth model that provides a snapshot of what a canopy may look like at a certain age;
2. a canopy discretization routine that describes the canopy in a three-dimensional manner applicable for use in the hydrology model;
3. a physically based, distributed hillslope/catchment hydrology model that simulates runoff on an event basis.

Figure 7.1 is a representation of the overall modelling scheme.

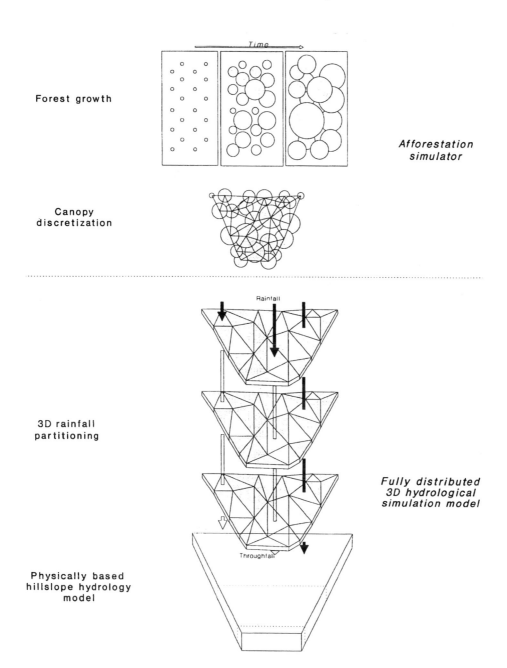

Figure 7.1 Schematic representation of the overall modelling scheme. The afforestation simulator operates independently of the hydrology model. The output from this is linked to the rainfall partitioning section of VSAS through a canopy discretization routine, dividing the canopy into a series of triangles which become the basic unit for throughfall

7.2.1 Forest Canopy Growth

The model treats each tree in a simplified manner. The shape of each tree is illustrated in Figure 7.2. For the simulations outlined here the forest canopy has been restricted to a monospecific coniferous forest although the growth of deciduous trees can be simulated using the alternative shape.

The model uses a logistic growth curve to simulate empirically potential tree growth. The logistic or autocatalytic curve has been used previously by Aikmann and Watkinson (1980) and Leps and Kindlmann (1987) as the basis for simulating potential growth in their respective plant growth models. It was also used by Ledig (1969) as a basis for a model simulating the increase in dry matter of pine seedlings. Hunt (1982) presents a table of 44 empirical applications of the logistic curve in plant ecology. Most of these have been in short-term studies (within a growth season) but Richards (1959) has indicated that the curve can be used for the complete growth of a plant, which is well supported by the modelling studies of Aikmann and Watkinson (1980) and Leps and Kindlmann (1987).

The logistic growth curve written as a differential equation is shown in equation 7.1.

$$\frac{dh}{dt} = h_t \cdot g \cdot \left(1 - \frac{h_t}{H}\right) \tag{7.1}$$

where h is tree height, t is time, g is growth ratio (see text), and H is maximum tree height.

The curve is asymptotic towards the maximum tree height. The main controlling parameter on curve shape is the growth ratio (g) which is an intrinsic rate of increase in size. There are limits to the range of g values. When $1.0 < g < 2.57$ the solution to the logistic equation becomes cyclic, and when $g > 2.57$ the solution exhibits chaotic behaviour (May 1974). A negative value of g is meaningless in the context of plant growth, therefore it is important that g falls between 0 and 1 to maintain a stable solution. By integrating equation 7.1, a value for g can be derived when the length of time it takes for a tree to reach a certain percentage of the maximum height is known; for example, if it takes 70 years for a tree to reach within 95% of its maximum height. This allows input values for g to be calculated given knowledge of possible maximum height and age to reach maximum height.

Equation 7.1 uses height as the predicted variable, but any plant growth variable can be predicted. There is a simple linear regression relationship between diameter at breast height (DBH) and tree height as verified by numerous studies (e.g. Helvey and Patric 1965; Ishikawa and Ito 1989), consequently DBH is the predicted variable from which tree height is later derived.

The value of g for each tree is drawn randomly from an assumed normal distribution with input mean and standard deviation. The stochastic nature of assigning g is assumed to simulate the difference in tree growth potential that cannot be described by inter-tree competition, e.g. genetic differences affecting mineral uptake or other growth factors. A normal distribution has been assumed as no evidence could be found to suggest otherwise.

Inter-tree competition is simulated by a series of zones of influence (ZOI) surrounding each tree. The ZOI of a tree is assumed to represent the zone in which it

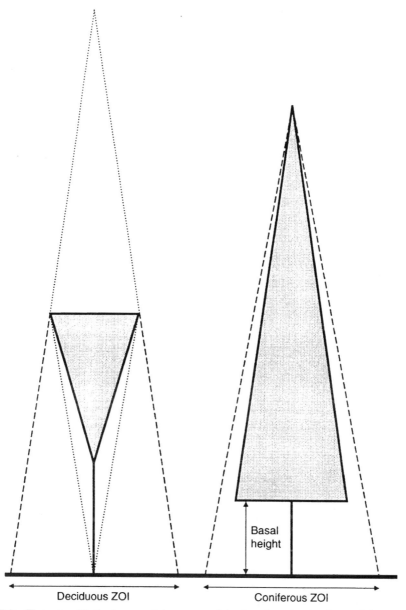

Figure 7.2 The simplified shape of trees used in the forest growth model and the calculation of zone of influence (ZOI) for each tree

is competing for resources. It is assumed that the major resource being fought for is light, i.e. water and minerals are uniformly available to all trees. This assumption has been made in previous modelling studies (e.g. Bella 1971; Ford and Diggle 1981; Gates 1982; Smith 1990) and although an obvious simplification of reality is not totally unreasonable for temperate environments where soil water is normally readily available to plants. Competition is two-sided but a large tree will necessarily affect a smaller tree to a greater degree than vice versa.

For coniferous trees the ZOI is assumed to be a set amount of the crown projected area, the actual proportion being input by the user (see Figure 7.2). For each tree the ZOI is first calculated then the amount that other trees interfere with the subject tree (η) is calculated as the total amount of overlap of ZOIs and the subject tree's crown projected area. The proportion of subject tree crown projected area overlapped by the ZOI of its neighbour is calculated as the overlap of two circles. η is used as a modifying factor on the growth ratio (g) of equation 7.1; as the amount of influence from neighbours increases the growth ratio decreases in size to a minimum possible amount.

When the tree growth and inter-tree competition is combined the initial DBH is input as either uniform for every tree (as for planted seedlings) or drawn randomly from a known distribution which is input. The growth rate is then randomly assigned to each tree and maintained throughout its lifetime.

The full equation for numerical solution is shown in equation 7.2. This is solved stepwise for each tree individually with a time increment of one year.

$$dbh_{t+1} = dbh_t \left(1 + g \left(1 - \frac{dbh_t}{dbh_{\max}} \right) \eta \right) \tag{7.2}$$

For each year the model sequentially counts through every tree, calculates the amount its neighbours infuence the subject tree, and then increases the subject tree DBH by an amount defined by equation 7.2.

In the model, trees die in order to maintain a maximum plot density as derived from Forestry Commission yield tables. The selection of which trees are to die depends on their competitive status (η). The η term relates to the amount of stress a tree suffers from inter-tree competition, therefore it seems reasonable that the trees under most stress will die. On the occasions where the input density data require that some trees should die but not enough trees suffer from competitive stress (e.g. at the start of the simulation before the trees start to interact, $\eta = 0$ for many trees) the trees are ranked according to their size and the smallest trees culled. The number of these size-related deaths is likely to be small so that it can be said that the primary process causing mortality is the competition for light.

When a tree dies it is removed from the simulation and the space it previously occupied is assumed vacant and available for other trees to grow into (although any tree crown expansion maintains a conical shape).

The largely empirical nature of the model reflects the use required of it. It has not been designed as a highly sophisticated forest growth model but rather as a simple feeding mechanism for VSAS. The model is an attempt to gain a distance-dependent, individual tree growth model so that the growth in canopy can be simulated in a three-dimensional manner.

7.2.2 Canopy Discretization

The individual position and size of each tree at a certain age is taken from the pre-processing forest growth model described above. The canopy discretization routine takes this tree position and size output and calculates the canopy parameters required by the three-dimensional rainfall partitioning section of VSAS.

The first step is to perform a tesselation, creating a series of triangles with a tree at the vertex of each. These triangles then become the basic unit for throughfall and interception. Each triangle is divided into a series of layers based upon the leaf area index (LAI) of each triangle. The LAI is calculated from a relationship found by Halldin (1985). This deterministically links the leaf area of a pine tree to its size (see Halldin 1985, p. 53). As area and volume of tree is known, the LAI can be calculated easily from the leaf area. The LAI of each triangle is calculated as the average of the LAIs for the trees contributing to the triangle. Secondly, the number of layers in the triangle is found by calculating the probability that a raindrop falling onto a triangle will be intercepted by n number of intercepting layers. The probability is calculated using a discrete Poisson distribution.

The probability of a triangle layer having n layers is shown in equations 7.3–7.5

$$P'_n = \begin{cases} x - (1-x)P_n & (i = j) \\ (1-x)P_n & (i \neq j) \end{cases} \tag{7.3}$$

where

$$x = \frac{P_n}{1 - P_n}\left(\frac{1}{\gamma} - 1\right) \tag{7.4}$$

and

$$P_n = \frac{L^n e^{-L}}{n!} \tag{7.5}$$

P_n is the probability of a raindrop encountering n layers, L is the leaf area index of a triangle, i is the mean number of layers, j is a discrete value, and γ is the weighting factor.

The inclusion of a weighting factor allows a degree of manipulation for the Poisson distribution so that there can be a clumping of layers or an even distribution (see Wu et al. 1987; Davie and Durocher; in prep a). Halldin (1985) found that the distribution of leaf area through a pine canopy was normal so a γ value of 0.55 was used in these simulations. This gives a near normal distribution, i.e. more rainfall will be intercepted in the middle of the verticle profile than at the top or bottom.

For each tree a certain proportion of the crown is assigned as generating stemflow. The actual proportion for each tree is generated randomly from a log normal distribution following the work of Durocher (1990). Johnson (1991) presents evidence from a collation of field studies to show that the stemflow contribution to below canopy rainfall does change with canopy age. This is because an individual tree structure changes as it grows, the proportion of leaf area to branch or trunk area alters, although the actual amount is difficult to assess. The data presented by Johnson (1991) suggest that the relationship is not linear, there being a higher percentage of stemflow in younger trees than in the older trees. The stochastic allocation of stemflow proportion used in this modelling scheme occurs once the final tree size at a certain age is known, i.e. at the end of a forest growth simulation. This means that the area of stemflow contribution will be larger for an older tree but the proportion of total crown surface area remains static. It is quite possible that the relationship between stemflow amount and tree age will be non-linear as the growth of each tree is non-linear. A large tree will definitely produce more stemflow than a smaller tree.

Within the stemflow proportion of the crown, a series of layers is generated in exactly the same manner as for the triangle layers. In this case the stem area index (SAI) is used in place of the LAI. The stem area of a tree is assumed to be one-twentieth of the leaf area, following the work of Halldin (1985).

7.2.3 VSAS

For a full description of the VSAS model, see Troendle (1985) and Bernier (1985), although the version used here does have a different rainfall partitioning routine from that described in the above references.

For calculating the soil mantle geometry VSAS requires that a catchment be divided into a series of polygons called *segments* by the user. This is achieved using a topographical map, and any *a priori* knowledge of flow paths within the catchment. A segment is an autonomous hydrological unit, stretching from the base of a slope (normally a channel) to the watershed boundary. It is assumed that the only water exiting from a segment occurs at the slope base by flowing into a stream channel. This requires the intersegment boundaries (normally a ridge or valley bottom), watershed boundary, and soil base to be impermeable. The side boundaries need not be parallel to each other, so a segment can take a rectangular, square, convergent, divergent, or skew form. This is a similar topographic representation to the IHDM4 model (e.g. Calver 1988); see Figure 7.3 for a comparison of VSAS with the SHE model.

(a)   SHE    (b)   VSAS

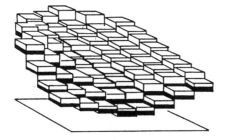 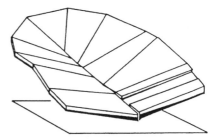

Figure 7.3 The catchment discretization used in VSAS as compared to the square grid system of the SHE model (e.g. Abbott *et al.* 1986). Each hillslope segment is treated as a separate hydrological unit (from Whitelaw 1988)

Each segment can be divided lengthways into an odd number of *subsegments* (maximum of five). These are effectively smaller versions of segments except that water is allowed to flow laterally between subsegments.

Each subsegment is divided slopewise into a series of *increments*. The division is performed in a logistic fashion. The number of slopewise divisions is decided by the user, normally depending on the detail required at the segment base. The logistic nature of this equation divides the slope into smaller increments at the stream side, which increase in size upslope towards the watershed boundary. This gives greater model sensitivity in the region nearest the stream which the variable source area concept suggests is the most important for runoff generation. At each increment centre point, the surface elevation, distance to stream, and soil depth to the impermeable bedrock is input. These values are obtained from a combination of topographical map and field investigation.

Each increment is divided with depth into a series of soil *elements*. The number and depth of these elements is uniform for each segment, the actual values being dependent on the soil heterogeneity. This is the basic unit of VSAS. All soil hydrological properties are assigned to a point in the centre of the element and are assumed uniform throughout an element. The element centre point is the solution point for the VSAS block centred finite-difference scheme, with soil water flow being computed in three dimensions (non-simultaneously).

The linkage between the element centres is the flow line between centre points along which it is assumed all subsurface flow moves toward the stream (see Figure 7.4).

Each subsegment can be divided into any number of hydrological *zones*. Zones are regions with distinct soil hydrology properties that stretch across subsegment width. They are delimited along the line running through the element centres, each element being assigned hydrological properties according to which zone the element centre is contained within.

A suction-moisture curve for each soil zone is input (hysteresis is assumed negligible), from which the Millington–Quirk method (Millington and Quirk 1961) calculates a table of unsaturated hydraulic conductivities for each soil zone. The value of saturated moisture content (porosity), saturated hydraulic conductivity, and suction moisture curve points is input for each soil zone along with the standard deviations of each value. The actual value of these parameters for each element is then stochastically derived prior to simulation, drawing the values randomly from a normal distribution of defined mean and standard deviation. This is to try to account for the heterogeneity in soil physical properties that is known to occur in the field.

VSAS operates under a system whereby each segment is considered an entirely separate hydrological unit. Rainfall is read in as hourly values prior to the simulation starting and the model computes the flows into and out of each segment in turn for the whole storm. The hourly rainfall is divided into even increments, the size of the increments being an input value, i.e. setting the internal timestep.

All of the process representations are calculated at this internal timestep except for channel flow which is lagged for every hour. Water passing through the canopy is calculated as a volume every timestep and then transferred to a depth per timestep in order to account for the difference in surface area between the canopy and the segment.

(a)

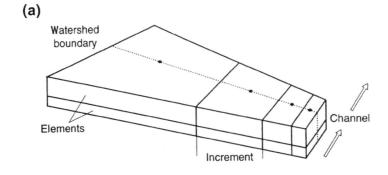

(b)

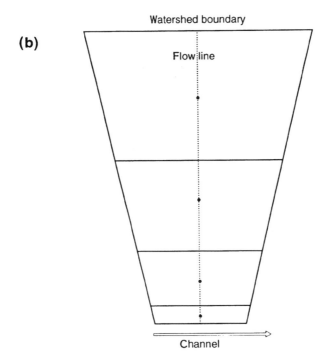

Figure 7.4 Discretization with a VSAS segment: (a) oblique view; (b) plan view

The process representations in this section are subdivided into four subsections: rainfall partitioning, soil water flow, overland flow and channel flow.

The rainfall partitioning section of VSAS is largely based upon the Rutter model (Rutter *et al.* 1971) for canopy interception. The major difference is that the scale of computation has been changed from canopy to leaf and stem scale to take advantage of the three-dimensional canopy generated as part of the modelling scheme.

Rainfall partitioning can be divided into four individual parts that are calculated separately within VSAS. These are

1. direct throughfall: rainfall falling between tree crowns to the soil surface;
2. free throughfall: rainfall draining from a canopy layer and falling directly to the soil surface, i.e. not draining to the next layer but falling through gaps within it. N.B. The free throughfall occurring through the top layer of the canopy is effectively direct throughfall but is kept separate because they are calculated in a different manner;
3. indirect throughfall: the remainder of rainfall, after direct and free throughfall, that drains sequentially down through the canopy layers until it reaches the soil surface. The water stored (storage) on each layer is available for evaporation;
4. stemflow.

Each of these parts are calculated separately within a timestep and the volumes reaching the soil surface through each of them is totalled as net rainfall below the canopy. For a more detailed description of the rainfall partitioning see Davie (1993) and Davie and Durocher (in prep a).

VSAS accounts for both saturated and Hortonian overland flow. The impermeable surface area for each subsegment is input to calculate the proportion of rainfall that flows to the channel as Hortonian overland flow. It is assumed that all water falling on this area does not infiltrate and flows directly to the channel, reaching it within a simulation timestep (frequently five minutes). The volume of rainfall routed in this manner is subtracted from the original volume, before being added to the surface elements.

It is assumed that all the remaining rainfall infiltrates into the surface elements and is available for soil water flow. If the volume of water entering a surface element (either rainfall or flow from an adjacent element) is too great, the excess water is routed downslope, filling any unsaturated surface elements until the channel is reached. The final amount of this saturated overland flow is added to the direct runoff and subsurface contributions to channel flow.

The soil water flow is the most significant part of the original VSAS structure. It is assumed that all rainfall falling onto an element infiltrates the soil surface (apart from a predesignated amount to account for any impervious surfaces; see above) and is then accounted for by either subsurface soil water flow, or saturated overland flow if the input volume is too great for the element. The rate of subsurface flow is calculated using the Richards generalization of Darcy's law for unsaturated flow and Darcy's law for saturated flow. This is solved sequentially in three dimensions (vertically between soil elements, horizontally between soil elements, and laterally between subsegments) using an explicit finite-difference scheme.

To calculate the channel flow the area of channel within a subsegment is input and any rainfall falling on this is treated in exactly the same manner as Hortonian overland flow. Channel flows are accumulated for each segment per hour, and are lagged according to the estimated time of travel from each segment to the basin outlet (input value). This involves passing a proportion of a given hours flow to the following hour. The flows from each segment are accumulated to give the final outflow hydrograph.

In summary VSAS is a physically based distributed hydrological model. It is a storm event simulator that has been used previously in small catchments ranging from $0.38 \, km^2$ (Troendle 1985) to $1.88 \, km^2$ (Fawcett *et al.* 1995). It has routines to simulate the main hydrological processes (except snowmelt) to differing degrees of sophistication. Originally this difference in emphasis on processes was due to lack of computational power to run the model. With modern-day computing this is less of a problem and it would be possible to have fully physically based representations of all processes. However, given the criticism of the ability of physically based, distributed models in recent years (e.g. Beven 1989; Grayson *et al.* 1992) there seems little worth in pursuing this line. Instead VSAS offers an opportunity to specialize a model towards different hydrological processes because they are known to be important to the issue being investigated. In this model version the computing emphasis is on soil water flow through the topography (important for runoff generation, particularly through saturated overland flow) and rainfall partitioning (important for forest canopy differentiation).

The combination of a forest growth model that acts as a pre-processor for VSAS provides an opportunity to investigate the effects of afforestation on hillslope hydrology. As with any modelling scheme, there are assumptions and simplifications that mean it will never be the perfect simulator of a natural process. Most notably in the scheme used here there is no allowance made for three factors that may be important in altering hillslope hydrology with afforestation:

1. the method of tree planting, particularly when the ridge and furrow technique is used;
2. the change in soil structure as the trees grow – it might be expected that this will change through the growth of roots and the extra build-up of organic litter;
3. the change in boundary layer conductance as trees grow.

Robinson (1986) investigated the effects of drainage (ridge and furrow) and early afforestation on an upland catchment in the UK. The results of drainage were extremely marked. The omission of catchment drainage, prior to tree planting, in the modelling scheme used here is a significant deterrent to this type of modelling scheme being used in an applied predictive mode for a particular catchment. It is still possible, however, to use the model in a series of hypothetical situations that ignore canopy drainage in order to investigate the effects of afforestation in a generalist sense.

It is very difficult to quantify the effects of afforestation on soil hydraulic properties, due to the extreme heterogeneity of these properties. Previous studies have investigated the impact of afforestation on soil properties such as soil profiles and some physical properties (Pyatt and Craven 1979) and soil chemistry (Anderson *et al.* 1993). These studies indicate that soil properties do change with the growth of a forest canopy. The lack of quantified data in this area led to its omission from the modelling scheme.

McNaughton and Jarvis (1983) summarize evidence on the importance of boundary layer conductance above a canopy for the transfer of water vapour from forest to atmosphere above. The aerodynamic resistance term used in the Penman–Monteith equation will alter as a forest grows, which will affect the potential evaporation rate. This has not been accounted for in this model. The use of the model for storm event simulations rather than longer-term predictions lessens the importance of this omission, although it may still be significant.

7.3   MODEL SIMULATIONS

VSAS has been tested previously (Troendle 1985; Bernier 1985; Whitelaw 1988; Prevost et al. 1990; Fawcett et al. 1995) and shown to be a reasonable simulator of catchment runoff. The simulations presented here aim to test to new parts of the VSAS scheme (forest growth and rainfall partitioning) and in particular to investigate the effects of afforestation on storm runoff. The testing does not validate the modelling scheme in full (a task that seems unlikely to be achieved without a prohibitively expensive field study) and therefore the results cannot be interpreted as presenting absolute truths about the affects of afforestation on stormflow hydrology. The aim of this modelling study is to highlight areas of apparent importance that may warrant further investigation.

7.3.1   Simulations in the Tanllwyth Catchment

The Tanllwyth catchment is a small (0.92 km^2) headwater tributary to the Severn River in Mid Wales. The location and shape of the catchment is shown in Figures 7.5 and 7.6. The catchment is monitored hydrologically by the Institute of Hydrology as a subcatchment within the Plynlimon experiment. The land is owned by the Forestry Commission, constituting part of the Hafren forest.

The Tanllwyth has a flume that provides hourly streamflow data for the catchment. Very close to the flume site there is an automatic weather station which provides the main meteorological data. The rainfall that is measured here is averaged for the whole of the Tanllwyth using earlier studies of rainfall distribution within the Severn catchment, before being entered into the Institute of Hydrology data bank. Also available from this site are potential evaporation estimates at hourly intervals, calculated using the Penman–Monteith equation. This has the advantage of obtaining a potential evaporation estimate without undergoing the calculations of the Penman–Monteith equation in full (as in Davie and Durocher; in prep b).

The length of hydrological record at Tanllwyth is less than the period of afforestation, so storms were selected from as much of the record as possible. The storms were selected as a series of similar storms (in peak rainfall size). These sets of data were obtained by searching through the daily rainfall records for the Tanllwyth to find the necessary storms and then extracting the rainfall, runoff and evaporation data for the hourly timestep ready for direct usage in VSAS. The three events chosen for initial detailed investigation are summer storms with an estimated stormflow

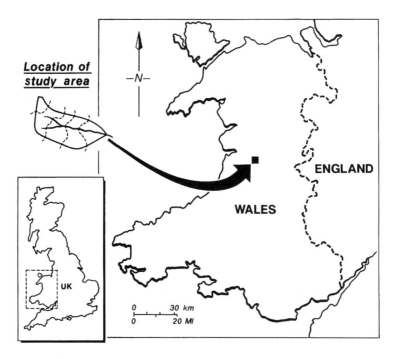

Figure 7.5   Location of the Tanllwyth catchment in the UK

return period of 7–10 years. The reasoning for the choice of summer storms is that snowmelt contribution is not a contributory factor to the runoff.

After the simulations of these relatively simple (i.e. single or double peak) hydrographs, two further storms were selected to test the ability of the modelling scheme in more extreme conditions. These storms were very large complex events (an estimated return period of at least 200 years) occurring in February (with no measured snowfall) and September.

In order to move away from the approach that combines validation with calibration, as many input data as possible were derived from several sources. These included field visits to the catchment, previous studies in the catchment, and other derived sources. By using these independently derived input data the model is run without attempting to fit the output to the observed. There are two exceptions to this: some of the forest growth parameters had to be obtained through calibration; and the initial soil moisture conditions were obtained through draining the catchment until the flows were close to the baseflows prior to each storm.

The first three summer storms simulated occurred in July 1976, July 1982 and July 1989; consequently the forest growth model simulated the growth of a 28, 34 and 41 year canopy respectively (planting was carried out in 1948). The second set of storms (larger and more complex) both occurred in February 1988 and September 1988, so a forest growth period of 40 years was used.

The storm in July 1976 (Figure 7.7) produces a triple peak hydrograph caused by four significant rainfall events on a dry catchment. The model predicts the form of

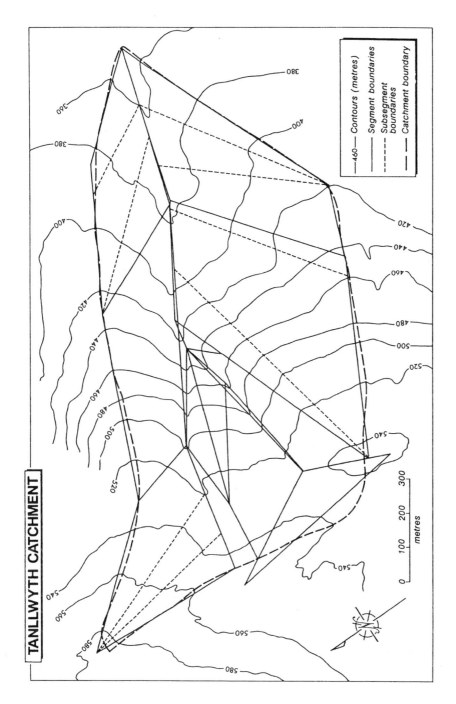

Figure 7.6 Subdivision of the Tanllwyth catchment into segments and subsegments for VSAS simulations

response to a reasonable extent (except for predicting a quadruple rather than triple response) but in all peaks the timing of maximum flow is too early. The shape of the rising and falling limbs of the hydrograph is reasonably close to the observed values although the volume of baseflow following the storm is considerably under-predicted. The model does predict the second storm peak being larger than the first (despite a less intense rainfall) but inaccurately shows the third peak as the largest and shows up a fourth small peak that was not in the observed values. The volume of streamflow for the entire simulation (including baseflow prior to and after the stormflow peaks) was under-predicted by 27%.

The July 1982 storm produces a single peak hydrograph after two significant, but barely separated, rainfall events (Figure 7.8). The response predicted by VSAS corresponds well with the observed hydrograph in timing and shape although the peak volume is slightly over-predicted. The model also predicts a more defined initial peak that is not evident in the observed values. The falling limb of the predicted hydrograph is steeper than that observed but the volume of baseflow after the event is well predicted. The volume of streamflow for the entire simulation (including baseflow prior to and after the stormflow peaks) was over-predicted by 11%.

The storm in July 1989 produces a single peak hydrograph from two significant and several minor rainfall events (Figure 7.9). VSAS predicts a greater response to all of the rainfall events than was observed but the timing of the main event is fairly well predicted. The main problem with the simulation is the prediction of smaller peaks either side of the main peak, which were not in the observed data set. The baseflow following the storm events is well predicted. The volume of streamflow predicted for the entire simulation (including baseflow prior to and after the stormflow peaks) was over-predicted by 15%.

In the more complex storm event occurring in February 1988 (Figure 7.10) the model provides a reasonable prediction except for the recession limbs of the multiple peaks within the main storm peak. This is especially marked in the recession limb and baseflow at the end of the storm event. The actual size and timing of storm peaks is quite well predicted but the model seems to be predicting much less soil water flow than in the observed hydrographs, hence the poor correspondence of recession limbs. The volume of streamflow for the entire simulation (including baseflow prior to and after the stormflow peaks) was under-predicted by 17%.

The results for the large storm event in September 1988 (see Figure 7.11) provide the worst simulation of these model runs. The model predictions are wrong in the size of the peak flows, and there is also difficulty with the timing, especially in the later states of the storm. The recession limb and baseflow after each peak are considerably different from the observed values. The volume of streamflow for the entire simulation (including baseflow prior to and after the stormflow peaks) was under-predicted by 33%, reflecting the poor prediction. Although this was poorly estimated by the model the September 1988 storm is the largest in the Tanllwyth streamflow record and is an extremely complex event that is likely to severely test the capabilities of any numerical model.

The storm hydrographs simulated here show that the modelling scheme is reasonable at reproducing the main features of simple storm hydrographs but it has

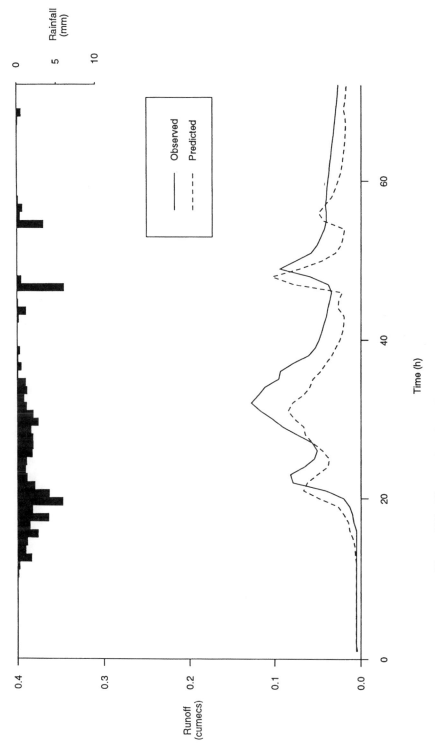

Figure 7.7 Observed and predicted stormflows for July 1976 storm event

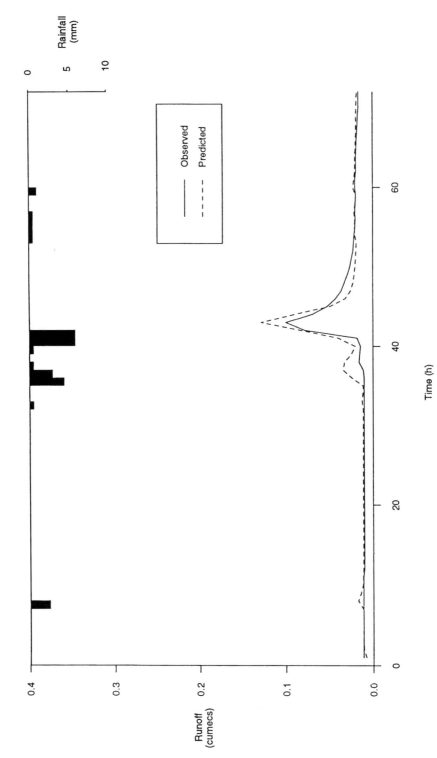

Figure 7.8 Observed and predicted stormflows for July 1982 storm event

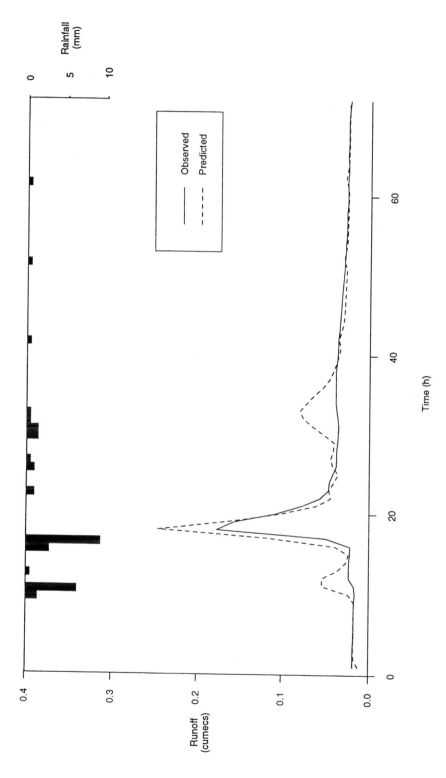

Figure 7.9 Observed and predicted stormflows for July 1989 storm event

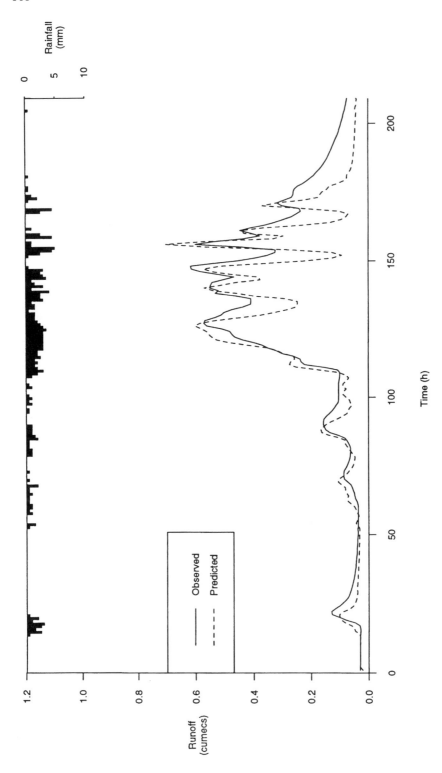

Figure 7.10 Observed and predicted stormflows for February 1988 storm event

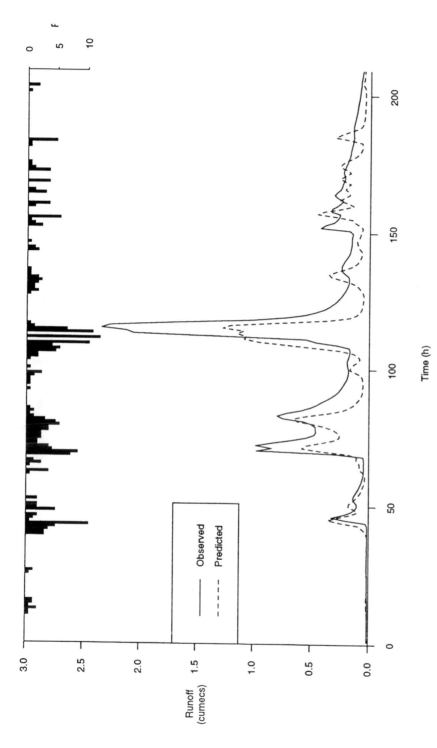

Figure 7.11 Observed and predicted stormflows for September 1988 storm event

much more difficulty when simulating complex events such as in the September 1988 examples.

In order to investigate the validity of the scheme including forest growth it is necessary to look at the three hydrographs that represent similar storms over a 13 year period (i.e. 1976, 1982 and 1989). If the model is over-predicting the effects of a forest then it might be expected that the 1989 event would be under-predicted, in comparison to the 1976 event, and vice versa for an under-prediction. A comparison of Figures 7.7 and 7.9 shows that the predicted response from the 1989 storm is over-predicted whereas the 1976 event is under-predicted, suggesting an under-prediction of the effects of a forest. This suggestion is by no means conclusive as only three storms have been compared and the length of record is not particularly long.

The main error in the VSAS predictions has been in the receding limb of the hydrograph and subsequent baseflow. This probably reflects the fact that the only soil water flow considered is in the soil matrix, although the range of saturated hydraulic conductivities (K_{sat}) values chosen were deliberately set high to allow for fairly rapid soil matrix flow. The lack of any soil water flow other than matrix flow is a severe restriction to the modelling effort, and although the K_{sat} values can be adjusted to allow for more rapid flow than could be reasonably expected in the matrix, this denigrates the physical basis of the study.

If the estimates of K_{sat} are reasonable for all soil water flow and the rainfall partitioning routine is accurately predicting interception loss then the storm response for a non-forested catchment could be expected to be larger than for the afforested catchment. Figure 7.12 shows the storm response during the 1976 event assuming the catchment was totally non-forested. This prediction has a considerably flashier response than the forested catchment and the baseflow after the peak is under-predicted, suggesting that K_{sat} estimates are too low in this case. There are two conclusions that could be drawn from this and no way of distinguishing if either is correct. The first conclusion is that the forest canopy is having a very large effect on catchment response. The second conclusion requires the assumption that K_{sat} does not change with forest growth, in which case it is possible that the input K_{sat} values are too low and that the rainfall partitioning routine may be overestimating the interception loss to produce the reasonably good estimates shown in Figures 7.7–7.10.

As there is no way of knowing what the observed rainfall partitioning totals were during the simulated storms, all that can be done is look at the total figures and see if these appear reasonable (see Table 7.1). Zinke (1967) reports that the amount of interception loss in conifers is commonly between 20 and 40% (10–20% for hardwoods) and Johnson (1991) reports values of 28–49% for sites in upland Britain (25% for Plynlimon). Johnson (1991) also reports stemflow proportions of 2–39% for sites in upland Britain (18% for Plynlimon). The values shown in Table 7.1 span the range of the other reported results and cannot be considered unreasonable. The difficulty with comparing these results is that the reported values are predominantly average yearly values (but not all) whereas the VSAS results are on a storm basis. Gash (1979) and Gash et al. (1980) indicate that interception loss of between 30 and 45% can be expected from conifer plantations in the UK. Note that the variation in throughfall percentages shown in Table 7.1 is a result of the different canopy ages

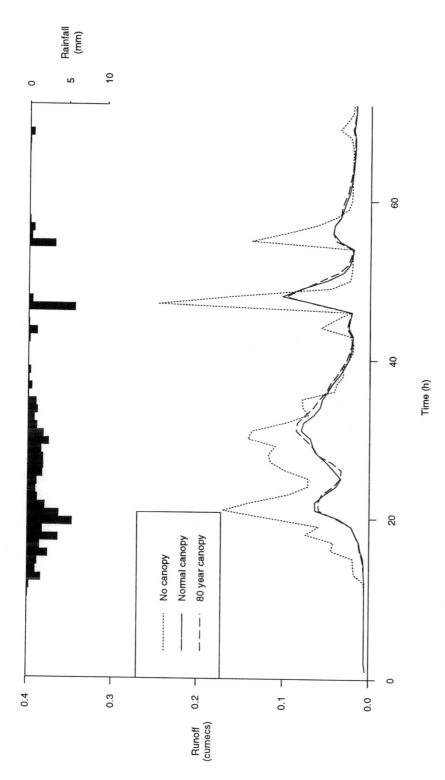

Figure 7.12 The July 1976 storm event simulated with different ages of canopy growth

Table 7.1 Rainfall partitioning percentages of above canopy rainfall for the Tanllwyth simulations

Storm	Throughfall (%)	Stemflow (%)	Interception loss (%)
July 1976	64	6	30
July 1982	51	7	42
July 1989	54	8	38
Feb. 1988	74	8	18
Sept. 1988	70	8	22

plus (or minus) the different meteorological conditions occurring at the time, which affects the potential evaporation. It is noticeable that there was less interception loss during the February 1988 event, which might be expected for a winter storm event.

Given that the simulated rainfall partitioning appears to be within reasonable limits then the model is predicting that the addition of a forest is having a severe effect on the storm events for the Tanllwyth.

The results presented above have indicated that VSAS has some capability at predicting stormflows on an afforested catchment when the storm event involves a single or double hydrograph peak but the capability declines as the storm events become larger and more complex. This does very little to investigate the implications of long-term land-use change. In order to achieve this the modelling scheme has been run on a series of hypothetical scenarios to see how the predictions change given a different set of conditions.

In order to look at the effect that canopy age has on the simulated hydrographs all five storms were simulated using no canopy cover, their normal canopy cover (i.e. 28, 34, 41, 40 and 40 years) and an 80-year-old forest to represent a fully mature forest stand. The results of these simulations showed an interesting point which is illustrated by the July 1976 event in Figure 7.12. There is a large difference in storm peak volumes between no canopy and normal canopy but very little distinction between the normal and 80-year-old forest cover. These results suggest that the canopy is having its most major influence on storm hydrographs prior to 28 years of growth.

In order to test this further and to see if the modelling scheme is able to distinguish any canopy age threshold, the forest growth model was run to simulate 10, 20 and 34 year canopies and the three summer storms simulated using these parameters. The hydrographs show very similar results. The simulations with no and 10-year-old canopy are almost identical while the 20-year-old canopy is approximately half-way between the extremes. Given that in Figure 7.12 the normal and 80-year-old canopies were very similar, the 34-year-old canopy can be said to represent a mature canopy and the major modification in storm hydrographs takes place between the tenth and thirtieth year of forest-cover growth.

By looking at various forest growth model outputs an attempt can be made to see which part of the model is having such a critical effect on the storm hydrographs.

Very broadly speaking, the forest growth model performs two functions: growing the trees in the horizontal dimension towards each other, thereby decreasing the volume of direct throughfall; and growing the trees in the vertical dimension, thereby increasing the number of rainfall intercepting layers in each tree, and hence increasing the amount of indirect throughfall. To assess these two functions the percentage canopy cover over the plot and the average number of intercepting layers per tree were output from the forest growth simulations. The canopy cover is an estimation calculated by totalling the crown area of each tree in the plot and dividing by the total plot area. This does not account for overlap between trees hence it being only an estimate. The average number of intercepting layers per tree is analogous to the average leaf area index (LAI) for the simulated forest.

The results of these are plotted in Figure 7.13. The factor that changes most between 10 and 30 years of growth is the percentage canopy cover. The percentage canopy cover increases from 10% after 13 years to canopy closure being predicted at 24 years of forest growth, whilst the LAI continues to increase throughout the simulation. There is a large increase in canopy layers between 10 and 30 years (one to eight layers) but the number of layers continues to increase beyond year 30 to a maximum of 12 in year 80. When this result is tied in with Figure 7.12 it suggests that canopy closure is having the greatest effect on the storm hydrographs and that the increase in intercepting layers is less important. As a canopy closes, the proportion of above-canopy rainfall reaching the soil surface as direct throughfall decreases and a certain proportion of the rainfall becomes indirect throughfall. As the number of intercepting layers increases, the time taken for the indirect throughfall to reach the surface increases thereby delaying the impact of the storm rainfall. This result suggests that the amount of direct throughfall is more important than the number of intercepting layers for the indirect throughfall. This result is slightly surprising as the greater the number of intercepting layers, the greater the amount of potential interception loss and therefore the less rainfall could be expected to reach the soil surface.

This is not to say that the increase in the number of canopy interception layers is insignificant, but rather that the canopy closure element appears more important. In an earlier version of this model the forest growth model significantly overestimated the number of canopy layers and still the same pattern emerged. In Davie (1993) the total number of layers predicted by the model was 40 layers after 80 years. This appears to be a gross overprediction when compared to similar estimates. Leyton et al. (1967) report a LAI of 15 for a mature Norway spruce plantation; Ford (1982) reports a LAI of 7.5 for a mature Sitka spruce plantation. A LAI of 15 does not necessarily mean that there are 15 layers to the canopy, therefore a direct comparison cannot be made but it seems fair to assume that if there is 15 times the area of leaf as there is of ground surface then the number of canopy layers will be closer to 15 than 40. In the earlier version of the model the storm hydrograph barely changed after canopy closure in spite of the apparent overestimation of number of intercepting canopy layers. The error in the earlier version of model has been corrected for the results shown here.

There is no record kept of when canopy closure did occur in the Tanllwyth so the exact timing of this result cannot be validated. The important feature of this result is that the modelling scheme predicts that the amount of canopy cover is the most

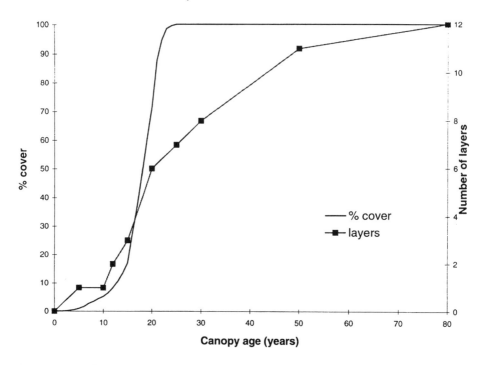

Figure 7.13 The percentage canopy cover and average number of canopy layers calculated by the model, plotted against time. The apparent lack of smoothness in the number of canopy layers line (at 10 years) is due to it always being a whole number

important governing factor in altering a storm hydrograph through rainfall partitioning and that canopy closure is a critical feature to ascertain in the forest growth simulation.

7.3.2 Hypothetical Hillslope Scenarios

The simulations on the Tanllwyth catchment have suggested that canopy closure is an extremely important influence on a storm hydrograph. In order to test this further a form of sensitivity analysis has been carried out. The aim of this sensitivity analysis is not to test the model as such, but to use the model as a controlled experiment so that different hydrological processes could be isolated and investigated and the importance of canopy closure tested under a range of conditions. The parameters were varied within a range which might be expected in the field.

For the sensitivity analysis a single hypothetical segment was used and several factors varied. These factors were topographical parameters, soil hydrology and forest age. The remainder of conditions were kept uniform throughout the simulations. The area of segment was kept at 10 ha throughout the simulations. Several storm sizes were used; the main analysis was done using the July 1982 storm shown above, although several hypothetical storms of larger and smaller size were also used.

The topographical parameters altered were to create conditions of

1. topographic convergence in plan; similar to a valley head;
2. a straight rectangular segment;
3. topographic divergence in plan, similar to an interfluve situation;
4. concavity in cross-section;
5. convexity in cross-section;
6. slope angle variation (4°, 8°, 12°, 16° and 20°).

All of these were varied between each other so that simulations were performed for all versions of slope and shape.

The soil hydrological condition altered was the saturated hydraulic conductivities of the soil in each segment. The three values chosen are designated by Rawls *et al.* (1982) to fall within the range of a clay loam (1×10^{-6} m/s); sand loam (1×10^{-5} m/s); and sand (1×10^{-4} m/s). Before each simulation the segments were "drained" from saturation by running the model without any rainfall input for 72 h to reach some kind of drainage equilibrium. This means that the initial soil moisture conditions for each model run were not the same.

The forest age was varied through snapshots of a three-dimensional canopy simulated by the forest growth pre-processing model. The ages simulated were 5, 10, 15, 20, 25, 30 and 50 years for a forest planted with a tree spacing of 1.4 m. This is not the same forest as was simulated for the Tanllwyth catchment, although it is very similar to part of the Tanllwyth simulation.

The model output analysed was both runoff at the bottom of the segment (a mixture of overland flow and soil water flow) and the soil moisture conditions at the end of each simulation. In total, 4050 model runs were carried out to simulate as wide a range of conditions as possible.

An initial qualitative assessment of the hydrographs and the final soil moisture conditions was carried out. This involved plotting hydrographs together in different combinations to assess the relative importance of each varied parameter (see Davie (1993) for the more detailed description of this). Not all of the hydrographs were plotted; for example, only three canopy ages were considered: pre-canopy closure, near the time of canopy closure and a mature stand.

For all storms it could be seen that the variation in saturated hydraulic conductivity had the largest influence. This is not altogether surprising given that the three soils varied represent a tenfold increase/decrease in hydraulic conductivity values between them and the importance that VSAS places on soil water flow calculation. Within the different soil conditions the importance of canopy age was more important than slope/convergence/convexity in the clay-loam situation but less important than in sand conditions. In the sand-loam conditions the differences could not be separated out into a straight ranking. This does not mean that canopy age had no affect on storm hydrographs in sand, but that topographic effects had a greater influence. These results suggest that canopy age is not always as important a factor as some other attributes known to control hillslope runoff but under certain conditions (notably low-permeability soils) it is very influential.

The qualitative assessment of the hydrographs does not consider the issue of the effect of canopy closure, purely whether for instance under certain conditions variation in topographic convergence of the slope appeared to affect the hydrograph more than variation in canopy age. To investigate the importance of canopy closure the sum of squares statistic was calculated for comparison between simulations. This is the total of the difference in runoff at each hour from the simulated runoff for a sand-loam, rectangular in plan, straight in cross-section, segment with no forest canopy. This can then be plotted against canopy age to see if the effect is significant during particular periods of canopy growth. In all of the examples shown here the storm was the same as the July 1982 storm shown in Figure 7.8.

Figure 7.14 shows the sum-of-squares difference for sand-loam soil conditions, with different lines representing the different states of topographic convergence. All of these are with a segment straight in plan and sloping at 12°. The similarity in values between the planar (rectangular) segment and the convergent segment when compared to the divergent segment reflects the fact that the divergent segment produces very little runoff. In looking at the affects of canopy change on different scenarios it is more important to look at the shape of the line rather than the comparison between the lines. In this modelled scenario there is a very gradual increase in difference until around 12 years, followed by rapid change and then a

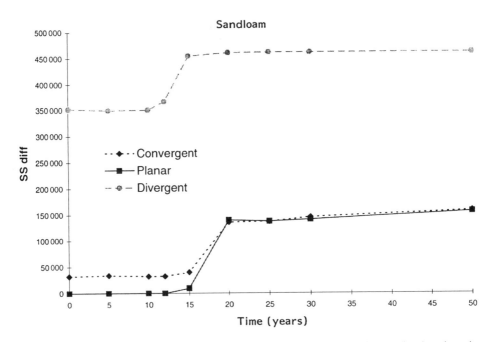

Figure 7.14 The sums of square difference between hillslope flows simulated under varying topographic plan and canopy age conditions. The simulated scenarios are all with sand-loam soil conditions. The sums of squares statistic calculates the difference in flow between a scenario and the values for a rectangular in plan, straight in cross-section, hillslope, with sand-loam soil and sloping at 12°

flattening off and gradual increase after 20 years. This suggests, as with the
Tanllwyth hypothetical simulations, that the greatest change coincides with the
period of rapid canopy closure. The forest growth used in the simulation is very
similar to that shown in Figure 7.13, with canopy closure occurring at around 20
years but the number of layers in the canopy continuing to increase throughout the
simulation. A very similar response can be seen when the maximum flow values are
used as a comparative statistic (see Figure 7.15). The greatest change in maximum
flow amounts occurs between 10 and 15 years of canopy growth. At other times there
is some change but it is very small in comparison to the apparent change during the
period of canopy closure.

The results from the clay-loam simulations were very similar to those of the sand
loam in Figure 7.14, so are not presented here. The sand conditions are shown in
Figure 7.16. It can be seen that the canopy age has virtually no effect on the
divergent segment although it does on the planar and the convergent segments. This
is the only situation of the modelled scenarios where canopy age was not significantly
affecting the storm runoff and reflects the lack of any runoff from a sandy divergent
slope. On the convergent and rectangular (planar) slopes the greatest effect is again
during the period between 10 and 20 years of canopy growth.

The variation in cross-sectional convexity produced very similar results to the
topographic convergence. The results from clay loam are shown as an example here
(Figure 7.17). The convex slope is significantly different from the others in all the soil
types and all of the scenarios had the greatest change between 15 and 20 years of

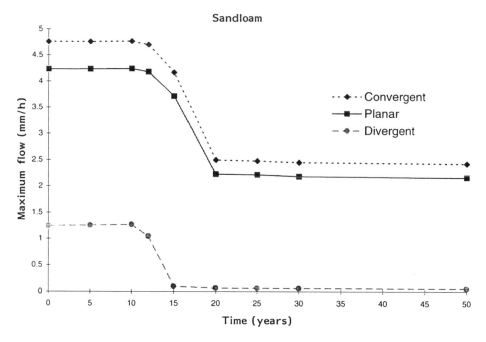

Figure 7.15   The maximum stormflow values from each simulated scenario in Figure 7.14

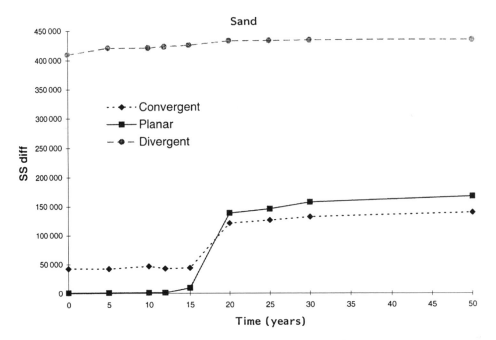

Figure 7.16 The sums of square difference between hillslope flows simulated under varying topographic plan and canopy age conditions. The simulated scenarios are all with sandy soil conditions

canopy growth. The results using peak runoff statistics show the same trends as in Figure 7.15.

The variation in slope angle again showed similar results to the above and only the sand conditions are illustrated here (Figure 7.18). In the qualitative analysis of hydrographs slope was found to be more important than canopy growth on a sandy soil. Figure 7.18 does not compare the two but it is evident from the difference in sum-of-squares values that the slope is having a major effect on the output. It is also evident that canopy age is having its greatest effect during the period between 15 and 20 years.

In almost all of the situations simulated the greatest change in the sum-of-squares (i.e. total difference) statistic and the peak runoff values occurred between 15 and 20 years. As can be seen in Figure 7.13 this is the period of canopy closure. This confirms the findings from the Tanllwyth simulations that canopy closure appears to be the largest forest growth factor affecting storm runoff and extends it to a range of different possible situations.

7.4 DISCUSSION

The modelling scheme used here provides a method of investigating land-use change and the growth of forests in particular, in a controlled environment. The

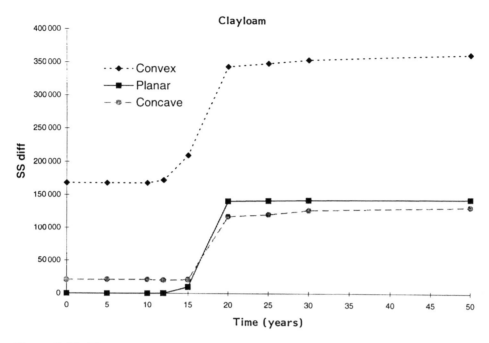

Figure 7.17 The sums of square difference between hillslope flows simulated under varying topographic cross section and canopy age conditions. The simulated scenarios are all with clay-loam soil conditions

simulations of the Tanllwyth catchment and the further modelling of hypothetical scenarios has indicated that throughout the growth of a forest any change in storm runoff hydrographs is not gradual. The greatest change occurs during the period of canopy closure. There are often significant changes in storm flows before and after this period but the greatest change appears to be during the canopy closure period. This suggests that any alteration afforestation makes on stormflow is not a gradual process throughout the period of forest growth but a rapid change during a relatively short period of time. This is not altogether surprising as canopy interception and evaporative loss (which would be more affected by the gradual change in canopy layers than the rapid change in canopy cover) is often assumed to be a minor process during storm events. It might be expected that under conditions of little soil water flow (e.g. the low-permeability clay-loam soil) the water falling directly through canopy gaps (which is what decreases rapidly during canopy closure) would quickly become storm runoff. However, the simulations have included a range of conditions where there is considerable infiltration and soil matrix flow. In these situations it might be expected that canopy gaps are not so important, but the modelling scheme used here predicts that the influence of canopy closure is still large.

Any modelling scheme is an imperfect representation of the complex real world and therefore predictions made from hypothetical scenarios cannot be relied upon as

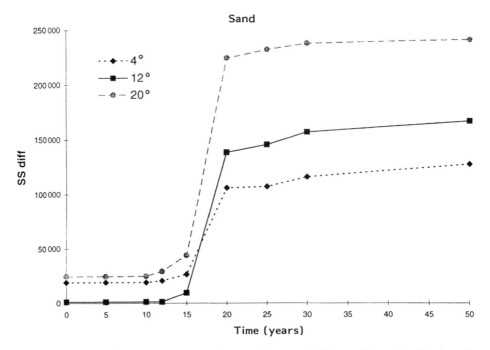

Figure 7.18 The sums of square difference between hillslope flows simulated under varying slope angle and canopy age conditions. The simulated scenarios are all with sandy soil conditions

absolutely accurate. The main assumptions of the model used here and the implications for the results are summarized below.

1. An assumption that the discretization of runoff and forest hydrology processes is correct. Incorrect discretization may lead to overemphasis on certain processes within the model and is a possible explanation for the results presented here. It is very difficult to verify whether a correct spatial discretization is being used as testing of the complete modelling scheme will not necessarily highlight this.
2. An assumption that all runoff is generated as saturated overland flow and all subsurface flow is in a Darcian manner. The simulation of saturated overland flow is a reasonable assumption for a humid mid-latitude forested environment. The absence of any mechanism to model rapid throughflow (e.g. pipe networks) is a restraint on the model. This may influence the results in the generation of extra overland flow as the soil will not be able to simulate water moving quickly through the matrix. This may intensify the importance of direct throughfall as it will be quickly routed to overland flow.
3. An assumption that the LAI is proportional to the number of intercepting canopy layers and which can be described by a modified Poisson distribution. This is the basis for the vertical discretization within the forest canopy. Incorrect usage of this relationship may influence the importance of vertical growth in the canopy.

This is an area that requires further investigation before the full implications of these results can be assessed.

4. An assumption that the soil structure does not change during the growth of the forest. This is probably an incorrect assumption but no directly quantifiable relationship could be established. It seems reasonable that the soil would store more water and allow water to pass through it more easily as a forest grows. It is not possible to say what effects this may have on the results presented here as the rate of change has not been studied.

Konikow and Bredehoeft (1992) suggest that models are a means for providing critical analysis and go on to say: "They are a means to organize our thinking, test ideas for their reasonableness, and indicate which are sensitive parameters. They point the way for further investigation" (Konikow and Bredehoeft 1992, p. 82). The results from this modelling study suggest that a further investigation needs to be carried out to fully understand whether it is purely a factor of the modelling scheme, or that in "real life" canopy closure is the single most important period in a forest canopy development.

Previous work in this area has tended to concentrate on interception loss with forest age. McNaughton and Jarvis (1983) quote the work of Langford and O'Shaughnessy (1980) as showing an increase in interception loss from seedling until canopy closure. Ford and Deans (1978) suggest that interception loss may decrease as conifers age due to changes in the angles of branching. Work by Teklehaimanot *et al.* (1991) has shown that average annual interception loss increases markedly for a decrease in tree spacing. This was attributed to the difference in boundary layer conductance.

The modelling work shown here has concentrated on storm events and ignored the change in boundary layer conductance. It has been concerned with hillslope storm runoff rather than interception amounts but the results appear to correspond to a certain degree with Teklehaimanot *et al.* (1991). It is difficult to imagine being able to isolate out a variable such as canopy growth (and therefore period of closure) from a streamflow record to analyse in an analogous way to this modelling study. A possible method of field investigation into the importance of canopy closure on storm runoff would be with a large-scale lysimeter, although in the past these have been used more for inception studies than hillslope (or plot) runoff. The results from this modelling study are a long way short of conclusive. Further investigation seems worthwhile into whether afforestation produces a gradual change in stormflow or is much more sporadic.

ACKNOWLEDGEMENTS

Much of this work was carried out while the author was supported by a scholarship from the University of Bristol. Thanks are due to staff at the Institute of Hydrology, Plynlimon Office, who were extremely helpful in providing meteorological and hydrological data from the Tanllwyth catchment.

REFERENCES

Abbott, M.B., Bathurst, J.C., Cunge, J.A., O'Connell, P.E. and Rasmussen, J. 1986. An introduction to the European Hydrological System – Système Hydrologique Européen, "SHE", 2: Structure of a physically based, distributed modelling system. *Journal of Hydrology*, **87**, 61–77.

Aikmann, D.P. and Watkinson, A.R. 1980. A model for growth and self thinning in even-aged monocultures of plants. *Annals of Botany*, **45**, 419–427.

Anderson, H.A., Miller, J.D., Gauld, J.H., Hepburn, A. and Stewart, M 1993. Some effects of 50 years of afforestation on soils in the Kirkton Glen, Balquhidder. *Journal of Hydrology*, **145**, 439–451.

Bella, I.E. 1971. A new competition model for individual trees. *Forest Science*, **17**, 364–372.

Bernier, P.Y. 1982. VSAS2: A revised source area simulator for small forested areas. *PhD Thesis*, University of Georgia.

Bernier, P.Y. 1985. Variable source areas and stormflow generation: an update of the concept and simulation effort. *Journal of Hydrology*, **79**, 195–213.

Beven, K. 1989. Changing ideas in hydrology – the case of physically-based models. *Journal of Hydrology*, **105**, 157–172.

Bosch, J.M. and Hewlett, J.D. 1982. A review of catchment experiments to determine the effect of vegetation charges on the water yield and evapotranspiration. *Journal of Hydrology*, **58**, 5–23.

Burt, T.P. and Butcher, D.P. 1985. On the generation of delayed peaks in the stream discharge. *Journal of Hydrology*, **78**, 361–378.

Calver, A. 1988. Calibration, sensitivity, and validation of a physically-based rainfall–runoff model. *Journal of Hydrology*, **103**, 409–417.

Davie, T.J.A. 1993. Modelling the effect of vegetation change on stormflow hydrology. *PhD Thesis*, University of Bristol.

Davie, T.J.A. and Durocher, M.D. (in prep a). A model to consider the spatial variability of rainfall partitioning within deciduous canopy. I. Model description. *Hydrological Processes*.

Davie, T.J.A. and Durocher, M.G. (in prep b). A model to consider the spatial variability of rainfall partitioning within deciduous canopy. II. Model parameterisation and testing. *Hydrological Processes*.

Durocher, M.G. 1990. Monitoring spatial variability of forest interception. *Hydrological Processes*, **4**, 215–229.

Fawcett, K.R., Anderson, M.G., Bates, P.D., Jordan, J.-P. and Bathurst, J.C. 1995. The importance of internal validation in the assessment of physically based distributed models. *Transactions of the Institute of British Geographers*, **20**, 248–265.

Ford, E.D. 1982. High productivity in a polestage Sitka spruce stand and its relation to canopy structure and development. *Forestry*, **55**, 1–7.

Ford, E.D. and Deans, J.D. 1978. The effects of canopy structure on stemflow, throughfall and interception loss in a young sitka spruce plantation. *Journal of Applied Ecology*, **15**, 905–917.

Ford, E.D. and Diggle, P.J. 1981. Competition for light in a plant monoculture modelled as a spatial stochastic process. *Annals of Botany*, **48**, 481–500.

Gash, J.H.C. 1979. An analytical model of rainfall interception by forests. *Quarterly Journal of the Royal Meteorological Society*, **105**, 43–55.

Gash, J.H.C., Wright, I.R. and Lloyd, C.R. 1980. Comparitive estimates of interception loss from three coniferous forests in Great Britain. *Journal of Hydrology*, **48**, 89–105.

Gates, D.J. 1982. Competition and skewness in plantations. *Journal of Theoretical Biology*, **94**, 909–922.

Grayson, R.B., Moore, I.D. and McMahon, T.A. 1992. Physically based hydrologic modeling. 2. Is the concept realistic? *Water Resources Research*, **26**, 2659–2666.

Halldin, S. 1985. Leaf and bark area distribution in a pine forest. In B.A. Hutchinson and B.B. Hicks (eds), *The Forest–Atmosphere Interaction*. D. Reidel, Dordrecht.

Helvey, J.D. and Patric, J.M. 1965. Canopy and litter interception of rainfall by hardwoods of eastern United States. *Water Resources Research*, 1, 193–206.

Hewlett, J.D. and Hibbert, A.R. 1967. Factors affecting the response of small watersheds to precipitation in humid areas. In W.E. Sopper and H.W. Lull (eds), *Forest Hydrology*. Pergamon Press, Oxford.

Hewlett, J.D. and Nutter, W.L. 1970. The varying source areas of stream flow from upland basins. Symposium on *Interdisciplinary Aspects of Watershed Management*. American Society of Civil Engineers, Bozeman, Montana, pp. 65–83.

Hewlett, J.D. and Troendle, C.A. 1975. Non-point and diffused water sources: a variable source area problem. Conference on *Watershed Management*, American Society of Civil Engineers, Logan, Utah, pp. 21–46.

Hunt, R. 1982. *Plant Growth Curves*. Edward Arnold, London.

Ishikawa, Y. and Ito, K. 1989. DBH, height, and crown radius of some component species of Nopporo National Forest, Central Hokkaido, Japan. *Journal of Graduate School of Environmental Science, Hokkaido University, Sapporo*, 12, 117–138.

Johnson, R.C. 1991. *Effects of Upland Afforestation on Water Resources. The Balquhidder Experiment 1981–1991*. Institute of Hydrology Report No. 116.

Kirby, C., Newson, M.D. and Gilman, K. (eds) 1991. *Plynlimon Research: The First Two Decades*. Institute of Hydrology Report No. 109.

Konikow, L.F. and Bredehoeft, J.D. 1992. Groundwater models cannot be validated. *Advances in Water Resources*, 15, 75–83.

Langford, K.J. and O'Shaughnessy, P.J. 1980. Second progress report Corranderrk. Report No. MMBW-W-0010, Melbourne and Metropolitan Board of Works, Victoria, Australia.

Ledig, F.T. 1969. A growth model for tree seedlings based on the rate of photosynthesis and the distribution of photosynthate. *Photosynthetica*, 3, 263–275.

Leps, J. and Kindlmann, P. 1987. Models of the development of spatial pattern of an even-aged plant population over time. *Ecological Modelling*, 39, 45–57.

Leyton, L., Reynolds, E.R.C. and Thompson, F.B. 1967. Rainfall interception in forest and moorland. In W.E. Sopper and H.W. Lull (eds), *International Symposium on Forest Hydrology*. Pergamon, Oxford.

May, R.M. 1974. Biological populations with non-overlapping generations: stable points, stable cycles, and chaos. *Science*, 186, 645–647.

McNaughton, K.G. and Jarvis, P.G. 1983. Predicting the effects of vegetation changes on transpiration and evaporation. In T.T. Kozlowski (eds), *Water Deficits and Plant Growth*. Academic Press, New York.

Millington, R.J. and Quirk, J.P. 1961. Permeability of porous solids. *Transactions of the Faraday Society*, 57, 1200–1207.

Prevost, M., Barry, R., Stein, J. and Plamonden, A.P. 1990. Snowmelt runoff modelling in a Balsam Fir Forest with a variable source area simulator. *Water Resources Research*, 26, 1067–1077.

Pyatt, D.G. and Craven, M.M. 1979. Soil changes under even-aged plantations. Conference on *The Ecology of Even-Aged Forest Plantations*, Edinburgh. Institute of Terrestrial Ecology, pp. 369–396.

Rawls, W.J., Brakensiek, D.L. and Saxton, K.E. 1982. Estimation of soil water properties. *Transactions of the American Society of Agricultural Engineers*, 25, 1316–1320.

Richards, F.J. 1959. A flexible growth function for empirical use. *Journal of Experimental Botany*, 10, 290–300.

Robinson, M. 1986. Changes in catchment runoff following drainage and afforestation. *Journal of Hydrology*, 86, 71–84.

Rutter, A.J., Kershaw, K.A., Robins, P.C. and Morton, A.J. 1971. A predictive model of rainfall interception in forests. I. Derivation of the model from observations in a plantation of Corsican pine. *Agricultural Meteorology*, 9, 367–384.

Smith, W.R. 1990. The static geometric modeling of three-dimensional crown competition. In R.K. Dixon, R.S. Meldahl, G.A. Ruark and W.G. Warren, (eds), *Process Modeling of Forest Growth Responses to Environmental Stress*. Timber Press, Oregon.

Son, I. 1991. Modelling the hydrological effects of land-use change in a small catchment. *PhD Thesis*, University of Southampton.

Teklehaimanot, Z., Jarvis, P.G. and Ledger, D.C. 1991. Rainfall interception and boundary layer conductance in relation to tree spacing. *Journal of Hydrology*, **123**, 161–278.

Troendle, C.A. 1979. A variable source area model for stormflow prediction on first order forested watershed. *PhD Thesis*, University of Georgia.

Troendle, C.A. 1985. Variable source areas models. In M.G. Anderson and T.P. Burt (eds), *Hydrological Forecasting*. John Wiley & Sons, Chichester.

Whitelaw, A.S. 1988. Hydrological modelling using variable source areas. *PhD Thesis*, University of Bristol.

Wu, H., Malafant, K.M., Penridge, L.K., Sharpe, P.J.H. and Walker, J. 1987. Simulation of two dimensional point patterns: applications of lattice framework approach. *Ecological Modelling*, **38**, 299–308.

Zinke, P.J. 1967. Forest inception studies in the United States. In W.E. Sopper and H.W. Lull (eds), *Forest Hydrology*, Pergamon Press, Oxford.

Section 3

HILLSLOPE SOLUTE PROCESSES

8 Movement of Water and Solutes from Agricultural Land: The Effects of Artificial Drainage

A. C. ARMSTRONG and G. L. HARRIS
ADAS Land Research Centre, UK

8.1 INTRODUCTION

For much of lowland Europe, the dominant feature of most land is its use by agriculture. Prediction of the hydrological response of such land thus requires information about the nature of the agricultural land use, and the way it interacts with the soil hydrological system. The whole of agriculture is a deliberate intervention in the natural soil–water–vegetation system, in which the natural vegetation is replaced by a desired artificial vegetation. These processes, involving the periodic disturbance of the soil and the manipulation of soil structure (cultivations), the choice, placing and removal of specific plants, and the direct manipulation of soil hydrological conditions (drainage and irrigation), all have an enormous impact both on the agricultural land itself, and on the surrounding environment. Agriculture therefore represents an enormously complex and varied range of activities. This chapter reviews much recent research into the impact of just one component of agricultural activity: that of artificial drainage on both the hydrology and the movement of solutes to the environment. Although this review will consider the effects of this one agricultural operation, it must be emphasized that it is only one of many such operations which may have effects on site hydrology.

The emphasis on the effects of drainage on clay soils is largely a reflection of the soils in England and Wales, where the vast majority of field drainage is undertaken to control surface wetness in soils with impermeable horizons close to the surface (Thomasson 1981). This review does not, therefore, comment on the use of drainage systems in deep permeable soils with regional groundwater. The majority of the field results quoted come from three major experiments at Cockle Park, Northumberland (Armstrong 1984, 1986a), North Wyke, Devon (Armstrong and Garwood 1991) and Brimstone Farm, Oxfordshire (Cannell *et al.* 1984; Harris *et al.* 1984, 1993a; Catt 1991). These three sites are all heavy clay soils, and were established to be representative of the range of agroclimatic types in the UK where drainage systems rely on mole drainage. Table 8.1 indicates some of the main characteristics of the sites, their soils, and their drainage systems.

Advances in Hillslope Processes, Volume 1. Edited by M. G. Anderson and S. M. Brooks.
© 1996 John Wiley & Sons Ltd.

Table 8.1 Characteristics of the three experimental sites

	Brimstone Farm	North Wyke	Cockle Park
Location (grid ref)	SU 248 947	SX 650 995	NZ 188 916
Soil series	Denchworth	Hallsworth	Dunkeswick
Soil type	pelo-stagnogley	pelo-stagnogley	Typical stagnogley
Topsoil texture	clay	clay loam	clay loam
Subsoil texture	clay	clay	clay
Topsoil clay %	54	37	
Subsoil clay %	62	40	35
Slope	2°	10°	3°
Main cropping	winter cereals	grass	cereals/grass
Main drainage			
Spacing	46 m	30 m	40 m
Depth	0.90 m	0.85 m	0.85 m
Secondary drainage			
	mole drains	mole drains	mole drains
Spacing	2.0 m	2.0 m	2.0 m
Depth	0.60 m	0.55 m	0.55 m

However, before discussing the results from the experiments in detail, the techniques of drainage, and its role in the economy of the farming system, is first described.

8.2 DRAINAGE: TECHNIQUES AND METHODS

The installation of drainage pipes, involving the use of specialized machinery, is expensive, and typical costs are of the order of £1000–2000 per hectare (see Armstrong (1979, 1981) for data on areas, practices and costs). However, at a time when government policy was one of increased agricultural production (MAFF 1975) drainage made economic sense: the costs involved were recouped by an increase in productivity (Trafford *et al.* 1977). Consequently, until 1984, drainage was supported by a system of government grants. During the decade 1971–1981, approximately one million hectares of agricultural land was drained, i.e. approximately 10% of the total agricultural area (Robinson and Armstrong 1988), but since then the rate of drainage activity has dropped to low levels. Although land is no longer being newly drained, existing drainage systems are being maintained, and good drainage remains an essential requirement for the continued utilization of heavy soils.

Artificial drainage of the soil involves the placing of outlets for water in the subsoil, to supplement the natural downward flux of water. Typically this is achieved by installing pipe drainage systems, which provide a free outfall for water below the water-table, which is thus lowered. A typical drainage system is installed as a series of parallel pipes, laid out in a regular pattern, usually a regular "grid-iron" or in a "herringbone" arrangement. Where wetness is due to the eruption of spring water in a field, less regular drainage patterns are imposed. In the UK, drainage systems are installed typically at depths between 0.85 and 1.4 m, and at spacings between 12 and 40 m, the exact dimensions (the "drainage design") being a function of the permeability of the soil and the level of water-table control required.

8.2.1 Mole Drainage

Heavy clay soils present exceptional problems for drainage. In these soils, the rate of water movement through the undisturbed soil is so slow that installation of pipe systems does not provide a sufficient improvement for normal agricultural usage. In these soils, then, a system of *mole drainage* is commonly practised in UK, although it is uncommon elsewhere (see e.g. Spoor 1994). Mole drains are introduced into the soil by the operation of pulling a mole plough through the soil, which consists of a single vertical blade, at the base of which is a cylindrical bullet, behind which is trailed a roughly spherical "expander". The vertical blade introduces fissures and cracks into the soil, while the soil round the bullet is deformed plastically to produce a continuous circular channel. The expander, following behind the bullet, consolidates the walls of the channel. The channel that is so formed will remain in the soil for up to 10 years. Mole drains are drawn at horizontal spacings typically around 2 m, and depths around 550 mm below the ground surface. Because it involves both the introduction of fissures into the soil and the provision of channels, mole drainage both increases the effective conductivity of the soil, and provides frequent cheap drains to provide outlets for the water. Mole drainage has thus been advocated as the best solution for the drainage of clay soils (Trafford and Massey 1975).

8.3 DRAINAGE THEORY

The theory of water movement to drains is derived from an understanding of the physics of soil-water flow processes that was developed largely from experimental studies of simple uniform porous materials and is based on Darcy's Law for saturated flow, and the Richards' equation for unsaturated flow (see e.g. Childs 1969; Hillel 1980; Marshall and Holmes 1988). Because systematic drainage provides a regular situation, a theoretical analysis can take, as a simplification, a segment of soil between two parallel drains, and assume two-dimensional flow. Again, because drainage is concerned with movement of water below the water-table, the unsaturated portion of water movement above the water-table can, at least as a first approximation, be ignored. Using the Dupuit–Forchheimer assumptions (Youngs 1990; Bos 1994) that the rate of water movement is proportional to the gradient of the free water surface (the water-table), we can describe the rate of movement through the soil as:

$$Q = K\frac{\mathrm{d}h}{\mathrm{d}x} \tag{8.1}$$

where Q is the discharge, h is the water-table height, x the horizontal direction, and K the hydraulic conductivity. For parallel drains intersecting the whole soil mass, with drains at spacing S, and with the water-table height at mid-drain spacing, H_m (the highest point), and the water level in the ditch being H_o we can simply write (Bos 1994):

$$Q = \frac{K}{S}(H_m^2 - H_o^2) \tag{8.2}$$

The Dupuit–Forchheimer theory assumes that all water flow is parallel, an approximation that is reasonable for open ditches. However, for drain pipes, the flow is forced to converge, resulting in additional resistance to flow. The commonly used analysis of Hooghoudt (described in Smedema and Rycroft 1983; Ritzema 1994) takes this into account by replacing the real depth of soil between the pipe and the base of the profile, D by an "effective depth", d, so that the flux due to water-table at height h above drains at half-spacing L $(L = S/2)$ is given by:

$$Q = \frac{8Kdh + 4Kh^2}{L^2} \tag{8.3}$$

Ritzema quotes formulae for calculating d from the real depth of soil D, and other parameters of the drainage system. This well-known Hooghoudt drainage equation, or one of its alternatives (see the review by Lovell and Youngs 1984), which relates the height of the water-table at mid-drain spacing, the flux through the system, and the system design parameters, forms the basis for both studies of drainage hydrology and for the design of drainage systems.

Similar drainage equations can be derived from other assumptions, particularly for non-uniform soils. The Ernst analysis (given for example in Ritzema 1994) deals with layered soils, and Youngs (1965) presents analysis for soils in which the hydraulic conductivity varies systematically with depth; results which have been shown to offer practical utility by Armstrong et al. (1991b).

8.3.1 Drainage Design

Drainage design typically requires the choice of drain depth and spacing. The flux, Q, is chosen by reference to the climatic data for the site in question. The maximum permitted water-table height is defined by the crop requirement as a depth of water-table from the surface. It then becomes possible to invert the Hooghoudt equation and so solve for the necessary drain spacing. In practice, the situation using the Hooghoudt equation is less simple because the calculation of the effective depth, d, requires a value for the spacing. However, the process of simple trial and error is easily programmed, or else nomographs can be used (see Ritzema (1994) and also van der Molen and Wesseling (1991) for a closed solution and a relevant computer program).

Drainage design procedures for the UK (see e.g. Castle et al. 1984) adopt the use of climatic data to estimate the flux term, Q. The choice of a suitable value for this term is based on the degree of risk associated with short-term waterlogging of the crop, which is expressed as the return period for the rainfall event. Duration of the design event is five days for normal pipe systems and one day for mole drainage systems. Data suitable for this purpose were compiled by Smith and Trafford (1976), dividing England and Wales into 51 "agroclimatic areas". Choice of suitable parameters for the pipe drainage system to accept the flows is then achieved following the procedures outlined in MAFF (1982) and also detailed in Castle *et al.* (1984).

8.3.2 Macropore Soils

A major shortcoming of conventional soil physical analyses is that they normally assume that the soil is a uniform homogenous medium. No soil meets this assumption absolutely, but it is often a reasonable working assumption. However, the assumption breaks down where the soil contains routes for rapid movement. The phenomenon of preferential or bypass flow in soils with macropores has been the subject of much recent research, although it was recorded as early as 1882 by Lawes *et al.* Frameworks for the study of rapid flows, in which the essential conceptual model is partition of the flow of water and solutes into two distinct but interacting components, normally termed macro- and micropores have been provided by, *inter alia*, Bouma and Dekker (1978), Bouma (1981), Beven and Germann (1982), Germann and Beven (1985) and Gerke and van Genuchten, (1993). More recent work (e.g. Gish *et al.* 1991; Chen and Wagenet 1992a,b) has given considerable attention to preferential flow processes in macropores in respect of water and its associated solutes draining to groundwater and surface-water resources.

Conventional drainage theory cannot adequately predict the movement of water through such soils. Clay soils frequently have matrix hydraulic conductivities less than 10 mm/day, yet have very rapid hydrological responses. Consequently, the prediction of flows through these soils must rely on more complex models, which are no longer amenable to simple manipulation in the same way as the Hooghoudt analysis. Models such as MACRO (Jarvis 1993) and CRACK (Jarvis and Leeds-Harrison 1987; Armstrong *et al.* 1995b) have thus been developed and used successfully to predict the movement of water and solutes in such soils.

8.4 THE ECONOMIC CONTEXT OF DRAINAGE ACTIVITY

The aim of drainage is to improve the profitability of agriculture. However, despite its widespread acceptance, even enshrined in an old dictum that "drainage is the foundation of good husbandry", experimental evidence to support the assertion that drainage increases yields is remarkably sparse. Part of the reason may be that most experimental studies have attempted to establish the effects of drainage by use of paired drained and undrained plots with common cultivation dates, in which the major benefit of drainage, i.e. the greater ease of flexibility of cultivation, is not recorded.

Consequently, only a relatively small number of studies have shown a positive effect of drainage on arable crop yield. The adverse effects of waterlogging on crop yield have been documented (e.g. by Trafford 1972, 1974). Early work at Drayton, Layer Breton and Brooksby (Armstrong 1978) showed an average yield benefit of around 10%, and also indicated an increased efficiency in the use of applied fertilizer nitrogen (Armstrong, 1980). Similarly yield effects directly attributable to drainage were reported from the Brimstone Farm experiment (Christian 1991). The details of the mechanisms involved have been elucidated using lysimeter installations by Cannell *et al.* (1980a,b) and Belford *et al.* (1985), which demonstrated that waterlogging of seeds during germination leads to a complete loss of the crop.

However, the effects of drainage are at least as great in terms of the workability benefit. The review of all available experimental data on drainage effects on crop yields in England and Wales, undertaken by Armstrong *et al.* (1988), showed that, in line with the earlier studies, whereas benefits are generally around 10%, in exceptional circumstances drainage was essential for successful cultivation, and without it no satisfactory crop could be established at all. Nearly every case showed the failed crop on undrained land to be the consequence of limitations on accessibility, and the inability to either cultivate or sow the crop. However, even without this major impact, the impact of drainage on increasing the period that the land is workable (up to three weeks in spring and autumn; Armstrong, 1986b) can have major economic effects (Armstrong 1987).

The economic value of the drainage of grasslands is even more difficult to establish, as the primary effect of drainage on the economy of grassland systems is mediated through its impact on the length of the grazing season, and through the reduction of poaching damage to swards by grazing stock (Patto *et al.* 1978; Davies and Armstrong 1986). Studies at North Wyke, Devon (Armstrong and Garwood 1991; Tyson *et al.* 1992) have shown that the effect of drainage of wet grassland is to increase the yield of grass at the start and the end of the grazing season, but to reduce the yield during the mid-summer. The overall increase in grass yield was thus only 7%, but because of the longer grazing season gave a benefit of 11% in animal production terms. Economic analysis of the results (Armstrong *et al.* 1992), however, shows that when short-term accounting and high interest rates apply, then these benefits do not outweigh the costs of installing drainage.

Drainage is now perceived as a contributor to environmental pollution, and so is no longer actively encouraged. This situation is common in Europe, such that new drainage of new land is discouraged or even prohibited (see e.g. van Hoorn 1990; Kraus 1990). However, where arable cultivation is continued on land prone to waterlogging, particularly heavy clay soils, then continued maintenance of existing drainage systems will be necessary to ensure adequate yields.

8.5 DRAINAGE EFFECTS: IMPACT ON WATER-TABLES

The immediate effect of drainage should be to lower the water-table, if it is above the drains. Experimental verification of this effect is reported in the national survey of drained sites reported by Armstrong *et al.* (1988). This survey reviewed over 400 site-years of soil-water regime data, measured using open auger holes to record the water-table position. The results, however, showed a wide variation in the magnitude of the effect. Although, in the vast majority of the sites examined, the result of the drainage was to reduce the period of winter waterlogging, for a significant number of sites (around 20% of those examined) no such benefit was discernible. Although a number of these could be explained as experimental problems in identifying the drained/undrained treatment pair, there still remained a proportion (around 10% of the sites examined) where the drainage systems that were installed did not appear to achieve their aim.

Much detailed study of the water-table effect on soil-water regime undertaken on ADAS experimental sites, such as those reviewed by Arrowsmith *et al.* (1989), has shown that well-designed and carefully installed drainage systems can achieve their primary aim of reducing the water-table during the field capacity period, thus reducing the duration of waterlogging, with its detrimental effects on crop performance. Tables 8.2 and 8.3 record the mean water-table levels for the experiments at Brimstone Farm and North Wyke respectively. These data show that the drainage effect can be maintained over a number of years. The data from North Wyke, however, also show that despite considerable care in the installation, there is still a high residual variability, as shown in the differences between plots.

Table 8.2 Mean water-tables (centimetres below ground level, January to March), Brimstone Farm (after Harris *et al.* 1993a)

	1979	1980	1981	1982	1983	1984	1985	1986	1987	1988	Mean
Undrained, ploughed	36	21	4	11	20	28	20	28	17	21	20
Undrained, direct drilled	52	15	11	11	36	27	11	26	16	25	23
Drained, ploughed	44	46	31	28	60	73	38	49	42	47	46
Drained, direct drilled	38	47	34	29	58	56	32	45	33	36	41

Table 8.3 Mean annual depth to water-table for each of the plots at North Wyke (centimetres below ground level)

Block	Treat	Plot	Status	1984/5	1985/6	1986/7	1987/8	1988/9	1989/90	Means
A	High N	1	Drained	64	60	59	68	74	81	68
A	Reseed	4	Drained	54	42	43	54	77	72	57
A	Low N	5	Drained	62	62	68	74	72	80	70
B	Low N	7	Drained	50	36	38	46	34	64	45
B	High N	10	Drained	58	49	49	67	68	80	62
B	Reseed	12	Drained	63	54	61	63	65	78	64
Means				59	51	53	62	65	76	61
A	Low N	2	Undrained	43	30	41	41	33	61	41
A	Reseed	3	Undrained	31	18	25	28	30	53	31
A	High N	6	Undrained	45	41	44	48	42	66	48
B	High N	8	Undrained	39	31	37	37	34	57	39
B	Low N	9	Undrained	50	34	39	44	41	64	45
B	Reseed	11	Undrained	33	19	25	28	27	45	30
Means				40	29	35	38	35	57	39

Much more diagnostic, in terms of understanding the hydrological impact of drainage, is the short-term behaviour of the water-table, as it responds to an input event in the form of rainfall. The typical response for a mole-drained clay site, as described by Trafford and Rycroft (1973) for example, shows a rapid rise in water-table accompanied by a rapid rise in the rate of drain discharge, followed by a rapid fall on the cessation of rainfall. A typical example is shown in Figure 8.1. This characteristic form implies both a low drainable porosity and rapid transmission of water to the drainage system, despite the apparently low hydraulic conductivity of such soils, and so indicates the importance of rapid flow through the system of macropores within the soil. By contrast, where the soils are more permeable, and flow is dominated by normal matrix flow through the whole of the soil (not just the system of macropores), the response in water-table terms is much slower.

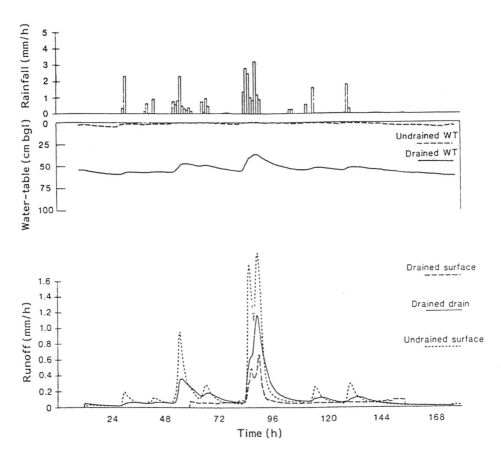

Figure 8.1 Typical water-table and drainflow response to a rainfall event, North Wyke (after Armstrong and Garwood 1991)

Armstrong (1988) has shown how conventional linear response functions (such as the familiar Unit Hydrographs) can be derived from water-table data, and used to characterize the behaviour of the soil-water system. If the water-table is initially at rest at some rest level, H_r, then the response of the water-table to a sequence of rainfall inputs at successive time (T) can be predicted as:

$$H(T) - H_r = \int_0^{tm} R(t)P(T - \tau) \tag{8.4}$$

The response functions, $R(t)$ of length t_m can be considered as an idealized, objectively generalized, response to a rainfall event. A method for deriving the discrete form of these functions for multiple events has been presented by Bruen and Doodge (1984).

Armstrong (1988) presented water-table response functions for contrasting clay and deep silt soils. Further examples (Armstrong et al. 1991a) from the Cockle Park drainage experimental site (Arrowsmith et al. 1989) showed contrasting response functions for different types of drainage at a single site. Here three drainage systems had been installed:

- a conventional system using pipe drains alone at 12 m spacings and 85 cm depth;
- mole drains at 2 m spacing, 55 cm depth over pipe collectors at 40 m spacing;
- winged subsoiling, 35 cm depth and 600 mm spacing, over drains at 20 m spacing.

The water-table response functions for these three systems were identified and showed characteristically different forms (Figure 8.2). The three response functions are different in form, and reflect different efficiencies of the three drainage systems.

The response function for the system using drains alone (Figure 8.2(a)), which starts at a rest level of only 26 cm below the surface, shows an almost instantaneous rise, followed by a long period of decline. The large rise in water-table, to a level close to the soil surface, coupled with a long decline, indicates that this drainage system is not very efficient, as it fails to lower the water table below the topsoil–subsoil interface.

By contrast, the mole-drained plot (Figure 8.2(b)) shows the water-table rising from a returning lower rest level, close to the mole drain depth at 55 cm. The rise to peak is slower than for the drains alone plot, but falls back to rest level more rapidly, reflecting the more efficient drainage offered by this system.

The winged subsoiling had greatly increased the porosity of the top 40 cm of the soil, so whereas it also failed to lower the water-table below that depth, the water-table rise in response to rainfall is very much lower, and the response is much slower (Figure 8.2(c)). By increasing the porosity, subsoiling had decreased the water-table rise in response to rainfall, but because it did not leave a channel behind for the water to leave the soil, the subsequent water-table fall is slower.

These observations indicate that different soil management treatments can have different hydrological effects. If subsoiling is carried out as a routine cultivation (as is common in some areas) then it can have a major impact on hydrological response. The water-table regime achieved by different approaches to drainage on the same site can

196

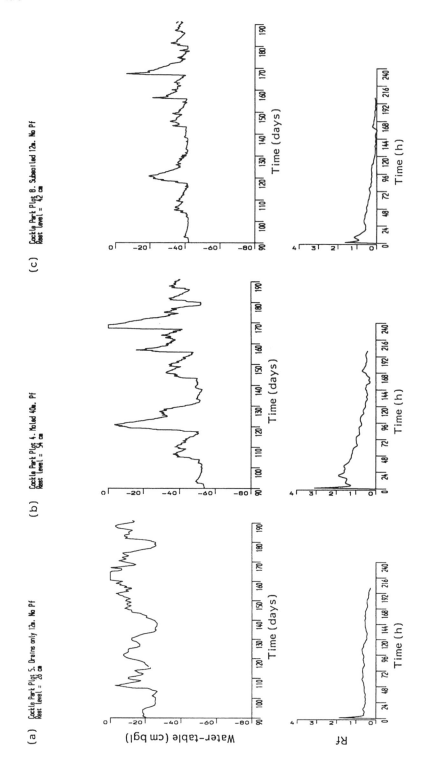

Figure 8.2 Water-table response functions for three different types of drainage at the same site, Cockle Park, Northumberland

lead to different soil-water regimes. It is thus important in hydrological studies to understand the nature of the drainage system in operation in any location.

8.6 DRAINAGE: IMPACT ON RUNOFF

Despite the impact on water-tables, drainage appears to have only a small impact on water balances. Studies at a number of sites have shown that flows through the drainage systems are typically very similar to the surface flows observed on paired undrained areas.

Components of the water balance for the drainage experiments at Cockle Park, North Wyke, and Brimstone Farm are shown in Tables 8.4, 8.5 and 8.6 respectively. At all three sites, the immediate effect is a reduction in the amount of surface or near-

Table 8.4 Mean annual components (mm) of the water balances, and the percentage of rainfall appearing as runoff during the December to March period at Cockle Park

	UDS[a]	DS[b]	DD[c]	Rain
Oct. 1979–Apr. 1980	224	47	233	467
Oct. 1980–June 1981	315	35	239	571
Means	270	41	236	519
Dec. 1979–Mar. 1980	195	44	199	290
Dec. 1980–Mar. 1981	225	27	167	251
Means	210	35	183	270
% of rain	78	13	68	

[a]UDS: undrained surface and near-surface flows.
[b]DS: drained surface and near surface.
[c]DD: drained drainflows.

Table 8.5 Mean annual components (mm) of the water balances, and the percentage of rainfall appearing as runoff during the December to March period at Brimstone Farm

	Ploughed			Direct drilled		
	UDS[a]	DS[b]	DD[c]	UDS[a]	DS[b]	DD[c]
1980/1	68	5	74	56	14	80
1981/2	73	7	84	83	25	62
1982/3	73	3	94	89	21	77
1983/4	19	10	82	57	3	95
1984/5	30	7	91	69	3	94
1985/6	54	3	96	81	11	81
1986/7	77	3	94	79	3	93
1987/8	83	3	91	93	20	78
Mean	60	5	88	76	13	83

[a]UDS: undrained surface and near-surface flows.
[b]DS: drained surface and near surface.
[c]DD: drained drainflows.

Table 8.6 Mean annual components (mm) of the water balances, and the percentage of rainfall appearing as runoff during the December to March period at North Wyke

	UDS[a]	DS[b]	DD[c]	Rain	HER
Dec.–Mar.	307	44	315	496	452
% of rain	62	9	64		

[a]UDS: undrained surface and near-surface flows.
[b]DS: drained surface and near surface.
[c]DD: drained drainflows.

surface runoff. Interception of surface flows by shallow drainage systems (as at Cockle Park and North Wyke) or by gutters and gravel interflow collectors (as at Brimstone Farm) both collect water moving close to the soil surface. On the undrained plots, this near-surface flow is the dominant mode of water movement, and only small components of runoff can be inferred to be moving below this depth (though they cannot actually be measured). By contrast on the drained plots, this near-surface component of runoff is virtually eliminated, and is observed only for a few hours during the peaks of large rainfall events (in excess of 20 mm/day). On drained land, amounts of discharge very similar to the near-surface runoff from undrained plots, are diverted to the drainage system, so that the total loss of water from the sites is normally similar to that from undrained land. The observations of Hallard and Armstrong (1992) and Goss et al. (1983) have emphasized the importance of near-surface or surface water in generating mole drainage flow. However, because the water balance is inevitably incomplete on the undrained land (as any attempt to capture deeper flows will introduce a drainage function which invalidates the undrained plots as controls for all other studies), it is concluded that the major impact of drainage is to affect only the route of water loss, not the total quantity.

The discharge data also show considerable variability in the relative contributions year to year, which suggests that the exact balance is influenced by factors other than the presence or absence of drainage, including both the pattern of weather and the effects of cultivations and other management treatments. The effects of cultivations were observed at both North Wyke and Brimstone Farm.

At North Wyke (Table 8.7), which remained in grass throughout the experiment, several of the plots were initially reseeded, after ploughing. This operation had quite dramatic effects, particularly in the first year after the ploughing, in which the near-surface runoff from the drained plots, through the recently ploughed zone, was some two to three times greater than the runoff through the drainage system, so completely reversing, for a single year, the observation that on the drained plots the drainage systems were the main route for water loss (Armstrong and Garwood 1991).

8.7 DRAINAGE EFFECTS: SHORT-TERM HYDROLOGY

As with water-table data, the most immediate effects of drainage are to be found in the short-term behaviour of the site, and in particular the effect on the rates of runoff

Table 8.7 Summary of the annual flow components (mm) for the plots at North Wyke

Block	Plot	Treat	Drainage[a]	Flow	1982/3	1983/4	1984/5	1985/6	1986/7	1987/8	1988/9	1989/90	1990/1	Mean
A	1	High N	D	Surface	—	15	18	10	133	14	131	6	11	42
A	1	High N	D	Drain	—	455	483	266	489	340	366	368	148	365
A	2	Low N	UD	Surface	—	482	425	305	427	398	357	365	363	390
A	3	Reseed	UD	Surface	—	419	498	315	431	384	385	414	351	400
A	4	Reseed	D	Surface	—	29	15	28	118	39	143	40	45	57
A	4	Reseed	D	Drain	—	409	414	222	345	458	240	289	361	342
A	5	Low N	D	Surface	—	8	5	4	32	4	24	8	8	12
A	5	Low N	D	Drain	—	415	438	249	411	407	320	359	354	369
A	6	High N	UD	Surface	—	187	319	191	246	235	338	224	—	249
A	7	Low N	D	Surface	—	26	13	36	151	30	76	53	178	70
A	7	Low N	D	Drain	—	523	537	177	—	345	351	330	402	387
B	8	High N	UD	Surface	108	387	435	314	631	496	399	—	328	406
B	9	Low N	UD	Surface	—	428	427	292	578	15	331	562	353	434
B	10	High N	D	Surface	43	20	23	31	33	15	52	65	27	33
B	10	High N	D	Drain	207	433	511	207	509	357	396	388	319	390
B	11	Reseed	UD	Surface	241	299	345	306	603	467	219	213	236	336
B	12	Reseed	D	Surface	81	25	18	79	126	18	61	24	15	46
B	12	Reseed	D	Drain	252	403	348	171	479	341	286	318	—	335

—, missing or unrecorded data.
[a]D=Drained, UD=undrained.

from a site. A long running debate has focused on the role of drainage on the generation of runoff, going back to a classic debate introduced by Bailey Denton (1862), and more recently reviewed by Robinson (1990).

The experimental data shown in Figure 8.1 identify the nature of the problem. On an undrained clay site, the dominant process leading to the generation of rapid runoff is infiltration-limited surface flow. If the water-table is close to the surface, then incident rainfall cannot infiltrate, and so becomes surface runoff. The speed with which this runoff reaches the receiving water course will depend on the nature of the surface, in particular whether it is rough (as in a recently ploughed field) or smooth (as in a recently rolled field), or whether tramlines are drawn across or down the slope. Management of the surface thus plays a major role in determining the rate at which runoff reaches the receiving water course. If, however, the undrained land is not waterlogged to the surface, then incident rainfall can infiltrate, and it will then move slowly through the soil to water courses as return or subsurface flow. Runoff from undrained land may thus be either slow or fast, depending on the nature of the soil, and its management.

The rate of runoff from drained land is generally less variable, as the water always has direct routes for movement to the water courses via the drains, although only if secondary drainage systems are to a sufficient standard. Direct comparison between drained and undrained areas will thus depend as much on the nature of the undrained condition as it does on the efficiency of the drainage system.

A detailed examination of these effects has been reported for the Brimstone Farm site by Harris *et al.* (1993a). This study also included an analysis of the effects of differential cultivation regimes. Initially, direct drilling (with burning of straw residues) was adopted as one treatment, but this was replaced by a minimum tillage regime. The experiment thus identified the effects of both drainage and cultivation from replicated isolated plots. From each of the four treatments, the maximum flow in each year was identified, and the flow growth curve identified. The resultant four flood growth curves (Figure 8.3) illustrate the effects of the two treatments.

These results suggest that the effect of the cultivation treatment has a greater magnitude than the drainage, but the effects of the two are roughly additive. Drained land, compared to the undrained land with the same cultivation treatment, generated lower runoff peaks, and thus the immediate effect of drainage is to reduce the flood risk. However, the effect of the change from ploughed to reduced cultivations is to increase the annual flow maximum.

Similar results were also observed on the North Wyke site (Figure 8.4). There is a clear indication that maximum flow rates are higher on the undrained land than on the drained, reflecting the dominance of overland flow on the undrained areas. However, there is also a clear indication that reseeding had affected the response of the undrained plots, with the reseeded plots having a lower peak flow response than the undisturbed plots. This effect is probably due to an increased throughflow component in the disturbed plots. On the drained areas, however, where the dominant mode of water movement remains through the soil to the mole drains, this effect is not repeated, and the two treatments cannot be easily separated. However, these data confirm the observation from Brimstone Farm that management options can have effects similar to the impact of drainage itself.

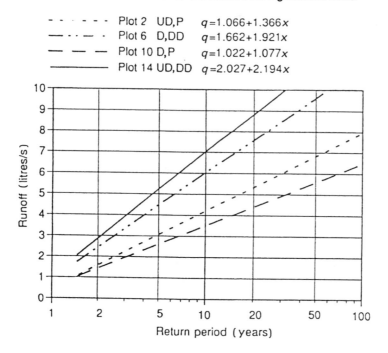

Figure 8.3 Annual maxima flow against return period for each cultivation and drainage treatment (after Harris *et al.* 1993a). (D=Drained; UD=undrained; P=ploughed; DD=direct drilled)

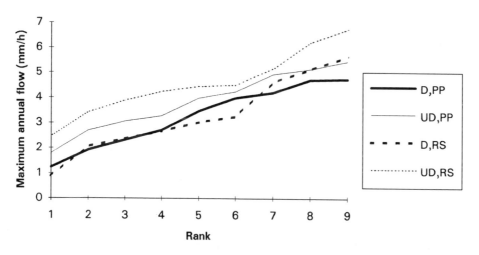

Figure 8.4 Annual maximum flows plotted against rank position, North Wyke. (D=Drained; UD=undrained; PP=permanent pasture; RS=reseed)

8.8 DRAINAGE AND SOLUTE MOVEMENT: NITRATE

Water moving through the soil to the drains also contains dissolved solutes. These solutes are now a major area of concern, as the need to meet quality standards for surface water is becoming increasingly a goal of agricultural management, particularly in the light of the EC Drinking Water Directive (Anon. 1980). Of the solutes most of concern, nitrate and pesticides have been studied the most intensively, and these studies are reviewed here.

The losses of nitrate through drainage systems have been the object of many studies and reviews, dating back to the classic work of Lawes *et al.* (1882). Much of the material has been reviewed by, for example, Addiscott *et al.* (1991) and Anon. (1986). The observation that nitrate levels in UK rivers have risen steadily since the Second World War to the middle 1970s, in step with the intensification of agriculture, and in particular with the increase in the use of nitrogen fertilizer, has led to an immediate correlation between the two, but the mechanisms for the link are far from clear (Davies and Sylvester-Bradley 1995). The assumption that over-use of nitrogen leads directly to leaching is too simplistic, and the process is one in which the use of fertilizer nitrogen enhances mineralization of soil nitrogen which is then available to be leached (Addiscott *et al.* 1991).

Leaching of nitrate through drainage waters has been considered both as a loss of a nutrient to agriculture, and as a pollutant. Traditionally, two major observations have been identified to identify the influence of drainage on nitrate leaching.

- First, waterlogged soil is anaerobic, and in anaerobic soil, nitrate is subject to bacterial action which leads to denitrification. In anaerobic soils, then, nitrate is lost by degradation and gaseous diffusion. By contrast, aerobic soils are suitable to bacteriological activity that can break down organic matter into soluble products, the process of mineralization. It might thus be anticipated that drained soils will have higher nitrate concentrations.
- Secondly, leaching is the process whereby water moving through the soil picks up solutes and moves them to depth, or to the drainage system. Again, because drainage enhances movement through the soil, it might be expected that the leaching loss from drainage water is higher than the equivalent loss from surface runoff which does not interact with the soil to the same degree.

Both these prior arguments suggest that drainage waters may constitute a major route for nitrate loss.

The role of drainage systems in the loss of nitrate has been the subject of intensive study at the Brimstone Farm experimental site (Dowdell *et al.* 1987; Howse and Catt 1991). Here, drainflow, interflow and surface flow have been sampled analysed for nitrate concentrates for an extended period. These results, when broken down into flow type and cultivation type (Table 8.8), show first that the dominant route for nitrate movement is through the drainage systems. Losses in surface water, whether from drained or undrained plots, were small. However, the same results also show an average of difference of 21% between ploughed and direct drilled land, against indicating that the impact of land management practices can be large. Goss *et al.*

Table 8.8 Losses of nitrate-N by surface flow, interflow and drainflow, for the harvest years 1985–1988, Brimstone Farm

			1985	1986	1987	1988	Mean
Drained	Surface	Plough	–	–	0.3	0.3	0.3
		Direct drill	–	–	0.4	0.5	0.5
	Interflow	Plough	–	–	2.0	0.4	1.2
		Direct drill	–	–	0.7	0.3	0.5
	Drainflow	Plough	48.6	50.9	32.5	17.9	37.5
		Direct drill	39.9	46.6	27.5	11.3	31.3
Undrained	Surface	Plough	–	–	1.3	1.1	1.2
		Direct drill	–	–	4.7	2.4	3.2
	Interflow	Plough	–	–	2.2	0.8	1.5
		Direct drill	–	–	4.4	1.0	2.7

(1988) suggest these are the consequence of the enhanced mineralization in ploughed soil providing a larger pool of nitrate in autumn.

Patterns of nitrate concentrations in the drainage waters from Brimstone Farm and other sites have also been reported by Howse and Catt (1991) and Rose *et al.* (1991). These show that the overall pattern is for high levels in the autumn, declining through the winter. Levels may rise again during the spring after the application of the fertilizer. Particularly important is the first drainage event of the winter period, which frequently has very high nitrate levels, in excess of the EC directive for drinking water. The patterns of nitrate leaching from fields are also repeated at the small catchment scale. Results presented by Rose *et al.* (1991) (Figure 8.5), show that the pattern of an initially high nitrate concentration at the start of the winter

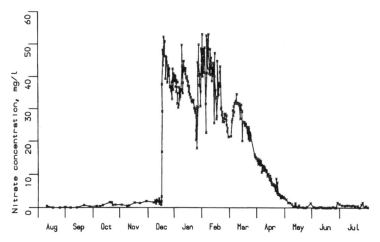

Figure 8.5 Typical nitrate concentrations in drainage waters from a small agricultural catchment, Swavesey, Cambridgeshire, 1989–90 (after Rose *et al.* 1991)

followed by a slow decline during the winter period, is typical of small surface water agricultural catchments.

Short-term patterns of nitrate concentrations from the cracked clay soils of Brimstone Farm are also reported by Rose *et al.* (1991) who demonstrate that during the winter period each rainfall event leads to a corresponding drainflow event, and that the nitrate concentrations show dilution during the peak of discharge, and some recovery in-between (Figure 8.6). The dilution of nitrate concentrations is thought to be associated with the predominance of bypass flow, which passes rapidly from the surface to the drains via the network of cracks and macropores, and in which the supply of nitrate ions is limited by the rate of diffusion out of the peds (Armstrong *et al.* 1995b).

Although nitrate leaching losses have been traditionally associated with arable agriculture (Davies and Sylvester-Bradley 1995), work presented by Garwood (1988) in which he presented a nitrogen budget for the grassland site at North Wyke demonstrated that significant losses of nitrogen can also occur from grassland, particularly when the swards are grazed (Table 8.9). These results show significant losses of nitrogen from the drained areas, in the drainage waters, reflecting the higher concentrations of nitrate found in water moving through the drains as opposed to the surface waters, whether it be from drained or undrained plots. The consequence is that the rate of nitrogen loss from the drained plots is some two to three times higher from drained land than from undrained. This effect is probably due to the combination in drained land of enhanced mineralization, coupled with the diversion of water through the soil. The pattern of leaching loss is broadly similar to that observed in arable areas by Rose *et al.* (1991), throughout the drainage period, starting with high concentrations in the autumn, declining steadily throughout the

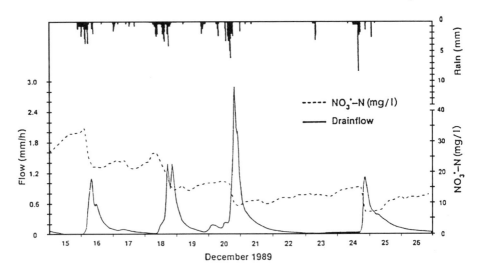

Figure 8.6   Drainage discharge and nitrate concentrations in drain water, Brimstone Farm, December 1989 (after Armstrong and Burt 1993)

Table 8.9 Annual nitrogen balance for drained and undrained permanent grass and reseeded perennial ryegrass swards at North Wyke (all quantities are in kilograms of nitrogen per hectare) (after Garwood 1988)

| | Permanent grass | | | | Reseed | |
	UD	D	UD	D	UD	D
N input	200	200	400	400	400	400
Animal output (cattle LWG)	24	27	28	30	27	34
Leached	20	56	48	187	24	74
Denitrified	90	56	111	84	115	81
Ammonia loss	30*	30	60*	69	101	113
Storage in soil organic matter	72	64	48	45	137	128
Total N accounted for	236	233	295	415	404	430

year (Scholefield *et al.* 1993). Perhaps the most striking result from the North Wyke data, however, is the contrast between the levels of leaching depending on the management regime. Losses were particularly high from those swards receiving high levels of nitrogen inputs.

8.9 DRAINAGE AND PESTICIDE MOVEMENT

Increasing concern over concentration and the load of pesticides in drainage water, again expressed in the EC Drinking Water Directive (Anon. 1980), has led to the study of patterns of movement of pesticides to surface waters. Initially, concern focused on the leaching of pesticides from sandy soils into underlying aquifers, but increasingly it has been realized that rapid movement of waters from the surface of cracked clay soils can generate flows which may contain pesticides and other pollutants. Because the majority of pesticides are applied to, and remain close to, the soil surface, water from the soil surface in particular is at risk of picking up significant quantities of pesticide. A particularly vulnerable situation is the application of autumn pesticides to cracked clay soils. With cracks still open in autumn, bypass flow can easily and rapidly transport water, and its associated dissolved load, from the soil surface to the drains.

Data presented by Harris *et al.* (1994) show that the first significant rainfall event leading to drainflow may contain pesticide concentrations above the EC limit. Because the major mode of flow is bypass flow, this first drainflow event may occur before the soil returns to true field capacity. Macropore flow events may be initiated while the centre of soil aggregates is still below saturation. As with nitrate, the short-term concentrations in the field experiment showed a dilution effect, reflecting the dominance of preferential flow. Modelling of this process requires the explicit consideration of the macropore element of flow (Armstrong *et al.* 1995a).

Because most of the pesticide stays at or close to the soil surface (Nicholls *et al.* 1993; Harris *et al.* 1994), surface runoff from either drained or undrained land may contain high pesticide concentrations at either the field scale (Harris *et al.* 1993b) or the small catchment scale (Harris *et al.* 1992). It is possible that, by increasing the soil-water contact time for pesticide held in the near-surface horizons, that drainage may reduce the concentration of pesticide reaching surface waters.

However, the experimental studies are derived from small areas, and the effects within larger catchments will include the influence of multiple water sources, with the potential for mixing of many different water qualities. Because of mixing, and possible readsorption and further degradation, concentrations of pesticides in drainage water are generally observed only some distance from potable water supplies.

8.10 CONCLUSIONS

This review has identified drainage as a major component of land management for agricultural operations. However, drainage remains only one of the components, and its effects can be either supplemented or negated by other options. It is thus suggested that drainage should not be seen as a single operation, but as a component of the total land management. Often drainage can then be seen as a stage in the intensification of land usage, and its effects need to be seen within the context of a total shift of hydrological and agricultural operations. Nevertheless, several specific generalizations can be made about drainage as an operation.

- Drainage, if effective, lowers the water-table, which in turn reduces the length of the period of waterlogging. This can be translated into higher crop yields, and greater flexibility of agricultural management.
- Drainage has only slight effects on the overall water balance of a site. However, it does provide rapid routes for water movement through the soil. Where the soil is previously saturated, drainage will alter the predominant mode of flow generation from surface to subsurface flow, and will therefore reduce the speed of peak flow generation. However, where the soil is previously unsaturated, then drainage will speed the rate of water movement through the soil, and so increase the speed of flow peaks.
- On the cracked clay soil at Brimstone Farm, the effects of differential cultivation techniques, particularly the contrast between reduced cultivations and ploughing, can have effects equal in magnitude to the drainage effect.
- Drainage remains a major route for the loss of solutes from agricultural land. It is observed that drained land loses significantly more nitrate than undrained land. However, this effect is more probably the result of the better soil conditions consequent upon the drainage, leading to enhanced mineralization, as much as to the direct hydrological effect.

• Pesticide residues my also move to drainage systems, particularly on cracked clay soils, where preferential bypass flow will carry surface waters to drainage systems without major possibility for readsorption.

The study of the mechanisms involved is, however, only the first stage of the process. The challenge is now to convert that understanding into practical agricultural management systems that can control the hydrology of site so as to minimize the impact of agricultural activity on its surrounding environment.

ACKNOWLEDGEMENTS

The authors of this chapter gratefully acknowledge the help of their many colleagues, too numerous to mention individually, who contributed to the work mentioned in this chapter. The majority of this work has been funded by the Ministry of Agriculture, Fisheries and Food, and this funding is gratefully acknowledged.

NOTE

Several of the references quoted in the paper emanate from the former ADAS Field Drainage Experimental Unit of the Ministry of Agriculture, Fisheries and Food, at an address in Trumpington. These publications are no longer available from this address, and readers wishing to follow up these publications should contact the authors of this chapter.

REFERENCES

Addiscott, T.M., Whitmore, A.P. and Powlson, D.S. 1991. *Farming, Fertilisers and the Nitrate Problem*. CAB International, Oxford.

Anon. 1980. EC Council directive relating to the quality of water intended for human consumption (80/778/EC). *Off.J.Eur.Comm.*, **L229**, 11–29.

Anon. 1986 *Nitrate in Water*. A report by the Nitrate Co-ordination Group, Department of the Environment Central Directorate of Environmental Protection, Pollution Paper 26, HMSO, London.

Armstrong, A.C. 1978. The effect of drainage treatments on cereal yields: results from experiments on clay lands. *Journal of Agricultural Science*, Cambridge, **91**, 229–235.

Armstrong, A.C. 1979. *A Digest of Drainage Statistics*. Technical Report 78/7, ADAS Field Drainage Experimental Unit, Trumpington.

Armstrong, A.C. 1980. The interaction of drainage and the response of winter wheat to nitrogen fertilisers: some preliminary results. *Journal of Agricultural Science*, Cambridge, **95**, 229–231.

Armstrong, A.C. 1981. *Drainage Statistics 1978–80*. Technical Report 80/1, ADAS Field Drainage Experimental Unit, Trumpington.

Armstrong, A.C. 1984. The hydrology and water quality of a drained clay catchment, Cockle Park, Northumberland. In T.P. Burt and D.E. Walling (eds), *Catchment Experiments in Fluvial Geomorphology*. Geo Books, Norwich, pp. 153–168.

Armstrong, A.C. 1986a. Mole drainage of a Hallsworth series soil. *Soil Use and Management*, **2**, 54–58.

Armstrong, A.C. 1986b. Drainage Benefits to land workability. In K.V.H. Smith and D.W. Rycroft (eds), *Hydraulic Design in Water Resources Engineering: Land Drainage*. Springer-Verlag, Berlin, pp. 589–598.

Armstrong, A.C. 1987. Soil water management inputs to crop simulation models. In J. Feyen (ed.), *Agriculture: Simulation Models for Cropping Systems in Relation to Water Management*. EUR 10869, pp. 91–104.

Armstrong, A.C. 1988. Modelling rainfall/watertable relations using linear response functions. *Hydrological Processes*, **2**, 383–389.

Armstrong, A.C. and Garwood, E.A. 1991. Hydrological consequences of artificial drainage of grassland. *Hydrological Processes*, **5**, 157–174.

Armstrong A.C., Rands, J.G. and Castle, D.A. 1988. Drainage benefits: watertable control, workability and crop yields. *Agricultural Water Management*, **14**, 43–52.

Armstrong, A.C., Castle, D.A. and Harris, G.L. 1991a. Use of response functions to characterise the efficiency of drainage installations. Paper presented to XVIth General Assembly of the European Geophysical Society, Wiesbaden, Germany, April 1991. Abstract in *Annales Geophysicae*, Supplement to Vol. 9, p. 517.

Armstrong, A.C., Youngs, E.G. and Arrowsmith, R. 1991b. Modelling water-table movement in drained soils with depth-dependent hydraulic conductivity. *Agricultural Water Management*, **20**, 101–108.

Armstrong, A.C., Castle, D.A. and Tyson, K.C. 1992. Economic evaluation of drainage benefits from grassland. *Farm Management*, **8**, 186–193.

Armstrong, A.C. and Burt, T.P. 1993. Nitrate losses from agricultural land. In T.P. Burt, A.L. Heathwaite and S.T. Trudgill (eds), *Nitrate: Processes Patterns and Management*. John Wiley & Sons, Chichester, pp. 239–267

Armstrong, A.C., Portwood, A.M., Harris, G.L., Catt, J.A., Howse, K.R., Leeds-Harrison, P.B. and Mason, J.D. 1995a. Mechanistic modelling of pesticide leaching from cracking clay soils. In A. Walker, R. Allen, S.W. Bailey, A.M. Blair, C.D. Brown, P. Günther, C.R. Leake and P.H. Nicholls (eds), BCPC Monograph No. 62, pp. 181–186.

Armstrong, A.C., Addiscott, T.M. and Leeds-Harrison, P.B. 1995b. Methods for modelling solute movement in structured soils. In S.T. Trudgill (ed.), *Solute Movement in Catchment Systems*. John Wiley & Sons, Chichester, pp. 133–161.

Arrowsmith, R.A., Armstrong, A.C. and Harris, G.L. 1989. The role of sub-surface drainage in the generation of runoff from sloping clay land under different agricultural management regimes. *Second National Hydrology Symposium*, British Hydrological Society, pp. 1.23–1.36.

Bailey Denton, J. 1862. On the discharge from underdrainage and its effects on the arterial channels and outfalls of the country. *Proceedings, Institute of Civil Engineers*, **21**, 48–130.

Belford, R.K., Cannell, R.Q. and Thomson, R.J. 1985. Effects of single and multiple waterloggings on the growth and yield of winter wheat on a clay soil. *Journal of the Science of Food and Agriculture*, **36**, 142–156.

Beven, K.J. and Germann, P.F. 1982. Macropores and water flow in soils. *Water Resources Research*, **18**, 311–325.

Bos, M.G. 1994. Basics of groundwater flow. In H.P. Ritzema (ed.), *Drainage Principles and Applications*. ILRI Publication 16, Second Edition, International Institute for Land Reclamation and Improvement, Wageningen, The Netherlands, pp. 252–259.

Bouma, J. 1981. Soil morphology and preferential flow along macropores. *Agricultural Water Management*, **3**, 235–250.

Bouma, J. and Dekker, L.W. 1978. A case study on the infiltration into dry clay soil. I Morphological observations. *Geoderma*, **20**, 27–40.

Bruen, M. and Doodge, J.C. 1984. An efficient and robust method for estimating unit hydrograph ordinates. *Journal of Hydrology*, **70**, 1–24.

Cannell, R.Q., Belford, R.K., Gales, K. and Dennis, C.W. 1980a. A lysimeter system used to study the effect of transient waterlogging on crop growth and yield. *Journal of the Science of Food and Agriculture*, **31**, 105–116.

Cannell, R.Q., Belford, R.K., Gales, K., Dennis, C.W. and Prew, R.D. Effects of waterlogging at different stages of development on the growth and yield of Winter Wheat. *Journal of the Science of Food and Agriculture*, **31**, 117–132.

Cannell, R.Q., Goss, M.J., Harris, G.L., Jarvis, M.G., Douglas, J.T., Howse, K.R. and Le Grice, S. 1984. A study of mole drainage with simplified cultivation for autumn-sown crops on a clay soil 1. Background, experiment and site details, drainage systems, measurements of drainflow, and summary of results, 1978–80. *Journal of Agricultural Science, Cambridge*, **102**, 539–559.

Castle, D.A., McCunnell, J. and Tring, I.M. (eds) 1984. *Field Drainage: Principles and Practice*. Batsford Academic, London.

Catt, J.A. (ed.) 1991. *The Brimstone Experiment: Proceedings of a Conference on Collaborative Work by ADAS Field Drainage Experimental Unit and Rothamsted Experimental Station*. Lawes Agricultural Trust, Harpenden.

Chen, C. and Wagenet, R.J. 1992a. Simulation of water and chemicals in macropore soils. Part 1. Representation of the equivalent macropore influence and its effect on soil water flow. *Journal of Hydrology*, **130**, 105–126.

Chen, C. and Wagenet, R.J. 1992b. Simulation of water and chemicals in macropore soils. Part 2. Application of linear filter theory. *Journal of Hydrology*, **130**, 127–149.

Childs, E.C. 1969. *An Introduction to the Physical Basis of Soil Water Phenomena*. Wiley-Interscience, London.

Christian, D.G. 1991. Crop yield and uptake of nitrogen. In J.A. Catt (ed.), *The Brimstone Experiment: Proceedings of a Conference on Collaborative Work by ADAS Field Drainage Experimental Unit and Rothamsted Experimental Station*. Lawes Agricultural Trust, Harpenden, pp. 41–53.

Davies, B.D. and Sylvester-Bradley, R. 1995. The contribution of fertiliser Nitrogen to leachable nitrogen in the UK: a review. *Journal of the Science of Food and Agriculture*, **68**, 399–406.

Davies, P.A. and Armstrong, A.C. 1986. Field measurements of grassland poaching. *Journal of Agricultural Science*, Cambridge, **106**, 67–73.

Dowdell, R.J., Colbourn, P. and Cannell, R.Q. 1987. A study of mole drainage with simplified cultivation for autumn sown crops on a clay soil. 4. Losses on nitrate-N in surface run-off and drain water. *Soil and Tillage Research*, **9**, 317–331.

Garwood, E.A. 1988. Water deficiency and excess in grassland. In R.J. Wilkins (ed.), *Nitrogen and Water Use by Grassland*. AFRC Institute for Grassland and Animal Production, Hurley, pp. 24–41.

Gerke, H.H. and van Genuchten, M.T. 1993. A dual-porosity model for simulating the preferential movement of water and solutes in structured porous media. *Water Resourcs Research*, **29**, 305–319.

Germann, P.F. and Beven, K. 1985. Kinematic wave approximation to infiltration into soils with sorbing macropores. *Water Resources Research*, **21**, 990–996.

Gish, T.J., Shirmohammadi, A. and Helling, C.S. 1991. Modelling preferential movement of agricultural chemicals. In T.J. Gish and A. Shirmohammadi (eds), *Preferential Flow. Proceedings*. ASAE National Symposium, Chicago, December 1991, pp. 214–222.

Goss, M.J., Harris, G.L. and Howse, K.R. 1983. Functioning of mole channels in a clay soil. *Agricultural Water Management*, **6**, 27–30.

Goss, M.J., Colbourn, P., Harris, G.L. and Howse, K.R. 1988. Leaching of nitrogen under autumn-sown crops and the effects of tillage. In D.S. Jenkinson and K.A. Smith (eds), *Nitrogen Efficiency in Agricultural Soils*. Elsevier Applied Science, London & New York, pp. 269–282.

Hallard, M. and Armstrong, A.C. 1992. Observations of water movement to and within mole drainage channels. *Journal of Agricultural Engineering Research*, **52**, 309–315.

Harris, G.L., Goss, M.J., Dowdell, R.J., Howse, K.R. and Morgan, P. 1984. A study of mole drainage with simplified cultivation for autumn sown crops on a clay soil. 2. Soil water regimes, water balances and nutrient loss in drain water. *Journal of Agricultural Science, Cambridge*, **102**, 561–581.

Harris, G.L., Rose, S.C., Muscutt, A.D. and Mason, D.J. 1992. Changes in hydrology, pesticides and soil-nitrogen levels under set-aside. In *Set-Asside*. British Crop Protection Council, Monograph 50, pp. 35–40.

Harris, G.L., Howse, K.R. and Pepper, T.J. 1993a. Effects of moling and cultivation on soil-water and runoff from a drained clay soil. *Agricultural Water Management*, **231**, 161–180.

Harris, G.L., Hodgkinson, R.A., Brown, C., Rose, D.A., Mason, D.J. and Catt, J.A. 1993b. The influence of rainfall patterns and soils on losses of isoproturon to surface waters. *Proceedings, Crop Protection in Northern Britain*, 247–252.

Harris, G.L., Nicholls, P.H., Bailey, S.W., Howse, K.R., and Mason, D.J. 1994. Factors influencing the loss of pesticides in drainage from a cracking clay soil. *Journal of Hydrology*, **159**, 235–253.

Hillel, D. 1980. *Fundamentals of Soil Physics*. Academic Press, New York.

Howse, K.R. and Catt, J.A. 1991. Nitrate leaching. In J.A. Catt (ed.) 1991. *The Brimstone Experiment: Proceedings of a Conference on Collaborative Work by ADAS Field Drainage Experimental Unit and Rothamsted Experimental Station*. Lawes Agricultural Trust, Harpenden, pp. 27–40.

Jarvis, N.J. 1993. *The MACRO Model (Version 3.1). Technical Description and Sample Simulations*. Swedish University of Agricultural Sciences, Department of Soil Science, Reports and Dissertations No. 19, Uppsala, Sweden.

Jarvis, N.J. and Leeds-Harrison, P.B. 1987. Modelling water movement in drained clay soil. I. Description of the model, sample output and sensitivity analysis. *Journal of Soil Science*, **38**, 499–509.

Kraus, W. 1990. Hydraulique agricole et environnement en Republique Federale D'Allemagne. In M.-P. Arlot (ed.), *Actes de Journeés "Hydraulique Agricole et Environnement"*. AFEID, Antony, France, pp. 155–159.

Lawes, J.B., Gilbert, J.H. and Warrington, R. 1882. *On the Amount and Composition of the Rain and Drainage Water Collected at Rothamsted*. Williams, Clowes & Sons, London.

Lovell, C.J. and Youngs, E.G. 1984. A comparison of steady-state land-drainage equations. *Agricultural Water Management*, **9**, 1–21.

MAFF 1975. *Food from Our Own Resources*, HMSO, London, Cmnd. 6020.

MAFF 1982. *The Design of Pipe Drainage Systems*. Ministry of Agriculture, Fisheries and Food, Reference Book 345, HMSO, London.

Marshall, T.J. and Holmes, J.W. 1988. *Soil Physics*, 2nd edn. Cambridge University Press, Cambridge.

Nicholls, P.H., Evans, A.A., Bromilow, R.H., Howse, K.R., Harris, G.L., Rose, S.C., Pepper, T.J. and Mason, D.J. 1993. Persistence and leaching of isoproturon and mecoprop in the Brimstone Farm plots. In *Weeds*, Brighton Crop Protection Conference, pp. 849–854.

Patto, P.M., Clement, C.R. and Forbes, T.J. 1978. *Grassland Poaching in England and Wales*. Permanent Grassland Studies 2, GRI–ADAS Joint Permanent Pasture Group, Grassland Research Institute, Hurley.

Ritzema, H.P. (ed.) 1994. *Drainage Principles and Applications*, 2nd edn. ILRI Publication 16, International Institute for Land Reclamation and Improvement, Wageningen, The Netherlands.

Robinson, M. 1990. *Impact of Improved Land Drainage on River Flows*. Institute of Hydrology Report no. 113, Wallingford.

Robinson, M. and Armstrong A.C. 1988. The extent of agricultural field drainage in England and Wales, 1971–80. *Transactions, Institute of British Geographers*, New Series, **13**, 19–28.

Rose, S.C., Harris, G.L., Armstrong, A.C., Williams, J.R., Howse, K.R. and Tranter, N. 1991. The leaching of agrochemicals under different agricultural land uses and its effect on water quality. In *Sediment and Stream Water Quality in a Changing Environment: Trends and Explanations*. IAHS Publication, **203**, 249–257.

Scholefield, D., Tyson, K.C., Garwood, E.A., Armstrong, E.A., Hawkins, J. and Stone, A.C. 1993. Nitrate leaching from grazed grassland lysimeters: effects of fertiliser input, field drainage, age of sward and patterns of weather. *Journal of Soil Science*, **44**, 601–613.

Smedema, L.K. and Rycroft, D.W. 1983. *Land Drainage: Planning and Design of Agricultural Drainage Systems*. Batsford Academic, London.

Smith, L.P. and Trafford, B.D. 1976. *Climate and Drainage*. Ministry of Agriculture, Fisheries and Food, Technical Bulletin 34, HMSO, London.

Spoor, G. 1994. Mole drainage. In H.P. Ritzema (ed.), *Drainage Principles and Applications*, 2nd edn. ILRI Publication 16, International Institute for Land Reclamation and Improvement, Wageningen, The Netherlands, pp. 913–927.

Thomasson, A.J. 1981. The distribution and properties of British soils in relation to land drainage. In M.J. Gardiner (ed.), *Land Drainage: A Seminar in the EC Program of Coordination of Research on Land Use and Rural Resources*. Balkema, Rotterdam, pp. 3–10.

Trafford, B.D. 1972. *The Evidence in Literature for Increased Yield due to Field Drainage*. Technical Bulletin 72/5, ADAS Field Drainage Experimental Unit, Trumpington.

Trafford, B.D. 1974. *The Effect of Waterlogging on the Emergence of Cereals*. Technical Bulletin 74/3, ADAS Field Drainage Experimental Unit, Trumpington.

Trafford, B.D. and Massey, W. 1975. *A Design Philosophy for Heavy Soils*. Technical Bulletin 75/5, ADAS Field Drainage Experimental Unit, Trumpington.

Trafford, B.D. and Rycroft, D.W. 1973. Observations on the soil-water regimes in a drained clay soil. *Journal of Soil Science*, **24**, 380–391.

Trafford, B.D., Mitchell, W., Homan, D.K. and Hunter, J.M. 1977. *Drainage in the Economy of the Farm*. MAFF Technical Management Note No 3. ADAS, MAFF, London.

Tyson, K.C., Garwood, E.A., Armstrong, A.C. and Scholefield, D. 1992. Effects of field drainage on the growth of herbage and the liveweight gain of grazing cattle. *Grass and Forage Science*, **47**, 290–301.

Van der Molen, W.H. and Wesseling, J. 1991. A solution in closed form and a series solution to replace the tables for the thickness of the equivalent layer in Hooghoudt's drain spacing formula. *Agricultural Water Management*, **19**, 1–16.

Van Hoorn, J.W. 1990. Hydraulique agricole et environnement aux Pays-Bas. In M.-P. Arlot (ed.), *Actes de Journeés "Hydraulique Agricole et Environnement"*. AFEID, Antony, France, pp. 141–153.

Youngs, E.G. 1965. Horizontal seepage through unconfined aquifers with hydraulic conductivity varying with depth. *Journal of Hydrology*, **3**, 283–296.

Youngs, E.G. 1990. An examination of computed steady-state water-table heights in unconfined aquifers: Dupuit-Forchheimer estimates and exact analytical results. *Journal of Hydrology*, **119**, 201–214.

9 Nitrogen and Phosphorus Flows from Agricultural Hillslopes

K. W. T. GOULDING,[1] L. S. MATCHETT,[2] G. HECKRATH,[1] C. P. WEBSTER,[1] P. C. BROOKES[1] and T. P. BURT[2]

[1] *Soil Science Department, IACR-Rothamsted, Harpenden, UK*
[2] *School of Geography, Oxford University, UK*

9.1 INTRODUCTION

9.1.1 Nitrogen and Phosphorus in the Aquatic System

The pollution of surface and groundwaters by nitrate and phosphate is a continuing problem and the subject of much debate (e.g. Tunney *et al.* 1995). Excess nitrate and phosphate within the aquatic system both constitute an environmental threat via their promotion of eutrophication. Indirect health concerns from elevated nitrate and phosphate concentrations in water arise from the secretion of toxins by the prolific algal growth associated with eutrophication. Direct health concerns focus on the incidence of methaemoglobinaemia at concentrations > 100 mg N/l and nitrate has been regarded as a possible cause of stomach cancer, but evidence suggests that the latter is unlikely (Addiscott *et al.* 1991). The reduction of nitrate and phosphate inputs to surface and groundwaters from agricultural sources is fundamental to alleviating nutrient pollution. The integration of strips of uncultivated land at the interface of terrestrial–aquatic systems to operate as riparian buffers to pollutants moving off hillslopes is a promising first-line of defence.

9.1.2 Regulatory Chemical and Biological Processes

This chapter considers some of the chemical and biological processes that regulate the flows and losses of nitrogen (N) and phosphorus (P) within hillslope and floodplain ecosystems. The chemical processes that dominate N and P cycling are dissolution (of fertilizers), precipitation (of P compounds), adsorption and desorption (of ammonium ions, NH_4^+, onto clays and of P forms onto clays, oxides, hydroxides and organic matter), and ion exchange (of NH_4^+ and P forms). Mineralization (the conversion of organic N, P and sulphur, S, to inorganic forms) and immobilization (the reverse) are the biological processes of greatest influence on N and P, especially in natural and semi-natural ecosystems. For N, mineralization involves two steps: ammonification (organic N to NH_4^+) and nitrification (NH_4^+ to NO_3^-).

Advances in Hillslope Processes, Volume 1. Edited by M. G. Anderson and S. M. Brooks.

The dominant fate of N and P within the soil is transport via catchment drainage and ultimately the river system if its passage is uninterrupted (Ritter 1988; Heathwaite *et al.* 1993; Tunney *et al.* 1995). Phosphorus is removed from percolating waters by plant uptake, by adsorption by soil and by the precipitation of insoluble phosphates (Sharpley and Menzel 1987). Of these, only plant uptake can continually remove P, and then only if the plants are harvested; otherwise uptake will reach a maximum, as will adsorption and precipitation, and then P will begin to leach again. Dissolved nitrate may leach directly to the aquatic system, be assimilated by vegetation or the soil microbial biomass, or undergo dissimilatory reduction to ammonium or to a nitrogenous gas via the process of denitrification (Addiscott *et al.* 1991; Armstrong and Burt 1993). Denitrification is the reduction of nitrate (NO_3^-) to dinitrogen gas (N_2), via nitrite (NO_2^-), nitric oxide (NO) and nitrous oxide (N_2O) by micro-organisms under anoxic conditions (Knowles 1982; Vinten and Smith 1993). The penultimate product, N_2O, is a greenhouse gas, and hence a potential atmospheric pollutant. Under certain conditions denitrification fails to reach completion and N_2O forms the major product of denitrification. If the process goes to completion, i.e. all of the product is N_2, the denitrification process is benign; indeed it closes the nitrogen cycle and provides an environmentally acceptable way by which nitrate can harmlessly and continuously be removed from the terrestrial system. Much effort is therefore being made to understand the factors regulating denitrification and to gain an insight as to how the process can best be managed in order to maximize nitrate reduction without exacerbating the production of N_2O.

The make-up, density and distribution of the microbial population of a soil, and the rate and products of chemical and biological reactions such as denitrification, are governed by temporally and spatially variable factors that operate over a range of scales. These factors are soil type (clay content, mineralogy, organic matter content), climate (temperature, moisture), hydrology, and management (grazing, cutting, fertilizers, irrigation). The effects of these are briefly described below, with references leading to a more complete literature.

9.1.3 Soil

Different soil types will form over time along a hillslope–floodplain catena as a result of various processes (Gerrard 1981; Burt and Trudgill 1985), notably through leaching and the erosion of fine materials (clay and organic matter, OM) from the slope, with accumulation in the valley. A generalized hillslope which commonly occurs in Britain is of topslope podsols in well-drained conditions, midslope brown earths and footslope poorly drained gleys. The timescale for these processes is at least decades but, more realistically, centuries. We can therefore safely assume that soil type remains constant during the lifetime of a research project of a few years. Soil is an open system (Gerrard 1981; Addiscott 1995). It receives and loses material and energy at its boundaries, e.g. at the top and bottom of hillslopes. So, when considering a soil system we need to define its boundaries, inputs and outputs, storage capacities, flow patterns, and the transformations that occur. For a riparian buffer zone, the boundry concept requires three-dimensional consideration. The interactive zone must not be considered as a superficial phenomenon, more as a 'reactor volume' within the soil body. Vertical boundaries are the watercourse and the top of the slope, the soil surface and the maximum depth of microbial activity or subsurface flow. Horizontally, side boundaries are more difficult to delimit. The horizontal component to the riparian soil system is best assayed via establishing the down-valley vector of the drainage lines.

9.1.4 Climate

On long slopes, it will be colder at the top than at the base by about 1°C per 200 m vertical change. More importantly, frost may occur on only a part of the slope, for example in the valley when cold air is trapped below warm air. The relevance of this is that the rate of biological processes often doubles with a 10°C increase in temperature, and freezing and thawing stimulate many chemical and biological processes such as adsorption and mineralization. Precipitation can vary widely over the slope: orographic cloud can result in much wetter conditions at the top of the slope, and seeder/feeder mechanisms can result in much more acid, N or S deposition at the top of the slope (Fowler *et al.* 1988).

9.1.5 Hydrology

Drainage patterns on a hillslope have been described by Gerrard (1981) and Burt (1986) and are closely associated with soil catena development. Generally one would expect leaching at the top of the slope, surface flow (and erosion) and throughflow and leaching on the slope, and waterlogging and leaching at the base of the slope. In terms of hydrology, one needs to assess the relative importance of surface and subsurface flow (and thus infiltration rate), and the nature and direction of all the flows. It is often erroneously assumed that water follows a simple transect down the hillslope, perpendicular to the river. However, this is too simplistic and can lead to misinterpretation of data if not verified. The role of both the hillslope gradient and the down-valley gradient must be accounted for when establishing the flow paths of a hillslope system (Haycock and Burt 1993).

9.1.6 Management

Management practices exert a very strong controlling influence upon the agricultural hillslope system and beyond through fertilizer and manure application, grazing and cultivation-enhanced erosion (Armstrong and Burt 1993; Vinten and Smith 1993). The interactions between these imposed changes and those inherent to the system generate environmental threats. For example, heavy rain coupled with overstocking and compaction, can result in the erosion of bare soil and the leaching of fertilizers or manures. Truly natural ecosystems are not managed and rarely export excessive N and P from their terrestrial to aquatic component; indeed these are much more likely to be N and/or P limited. However, excessive atmospheric deposition of N and the ensuing acidity increase may unbalance a natural hillslope ecosystem and lead to losses to drainage waters (UKRGIAN 1994).

We shall discuss the relative importance of some of these processes and their controlling factors in determining N and P losses from hillslopes, using an experiment at Cuddesdon on the River Thame in Oxfordshire, and the 152-year old Broadbalk Continuous Wheat Experiment at Rothamsted Experimental Station in Hertfordshire. The former focuses on nitrogen flows, the latter on many aspects of nutrient cycling. Complimentary results from these experiments show how hydrology is the dominant controlling factor in N and P losses at these two sites, limiting the mediation of losses by chemical and biological processes.

9.2 EXPERIMENTAL METHODS

9.2.1 Nitrogen Flows in a Riparian Zone

9.2.1.1 Background

Denitrification is thought to be a key process operating in riparian soils, removing nitrate from waters moving through them from the terrestrial system upslope to the aquatic system downslope. There is much evidence of a disappearance of nitrate as water passes through such zones (Howard-Williams and Downes 1993; Haycock and Burt 1993), resulting in the concept of such riparian strips operating as buffers for nitrate pollution. This missing nitrate is commonly attributed to denitrification but direct quantification is frequently lacking and possible alternative fates neglected. In 1993 we began to study the movement of nitrate through the hillslope and floodplain section of a riparian zone near Cuddesdon in Oxfordshire. The focus was placed upon quantifying and understanding the relationship between groundwater nitrate fluxes and the process of denitrification. The measurements entailed a detailed monitoring of water and nitrate flows, actual and potential denitrification and associated controlling factors such as temperature, soil moisture status and soil carbon and inorganic nitrogen.

9.2.1.2 Site and Methods

The riparian zone (Figure 9.1) constituted the floodplain and backslope of a section of the River Thame. The 100 m wide strip consists of unimproved pasture for summer grazing and a narrow section of scrubland (small bushes and rough grasses). Above lies several hundred metres of intensively farmed arable land. The site appears to be a perfect buffer zone, where waterlogged silty-clay soil in the valley (which includes an oxbow) could be denitrifying nitrate from water moving through it, hence reducing nitrate input to the River Thame.

The site was instrumented with a grid network of dipwell piezometers which allowed weekly measurements of the water-table and the sampling of groundwater, with analysis being made for nitrate (NO_3-N), nitrite (NO_2-N) and ammonium (NH_4-N) (Haycock and Burt 1993). Dominant flow paths were determined and nitrogen fluxes studied. A transect was established across the riparian strip, along which both intact cores and loose soil were collected from the top 10 cm each month over a period of one year, from April 1994 to May 1995 (Webster and Goulding 1989; Goulding et al. 1995). Measurements were restricted to the top

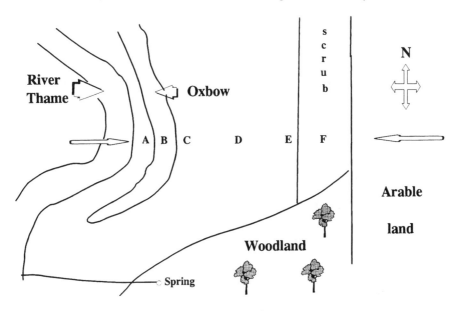

Figure 9.1 A sketch map of the buffer zone at Cuddesdon, Oxfordshire, showing the transect through Zones A–F along which samples were collected

10 cm of soil because previous research has shown that most, if not all, denitrification activity in grassland occurs in this layer (Jarvis *et al.* 1991). Samples were taken at six points, chosen on the basis of topographical, soil and management characteristics of the buffer zone, shown in Figure 9.1: riverbank (A), oxbow (B), midslope (C), footslope (D), topslope (E), scrub (F). Denitrification was quantified in the field using (i) a 24-hour paired intact core incubation, with and without acetylene-inhibition, and (ii) a grid of cover boxes (Webster and Goulding 1989). The latter approach proved unsuccessful, however, due to the lack of sensitivity of this method and practical problems of flooding, freezing and curious, destructive grazing cattle!

In the laboratory, loose soil collected from around the cores was used to assess the *maximum potential* for denitrification and the production of N_2O, using the denitrifying enzyme assay (DEA) technique of Smith and Tiedje (1979). This involved slurrying the soil in a flask with excess nitrate and carbon under anaerobic conditions (a $N_2 + 10\%$ acetylene atmosphere) (Webster and Goulding 1989). Acetylene inhibits the reduction of N_2O to N_2 and so all of the product of denitrification is N_2O, which can be measured by gas chromatography. Chloramphenicol was used in a further set of incubations to inhibit *de novo* enzyme synthesis (i.e. a growth in the numbers of denitrifiers and therefore the potential to denitrify; Dendooven *et al.* 1994) and thus to provide a laboratory measure of the *actual denitrification* activity. These measurements provided a full picture of the variation in space and time of flows of N and of denitrification and its controlling factors across the buffer zone.

9.2.2 Phosphorus Losses from the Broadbalk Experiment

9.2.2.1 Background

An accumulation of P in agricultural soils, resulting from fertilizer, organic manure and sewage sludge applications, has been observed in several countries in the EU (Sibbesen 1989). This could enhance the potential for P losses from soil to water. Because of the high P fixation capacity of most soils, the movement of P through the soil profile into drainage water has generally been considered of little importance (Sharpley and Menzel 1987). However, the increased concentrations of P in surface and sea waters is not solely the result of sewage effluent: agriculture has also made a contribution (Burt and Heathwaite 1993; Heckrath et al. 1995; Tunney et al. 1995). Although P is strongly held in soil, surface runoff and the erosion of colloidal material (clays and OM) rich in P does result in the movement of P from land to waters. P also moves from enriched surface soil through the subsoil to drains in mobile forms which are not adsorbed, such as in the bodies of the microbial biomass. Finally, the preferential flow of water from surface soil through cracks and channels, where contact with the subsoil is brief and limited, can carry dissolved and suspended P to drains and thus to surface waters. The 152-year-old Broadbalk Continuous Wheat Experiment at Rothamsted Experimental Station, near St Albans in Hertfordshire, 40 km north of London, is an excellent experimental site on which to examine P movement.

9.2.2.2 Site and Methods

Established in 1843, the experiment compares the effects of different types and amounts of fertilizers on the growth and yield of winter wheat. The Broadbalk soil is a silty clay loam of the Batcombe Series overlying clay-with-flints; the site is gently sloping with a maximum gradient of $2°$ (Avery and Bullock 1969). The experiment consists of 20 plots which have pipe drains installed along the middle of each plot to a depth of $c.$ 65 cm. The original drains, inserted in 1849, were replaced in 1993; this had no effect on P leaching. For 152 years the plots have received either no P, P as superphosphate at a rate of 35 kg P/ha/year or 17 kg P/ha/year, or P in farmyard manure at a rate of 40 kg P/ha/year. Drainage water was collected five times between October 1992 and February 1994 and analysed for total P (TP), total dissolved P (TDP), total particulate P (TPP), dissolved organic P (DOP) and molybdate reactive P (MRP). The P extracted with 0.5 M $NaHCO_3$ (Olsen P) was determined in air-dried soil from the plough layer (0–23 cm) as the most commonly used (in the UK) measure of available P in soil.

  This chapter takes the results from these two experiments of nutrient flows and transformations and shows how the physico-chemical and biological processes that might control and reduce nutrient losses from these two systems are largely controlled by the hydrological regime. The results illustrate that a recognition of this is vital to the effective management of these systems for optimum utilization of nutrients and minimal losses.

9.3 RESULTS AND DISCUSSION

9.3.1 Nitrogen Flows in a Riparian Buffer Zone

9.3.1.1 *Inorganic N in the Soil*

The interaction of land use, soil type and hydrology, and its impact on the physico-chemical character and microbial composition of the soil is well seen at the Cuddesdon site. The zone certainly follows the typical pattern of soil type, described in the Introduction. Zone A nearest the River Thame has the highest proportion of silt and clay, with coarser silt and sand increasing towards the backslope, and silt and sand fractions dominating the topslope and scrub soil profiles. The percentage of sand, silt and clay will affect the infiltration and water-holding capacity of the soil and its temperature fluctuations. Of special relevance is the presence, about 10 m from the river, of the clay-rich sediment of the oxbow, Zone B (63% of the particles $<2\,\mu m$). The dense microporous nature of the soil here resulted in a substantially greater moisture content than that elsewhere in the buffer zone. The annual mean Water-Filled Pore Space (WFPS) of the surface (0–10 cm) soil within the oxbow was 99%, whereas the zones either side, A and C, had WFPS averages of 75% and 80% respectively. However, the denser clay also produces a soil lens with a lower infiltration capacity and transmissivity than that of the remainder of the site. In times of flooding, water is diverted either side, into zones A and C.

Soil variation has a significant impact on soil N; how is described later. Figure 9.2 shows the average inorganic N status of the soil during the study year; seasonal variation was surprisingly small, with a clear zonal pattern throughout the year. The average NO_3-N value for the site is 12.5 mg/kg dry soil, but this masks the split in the transect between the nitrate-rich floodplain (Zones A–D, average value of 14.7 mg/kg dry soil) and the nitrate-impoverished hillslope (Zones E–F, average value 1.72 mg/kg dry soil). The prominent NO_3-N maximum of 27.9 mg/kg dry soil is associated with the midslope (Zone C). The NH_4-N and NO_2-N concentrations are generally small, with annual mean values of 3.11 and 0.32 mg/kg dry soil, respectively. They too have maximum concentrations of 8.06 mg/kg dry soil and 1.41 mg/kg dry soil, respectively, in the midslope. So, we see a clear zonal pattern in the distribution of N.

9.3.1.2 *Potential of the Riparian System to Denitrify*

The *potential denitrification activity* was established by measuring DEA. Figure 9.2 shows the averages in potential denitrification along the transect. The zonal pattern is the same as that for inorganic N in the soil and standard errors on the data are small, as would be expected for such a carefully controlled laboratory incubation. The main points to note are (i) there was relatively little denitrifier activity within the hillslope (4.15 mg N/kg dry soil/day, *c.* 5 kg N/ha/day) compared with that of the floodplain (60.3 mg N/kg dry soil/day, *c.* 65 kg N/ha/day); (ii) there was less potential for denitrification in the oxbow than in the adjacent zones, 40.0 mg N/kg dry soil/day, and the corresponding data for CO_2-C emissions (not shown) suggests the zone as a whole to be microbially impoverished; (iii) the midslope of the floodplain (Zone

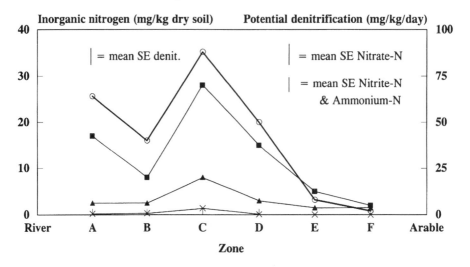

Figure 9.2 Mean annual concentrations of inorganic nitrogen, and mean annual potential denitrification in the surface (0–10 cm) soil along the transect at the Cuddesdon Mill buffer zone, shown as nitrate-N (–■–), ammonium-N (–▲–) and nitrite-N (–×–) in mg/kg dry soil, and N denitrified (–○–) in mg N/kg dry soil per day

C) is potentially the most active site for denitrification within the riparian zone, with a maximum rate of 87.0 mg N/kg dry soil/day, *c.* 100 kg N/ha/day. Taking the width of each zone and summing the contribution of each, the potential denitrification for a 100 m length of the buffer zone is *c.* 50 kg N/day. This illustrates the importance of such areas for alleviating water pollution, if they can be managed to their full potential.

A direct comparison of inorganic N and potential denitrification in Figure 9.2 shows the strong determining influence of the soil's inorganic N status, most notably that of nitrate, on its potential to denitrify. Previous research has suggested that available carbon is the main determining factor (Webster and Goulding 1989) but, where carbon supplies are adequate, N becomes the main limiting factor (Jarvis *et al.* 1991). This is discussed further below.

9.3.1.3 Denitrification in the Field

Figure 9.3 shows the mean annual field denitrification activity, measured by the slurry and core incubation techniques. The two techniques give very different rates of denitrification, as noted in previous work (Webster and Goulding 1989). Although the slurrying technique is a useful indicator of field rates, it still presents the denitrifying organisms in the slurried soil with ideal conditions in terms of a completely wetted and well-mixed soil in a fully anaerobic environment. Rates are therefore still artificially enhanced over true field rates. By contrast, field-incubated cores have been found to give rates of denitrification of the same order of magnitude as other techniques used to measure denitrification in the field (Webster and

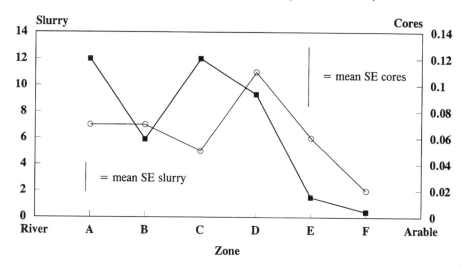

Figure 9.3 Mean annual field denitrification (mg N/kg dry soil per day) along the transect at the Cuddesdon Mill buffer zone: —■—, measured using a slurrying technique; —O—, measured using incubated cores

Goulding 1989), with very good agreement in some cases (Ryden *et al.* 1987). A further consideration, however, is the large standard errors that result from core incubation techniques, resulting from the large spatial variation of denitrification in the field and the relatively small surface area of the cores. Standard errors for the slurrying technique were much smaller and, obviously, similar to those for potential denitrification because the disturbed soil was taken from a wider area than a core. In estimating field rates of denitrification one therefore has to balance accuracy (the degree of conformity to the true value) with precision (the degree of agreement of repeated independent measurements with the mean) (Milner 1992).

Both techniques show approximately the same pattern of denitrification, and the pattern is the same as that of potential denitrification and soil N, because it is determined by the same factors. Maximum rates of denitrification were measured in the floodplain: *c.* 10 kg N/ha/day by slurrying, 0.07 kg N/ha/day with cores; minimum rates were measured on the hillslope: *c.* 0.05 kg N/ha/day with the slurrying technique, 0.01 kg N/ha/day with cores. The oxbow "trough" is again evident in the slurrying results but not in the cores. Calculating the total contribution of the buffer zone to denitrification as before, the average N_2O-N flux for a 100 m length of buffer strip is *c.* 10 kg N/day using the slurrying technique, 0.1 kg N/day using the cores.

9.3.1.4 Control of Mineral Nitrogen and Denitrification by Water Flow Patterns

Although the absolute values of mineral N and of denitrification varied each month, their relative value did not. The pattern shown in Figure 9.2 was consistent throughout the study year and determined by the hydrological, chemical and biological characteristics of each zone. The six zones are differently affected by the River Thame and catchment runoff which, together with land use, largely governed the physical, chemical and biological nature of the riparian strip.

The topslope (E) and scrub zones (F) are beyond the direct impact of the river system and their character is dictated by land use and hillslope fluxes. The remainder of the transect is affected by both the river and hillslope processes to varying degrees. The grid of piezometers showed a confluence of flow lines in the footslope (D), further illustrated by the presence of occasional seeps, with the hillslope saturated wedge spreading out from here. The maximum extent of flooding is set by the backslope, but full inundation was not always achieved. Greater persistence of flood water and rapidly fluctuating water-tables were evident within the middle of the floodplain (C). Zones C and D therefore form a dynamic fringe between the terrestrial and aquatic components of the riparian zone, and a fringe of detritus was apparent when the flood water retreated.

The meeting of floodwaters and hillslope drainage will also result in a confluence of sediment and nutrient fluxes. Zone A, and in winter Zone B, are primarily under the influence of the river, the extent varying seasonally. The N status of the hillslope section is solely controlled by runoff from the agricultural land above it, by leaching and by direct inputs from vegetation and precipitation. The lower riparian zone is subject to both riverine and hillslope inputs of water, sediment and nutrient deposition, chiefly in winter, and on inputs from grazing cattle in the summer. Our data showed that the flooding of Zones A–D tended to increase the nitrate levels mainly in Zones C and D, where standing pools of water from the river caused prolonged infiltration of water relatively rich in nitrate, and where detritus was deposited which yielded N upon mineralization.

The oxbow had a low infiltration capacity compared to the adjacent zones because of its larger clay content and mean water-filled pore space, as described above. In addition, the measured hydraulic gradients indicated that the relatively nitrate-rich flood water (mean NO_3-N concentration 8.0 ± 1.0 mg/l) is diverted to either side of this zone, especially into the midslope, Zone C. Bank seepage from the river into Zone A explains its relatively higher nitrate status (mean inorganic N content 19.9 mg/kg dry soil) and explains why there is no comparable rise in the NH_4-N and NO_2-N fractions. In the case of the midslope zone, the prolonged receipt of both directly infiltrating flood water and that diverted from the oxbow, the convergence of hillslope drainage lines forming a saturated wedge extending down the transect, and "ponding up" in front of the oxbow all contribute to the high N status of this zone.

The cattle were confined to Zones A–E by a fence at the edge of the scrub, and weekly observation of grazing behaviour and the distribution of excretal returns showed that they preferred to graze Zones A–D. The urea input from the cattle forms ammonia through hydrolysis, which is toxic to *nitrobacter*, briefly halting the nitrification process at $NO_2{}^-$. This would explain the close positive relationship between the soil NH_4-N and NO_2-N concentrations.

Thus we have a relatively dry, nitrogen-impoverished topslope where little denitrification occurs, a wet nitrogen-rich midslope where denitrification rate reaches a maximum, a wet but nitrogen and microbially-impoverished oxbow, and a very wet, nitrogen-rich and strongly denitrifying riverbank. This explains at least in part, the difference in the N status of the soil of the upper and lower sections.

9.3.1.5 Nitrate Loss other than by Denitrification

The addition of extra NO_3-N has a large positive impact on the activity of denitrifiers in the buffer zone: the average N_2O-N flux rate for the site with the NO_3-N amendment was 103 mg N/kg/day compared to the 40.6 mg N/kg/day without additional NO_3-N. This emphasizes that N is the limiting factor in the buffer zone. Conversely, the overall response of the system to excess carbon was a *decrease* in denitrification from 40.6 to 33.4 mg N/kg/day. This is not what would be expected: extra carbon should enhance denitrification (Webster and Goulding 1989). Respiration (estimated by CO_2-C emission) of the soil microbial biomass increased by 108% when carbon was added, so the decrease in N_2O evolution is not due to a decline in the population of denitrifiers and thus denitrification. The presence of chloramphenicol, which prevents the growth of the microbial population, rules out any assimilation of nitrate into the bodies of an expanding population of organisms. What is more likely is a redistribution of N within the soil system.

There are three possible pathways for this redistribution: (i) N_2O was produced by denitrification but immediately reused by denitrifiers, i.e. complete denitrification to N_2 occurs despite the presence of acetylene (this is very unlikely in the laboratory experiments as conducted because acetylene is always an efficient inhibitor), (ii) the relative abundance of readily available carbon, compared to the availability of inorganic N, caused a net immobilization of N; (iii) the large C/N ratio resulted in the dominance of dissimilatory nitrate reduction to ammonium (DNRA) over denitrification (Tiedje *et al.* 1979). The response to added N and other research currently in progress (Ellis and Dendooven, pers. comm.) suggests that the N is immobilized and slowly remobilized later.

9.3.1.6 The Riparian System as a Nitrate Filter

Weekly sampling of the River Thame and the spring emanating from a wooded area bordering the study site produced 18 month average NO_3-N concentrations of 7.70 mg/l and 25.9 mg/l, respectively. It was presumed that the spring provided a representative sample of the water contained within the Portland limestone aquifer and thus of the water prior to its passage through the riparian zone. However, this relatively nitrate-rich water was not detected entering the riparian zone. The average NO_3-N concentration fell from 3.42 mg/l at the topslope to 0.84 mg/l at 15 m into the riparian strip – a 75% decrease in concentration. The surprisingly small nitrate concentration at the upper edge of the riparian zone transect suggests that either (i) the sampling regime does not extend far enough back up the slope to detect the nitrate-rich water sampled in the spring, and that dilution, transformation and uptake of the nitrate has occurred before it reached the riparian zone, or (ii) the nitrate-rich water may bypass the riparian zone, entering the river via a gravel conduit, upwelling as base-flow, or during the winter months as overland flow. Whatever the explanation the outcome is that the denitrifying buffering activity within the riparian zone is largely unused. Maximizing the denitrification activity of this site by using substrate amendments would be pointless unless the hydrology was altered fundamentally to direct the nitrate-rich water into it.

9.3.2 Phosphorus Losses from the Broadbalk Experiment

The research at Cuddesdon showed that hydrology can reduce the potential of processes such as denitrification to reduce water pollution by nitrate. Complementary data from the Broadbalk Experiment show that hydrology can have a major effect on P losses. Phosphorus in drainage water collected from the Broadbalk Experiment and the relative proportions of the different P fractions varied substantially between the five drainage events and between the different plots (Figure 9.4). MRP was the largest fraction in the drainage water, ranging from 65% to 85% of the total P. Dissolved organic P accounted for less than 5% of TP, on average. Compared to literature data from heavy clay soils (e.g. Sharpley and Menzel 1987), concentrations of TP in drainage water were large, exceeding 1 mg/l on several plots at most drainage events.

The amount of P in the soil, measured by Olsen's method as available to plants, ranged from 7 to 90 mg/kg soil and resulted from different P applications and, especially, different crop offtakes of P in response to different rates of N fertilizer application. The concentrations of TP (and MRP) in drainage water from plots with less than about 60 mg Olsen P/kg soil were very small, <0.2 mg/l. There was a rapid increase of TP in drainage coming from plots containing >60 mg Olsen P/kg soil up to the maximum concentration of Olsen P (Figure 9.4). A simple split-line model

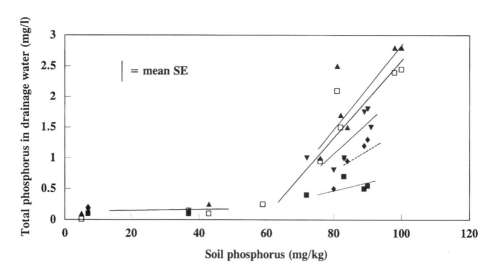

Figure 9.4 Concentrations of total phosphorus (as mg/l) in drainage water from the Broadbalk Experiment measured on five occasions in October (—■—) and November (—▼—) 1992, June 1993 (—◆—), and January (—□—) and February (—▲—) 1994, and plotted against available phosphorus in the soil (mg/kg) as measured by Olsen's Method using 0.5 M sodium bicarbonate

described this relationship well for all five drainage events. The initial slopes were not significantly different from zero below the change point which the model predicted to be 56 mg Olsen P/kg soil. Above this, the slopes differed significantly ($p < 0.01$) reflecting the variation in soil and weather conditions between the five events. This implies that, up to 56 mg Olsen P/kg soil, P was strongly retained in the surface, plough layer; losses were small and were unrelated to soil conditions or the weather. Beyond this change point, P losses in drainage water were very closely related to the P in the soil of the plough layer and to the weather. Phosphorus loadings were $\leqslant 0.5$ kg/ha from the plots below the change point, rising to 3–4 kg/ha at Olsen P values at 80–90 mg/kg. Losses of 3–4 kg P/ha are insignificant in agronomic terms but would make a major input to P in fresh waters.

The sharp increase in P leached beyond the change point could imply a shift towards more P being held on low-energy sites in the surface soil (Holford and Mattingly 1976). For example, Lookman (1995) found a similar split-line pattern relating water-extractable P from the plough layer of a range of acid sandy soils to their degree of P saturation. A build-up of calcium-bound Apatite P in P-saturated soils was believed to represent the more mobile P fraction in those soils. However, saturation of the surface soil and the leaching of excess does not explain why this P reached the drains and was not adsorbed in the very P-deficient subsoil. The mechanisms involved could either be transport of P in forms which make it less susceptible to sorption (e.g. colloidal or organically associated P) or, more likely, preferential or bypass flow. Such preferential flow was noted on Broadbalk for nitrate leaching (Goulding and Webster 1992), so we know that cracks and channels exist. Presumably the walls of the cracks and channels through which the water moves have become saturated with P over the 150 years of the experiment, and so any P leached or mobilized is not adsorbed. Research using tracers is currently in progress to check this.

9.4 CONCLUSIONS

At the Cuddesdon site, denitrification certainly occurs and there is the potential for much nitrate to be removed from water by this process. However, most of the water draining from the arable land does not flow through the riparian soil system. The "buffer zone" therefore does not buffer the river very much, if at all. Whilst denitrification is an important process in riparian buffer zones there is now increasing evidence that nitrate within these zones may follow more complex pathways, notably immobilization or DNRA prior to later mineralization, followed by leaching, denitrification, or vegetative or microbial uptake. A much better understanding of the processes controlling N flows and transformations is needed for an effective design and implementation of riparian areas as buffers against aquatic nitrate pollution. On the Broadbalk Experiment, preferential flow allows P to pass through a soil that adsorbs P very strongly and to enter drains (and by inference, surface waters) in comparatively large concentrations, certainly large enough to cause eutrophication. In both studies the drainage and runoff pathways are paramount in determining N and P flows. Our conclusion must be, then, that whilst chemical and biological processes within the soil system directly regulate nutrient fluxes from the hillslope to the river system, an understanding of the prevailing hydrological regime is a necessary prerequisite to their study and to the advancement of management strategies for buffer zones.

ACKNOWLEDGEMENTS

The research was funded by the Ministry of Agriculture, Fisheries and Food, that on nitrogen flows in Oxfordshire via an Open Contract (OC9114). IACR receives grant-aided support from the Biotechnology and Biological Sciences Research Council.

REFERENCES

Addiscott, T.M. 1995. Entropy and sustainability. *European Journal of Soil Science*, **46**, 161–168.

Addiscott, T.M., Whitmore, A.P. and Powlson, D.S. 1991. *Farming, Fertilisers and the Nitrate Problem*. CAB International, Wallingford.

Armstrong, A.C. and Burt, T.P. 1993. Nitrate losses from agricultural land. In T.P. Burt, A.L. Heathwaite and S.T. Trudgill (eds), *Nitrate. Processes, Patterns and Management*. John Wiley & Sons, Chichester, pp. 239–267.

Avery, B.W. and Bullock, P. 1969. The soils of Broadbalk. Morphology and classification of Broadbalk soils. *Rothamsted Report for 1968*, Rothamsted Experimental Station, Harpenden, pp. 63–81.

Burt, T.P. 1986. Slopes and slope processes. *Progress in Physical Geography*, **10**, 547–562.

Burt, T.P. and Heathwaite, A.L. 1993. A scale-dependent approach to the study of nutrient export from basins. *Institute of Hydrology Report 120*, 82–92.

Burt, T.P. and Trudgill, S.T. 1985. Soil properties, slope hydrology and spatial patterns of chemical denudation. In K.S. Richards, R.R. Arnett and S. Ellis (eds), *Geomorphology and Soils*. George Allen & Unwin, London, pp. 13–36.

Dendooven, L., Splatt, P. and Anderson, J.M. 1994. The use of chloramphenicol in the study of the denitrification process: some side-effects. *Soil Biology and Biochemistry*, **26**, 925–927.

Fowler, D., Cape, J.N., Leith, I.D., Choularton, T.W., Gay, M.J. and Jones, A. 1988. The influence of altitude on rainfall composition at Great Dunn Fell. *Atmospheric Environment*, **22**, 1355–1362.

Gerrard, A.J. 1981. *Soils and Landforms*. George Allen & Unwin, London.

Goulding, K.W.T. and Webster, C.P. 1992. Methods for measuring nitrate leaching. *Aspects of Applied Biology*, **30**, 63–70.

Goulding, K.W.T., Hütsch, B.W., Webster, C.P., Willison, T.W. and Powlson, D.S. 1995. The effect of agriculture on methane oxidation in soil. *Philosophical Transactions of the Royal Society of London*, **351A**, 313–325.

Haycock, N.E. and Burt, T.P. 1993. Role of floodplain sediments in reducing the nitrate concentration of subsurface run-off: a case study in the Cotswolds, UK. *Hydrological Processes*, **7**, 287–295.

Heathwaite, A.L., Burt, T.P. and Trudgill, S.T. 1993. Overview – The nitrate issue. In T.P. Burt, A.L. Heathwaite and S.T. Trudgill (eds), *Nitrate. Processes, Patterns and Management*. John Wiley & Sons, Chichester, pp. 3–21.

Heckrath, G., Brookes, P.C., Poulton, P.R. and Goulding, K.W.T. 1995. Phosphorus leaching from soils containing different P concentrations in the Broadbalk Experiment. *Journal of Environmental Quality*, **24**, 904–910.

Holford, I.C.R. and Mattingly, G.E.G. 1976. A model for the behaviour of labile phosphate in soil. *Plant and Soil*, **44**, 219–229.

Howard-Williams, C. and Downes, M.T. 1993. Nitrogen cycling in wetlands. In T.P. Burt, A.L. Heathwaite and S.T. Trudgill (eds), *Nitrate. Processes, Patterns and Management*. John Wiley & Sons, Chichester, pp. 141–167.

Jarvis, S.C., Barraclough, D., Williams, J. and Rook, A.J. 1991. Patterns of denitrification loss from grazed grassland: Effects of nitrogen fertilizer inputs at different sites. *Plant and Soil*, **131**, 77–88.

Knowles, R. 1982. Denitrification. *Microbiological Reviews*, **46**, 43–70.

Lookman, R. 1995. Phosphorus chemistry in excessively fertilized soils. *PhD thesis*, Katholieke Universiteit Leuven, Leuven, Belgium.

Milner, O.L. 1992. *Successful Management of the Analytical Laboratory*. Lewis Publishers, Boca Raton, USA.

Ritter, W.F. 1988. Reducing impacts of nonpoint source pollution from agriculture. A review. *Journal of Environmental Science and Health*, **23A**, 645–667.

Ryden, J.C., Skinner, J.H. and Nixon, D.J. 1987. Soil core incubation system for the field measurement of denitrification using acetylene inhibition. *Soil Biology and Biochemistry*, **19**, 753–757.

Sharpley, A.N. and Menzel, R.G. 1987. The impact of soil and fertilizer phosphorus on the environment. *Advances in Agronomy*, **41**, 297–324.

Sibbesen, E. 1989. Phosphorus cycling in intensive agriculture with special reference to countries in the temperate zone of Western Europe. In H. Tiessen (ed.), *Phosphorus Cycles in Terrestrial and Aquatic Ecosystems*. University of Saskatchewan, Saskatoon, Canada, pp. 112–122.

Smith, M.S. and Tiedje, J.M. 1979. Phases of denitrification following oxygen depletion in soil. *Soil Biology and Biochemistry*, **11**, 261–267.

Tiedje, J.M., Sexstone, A.J., Myrold, D.D. and Robinson, J.A. 1979. Denitrification: ecological niches, competition and survival. *Antonie van Leeuwenhoek*, **48**, 569–583.

Tunney, H., Carton, O.T., Brookes, P., Johnston, A.E., Foy, R.H., Morgan, A., Sharpley, A.N. and Pionke, H.B. 1995. *Phosphorus Loss to Water from Agriculture*. TEAGASC, Johnstown Castle, Wexford.

UKRGIAN 1994. *Impacts of Nitrogen Deposition in Terrestrial Ecosystems*. United Kingdom Review Group on the Impacts on Nitrogen Deposition, Department of the Environment, London.

Vinten, A.J.A. and Smith, K.A. 1993. Nitrogen cycling in agricultural soils. In T.P. Burt, A.L. Heathwaite and S.T. Trudgill (eds), *Nitrate. Processes, Patterns and Management*. John Wiley & Sons, Chichester, pp. 141–167.

Webster, C.P. and Goulding, K.W.T. 1989. Influence of soil carbon content on denitrification from fallow land during autumn. *Journal of the Science of Food and Agriculture*, **49**, 131–142.

10 Phosphorus Fractionation in Hillslope Hydrological Pathways Contributing to Agricultural Runoff

R. M. DILS and A. L. HEATHWAITE
Department of Geography, University of Sheffield, UK

10.1 INTRODUCTION

The form of phosphorus (P) in runoff from agricultural land is largely dependent on the P source, flow pathways and transformations during transport. Hillslope hydrological pathways determine rainfall–soil–runoff interactions and therefore the extent to which newly infiltrating water is modified before it reaches the drainage network. Consequently, different hydrological pathways are characterized by distinct P forms. As P fractions vary in their bioavailability, the relative contribution from different pathways will influence the overall runoff composition and its impact on the receiving stream. Phosphorus is essential for maintaining biological productivity in terrestrial and aquatic systems. In natural freshwater ecosystems orthophosphate concentrations may be limiting to primary productivity but the artificially elevated concentrations of P present in agricultural runoff can enrich the low natural background levels. Excessive growths of algae and macrophytes are the result of this accelerated eutrophication process (Sharpley and Menzel 1987).

Many studies of P transport in agricultural runoff have focused on catchment-scale losses, with annual loadings estimated from the cumulative product of monthly stream P concentrations and discharge (Burwell *et al.* 1977; Heathwaite *et al.* 1990; Sharpley and Smith 1990). Such investigations permit comparisons of the extent of water quality deterioration as a function of land use but they provide limited information regarding the relative importance of different stormflow pathways responsible for P transport. At the soil profile and experimental plot scales, soil–nutrient–runoff interactions have been studied by controlled simulation rainfall–runoff experiments using soil boxes and monolith lysimeters. The validity of extrapolating laboratory results to the field scale is questionable as independent variables and generalizations are scale-specific (Young and Mutchler 1976), and using reductionist theories to predict large-scale behaviour of systems from an aggregation on small-scale results is not necessarily valid (Armstrong and Burt 1993). Hillslope and small catchment-scale studies have been identified as forming the link between laboratory- and catchment-scale investigations (Young and

Advances in Hillslope Processes, Volume 1. Edited by M. G. Anderson and S. M. Brooks.
© 1996 John Wiley & Sons Ltd.

Mutchler 1976). Presently, the interpolation and integration of hydrochemical data obtained from different spatial scales is limited.

The present study examines P fractions in surface and subsurface hydrological pathways at plot, hillslope and small catchment scales. The objective of the study was to use a nested approach to investigate the hydrochemical dynamics of P in agricultural runoff at integrated spatial scales.

10.2 BACKGROUND

10.2.1 Agricultural Activities as a Source of Excess P in Freshwaters

Since the 1940s, agricultural intensification has resulted in an increase in the area under cereals and temporary grass at the expense of permanent pasture and rough grazing (National Rivers Authority 1992a). These land-use changes have reduced soil organic matter and nutrient levels in areas where the application of livestock-generated wastes are restricted. Thus inorganic fertilizers with specific chemical compositions are applied to ensure crop nutrient demands are satisfied. The superior attributes of manufactured fertilizers has rendered the recycling of organic wastes uneconomical and consequently its storage and disposal is a problem. The waste is kept in slurry stores and disposed of by spreading on the land. Insufficient storage capacity may force a farmer to spread slurry on "high risk" fields (those susceptible to waterlogging, recently drained or severely compacted) during months when runoff losses will be high.

During the 1980s, the contribution of agriculture to the deteriorating quality of freshwaters was substantial, constituting 12% of all pollution incidents reported to the National Rivers Authority (Ministry of Agriculture, Fisheries and Food 1991) and between 24 and 71% of the total P load to rivers in Europe (Vighi and Chiaudani 1987). Pollutants from agriculture are extremely diverse in their origin. Point sources include runoff from temporary manure heaps, leakage of effluent from slurry or silage stores and overflow from dirty water storage units. Diffuse sources are more difficult to identify and include pollutants present in runoff from agricultural land and discharge from field drains.

With the implementation of complex regulations, EC directives, codes of practice and similar legislation, farmers are being actively encouraged to adopt agricultural management practices that minimize nutrient losses to inland waters (National Rivers Authority 1992b). Preventative legislation has proved to be the most effective method of protecting water resources with initiatives including catchment land management planning, buffer zones and the development of risk maps. Such strategies require an improved knowledge of

- the quantity and composition of diffuse and point *sources* in the catchment;
- P fractionation within hydrological *pathways* that transport P from terrestrial to aquatic environments;
- the *impact* of runoff on the receiving water body.

Setting P standards for rivers to meet specific water quality objectives is complex due to the dynamic and individual nature of river systems. Currently, insufficient data exist to introduce effective management schemes to protect natural waters from P pollution, especially from diffuse agricultural sources.

10.2.2 P Forms in Aquatic Systems

Phosphorus is geochemically classified as the eleventh most abundant element, constituting 0.1% of the earth's crust by weight (Holtan *et al.* 1988) and between 0.022% and 0.83% of soil mass (Persson and Jansson 1988). It occurs naturally in numerous forms which are broadly distinguished by chemical type as inorganic or organic and by physical state as particulate or dissolved (Olsen 1967). Identification of specific forms and discrimination between the individual fractions is analytically problematical and consequently conflicting terminology is found in the literature.

An arbitrary distinction between dissolved (DP) and particulate P (PP) is commonly used whereby dissolved P is operationally defined as that fraction which passes through a $0.45\mu m$ cellulose membrane filter (Broberg and Pettersson 1988). In the soluble phase, most inorganic and several organic compounds are considered *bioavailable* and therefore immediately available for biological utilization (Wild and Oke 1966). In the particulate form, labile P provides a long-term source of P as it becomes available following dissolution or desorption (Logan 1982). Eutrophication control programmes and similar research has tended to quantify the soluble inorganic P fraction and total P (TP), but the bioavailable portion cannot be precisely determined from these indices. Consequently it is essential to quantify all P fractions present in aqueous samples.

10.2.3 P Forms and Transformations in Soil Systems

The soil P cycle is well documented in the literature (see Kofoed 1984; Brady 1990) so only the key aspects relevant to soil–runoff interactions are summarized below. The ability of P to sorb onto clay particles, organic matter and hydrated non-cyrstalline iron and aluminium oxides, combined with the low solubility of primary and secondary phosphates, results in low concentrations of phosphate ions in soil solution. This minimizes the likelihood of P being lost in percolating water but maximizes the potential for losses associated with eroded material in overland flow. P transformations in the soil determine the relative amounts present in dissolved and particulate forms. In brief, P is removed from soil solution by physical and chemical processes including adsorption, immobilization (chemisorption) and precipitation; desorption, mobilization and solubilization convert P into more soluble forms (Finkl and Simonson 1979). The main microbiological transformations are mineralization, solubilization and immobilization. The physical or chemical adsorption–desorption reactions occur very rapidly in soils, complete within minutes to hours, whereas the slower biological immobilization–mobilization transformations require days or months to reach "equilibrium", with true equilibrium rarely being attained (Finkl and Simonson 1979).

10.2.4 Hillslope Hydrological Pathways

During a rainfall event there are a number of hillslope hydrological pathways which water may take. These can be broadly classified as surface, unsaturated subsurface and groundwater runoff (Ryden *et al.* 1973). The pathway followed will determine the speed with which the water reaches the stream and hence the timing of the hydrochemical response to the storm (Heathwaite 1993). Infiltration-excess surface runoff occurs soon after the onset of rainfall and moves water directly from the hillslope to the stream, minimizing the possibility of P forms originally present in the rainwater undergoing transformations. Unsaturated subsurface runoff generally takes slightly longer to reach the stream (12–48 h) and is rarely identified in the hydrograph peak but the slow response means this pathway may contribute to streamflow for several days after the rainfall event (Whipkey and Kirkby 1990). However, unsaturated subsurface throughflow may reach the drainage network within hours of rainfall if flow is through cracks, fissures and macropores therefore bypassing the soil matrix (Anderson and Burt 1990). Groundwater runoff is considerably delayed, taking 2–5 days to reach the stream, the lag time being dependent upon the rate of vertical percolation and the subsequent lateral transport rate (Weyman 1973; Whipkey and Kirkby 1990). Large delayed discharge peaks are also the result of lateral subsurface runoff converging in hillslope spurs, hollows and interfluve plateaux as found in the Bicknoller Combe catchment, Somerset, England (Burt 1978).

Surface runoff transports both dissolved and particulate P to surface waters although the particulate phase, associated with eroded soil particles accounts for the majority of the total P transported (Burwell *et al.* 1975; Sharpley and Syers 1979). This is because soil erosion is a preferential process in which clay-sized particles and light organic and colloidal materials are dominantly transported due to their small size and light weight (Sharpley and Menzel 1987). These materials contain relatively high quantities of P so overland flow transports P-rich particulate matter. A number of studies have investigated the influence of land use and management practices on concentrations, forms and amounts of P transported in surface runoff. For example, P transported in the particulate form accounted for more than 95% of the annual P lost in surface runoff from plot-scale experiments in Minnesota and four of the five cropping practices (Burwell *et al.* 1975). Detailed summaries of these studies are presented in review papers by Sharpley and Menzel (1987) and Sharpley and Halvorson (1994). Attempts to quantify P losses in overland flow are complicated by slope heterogeneity creating differences in the energy available for transport. Further, several studies have acknowledged that transformations between the various P forms occur during overland transport; in particular, soluble P can be re-adsorbed by soil material during transit (Burwell *et al.* 1975; Sharpley and Syers 1979; Sharpley *et al.* 1981).

Unsaturated subsurface runoff can contribute considerably to the transport of soluble P in surface runoff (Ahuja 1986). Soluble P at or near the soil surface is transferred to overland flow by diffusion of P retained in the soil pores and dissolution or desorption of P present in the solid phase (Ahuja 1986). This transfer occurs in a thin (1–3 mm) surface layer of soil termed the effective depth of interaction (EDI) (Logan 1982). The extent of interaction is non-uniform and decreases exponentially with soil depth (Ahuja 1986).

Soil structure has an important influence on P transport and transformations in storm runoff through controlling the contact time between the new water, the soil and the pre-existing soil water (Wilson *et al.* 1991). In unsaturated, structured soils, the presence of macropores means that infiltrating water rapidly flows through the soil bypassing a large part of the soil matrix. This produces irregular, non-Darcian hydrochemical flow patterns in the soil because high flow velocities minimize contact time; thus non-equilibrium, low solute concentrations are prevalent (Wild and Babiker 1976; Bouma *et al.* 1980). This is in contrast to matrix flow in which micropores remain water-filled for extended periods of time, allowing a chemical equilibrium between the pore surface and the soil solution to develop. Flow may temporarily occur in both domains as macropores can actively carry water before the soil matrix is fully saturated (Germann 1990).

Artifical drainage pipes can be considered as large macropores which exhibit greater stability than natural pipes and macropores. As the pipes are large and well connected, intercepted water moves rapidly from the source area to the stream outlet with minimal opportunity for transformation or uptake by P-deficient subsoil. Consequently, P concentrations and loads are greater in water transported via tile drains than in natural subsurface pathways. Artificial drains are fed by saturated throughflow under wet conditions and by freely-draining water from the unsaturated zone during periods when the water-table drops below the pipeline. The hydrological effect of agricultural drainage on river flow is to suppress and delay peak discharge due to an increased moisture storage capacity and a decrease in surface runoff contributions respectively (Newson and Robinson 1983; Armstrong 1984).

P concentrations in *groundwater* are generally low because phosphate anions are tightly bound in the soil and P in percolating waters is sorbed by P-deficient subsoils. Consequently, the contribution of groundwater P to total runoff is usually assumed to be minimal. Organic soils are an exception owing to low iron and aluminium oxide fractions, as are sandy soils which are characterized by rapid rates of water percolation and low P sorption capacities. P leaching from sandy and sandy loam soils has been recorded under both laboratory (Logan and McLean 1973) and field conditions (Reddy *et al.* 1978). Recent research in the Netherlands suggested that the P content of parent material influenced P concentrations in groundwater, in particular P-rich peat and marine clay soils (Breeuwsma 1995). This study also identified areas where the water-table rose and contacted P-enriched soil layers, generating high P concentrations in groundwater. The majority of P in groundwater runoff is expected to be in the soluble form, although preferential pathways including macropores and artificial drainage systems may transfer colloidal and particulate P to the saturated zone.

10.3 METHODOLOGY

10.3.1 Site

A small (120 ha) mixed agricultural catchment near Ashby-de-la-Zouch, Leicestershire, has been selected for this study (UK National Grid Reference: SK 356 197).

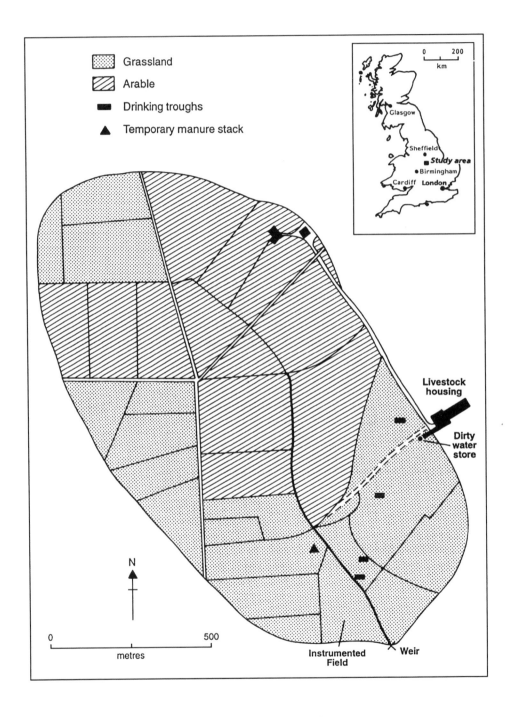

Figure 10.1 Pistern Hill catchment

This catchment was chosen because: it is first order, so that stream P loadings are exclusively from catchment sources; it only has agriculturally derived P inputs; and it is representative of intensive farming practices in the Midlands region. The catchment is composed of mixed arable fields to the north and temporary grass (re-seeded every 5 years) to the south (Figure 10.1). A herd of 180 dairy cows graze the southern grassland fields in rotation from April to October and are kept in winter housing for the remainder of the year. The soils are freely-draining fine loams overlying impermeable clays of the Salop Series. A perched water-table is present at the intersection of the permeable B horizon and the impermeable clay C horizon at a depth varying from 70 cm adjacent to the stream to a maximum of 2 m at the catchment boundary. Geologically, the catchment is dominated by Triassic Keuper Sandstone overlying Coal Measures. The fields are underlaid by permanent tile drains at 0.7 m depth to reduce the spatial extent and duration of waterlogging, thereby increasing the length of the grazing season. In most fields primary drains run parallel to the stream and feed into secondary drains which terminate at the stream bank.

10.3.2 Instrumentation

At the catchment outlet, a calibrated weir in conjunction with an Ott water-level recorder were installed to continually monitor stream discharge, and an automatic Rock and Taylor water sampler was set to sample runoff that occurred at stream discharges exceeding baseflow. The position of the float switch to activate the sampler and the height of the sampling tube inlet were seasonally adjusted depending on baseflow depth. To examine P transport in hillslope hydrological pathways, a series of overland flow troughs, macropore water samplers and hydrological instruments (dipwells and piezometers) were installed in banks at top-, mid- and bottom-slope positions along two hillslope transects in a grassland field adjacent to the catchment outlet (Figure 10.2). A gutter, 1.5 m in width, was installed at the base of a 5.5 m^2 hydrologically isolated area to collect overland flow. The gutter was covered to prevent contamination by direct rainfall. Macropore samplers, similar in design to those used by Simmons and Baker (1993), were installed in the field at an angle of 45° to intercept free-flowing water at three soil depths (0–15 cm, 15–30 cm and 30–45 cm). Water-table depth was measured in dipwells (perforated PVC tubing 2 m in length, 32 mm in diameter, covered with a fine-mesh nylon screen) installed to a depth of 2 m with a gravel backfill to ensure good hydraulic contact. Piezometers were installed in nests consisting of four PVC pipes, 22 mm in diameter, perforated near the base and covered with a nylon mesh (50, 100, 150 and 200 cm length). At the smallest scale, three hydrologically isolated experimental plots (10 m × 1.5 m) were set up adjacent to the hillslope transect on a 10° slope (Figure 10.2). The plots were instrumented in an analogous arrangement to the hillslope transect but on a smaller spatial scale.

10.3.3 Sampling Strategy

A significant portion of annual P loss occurs during a few large storm events (Johnson *et al.* 1976; Heathwaite *et al.* 1989; Kronvang 1992). Hence the research

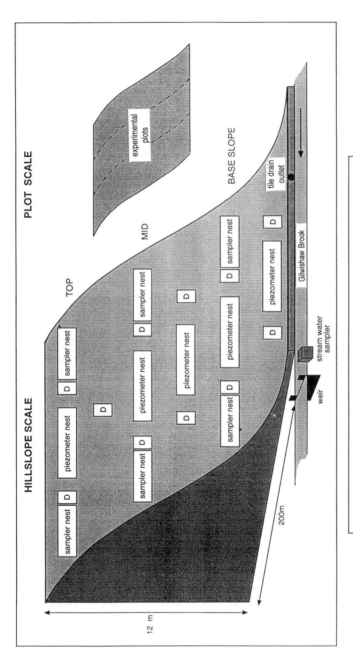

Figure 10.2 Layout of sampling and hydrological instrumentation at hillslope and experimental plot scales

Table 10.1 Supplementary data used in the study

	Parameter
Soil physical properties	Infiltration capacity, bulk density, moisture content, particle size
Soil chemical properties	pH, water-extractable P, Olsen's-extractable P, iron oxide-strip P, TP, extractable ions (Ca^{2+}, Al^{3+}, Fe^{2+}, Mn^{2+})
Hydrology	Water-table depth, saturated hydraulic conductivity

described here combined a regular sampling strategy to measure the baseline water quality together with intense sampling during storm events to monitor short-term changes during and following events. Stream-water samples were manually taken on a weekly basis and during storm events, automatic samplers were triggered to sample at fixed time intervals.

To support the interpretation of water sample results, additional data regarding the catchment pedological and hydrological characteristics were periodically obtained (Table 10.1). Water-table depth was measured on a weekly basis whereas the other parameters presented in Table 10.1 were measured every three months.

10.3.4 Laboratory Procedures

Samples were collected in iodized 60-ml polyethylene sampling bottles, an aliquot filtered on returning to the laboratory using a pre-washed 0.45 μm membrane filter paper and stored in the dark at 4°C prior to analysis. The soluble inorganic component was analysed within 24 h of collection by the colorimetric technique of Murphy and Riley (1962) using a UNICAM UV/VIS Spectrometer with a detection limit of 1 μg P/l. This molybdate-reactive P (MRP) is considered to be analogous to the soluble inorganic P fraction, although the acidic conditions may hydrolyse some condensed phosphates and high molecular weight organic P compounds (Rigler 1968). Bioavailable (algal-available) P was determined on unfiltered samples using iron-oxide impregnated filter paper strips (Sharpley 1993; Robinson *et al.* 1994). This method proved to be accurate (101% recovery of inorganic standards), precise (standard deviation = 4 μg P/l per strip (n = 25)), have a low detection limit (5 μg P/l) and be interference-free. Total P and total dissolved P were determined on unfiltered and filtered aliquots respectively using an alkaline persulphate microwave digestion

Table 10.2 Terminology and analytical method of determination

P fraction	Acronym	Pre-treatment	Determination
Molybdate-reactive P	MRP	Filtration (0.45 μm)	Colorimetry
Dissolved unreactive P	DUP		Difference (TDP − MRP)
Total dissolved P	TDP	Filtration (0.45 μm)	Persulphate digestion
Particulate P	PP		Difference (TP − TDP)
Total P	TP	None	Persulphate digestion

technique with a detection limit of $12\,\mu g$ P/l, carried out using a CEM MDS-81D Microwave Digestive System (Johnes and Heathwaite 1992).

Table 10.2 summarizes the terminology and analytical methods used when referring to P fractions in water samples.

10.4   RESULTS

The results presented in this chapter are for the period September 1994 to April 1995.

10.4.1   Point Sources of P

Within the catchment, several point sources of pollution were identified including livestock housing facilities, dirty water stores, temporary manure stacks and drinking troughs around which cattle congregated (Figure 10.1). TP concentrations averaged between 2.21 and 3.42 mg P/l (Table 10.3), of which the dissolved unreactive P fraction was dominant due to the faecal and urinal origins of the material (Cooke 1986). Runoff from cattle housing buildings and overflow from the dirty water store was transported as overland flow down a farm track directly to the stream, a distance of approximately 200 m (Figure 10.1). A 30 m grass strip adjacent to the stream acted as a buffer zone to reduce the polluting potential of the runoff. Inadequate storage facilities during the winter period meant that temporary manure piles were created in the grassland fields approximately 60 m from the stream. In February 1995, a TDP concentration of 1.87 mg P/l was measured in runoff from a slurry pile located in one of the grassland fields, which gives an indication of the pollution risk posed by such heaps. Drinking troughs and gateways were also important point sources of P due to the combination of soil compaction from repetitive trampling by cattle and high nutrient concentrations from excrement. Throughout the year troughs were surrounded by ponded water which was intermittently lost via infiltration–excess overland flow during intense rainfall events.

Table 10.3   P composition of selected point sources in catchment (mg P/l) (standard error values in parentheses)

P source	n	MRP	DUP	TDP	PP	TP	DUP/TP
Runoff from cowsheds	3	0.71	2.32	3.03	0.13	3.16	73.4
		(0.12)		(0.08)		(0.11)	
Runoff from dirty water store	3	1.27	2.14	3.41	0.01	3.42	62.6
		(0.28)		(0.46)		(0.30)	
Runoff from manure heap	2	0.32	1.55	1.87	0.34	2.21	70.1
		(0.08)		(0.06)		(0.11)	
Ponded water by drinking trough	2	0.58	1.55	2.13	0.27	2.40	64.6
		(0.15)		(0.09)		(0.29)	

10.4.2 Diffuse Sources of P

P in rainfall, leachate from plant material (Sharpley 1981) and desorption and dissolution of soil P have been identified as the main naturally derived diffuse sources of P (Newman 1995). In the study catchment, P originating in the topsoil (0–30 cm) may be the most important diffuse source of P contributing to stream load. Mean TP present in surface soil measured 116.9 μg P/g, compared to only 41.0 μg p/g in the C horizon at 150 cm (below the water-table). Of greater environmental significance was the water-extractable fraction (WEP), a surrogate index for the amount of P that can be lost to percolating waters. WEP measured 0.31 μg P/g in the topsoil and 0.03 μg P/g at 150 cm, suggesting that only a fraction of the TP present maybe liable to loss through leaching in saturated subsurface waters. At a depth of 80 cm, where the tile drains are situated, P concentrations were notably higher than the average C horizon values, 88.0 μg P/g for TP and 0.11 μg P/g for WEP. This may indicate preferential flow as a result of artificial drainage.

10.4.3 Catchment-Scale Pathways

The catchment-scale results presented in this chapter are based on two storm events that occurred during the sampling period from September 1994 to April 1995. Peak stormflow concentrations of P exceeded baseflow concentrations by at least two orders of magnitude, confirming the importance of intense rainfall events to the transport of P.

The stream response at the catchment outlet to a 38 h, 54.0 mm storm commencing on 15 September 1994 is shown in Figure 10.3. This event (Storm 1) was the first major rainfall event of the autumn and was characterized by three distinct stages. For the initial 14 h following the onset of rainfall, the stream exhibited a suppressed discharge response due to the low antecedent moisture conditions. Subsequent rain wet-up the catchment and the stream became progressively more responsive, with the peak discharge being recorded 19 h after the start of rainfall. Diminishing rainfall intensities after 24 h resulted in the recession of stream discharge although delayed subsurface throughflow continued to contribute to stream discharge. After 30 h, approximately 15% of the rainfall had been transported to the catchment outlet, suggesting considerable adsorption into the soil profile and runoff in delayed hydrological pathways. Hourly stream sampling was started at the onset of rainfall (10:00 hours) and continued for 30 h.

TP concentrations rapidly increased from 0.07 mg P/l to a peak of 1.98 mg P/l after eight hours of rainfall and then gradually decreased to 0.77 mg P/l after a further 13 h (Figure 10.3). A slight recovery to 1.05 mg P/l occurred in response to the TDP peak but this peak was temporary and was followed by a decline to the minimum concentration of 0.49 mg P/l. In general, the TP chemograph was controlled by the MRP fraction. Conversely, particulate P concentrations were low and in relative terms only contributed between 3 and 37% to TP. The larger PP contributions coincided with periods of heavy rainfall when P-associated soil particles could be transported in surface and subsurface pathways.

After 8 h of rainfall, the TDP concentration had almost tripled, increasing from a pre-storm value of 0.06 mg P/l to 1.74 mg P/l. This probably reflected terrestrial

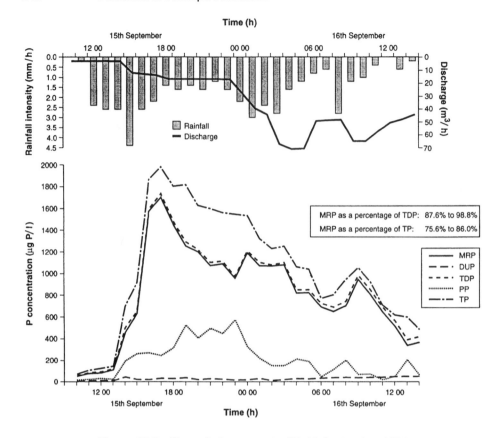

Figure 10.3 Storm 1 chemograph, 15–16 September 1994.

inputs rapidly transported in surface and subsurface runoff reaching the catchment outlet. TDP concentrations decreased as the storm period continued except during short, high-intensity rainfall periods (3.0 and 2.8 mm/h) which produced secondary discharge and TDP peaks. Inorganic P dominated the soluble fraction (87–99%) and as a consequence, the TDP and MRP chemographs were closely matched. This inorganic P may represent soil P released by mineralization during the summer, which, in the absence of water moving through the system, accumulated in the solid phase in the topsoil (Jordan and Smith 1985). Dissolution or desorption of this P, in addition to P released from the groundcover, probably resulted in the transport of MRP in overland and percolating waters. Although in absolute terms the organic fraction was minimal, in relative terms its contribution gradually increased as the storm continued, from approximately 1% (17:00 hours) to a maximum of 12% at the end of the storm, indicating that organic P was mobile but transported in slow or delayed hydrological pathways.

In a similar way to TDP, MRP increased in concentration from 0.05 mg P/l at the start of sampling to a maximum value of 1.70 mg P/l 8 h after the storm began

(Figure 10.3). The mobility of MRP can explain this rapid response and suggests that the P source was in close proximity to the stream, either of aquatic origin such as desorption from sediments, or of terrestrial origin, transported as overland flow or in rapid subsurface pathways. Two smaller peaks of 1.19 mg P/l and 0.95 mg P/l were later recorded, coincident with the short, high-intensity rainfall periods (Figure 10.3). After the third peak, the MRP concentration rapidly decreased to a minimum of 0.37 mg P/l. Storm chemographs for the Slapton catchment, Devon, UK, showed a similar MRP response with coincident MRP and discharge peaks succeeded by rapid recession limbs for both variables (Heathwaite *et al.* 1989). Overall, MRP concentrations were exceptionally high, ranging from 0.05 to 1.70 mg P/l, compared to spring and summer low baseflow which averaged 0.035 and 0.010 mg P/l respectively.

For dissolved P fractions, concentrations generally decreased as Storm 1 proceeded due to dilution effects, but overall loadings increased due to the overriding influence of discharge. These results are consistent with those of Houston and Brooker (1981) who found MRP concentrations in the River Frome, UK, decreased significantly with increased river discharge during flood events ($r = 0.741$). The chemograph resulting from Storm 1 was dominated by the MRP fraction responses and consequently TDP greatly exceeded PP. The delayed PP peak found during Storm 1 appeared to be at the expense of MRP concentrations (Figure 10.3). This may be due to the adsorption of MRP onto particulate matter transported in overland flow (Sharpley *et al.* 1981) or may represent the delayed contribution from tile drains.

Storm 2 occurred between the 2nd and 3rd of October 1994, when a total of 19.4 mm of rain was recorded at the study site in a 24-h period. The stream showed a slightly delayed but very rapid response to the storm, the maximum discharge of 0.22 m^3/s being recorded 6 h after the onset of rainfall (Figure 10.4). Unlike Storm 1, the stream remained elevated for 72 h, indicating that delayed subsurface or groundwater runoff was sustaining the high stream discharge. Stream sampling was started at 10:00 hours on 2nd October, at the onset of rainfall and coincident with the start of the rising limb of the stream hydrograph. Samples were taken every 30 min for a 24-h period.

The TP chemograph for Storm 2 peaked at 1.31 mg P/l 5 h after the onset of rainfall and gradually receded to 0.20 mg P/l after 20 h (Figure 10.4). This peak concentration was considerably lower than the 1.98 mg/l recorded during Storm 1 due to a lower rainfall total, the dominance of different hydrological pathways and possibly a depleted P store from "flushing out" by intervening rainfall events. A secondary peak of 0.50 mg P/l occurred after 23 h with the onset of the second rainfall period.

The most distinctive features of the TP chemograph were the dominance of the PP fraction, constituting between 48% (10:00 hours) and 88% (13:00 hours) of TP and, conversely, the insignificance of the soluble fraction (Figure 10.4). PP concentrations rose rapidly to a peak of 1.04 mg P/l 5 h after the start of the storm, then gradually receded to a minimum of 0.12 mg P/l after 21 h. A second delayed peak of 0.27 mg P/l was measured shortly after 5.0 mm of rainfall. In absolute terms the maximum PP concentration was greater than that measured in Storm 1, the values being 1.04 and 0.53 mg P/l respectively. Low PP concentrations during baseflow periods and high

PP concentrations during storm periods have also been observed at the catchment scale in Ireland (Stevens and Smith 1978), Denmark (Kronvang 1992) and Canada (Ng *et al.* 1993). A possible source of the PP was saturation excess overland flow. A saturated wedge developed at the hillslope base adjacent to the stream and the P-rich topsoil interacted with the overlying water to provide a major diffuse source of P. These transient runoff producing zones have been identified as important source areas for suspended solids and P in several previous studies including Duda and Finan (1983) and Heathwaite *et al.* (1989). Additionally, P-associated soil particles transported in overland flow, or through erosion, accumulated at the hillslope base so that when the area became saturated these particles were rapidly moved overland and into the stream. Alternatively, disturbance of stream sediments by fast-flowing water may have contributed to the particulate fraction (Dorioz *et al.* 1989; House *et al.* 1995) or the dispersion of clay particles by percolating water (Sharpley and Withers 1994).

Although TP was always dominated by the particulate fraction during Storm 2, the dissolved fraction increased in relative importance as the storm continued, indicating that slower subsurface hydrological pathways, dominated by dissolved P compounds, were used to transport soluble P. An alternative hypothesis is that P was desorbed from particulate matter during transit in either the terrestrial or the aquatic part of the system. The TDP chemograph exhibited two peaks of $0.30\,mg\,P/l$ and $0.25\,mg\,P/l$ after 6 and 23 h respectively, the first in response to the increase in discharge and the second, to a small rainfall event (Figure 10.4). Throughout the storm, MRP and TDP chemographs were similar as MRP dominated the TDP fraction, constituting between 60% (06:00 hours) and 96% (13:30 hours). Compared to Storm 1 when the maximum TDP concentration was $1.74\,mg\,P/l$ and the soluble fraction dominated TP, Storm 2 TDP concentrations were extremely low and formed a very insignificant part of the total P present. Similarly, MRP concentrations were extremely low, with the maximum of $0.26\,mg\,P/l$, compared to $1.74\,mg\,P/l$ in Storm 1. This may be due to the flushing out and dilution of the soluble fraction during the wet period between the two events.

For the initial 3 h of the rising limb of the stream hydrograph, stream P loadings were positively influenced by the rapid change in discharge. After the peak discharge, the stable, elevated stream height meant that loading values were influenced by concentration rather than discharge, chemographs for concentration and load fluctuating simultaneously.

10.4.4   Hillslope- and Plot-Scale Pathways

P fractions present in overland, unsaturated subsurface and groundwater hydrological pathways were investigated at the experimental plot and hillslope transect scales (Figure 10.2) to identify the dominant pathways transporting P. During several monitored winter storm events ($n = 5$), the concentration of P in overland flow was high, MRP averaging $1.80\,mg\,P/l$ and TP, $2.48\,mg\,P/l$. In all samples, the soluble fraction was dominated by the inorganic component (61–98%). Both hillslope transects showed a downslope increase in the TP concentration due to the accumulation of upslope contributions.

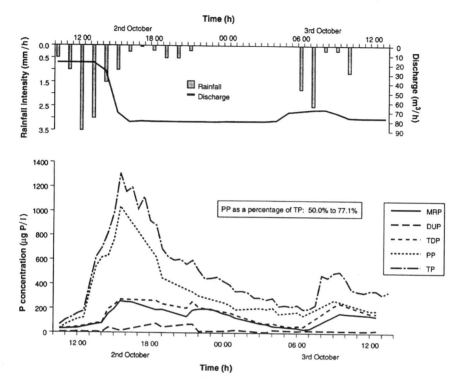

Figure 10.4 Storm 2 chemograph, 2–3 October 1994.

Grab samples from all piezometers indicated that for the dissolved fraction, as the piezometer depth increased (50, 100, 150 and 200 cm), absolute concentrations of MRP, DUP and TDP decreased. This relationship was expected given that soil analysis showed a clear decrease in water-extractable and TP with depth. MRP concentrations ranged from 18 μg P/l at 50 cm depth to 1 μg P/l at 200 cm, compared to 143 μg P/l in the stream. The ratio of MRP:DUP ranged from 0.02 to 0.30, indicating that unreactive P dominates the soluble phase in the saturated zone possibly as a result of anaerobic conditions preventing mineralization of organic compounds. These data confirm that subsurface throughflow has a very low dissolved P content. Extremely high particulate P concentrations were measured in all piezometers with a maximum of 1.93 mg P/l at 100 cm and a minimum of 0.05 mg P/l recorded at 200 cm. As a consequence, TP concentrations largely reflected PP fluctuations and showed a similar negative relationship with profile depth. TP decreased from 2.00 mg P/l at 100 cm to 0.10 mg P/l at 200 cm, of which PP contributed 97 and 53% respectively.

Grab samples collected from the plot experiments during winter (December to February) indicated the importance of zero-tension (macropore) pathways in the transport of phosphates under unsaturated conditions. The MRP fraction exhibited considerable inter- and intra-plot variation, from a minimum of 1 μg P/l to a maximum of 478 μg P/l. However, the majority of this variation can be explained by the sampling depth as MRP concentrations generally decreased with depth. Alternatively, spatial variability may be caused by enriched surface P concentrations resulting from isolated dung pats. The most important finding was the large quantity of PP present, which confirmed that subsurface pathways were extremely important in the transport of phosphates during storm events. PP exceeded 1 mg P/l in a third of the samples and the particulate fraction consistently dominated the TP present.

Underground tile drains provided an efficient route for the transport of water from the terrestrial to the aquatic environment particularly during storm events. Under normal baseflow conditions, P concentrations in the outfall from a drain terminating near the catchment outlet were lower than in the receiving stream. For example, in January 1995, the TP concentration in drain-water was 0.06 mg P/l, compared to 0.29 mg P/l in the stream. Similar results were found at the Saxmundham Experimental Station, East Suffolk, UK, where concentrations in land drains were 0.025 mg P/l compared to 0.5 mg P/l in the receiving stream (Cooke and Williams 1972). Drain outflow samples generally contained a larger concentration of soluble rather than particulate P, consistent with the results of Sharpley and Syers (1979) and Hergert *et al.* (1981). Yet during storm events, TP concentrations approaching 1 mg P/l were measured. For example, on 15 September 1994, a P concentration of 0.29 mg P/l was measured, of which 98% was in the soluble form. Similarly, the recession limb of an 11.5 mm event in February 1995 produced a TP peak of 0.39 mg P/l and was followed by a gradual decline to 0.16 mg P/l within 8 h. In March 1995, the recession limb of a 10 mm storm had a peak TP concentration of 0.97 mg P/ l, of which 80% was in the particulate form. The high concentration of P rapidly declined, and within 5 h had dropped to 0.04 mg P/l. A grab sample taken the following week contained only 0.02 mg TP/l and 0.01 mg PP/l. In summary, during dry periods, accelerated subsurface flow through tile drains is minimal and low in P, but during storm conditions this pathway becomes important in transporting all forms of P.

10.5   DISCUSSION

10.5.1   Storm Events

This study confirmed that the major portion of annual P exported from a mixed agricultural catchment occurs during intense, infrequent storm events. Similar results have been observed for grass (Jordan and Smith 1985) and arable catchments (Owens *et al.* 1991; Kronvang 1992). During baseflow conditions, P loads were fairly low and stable due to groundwater sources dominating streamflow. Conversely, during storm events, a greater number of source areas and hydrological pathways were activated, discharge increased by several orders of magnitude, and stored P was mobilized. For example, a typical autumn rainfall event (Storm 2) increased TP loads from a pre-storm value of 1.2 g/h to a maximum value of 96.5 g/h. Discharge was found to be an important factor influencing P loads although maximum P concentrations generally occurred before peak discharge.

The contrast between high soluble-P losses in the first autumnal storm (Storm 1) and high particulate-P losses in subsequent events (Storm 2) was attributed to a "flushing out" of mineralized P accumulated during the dry summer period by the first substantial rain of the autumn period (Jordan and Smith 1985). Urine and faeces deposited on the ground surface undergo mineralization and immobilization, producing a store of labile inorganic P in the solid phase. Residual soil P rapidly undergoes desorption on contact with infiltrating water and the mobilized P is transported in the aqueous phase in subsurface pathways. Further physical, chemical or biological transformations may occur before the water reaches the stream channel (Ryden *et al.* 1973). Unsaturated subsurface throughflow via tile drains and soil macropores were identified as important pathways operating during this event (Storm 1). Storm 2 was characterized by lower overall P concentrations and dominance by the particulate fraction. Infiltration- and saturation-excess overland flow, tile drain runoff and groundwater runoff were noted for transporting P from the terrestrial to the aquatic environment.

10.5.2 Pathways

The early concentration peak recorded during storm events in response to an increase in rainfall or stream discharge probably represented P transported in infiltration–excess overland runoff from partial source areas in close proximity to the stream channel and rapid shallow subsurface pathways particularly through macropores, as well as stream sediment-derived P. Subsurface runoff water possibly became enriched in dissolved P diffused from soil pores and from labile soil P desorbed on contact with the percolating water. The flushing out of pre-event water and soil-derived P continued until exhaustion or dilution by P-poor event water became overwhelming. Unsaturated subsurface throughflow via tile drains was a delayed but quantitatively important pathway for PP during the early storm period and for soluble P as the storm progressed. A combination of high discharge and P-deficient subsoils in the saturated zone produced low concentrations in the final stages of the hydrograph recession limb.

For the study catchment, the hydrological pathways responsible for the transport of P from the terrestrial to the aquatic environment largely depended upon the antecedent moisture conditions, rainfall characteristics and the recurrence interval between events. For example, the low water-table depth preceding autumn recharge produced high infiltration and low runoff rates during the first storm, hence the transport of PP in overland flow was minimal. After autumnal water-table recharge had occurred, soil infiltration capacity decreased, erosional and overland flow losses increased and variable sources areas adjacent to the stream became activated.

In the majority of previous research, P losses through subsurface pathways have been assumed to be minimal, particularly regarding the particulate fraction (Sharpley and Menzel 1987). However, the high particulate P concentrations measured in zero-tension samplers, tile drain outflow and piezometers in the present study emphasized the importance of subsurface transportation mechanisms. Considering the greater spatial and temporal extent of subsurface as opposed to overland flow, the high P concentrations but low overall P loads in overland flow were probably of secondary importance compared to the large P loads generated by subsurface transport.

In general, soluble P concentrations were low in soil solutions and consequently, subsurface drainage waters contained less soluble P than the receiving stream. Nevertheless, underground tile drains provided a rapid and quantitatively important route for soluble P transportation in the study catchment, particularly during storm events. The role of tile drains in transporting MRP from grassland fields is well documented (Foy and Withers 1995) but their contribution to stream PP concentrations appears controversial. Substantial losses of PP in drainflow have been found from a small (6 ha) intensively managed grassland catchment in Ireland (Jordan and Smith 1985). Conversely, P losses in effluent from 12 manure-amended 0.33-ha tile-drained grassland plots (Aurora, NY) collected over a 7-year period were dominated by the soluble inorganic P fraction and contained very low concentrations of sediment (Hergert *et al.* 1981). Plot experiments at Rothamsted, Herts, UK, have similarly shown that MRP is the dominant fraction in subsurface drainage water from wheat, with concentrations exceeding 1.5 mg P/l during a storm event (Heckrath *et al.* 1995). The results of the present study suggested that the ratio of soluble to particulate P is dependent upon rainfall and antecedent moisture conditions. During normal winter baseflow periods, drain discharge was dominated by the soluble fraction and concentrations of all fractions were low compared to the receiving stream. During the initial stages of stormflow, water percolating through the soil profile in rapid subsurface pathways was intercepted by tile drains, resulting in extremely high PP concentrations (exceeding 1 mg P/l). As the storm progressed, the relative importance of the soluble fraction compared to the particulate fraction increased, although in absolute terms both fractions decreased. The recession of the concentration chemograph may be the result of a dilution effect or an exhaustion of the P store.

The combination of a saturated wedge and soil compaction from farm traffic close to the stream channel generated a variable source area approximately 20 m in width on both sides of the stream channel. P-rich, shallow subsurface water appeared to converge and re-surface at the hillslope base and remain ponded on the ground surface in a dynamic equilibrium with the P-rich topsoil until the next rainfall event which transported it from the terrestrial to the aquatic environment. Grab samples taken in December 1994 indicated low TP concentrations in this surface water, measuring 0.15 mg P/l compared to 0.40 mg P/l in the adjacent stream. This was due to previous heavy rainfall, the surface water reflecting precipitation composition rather than that transported from upslope. Further samples taken in February 1995 supported this hypothesis as TP concentrations ranged from 0.80 mg P/l to 0.87 mg P/l of which the particulate fraction dominated.

10.6 CONCLUSIONS

The research reported in this chapter investigated the quantities and forms of P present in several hillslope hydrological pathways to clarify the interpretation of stream chemographs produced for a 120-ha mixed agricultural catchment during two intense rainfall events. An understanding of the sources and pathways responsible for the final form of P in runoff is essential as P fractions vary in their availability to aquatic biota. The presence of persistent algal blooms in the study catchment demonstrated the detrimental impact of agricultural activities on stream ecosystems and highlighted the importance of increasing the current understanding of P transport mechanisms.

The majority of studies investigating P losses in runoff have focused on either the lysimeter, plot or catchment scales. Catchment-scale research cannot accurately reveal the processes and hydrological flow pathways by which nutrients are transported. Conversely, field lysimeters and experimental plots cannot adequately represent catchment scale processes (Heathwaite *et al.* 1990). These drawbacks advocated the use of the integrated-scale approach adopted in this study. Here catchment-scale findings were supported by transect and plot-scale results. The smaller-scale instrumentation used here intercepted water as it moved towards the stream without restricting or disrupting the natural system.

The identification of point and diffuse sources of P within agricultural catchments is important for the implementation of management strategies aimed at reducing P losses. Concentrations exceeding 2 and 3 mg P/l of soluble organic P and TP respectively were measured at a few points within the catchment, including runoff from manure heaps, overflow from reception pits and around drinking troughs. As the distance between the point sources and the stream usually exceeded 50 m, P mobilized during rainfall events was sorbed on contact with P-deficient subsoil and was not therefore lost directly to the stream but stored until further transport or transformation. This suggested that the majority of P present in storm-derived stream discharge originated from diffuse sources within the catchment including rainfall, plant leachate, and the soil.

Storm events proved to be the most critical time periods for the transport of P as dormant hydrological pathways were activated and stored soil P was mobilized. The characteristics of individual events appeared to be a function of several factors including antecedent moisture conditions, rainfall characteristics and the recurrence interval between events, although chemographs were generally similar in appearance due to the overriding influence of discharge.

The identification and quantification of P fractions mobilized from point and diffuse agricultural sources is important if their impact on receiving water bodies is to be minimized. Management strategies aimed at reducing field-, farm- and catchment-scale losses of P need to take into consideration processes and pathways operating at smaller scales if the contribution of agriculture to deteriorating water quality is to be reduced.

ACKNOWLEDGEMENTS

This research was funded by the Ministry of Agriculture, Fisheries and Food, Award no. AE 8750. Thanks are extended to Prof. R. Dils for his constructive comments on the manuscript and the cartographic unit at Sheffield University for their help in preparing the illustrations. We would also like to thank the ADAS Hydrology Unit at Gleadthorpe for the weir installation and Mr John Smith, the host farmer for allowing catchment monitoring.

REFERENCES

Ahuja, L.R. 1986. Characterization and modelling of chemical transfer to runoff. *Advances in Soil Science*, **4**, 149–188.

Anderson, M.G. and Burt, T.P. 1990. Subsurface runoff. In Anderson, M.G. and T.P. Burt (eds), *Process Studies in Hillslope Hydrology*. Wiley, Chichester, pp. 365–400.

Armstrong, A.C. 1984. The hydrology and water quality of a drained clay catchment, Cockle Park, Northumberland. In T.P. Burt and D.E. Walling (eds), *Catchment Experiments in Fluvial Geomorphology*. Geo Books, Norwich, pp. 153–168.

Armstrong, A.C. and Burt, T.P. 1993. Nitrate losses from agricultural land. In T.P. Burt, A.L. Heathwaite and S.T. Trudgill (eds), *Nitrate: Processes, Patterns and Management*. Wiley, Chichester, pp. 239–267.

Bouma, J., Dekker, L.W. and Haans, J.C.F.M. 1980. Measurement of depth to water table in a heavy clay soil. *Soil Science*, **130**, 264–270.

Brady, N.C. 1990. *The Nature and Properties of Soils*, 10th edn. Macmillan, New York.

Breeuwsma, A., Reijerink, J.G.A. and Schoumans, O.F. 1995. Impact of manure on accumulation and leaching of phosphate in areas of intensive livestock farming. In K. Steele (ed.). *Animal Waste and the Land–Water Interface*. Lewis Publishers, pp. 239–249.

Broberg, O. and Pettersson, K. 1988. Particulate and dissolved phosphorus forms in freshwater: composition and analysis. *Hydrobiologia*, **170**, 61–90.

Burt, T.P. 1978. *Three Simple and Low-Cost Instruments for the Measurement of Soil Moisture Properties*. Huddersfield Polytechnic, Department of Geography and Geology Occasional Paper No. 6.

Burwell, R.E., Timmons, D.R. and Holt, R.F. 1975. Nutrient transport in surface runoff as influenced by soil cover and seasonal periods. *Soil Science Society of America Proceedings*, **39**, 523–528.

Burwell, R.E., Schuman, G.E., Heinemann, H.G. and Spomer, R.G. 1977. Nitrogen and phosphorus movement from agricultural watersheds. *Journal of Soil and Water Conservation*, **32**, 226–230.

Cooke, G.W. 1986. *Fertilizing for Maximum Yield*, 3rd edn. Collins Professional and Technical Books, London.

Cooke, G.W. and Williams, R.J.B. 1972. The phosphorus involved in agricultural systems and possibilities of its movement into natural water. *Water Research*, **7**, 19–33.

Dorioz, J.M., Pilleboue, E. and Ferhi, A. 1989. Phosphorus dynamics in watersheds: role of trapping processes in sediments. *Water Research*, **23**, 147–148.

Duda, A.M. and Finan, D.S. 1983. Influence of livestock on nonpoint source nutrient levels of streams. *Transactions of the American Society of Agricultural Engineers*, **26**, 1710–1716.

Finkl, C.W. Jr and Simonson, R.W. 1979. P Cycle. In R.W. Fairbridge and C.W. Finkl (Jr) (eds), *The Encyclopaedia of Soil Science, Part 1. Physics, Chemistry Biology, Fertility and Technology*. Dowden, Hutchinson and Ross, Pennsylvania, pp. 370–377.

Foy, R.H. and Withers, P.J.A. 1995. The contribution of agricultural phosphorus to eutrophication. *Proceedings of the Fertiliser Society*. No. 365.

Germann, P.F. 1990. Macropores and hydrological hillslope processes. In M.G. Anderson and T.P. Burt (eds), *Process Studies in Hillslope Hydrology*. Wiley, Chichester, pp. 327–364.

Heathwaite, A.L. 1993. *The Effect of Land Use on Phosphorus Losses*. Oral paper presented at the Phosphorus and the Environment Conference, 19 October 1993, organised by the Society of Chemical Industry, Agriculture and Environment Group, London.

Heathwaite, A.L., Burt, T.P. and Trudgill, S.T. 1989. Runoff, sediment, and solute delivery in

Heathwaite, A.L., Burt, T.P. and Trudgill, S.T. 1989. Runoff, sediment, and solute delivery in agriculture drainage basins: a scale-dependent approach. *Proceedings of the Baltimore Symposium, May 1989.* IAHS Publication No. 182, pp. 175–190.

Heathwaite, A.L., Burt, T.P. and Trudgill, S.T. 1990. The effect of land use on nitrogen, phosphorus and suspended sediment delivery to streams in a small catchment in southwest England. In J.B. Thornes (ed.). *Vegetation and Erosion.* Wiley, Chichester, pp. 167–179.

Heckrath, G., Brookes, P.C., Poulton, P.R. and Goulding, K.W.T. 1995. Phosphorus leaching from soils containing different P concentrations in the Broadbalk experiment. *Journal of Environmental Quality,* **24,** 904–910.

Hergert, G.W., Klausner, S.D., Bouldin, D.R. and Zwerman, P.J. 1981. Effects of dairy manure on phosphorus concentrations and losses in tile effluent. *Journal of Environmental Quality,* **10,** 345–349.

Holtan, H., Kamp-Nielsen, L. and Stuanes, A.O. 1988. Phosphorus in soil, water and sediment: an overview. *Hydrobiologia,* **170,** 19–35.

House, W.A., Denison, F.H., Smith, S.T. and Armitage, P.D. 1995. An investigation of the effects of water velocity on inorganic phosphorus influx to a sediment. *Environmental Pollution,* **89,** 263–271.

Houston, J.A. and Brooker, M.P. 1981. A comparison of nutrient sources and behaviour in two lowland subcatchments of the River Wye. *Water Research,* **15,** 49–57.

Johnes, P.J. and Heathwaite, A.L. 1992. A procedure for the simultaneous determination of total nitrogen and total phosphorus in freshwater samples using persulphate microwave digestion. *Water Research,* **26,** 1281–1287.

Johnson, A.H., Bouldin, D.R., Goyette, E.A. and Hedges, A.M. 1976. Phosphorus loss by stream transport from a rural watershed: quantities, processes and sources. *Journal of Environmental Quality,* **5,** 148–157.

Jordan, C. and Smith, R.V. 1985. Factors affecting leaching of nutrients from an intensively managed grassland in County Antrim, Northern Ireland. *Journal of Environmental Management,* **20,** 1–15.

Kofoed, A.D. 1984. Pathways of nitrate and phosphate to ground and surface waters. In F.P.W. Winteringham (ed.), *Environment and Chemicals in Agriculture.* Proceedings of a Symposium held in Dublin, 15–17 October 1984. Elsevier Applied Science Publishers, London, pp. 27–69.

Kronvang, B. 1992. The export of particulate matter, particulate phosphorus and dissolved phosphorus from two agricultural river basins: implications on estimating the non-point phosphorus load. *Water Research,* **26,** 1347–1358.

Logan, T.J. 1982. Mechanisms for release of sediment-bound phase phosphate to water and the effects of agricultural land management on fluvial transport of particulate and dissolved phosphate. *Hydrobiologia,* **92,** 519–530.

Logan, T.J. and McLean, E.O. 1973. Nature of phosphorus retention and adsorption with depth in soil columns. *Soil Science Society of America Proceedings,* **37,** 351–355.

Ministry of Agriculture, Fisheries and Food 1991. *Code of Good Agricultural Practice for the Protection of Water.* MAFF Publications, London.

Murphy, J. and Riley, J.P. 1962. A modified single solution method for the determination of phosphate in natural waters. *Analytica chim. Acta,* **27,** 31–36.

National Rivers Authority, 1992a. *Water Pollution Incidents in England and Wales – 1990.* Water Quality Series Report No. 7.

National Rivers Authority, 1992b. *The Influence of Agriculture on the Quality of Natural Waters in England and Wales, A Report by the National Rivers Authority.* Water Quality Series, No. 6.

Newman, E.I. 1995. Phosphorus inputs to terrestrial ecosystems. *Journal of Ecology,* **83,** 713–726.

Newson, M.D. and Robinson, H. 1983. Effects of agricultural drainage on upland streamflow. Case studies in Mid-Wales. *Journal of Environmental Management,* **17,** 333–348.

Ng, H.Y.F., Mayer, T. and Marsalek, J. 1993. Phosphorus transport in runoff from a small agricultural watershed. *Water, Science and Technology,* **28,** 451–460.

Olsen, S.R. 1967. Recent trends in the determination of orthophosphate in water. In H.L. Golterman and R.S.N.V. Clymo. *Chemical Environment in the Aquatic Habitat.* Noord-Hollandsche uitgevers Maatschappij, Amsterdam, pp. 663–105.

Owens, L.B., Edwards, W.M. and Van Keuren, R.W. 1991. Baseflow and stormflow transport of nutrients from mixed agricultural watersheds. *Journal of Environmental Quality,* **20**, 407–414.

Persson, G. and Jansson, M. (Eds) 1988. Phosphorus in freshwater ecosystems – Proceedings of a symposium held in Uppsala, Sweden, 25–28 September 1985. *Hydrobiologia,* **170**.

Reddy, G.Y., McLean, E.O., Hoyt, G.D. and Logan, T.J. 1978. Effects of soil, cover crop, and nutrient source on amounts and forms of phosphorus movement under simulated rainfall conditions. *Journal of Environmental Quality,* **7**, 50–54.

Rigler, F.H. 1968. Further observations inconsistent with the hypothesis that the molybdenum blue method measures orthophosphate in lake water. *Limnological Oceanography,* **13**, 7–13.

Robinson, J.S., Sharpley, A.N. and Smith, S.J. 1994. Development of a method to determine Bioavailable P loss in agricultural runoff. *Agriculture, Ecosystems and Environment,* **47**, 287–297.

Ryden, J.C., Syers, J.K. and Harris, R.F. 1973. Phosphorus in runoff and streams. *Advances in Agronomy,* **25**, 1–45.

Sharpley, A.N. 1981. The contribution of phosphorus leached from crop canopy to losses in surface runoff. *Journal of Environmental Quality,* **10**, 160–165.

Sharpley, A.N. 1993. An innovative approach to estimate bioavailable phosphorus in agricultural runoff using iron-oxide impregnated paper. *Journal of Environmental Quality,* **22**, 597–601.

Sharpley, A.N. and Halvorson, A.D. 1994. The management of soil phosphorus availability and its impact on surface water quality. In R. Lal and B.A. Stewart (eds), *Soil Processes and Water Quality.* Advances in Soil Science, Lewis Publishers, pp. 7–90.

Sharpley, A.N. and Menzel, R.G. 1987. The impact of soil and fertilizer on the environment. *Advances in Agronomy,* **41**, 297–324.

Sharpley, A.N. and Smith, S.J. 1990. Phosphorus transport in agricultural runoff: the role of soil erosion. In J. Boardman, I.D.L. Foster and J.A. Dearing (eds), *Soil Erosion on Agricultural Land.* Wiley, Chichester, pp. 351–366.

Sharpley, A.N. and Syers, J.K. 1979. Phosphorus inputs into a stream draining an agricultural watershed. II. *Water, Air and Soil Pollution,* **11**, 417–428.

Sharpley, A.N. and Withers, P.J.A. 1994. The environmentally-sound management of agricultural phosphorus. *Fertilizer Research,* **39**, 133–146.

Sharpley, A.N., Menzel, R.G., Smith, S.J., Rhoades, E.D. and Olness, A.E. 1981. The sorption of soluble phosphorus by soil material during transport in runoff from cropped and grassed watersheds. *Journal of Environmental Quality,* **10**, 211–215.

Simmons, K.E. and Baker, D.E. 1993. A zero-tension sampler for the collection of soil water in macropore systems. *Journal of Environmental Quality,* **22**, 207–212.

Stevens, R.J. and Smith, R.V. 1978. A comparison of discrete and intensive sampling for measuring the loads of nitrogen and phosphorus in the River Main, County Antrim. *Water Research,* **12**, 823–830.

Vighi, M. and Chiaudani, G. 1987. Eutrophication in Europe: the role of agricultural activities. In E. Hodgson (ed.), *Reviews in Environmental Toxicology 3.* Elsevier, pp. 213–257.

Weyman, D.R. 1973. Measurements of the downslope flow of water in a soil. *Journal of Hydrology,* **20**, 267–288.

Whipkey, R.Z. and Kirkby, M.J. 1990. In M.J. Kirkby (ed.), *Hillslope Hydrology.* Wiley, Chichester, pp. 121–144.

Wild, A. and Babiker, I.A. 1976. The asymmetric leaching pattern of nitrate and chloride in a loamy sand under field conditions. *Journal of Soil Science,* **27**, 460–466.

Wild, A. and Oke, O.L. 1966. Organic phosphate compounds in calcium chloride extracts of soils: identification and availability to plants. *Journal of Soil Science,* **17**, 356–371.

Wilson, G.V., Jardine, D.M., Luxmore, R.J., Zelanzy, L.W., Lietzke, D.A. and Todd, D.E. 1991. Hydrogeochemical processes controlling subsurface transport from an upper subcatchment of Walker Branch watershed during storm events. 1. Hydrologic transport processes. *Journal of Hydrology*, **123**, 297–316.

Young, R.A. and Mutchler, C.K. 1976. Pollution potential of manure spread on frozen ground. *Journal of Environmental Quality*, **5**, 174–179.

11 Nutrient Cycling in Upland Catchment Areas: The Significance of Organic Forms of N and P

A. C. EDWARDS, M. D. RON VAZ,[1] S. PORTER[2] and L. EBBS

Macaulay Land Use Research Institute, Craigiebuckler, Aberdeen, UK

11.1 INTRODUCTION

Upland areas are located predominantly in the northern and western regions of the UK. They can be broadly defined as land over 300 m above sea level, and represent approximately 20% of the total land area (Harrison 1974). As hydrological source areas, uplands are extremely important, supplying 40% of the public water supply in Britain. In addition, their distinctive flow and chemical characteristics can greatly influence regions downstream.

Typically, upland soils are acidic, highly organic and nutrient-poor, properties which give rise to oligotrophic drainage waters containing low concentrations of solutes. The close link between terrestrial and aquatic ecosystems means that in upland areas water chemistry is particularly sensitive to any perturbations that may disrupt nutrient cycling processes. There is concern, for example, over the consequences of increasing atmospheric N deposition in terms of vegetation community structure (Lee and Stewart 1978) and major changes in land use such as afforestation. A clear understanding of the effects on soil processes that influence nutrient availability and interactions with the hydrological cycle is essential for the long-term sustainability of upland ecosystems.

The central role that organic matter plays in soil nutrient cycling is well established and large quantities ($> 200\,000$ Mg C/km^2) of carbon can accumulate in upland soils (Howard *et al.* 1995), often over relatively short timescales (Billett *et al.* 1990). Both particulate (POC) and dissolved (DOC) carbon are significant components in aquatic systems (Aiken 1985), participating in numerous physical, chemical and biological reactions (Vaughan *et al.* 1993). The nutrient content of surface organic horizons can be substantial; Williams (1992), for example, reported a value of 700 kg N/ha for a forest soil. Consequently upland soils have the potential to act as a major potential source of soluble nutrients. While substantial information exists on the organic carbon contents of surface waters (Hope *et al.* 1994), only limited

Present address: [1]Instituto De Recursos Naturales y Agrobiologia De Sevilla, Sevilla, Spain. [2]Department Plant and Soil Science, Aberdeen University, Aberdeen, UK.

consideration has been given to quantifying the significance of any associated nutrients such as nitrogen (N), phosphorus (P) and sulphur (S) (Williams and Edwards 1993). This situation can be partly attributed to analytical constraints, with it being much more common to determine only inorganic forms of the elements (Heathwaite 1993). This can result in a considerable underestimation of total nutrient fluxes and, therefore, in inaccuracies when calculating input/output budgets.

Soluble organic compounds are structurally complex with very diverse origins. The N, P and S can either be present as an integral part of a large organic molecule or as discrete compounds that may be attached in some way to the larger carrier molecule. Organically associated N, P and S can contribute significantly to the total amounts present (Qualls *et al.* 1991) and as such represent key components of the respective nutrient cycles in both terrestrial and aquatic ecosystems. Organic forms can be either a direct (Chapin *et al.* 1993) or an indirect (after chemical hydrolysis) source of nutrients for plant uptake. There is growing evidence that changes in atmospheric deposition of N in particular (Yesmin *et al.* 1995) and patterns of land use (Stevens and Wannop 1987) can have a direct influence on the amounts of soluble organic nutrients.

In this chapter we compare the total amounts and chemical composition of N and P in soil solution from a range of sites located in northeast Scotland. The soils are all developed from the same parent material but they support different vegetation covers. The relative importance of soluble organic N and P is discussed, particularly in terms of their potential to be preferentially leached compared to inorganic forms. The influence land use can play in determining the total N concentration and the N species present in river water is investigated.

11.2 FUNDAMENTAL METHODS

11.2.1 Collection of Samples

Soil samples were collected from two different locations. The bracken (*Pteridium aquilinum*) and heather (*Calluna vulgaris*) sites were adjacent to each other at Glendye (NGR NO 647846), a subcatchment of the River Dee. Soil cores (nine for each vegetation cover) were collected using a 7 cm diameter auger to a depth of 15 cm. A second series of samples was collected from a catenal sequence at Brimmondhill, Craibstone (NGR NJ 855 100), Aberdeen. Three vegetation covers were sampled: heather moorland, permanent grassland and temporary grassland. Full details of the management history are given in Arowolo *et al.* (1994). All soils are freely draining iron podsols from the Countesswells Association, derived from granitic-type parent materials (Glentworth and Muir 1963). Soil samples were coarse sieved (<6 mm) and analysed field moist. Unless stated otherwise, all results are expressed on an air-dry basis and are the means of at least three replicate extractions.

Samples of water were collected at various locations along the main channel of the rivers Dee, Don and Ythan and a selection of their main tributaries. The data are means of eight samples collected over a two-month period (August and September 1994) on a weekly basis and included a range of flow conditions. The river catchment areas are contrasting in terms of their land use, which varies from being predominantly semi-natural moorland/forestry in the Dee to the agriculturally dominated Ythan catchment (MacDonald *et al.* 1994). Water samples were filtered through 0.45 μm membrane filters immediately on return to the laboratory, with all analysis being carried out within two days of sample collection.

11.2.2 Soil Extractions

Soil water extractions were prepared from 4 g of fresh soil and 40 ml of deionized water, shaken for 1 h, and then centrifuged at 2000 g for 35 min. The samples were filtered initially through pre-washed Whatman No. 42 and then through 0.45 μm membrane filters.

11.2.2.1 Potassium sulphate

Fifty millilitres of 0.5 K_2SO_4 was added to 10 g of fresh soil and shaken end-over-end for 2 h. The mixture was filtered through Whatman No. 42 filters, and the soil residue washed through with a further 50 ml of extractant. Finally, samples were filtered through 0.45 μm membrane filters, made up to 100 ml with 0.5 K_2SO_4.

All results are expressed on an oven-dry basis. Soil moisture contents were determined by drying 10 g of fresh soil at 105°C for 24 h.

11.2.3 Chemical Analysis

11.2.3.1 Phosphorus

Inorganic-P (MRP) in both water and K_2SO_4 extracts was determined by the colorimetric molybdenum blue method (Murphy and Riley 1962). Total-P was determined following the persulphate oxidation procedure of Williams *et al.* (1995). Dissolved organic phosphorus (DOP) was determined by an automated (10 samples/h) photo-oxidation procedure (Ron Vaz *et al.* 1992). Dissolved condensed phosphorus (DCP) was determined by difference, DCP = TDP − (MRP + DOP).

11.2.3.2 Nitrogen

Ammonium and nitrate concentrations in waters, extracts and oxidized solutions were determined colorimetrically using a Technicon TRAACS auto-analyser system. Total-N in solutions was analysed as nitrate after oxidation with alkaline potassium persulphate (Williams *et al.* 1995). Organic-N in water extracts was assumed to be the difference between total-N and ($NH_4 - N + NO_3 - N$).

Dissolved organic carbon (DOC) was determined on acidified samples using a TOCsin II system (Phase Separation Ltd).

11.2.3.3 Statistical Analysis

Analysis of variance was performed on the mean concentration for each soluble fraction using Genstat 5 (1990). All least significant differences were calculated at the probability level $P < 0.05$.

Table 11.1 Total organic carbon (TOC) and nitrogen (N_t) concentrations ($\mu g/g$) in water extracts from soil collected at two depths under bracken and heather with standard errors in parentheses

Depth	TOC	N_t
0–7 cm		
Bracken	354(42.8)[a]	26.2(2.45)[a]
Heather	274(21.6)[a]	15.6(0.87)[b]
7–15 cm		
Bracken	n/d	10.3(1.50)[a]
Heather	n/d	8.5(1.58)[a]

n/d=not determined.
Within each column, numbers followed by a different letter are significantly different for individual depths ($p < 0.05$).

11.3 DISCUSSION

11.3.1 Extractable Soil Nitrogen

The total amounts of C and N in water extracts of soil collected from the two adjacent bracken and heather vegetation sites are shown in Table 11.1. Concentrations of N were significantly greater under bracken than heather for topsoil samples (0–7 cm) and a marked decline occurred in the lower (7–15 cm) soil horizon. Only a small proportion (<20%) of the total soluble N was present in an inorganic form, mainly as nitrate (Figure 11.1). The C to organic N ratios for bracken and heather

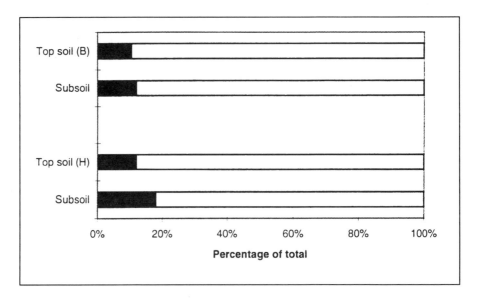

Figure 11.1 Relative composition of the N present in water extracts of soil from areas of bracken (B) and heather (H) expressed as a percentage of the total present. Nitrate (■) and organic N (□)

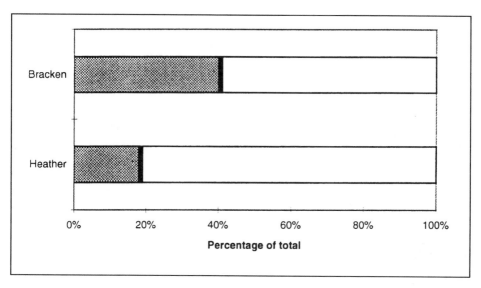

Figure 11.2 Relative composition of the N present in potassium sulphate extracts of soil from areas of bracken and heather expressed as a percentage of the total present ammonium (▨), nitrate (■) and organic N (☐)

were 15:1 and 20:1, respectively. Extraction of the topsoil samples with potassium sulphate, which includes exchangeable NH_4-N, increased the amounts of total N recovered (180 and 126 $\mu g\,N/g$ for bracken and heather, respectively). Organic N remained the dominant form present (Figure 11.2).

11.3.2 Extractable Soil Phosphorus

The concentration of total P and the relative contribution organic and inorganic fractions make to soil-water extracts from a catenal transect are shown in Table 11.2.

Table 11.2 Concentration (μg P/g) of water-extractable total P (TDP) and included fractions for soil collected under three vegetation covers (values are the mean of five replicates)

	TDP	MRP	DOP	DCP
Topsoil				
Heather moorland	1.28[c]	0.34[c] (27)	0.57[c]	0.37[a]
Permanent grassland	4.11[b]	1.20[b] (29)	2.82[a]	0.09[a]
Temporary grassland	13.05[a]	11.74[a] (90)	1.31[b]	0.0[a]
Subsoil				
Heather moorland	1.06[b]	0.23[b] (22)	0.72[b]	0.11[b]
Permanent grassland	0.74[c]	0.15[b] (20)	0.41[c]	0.18[b]
Temporary grassland	3.00[a]	1.44[a] (48)	1.61[a]	0.05[b]

Numbers in parentheses indicate the percentage contribution DRP makes to TDP.
Within each column, numbers followed by a different letter are significantly different ($p < 0.05$).

Table 11.3 Concentrations of total nitrogen (μg N/ml) and the relative contribution of organic N for various headwater tributaries of the rivers Dee and Don, northeast Scotland (mean of eight individual samples with the range of values in parentheses)

	Total N	% Organic-N	C:organic-N
Dee			
Girnock	0.175 (0.09–0.37)	93	30
Gaim	0.174 (0.095–0.325)	94	18
Muick	0.316 (0.18–0.61)	66	24
Tanner	0.244 (0.1–0.51)	95	19
Don			
Ernan	0.291 (0.12–0.63)	80	21
Nochty	0.243 (0.155–0.325)	61	21
Buchat	0.814 (0.65–1.03)	84	10
Kindie	0.99 (0.85–1.2)	24	22

Large significant differences in the amount of TDP extracted from each vegetation cover were observed for both top and subsoil samples. The highest TDP concentrations were associated with the agriculturally improved temporary grass. The P fractionation also differed markedly: while inorganic P (DRP) constituted 90% of the TDP present in the topsoil from temporary grass, organic forms dominated the moorland and permanent grass sites (Table 11.2). While amounts of TDP in subsoils were less than those of the respective topsoils, the contribution made by DOP was greater.

11.3.3 Composition of River Water Samples

Concentrations of total N in various headwater tributaries of the rivers Dee and Don are shown in Table 11.3. Total N varied considerably both between sites and during the sampling period but generally the concentrations were low. In all but one of the rivers organic N contributed more than 60% and in three cases more than 90% of the total N present. The C to organic N ratio varied between 10 : 1 and 30 : 1.

The change in concentrations of total N and relative contributions of organic and inorganic N for samples collected at locations down the Dee, Don and Ythan are shown in Figure 11.3(a) and (b). Generally, the upland semi-natural headwaters of the Dee and Don have relatively low total N concentrations, with organic N being the major component. The relative significance of the inorganic N (as nitrate) increases downstream, as does total N, closely reflecting changes in land use (MacDonald *et al.* 1994). The pattern in the agriculturally dominated Ythan is rather different in that it displays a constant total N concentration and is mainly in the form of nitrate.

11.3.4 Implications of Soluble Organic Nutrients to Upland Ecosystems

Organic forms of N and P have been shown to contribute significantly to the total amounts of soluble nutrients present in a range of semi-natural and managed soils.

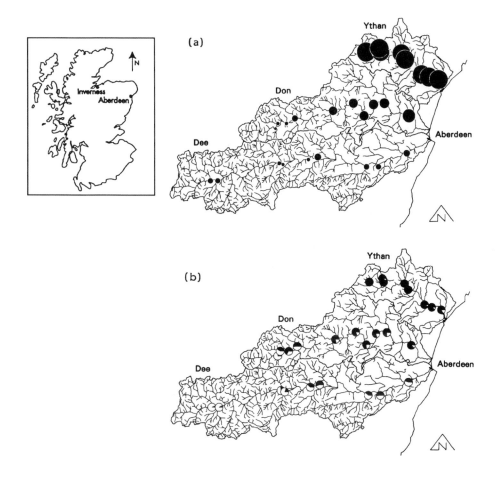

Figure 11.3 (a) Total N content of river water where the circle size is related to concentration (e.g. total N (●)=2 mg/l), and (b) the relative proportions of inorganic (black) and organic (open) N at various locations down the Rivers Dee, Don and Ythan

This observation has a number of important implications, particularly for upland catchments where in general the amount of available nutrients is low. First, these organic forms represent a potential source of readily available plant nutrients and, secondly, they enhance the transfer of N, P and S through the vegetation and soil horizons and potentially into drainage waters. The sorption properties and therefore mobility of neutral organic compared to charged inorganic species are very different (McKercher and Anderson 1989). The dominant form of inorganic species (e.g. NH_4, SO_4 and PO_4) which are present in acidic soils tend to be more strongly adsorbed and therefore retained than the respective organic species. As a consequence the leaching of soluble organic forms can be significant.

It is useful to subdivide a soil profile into an organic-rich zone in which biological processes dominate and a lower mineral zone where mineral weathering processes predominate (Ugolini *et al.* 1977). Large changes in the composition of soil water can occur as it drains through the profile with the "products" from one soil horizon forming the "reactants" for subsequent horizons. This situation has been well demonstrated by Qualls *et al.* (1991) for a deciduous forest in the Southern Appalachians of North Carolina, USA. They showed that the forest floor was an abundant source of N and P for the underlying mineral soil in the form of dissolved, and possibly particulate, organic matter. A similar conclusion was reached by Anderson *et al.* (1991) for two forested hillslopes in central Scotland, who also noted a marked seasonal pattern in organic N production.

It is important to include this organic component in any "catchment type" input/output budgets as considerable amounts of terrestrially derived organic N can reach drainage waters (Williams and Young 1994). This information is urgently required (INDITE 1994), particularly for upland catchments where losses of the total N in surface waters could currently be underestimated. If the average C : N of 20 : 1 from this study is used with published values of organic C losses (Hope *et al.* 1994) which are in the range of 50 kg C/ha, an annual figure of 2–3 kg/ha organic N can be expected. This value, while perhaps not significant for lowland agricultural catchments, where nitrate tends to dominate and losses are of the order of 20–30 kg N/ha, represents a major component of the N cycling within upland areas (Edwards *et al.* 1985; Lepisto *et al.* 1995). Similar conclusions can be made for the role of organically associated P, although these are more difficult to demonstrate clearly because of the rapid sorption reactions which take place with soil and river sediment.

A comparable situation exists when determining the likely impacts of atmospheric sulphur deposition and assessing the risk of acidification. The major emphasis has again tended to be placed upon catchment input/output balances where only inorganic S has been included. The significant contribution (up to 50% of the total S) that organic S can make to soil solution extracted from a comparable range of soils as those used in the present study has been demonstrated by Arowolo *et al.* (1994). The analysis of a range of river and lake samples indicated that 30% of the total S present was in a non-sulphate form (Edwards *et al.* 1992).

Recent studies have indicated that soil nutrient cycles, and in particular the part involving soluble organic components, are sensitive to various types of disruption. Stevens and Wannop (1987) and Yesmin *et al.* (1995), for example, demonstrated that clear-felling and increased atmospheric N deposition respectively caused an increase in the leaching of organic N. The possible implications this situation may pose for surface waters and in particular eutrophication have not yet been assessed adequately (Lepisto *et al.* 1995).

ACKNOWLEDGEMENTS

We thank Mr J. A. M. Ross for soil sampling and the Scottish Office Agricultural and Environment Fisheries Department for financial support.

REFERENCES

Aiken, G.R. 1985. Isolation and concentration techniques for aquatic humic substances in soil, sediment and water. In G.R. Aiken, D.M. McKnight, R.L. Wershaw and P. MacCarthy (eds), *Humic Substances in Soil Sediment and Water*. J. Wiley and Sons, Chichester.

Anderson, H.A., Stewart, M., Miller, J.D. and Hepburn, A. 1991. Organic nitrogen in soils and associated surface waters. In W.S. Wilson (ed.), *Advances in Soil Organic Matter Research. The Impact on Agriculture and the Environment*. Royal Society of Chemistry, Redwood Press, Melksham, Wiltshire.

Arowolo, T.A., Cresser, M.S. and Edwards, A.C. 1994. Impact of land use and soil type on the contribution of sulphate to total sulphur in drainage waters from upland soils. *Science Total Environment*, **158**, 139–146.

Billett, M.F., FitzPatrick, E.A. and Cresser, M.S. 1990. Changes in carbon and nitrogen status of forest soil organic horizons between 1949/50 and 1987. *Environmental Pollution*, **66**, 67–79.

Chapin, F.S., Moilanan, L. and Keilland, K. 1993. Preferential use of organic nitrogen for growth by a non-mycorrhizal arctic sedge. *Nature*, **361**, 150–153.

Edwards, A.C., Creasey, J.C. and Cresser, M.S. 1985. Factors influencing nitrogen inputs and outputs in two Scottish upland catchments. *Soil Use and Management*, **1**, 83–87.

Edwards, A.C., Ferrier, R.C. and Miller, J.D. 1992. The contribution of sulphate to total sulphur in a range of natural water samples. *Hydrological Sciences Journal*, **37**, 277–283.

Genstat 5. 1990. *Reference Manual*. Clarendon Press, Oxford.

Glentworth, R. and Muir, J.W. 1963. *The Soils of the Country Round Aberdeen, Inverurie and Fraserburgh*. Soil Survey of Scotland. MLURI, Aberdeen.

Harrison, S.J. 1974. Problems in the measurement and evaluation of the climatic resources of upland Britain. In J.S. Taylor (ed.), *Climatic Resources and Economic Activity*. David and Charles, Newton Abbott, UK.

Heathwaite, A.L. 1993. The impact of agriculture on dissolved nitrogen and phosphorus cycling in temperate ecosystems. *Chemistry and Ecology*, **8**, 217–231.

Hope, D., Billett, M.F. and Cresser, M.S. 1994. A review of the export of carbon in river water: fluxes and processes. *Environmental Pollution*, **84**, 301–324.

Howard, P.J.A., Loveland, P.J., Bradley, R.I., Dry, F.T., Howard, D.M. and Howard, D.C. 1995. The carbon content of soil and its geographical distribution in Great Britain. *Soil Use and Management*, **11**, 9–15.

INDITE. 1994. *Impacts of Nitrogen Deposition in Terrestrial Ecosystems*. UK Review Group on Impacts of Atmospheric Nitrogen, Department of the Environment.

Lee, J.A. and Stewart, G.R. 1978. Ecological aspects of nitrogen assimilation. In H.W. Woolhouse (ed.), *Advances in Botanical Research*, Vol. 6. Academic Press, pp. 2–43.

Lepisto, A., Andersson, L., Arheimer, B. and Sundblad, K. 1995. Influence of catchment characteristics, forestry activities and deposition on nitrogen export from small forested catchments. *Water, Air and Soil Pollution*, **84**, 81–102.

MacDonald, A.M., Edwards, A.C., Pugh, K.B. and Balls, P.W. 1994. The impact of land use on nutrient transport into and through three rivers in the north east of Scotland. In C. Kirby and W.R. While (eds), *Integrated River Basin Development*, Wiley, pp. 201–214.

McKercher, R.B. and Anderson, G. 1989. Organic phosphate sorption by neutral and basic soils. *Communications in Soil Science and Plant Analysis*, **20**, 723–732.

Murphy, J. and Riley, J.P. 1962. A modified single solution method for the determination of phosphate in natural waters. *Analytica Chimica Acta*, **27**, 31–36.

Qualls, R.G., Haines, B.L. and Swank, W.T. 1991. Fluxes of dissolved organic nutrients and humic substances in a deciduous forest. *Ecology*, **21**, 254–266.

Ron Vaz, M.D., Edwards, A.C., Shand, C.A. and Cresser, M.S. 1992. Determination of dissolved organic phosphorus in soil solutions by an improved automated photo-oxidation procedure. *Talanta*, **39**, 1479–1487.

Stevens, P.A. and Wannop, C.P. 1987. Dissolved organic nitrogen and nitrate in an acid forest soils. *Plant and Soil*, **102**, 137–139.

Ugolini, F.C., Minden, R., Davidson, H. and Zachara, J. 1977. An example of soil processes in the *Aqbies arnabilis* zone of central cascades, Washington. *Soil Science*, **124**, 291–302.

Vaughan, D., Lumsdon, D.G. and Linehan, D.J. 1993. Influence of dissolved organic matter on the bio-availability and toxicity of metals in soils and aquatic systems. *Chemistry and Ecology*, **8**, 185–201.

Williams, B.L. 1992. Nitrogen dynamics in humus and soil beneath Sitka spruce (*Picea sitchensis* (Bong.) Carr.) planted in pure stands and in mixture with Scots pine (*Pinus sylvestris* L.). *Plant and Soil*, **144**, 77–84.

Williams, B.L. and Edwards, A.C. 1993. Processes influencing dissolved organic nitrogen, phosphorus and sulphur in soils. *Chemistry and Ecology*, **8**, 203–215.

Williams, B.L. and Young, M. 1994. Nutrient fluxes in runoff on reseed blanked bog, limed and fertilized with urea, phosphorus and potassium. *Soil Use and Management*, **10**, 173–180.

Williams, B.L., Shand, C.A., Hill, M., O'Hara, C., Smith, S. and Young, M.E. 1995. A procedure for the simultaneous oxidation of total soluble N and P in extracts of fresh and fumigated soils. *Communications in Soil Science and Plant Analysis*, **26**, 95–106.

Yesmin, L., Gammack, S.M., Sanger, L.J. and Cresser, M.S. 1995. Impact of atmospheric N deposition on inorganic- and organic-N outputs in water draining from peat. *Science Total Environment*, **166**, 201–209.

12 Identification of Critical Source Areas for Phosphorus Export from Agricultural Catchments

W. J. GBUREK, A. N. SHARPLEY and H. B. PIONKE
Pasture Systems and Watershed Management Research Laboratory, USDA-ARS, Pennsylvania, USA

12.1 INTRODUCTION

In areas of intensive livestock production, continuing inputs of phosphorus (P) from fertilizers and manure to agricultural systems often exceeds its output in crop and animal produce. This situation can increase P loss in runoff from the land surface, which may, in turn, promote eutrophication of freshwater bodies (Sharpley *et al.* 1994). Loss of P from a catchment is controlled primarily by the interaction of its source factors of rate, timing, method, type of P applied and soil P content, with its transport factors of runoff and erosion. Thus, efforts to minimize P loss generally involve reducing runoff and erosion by conservation tillage, buffer strips, cover crops, and contour farming, as well as minimizing build-up and availability of surface soil P through the use of soil test P recommendations to guide fertilizer and manure applications.

The dynamic interactions between soil P and runoff water which control P levels in runoff at the point or plot scale are relatively well understood (Ryden *et al.* 1973; Sharpley *et al.* 1994). Extension of this knowledge to multi-field or catchment scales is tenuous but critical to the issue of nutrient management, since we are beginning to recognize that spatially variable P sources, temporary storages, sinks and transport processes are linked by the catchment-scale flow system. Further, a comprehensive P management strategy should address *downgradient* water quality impacts. Such a strategy must integrate effects at the local scale where specific management practices are implemented (i.e. the field), with the scale of the logical management unit (i.e. the farm), and finally with the larger scales at which results of the strategy are evaluated (i.e. the catchment).

Finally, specific control measures implemented within an overall management strategy will reduce P losses most effectively if they are targeted to "critical source-areas" (CSAs), i.e. specific identifiable areas within a catchment that are most vulnerable to P loss in runoff (Heatwole *et al.* 1987; Prato and Wu 1991). These CSAs are a special condition of the nitrate critical zone defined by Pionke and Lowrance (1991) as "a bounded area or volume within which one or a set of related

Advances in Hillslope Processes, Volume 1. Edited by M. G. Anderson and S. M. Brooks.

processes dominate to provide excessive production (source), permanent removal (sink), detention (storage) or dilution of nitrate." Catchment hydrology provides the basis for definition of these CSAs.

12.2   THE HYDROLOGIC BASIS FOR PHOSPHORUS TRANSPORT

Catchment hydrology was founded on the concept of defining catchment-scale response to inputs. Because water quantity, not quality, was originally of concern, hydrologists traditionally gave minimal consideration to pathways of water movement when quantifying these responses. However, as our concern with water quality began in the late 1960s, we began to consider an alternative view. Quantifying P transport within and from a mixed-land-use catchment requires knowledge of pathways of water movement, not simply catchment response. Water can translocate P from specific source-areas of application (typically on or within the soil zone), to or through zones of reaction and sinks within the catchment (in either the surface or subsurface), and finally to positions where it is removed from the catchment (generally by streamflow but possibly via subsurface flow). Thus, the key to understanding P transport within and from a catchment is a detailed understanding of pathways of flow (primarily surface runoff), not simply responses of all individual fields within the catchment.

Surface runoff was originally viewed as a soil-controlled phenomenon occurring more or less uniformly over the entire catchment. Techniques developed to predict storm runoff (e.g. the Curve Number) were based on this assumption. More recently, however, partial-area hydrology, which evolved to variable-source-area (VSA) hydrology, has become accepted as a descriptor of catchment response to precipitation, especially under humid-climate conditions (Ward 1984). The basic premise of VSA hydrology is that there is a dynamic contributing subcatchment within the topographically defined catchment; it expands and contracts seasonally, and also during a storm, as a function of precipitation, soil type, topography, and groundwater level and moisture status. Runoff from the VSA is dominated by saturation overland flow and rapidly responding subsurface flow. The remainder of the catchment provides only limited runoff, with infiltration and groundwater recharge being the dominant hydrologic processes.

In the most simple sense, the intersection of surface runoff source-areas within a catchment with areas of fertilizer application, grazing and/or manure disposal producing excess P available for transport creates the CSA's controlling P export. Thus, techniques or tools must be developed which identify these phosphorus CSAs if we want to implement cost-effective remedial strategies. The tools may be developed by field studies or computer simulation techniques. However, assessment by field studies is time-consuming, costly and labour intensive, while application of simulation models generally requires extremely detailed soil and management information.

Because of this, a Phosphorus Index (PI) was conceived and developed as a simple tool to quantify the relative importance of source and transport factors controlling P loss in runoff and to rank P loss vulnerability of individual sites (Lemunyon and Gilbert 1993). To date, the index has received limited testing and validation (e.g. Stevens *et al.* 1993; Sharpley 1995), so further studies related to the PI are necessary. The focus of this chapter will be on application of the PI to unit-source areas and investigation of its utility at the small mixed-land-use catchment scale, a step above field and farm scales. Goals will be

1. to evaluate the PI as applied to a variety of small unit-source areas;
2. to identify source-areas of potential P loss within a small mixed-land-use catchment using the PI; and
3. to compare these source-areas to a simulation-based delineation of phosphorus CSAs within the same catchment.

We will consider unit-source areas within Oklahoma and Texas, and an agricultural hill-land catchment in east-central Pennsylvania, USA.

12.3   THE PHOSPHORUS INDEX

Lemunyon and Gilbert (1993) developed the P indexing system shown as Table 12.1 to rank sites according to their vulnerability to P loss in runoff. The PI rates source (expressed as soil test P and fertilizer and manure inputs) and transport (runoff and erosion potential) factors of a site to provide the numerical ranking. Each site characteristic affecting P loss is assigned a weighting based on the assumption that certain characteristics have relatively greater effects on P loss than others, and a rating ranging from none to very high. The rating value for each characteristic is selected from Table 12.1, multiplied by the appropriate weighting factor, and the weighted values of all site characteristics are then totalled to quantify overall site vulnerability to P loss.

Data to rate the site characteristics are generally readily available. However, some soil and organic material testing may be required to determine their loss rating levels; this soil and material analysis is essential to reliable site assessment. In the original index, soil test P was not quantified, but rather simply categorized as none, low, medium, high and very high (Lemunyon and Gilbert 1993). It was intended that the appropriate ranges in soil test P values be established regionally, dependent on area conditions, proximity of P-sensitive waters, and agro-economic priorities.

The PI was envisioned to provide field staff, catchment planners and land managers with a tool to assess the potential risk for P movement to surface waters from a variety of landforms and management practices. The ranking of PIs of individual sites identifies where the risk of P movement (field losses) is highest. Following such an analysis, it may be apparent that one or more site parameters are influencing the index disproportionately; the parameters identified can serve as the basis for implementation of remedial soil and water conservation practices and land management techniques.

Table 12.1 The Phosphorous Index: weighted rating calculations and site vulnerability evaluation (adapted from Lemunyon and Gilbert 1993)

Site characteristic	Weight	Phosphorous loss rating (value)				
		None (0)	Low (1)	Medium (2)	High (4)	Very high (8)
Soil erosion	1.5	Not applicable	<10 Mg/ha	10–20 Mg/ha	20–30 Mg/ha	>30 Mg/ha
Runoff class	0.5	Negligible	Very low or low	Medium	High	Very high
Soil P test	1	Not applicable	Low	Medium	High	Excessive
P fertilizer application rate	0.75	None applied	1–15 kg P/ha	16–45 kg P/ha	46–75 kg P/ha	>76 kg P/ha
P fertilizer application method	0.5	None applied	Placed with planter deeper than 5 cm	Incorporated immediately before crop	Incorporated >3 months before crop or surface applied <3 months before crop	Surface applied >3 months before crop
Organic P source application rate	1	None applied	1–15 kg P/ha	16–30 kg P/ha	31–45 kg P/ha	>46 kg P/ha
Organic P source application method	1	None	Injected deeper than 5 cm	Incorporated immediately before crop	Incorporated >3 months before crop or surface applied <3 months before crop	Surface applied to pasture, or >3 months before crop

Total of weighted rating values	Site vulnerability
<8	Low
8–14	Medium
15–32	High
>32	Very high

12.4 METHODS

12.4.1 Unit-Source Areas

The PI was evaluated for 30 unfertilized and P-fertilized, grassed and cropped unit-source areas (small, single land use and management) within the Southern Plains of Oklahoma and Texas, USA. The areas and their measured losses have been previously reported by Chichester and Richardson (1992), Sharpley *et al.* (1985,

Table 12.2   Unit-source area characteristics for 1976 to 1992

Site	Study period	Land use	Tillage	Area (ha)	Slope (%)	Soil P[a] (mg/kg)	Fert. P added (kg/ ha/year)
Bushland, TX							
B10A	1984–1992	Wheat–sorghum– fallow rotation	No till	3.0	1.0	25	0
B11A	1984–1992		No till	3.0	1.0	31	0
B12A	1984–1992		No till	3.0	1.0	24	0
B10B	1984–1992	Wheat–sorghum– fallow rotation	Sweeps, mulch tread	2.6	1.0	27	0
B11B	1984–1992		Sweeps, mulch tread	2.6	1.0	29	0
B12B	1984–1992		Sweeps, mulch tread	2.6	1.0	23	0
El Reno, OK							
E1	1977–1992	Native grass	None	1.6	2.6	13	0
E2	1977–1992	Native grass	None	1.6	2.9	15	0
E3	1977–1992	Native grass	None	1.6	3.2	14	0
E4	1977–1992	Native grass	None	1.6	3.6	15	0
E5	1984–1992	Wheat–sorghum	Moldboard, disk	1.6	3.5	22	12
E6	1979–1992	Wheat	Moldboard, disk	1.6	2.9	32	12
E7	1984–1992	Wheat	No till	1.6	2.9	38	13
E8	1979–1992	Wheat	Sweeps, disk	1.6	2.7	21	13
Ft Cobb, OK							
C1	1982–1992	Peanuts–grain– sorghum rotation	Moldboard, disk	2.6	2.0	26	19
C2	1982–1992		Moldboard, disk	2.1	2.0	23	18
Riesel, TX							
Y	1976–1982	Mixed grass, cotton, oats, sorghum rotation	Disk	125.0	3.0	16	28
Y2	1976–1982		Disk	52.8	3.0	15	24
Y6	1976–1982	Cotton–oats– sorghum rotation	Disk	6.1	3.0	22	20
Y8	1976–1982		Disk	7.8	2.0	18	15
Y10	1976–1982		Disk	7.0	2.0	20	22
Y14	1976–1982	Klein grass	None	2.2	1.0	8	0
W10	1976–1982	Ctl. Bermudagrass	None	8.5	2.0	7	0
SW11	1976–1982	Wtgrn. Hardinggrass	None	1.1	1.0	4	0
Woodward, OK							
W1	1977–1992	Native grass	None	4.8	7.0	14	0
W2	1977–1992	Native grass	None	5.6	8.2	15	0
W3	1979–1986	Wheat	Sweeps, disk	2.7	8.6	29	23
W3	1987–1992	Native grass	None			22	23[b]
W4	1982–1986	Wheat	No till	2.9	7.4	41	23
W4	1987–1992	Native grass	None			30	23[b]

[a]Average soil P content during the study represented by Mehlich-3 extractable P.
[b]Native grasses received P fertilizer year of establishment only.

1991, 1992), and Smith *et al.* (1991a), so only a summary is given here. Generalized characteristics of the unit-source areas are shown in Table 12.2. These areas represent the variety of soils (fine sandy loams to clays), agricultural land uses, and P management for the Southern Plains of the USA. Surface soil samples (0–15 cm) were collected from each area to determine soil test P using the Mehlich 3 method (Mehlich 1984); this was then used to recommend fertilizer P applications rates. Fertilizer P was broadcast at autumn planting for both conventional and no-till wheat at El Reno and Woodward, and during harrowing in March at Ft Cobb. At the Riesel sites, P was applied in March prior to planting and disked to incorporate it.

Catchment runoff was measured using precalibrated flumes equipped with water-level recorders; 5 to 15 water quality samples were collected automatically as a function of stage during each runoff event. Samples were composited in proportion to flow rate to provide a single representative sample. All runoff samples were refrigerated at 4°C between collection and analysis, usually only a matter of days. Total P was determined by perchloric acid digestion of unfiltered samples (O'Connor and Syers 1975). Suspended sediment concentration of runoff was determined in duplicate, as the difference in weight of 250-ml aliquots of unfiltered and filtered (0.45 μm) runoff after evaporation (105°C) to dryness.

12.4.2 The Mixed-Land-Use Catchment

The PI was also evaluated for all individual fields within the 26-ha mixed-land-use Brown Catchment located in east-central Pennsylvania, USA. This area can be considered typical of small, first-order, upland agricultural catchments in the northeastern USA, and is located within the Susquehanna River Basin. Climate is temperate and humid; precipitation is approximately 1100 mm/year and streamflow about 450 mm/year. Soils on the catchment are all stony silt loams ranging in depth from 25 to 120 cm, but are hydrologically similar. Land use is almost entirely agricultural (Figure 12.1). Field geometry and associated land use was determined from a detailed survey conducted by the ARS Field Staff. There is a small amount of forest, and an unmanaged grass strip borders the lower portion of the stream channel. The typical crop rotations are variations of corn–small grain–hay, and about 35% of the catchment is in corn in any year. Bray P concentration of the soils, based on a 1-cm depth sample, ranged from 6 mg/kg under permanent grass and woods to 36 mg/kg in the intensively cropped areas. Individual field characteristics are summarized in Table 12.3.

Topographic data used in the study were derived from a USGS topographic map augmented by a local survey. Soil distribution and hydrologic characteristics were determined from a Natural Resource Conservation Service (formerly Soil Conservation Service, USDA) soil survey. Methods of hydrologic data collection (climate, rainfall and runoff) and associated sediment and water-quality sampling and analysis techniques have been presented in a number of references (e.g. Pionke *et al.* 1988, 1995: Zollweg *et al.* 1995); all follow commonly accepted methods and thus are not given here.

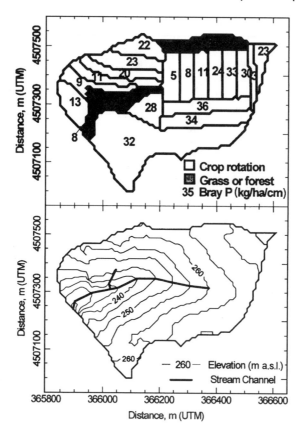

Figure 12.1   Brown catchment field distribution, generalized land use, and topography

Methodology for simulating the runoff, sediment and P source-areas of a catchment has been developed by modifying a VSA-based storm runoff model integrated within the GRASS GIS framework. The Soil Moisture-based Runoff Model (SMoRMod) (Zollweg 1994) has been shown to successfully simulate both the long-term daily hydrograph and storm hydrographs for small to large non-winter storms on small (<200 ha) New York and Pennsylvania catchments (Zollweg *et al.* 1995). SMoRMod is a physically based and spatially distributed model of catchment processes, using climatic variables as input, and requiring topography, land use, soils distribution and geology as GIS data layers. It divides the catchment into homogeneous small rectangular cells within which calculations are performed, and includes the hydrologic processes of infiltration, soil moisture redistribution, groundwater flow and surface runoff generation. Surface runoff generated is translated through the channel system to the catchment outlet to simulate formation of the storm runoff hydrograph. Erosion can be computed for each cell using the Universal Soil Loss Equation (Wischmeier and Smith 1978), and then transferred to the storm hydrograph directly using SMoRMod. Cells with no surface runoff are

Table 12.3 Brown catchment field characteristics and PI rankings

Field no.[a]	Land use	Area (ha)	Slope (%)	USLE (kg/ha/year)	Soil P[b] (mg/kg)	Fertilizer P (kg/ha/year)	P index rating
335	rotation	1.33	9	4230	9	17	5.5
337	rotation	2.67	11	13931	11	18	8
342	rotation	2.07	16	45260	20	23	18
346	woodland	4.03	6	440	6	0	2
347	rotation	2.01	16	30173	23	21	12
354	rotation	1.70	16	15087	23	18	9
358	rotation	0.87	18	18734	9	16	7
362	rotation	0.67	16	76988	13	24	17
363	rotation	2.01	12	20160	22	21	9
383	rotation	2.62	12	3024	28	17	7.5
364	rotation	0.43	15	11635	13	18	8
378	rotation	0.06	18	75714	8	24	16
372	rotation	1.73	10	1421	5	0	2.5
373	rotation	1.63	10	14210	8	24	7.5
375	rotation	2.16	9	13375	11	16	8
376	woodland	1.63	18	3029	6	0	2
377	rotation	2.06	9	6688	24	19	7.5
379	rotation	2.26	6	635	33	13	7.25
381	rotation	1.96	5	505	30	13	7.25
384	woodland	1.11	15	1939	6	0	2
385	rotation	2.45	8	9542	36	16	7.5
386	rotation	3.02	8	12319	34	19	9
500	woodland	0.48	3	100	23	0	4
501	rotation	0.97	3	612	23	0	4
504	rotation	0.27	3	1666	23	16	7.5
505	rotation	0.23	3	1666	23	16	7.5
506	rotation	0.47	3	2041	23	16	7.5
507	rotation	0.13	3	1666	23	16	7.5
508	rotation	0.13	3	1178	23	16	7.5
509	rotation	0.04	3	1414	23	16	7.5
510	rotation	0.00	3	1000	23	16	7.5
602	rotation	9.41	15	90022	32	16	18
603	rotation	5.03	4	5725	32	16	7.5
999	grass	4.71	6	314	6	0	7

[a]Field number is identification within the mapping scheme for land-use description.
[b]Bray P is kg/ha within top 1 cm of soil.

designated as noncontributing. USLE parameters are determined directly from maps and available data.

For simulating P output from the catchment, an additional data layer representing P availability over the catchment is required. SmoRMod-P can then combine P input from the land surface with the generated surface runoff and erosion values and route this mix to the catchment outlet using the same time-area approach used to produce the storm hydrograph. Dissolved P (DP, soluble only) in surface runoff was computed from Bray P for both cropland and forest/grassland soils (Figure 12.1) using the linear relationships presented in Daniel *et al.* (1994). These simplistic relationships provide a representation of the soil/runoff interaction appropriate to the initial stage of modelling being proposed. Subsurface flow, simulated by the long-term hydrologic sub-model of SMoRMod-P, dilutes storm runoff in the stream with an assigned concentration of 0.007 mg/l DP (Pionke *et al.* 1988). The DP simulation methodology was calibrated and tested against measured runoff and P response to storms on the Brown catchment. Initial simulation results compared favourably to observed P output data from a wide range of storms (Zollweg *et al.* 1995).

Sediment bioavailable P (BAP) was based on the enrichment ratio (ER) defined by Sharpley (1985):

$$\ln(ER) = 1.21 - 0.16 \ln \text{ (soil loss)}$$

and calculated directly as $ER \times Bray P \times soil loss$. Total bioavailable P delivered to the catchment outlet is then the sum of sediment BAP and DP.

12.5   RESULTS

12.5.1   Unit-Source Areas

When examining the basic data collected on the unit-source areas as shown in Table 12.4, average annual total P loss is found to be strongly related to sediment yield ($r = 0.95$; $p < 0.1$), but runoff alone is not related ($p > 0.05$) to total P loss (Figure 12.2). Thus, runoff production, through its effects on subsequent erosion, appears to be the dominant transport process controlling P loss from these areas.

The PI was first applied to all unit-source areas using runoff and erosion estimates by Curve Number and USLE techniques, respectively, as suggested in the original index formulation. When used in this way, only measurement of soil test P is required; the other index characteristics are obtained from soil survey information and site inspection. For this part of the study, soil test P ratings were based on potential crop response and fertilizer recommendations (Johnson and Tucker 1980). The very high category (> 65 mg/kg Mehlich 3 P) represents the level at which no yield response to fertilizer P is expected in the study area. PI values determined in this way are shown in Table 12.3 as "Original".

To better evaluate the importance of accurately estimating runoff and erosion when assessing potential for P loss in runoff, the PI was again applied to the unit-source areas, this time using measured, rather than estimated, values of runoff and sediment. For each area, calculation of the PI was optimized to measured P loss by including measured annual runoff and erosion (from Table 12.4) along with soil test P values and amount of fertilizer applied from Table 12.2. Details of this technique are presented in Sharpley (1995); the resulting PI values are shown in Table 12.4 as "Measured".

Vulnerability to P loss determined by the Original and the Measured PIs was related to total P loss as shown in Figure 12.3. The close relationship between PI rating by both techniques and total P loss,

$$\text{measured PI} = 11.73 + 9.09 \log \text{ (total P loss)}, \ r = 0.89$$

Table 12.4 Mean annual sediment, runoff and total P loss and P index rating value as a function of unit-source area management[a]

Site	Sediment (kg/ha/ year)	USLE (kg/ ha/year)	Runoff (cm)	Total P (g/ha/ year)	P index rating[b] Measured	Original	Sediment based	Runoff based
Grass								
E1	39	1258	11.6	210	5.3	3.0	3.0	5.3
E2	27	1422	9.5	210	5.1	3.0	3.0	5.1
E3	35	1567	10.0	195	5.0	3.0	2.9	5.1
E4	40	1770	9.0	283	5.5	3.0	2.8	4.7
W1	33	2113	0.6	27	2.5	3.0	2.4	3.2
W2	465	3284	1.3	202	4.1	3.0	3.9	3.2
W3	115	2930	1.4	122	4.3	4.5	4.7	4.9
W4	153	2412	1.6	99	4.6	4.5	5.1	4.7
Y14	603	477	15.2	345	4.6	3.0	4.3	4.7
W10	74	1041	11.2	217	4.3	3.0	3.3	4.0
SW11	870	401	13.1	516	4.6	3.0	3.3	4.3
Average	223a	1700a	7.7ab	221a	4.5a	3.3a	3.5a	4.5a
No till								
B10A	213	1381	2.6	224	5.8	4.5	5.8	4.4
B11A	440	1381	3.2	382	7.5	4.5	7.3	4.7
B112A	911	1381	5.5	634	8.5	4.5	7.7	5.3
E7	375	7112	10.3	1128	11.3	6.8	8.5	8.8
W4	883	8041	6.5	1262	12.9	9.5	12.2	10.7
Average	564b	3589ab	5.6ab	726b	9.2ab	6.0ab	8.3bc	6.8ab
Reduced till								
B10B	617	1332	1.7	423	5.6	4.5	6.0	4.1
B11B	790	1332	2.3	438	7.6	4.5	7.7	4.4
B12B	2418	1332	2.0	541	8.0	4.5	8.3	4.3
Average	1275b	1332a	2.0a	467b	7.1a	4.5a	7.3b	4.2a
Conventional till								
E5	2250	8589	10.5	1534	12.6	6.8	10.1	8.7
E6	5341	7112	9.8	2967	16.9	6.8	11.6	8.9
E8	2876	6649	8.6	1945	14.9	6.8	10.1	8.8
W3	11738	14650	4.3	4617	16.4	9.0	12.0	9.9
Y	1599	8740	11.4	671	13.9	7.5	11.6	9.0
Y2	901	7046	12.6	860	13.4	7.5	11.6	8.7
Y6	2083	6846	8.3	1105	14.9	8.5	13.5	9.5
Y8	1363	5095	9.6	883	13.7	7.5	12.0	8.7
Y10	4015	4959	14.4	2458	16.5	7.5	13.0	9.0
C1	14351	12444	12.2	4309	21.4	9.0	16.9	11.4
C2	19016	11797	10.3	5961	20.9	9.0	16.4	11.3
Average	5867c	8539b	10.2b	2528c	16.0b	7.8bc	12.6c	9.4b

[a]Means followed by different letters are significantly different ($p < 0.05$) between specific management types as determined by analysis of variance.
[b]Measured index used measured runoff and erosion; original index used curve number and USLE estimates of runoff and erosion; sediment-based index used measured erosion and curve number runoff; runoff-based index used measured runoff and ULSE erosion.

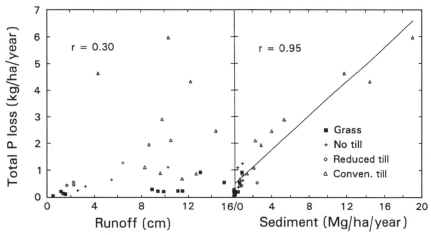

Figure 12.2   Relationship between total P and runoff and sediment discharge from the unit-source areas in Oklahoma and Texas

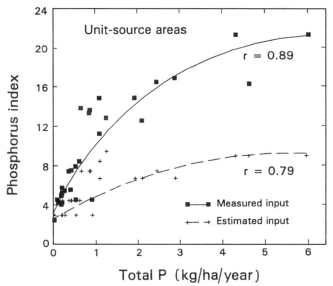

Figure 12.3   Relationship between P index rating of vulnerability to P loss in runoff from the unit-source areas using measured and estimated runoff (curve number) and erosion (USLE) inputs and measured total P loss in runoff

original PI $= 6.22 + 3.31$ log (total P loss), $r = 0.79$

indicates that the PI technique can give reliable estimates of site vulnerability to P loss in runoff from areas of widely differing management. Measured values make the estimates somewhat more consistent with observed P losses.

Two additional PIs were developed for each site. One was optimized using measured sediment values and estimated runoff (sediment-based; Table 12.4), while the other was optimized based on measured runoff values and estimated sediment (runoff-based). When measured runoff values were used with estimated sediment, resulting PI values for the grass and no-till unit-source areas were not significantly different ($p < 0.05$) from PI values determined using both measured runoff and erosion. However, when measured sediment and estimated runoff was used to develop PIs for these same areas, PI values were consistently less than those based on both measured values. In contrast, for the reduced till and conventional till areas, PI values using measured runoff along with USLE erosion (runoff-based; Table 12.4) were consistently less than those based on all measured values. PIs developed from measured sediment and estimated runoff more closely matched those values derived from measured runoff *and* sediment. The greater influence of erosion than runoff on P-loss vulnerability for tilled areas, and of runoff than erosion for the grassed and no-till areas is consistent with the relative amounts of dissolved and sediment-bound P transported in runoff from these areas. Over the 16 years studied, approximately 60% of the P transported from the grassed and no till areas was dissolved in runoff (Sharpley *et al.* 1992). In contrast, 90% of P lost from the reduced and conventionally tilled areas was sediment bound.

12.5.2   The Mixed-Land-Use Catchment

Clearly, runoff generation and erosion are critical factors controlling site vulnerability to P loss from the unit-source areas, and we might expect this finding to be extrapolated to other unit-source areas, either alone or within a mixed-land-use catchment. However, as previously mentioned, the impact of management strategies applied to individual fields will likely be evaluated at the catchment scale. As site hydrology and management increase in complexity, the runoff and erosion processes controlling P loss become more dynamic, interactive, and spatially and temporally variable. Thus, it is essential that we be able to integrate the output of individual source-area vulnerabilities within catchment-scale flow and sediment transport systems. This problem is considered in the following section.

Figure 12.4 shows the distribution of Bray P over the Brown catchment on a field-by-field basis, USLE values, and finally, the PI for every field. Table 12.3 and Figure 12.4 show that the PIs of individual fields within the Brown catchment are strongly influenced by both the Bray P and the USLE values. Runoff classes are almost the same over the watershed because of the hydrologic similarity of the soils, and likewise, fertilization rates do not vary much, except between cropped and non-cropped areas. In total, this would infer that management options should be focused on reducing erosion and controlling P build-up in those fields with high Bray P. However, as discussed in the unit-source section, flow components important to the question of P export from a catchment are both surface runoff and its associated water-borne sediment. VSA hydrology implies that these do not occur uniformly over the catchment, so runoff determined via runoff classes and erosion from the USLE technique may not tell the entire story. That is why we resort to use of a model such as SmoRMod-P to determine source-areas of runoff, sediment and associated P loss from the catchment for comparison to individual field-based PI values.

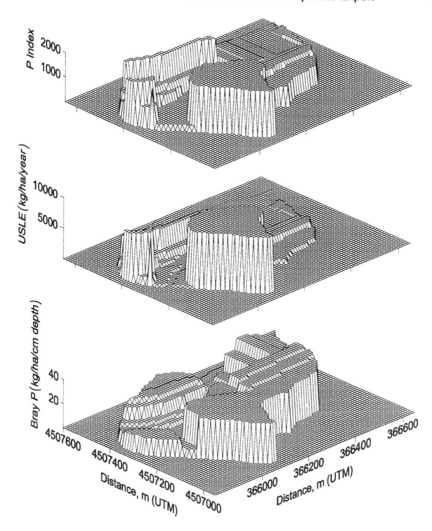

Figure 12.4   P Index, USLE, and Bray P distributions for the Brown catchment

Surface runoff, erosion, DP and sediment BAP source-areas were delineated over the Brown catchment by applying SMorMod-P to a 21-mm storm from April 1992 (Figures 12.5 and 12.6). For a different size storm under different initial conditions, runoff yields will change, and the areal extent of the surface runoff response might expand (larger storm) or contract (smaller storm); however, based on the results of field research recently begun on the Brown catchment, the general patterns of surface runoff, erosion, and P loss are not expected to vary significantly from storm to storm. These erosion and sediment BAP simulations have not yet been rigorously compared against observed data and, thus, are used here only as examples.

Results show that surface runoff and erosion are generated only from very limited areas within the Brown catchment (Figure 12.5). Most surface runoff occurs near the

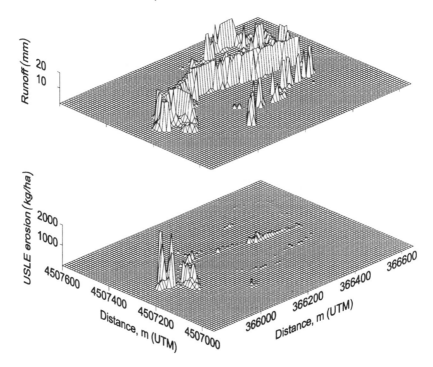

Figure 12.5 Surface runoff and erosion (USLE) simulated for the April 1995 storm on the Brown catchment

stream channel with 98% of the volume being produced by about 14% of the catchment area. Some surface runoff is produced from areas located along slope breaks that parallel the channel. Basically this is a VSA-driven hydrologic system where the location of surface runoff is controlled by catchment (water-table position) rather than soil (infiltration) related processes. The erosion source-areas are even more limited, located primarily in the lower portions of the catchment where a cropped field intersects the hydrologically active area. Very little erosion occurs elsewhere within the surface runoff source-areas because of the excellent cover; contrast this to the plot of USLE values over the watershed (Figure 12.4).

The source-areas of surface runoff and erosion establish the outer boundaries for the source-areas of DP and sediment BAP loss (Figure 12.6). Within these limits, most of the DP is lost from cropland where Bray P concentrations in soil are highest, near the upper channel and along the hydrologically active slope break. In contrast, the much lower Bray P values associated with the permanently grassed area around the lower channel counteract the effect of the high surface runoff volumes on DP export. Sources of sediment BAP are even more localized, occurring primarily in cropland near the lower channel because of the high erosion rates. When the DP and sediment BAP are combined as total BAP loss (Figure 12.6), there are only two areas that contribute primarily to this loss. A limited zone along the lower channel reach provides the sediment-based BAP, while a similar zone along the upper end of the

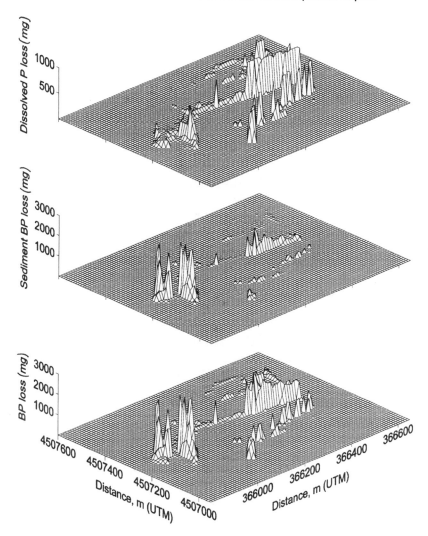

Figure 12.6   Dissolved P loss, sediment bioavailable P loss, and total bioavailable P loss simulated for the April 1992 storm on the Brown catchment

channel provides the DP. Based on the simulations, DP (68 g) and sediment BAP (77 g) losses are about equal. In terms of area, 98% of the sediment BAP loss is from 6% of the catchment, and most of the DP loss is from 11% of the catchment. In total, about 20–30% of the BAP is produced by approximately 1% of the catchment area.

This analysis shows that a very small portion of the catchment is the source of most of the exported BAP. It also shows where these BAP losses are concentrated within the catchment. The findings are at odds with application of the PI to the individual fields over the catchment. Certainly where there are high BAP values indicated by the simulation approach, there are relatively high PI values in Figure 12.5. However, the CSAs of runoff and erosion mitigate the higher PI values over much of the catchment, only "activating" these high PI values in limited portions of the catchment. With the simulation-based approach, it is possible to better target research, monitoring, and remediation programmes, and more realistically explore the impact of alternative land use and P management options.

12.6 DISCUSSION

As originally developed, the PI is to be applied in field situations using estimated runoff and sediment yield. Like any simulation, reliability of its output is dependent on both representativeness of the model formulation and accuracy of inputs. In the unit-source portion of this study, observed runoff and sediment yield were used to facilitate a better test of the P index by eliminating the shortcomings in runoff or sediment yield estimation. However, the PI applied to these areas in its original format also did a good job of ranking site vulnerability to P loss. Further, the range of soil test P categories was based on crop yield response in the geographic area of concern, so the magnitude of site vulnerability indicated by the PI depends on the range of soil test P values used. This implies that reliable soil test P values for each category be established for specific areas based on regional agronomic, economic and environmental concerns. Clearly, soil test P values must be categorized at regional levels to ensure index flexibility in accounting for the fact that P-related water quality problems occur in specific areas rather than being nation-wide.

Currently, the index assesses specific site vulnerability to P loss using annual erosion and runoff losses and average soil test P values. This approach may not accurately represent site vulnerability in specific cases. For example, up to 90% of annual soil, runoff and P loss can occur in only one or two intensive storms during a year (Edwards and Owens 1991; Smith et al. 1991b). If climatic, topographic and agricultural management factors are such that these losses occur during 15–20 runoff events spread over a year, site vulnerability to P loss would not be as great as similar losses from only one storm.

In addition, modification of the index to separately address DP may be warranted in situations with high soil P. For example, even if erosion is minimal, elevated surface soil P can support total P losses in runoff in excess of 5 kg/ha, of which over 95% is DP (Edwards and Daniel 1993). Further, as DP is immediately available for algal uptake, the loss of total P predominantly as DP will have a greater impact on the biological productivity of P-sensitive surface waters than will particulate or sediment-bound P. Thus, the index might be modified to describe the primary form of P transported in runoff as well as the total amount.

In areas where surface waters have been determined to be biologically sensitive to P inputs, field-scale vulnerability to P loss may be used to assess farm management options in terms of soil testing, soil conservation and P management (Sims 1993). As vulnerability increases from low to very high, the frequency of soil P testing should increase from every three years to every year, and estimates of the time required to deplete soil P to optimum levels should be made. Basic soil conservation measures should also be encouraged, but as site vulnerability approaches medium or greater levels, more extensive measures, such as reduced tillage, buffer strips, grassed waterways, and the use of high P demand crops, should be considered. Any changes in P management that may affect P loss in runoff should be considered before implementation. Also, development of a long-term P management plan should be encouraged for sites of high vulnerability to P loss and be required to be implemented on sites of very high vulnerability (Sims 1993).

The results of the unit-source study show that the PI may be a valuable tool to identify individual sites *within* a catchment that are *potential* sources of P vulnerable to loss in runoff. Based on this, it is assumed that the PI will identify and reliably assess management options available to land users that will allow them flexibility in targeting control strategies that minimize P loss, while maintaining crop productivity. However, the analyses done within context of the small mixed-land-use catchment illustrates the potential problems with determining management options by applying the PI to individual fields within the catchment or mixed-land-use scale, if reduction of P loss from the entire catchment is the concern. While it may have been the original intent that the PI not be applied in this way, there is little doubt that if it is shown to be successful when applied to individual fields, attempts will be made to extend it to mixed-land-use conditions within small catchments. This same scenario has occurred in the cases of tools with similar limitations, the Curve Number runoff prediction technique and the USLE. Thus, results of examination of the PI in context of the catchment are instructive.

Typically, small identifiable areas within the catchment will be the source of most BAP exported. The base control for these BAP critical source-areas is where surface runoff and erosion occur. In VSA-controlled catchments, the location of surface runoff source-areas is predictable and follows a set pattern, even when hydrologic conditions change. If BAP loss from the catchment is a major issue, the CSAs of runoff, erosion and P loss can be identified and targeted for remediation. But such identification will require modification of the PI to better reflect actual, rather than potential, runoff and erosion. Finally, knowing where surface runoff is least likely to occur can also be important for development of land-management alternatives. For instance, these areas might be the best areas for land-based disposal of manure, although enhanced loading of N in these areas must also be considered.

Analogous to identification of these CSAs, there may also be critical time or storm periods where relatively little time or few storms account for most of the BAP lost from the catchment. Using nine years of data from the larger 7.4-km^2 Mahantango Creek Research catchment, which contains the Brown catchment, Pionke *et al.* (1995) found that nearly 70% of the DP exported occurred during the 10% of the time dominated by the larger runoff events. This increased to about 90% BAP exported when sediment BAP was included. Sixty-two storm events, averaging seven per year, were the dominant controls on BAP loss, and the remedial implications of such a time period or event class being critical are several. First, we can probably ignore base flow and the sources of small storm events, which together export most water, but very little of the BAP. Secondly, we may be able to develop design or index storms to represent the range in size, intensity and initial conditions of these larger controlling events. The index storms can be used to do simulations to establish the location and dimensions of CSAs under these conditions, and perhaps be used as a basis to develop modifications to the PI technique.

In summary, relatively few processes and parameters control BAP export at the catchment scale, and even fewer are efficiently or readily manageable. The amount and location of surface runoff is not typically nor readily managed on site. Erosion is more manageable, particularly by manipulating cover directly or indirectly through a conservation practice, such as modified tillage. Dissolved P export can be managed though, by controlling P fertility level in surface runoff zones, and sediment BP is managed by controlling erosion and/or P fertility level in the primary erosion zones. Thus, if a remedial control strategy was needed for the Brown catchment, it would be to reduce the P soil fertility levels in the surface runoff source-areas.

12.7 SUMMARY

We remain far removed from development of comprehensive fertilizer management and animal manure application strategies which account for P loss from fields and/or catchments by considering all hydrologic implications. Modelling tools and field data are simply not available to integrate all aspects of the hydrologic cycle from the flow perspective alone, much less from that of water quality. However, we can draw conclusions based on what has been found.

Application of the PI to the unit-source areas shows that the index has potential to identify vulnerability to P loss when comparing site against site, providing one is concerned simply with *ranking* outputs from the individual sites. However, within a mixed-land-use catchment, hydrologic implications are that to control P loss, we must control P application and soil P build-up primarily in the runoff- and erosion-producing zones. These are typically the near-stream zones for the Brown catchment considered, and are not readily reflected by application of the PI in its present form. Levels in the soil at greater distances from the stream are of less concern as there is less chance of runoff occurring at these sites to move the P to the stream. Thus, the most obvious management option for P is to use less rigorous guidelines for fertilizer or manure P application on landscape positions at distance from the stream where runoff potential is low compared to the most vulnerable sites adjacent to the channel.

Whether in the unit-source or mixed-land-use catchment context, P management strategies must account for interactions between areas of nutrient availability over the landscape and areas of potential surface runoff and associated sediment transport. We may apply a tool as simple as the PI in its present form when comparing field to field in a more straightforward situation such as the unit-source areas. However, when concerned with a mixed-land-use catchment, such as the Brown catchment, we must either resort to other more sophisticated techniques, or modify the Phosphorus Index to provide an accurate portrayal of the hydrologic flow system controlling P loss from the catchment.

ACKNOWLEDGEMENTS

Contribution from the Pasture Systems and Watershed Management Research Laboratory, US Department of Agriculture, Agricultural Research Service, in co-operation with the

Pennsylvania Agricultural Experiment Station, the Pennsylvania State University, University Park, Pennsylvania, USA.

REFERENCES

Chichester, F.W. and Richardson, C.W. 1992. Sediment and nutrient loss from clay soils as affected by tillage. *J. Environ. Qual.*, **21**, 587–590.

Daniel, T.C., Sharpley, A.N., Edwards, D.R., Wedepohl, R. and Lemunyon, J.L. 1994. Minimizing surface water eutrophication from agriculture by phosphorus management. In Nutrient Management, supplement to *J. Soil and Water Conser.*, **49**, 30–38.

Edwards, D.R. and Daniel, T.C. 1993. Effects of poultry litter application rate and rainfall intensity on quality of runoff from fescuegrass plots. *J. Environ. Qual.*, **22**, 361–365.

Edwards, W.M. and Owens, L.B. 1991. Large storm effects on total soil erosion. *J. Soil Water Conserv.*, **46**, 75–77.

Heatwole, C.D., Bottcher, A.B. and Baldwin, L.B. 1987. Modeling cost-effectiveness of agricultural nonpoint pollution abatement programs in two Florida basins. *Water Res. Bull.*, **23**, 127–131.

Johnson, G. and Tucker, B. 1980. *Oklahoma State University Soil Test Calibrations*. Okla. State Univ. Ext. Serv. Fact Sheet 2225, Oklahoma State University, Stillwater, OK.

Lemunyon, J.L. and Gilbert, R.G. 1993. The concept and need for a phosphorus assessment tool. *J. Prod. Agric.*, **6**, 483–496.

Mehlich, A. 1984. Mehlich 3 soil test extractant: A modification of Mehlich 2 extractant. Commun. *Soil Sci. Plant Anal.*, **15**, 1409–1416.

O'Connor, P.W. and Syers, J.K. 1975. Comparison of methods for the determination of total phosphorus in waters containing particulate material. *J. Environ. Qual.*, **4**, 347–350.

Pionke, H.B. and Lowrance, R.R. 1991. Fate of nitrate in subsurface waters. In R.F. Follett, D.R. Keeney and R.M. Cruse (eds), *Managing Nitrogen for Groundwater Quality and Farm Profitability*. Soil Sci. Soc. Am. Special Publication, pp. 237–257.

Pionke, H.B., Hoover, J.R., Schnabel, R.R., Gburek, W.J., Urban, J.B. and Rogowski, A.S. 1988. Chemical–hydrologic interactions in the near-stream zone. *Water Resour. Res.*, **24**, 1101–1110.

Pionke, H.B., Gburek, W.J., Sharpley, A.N. and Schnabel, R.R. 1996. Flow and nutrient export patterns for an agricultural hill-land watershed. *Water Resour. Res.*, **32**, 1795–1804.

Prato, T. and Wu, S. 1991. Erosion, sediment, and economic effects of conservation compliance in an agricultural watershed. *J. Soil Water Conserv.*, **46**, 211–214.

Ryden, J.C., Syers, J.K. and Harris, R.F. 1973. Phosphorus in runoff and streams. *Adv. Agron.*, **25**, 1–45.

Sharpley, A.N. 1985. The selective erosion of plant nutrients in runoff. *Soil Sci. Soc. Am. J.*, **49**, 1527–1534.

Sharpley, A.N. 1995. Identifying sites vulnerable to phosphorus loss in agricultural runoff. *J. Environ. Qual.*, **24**, 947–951.

Sharpley, A.N., Smith, S.J., Berg, W.A. and Williams, J.R. 1985. Nutrient runoff losses predicted by annual and monthly soil sampling. *J. Environ. Qual.*, **14**, 354–260.

Sharpley, A.N., Smith, S.J., Williams, J.R., Jones, O.R. and Coleman, G.A. 1991. Water quality impacts associated with sorghum culture in the Southern Plains. *J. Environ. Qual.*, **20**, 239–244.

Sharpley, A.N., Smith, S.J., Jones, O.R., Berg, W.A. and Coleman, G.A. 1992. The transport of bioavailable phosphorus in agricultural runoff. *J. Environ. Qual.*, **21**, 30–35.

Sharpley, A.N., Chapra, S.C., Wedepohl, R., Sims, J.T., Daniel, T.C. and Reddy, K.R. 1994. Managing agricultural phosphorus for the protection of surface waters: Issues and options. *J. Environ. Qual.*, **23**, 437–451.

Sims, J.T. 1993. *The Phosphorus Index: A Phosphorus Management Strategy for Delaware's Agricultural Soils*. Dept. Plant and Soil Science, Univ. Delaware, Newark, DE.

Smith, S.J., Sharpley, A.N., Naney, J.W., Berg, W.A. and Jones, O.R. 1991a. Water quality impacts associated with wheat culture in the Southern Plains. *J. Environ. Qual.*, **20**, 244–249.

Smith, S.J., Sharpley, A.N., Williams, J.R., Berg, W.A. and Coleman, G.A. 1991b. Sediment-nutrient transport during severe storms. In S.S. Fan and Y.H. Kuo (eds), *Fifth Interagency Sedimentation Conf.*, March 1991, Las Vegas NV. Federal Energy Regulatory Commission, Washington, DC, pp. 48–55.

Stevens, R.G., Sobecki, T.M. and Spofford, T.L. 1993. Using the phosphorus assessment tool in the field. *J. Prod. Agric.*, **6**, 487–492.

Ward, R.C. 1984. On the response to precipitation of headwater streams in humid areas. *J. Hydrol.*, **74**, 171–189.

Wischmeier, W.H. and Smith, D.D. 1978. *Predicting Rainfall Erosion Losses – A Guide to Conservation Planning*. US Dept. Agr. Handbook 537.

Zollweg, J.A. 1994. Effective use of geographic information systems for rainfall–runoff modeling. *PhD Dissertation*, Cornell Univ., Ithaca, NY.

Zollweg, J.A., Gburek, W.J., Pionke, H.B. and Sharpley, A.N. 1995. GIS-based delineation of source areas of phosphorus within agricultural watersheds of the northeastern USA. *Modelling and Management of Sustainable Basin-Scale Water Resource Systems*, IAHS Publication No. 231, pp. 31–39.

13 Pathways and Forms of Phosphorus Losses from Grazed Grassland Hillslopes

P. M. HAYGARTH and S. C. JARVIS
Institute of Grassland and Environmental Research, Okehampton, UK

13.1 INTRODUCTION

It is a commonly held assumption that intensive agricultural land use causes increased discharge of phosphorus (P) into surface water systems, yet this is based on a small amount of substantive, published information (e.g. Roberts *et al.* 1986; Sharpley and Smith 1989; Heathwaite *et al.* 1990; Kronvang 1990). Grassland agricultural systems typically receive inputs of 10–40 kg/ha P per year, and this input is enhanced by recycling in sheep and cattle dung and returns of farmyard manure, slurry and dirty waters generated by cattle systems (e.g. Jordan and Smith 1985). Grasslands cover over 70% of the UK land area (Waters 1994), and despite the fact that grasslands easily lend themselves to studies at a hillslope scale, often occurring on sloping ground and in areas of higher and intense rainfall, the processes, forms and pathways controlling P movement are not well understood.

Many factors play a role in influencing the loss of P from grassland soils. A non-exhaustive list may be

1. drainage status (natural and artificial) (e.g. Armstrong and Garwood 1991);
2. fertilizer management, particularly amount and timing (e.g. Haygarth and Jarvis 1996);
3. farm waste management and, in particular, timing of amendments (e.g. Jordan and Smith 1985);
4. intensity and timing of grazing (e.g. after Harrison 1985);
5. amount and intensity of rainfall (e.g. Haygarth and Jarvis 1996);
6. soil texture, contributing to the soil hydrological classification (e.g. Boorman *et al.* 1995);
7. soil P status and sorption characteristics (e.g. Sharpley and Smith 1989).

Additional substantive data on any of these would contribute significantly to understanding the problem, and in this chapter we describe new information which contributes to the first three. The work was based on an experimental design known as The Rowden Experiment, comprising 14 1-ha grazed grassland lysimeters,

Advances in Hillslope Processes, Volume 1. Edited by M. G. Anderson and S. M. Brooks.

managed in order to study the effect of soil drainage, coincident with regular fertilizer, slurry and animal management practices on nutrient transfer.

13.2 METHODOLOGY

The Rowden drainage experiment (at North Wyke, Devon, southwest England, 7 km north of Dartmoor, SX 650995) involves a number of nitrogen and drainage treatments, and was established in 1982 on old, unimproved pasture, with sloping land (5% to 10%) (Scholefield et al. 1993). The soil is a clayey non-calcareous pelostagnogley of the Hallsworth Series (USDA typic haplaquept, FAO dystric gleysol), overlying clay shales of the Crackington Formation. The soil has been classified (Boorman et al. 1995) by the hydrology of soil types (HOST) as class 24 (model J) (base flow index 0.311), representing the single most common hydrologic soil type in England, Scotland and Wales at 13.9% of the land area. The lysimeters are each approximately 1 ha, and each is hydrologically isolated from its neighbour.

There are 14 lysimeters; seven have artificial drainage and seven are undrained (Figure 13.1). Only a single hydrological pathway is sampled from the undrained lysimeters, being that which is collected in gravel-filled ditches to 30 cm at the downslope boundaries of each plot, thus incorporating surface plus lateral flow to 30 cm depth. Drained lysimeters are regularly mole drained, with pipe drains at 85 cm depth (Armstrong and Garwood 1991). Two pathways are sampled from the drained lysimeters: that collected in gravel-filled ditches to 30 cm (as in the undrained plot) and that collected from the pipe drains at 85 cm. Therefore, there were 21 collection points in total, and drainage waters were channelled through continuous level recording reservoirs and then through V-notch weirs, where discharges were continuously gauged (described by Talman 1983; Armstrong and Garwood 1991).

On all the lysimeters, the grass swards (dominated by Lolium perenne L.) were grazed with 4 steers/ha during April–October. Although the nutrient inputs to the lysimeters were managed for N studies (Scholefield et al. 1993), P inputs were also incorporated in the form of triple super phosphate (TSP) fertilizer, recycling from cattle excreta during grazing and in some cases from slurry additions. Twelve out of 14 of the lysimeters received 16 kg P in a single application during May, and slurry applications (equivalent to a mean annual application of 16 kg total P/ha). These P inputs are in accordance with standard management practices for grassland soils in the UK. The remaining two lysimeters (one drained, one undrained) were grazed but received no fertilizer or slurry amendments. Total P concentration of the lysimeter soils was c. 540 mg/kg (0–20 cm depth). Olsen's bicarbonate extractable P was determined in the 0–20 cm soil horizon and had a mean value of 5.3 ($\pm$0.6) mg/kg on all the fertilized lysimeters, and 4.0 ($\pm$1.3) mg/kg where there had been no P additions. Both of these values are small compared to other UK agricultural soils, although it is acknowledged that the value may have been larger if the sampling represented the 0–7.5 cm horizon following more conventional sampling methods for grassland (Ministry of Agriculture, Fisheries and Food 1984).

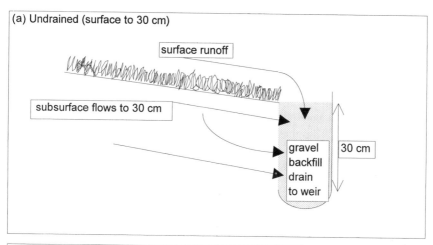

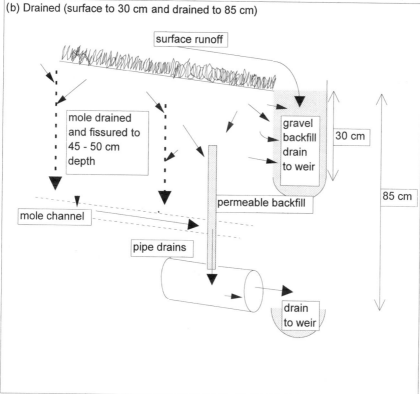

Figure 13.1 Schematic illustration of water pathways sampled from the Rowden experiment (Haygarth and Jarvis, unpublished)

During 1993 there were 193 days when field capacity was exceeded, and an excess discharge of 890 mm was generated from a total annual rainfall of 1310 mm. The overall magnitude of water flows from the undrained lysimeters (i.e. surface to 30 cm pathway only) were approximately equal to those from the drained lysimeters (i.e. surface to 30 cm + drained to 85 cm). Samples were collected and analysed for MRP on 10 occasions from all the 21 drain pathways (14 lysimeters), spanning a range of discharge conditions (incorporating base and storm flows) from 0 to 4.5 l/s, as measured on the weirs and for all pathways. During 8 of the 10 events, samples were analysed for TP and during 6 of the 10 occasions the samples underwent full fractionation of all the phosphorus species. The sampling strategy aimed to capture a range of flow events throughout the year, but was also timed to capture discharges which followed surface amendments of slurry or fertilizer. After sampling, waters were rapidly transferred to the laboratory in 150-ml polyethylene vessels and, following shaking, subjected to a series of filtrations and analytical determinations which separated the operationally defined species, the sum of which comprised the total phosphorus fraction, as summarized in Figure 13.2 (see also list of abbreviations). For filtration, 0.45 μm Millepore HA composite (cellulose–nitrate/acetate) membrane filters were used, which had been preconditioned by washing with 20 ml of deionized water, followed by 20 ml of sample, both of which were discarded. A 50 ml sample was then passed through the filter and the filtrate retained for TDP and MRP determination. All plastic and glassware which came into contact with the sample had been soaked for 16 h in 10% v/v sulphuric acid and rinsed six times with deionized water. The preparation protocol was that defined previously by Haygarth et al. (1995) as optimal in order to minimize transformations between labile species during sample storage.

MRP and UMRP were determined with a Tecator 5020 flow injection analyser with an autosampler (Tecator Ltd., Method Application ASN 60–03/83) using a molybdenum blue reaction. Colour development was determined at 690 nm wavelength, following reduction with stannous chloride and reaction with ammonium molybdate. The instrumental limit of detection for this method was 10 μg/l. TP and TDP were determined on the unfiltered waters, after a method optimized from Eisenreich et al. (1975). A 20 ml unfiltered sample was pipetted into a 60 ml Teflon vessel, and mixed with 0.15 g of $K_2S_2O_8$ and 1.0 ml of 1.01 N H_2SO_4, and, with the vessel lid untightened, the sample was autoclaved at 121°C and 1.4 bar for 1 h. Colour was determined spectrophotometrically at 880 nm after formation of the molybdenum blue complex in a reaction using a blend of acid–antimony–molybdate reagent and L-ascorbic acid, optimized from the Murphy and Riley (1962) procedure. Calibration standards were treated as samples to compensate for the digestion matrix. For all methods of determination, calibration was over the range 0–450 μg/l, using six standards which were prepared on each day of analysis. Field blanks were prepared during every sample event and blank sample vessels were taken through the entire protocol, with appropriate washing, sampling, filtering and determination. Quality control was maintained by using routine determinations of unknown amounts of MRP and TP, both as part of an in-house QC programme and a wider sample exchange scheme with other laboratories in the UK. Only those analytical batches where QC samples fell within ± 5% of the known value were accepted.

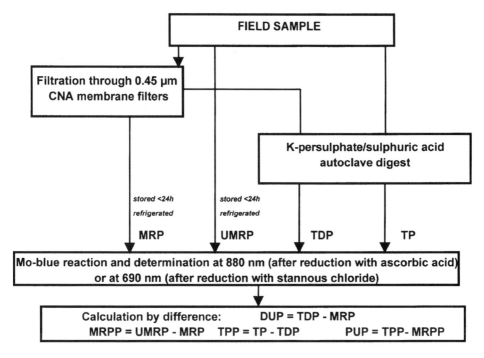

Figure 13.2 Operationally defined procedure for P fractionation in discharge waters (● also see list of abbreviations) (Haygarth and Jarvis, unpublished)

13.3 RESULTS AND DISCUSSION

13.3.1 Effect of Drainage on Pathways and Forms of P Loss

Complete fractionation of P was undertaken on six events and a summary of the data is presented in Table 13.1 (Haygarth and Jarvis, preliminary unpublished data). Standard errors, which reflect true field inter-lysimeter variability, are high and this should be taken into consideration in data interpretation. This relatively high degree of field variability clearly contributed to the small number of statistically significant differences that were found: only the MRPP and the PUP fractions were significantly different ($p < 0.05$) when comparing the drained and undrained surface to 30 cm pathways. Another limitation is that six to ten is only a small number of events, reflecting a limitation of the sampling design used. As a result the data may reflect bias, for example towards those events which follow fertilizer and slurry amendments. It should also be noted that these concentration data do not reflect water discharge and thus loadings of P (loadings are not addressed in this chapter).

Concentrations of P in the drained (surface to 30 cm) pathway were consistently higher than the other pathways for all forms of P, with a mean loss of 232 μg/l TP being determined. In drained plots the majority of water passed into mole and

Table 13.1 Concentration of P species in different drainage pathways from Rowden grassland lysimeters, mean for seven lysimeters over six events during 1994 (μg/l) (Haygarth and Jarvis, unpublished)

		MRP	DUP	TDP	MRPP	PUP	TPP	TP
Undrained soil	mean	60	44	105	23	24	47	152
surface to	range	10–1296	10–447	18–1743	10–63	10–81	10–104	26–1773
30 cm	standard error	32	11	42	2	4	5	42
Drained soil	mean	116	43	160	39	34	72	232
surface to	range	10–597	10–179	10–308	10–130	10–259	10–308	10–892
30 cm	standard error	37	8	43	7	12	15	54
30 cm to	mean	41	35	76	26	30	56	132
85 cm	range	10–304	10–165	10–355	10–145	10–196	10–253	10–605
	standard error	9	6	13	5	8	12	24

Surface to 30 cm pathways were compared using a t-test on drained and undrained plots. Concentrations in the drained plot were significantly higher for TPP (p=0.036, t=1.822, one tailed hypothesis only) and MRPP was significantly different between the drained and undrained (p=0.013, t=2.5, two tailed hypothesis).

drainage pipes, thus a relatively small proportion of water remained for discharge by the surface pathways. This apparently resulted in higher concentrations removed in the surface to 30 cm pathway, presumably as a combined result of (i) high surface soil concentrations of P and (ii) small volumes of water leaving by the surface pathways (as a result of drainage).

Although the highest mean concentrations were recorded on the surface pathways of the drained plots, the highest *maximum* values were recorded in the surface to 30 cm pathway on the *undrained* lysimeters, where concentrations were up to 1773 μg TP/l in the extreme. Clearly, where land is not drained there is a greater *range* of concentrations, perhaps reflecting the importance of single major events, such as surface fertilizer or slurry removal during intense rainfall shortly after applications (see the following section).

In contrast, the mean and maximum concentrations of TP were lower when P was removed from drained land (to 85 cm depth). Although drainage clearly creates an additional pathway for P loss, the opportunity for entrainment of colloidal and particulate P, or dissolution of soluble P and inorganic and organic P forms may be relatively restricted when water passes directly to drains. These processes are probably further avoided because of the speed of water flux, which is probably very rapid as mole drainage enhances preferential pathways of water movement.

Figure 13.3 illustrates the relative proportions of the P fractions in the three pathways. The ratio of particulate : dissolved P remained equal in both undrained and drained (surface to 30 cm) pathways, with the dissolved fraction contributing

69% of total P, predominantly in inorganic form as MRP. In the drained (to 85 cm) pathway, the dissolved fraction was a smaller proportion of the total P, and the dissolved organic components comprised a greater proportion of the total and total disolved P. This is supporting evidence that dissolution of inorganic P was restricted when water travels rapidly to the mole drains, and suggests that water flow along these specific pathways has the effect of (i) reducing potential contact with soil surfaces (thus reducing dissolution of inorganic P), and (ii) decreasing contact time and the possibility for desorption equilibria to occur. Thus, although organic P is relatively dominant in the drained pathway, we suggest that this reflects a deficient source of inorganic P (relative to surface pathways), rather than an increased input of organic forms *per se*.

Inorganic forms of P (i.e. MRP and MRPP) clearly dominated the forms of P in the surface pathways, but it would be incorrect to disregard the importance of organic species. The organic forms of P (i.e. DUP and PUP) constituted 50% of the loss through the drained (to 85 cm) pathway. Even in the surface pathways the organic forms represented at least 33% of TP loss. Transfer of organic P compounds warrants further investigation, as the mechanisms of release from solid to solution phase, and subsequent removal pathways are poorly understood.

13.3.2 Storm Magnitude and Application of Fertilizer and Slurry

It has already been suggested that "extreme" storm events play an important role in the removal of P from grassland hillslopes, particularly on the undrained soil, where local hydrological conditions (i.e. impermeable clay in the B horizon) prevent water movement by any alternative other than by the surface to 30 cm pathway. This work is considered in detail by Haygarth and Jarvis (1996), and is summarized in Figures 13.4 and 13.5. Figure 13.4 is a histogram of the frequency of total P concentration from the undrained plots (surface to 30 cm pathway), for seven lysimeter plots over eight events during 1994. Notice that on the right-hand tail of the distribution, the concentration categories increased at an accelerating rate: the data were highly skewed with a skewness coefficient of 6.5. There were, however, a number of low-frequency but high-magnitude events which contributed high concentrations of P removal.

The mechanisms that caused such events are related to the coincidence of high rainfall with the surface application of TSP fertilizer or slurry. Figure 13.5 illustrates a scatter of P concentrations and surface discharges on the undrained plots (surface to 30 cm pathway), and demonstrates the significance of a "single" dominant event: 25.2 mm rainfall in the 48 h prior to 0900 on 17th May. On 10th May, 16 kg P/ha was added as TSP fertilizer ("typical" for grassland management), in dry conditions when a soil moisture deficit of 38 mm had accumulated. In the period 14th–18th May, over 50 mm of rain fell, with a resulting equivalent drainage of over 20 mm. This event resulted in the highest P losses recorded during the year. The maximum and mean recorded TP loads during this event were 18.5 and 4.6 g/h respectively, and although data were not gathered for the duration of the storm, it has been suggested that a significant proportion of the fertilizer added on 10th May was removed during this single event (Haygarth and Jarvis 1996).

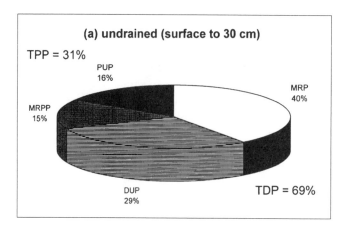

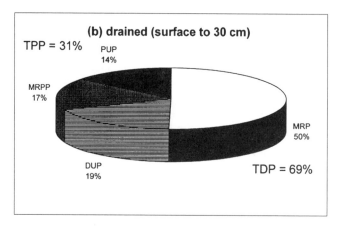

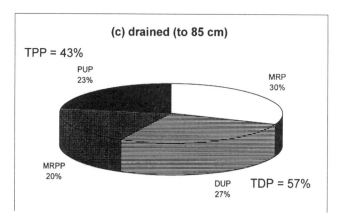

Figure 13.3 Forms of phosphorus determined in (a) undrained lysimeter, surface to 30 cm; (b) drained lysimeter, surface to 30 cm, and (c) drained lysimeter, surface to 85 cm export pathways, mean for seven lysimeter plots over six events during 1994 (Haygarth and Jarvis, unpublished)

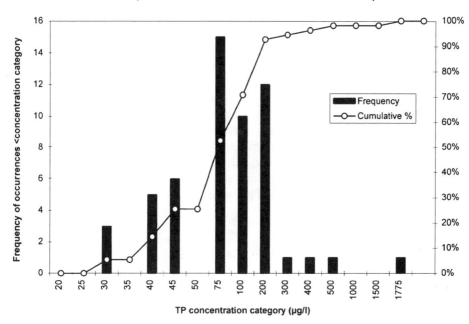

Figure 13.4 Histogram illustrating the frequency of total P concentration in the undrained lysimeters (surface to 30 cm pathway), for seven lysimeters over eight events during 1994

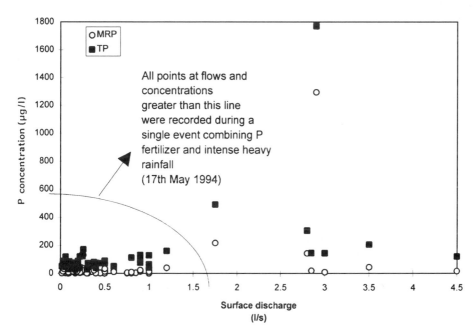

Figure 13.5 Scatter of P concentration and surface discharge on the undrained lysimeters (surface to 30 cm pathway), illustrating the significance of a "single" dominant event on the range of data acquired (after Haygarth and Jarvis 1996)

Similarly, Harris *et al.* (1995) also reported data from the Rowden Experiment from a single event during 1995, and found that concentrations of P in runoff were significantly elevated when heavy rainfall followed surface slurry application (Figure 13.6). On the 25th April 1995 an intense storm occurred approximately 12 h after the addition of 11 tonnes of surface spread slurry to two drained and two undrained lysimeters. Concentrations of MRP in drainage water on these plots peaked at 267 μg/l compared to 30 μg/l on plots which had not received slurry. MRP in the surface pathways showed a similar order of magnitude difference. Concentrations of UMRP were greater in surface runoff than in drainage waters, i.e. over 400 μg/l (compared with 30 μg/l where plots received no slurry). This is supporting evidence for preferential transport of inorganic particulate bound material in surface runoff, which does not occur to the same extent in subsurface movements.

13.4 CONCLUSIONS AND IMPLICATIONS

Drainage of grassland affects the forms and pathways of P removal from grassland hillslopes. Mole and pipe draining clearly encouraged downward movement of water, but as a result the concentrations of P removed in the surface pathways were higher. In both drained and undrained land, nearly 70% of the P removed was in dissolved, mainly inorganic, forms and was therefore potentially bioavailable to algae. However, even dissolved unreactive (organic) forms of P are potentially bioavailable, by the action of alkaline phosphatase enzymes which hydrolyse organic P compounds (Shan *et al.* 1993).

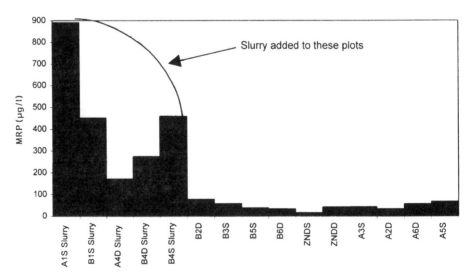

Figure 13.6 Increased concentration of MRP determined on unfiltered discharge waters from a range of plots on the Rowden Experiment, when heavy rainfall was preceded by slurry amendment (after Harris *et al.* 1995)

Whereas surface removal pathways were dominated by inorganic forms of P, when P was removed via mole and pipe drains, the relative proportions of organic and particulate P were increased. This suggests that dissolution, desorption and hydrolysis of P were dominant processes in surface horizons, presumably where soil P concentrations were greatest. This may involve simple desorption of P from soil into water and also removal of P from leaf surfaces, microbial exudates and algal films in wet conditions and entrainment of "soluble" forms of fertilizer, slurry and animal excretal residues. Conversely, drained pathways have a greater amount of organic P, perhaps because dissolution and hydrolysis of inorganic material does not occur to a high degree when water moves rapidly down preferential drain pathways. There is little mechanistic evidence to suggest how and why organic forms find their way into solution, but interactions with the sides of the pores leading into the mole channel must be important. There is little doubt that single storm events are important, particularly in surface pathways on undrained land. Surface spreading of fertilizer and slurry can have catastrophic effects if they coincide closely with intense rainfall: P exports of over 18 g/ha were determined during one event. Conversion of such data to annual export coefficients is difficult and we have not attempted it for all pathways in this chapter. However, based on the undrained lysimeters, Haygarth and Jarvis (1996) speculated that around 3 kg P/ha were removed by the surface to 30 cm pathway, but emphasize the pitfalls in calculating and extrapolating to an annual export figure. The apparent importance of low-frequency high-magnitude single storm events suggests that the only definitive way to determine hillslope-scale export coefficients is to monitor continually. However, even if this could be successfully achieved, the problems of spatial, as well as temporal inter-lysimeter variability can undermine confidence of export coefficient data currently being generated from grassland. Before export coefficients can be produced and used effectively, we suggest that future research should concentrate on improving understanding of *mechanisms* and *processes* of P transformations within the hillslope.

ACRONYMS

MRP	Molybdate reactive phosphorus ($<0.45\,\mu$m)
DUP	Dissolved unreactive phosphorus ($<0.45\,\mu$m)
TDP	Total dissolved phosphorus ($<0.45\,\mu$m)
UMRP	Unfiltered molybdate reactive phosphorus
MRPP	Molybdate reactive particulate phosphorus ($>0.45\,\mu$m)
PUP	Particulate unreactive phosphorus ($>0.45\,\mu$m)
TPP	Total particulate phosphorus ($>0.45\,\mu$m)
TP	Total phosphorus

ACKNOWLEDGEMENTS

The authors are grateful to Craig Ashby, Andrew Bristow (IGER North Wyke), Robert Harris (University of Sheffield and IGER North Wyke) and Tim Harrod (SSLRC North Wyke) for help and discussion. The work was funded by the Ministry of Agriculture, Fisheries and Food, London. IGER is supported by the BBSRC.

REFERENCES

Armstrong, A.C. and Garwood, E.A. 1991. Hydrological consequences of artificial drainage of grassland. *Hydrological Processes*, **5**, 305–314.

Boorman, D.B., Hollis, J.M. and Lilly, A. 1995. Hydrology of soil types: a hydrologically-based classification of the soils of the United Kingdom. IH Report No. 126, Institute of Hydrology, Natural Environment Research Council, Wallingford.

Eisenreich, S.J., Bannerman, R.T. and Armstrong, D.E. 1975. A simplified phosphorus analytical technique. *Environmental Letters*, **9**, 45–53.

Harris, R.A., Heathwaite, A.L. and Haygarth, P.M. 1995. High temporal resolution sampling of P exported from grassland soil during a storm, and the impact of slurry additions. *Proceedings of the International Workshop on Phosphorus Loss to Water from Agriculture*, TEAGASC, Johnstown Castle, Wexford, Ireland, 26–29 September 1995, pp. 25–26.

Harrison, A.F. 1985. Effects of environment and management on phosphorus cycling in terrestrial ecosystems. *Journal of Environmental Management*, **20**, 163–179.

Haygarth, P.M. and Jarvis, S.C. 1996. Soil derived phosphorus in surface runoff from grazed grassland lysimeters. *Water Research*, in press.

Haygarth, P.M., Ashby, C.D. and Jarvis, S.C. 1995. Short term changes in the molybdate reactive phosphorus of stored soil waters. *Journal of Environmental Quality*, **24**(6), 1133–1140.

Heathwaite, A.L., Burt, T.P. and Trudgill, S.T. 1990. The effect of land use on nitrogen, phosphorus and suspended sediment delivery to streams in a small catchment in southwest England. In J. Boardman, L.D.L. Foster and J.A. Dearing (eds), *Soil Erosion on Agricultural Land*. John Wiley & Sons, Chichester, pp. 161–177.

Jordan, C. and Smith, R.V. 1985. Factors affecting leaching of nutrients from an intensively managed grassland in County Antrim, Northern Ireland. *Journal of Environmental Management*, **20**, 1–5.

Kronvang, B. 1990. Sediment associated phosphorus transport from two intensively farmed catchment areas. In J. Boardman, L.D.L. Foster and J.A. Dearing (eds), *Soil Erosion on Agricultural Land*. John Wiley & Sons, Chichester, pp. 313–330.

Ministry of Agriculture, Fisheries and Food (1984). *Lime and Fertilizer Recommendations. No. 5: Grass and Forage Crops*. Booklet 2430, MAFF, London.

Murphy, J. and Riley, J.P. 1962. A modified single solution method for the determination of phosphate in natural waters. *Analytica chim. Acta*, **27**, 31–36.

Roberts, G., Hudson, J.A. and Blackie, J.R. 1986. Effect of upland pasture improvement on nutrient release in flows from a 'natural' lysimeter and a field drain. *Agricultural Water Management*, **11**, 231–245.

Scholefield, D., Tyson, K.T., Garwood, E.A., Armstrong, A.C., Hawkins, J. and Stone, A.C. 1993. Nitrate leaching from grazed grassland lysimeters: effects of fertilizer input, field drainage and patterns of weather. *Journal of Soil Science*, **44**, 601–614.

Shan, Y., McKelvie, I.D. and Hart, B.T. 1993. Characterisation of immobilised *Escherichia coli* phosphatase reactors in flow injection analysis. *Analytical Chemistry*, **65**, 3053–3060.

Sharpley, A.N. and Smith, S.J. 1989. Prediction of soluble phosphorus transport in agricultural runoff. *Journal of Environmental Quality*, **18**, 313–316.

Talman, A.J. 1983. A device for recording fluctuating water tables. *Journal of Agricultural Engineering Research*, **28**, 273–277.

Waters, G.R. 1994. Current government policy and existing instruments balance on the countryside. In R.J. Haggar and S. Peel (eds), *Grassland Management and Nature Conservation*. Proceedings of a joint meeting organized by the British Grassland Society and the British Ecological Society, Leeds University, 27–29 September 1993. The British Grassland Society, Occasional Symposium No. 28, Aberystwyth, pp. 3–9.

14 Modelling the Solute Uptake Component of Hillslope Hydrochemistry: Are Flow Times and Path Lengths Important during Mineral Dissolution?

STEPHEN TRUDGILL,* JASON BALL and BARRY RAWLINS
Department of Geography, University of Sheffield, UK

14.1 INTRODUCTION

The removal of material from hillslopes in solution is widely seen as a dominant land-forming process. Rapp (1960) concluded from his study of the Karkevagge area of Arctic Sweden that transport in solution was at least equal to the movement of sediment by earthslides and mudflows. Carson and Kirkby (1972) were able to assert that "chemical removal is a major form of hillslope erosion in temperate and humid tropical areas, of the same order of magnitude as all forms of mechanical erosion combined". In a recent review of rates of geomorphological processes, Goudie (1995) was able to present a broad databased picture: at a world scale, values for chemical denudation (as a percentage of total denudation) could be as high as 89% (the St Lawrence catchment) and, at a more detailed scale, for example in gauged catchments in Devon, UK, the mean value was 70%, with a range of 54–79%.

Despite the significance of chemical denudation, there are in fact relatively few models of the solutional processes at the hillslope scale when compared to either the catchment scale, as recently summarized by Trudgill (1995), or to the wealth of models on groundwater chemical evolution (e.g. Plummer *et al.* 1983; Domenico and Schwartz 1990; Glynn *et al.* 1990; Kenoyer and Bowser 1992a,b; Fetter 1994, and models such as HYDROGEOCHEM, SUTRA and NETPATH: Voss 1984; Plummer *et al.* 1994). Earlier geomorphological reviews indeed showed a much greater endeavour concerning hillslope hydrological and sediment transfer processes than solute processes (Kirkby 1978; Anderson 1988; Anderson and Burt 1990), with a range of models such as the simple demonstration LINEAR and more advanced topographically distributed TOPMODEL being available for hillslope hydrological processes (Beven and Kirkby 1979; Kirkby *et al.* 1993). A review of solute processes (Trudgill 1986b) showed a wealth of data on solutes in river systems (Walling and

---

Present address: Department of Geography, University of Cambridge, UK.

Advances in Hillslope Processes, Volume 1. Edited by M. G. Anderson and S. M. Brooks.
© 1996 John Wiley & Sons Ltd.

Webb 1986), but with the exceptions of Burt (1986), Crabtree (1986) and Kirkby (1986), there was little attempt to quantify and model the relevant downslope hydrochemical processes and their spatial distribution. It is also clear from Trudgill (1995) that much more effort has been made on the modelling of chemical weathering in an "acid rain" context than in a geomorphology context (e.g. SAFE and PROFILE; Sverdrup *et al.*, 1995). Given that hillslopes are a major source for solutes in fluvial systems and a major location for chemical weathering, modelling hillslope hydrochemistry can be seen as worthy of further attention.

The question asked in this chapter is whether flow times and path lengths (Quinn *et al.* 1991) are important for solute modelling at the hillslope scale. In mixing models (e.g. Sklash 1990; Kendall *et al.* 1995) it can often be assumed that there is a base flow component which has a stable value of high solute concentration because of high residence time; this then mixes with a rapid flow component of $\cong 0$ concentration. We consider the possibility that at the hillslope scale dissolution rates may be slow enough so that maximum, stable values may not be reached because the path length/residence time is not long enough (as recorded for some silicate minerals by Gíslason and Anórsson 1993). Here distinction must be made between rapid flow, which has no physical *opportunity* to take up solutes (overland flow and bypassing flow), and where the *contact time* of water in the soil is the limiting factor relative to a slow rate of dissolution. Conversely, rapid flows may have path lengths long enough for solute uptake from rapidly dissolving solids to be significant (comparable to the case for fine, soluble material in glacial streams: Fountain 1992; Brown *et al.* 1994). Thus, are flow rates in soils such that modelling should allow for a progressive solute uptake downslope or is chemical evolution along a path length irrelevant such that a stable maximum value can be assigned to specified unit areas of a hillslope?

There is a clear interest in how any spatial variations in solute uptake on hillslopes influences runoff water quality as well as the spatial distribution of chemical denudation rates. Spatial variations are also seen as key considerations in hydrological modelling (e.g. Beven 1995; Quinn and Beven 1993). Renewed interest in spatially distributed phenomena have also been prompted by developments in GIS (e.g. Moore *et al.* 1993a) with an emphasis on the linking of pattern to process (Moore *et al.* 1993b). Such trends permit a renewed interest in the spatial distribution of solutional processes. If this can be developed further in relation to the spatial variation in soil types (Billett and Cresser 1992), this is facilitated by a knowledge of downslope variations in soil type, including the concept of the catena (Dixon 1986; Furley 1968, 1971).

Our current knowledge relevant to the study of solutional processes on hillslopes includes both dynamic (temporal) and spatial aspects:

1. Small scale

- data on rates of mineral dissolution from laboratory studies, including carbonate and silicate minerals
- data and models on soil column/profile solute transport and leaching

2. Hillslope scale

- data on and models of hillslope hydrological processes, especially in relation to topography
- data on variations in soil type and soil characteristics downslope
- data on and models of solutional erosion rates on hillslopes
- long term models of hillslope evolution

3. Catchment and larger scale

- data and models of catchment solute processes, especially under acidification
- data on catchment solute losses, and hence on weathering rates from mass balance approaches
- data and models on the geochemical evolution of groundwater

Clearly processes and rates have to be scaled up under small-scale aspects and scaled down under large-scale aspects but the most promising link between the scales is that of soil type, with characteristic texture, porosity and mineralogy varying in relation to topography. Of especial use is the conclusion of Bluth and Kump (1994) that average river chemistries could be related to the distribution of soil associations within a catchment and the work of Sverdrup et al. (1995) who used the factors of soil reactive surface area, soil moisture and mineral reaction rate in modelling weathering in an acid rain context.

The earlier papers of Burt (1986), Crabtree (1986), Crabtree and Trudgill (1987), Trudgill et al. (1994a), Trudgill (1976, 1985, 1986a) directly related rates of weathering to soil type using weight-loss tablets (Trudgill 1975). The problem with this approach is that it relies upon empirical values derived from time-extensive fieldwork. If hillslope hydrochemical modelling is to be placed on a more generally applicable footing, some more "universal" approach is needed. Given that hillslope hydrological models already exist, and that the reviews of models of hillslope hydrochemical processes by Burt (1986), Crabtree (1986) and Kirkby (1986) suggest that it is indeed possible to model the spatial variation of solutional erosion, the aim of this chapter is to develop the modelling further by using several aspects of the existing approaches summarized in the list above, especially data on rates of mineral dissolution from laboratory studies, data on variations in soil type and soil characteristics downslope, and data on catchment solute losses.

14.2   MODELLING HILLSLOPE SOLUTE UPTAKE FROM MINERAL DISSOLUTION

14.2.1   Basic Assumptions

A basic assumption is that the more time water is resident in the soil, then the more it will tend to increase in solute concentration to a maximum value (Figure 14.1(a)). Path length down a hillslope can then be seen as some form of a substitute measure

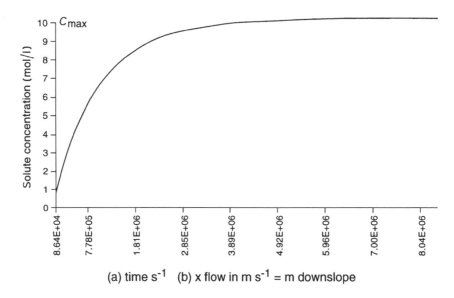

(a) time s^{-1} (b) x flow in m s^{-1} = m downslope

Figure 14.1   Solute uptake curve (a) over time; (b) substituting space (downslope distance) for time

for the time which water spends in the soil in contact with soluble material (Figure 14.1(b)) through the intervening variable of slow water flow velocity which controls contact time with soluble soil material (Lasaga *et al.* 1994). As water moves downslope at a given rate of distance per time, each downslope spatial unit inherits solute concentrations from a previous upslope unit just as a stationary body of water would inherit solute concentrations from a previous time step. The difference between a stationary body of water is that hillslope water will have moved a distance downslope and taken a given amount of time to do this under the contraints of hydraulic gradient and effective porosity. We are thus envisaging a pulse of water going downslope at a specified velocity.

Clearly, in seeing hillslope path length as some form of substitute measure for time, the words "some form" are of considerable importance since there will be many downslope changes, with additional water (and atmospheric solutes) from rainfall inputs, increases in discharge downslope, changes in porosity and in soil chemistry, all of which will need to be allowed for. However, the starting point of the modelling procedure is to evaluate the ranges of the rates of solute uptake over time and then to include the role of the rates of downslope water flow over the same time steps. Thus the formulation is that after the elapse of t seconds the chemical concentration in the flowing soil water should be C mol/l at distance m downslope given v, a water flow velocity in m/s up to a maximum concentration C_e at time t_e at slope position d_e.

True chemical equilibrium C_e is approached asymptotically and so for practical purposes it can be defined as C_{max} either as a 95% or 99% value of C_e or as an operational definition when rounding up of data within analytical accuracy gives C_e

or simply as the maximum recorded value in laboratory or field experimental work. The terms t_{max} and d_{max} can then also be used to define respective maximum times and downslope positions.

From a knowledge of slope length and flow rate the solute output to streams from that slope can be predicted. From a prediction of solute uptake rates and flow we can also compare predictions of d_{max} with λ, the slope length of interest and to which the calculations apply (Figure 14.2). λ could be an entire slope length from crest to stream but, given that slope conditions change, it is more likely to be a spatial unit which can be characterized as having internally relatively uniform conditions (such as a soil mapping unit and/or a topographic unit). If $d_{max} < \lambda$ (rapid rate of solute uptake per slope length or long slope; Figure 14.2(a)), then C_{max} can be used to characterize the output from that slope length/unit. If $d_{max} < \lambda$ and λ is the length of the last downslope unit adjacent to a stream, then only the characteristics of that unit need to be considered in order to predict hillslope solute output to that stream. If $d_{max} > \lambda$ (slow rate of solute uptake per slope length or short slope; Figure 14.2(b)) it is clear that contact time, flow rate and distance will be critical to hydrochemical modelling within that hillslope length or subunit of it.

The curve of solute uptake can be described by the equations (Mercado and Billings 1975; Wallach and Shabtai 1992; Ferguson et al. 1994):

$$dC/dt = k(C_{max} - C) \tag{14.1}$$

$$C = C_{max}(1 - e^{-kt}) \tag{14.2}$$

where C = solute concentration (mol/l); C_{max} = maximum concentration (mol/l); t = time (s); k = rate constant appropriate to the dissolving mineral and conditions.

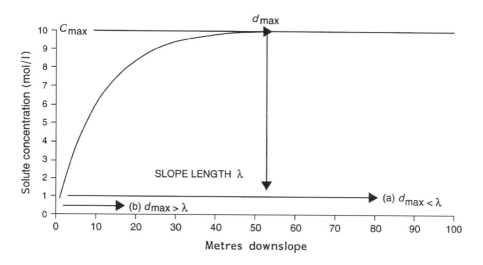

Figure 14.2 Solute uptake downslope to distance d_{max} at which C_{max} is attained and slope length, λ. (a) $d_{max} < \lambda$ (long slope length); (b) $d_{max} > \lambda$ (short slope length)

The expression in brackets in equation (14.2) reduces the amount of C_{max} that can occur in solution exponentially over time (e.g. for short times $C = C_{max} \times 0.2$ and for longer times $C = C_{max} \times 0.9$ until the time when C_{max} is reached when $C = C_{max} \times 1$).

Enrichment from an initial concentration to C_{max} is 63% complete by the time $t = 1/k$ and 95% complete by time $3t$ (Ferguson et al. 1994, p. 225). Therefore $k = 1/t$ at 63% enrichment and using 95% enrichment as an operational definition of equilibrium (which is in fact approached asymptotically) as used by Mercado and Billings (1975) at that stage $k \approx 3t$ and thus t_{max} can be defined as $3/k$ at C_{max} and so distance downslope d_{max} to reach C_{max} and t_{max} is:

$$d_{max} = 3v/k \qquad (14.3)$$

where $3/k$ = time to t_{max} (in s) and v = rate of movement of water (in m/s). d_{max} can be evaluated for a body of mobile water using a mean value of v or additively for the proportions of water flowing at different rates of v at any one point or successively for downslope changes.

In order to define spatial scales of interest, k and v per cubic metre and their variations with mineralogy, soil type and slope position have to be evaluated. The higher the value of k, the more quickly the C_{max} value is reached and thus, for the mobile water per cubic metre, the shorter the path length in metres will be to achieve that maximum value.

For porous media such as soils on hillslopes, k is increased not only with rate of mineral reaction but also the amount of reactive surface exposed to soil water per volume of soil (giving a shorter path length per metre of land surface as compared to a planar rock surface). The fraction of the total reactive surface which is comprised of any mineral under consideration is also critical. At soil moisture contents below saturation, the lower amount of water actually in contact with the maximum potential reactive surface may reduce the value of k. This, however, may be compensated for by the fact that low water contents are associated with slower rates of flow and thus more solid–solvent contact time. Similarly, high flow rates would tend to reduce the value of k but this may be compensated for by greater contact area and greater ease of chemical diffusion from small to the larger water-filled pores at higher moisture contents.

14.2.2 Defining Conditions of Interest: Water Flow Versus Reaction Rate

Water moves at a range of pore velocities at any one place and if there is free drainage they will tend to increase over time downslope as soil water content increases. The value of k can also vary considerably not only with mineral type and surface area but also with pH and the movement of water. In addition, in soils, rather than simply being an aggregation of dissolving minerals, processes such as cation exchange, anion production and weathering reactant concentration, adsorption, diffusion and dispersion will be involved. There are thus a large number of possible combinations of v and k and a wide range of factors influencing k to which the mineral dissolution rate is only a first approximation. It is therefore unlikely that this approach can provide a precise predictive tool without detailed calibration. In particular, mineral dissolution rates provide an indication of potential solute

Table 14.1 Ranges of v for soil types

| | Hydraulic conductivity | | | Porosity | |
	(cm/s)	(m/s)	v (m/s)	Mean	Loam (%)
Coarse sand					
mx.	1.00×10^{-1}	1.00×10^{-3}	$0.46\ 2.17 \times ^{-3}$		est.
mn	1.00×10^{-3}	1.00×10^{-5}	$0.31\ 3.23 \times 10^{-5}$	$1.10 \times 10^{-3}\ 0.05$	6.06×10^{-5}
Fine sand					
mx	1.00×10^{-3}	1.00×10^{-5}	$0.53\ 1.89 \times ^{-5}$		
mn	1.00×10^{-5}	1.00×10^{-7}	$0.26\ 3.85 \times 10^{-7}$	9.63×10^{-6}	0.55
Silt					
mx	1.00×10^{-4}	1.00×10^{-6}	$0.61\ 1.64 \times 10^{-6}$		
mn	1.00×10^{-6}	1.00×10^{-8}	$0.34\ 2.94 \times 10^{-8}$	8.34×10^{-7}	0.2
Clay					
mx	1.00×10^{-6}	1.00×10^{-8}	$0.6\ 1.67 \times 10^{-8}$		
mn	1.00×10^{-9}	1.00×10^{-11}	$0.34\ 2.94 \times 10^{-11}$	8.35×10^{-9}	0.3

production; actual concentrations in the short term are more likely to be related to cation exchange processes and biological factors. In this context, as discussed under Section 3 below, it is the rates which are derived from catchment output (which is the net result of all these processes) which are liable to be more realistic for hillslope modelling than laboratory mineral dissolution rates. However, while there are many complicating factors, many of the possible combinations of flow and solute uptake are either unlikely or not significant to our consideration. What is possible is to define probable combinations in order to define how significant the considerations might be. For example, given a hillslope path-length, assuming chemical equilibrium, or some maximum value, for base flow waters and no solute uptake in rapidly flowing storm waters, how wrong can we be?

If we define the probable ranges for mean pore water velocities for different soil textures (Table 14.1) and multiply these by t_{max} values derived from 3/probable k values (Ferguson *et al.* 1994), we are then able to define a finite range of interest in the combinations of these values (Table 14.2).

For the shaded areas <1 m, k values are so high that the water will have already reached C_{max} after 1 m of flow. Thus for all practical modelling purposes, the downslope evolution of soil water solute content can be ignored and C_{max} values can be used.

For the shaded areas lower in Table 14.2, the slope length would have to be greater than 100 km for C_{max} to be reached and few, if any, slope crests are that far from a stream. We can also rule these out if we note the very long timescales for these larger slope length values to be reached. If we are modelling hillslope flow, base flow residence times (Trudgill *et al.* 1983; Stewart and McDonnell 1991; Unnikrishna *et al.* 1995) are liable to be those below the order of magnitude of 10^1 days (the horizontal bar on the table) but there is decreasing interest down the chart if, for the moment, we are excluding long residence time groundwater. If we, then, set the residence time limits of the order of 1×10^1–10^2 days for base flow, we are only then

Table 14.2 Combinations of k and v to give d_{max}

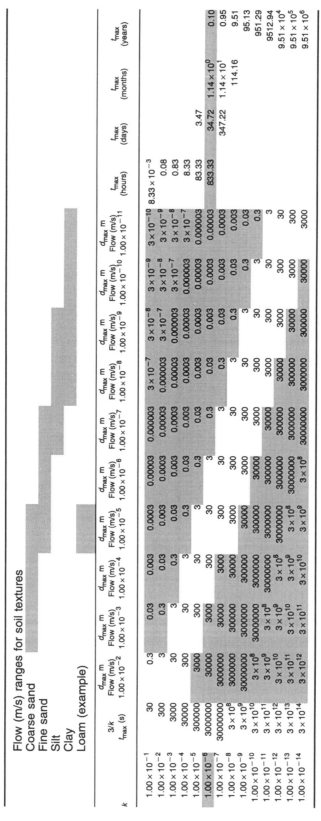

Flow (m/s) ranges for soil textures
Coarse sand
Fine sand
Silt
Clay
Loam (example)

k	$3/k$ t_{max} (s)	d_{max} m Flow (m/s) 1.00×10^{-2}	d_{max} m Flow (m/s) 1.00×10^{-3}	d_{max} m Flow (m/s) 1.00×10^{-4}	d_{max} m Flow (m/s) 1.00×10^{-5}	d_{max} m Flow (m/s) 1.00×10^{-6}	d_{max} m Flow (m/s) 1.00×10^{-7}	d_{max} m Flow (m/s) 1.00×10^{-8}	d_{max} m Flow (m/s) 1.00×10^{-9}	d_{max} m Flow (m/s) 1.00×10^{-10}	d_{max} m Flow (m/s) 1.00×10^{-11}	t_{max} (hours)	t_{max} (days)	t_{max} (months)	t_{max} (years)
1.00×10^{-1}	30	0.3	0.03	0.003	0.0003	0.00003	0.000003	3×10^{-7}	3×10^{-8}	3×10^{-9}	3×10^{-10}	8.33×10^{-3}			
1.00×10^{-2}	300	3	0.3	0.03	0.003	0.0003	0.00003	0.000003	3×10^{-7}	3×10^{-8}	3×10^{-9}	0.08			
1.00×10^{-3}	3000	30	3	0.3	0.03	0.003	0.0003	0.00003	0.000003	3×10^{-7}	3×10^{-8}	0.83			
1.00×10^{-4}	30000	300	30	3	0.3	0.03	0.003	0.0003	0.00003	0.000003	3×10^{-7}	8.33			
1.00×10^{-5}	300000	3000	300	30	3	0.3	0.03	0.003	0.0003	0.00003	0.000003	83.33	3.47		
1.00×10^{-6}	3000000	30000	3000	300	30	3	0.3	0.03	0.003	0.0003	0.00003	833.33	34.72	1.14×10^{0}	0.10
1.00×10^{-7}	30000000	3000000	30000	3000	300	30	3	0.3	0.03	0.003	0.0003		347.22	1.14×10^{1}	0.95
1.00×10^{-8}	3×10^{8}	3000000	300000	30000	3000	300	30	3	0.3	0.03	0.003			114.16	9.51
1.00×10^{-9}	3×10^{9}	30000000	3000000	300000	30000	3000	300	30	3	0.3	0.03				95.13
1.00×10^{-10}	3×10^{10}	3×10^{8}	30000000	3000000	300000	30000	3000	300	30	3	0.3				951.29
1.00×10^{-11}	3×10^{11}	3×10^{9}	3×10^{8}	30000000	3000000	300000	30000	3000	300	30	3				9512.94
1.00×10^{-12}	3×10^{12}	3×10^{10}	3×10^{9}	3×10^{8}	30000000	3000000	300000	30000	3000	300	30				9.51×10^{4}
1.00×10^{-13}	3×10^{13}	3×10^{11}	3×10^{10}	3×10^{9}	3×10^{8}	30000000	3000000	300000	30000	3000	300				9.51×10^{5}
1.00×10^{-14}	3×10^{14}	3×10^{12}	3×10^{11}	3×10^{10}	3×10^{9}	3×10^{8}	30000000	3000000	300000	30000	3000				9.51×10^{6}

Shaded areas exclude <1 m (top right, allows t_{max} to be reached) and > 100 km (bottom left, unlikely slope length). Above bar, likely storm flow values, below base flow values

concerned with k values of 1×10^{-6} to 1×10^{-7} and flow rates v higher than 1×10^{-7} m/s.

Above the horizontal bar are those values liable to apply only to storm flow and any rapid bypassing flow. If we set residence times for storm flow on a slope at ranging from 1 hour to 1 day, the values of interest of k range between 1×10^{-4} and 1×10^{-5} and of flow rates greater than 1×10^{-5}.

As k depends both on the rate of reaction and C_{max} as well as surface area, we cannot readily define the values of interest without specifying mineral type and soil type but given the porous nature of soils, the high reactive surface area per cubic metre of soil (or as per square metre of land surface for a given depth) the mineral reaction rates of interest will be some orders of magnitude lower than the k values.

14.2.3 Defining Probable Reaction Rates, Surface Areas, Porosities and Water Contents

There is considerable discussion about the range of possible mineral reaction rate values (Casey 1995), not least concerning the differences between field-derived and laboratory-derived values (Casey et al. 1993; Velbel 1993; the field values being lower, for example, by up to 200–300 times in the work of Swoboda-Colberg and Drever 1993). Given this uncertainty over actual values, it would be appropriate to first establish the sensitivity of predictions to a range of reaction ranges and thus we can assess whether, and if so to what extent, such differences are important.

The important factors in the dissolution of soil minerals (and which can be assessed per cubic metre of hillslope soil) are

1. mineral reaction rate k_r (mol/m^2)
2. reactive surface area A_w (m^2)
3. the fraction of that area which is of any mineral in question A_{wm} (% composition as 0.99 to 0.01)
4. pore volume of water actually in contact with that reactive surface area θ (cm^3/cm^3)
5. volume of water per cubic metre of soil at that θ in which solutes are concentrated (1)
6. rate of change of solute concentration dC/dt (mol/l/s) which decreases as solute concentrations C increase $= k(C_{\text{max}} - C)$
7. concentration for given time: $C = C_{\text{max}} (1 - e - kt)$ (mol/l)

None of these can be given unique and universally applicable values. Many are, however, mutually dependent; for example, decreasing soil moisture will lead to a reduced reactive surface area in contact with water, reduced volume for concentration and slower rate of water flow. Plausible boundary conditions can thus be deduced by considering the combined ranges of likely values (though high rate of flow could be combined with low contact area in pipes or macropores; Germann 1990; Velbel 1993).

The mineral reaction rates, k_r, of interest are rarely higher than 1×10^{-6} mol/m^2/s. For example, Svensson and Dreybrodt (1992) calculated that for 100 μm size calcite crystals at pH 4, with stirring and soluted products removed, the rate was 2.20×10^{-6}; field values in soils where calcite is present in larger rock particles at higher pH values, without stirring and with solute products accumulating to some extent are clearly liable to be lower than this rate. Many laboratory rates are evaluated at far from equilibrium and thus again will considerably exaggerate the rapidity of dissolution in the field (Berner 1981) except at the initial stages of contact of acid water with fresh mineral. As discussed further below, using such rates is liable to give an incorrectly short path length. Rates can drop as low as 1×10^{-14} or less (quartz is quoted as 4.1×10^{-14} by Drever 1988; Lasaga 1984; Lasaga et al. 1994). The values are pH dependent to a greater or lesser extent (e.g. Acker and Bricker 1992; Zulla and Billett 1994), as well as temperature dependent, and vary with grain size (e.g. Anbeek 1992) but for any one particular mineral there tends to be a finite range of published values, whatever the conditions of experimentation or the basis of the calculation. In general, higher C_{max} values are associated with higher reaction rates. We could thus usefully consider k_r rates from 1×10^{-6} to 1×10^{-14} mol/m²/s. For C_{max}, Berner (1981) cites a range of solubilities for minerals in pure water down to 3×10^{-7} mol/l for $KAlSi_3O_8$ and 6×10^{-7} for $NaAlSi_3O_8$; these will be higher where acid reactants are present (CO_2 and organic acids, as shown by Huang and Kiang 1972). A likely range is from 1×10^{-3} to 1×10^{-6} mol/l for C_{max}, which includes the ranges from the more soluble carbonate to the less soluble silicate minerals.

Reactive surface areas, A_w, in soils can be calculated in relation to particle size distribution (assuming that the particles are spherical) or by gas absorption. Surface area increases with decreasing particle size, clay size particles having the highest surface area per unit volume. Values of $10 \, m^2/g^1$ are quoted for kaolinite and chlorite, 20 for illite and an extreme value of 900 for allophane (Rowell 1994). At low bulk densities ($0.8 \, g/cm^3$) these convert to 8.00×10^6 to 1.6×10^7 and $7.20 \times 10^8 \, m^2/^3$. At maximum bulk densities around $1.4 \, g/cm^3$, these convert to 1.4×10^7, 2.8×10^7 and 1.26×10^9. These latter values would appear to set the level of the maximum possible. The extreme values are unlikely to be found, values of 10^7 being the most likely highest values, with lower orders of magnitude being possible as particle size increases, as with the inclusion of sand (and also with the inclusion of organic material). Sverdrup and Warfvinge (1988) calculated a range from 5×10^5 to 5×10^6 $m^2/^2$ to a depth of 0.5 m (1×10^6 to $1 \times 10^7 \, m^2/m^3$) on podsolic soils (Pačés (1983) estimated 5×10^3 to $5 \times 10^5 \, m^2/m^3$. This gives possible boundaries of 5×10^3 for lighter sand soils to 5×10^7 for clay soil, varying between these values with soil textural mixtures.

Using these values implies that all the reactive surface is composed of the mineral in question. If only 10% of the soil is composed of the mineral in question, then these values should be multiplied by 0.1 (A_{wm}) and so on, with possibly more precise adjustments if it is known that the mineral is confined to a particular size range.

Pore volumes in soils rarely exceed $0.6 \, cm^3/cm^3$ (except in peat soils) and extreme values for mineral soils are given as 0.69 (Rowell 1994). For saturated soils this maximum value could be taken as representing the maximum wetted area in fine-grained soils with small pores. For lower moisture values the range of reactive surfaces in contact with water must be decreased proportionally but at the dry end of the scale only: simply decreasing soil moisture in coarse-grained soils with large pores will lead to the centre of the pore emptying, with a film of water still being in contact with the edge of the pore. Larger pores have been identified as critical to solute transport (Rose et al. 1982; Harvey 1993; Armstrong et al. 1995) but a complicating factor is the low mobility of water films and that immobile water will be present in micropores through which solutes can diffuse; Brusseau (1993), however, felt that this process was negligible for fine-grained soils. Field capacity (48 h drainage after saturation) is often around 0.4–0.5. The boundary between mobile and retained water ranges around 0.25–0.35 (Clothier et al. 1992). The specific application of soil moisture values to influencing effective reactive surface can therefore be seen as not straightforward in this context. Velbel (1993) especially makes the point that not all the potential mineral surface is available for reaction because of limited mobile water contact. The greatest relevance of these values is that they can be used in Darcian equations to predict decreases in flow rate with lower moisture content. Water content is probably most important in terms of the influence of water content on dilutions and concentrations (i.e. C values in the soil water). Corresponding water contents in cubic decimetres (l) per cubic metre of soil would for the θ values above be 600, 400–500 and 250–350 respectively.

Rates of change of solute concentrations (dC/dt) can be described by equation (14.1) above.

Mean rates of flow of water in metres per second are already given in Table 14.1 and range from 1×10^{-3} to 1×10^{-11}, with a plausible value for a loam being 1×10^{-5}.

The procedure adopted is to take mean, central or otherwise plausible values in plausible combinations for a range of soils and to work out maximum and minimum end products given the likely conditions. The conditions are initially worked out for a cubic metre of soil and then the evolution of that water in that soil in successive downslope steps of $1\,m^2$ (a cubic metre under a square metre of hillslope) initially assuming all conditions are constant. Further refinement can involve the successive additions of water and changing downslope conditions, especially of soil texture, but the initial step involves sensitivity to values and conditions using plausible values for surface area and hydrological parameters, based on soil texture. The initial assumption is that all pores are full, i.e. the reactive surfaces are in contact with mobile water. Results show, in fact, that predictions are not especially sensitive to soil moisture decreases as while reducing soil moisture could be seen as reducing the amount of effective reactive surface, there would also be slower flows which allows more time for uptake per volume.

14.2.4 Modelling Probable Path Lengths for Probable Reaction Rates

The steps for a $1\,m^3$ of soil ($1\,m^2$ occupying 1 m path length) are as follows:

$$k = (k_r.A_w.\theta)/(C_{max} - C); \quad d_{max} = 3v/k; \quad C = C_{max}(1 - e^{-kt})$$

1. Define k_r (mol/m^2/s) (range of interest or mineral specific).
2. Define the area A_w (m^2/m^3) (specific to a soil texture).

3. $k_r \times A_w$ (mol/m³/s).
4. Define pore volume, θ, and thus (from cubic centimetres of water to cubic centimetres of soil, effectively a dimensionless factor) to 1 (dm³) m³ (specific to texture; maximum amount or in proportion to soil moisture).
5. $k_r \times A_w \times \theta$ (mol/l/s) = dC/dt.
6. Define C_{max} (range of interest or mineral solubilities).
7. Re-writing equation (14.1):
$$k = (dC/dt)/(C_{max} - C \text{ for time } t; t = 1)$$
8. t_{max} (m³) = $3/k$.
9. Define flow rate v (m/s) (specific to soil texture).
10. $d_{max} = (3/k).v$ (m).
11. Define C at time t using $C = C_{max} (1 - e - kt)$.
12. Define slope position d for time t as $v \times t$ (m) (d_{max} for t_{max}).

Options:

• Compare d_{max} with λ (slope length which maintains the above input characteristics of flow, surface area, etc.), $d_{max} > \lambda$ = go to third option; $d_{max} < \lambda$ use C_{max} value.
• Re-run for next unit with changed characteristics.
• Show m at time t against C for time t for downslope evolution of chemistry.

The above applies to conditions when rainfall is not occurring and does not allow for evaporation. Volumes from new rainfall/increased moisture can be added progressively downslope which will dilute C (allowing for input C_i) and increase flow rate.

Following Table 14.2, there is an initial assumption that clay soils are not of interest since there are low values for flow velocities and high values of surface area for reaction which means that $d_{max} \ll \lambda$, though this will not be true in cracking clay soils when bypassing conditions can occur.

In Table 14.3 the following combinations of conditions, based on the areas of interest defined in Table 14.2, are used for sand and silt textures and an example of a loam.

k_r: 1×10^{-6} to 1×10^{-14} mol/m² s in order of magnitude steps

C_{max}: 1×10^{-2} to 1×10^{-6} mol/1 in order of magnitude steps (10–0.001 mM/l)

(while these were used in every stepwise combination, the lower range of k is liable to apply to the lower range of C_{max}).

A_w: coarse sand 1×10^3; fine sand 10^4; silt 10^5; loam 10^6

k: coarse sand: 1×10^{-3}, 1×10^{-4}; fine sand: 1×10^{-4}; 1×10^{-5}; 1×10^{-6}; silt: 1×10^{-5}; 1×10^{-6}; 1×10^{-7}; loam: 1×10^{-5}

Results showed that for all the combinations considered in Table 14.3, the k_r values for the maximum k of interest from Table 14.2 (1×10^{-4}) was 1×10^{-6}. Higher values gave $d_{max} < \lambda$, whatever input data were used. This thus tends to define the maximum k_r of interest when studying realistic k_r rates (below) except for the possible cases of transient flow at high velocities (bypassing or overland flow). d_{max} is sensitive to the input data of C_{max}, k_r, A_w and v in order of magnitude terms. Raising C_{max}, k_r and A_w brings d_{max} upslope; raising flow values and decreasing A_w for the proportion of reactive mineral brings d_{max} downslope. If C_{max} is raised and k_r lowered for the same A_w and flow or if A_w is lowered and flow is lowered for the same C_{max} and k_r then d_{max} is the same. The calculation of d_{max} reduces to

$$d_{max} = 3[(C_{max}/k_r)/(0.002(A_w/v))] \tag{14.4}$$

In terms of path lengths, d_{max} can clearly range from one or two metres to several kilometres. One approach can be to plot downslope distance to c_{max} for given conditions (as in Figure 14.2). Another approach is to define a slope length of interest, which may be as short as 30 m in catchment headwaters or an order of magnitude longer in larger basins. A further approach is to consider the scales at which supporting information is already available. On soil maps units often appear in widths of from 100–200 m to 1–10 km, depending on the map scale. The width of such units could be taken as equivalent to λ when considering the input of soil data to a GIS. Taking this as a guide, there is then considerable interest in d_{max}, at a wide range of C_{max} values for the flow rates and surface areas given, for maximum k_r rates as follows:

$$d_{max} = 10^2 - 10^4 \, m$$

coarse sand: k_r values 1×10^{-6} mol/m/s

loam: 1×10^{-9}

fine sand: k_r values of 1×10^{-9}

silt soils: 1×10^{-10}

Values less than these will give progressively longer path lengths to C_{max} at d_{max} values of interest in hillslope modelling. These values assume a mineral being considered occupies 100% of A_w; they will be an order of magnitude higher if only 10% of A_w is of the particular mineral in question.

Given considerations of the likely ranges of reactive surface areas, porosity, soil moisture and water flow rates in soils, it is concluded that conditions can combine so that residence times and the attendant path lengths to maximum solute concentrations can be of the orders of magnitude relevant to hillslope path lengths. These conditions are, however, limited to coarse-textured soils with lower orders of magnitude of reactive surface (1×10^3 to 10^4 m^2/m^3) and relatively high flow rates $> 1 \times 10^{-6}$.

For minerals dissolving at rates greater than 1×10^{-6}, path length is irrelevant for flows slower than 1×10^{-3} m/s as maximum values could be attained within 1 m³ of soil. For many commonly occurring minerals with slower reaction rates, the combinations of conditions can be critical in that estimated distances to reach maximum concentrations may readily be greater or lesser than either hillslope length or the lengths of internally consistent subdivisions of hillslopes, but again only in coarse-textured soils.

This means that it is worth considering further the dynamics of solute uptake in hillslope hydrochemical modelling for porous, coarse-textured soils and other conditions associated with rapid flow. We might then make a working assumption that hillslope areal units of the order of 100 m will have stable characteristic soil water chemical values with rapidly dissolving solutes $>1 \times 10^{-6}$ mol/m²/s at flows $<1 \times 10^{-4}$ or 10^{-5} in fine-textured soils with reactive surface of 1×10^{4} or more.

This assumes that all the reactive surface is of the mineral in question, however, and if only 10% of the reactive surface is occupied by one mineral, for any one combination of reaction rate and flow, this increases the path length by an order of magnitude and shifts the critical flows down one order of magnitude to $<1 \times 10^{-5}$ to 10^{-6} or the critical reaction rate up one order of magnitude to make the working assumption of areal constancy appropriate.

The estimates of d_{max} in Table 14.3 are thus maximum values in relation to 100% A_w but they can be taken as an "indication of interest". They do not assess the importance of solute uptake in rapidly flowing bypassing water nor yet indicate the solute levels which might occur in very long residence time groundwater. They do, however, indicate that the approach is worth pursuing further in terms of being more realistic about rates of reaction and flow. This is especially so in terms of specifying the uncertainties pertaining to field reaction rates and the distributions of flow velocities and in evaluating the sensitivity of the predictions to the ranges of uncertainties.

14.3 MODELLING USING EXISTING VALUES

14.3.1 Information Available

Where drainage basin output is used as a basis for calculation, the net output losses can be distributed over the source area per unit time, making appropriate subtractions for other inputs such as atmospheric inputs and fertilizer additions if mineral weathering is of interest. The range of expressions vary but kg (or mol)/ha per year is commonly worked out by transferring mg (or mol)/l to mg (or mol)/s via a knowledge of l/s using a rating curve. In this exercise, the areal expression is in fact physically meaningless and is merely a mathematical distribution over the land surface area; it does not necessarily imply that the land surface is actually reacting at that rate. This would only be the case if the catchment were a non-porous planar surface: here the distributed rate would be applicable to the actual reactive surface. In soil-covered catchments the reactive surface is much greater as the reactions occur in a three-dimensional porous medium. A distinction thus has to be made between values per unit land area (A_{wc}) and per unit reactive surface area.

At the hillslope scale, what we are actually interested in is the spatial differentiation of bulk values derived from catchment output by internal allocations to the areal source components, rather than the mathematical distribution per unit area. Bounded plots can be set up and provided they are hydrologically isolated, identical measurements and calculations at a smaller scale can be made. Other measures (e.g. Genereux *et al.* 1993) have been effected from the use of inputs and outputs of stream reaches bracketed by upper and lower measurement points, the upper values being subtracted from the lower to assess the sideslope contribution (which again can be mathematically distributed over the catchment segment area). There has also been a considerable amount of soil column modelling and observation, though rather more on transport of solutes once in solution, e.g. for surface applied solutes (such as fertilizers), rather than for soil-derived solutes from mineral dissolution. Values can be expressed in mass or mol per unit water volume and per solid land surface area and/or volume. At the hillslope scale, column work obviously requires aggregation on some logical basis.

Weathering rates have been measured quite extensively in geomorphology, especially directly using such devices as micro-erosion meters, weight-loss rock tablets and calculations from known datum points. Usually they are point measurements. Values are often expressed in millimetres or grammes per year.

Mineral dissolution rates (k_r) has been measured extensively in geochemistry directly in the laboratory, usually with pure mineral phases under well-defined conditions of temperature, pH pCO_2 and turbulence (see, for example, Busenberg and Clemency 1976; White and Claassen 1979; Lin and Clemency 1981; Lasaga 1984; Colman and Dethier 1986; Drever 1988; Svennson and Dreybrodt 1992; Anbeek *et al.* 1994; Franke and Teschner-Steinhardt 1994; Lasaga *et al.* 1994).

14.3.2   Cautions in using Existing Data in Modelling

Situations where information is available at comparable bases are rare. Rates may apply only to specific sites or minerals phases. Many endeavours have, however, taken some known aspect of these three measurement types and predicted or calculated another aspect which is unknown but of interest. It is clear that all the values *can* be converted *mathematically* to one another (provided the appropriate parameters are known, for example conversions involving density, mass and volume) but also that the *validity* of doing so and the *interpretations* can be open to question.

The basic difficulties lie in the following aspects:

1. Measurements are often made at particular *scales* of time and space and it may be inappropriate to transfer them to other scales.
2. Measurements may be made under particular *conditions* (say of mineral compositions, pH, temperature, rainfall, etc.) and unless the effects of these conditions are calibrated for, it may be unrealistic to transfer the results from one scale to another, especially from the laboratory to the field where different *combinations* of conditions obtain and even more so when *further* conditions *also* apply, e.g. biological factors (Huang and Kiang 1972; Taylor and Velbel 1991; Drever 1994).

3. Rates, especially laboratory k_r rates, may be *mineral specific* and the solutes derived thus interpreted in relation to the specific minerals. Measurements of solute output from heterogeneous systems are not mineral specific but are *solute specific*. Inference is then involved in the application of solute data to mineral data (as, for example, the inference that Na in solution is derived from Na-feldspars or a knowledge of the composition of the mineral and the stochiometry of solution; Velbel 1985, 1992, 1995). The easiest assumption is of congruent solution where the solutes are present in solution in the same ratios as they are in the minerals (e.g. Pačés 1983), or defining the plausible combination of minerals present as used in NETPATH (Plummer *et al.* 1994).
4. Catchment rates per surface area have to be made three-dimensional by multiplying by reactive surface area.

The key element in the discussions is that differences can be observed between laboratory-derived rates and field-derived rates. Van Grinsven and van Riemsdijk (1992) insist that "any comparison of weathering rates measured in the laboratory and rates from field studies can be questioned", citing the issues of incomparable time and spatial scales and of the supplies of weathering reactants which they conclude "makes quantitative comparison of field rates and laboratory rates senseless". Berner (1978) felt that "actual dissolution rates cannot be accurately predicted from laboratory experiments, because experiments ordinarily fail to reproduce the composition and structure of natural mineral surfaces and poorly understood biological factors are usually ignored". Laboratory rates are thus precise but only accurate within their defined conditions. The field situation is more complex, but the benefit of catchment rates is that they are the net product of all naturally occurring processes. The difficulty with them is that they are uncalibrated for local variations within the catchment and an adjustment has to be made for A_w.

Taking the example of rates as 200–300 times less in the field than in the laboratory (Swoboda-Colberg and Drever 1993), Table 14.3 enables us to establish the basis for a sensitivity analysis. The table already shows the outcome if the rates are lowered by an order of magnitude which enables us to suggest simply that d_{max} derived from field values will be proportionately longer than those derived from laboratory rates.

14.3.3   Using Existing Data to Define the Minerals of Interest in Path Length Modelling

The reason for first defining relationships in Table 14.3 rather than first using published values of k_r and C_{max} should now be evident: the values for are context-dependent (pH, temperature and flow rate) and with different bases of calculation (e.g. far from equilibrium). It is not therefore simply a matter of plugging in one value for k_r and one for C_{max} into equation (14.4) but of examining published values for given conditions, predicting path lengths and examining the relevance of the observational conditions to the conditions in that path length.

If we are to use existing k_r rates, this is most easily effected in monominerallic rocks, such as calcite limestone. With silicate minerals, the relationships between

Table 14.3 Adjusting d_{max} for soil type

k_r (mol/ m²/s)	C_{max} (mol/l)	Sand k sa 10^3	k sa 10^4	1.00×10^{-3} CS d_{max}	1.00×10^{-4} CS d_{max}	1.00×10^{-4} FS d_{max}	1.00×10^{-5} FS d_{max}	1.00×10^{-6} FS d_{max}	Silt k sa 10^5	1.00×10^{-5} silt d_{max}	1.00×10^{-6} d_{max}
1.00×10^{-6}	1.00×10^{-2}	2.00×10^{-4}		15							
1.00×10^{-7}	1.00×10^{-2}	2.00×10^{-5}	2.00×10^{-4}	150							
1.00×10^{-3}	2.00×10^{-4}	2.00×10^{-6}		15	15						
1.00×10^{-8}	1.00×10^{-2}	2.00×10^{-6}	2.00×10^{-5}	150	150	150	15				
1.00×10^{-3}	2.00×10^{-5}	2.00×10^{-4}		15	15	15					
1.00×10^{-4}	2.00×10^{-4}	2.00×10^{-4}									
1.00×10^{-9}	1.00×10^{-2}	2.00×10^{-7}	2.00×10^{-6}	1500	1500	1500	150	15	2.00×10^{-6}	15	
1.00×10^{-3}	2.00×10^{-6}	2.00×10^{-6}		150	150	15	15				
1.00×10^{-4}	2.00×10^{-5}	2.00×10^{-5}		150	150	15					
1.00×10^{-5}	2.00×10^{-4}	2.00×10^{-4}		15	15						
1.00×10^{-10}	1.00×10^{-2}	2.00×10^{-8}	2.00×10^{-7}	1500	1500	1500	1500	50	2.00×10^{-7}	150	15
1.00×10^{-3}	2.00×10^{-7}	2.00×10^{-6}		1500	150	150	15	2.00×10^{-7}		15	
1.00×10^{-4}	2.00×10^{-6}	2.00×10^{-5}		150	150	15	15	2.00×10^{-6}			
1.00×10^{-5}	2.00×10^{-5}	2.00×10^{-4}		150	15	15					
1.00×10^{-6}	2.00×10^{-4}	2.00×10^{-4}		15	15						
1.00×10^{-11}	1.00×10^{-2}	2.00×10^{-9}	2.00×10^{-8}	1500	1500	1500	1500	1500	2.00×10^{-7}	150	15
1.00×10^{-3}	2.00×10^{-8}	2.00×10^{-7}		1500	150	150	15	2.00×10^{-6}		15	
1.00×10^{-4}	2.00×10^{-7}	2.00×10^{-6}		150	150	15	15				
1.00×10^{-5}	2.00×10^{-6}	2.00×10^{-5}		150	150	15					
1.00×10^{-6}	2.00×10^{-5}	2.00×10^{-4}		15	15						
1.00×10^{-12}	2.00×10^{-4}	2.00×10^{-4}									
1.00×10^{-3}	1.00×10^{-2}	2.00×10^{-10}	2.00×10^{-9}	1500	1500	1500	1500	1500	2.00×10^{-7}	150	15
1.00×10^{-4}	2.00×10^{-9}	2.00×10^{-8}		1500	150	150	150	2.00×10^{-6}		15	
1.00×10^{-5}	2.00×10^{-8}	2.00×10^{-7}		150	150	15	15				
1.00×10^{-6}	2.00×10^{-7}	2.00×10^{-6}		150	150	15	15				
	2.00×10^{-6}	2.00×10^{-5}		150	150						
	2.00×10^{-5}	2.00×10^{-4}									
1.00×10^{-13}	2.00×10^{-4}										
1.00×10^{-3}	1.00×10^{-2}	2.00×10^{-11}	2.00×10^{-10}	1500	1500	1500	1500	1500	2.00×10^{-7}	150	15
1.00×10^{-4}	2.00×10^{-10}	2.00×10^{-9}		1500	150	150	150	2.00×10^{-6}		15	
1.00×10^{-5}	2.00×10^{-9}	2.00×10^{-8}			150		15				
1.00×10^{-6}	2.00×10^{-8}	2.00×10^{-7}									
1.00×10^{-14}	2.00×10^{-7}										
1.00×10^{-4}	1.00×10^{-3}	2.00×10^{-10}	2.00×10^{-10}	1500	1500	1500	1500	1500	2.00×10^{-7}	150	15
1.00×10^{-5}	2.00×10^{-9}	2.00×10^{-9}			150	150	150	2.00×10^{-6}		15	
1.00×10^{-6}	2.00×10^{-8}	2.00×10^{-8}									
	2.00×10^{-7}	2.00×10^{-7}									

Table 14.4   Published values of k_r

Mineral	k_r (mol/m²/s)	Author	
Quartz	4.0738×10^{-14}	Lasaga *et al.* (1994)	Lab., pH 5, 25°C
Kaolinite	5.24807×10^{-14}		
Muscovite	8.51138×10^{-14}		
Epidote	2.45471×10^{-13}		
Microcline	3.16228×10^{-13}		
Prehnite	3.89045×10^{-13}		
Albite	5.49541×10^{-13}		
Sanidine	1×10^{-12}		
Gibbsite	3.54813×10^{-12}		
Enstatite	1×10^{-10}		
Diopside	7.07946×10^{-11}		
Forsterite	3.16228×10^{-10}		
Nepheline	2.81838×10^{-9}		
Anorthite	2.81838×10^{-9}		
Wollastonite	1.00×10^{-8}		
Quartz	4.10×10^{-14}	Lasaga (1984)	Lab: 25°C, pH 5
K-feldspar	1.70×10^{-12}		
Albite	1.20×10^{-11}		
Anorthite	6.60×10^{-9}		
Anorthite	3.20×10^{-13}	Busenberg and Clemency (1976)	Lab.
Orthoclase	5.50×10^{-13}		
Oligoclase	1.70×10^{-12}		
Albite	2.50×10^{-11}		
Muscovite	6.50×10^{-15}	Lin and Clemency (1981)	Lab.
Calcite min.	1.60×10^{-6}	Franke and Teschner-Steinhardt (1994)	Lab.
Calcite max.	2.20×10^{-6}		
Calcite min.	1.50×10^{-6}	Svensson and Dreybrodt (1992)	Lab.
Calcite max.	1.82×10^{-6}		
Oligoclase	3.60×10^{-14}	Pačés (1983)	Catchment (as Na)
Oligoclase	6.80×10^{-13}		
Oligoclase	8.90×10^{-13}	Velbel (1985)	Catchment

species in solution and mineral composition has to be carefully evaluated. Congruent dissolution cannot necessarily be assumed (where water composition reflects mineral composition in the same element ratios). C_{max} values could be used as a given solubility under defined conditions or as the highest observed field concentration of relevant solutes at a site (Pilgrim *et al.* 1979). For any one mineral with specific probable k_r and C_{max} values, the combination may be sought in Table 14.3 to define the potential interest in path length modelling.

Table 14.4 illustrates the published values of k_r for a variety of minerals. Laboratory measurements should clearly be regarded as probably too high to apply to the field but it helps us to define the minerals of interest when modelling at the hillslope scale, remembering that path lengths to d_{max} will be longer if the laboratory rates are too high.

Calcite is a relatively rapidly dissolving mineral and liable to show little path length change unless flows are above those considered in Table 14.3 or unless A_w is effectively lowered by lowering the $CaCO_3$ content of a soil.

Most feldspars appear to be within the zone of interest, clearly predictable as not reaching maximum values in hillslope path lengths unless flowing through soils of high A_{wm} at slow rates.

14.4 DISCUSSION: HILLSLOPE SOILS AND HILLSLOPE HYDROLOGY

Soil characteristics change markedly downslope, so the precise application of this approach will need careful calibration. The likely variables include downslope and soil horizon changes in porosity, surface area and reactive mineral present in those surface areas (effective A_w). Pačés (1983), for example, quotes the surface fraction formed by oligoclase as ranging from 0.14 to 0.19. It is most likely that decreasing proportions from 100% A_w and flow rate are liable to extend path lengths, making it highly likely that uptake considerations will be important for any one mineral (but not necessarily for any one *solute*).

Figure 14.3 illustrates the possibilities and the sensitivities to parameters: mineral of k_r 1×10^{-14} mol/m^2/s, $A_w = 1 \times 10^5$ m^2/m^3, $\theta = 0.5$, $C_{max} = 1 \times 10^{-5}$ and $v = 1 \times 10^{-5}$.

Line 1 assumes the above conditions are fulfilled. If λ is 50 m, each soil unit has a characteristic solute concentration: dissolution rates are so rapid that contact time is not significant and downslope progression is unimportant except in terms of the solubility of the minerals in each unit as the path length d_{max} required to reach $C_{max} < \lambda$.

Line 2 shows the effect of reducing the proportion of mineral in question to 10% of the reactive surface area. If λ is 50 m, $d_{max} > \lambda$, C_{max} is never reached and solute concentrations depend very much on solute uptake rates, flow rates and path length – 6.0×10^{-6} mol/l at $\lambda = 50$ m.

Line 3 suggests that additionally the moisture content increases downslope to a maximum at 80 m, the chief effect of which is dilute solute concentrations at the end of the 130 m slope from 8.0×10^{-6} to 7.0×10^{-6} mol/l.

Line 4 suggests that additionally the amount of reactive mineral increases in proportion downslope till it becomes the dominant mineral. Line 5 is as line 4 but the maximum value is 50%. Line 6 increases the A_w downslope. If the mineral was abundant, we might assume that above path lengths of 100 m, constant values could be reached.

Figure 14.4 again assumes an increase in surface area, proportion of mineral and soil moisture downslope but for a range of k_r values. For minerals dissolving at 1×10^{-8}, 10^{-10} and 10^{-12}, everything is irrelevant as C_{max} is rapidly reached.

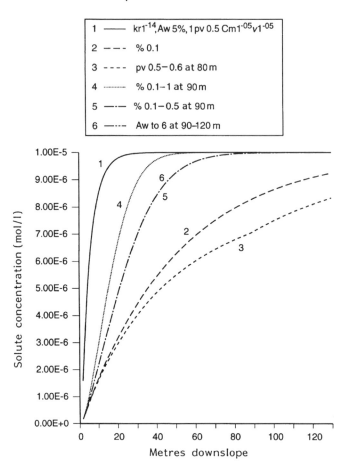

Figure 14.3 Solute concentrations downslope. Line 1: initial conditions kr 1×10^{-14}, Aw 5%, pore volume 0.5, C_{max} 1×10^{-5}, velocity 1×10^{-5}, and subsequent lines vary conditions of Aw, pore volumes showing the effect of soil conditions on downslope solute uptake

Conditions are critical for 1×10^{-14} and a slowly dissolving solute 1×10^{-16} is highly sensitive to stepped increases in reactive surface.

Figure 14.5 imagines three soil series occurring downslope with sharp boundaries between an upper, drier coarser-textured slope segment, a central loam segment and a lower clay segment. Path lengths are set at 50 m. Under the conditions set for $k_r = 1 \times 10^{-12}$ a constant output value can be assumed and the transition values rapidly adjust, mainly in terms of greater reactive surface area jumps.

Figure 14.6 uses plausible values for oligoclase (Pačés 1983; Huang and Kiang 1972) and partitions total cations in solution congruently with the composition of the mineral to predict solute concentration evolution and output at the slope foot. Figure 14.7 confirms that for rapid rates, the modelling of upslope conditions is irrelevant.

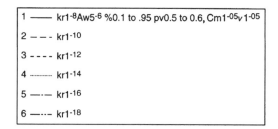

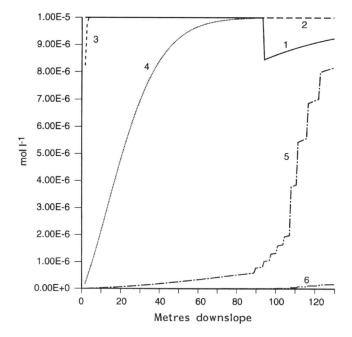

Figure 14.4 Effect of varying kr on downslope solute uptake with stepwise varying soil conditions. Solutes of rapid uptake (1×10^{-8}, 10, 12) reach C_{max} whatever the conditions; those slower than 1×10^{-16} show d_{max} distances longer than most slopes while those in-between show critical relationships between varying soil conditions and slope lengths

Figure 14.8 increases A_w and fractions, θ and p_v downslope but keeps C_{max} and v constant. As water content increases, d_{max} increases with t_{max}.

14.5 CONCLUSIONS

This chapter is concerned with conditions, probabilities and uncertainties. From a consideration of conditions relevant to hillslopes we are able to conclude that the dissolution rates of several commonly occurring minerals are of interest in hillslope solute modelling. Critical conditions of dissolution rate, C_{max}, surface area, soil

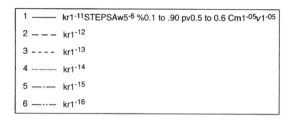

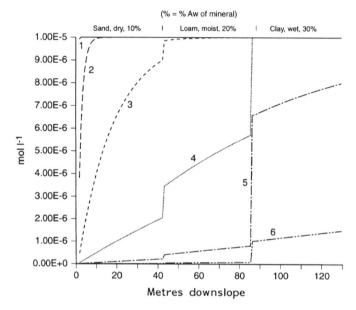

Figure 14.5 Solute uptake downslope with changing soil conditions appropriate to soil units of changing texture downslope

moisture and flow rate, can combine so that residence times and the attendant path lengths to C_{max} are of an order of magnitude or greater than those to be found on hillslopes. This is especially the case if published laboratory dissolution rates used are too high, increasing path lengths over which reactions take place. This means that it is worth considering the dynamics of solute uptake in hillslope hydrochemical modelling.

Despite the considerable uncertainties about the values used and processes envisaged, there would appear to be identifiable conditions:

1. For *rapidly dissolving solutes*, k_r greater than 1×10^{-6} mol/m^2/s path length is not important since whatever the hydrological or soil texture conditions (however rapid the flow or coarse-textured the soil with low A_w), they can reach C_{max} at d_{max} values of the order of 1×10^{-1} to 1×10^1 m.

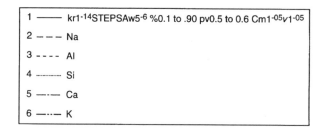

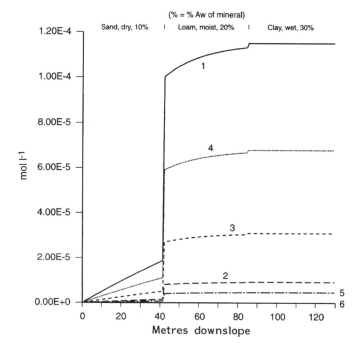

Figure 14.6 Partitioning of solute uptake appropriate to the congruent dissolution of oligoclase

2. For most solutes, however slowly they dissolve, path length is not important in fine-textured soils as *high* $A_w > 1 \times 10^5$ m^2/m^3 and *low v* $< 1 \times 10^{-5}$ m/s again means that they can reach C_{max} at d_{max} values of the order of 1×10^{-1} to 1×10^1 m.

3. Critical combinations are where path length is liable to be of the same order of hillslope length are solutes dissolving at rates of less than 1×10^{-6} mol/m^2/s in coarse-textured soils of $A_w > 10$ 1×10^4 m^2/m^3 and flows above 1×10^{-6} m/s.

4. Under realistic conditions it is likely that path length becomes more, not less important: this is especially true since (a) laboratory dissolution rates are widely seen as overestimates for field rates, and (b) as A_{wm} decreases and (c) if rapid flow pathways exists within the soil (1) and (2) become less true. This will be even more

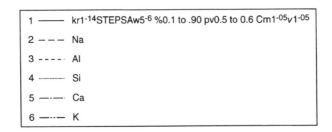

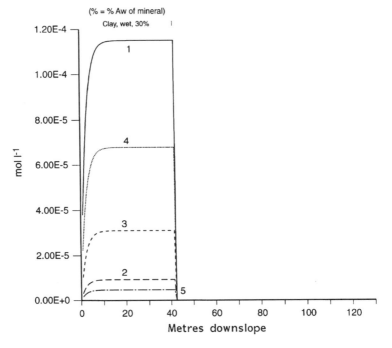

Oligoclase kr 8.9 x 10^{-13}, Cmax 1.15 x 10^{-4}, Aw 10+3,4,5, % 0.1,0.2,0.3, pv 0.3,0.4, 0.56, v 1 x 10^{-5}

Figure 14.7   As for Figure 14.6 but confirming that only the slope foot conditions need to be modelled

the case if soil hydrology acts to limit the actual contact of flowing water with the soil mineral material (Velbel 1993; Parnell 1993).

5. Countering (4) is the tendency for hillslope soils to have finer textures downslope, with higher A_w. Modelling of slope foot conditions is critical where this condition occurs and for rapidly dissolving solutes it could render hillslope path length irrelevant, rather the path length of the finer-textured condition will be important; here points (1) and (2) are liable to be important. The absence or presence of a slope foot alluvial deposit is thus critical to this consideration; however, if such an occurrence leads low permeability and thus to return overland flow so that further

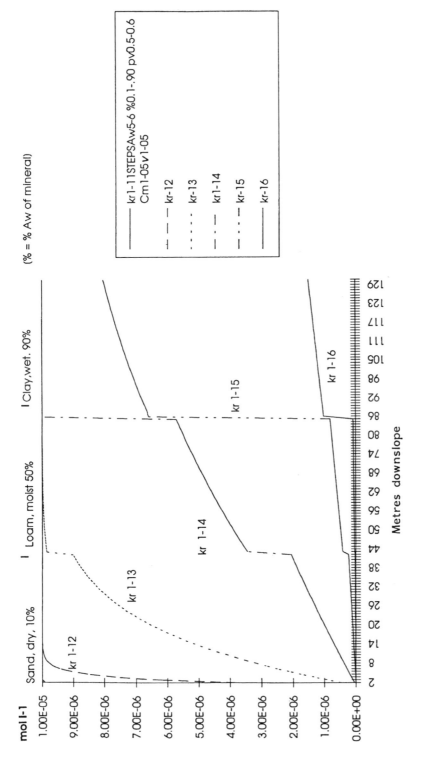

Figure 14.8 Increases in Aw and fractions, λ and pv downslope but keeps C_{max} and v constant. As water content increases d_{max} increases with t_{max}

contact with the soil does not happen, modelling upslope conditions under (3) and (4) then become important again.

6. Under many conditions A_{wm} is liable to be the controlling factor, together with the availability of reactants (especially pH) controlling C_{max}. Anion supply and exchange sites will also control concentrations. Since pH tends to increase downslope and A_{wm} tends to increase downslope, upslope conditions will be characterized by high k_r but low C_{max} and downslope by low k_r and high C_{max}. Thus, concentrations will tend to increase downslope not just in relation to solute uptake from dissolving minerals but also by virtue of greater reactive surface areas. The disposition of particle sizes, weatherable minerals and exchange sites are thus critical factors: increasing them decreases path length, bringing even the slowly dissolving minerals into the sphere of interest and making rapidly dissolving minerals more likely to have representative areal values rather than path lengths of evolution.

Ways ahead include using the existing hydrogeochemical models for groundwater chemical evolution and building them in to existing hillslope hydrological models. The difficulties lie in the adequate representation of the downslope variations in reactive surface areas and flows and in relating C_{max} for solutes derived from complex mineralogies. This is made easier when considering soils derived from one lithology but especially difficult in soils like those developed on glacial drift where heterogeneous admixtures of minerals exist. In these situations either the proportions of minerals present must be known or the alternative of back-calculating k values from output concentrations might be more viable (Ferguson *et al.* 1994; Trudgill *et al.* 1994b). Further papers will consider some of these developments.

ACKNOWLEDGEMENT

R. I. Ferguson for helpful comments.

REFERENCES

Acker, J.G. and Bricker, O.P. 1992. The influence of pH on biotite dissoluton and alteration kinetics at low temperature. *Geochimica et Cosmochimica Acta*, **56**, 3073–3092.

Anbeek, C. 1992. The dependence of dissolution rates on grain size for some fresh and weathered feldspars. *Geochimica et Cosmochimica Acta*, **56**, 3957–3970.

Anbeek, C., van Breemen, N., Meijer, E.L. and van der Plas, L. 1994. The dissolution of naturally weathered feldspar and quartz. *Geochimica et Cosmochimica Acta*, **58**, 4601–4613.

Anderson, M.G. (ed.) 1988. *Modelling Geomorphological Systems*. Wiley.

Anderson, M.G. and Burt, T.P. (eds). 1990. *Process Studies in Hillslope Hydrology*. Wiley.

Armstrong, A., Addiscott, T. and Leeds-Harrison, P. 1995. Methods for modelling solute movement in structured soils. In S.T. Trudgill (ed.), *Solute Modelling in Catchment Systems*. Wiley, pp. 133–161.

Berner, R.A. 1978. Rate control of mineral dissolution under earth surface conditions. *American Journal of Science*, **278**, 1234–1252.

Berner, R.A. 1981. Kinetics of weathering and diagenesis. In A.C. Lasaga and R.J. Kirkpatrick (eds), Kinetics of geochemical processes. *Reviews in Mineralogy*, **8**, 111–134.

Beven, K. 1995. Linking parameters across scales: subgrid parameterizations and scale dependent hydrological models. *Hydrological Processes*, **9**, 507–525.

Beven, K. and Kirkby, M.J. 1979. A physically based variable contributing area model of basin hydrology. *Hydrological Sciences Bulletin*, **24**, 43–69.

Billett, M.F. and Cresser, M.S. 1992. Predicting stream water quality using catchment and soil chemical characteristics. *Environmental Pollution*, **77**, 263–268.

Bluth, G.J. and Kump, L.R. 1994. Lithologic and climatologic controls of river chemistry. *Geochimica et Cosmochimica Acta*, **58**, 2341–2359.

Brown, G.H., Sharp, M.J., Tranter, M., Gurnell, A.M. and Nienow, P.W. 1994. Impact of post-mixing chemical reactions on the major ion chemistry of bulk meltwaters draining the Haut Glacier d'Arolla, Valais, Switzerland. *Hydrological Processes*, **8**, 465–480.

Brusseau, M.L. 1993. The influence of solute size, pore water velocity and intraparticle porosity on solute dispersion and transport in soil. *Water Resources Research*, **29**, 1071–1080.

Busenberg, E. and Clemency, C.V. 1976. The dissolution kinetics of feldspars at 25°C and 1 atm CO_2 partial pressure. *Geochimica et Cosmochimica Acta*, **40**, 41–49.

Burt, T.P. 1986. Runoff processes and solutional denudation rates on humid temperate hillslopes. In S.T. Trudgill (ed.), *Solute Processes*. Wiley, pp. 193–249.

Carson, M.A. and Kirkby, M.J. 1972. *Hillslope Form and Processes*. Cambridge.

Casey, W.H. 1995. Surface chemistry during the dissolution of oxide and silicate materials. In D.J. Vaughan and R.A.D. Pattrick (eds), *Mineral Surfaces*. Chapman and Hall.

Casey, W.H., Banfield, J.F., Westrich, H.R. and McLaughlin, L. 1993. What do dissolution experiments tell us about natural weathering? *Chemical Geology*, **105**, 1–15.

Clothier, B.E., Kirkham, M.B. and McLean, J.E. 1992. In situ measurement of the effective transport volume for solute moving through soil. *Soil Science Society of America, Journal*, **56**, 733–736.

Colman, S.M. and Dethier, D.P. 1986. *Rates of Chemical Weathering of Rocks and Minerals*. Academic Press.

Crabtree, R.W. 1986. Spatial distribution of solutional erosion. In S.T. Trudgill (ed.), *Solute Processes*, Wiley, pp. 329–361.

Crabtree, R.W. and Trudgill, S.T. 1987. Hillslope solute sources and solutional denudation on a Magnesian Limestone hillslope. *Transactions, Institute of British Geographers*, **12**, 97–106.

Dixon, J.C. 1986. Solute movement on hillslopes in the alpine environment of the Colorado Front Range. In A.D. Abrahams (ed.), *Hillslope Processes*. Allen and Unwin, pp. 139–159.

Domenico, P.A. and Schwartz, F.W. 1990. *Physical and Chemical Hydrogeology*. Wiley.

Drever, J.I. 1988. *The Geochemistry of Natural Waters*. Prentice Hall.

Drever, J.I. 1994. The effect of land plants on weathering rates of silicate minerals. *Geochimica et Cosmochimica Acta*, **58**, 2325–2332.

Ferguson, R.I., Trudgill, S.T. and Ball, J. 1994. Mixing and uptake of solutes in catchments: model development. *Journal of Hydrology*, **159**, 223–233.

Fetter, C.W. 1994. *Applied Hydrogeology*, 3rd edn. Macmillan.

Fountain, A.G. 1992. Subglacial water flow inferred from stream measurements at South Cascade Glacier, Washington, USA. *Journal of Glaciology*, **38**, 51–64.

Franke, W.A. and Teschner-Steinhardt, R. 1994. An experimental approach to the sequence of the stability of rock-forming minerals to chemical weathering. *Catena*, **21**, 279–290.

Furley, P.A. 1968. Soil formation and slope development. 2: The relationship between soil formation and gradient angle in the Oxford area. *Zeitschrift für Geomorphologie*, **NF12**, 25–42.

Furley, P.A. 1971. Relationship between slope form and soil properties developed over chalk parent materials. In D. Brunsden (ed.), *Slopes, Form and Process*. Institute of British Geographers, Special Publication 3, pp. 141–164.

Genereux, D.P., Hemond, H.F. and Mulholland, P.J. 1993. Spatial and temporal variability in streamflow generation on the West Fork of Walker Branch watershed. *Journal of Hydrology*, **142**, 137–166.

Germann, P.F. 1990. Macropores and hydrologic hillslope processes. In M.G. Anderson and T.P. Burt (eds), *Process Studies in Hillslope Hydrology*. Wiley, pp. 327–363.

Gíslason, S.R. and Anórsson, S. 1993. Dissoluton of primary basaltic minerals in natural waters: saturation state and kinetics. *Chemical Geology*, **105**, 117–135.

Glynn, P.D., Reardon, E.J., Plummer, L.N. and Busenberg, E. 1990. Reaction paths and equilibrium end-points in solid-solution aqueous-solution systems. *Geochimica et Cosmochimica Acta*, **54**, 267–282.

Goudie, A.S. 1995. *The Changing Earth: Rates of Geomorphological Processes*. Blackwell.

Harvey, J.W. 1993. Measurement of variation in soil solute tracer concentration across a range of effective pore sizes. *Water Resources Research*, **29**, 1831–1837.

Huang, W.H. and Kiang, W.C. 1972. Laboratory dissolution of plagioclase feldspars in water and organic acids at room temperature. *American Mineralogist*, **57**, 1849–1859.

Kendall, C., Sklash, M.G. and Bullen, T.D. 1995. Isotope tracers of water and solute sources in catchments. In S.T. Trudgill (ed.), *Solute Modelling in Catchment Systems*. Wiley, pp. 261–303.

Kenoyer, G.J. and Bowser, C.J. 1992a. Groundwater chemical evolution in a sandy silicate aquifer in northern Wisconsin. 1. Patterns and rates of change. *Water Resources Research*, **28**, 579–589.

Kenoyer, G.J. and Bowser, C.J. 1992b. Groundwater chemical evolution in a sandy silicate aquifer in northern Wisconsin. 1. Reaction modelling. *Water Resources Research*, **28**, 591–600.

Kirkby, M.J. (ed.) 1978. *Hillslope Hydrology*. Wiley.

Kirkby, M.J. 1986. Mathematical models for solutional development of landforms. In S.T. Trudgill (ed.), *Solute Processes*. Wiley, pp. 439–495.

Kirkby, M.J., Naden, P., Burt, T.P. and Butcher, D. 1993. *Computer Simulation in Physical Geography*. Wiley.

Lasaga, A.C. 1984. Chemical kinetics of water–rock interactions. *Journal of Geophysical Research*, **89**, 4009–4025.

Lasaga, A.C., Soler, J.M., Ganor, J., Burch, T.E. and Nagy, K.L. 1994. Chemical weathering rate laws and global geochemical cycles. *Geochimica et Cosmochimica Acta*, **58**, 2361–2386.

Lin, H.C. and Clemency, C.V. 1981. The kinetics of dissolution of muscovites at 25 °C and 1 atm CO_2 partial pressure. *Geochimica et Cosmochimica Acta*, **45**(4), 571–576

Mercado, A. and Billings, G.K. 1975. The kinetics of mineral dissolution in carbonate aquifers as a tool for hydrological investigations. 1. Concentration–time relationships. *Journal of Hydrology*, **24**, 303–331.

Moore, A.D., Turner, A.K., Wilson, J.P., Jenson, S.K. and Band, L.E. 1993a. GIS and land-surface–subsurface process modeling. In M.F. Goodchild, B.O. Parks and L.T. Steyaert (eds), *Environmental Modeling with GIS*. Oxford University Press, pp. 196–230.

Moore, I.D., Gessler, P.E., Nielsen, G.A. and Peterson, G.A. 1993b. Soil attribute prediction using terrain analysis. *Soil Science Society of America Journal*, **57**, 443–452.

Pačés, T. 1983. Rate constants of dissolution derived from the measurements of mass balance in hydrological catchments. *Geochimica et Cosmochimica Acta*, **47**, 1855–1864.

Parnell, R.A. 1993. Hydrologic control of chemical disequilibria in soil and surface waters, Sogndal, Norway. *Chemical Geology*, **105**, 101–115.

Pilgrim, D.H., Huff, D.D. and Steele, T.D. 1979. Use of specific conductance and contact time relationships for separating flow components in storm runoff. *Water Resources Research*, **15**, 329–339.

Plummer, L.N., Parkhurst, D.L. and Thorstenson, D.C. 1983. Development of reaction models for ground-water systems. *Geochimica et Cosmochimica Acta*, **47**, 665–686.

Plummer, L.N., Prestemon, E.C. and Parkhurst, D.L. 1994. An interactive code (NETPATH) for modeling *NET* geochemical reactions along a flow *PATH*. Version 2.0. US Geological Survey, Water-Resources Investigation Report 94-4169.

Quinn, P.F. and Beven, K.J. 1993. Spatial and temporal predictions of soil moisture dynamics, runoff, variable source areas and evapotranspiration for Plynlimon, mid-Wales. *Hydrological Processes*, **7**, 425–448.

Quinn, P.F., Beven, K., Chevallier, P. and Planchon, O. 1991. The prediction of hillslope flowpaths for distributed modelling using digital terrain models. *Hydrological Processes*, **5**, 59–79.

Rapp, A. 1960. Recent development of mountain slopes in Karkevagge and surroundings, northern Scandinavia. *Geografiska Annaler, A*, **42**, 65–200.

Rose, C.W., Chichester, F.W., Williams, J.R. and Ritchie, J.T. 1982. A contribution to simplified models of field solute transport. *Journal of Environmental Quality*, **11**, 146–155.

Rowell, D.L. 1994. *Soil Science: Methods and Applications*. Longman.

Sklash, M.G. 1990. Environmental isotope studies of storm and snowmelt generation. In M.G. Anderson and T.P. Burt (eds), *Process Studies in Hillslope Hydrology*. Wiley, pp. 401–435.

Stewart, M.K. and McDonnell, J. 1991. Modelling base flow soil water residence times from deuterium concentrations. *Water Resources Research*, **27**, 2681–2693.

Svensson, U. and Dreybrodt, W. 1992. Dissolution kinetics of natural calcite minerals in CO_2–water systems approaching calcite equilibrium. *Chemical Geology*, **100**, 129–145.

Sverdrup, H. and Warfvinge, P. 1988. Weathering of primary silicate minerals in the natural soil environment in relation to a chemical weathering model. *Water, Air and Soil Pollution*, **38**, 387–408.

Sverdrup, H., Alveteg, M., Langan, S. and Paces, T. 1995. Biogeochemical modelling of small catchments using PROFILE and SAFE. In S.T. Trudgill (ed.), *Solute Modelling in Catchment Systems*. Wiley, pp. 75–99.

Swoboda-Colberg, N.G. and Drever, J.I. 1993. Mineral dissolution rates in plot-scale field and laboratory experiments. *Chemical Geology*, **105**, 51–69.

Taylor, A.B. and Velbel, M.A. 1991. Geochemical mass balances and weathering rates in forested watersheds of the southern Blue Ridge. II. Effects of botanical uptake terms. *Geoderma*, **51**, 29–50.

Trudgill, S.T. 1975. Measurement of erosional weight loss of rock tablets. *British Geomorphological Research Group, Technical Bulletin*, **17**, 13–19.

Trudgill, S.T. 1976. The erosion of limestone under soil and the long-term stability of soil–vegetation systems on limestone. *Earth Surface Processes*, **1**, 31–41.

Trudgill, S.T. 1985. Field observations of limestone weathering and erosion in the Malham District, North Yorkshire. *Field Studies*, **6**, 201–236.

Trudgill, S.T. (ed.) 1986b. *Solute Processes*. Wiley.

Trudgill, S.T. 1986a. Limestone weathering under a soil cover and the evolution of limestone pavements, Malham District, North Yorkshire, UK. In K. Paterson and M.M. Sweeting (eds), *New Directions in Karst*. Geo Books, pp. 461–471.

Trudgill, S.T. (ed.) 1995. *Solute Modelling in Catchment Systems*. Wiley.

Trudgill, S.T., Pickles, A.M., Smettem, K.J. and Crabtree, R.W. 1983. Soil water residence time and solute uptake, 1: dye tracing and rainfall events. *Journal of Hydrology*, **60**, 257–279.

Trudgill, S.T., Crabtree, R.W., Ferguson, R.I., Ball, J. and Gent, R. 1994a. Ten year re-measurement of chemical denudation on a Magnesian Limestone Hillslope. *Earth Surface Processes and Landforms*, **19**, 109–114.

Trudgill, S.T., Ball, J. and Ferguson, R.I. 1994b. Excel modelling of hydrological systems. *Earth Surface Processes and Landforms*, **19**, 815–817.

Unnikrishna, P.V., McDonnell, J.J. and Stewart, M.K. 1995. Soil water isotopic residence time modelling. In S.T. Trudgill (ed.), *Solute Modelling in Catchment Systems*. Wiley, pp. 237–260.

Van Grinsven, J.J.M. and van Riemsdijk, W.H. 1992. Evaluation of batch and column techniques to measure weathering rates in soils. *Geoderma*, **52**, 411–457.

Velbel, M.A. 1985. Geochemical mass balances and weathering rates in forested watersheds of the southern Blue Ridge. *American Journal of Science*, **285**, 904–430.

Velbel, M.A. 1992. Geochemical mass balances and weathering rates in forested watersheds of the southern Blue Ridge. III. Cation budgets and the weathering rate of amphibole. *American Journal of Science*, **292**, 58–78.

Velbel, M.A. 1993. Constancy of silicate-mineral weathering rate ratios between natural and experimental weathering: implications for hydrologic control of differences in absolute rates. *Chemical Geology*, **105**, 89–98.

Velbel, M.A. 1995. Interactions of ecosystem processes and weathering processes. In Trudgill, S.T. (ed.), *Solute Modelling in Catchment Systems*, Wiley, Chichester, pp. 193–209.

Voss, C.I. 1984. *SUTRA – Saturated–Unsaturated Transport*. US Geological Survey, Reston, Virginia.

Wallach, R. and Shabtai, R. 1992. Surface contamination by soil chemicals: simulations for equilibrium and first-order kinetics. *Water Resources Research*, **28**, 167–173.

Walling, D.E. and Webb, B.W. 1986. Solutes in river systems. In S.T. Trudgill (ed.), *Solute Processes*. Wiley, pp. 251–327.

White, A.F. and Claassen, H.C. 1979. Dissolution of silicate rocks – applications to solute modelling. *American Chemical Society, Symposium Series*, **93**, 447–473.

Zulla, Y. and Billett, M.F. 1994. Long-term changes in chemical weathering rates between 1949–1950 and 1987 in forest soils from north-east Scotland. *European Journal of Soil Science*, **45**, 327–335.

Section 4

SOIL PROCESSES ON HILLSLOPES

15 The Influence of Slope on the Nature and Distribution of Soils and Plant Communities in the Central Brazilian *cerrado*

PETER A. FURLEY
Department of Geography, University of Edinburgh, UK

15.1 INTRODUCTION

Recent research has indicated the highly diverse nature of the flora of the Brazilian savannas (*cerrados*) and has begun to analyse the environmental and biotic controls over the vegetation (Ratter and Dargie 1992; Oliveira-Filho and Ratter 1995). The more xeromorphic features of *cerrado* vegetation were associated with soil properties as early as Waibel (1948) and Alvim and Araujo (1952) but the influence of topography and drainage has only been elaborated in detail over the past two decades (Furley 1992).

The aim of this research is to assess the extent of the interdependence between soil characteristics and plant communities over a sequence of slope-affected sites in the core *cerrado* area of central Brazil.

Although the observations reported here relate principally to the ecological reserve at Fazenda Água Limpa, situated approximately 18 km from the centre of Brasília (Figure 15.1), the distribution of the major soil groups and their relationship to vegetation formations is typical of much of the central plateau, an area covering approximately 2 million km².

15.2 IMPACT OF TOPOGRAPHY AND DRAINAGE ON HILLSLOPE PROCESSES AND THEIR EFFECT ON BIOTIC AND EDAPHIC DISTRIBUTIONS

The characteristic plant communities and soils can be represented in a model which illustrates their typical spatial location (Figure 15.2).

The uppermost land surfaces are composed of level or gently sloping, convex interfluves. Frequently they are capped by densely packed lenses of plinthitic gravel (Figure 15.3) derived from several cycles of weathering, transport and re-deposition, possibly involving landscape inversion in a complex geomorphological and pedological history. Such Latosols (Oxisols, US Soil Taxonomy) and Red-Yellow

Advances in Hillslope Processes, Volume 1. Edited by M. G. Anderson and S. M. Brooks.
© 1996 John Wiley & Sons Ltd.

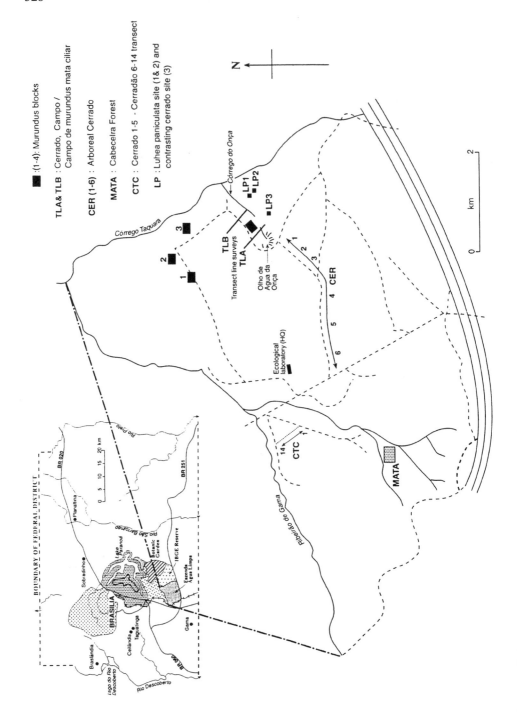

N

■ :(1-4): Murundus blocks

TLA& TLB : Cerrado, Campo /
Campo de murundus mata ciliar

CER (1-6) : Arboreal Cerrado

MATA : Cabeceira Forest

CTC : Cerrado 1-5 - Cerradão 6-14 transect

LP : Luhea paniculata site (1& 2) and
contrasting cerrado site (3)

Podsols (Ultisols) support woody savanna, where tree roots are able to draw from stored water throughout the intense dry season. At locations where deeper and finer-textured soils have developed, a dark red Latosol (Eutrorthox) can occur and provide favourable conditions for closed arboreal savanna which in turn affects the colour and surface soil properties. At other shallow or stony sites with Cambisols (Entisols and Inceptisols), a dry shrubby savanna frequently evolves. Downslope from the crest of such interfluves, the hillsides become transport slopes and the characteristics of the soil profiles vary according to drainage. In well-drained situations, a typical deep red oxidized profile develops (Red Latosols and Podsols), which is replaced by yellowish red soils where drainage is impeded, resulting in reduction of metallic oxides. Over such slopes, open woody savanna often gives way to shrubby grassland. The vegetation pattern is often interrupted towards the foot of slopes, where pockets of woody savanna signify the presence of indurated iron concretions (ironstone) at seepage points or the summits of earthmounds (*murundus*) which protrude above the surface and provide drier sites in the progressively wetter grassland (Oliveira-Filho and Furley 1990; Ponce and da Cunha 1993). Over the lower slopes and footslopes, gleyed, colluvial soils are widespread with buried profiles (Hydromorphic soils, Inceptisols and Entisols), supporting grasslands which lead down to the forest–savanna boundary and into the gallery forest (Ratter 1992; da Silva *et al.*, Chapter 20, this volume).

Each of these plant communities occupies a distinctive hydrological niche on the slope and appears to be associated with a characteristic suite of soil properties.

15.3 METHODOLOGY

No detailed soil survey exists for Fazenda Água Limpa. Some of the earliest general observations were made for agronomic purposes. The only systematic surveys have been those carried out by EMBRAPA (1978) for the whole of the Federal District at a scale of 1:100 000 and the concurrent Projeto Radambrasil surveys at a reconnaissance scale of 1:250 000 (MME/SG 1982). The most detailed soil investigations prior to the 1980s related to research sites set up to study plant ecology (Furley 1985). Subsequent studies have dealt mainly with aspects of soil–plant relationships and soil faunal influences (e.g. Haridasan 1982, 1985; Felfily and da Silva 1986, 1988, 1992). However, there remains a lack of systematic soil survey at an appropriate scale for examining small-scale variations in plant distribution (for instance at a scale of 1:10 000 or larger). The distribution of plant communities is better known and detailed surveys of the vegetation were made by means of both belt and point-centred quarter transects, as well as systematic observations involving compilation of species lists (Ratter 1980, 1991) (Table 15.1(a)). By way of comparison, the proportional areas in the nearby and much larger Parque Nacional de Brasília were calculated from LANDSAT TM imagery (Table 15.1(b)).

Figure 15.1 (*opposite*) Location of Fazenda Água Limpa showing the sampling sites

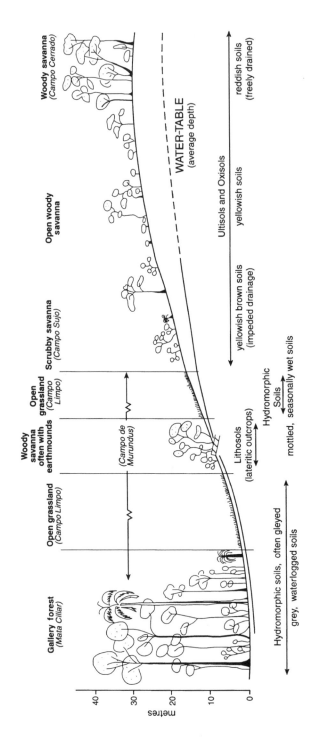

Figure 15.2 Model depicting the characteristic sequence of soils and vegetation with slope and drainage in the central *cerrado*

Figure 15.3   Convex interfluve sites showing the capping lens of plinthitic gravels

The soils have been investigated by means of over a 100 pits and trenches, selected
to represent variations in the vegetation both at a broad scale and at detailed sites of
particular interest (Table 15.2). Throughout the Federal District the distributional
pattern of vegetation formations is remarkably consistent and can be related to the
general model of soils, plants and slope illustrated in Figure 15.2. These relationships
are examined for each of the major types of vegetation. Variations in the water-table
were measured directly for two of the transects over a two-year period (Furley 1992,

Table 15.1   (a) Area occupied by different vegetation formations at Fazenda Água Limpa (total area 4062 ha). (b) Vegetation formations at the Parque Nacional de Brasília (total area 33 244 ha)

(a)

Vegetation formation	%
Dense arboreal savanna (cerradão)	0.2
Open woody cerrado	36.5
Grassland	
campo limpo	28.2
campo de murundus	4.4
Shrubby grassland (campo sujo)	20.1
Gallery forest/wet headwater catchments (mata ciliar/cabeceira)	9.9
Mesotrophic deciduous woodland	0.7

Based on Ratter (1980, 1991) and air photographs.

Vegetation formations at the Parque Nacional de Brasília (total area 33 244 ha)

(b)

Vegetation formation	%
Dense woody cerrado	29.7
Open shrubby savanna	22.6[a]
Gallery forest/wet catchment	22.3
Water	1.6
Rocky outcrops and gravel workings	9.1
Bare ground/urban	14.0[b]
Agriculture	0.7

[a]Probably underestimated.
[b]Probably overestimated.
Based on LANDSAT TM.

Fig. 6.5, p. 98), and taken from well records at several points on Fazenda Água Limpa over a five-year period.

15.4   THE MAJOR PLANT COMMUNITIES AND THEIR EDAPHIC CHARACTERISTICS

15.4.1   Cerradão

The area comprises a small, gently sloping patch of woodland forming an interfluve of variable tree density stretching across the upper slopes of the Riberão de Gama (Figure 15.4).

The trees reach 10–12 m tall and provide a crown cover which in places attains 100% but is generally much less, often due to gaps formed by treefall. Many trees, however, exceed this height and some specimens of *Bowdichia virgilioides* are

Table 15.2 Location and nature of research sites

Vegetation formation	Type of survey	No. of soil profiles	No. of plant sites
Cerradão	line transect	8	10 (25 × 10 m blocks) 1 × (50 × 50 m) block 2 PCQs (70 and 90 points)
Cerrado (open)	line transect	5	9 (25 × 10m) blocks; 1 × (50 × 50 m) block
Cerrado (dense)	line transect	6	10 m radius around each profile
Cerrado (medium/ open)	catena segment	14	10 m radius around each profile
Cerrado (medium)	catena segment	20	10 m radius around each profile
Campo de murundus	trenches	6	10 m radius around each profile
Gallery forest	blocks/catena segment	13	belt transects
Mesotrophic woodland	block	2	1 (50 × 50 m) block

particularly tall, attaining heights of nearly 20 m with an upright, narrow form. A number of large trees are common in this type of woodland, notably *Blepharocalyx suaveolens*, *Siphoneugena densiflora*, *Bowdichia virgilioides,*, *Emmotum nitens*, *Caryocar brasiliense*, and in places, *Copaifera langsdorfii* and *Simarouba versicolor*. There are also many big, venerable-looking specimens of *Qualea grandiflora*, *Dalbergia violacea* and *Vochysia thyrsoidea*. Some species of the characteristic large trees are occasional or rare, such as *Ocotea spixiana*, *Phoebe erythropus* and *Callisthene major*. Details of the species occurrence for trees and shrubs are given in Ratter (1980, 1991).

Eight profiles were examined with a further profile in the diffuse transition zone. All sites were tentatively classified as Dark Red Latosols (Latosolo Roxo or Terra Roxa Legitima – Eutrustox or Eutrorthox), having a thin litter layer (less than 2 mm), dark reddish brown colour (typically 5YR3/4 to 3/6) and sandy clay loam textures with moderately strong, cohesive structure in the surface horizons, becoming reddish brown (2.5YR4/6–4/8) with depth. This was accompanied by a slight textural increase and a greater angularity and blockiness of structure. The surface horizons contain abundant medium and fine roots, which continue throughout the profile, and numerous termites. Boundaries become diffuse below the surface 20 cm.

From the summary of mean values for the soil properties (Table 15.3), the clay-rich, mildly acidic, low nutrient status is evident with medium to high aluminium levels giving low *ki* values, indicating intensive weathering and leaching. There are appreciable amounts of surface organic matter, as reflected in the organic carbon percentages with well mineralized nitrogen giving characteristic C/N ratios of around

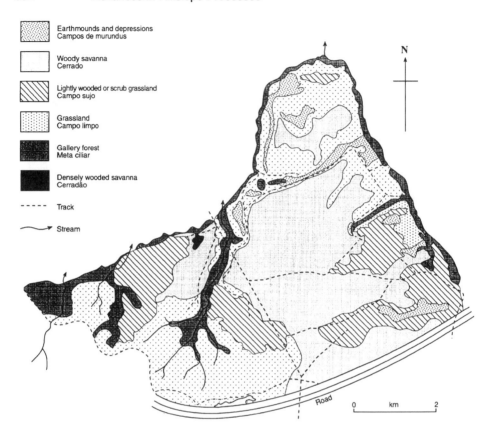

Figure 15.4 Vegetation formations of Fazenda Água Limpa (after Ratter 1980)

10. The rich colour and relatively high surface levels give a spurious impression of fertility.

Species growth and plant distribution are highly dynamic. On the one hand, it was observed that in undisturbed conditions, there is active plant growth and that the *cerrado–cerradão* ecotone becomes increasingly woody with time. On the other hand, periodic fires suppress and eliminate this tendency. It was observed that the *cerradão* appeared to be markedly more fire-resistant once the young trees had reached some 3 m in height and 10 cm in diameter (dbh). The implication is that, since the underlying soil profile properties do not vary significantly from the adjacent *cerrado* sites (see below), the edaphic factor has less influence here than the disturbing factors, principally fire. The surface soil organic levels and biomass are evidently related to undisturbed plant growth. Whereas the distinctive colour may reflect a mineralogy not determined here, it may be a function of the redox state of the metallic elements and therefore relate to water status in these well-drained, interfluvial sites.

Table 15.3 Summary of the soil characteristics of the *cerradão*

Means of seven profiles	Surface (0–20 cm)	Subsurface (20–40 cm)
Fractions of the total sample >2 mm (%)	1.1	1.3
Fractions of the fine earth sample (<2 mm) (%)		
sand (2–0.5 mm) (%)	7.5	7.5
silt (0.05–0.002 mm) (%)	13.6	10.6
clay (<0.002 mm) (%)	78.9	82.4
clay dispersible in water (%)	41.9	49.6
Degree of flocculation (%)	47.0	39.6
Acidity and nutrient status		
pH H_2O	5.0	5.3
pH KCl	4.0	4.5
Ca+Mg (cmol/kg)	0.28	0.12
K	0.13	0.03
Sum (S) Ca, Mg, K, Na	0.44	0.14
Al	1.41	0.35
H	9.61	5.20
base saturation (%)	3.9	2.4
100Al/(S)+Al	75.8	66.4
Organic properties		
P (assimilable) ppm	<0.5	<0.5
C organic (%)	3.02	1.72
N total (%)	0.30	0.19
C/N	10.1	9.6
Oxides		
SiO_2 (%)	14.1	14.3
Al_2O_3 (%)	31.4	32.0
Fe_2O_3 (%)	13.7	13.9
TiO_2 (%)	0.85	0.93
Weathering indices		
SiO_2/Al_2O_3 (ki)	0.77	0.76
SiO_2/R_2O_3 (kr)	0.60	0.60
Al_2O_3/Fe_2O_3	3.58	3.61

15.4.2 Open Woody Savanna (*Cerrado*)

Although the term *cerrado* is broadly synonymous with savanna, in its stricter sense it defines a vegetation formation lying over well-drained, gently inclined topography usually occupying the upper convex slopes and interfluves. The great variety of different plant associations reflects the diversity of environmental factors (Eiten 1972; Goodland and Pollard 1973; Oliveira Filho *et al.* 1989). There are numerous smaller trees, shrubs and palms under 3 m tall. The most abundant include *Byrsonima crassa* and *B. verbascifolia*, *Kielmeyer coriacea* and *K. speciosa*, *Oratea lexasperma*, *Qualea grandiflora* and *Q. parviflora*, but the species composition varies from location to location (Ratter and Dargie 1992). The ground layer is formed mainly by grasses, smaller shrubs and sedges, varying from a sparse to a dense ground cover. The present analysis summarizes variations observed over the three subdivisions based on tree density.

The *cerrado* studied was located in two areas some 2 km apart, typically found on gently sloping ground near the crests and over the interfluves. The vegetation varies in density but there are no areas with complete or almost complete crown cover and, in consequence, the ground vegetation is everywhere well-developed. The taller trees are generally 5–6 m tall but occasional individuals, particularly of *Eriotheca pubescens* and *Bowdichia virgilioides*, can reach much greater heights.

The soils are extremely varied over the full range of plant associations of the *cerrado*, although there are a number of common characteristics (Table 15.4). These shared features include good drainage, high clay content, a high degree of flocculation, low pH, low assimilable phosphorus, low base cation and base saturation levels, consistently high Al_2O_3 and significant Fe_2O_3 amounts which make up the predominantly kaolinite content (MME/SG 1982) of the clay fraction. The Al levels are so consistently high that most native *cerrado* plants are adapted and many are notable Al-accumulators, such as members of the Vochysiaceae (Haridasan

Table 15.4 Summary of the soil characteristics of the savanna (*cerrado* sensu lato)

Means of 45 profiles	Surface (0–20 cm)	Subsurface (20–40 cm)
Fractions of the total sample >2 mm (%)	21.0	41.0
Fractions of the fine earth sample (<2 mm) (%)		
sand (2–0.5 mm) (%)	14.0	13.0
silt (0.05–0.002 mm) (%)	25.0	19.0
clay (<0.002 mm) (%)	60.0	68.0
clay dispersible in water (%)	22.0	23.0
Degree of flocculation (%)	69.0	69.0
Acidity and nutrient status		
pH H_2O	4.5	4.9
pH KCl	4.0	4.6
Ca+Mg (cmol/kg)	0.58	0.14
K	0.19	0.08
Sum (S) Ca, Mg, K, Na	0.83	0.25
Al	1.13	0.42
H	9.58	5.05
base saturation (%)	6.7	4.5
100Al/(S)+Al	60.0	49.0
Organic properties		
P (assimilable) ppm	<0.50	<0.50
C organic (%)	2.70	1.46
N total (%)	0.23	0.14
C/N	12.0	11.0
Oxides		
SiO_2 (%)	8.31	8.24
Al_2O_3 (%)	29.63	30.64
Fe_2O_3 (%)	14.46	14.95
TiO_2 (%)	0.78	0.82
Weathering indices		
SiO_2/Al_2O_3 (ki)	0.50	0.55
SiO_2/R_2O_3 (kr)	0.38	0.38
Al_2O_3/Fe_2O_3	3.31	3.24

1982). The soils are therefore both dystrophic and allic, resulting from intensive weathering as demonstrated by the low *ki* and *kr* indices. Nearly all the surface soils have organic carbon percentages over 2 and appreciable levels in the subsurface horizons giving consistent C/N values of 10 to 14. Several local variations occur which have a marked affect on plant distribution. Soil characteristics which are especially variable include depth (affecting rooting), the sand and the coarser size fraction, varying degrees of plinthitization and hardening, and the subsurface water content affecting the oxidation states of the metallic elements. These properties generally fall within the broad parameters suggested by Lopes and Cox (1977a,b), Furley and Ratter (1988).

Within this broad characterization, there are several distinct trends in soil–plant community relationships. Since most of the area is gently concave–convex with broad interfluves, differences in topography, geomorphological evolution and drainage can be seen to have a determining effect on both the soil properties and plant communities. At the extreme end of the spectrum on shallow infertile soils with sparse, short grass- and shrub-dominated plant associations there is a much enhanced coarse fraction increasing to 75% of the total soil volume. This vegetation community is also associated with rocky outcrops (*campo rupestre*). The clay percentages are still significant (over 50% of the weight of the fine earth fraction). The proportion of this clay fraction dispersible in water remains constant at about a third. The degree of flocculation is high, generally over 70%. The pH values are low, at or under 4, with predictably low levels of base nutrients. The summation of the bases is frequently less than 1 cmol/kg. Exchangeable aluminium values rise to over 1 cmol/kg and form a more dominant part of the total acidity. Base saturation and available phosphorus are extremely low (less than 10% and 0.1 mg/100 g respectively). The organic complex is well mineralized but retains a reasonable organic carbon percentage with C/N ratios typically around 10. The SiO_2 levels are lower than for other *cerrado* sites (less than 5%) with high iron (over 30%) and aluminium (over 14%). The *ki* ($< c.\ 0.25$) and *kr* ($< c.\ 0.2$) values are depressed with lower Al/Fe ratios of less than 4.

These conditions are often associated with interfluves, which frequently have a dense packing of granular ironstone. On the other hand, further downslope, the soils more closely resemble the grassy (campo) sites (see below) with deeper colluvial profiles (Figure 15.5). The influence of the wet season and raised water-tables is more evident, especially in the subsurface horizons. The *cerrado* is therefore maintained over a range of soils, although always well drained within the main rooting zone, acidic and lacking in base nutrients. It is therefore typical of the convex summits sometimes with granular ironstone soils and of the deep, weathered soils typical of the upper slope sections of toposequences.

15.4.3 Grassland (*Campos*) and Grasslands with Earthmounds (*Campos de Murundus*)

These predominantly herbaceous areas are typically located towards the lower slopes of valleys below the *cerrado* and upslope of the gallery formations (Ratter 1980, 1991). Other shrubby and bushy grasslands (the dry hill *campos*) on the upper slopes

Figure 5.5 Footslope section showing a buried Red Latosol profile beneath colluvial deposits forming the present-day soil

and interfluves are more associated with the extreme *cerrado* conditions outlined earlier. Frequently, tracts of earthmounds cover the slopes with densities as great as 61 per hectare (Diniz *et al.* 1986). Although often related to termite activity, most of the murundus in the topographic situations examined here have been shown to be formed by differential erosion from seepage water associated with the water-table and with lateral surface erosion (Furley 1986). In these situations, the earthmounds are covered with typical *cerrado* flora whilst each discrete mound is surrounded by seasonally wet grassland. The *cerrado* vegetation and the mound soils will not be considered further here since they conform with the model already described.

The vegetation of the wet grassland varies from low, tufted species made up of short grasses, sedges and xyrids enlivened with Burmanniaceae, Lentibulariaceae, Gentianaceae, Eriocaulaceae and lycopods to waist-high thickets of grasses and sedges. The dry hill grasslands are often dominated by the same species as constitute the "herb layer' of the *cerrado* but contain important characteristic species as well, such as the sedges *Bulbostylis emmerichiae* T. Koyama, *Rhynchospora warmingii* Boeck and *R. Terminalis* (Nees) Steudal, which dominate considerable areas (Ratter 1980, 1991).

In contrast to the *cerrado* sites, the grasslands are underlain by darker-coloured soils lacking the coarser grain-size fractions (over 2 mm eps) and iron oxide levels, and containing high exchange acidity in the surface horizons. The organic levels are notably higher. The weathering ratios are less than in the *cerrado*, partly resulting from the low SiO_2 levels. The higher pH levels, base saturation and slight increase in clay content and cations, together with the organic concentrations in the subsurface soils, may be a reflection of the grass root densities and the ability of plants to concentrate materials in the rhizosphere (Table 15.5).

Thus variations in edaphic properties are related to the presence, absence or periodicity of water, together with the effectiveness of soil drainage. Consequently, on raised microrelief such as the earth mounds which characterize these areas, the summits are well-drained and typically have open woody vegetation underlain by Latosols, whereas the depressions contain wet grasslands typified by poor drainage and hydromorphic, highly gleyed and greatly reduced soils. The marked seasonality means that the tree species which form the gallery forests, are unable to survive during the dry season in the wet *campos* and, at the same time, these sites are too wet during the rainy season for the *cerrado* vegetation. In fact, our observations indicate that the gallery is advancing upslope and therefore demonstrate that the forest trees are able to adapt to the constraints posed by regular water supply during the dry season, at least over the lower parts of the slope (Ratter 1992).

15.4.4 Gallery Forests and Wet Headwater Catchment Forests (*Mata ciliar e Cabeceiras*)

Two sites were investigated which only partially represent the extensive gallery forests on the Fazenda, covering about 10% of the total (Ratter 1980; Felfily & da Silva 1992). This is however, less than half the relative proportion of gallery forest on the nearby Parque Nacional de Brasília (Table 15.1(b)).

The vegetation of the gallery forest of the Corrego da Onça, where most of the soil transects were sited, consists of swampy forest in areas closest to the stream and rather better drained tracts closer to the forest–savanna boundary. In general, the taller trees are 14–20 m in height and their crowns form a fairly dense canopy in the centre of the gallery but are sparser in the marginal areas. The commonest tall tree species in the damper parts of the gallery are *Pseudolmedia laevigata*, *Protium* spp.

Table 15.5 Summary of the soil characteristics of grasslands and earthmound grasslands (*campos* and *campos de murundus*)

Means of six profiles	Surface (0–20 cm)	Subsurface (20–40 cm)
Fractions of the total sample >2 mm (%)	4.0	2.0
Fractions of the fine earth sample (<2 mm) (%)		
sand (2–0.5 mm) (%)	21.0	20.0
silt (0.05–0.002 mm) (%)	23.0	17.0
clay (<0.002 mm) (%)	57.0	63.0
clay dispersible in water (%)	28.0	14.0
Degree of flocculation (%)	51.0	78.0
Acidity and nutrient status		
pH H_2O	4.9	5.5
pH KCl	4.2	5.6
Ca+Mg (cmol/kg)	0.18	0.15
K	0.10	0.04
Sum (S) Ca, Mg, K, Na	0.32	0.20
Al	1.05	0.02
H	9.85	3.28
base saturation (%)	2.5	9.0
100Al/(S)+Al	76.0	12.0
Organic properties		
P (assimilable) ppm	<0.10	<0.75
C organic (%)	4.29	1.18
N total (%)	0.33	0.12
C/N	8.0	10.0
Oxides		
SiO_2 (%)	5.60	4.22
Al_2O_3 (%)	33.00	40.03
Fe_2O_3 (%)	4.90	5.83
TiO_2 (%)	0.86	0.91
Weathering indices		
SiO_2/Al_2O_3 (ki)	0.28	0.18
SiO_2/R_2O_3 (kr)	0.26	0.17
Al_2O_3/Fe_2O_3	11.04	11.17

and *Calphyllum brasiliense*. In drier and marginal tracts, *Richeria obovata, Tapirira guianensis, Maprounea guianensis* and *Copaifera langsdorifii* are important. The other major tract of gallery forest is the large headwater catchment (cabeceira) of the Corrego Capitinga.

The poorly drained soils of the galleries, which are found over footslopes and valley floors, are clearly distinguished from those of the better-drained soils typical of the slopes and interfluves. They appeared to be mostly alluvial and partly colluvial depending upon position and the provenance of deposited sediment. The groundwater is close to the surface even during the dry season. The soils are characterized by low percentages of coarser size fractions, high silt contents, low amounts of water-dispersible clay, high degrees of flocculation and elevated cation values (especially exchangeable potassium). They are still relatively acid, with high exchangeable aluminium and exchange acidity, high phosphorus, low Fe_2O_3 and

Table 15.6 Summary of the soil characteristics of gallery forest (*mata ciliar*)

Means of 13 profiles	Surface (0–20 cm)	Subsurface (20–40 cm)
Fractions of the total sample >2 mm (%)	6.0	N/A (water-table)
Fractions of the fine earth sample (<2 mm) (%)		
sand (2–0.5 mm) (%)	8.0	
silt (0.05–0.002 mm) (%)	61.0	
clay (<0.002 mm) (%)	31.0	
clay dispersible in water (%)	14.0	
Degree of flocculation (%)	62.0	
Acidity and nutrient status		
pH H_2O	4.2	
pH KCl	3.5	
Ca+Mg (cmol/kg)	1.23	
K	0.44	
Sum (S) Ca, Mg, K, Na	1.70	
Al	6.41	
H	34.45	
base saturation (%)	4.0	
100Al/(S)+Al	80.0	
Organic properties	3.04	
P (assimilable) ppm	12.51	
C organic (%)	0.98	
N total (%)	13.0	
C/N		
Oxides		
SiO_2 (%)	10.87	
Al_2O_3 (%)	14.40	
Fe_2O_3 (%)	2.84	
TiO_2 (%)	n/a	
Weathering indices		
SiO_2/Al_2O_3 (ki)	1.50	
SiO_2/R_2O_3 (kr)	1.32	
Al_2O_3/Fe_2O_3	8.26	

with high organic levels including carbon and nitrogen. In terms of nutrient status, these soils are by far the most favourable for plant growth, despite the high aluminium levels (Table 15.6).

It is evident that water plays a dominant role in determining the distribution and, to some extent, the nature of the galleries. Variations in the soil water content have been shown to increase progressively towards the margins, particularly the upslope forest–savanna boundary on catenary transects (Furley 1992). The alluvial and colluvial character of the soils provides a slightly more favourable nutrient status although the exchange acidity levels are still very high and native plants have had to adapt to the aluminium levels. The high organic content reflects the increased biomass input (in comparison with other plant communities on the Reserve) as well as the lack of decomposition in the damp subsurface conditions which persist for most of the year. More detailed work has shown subtle differences within the galleries (see da Silva *et al.*, Chapter 20, this volume).

15.4.5 Mesotrophic Savannas (Mesotrophic facies *cerrado*)

The close relationship between deciduous tree growth and more favourable edaphic conditions has been noted in several publications (Ratter *et al.* 1973, 1977, 1978; Furley and Ratter 1988). The *cerrado* has been relatively recently subdivided into dystrophic and mesotrophic facies; the characteristic vegetation occurring on mesotrophic soils being described in a series of publications (for instance, Ratter 1971). The vegetation can be recognized by the presence of a number of indicator species, and they were found to occur in one small area of the Reserve.

The area enclosing the plant community carries an open *cerrado* with strongly developed ground vegetation with a well-drained soil. The community is located at midslope sites with good drainage. Trees of *Luhea paniculata* Mart., *Platypodium grandiflorum* Benth., *Pseudobombax longiflorum* (Mart. & Zucc.) A. Robyns, *P. tomentosum* (Mart. & Zucc.) A. Robyns and *Terminalia argentea* (Mart. & Zucc.) commonly reaching heights of 11 m. All these species, with the possible exception of *P. longiflorum*, are indicators of mesotrophic soils. *L. paniculata* and *Platypodium grandiflorum* have not been observed at other localities on the Reserve; the other species occur in *cerrado*-fringing gallery forest but only *Pseudobombax longiflorum* is frequent in such habitats.

Two soil profiles were examined in a small patch of this mesotrophic *cerrado* measuring less than 1 ha. The results are in striking contrast to the range of analyses for other vegetation formations on the Fazenda and contrast sharply with the surrounding dystrophic soils.

For the surface soils, the pH values are higher, the exchangeable cation levels are much greater (particularly calcium) and there are low exchangeable aluminium levels. The phosphorus levels are more favourable and the amounts of organic carbon are more typical of the densest gallery forest than open woody vegetation. The nitrogen is well mineralized in these oxic conditions giving C/N ratios of around 15. The high calcium and particularly magnesium levels are notable, with appreciable quantities of potassium (Table 15.7), giving high overall figures for the cations. Even rapid inspection is sufficient to pick up the distinctly mesotrophic character of the soils (Furley *et al.* 1984).

The influence of edaphic factors is perhaps most clearly indicated within this vegetation formation, not so much in terms of major structural differences in the flora, but in the subtlety and complexity of the relationship. The principal soil parameters seem to be calcium and magnesium although increased potassium levels are also evident. For the surface horizons, the figures for calcium and magnesium are around 40 times greater than the levels in the surrounding *cerrado*, although they drop sharply in the subsurface. The base cations and pH levels are also distinctly higher than the equivalent means for *cerrado* surface soils. Together with the lower exchange acidity, the chemical environment is clearly more favourable to plant growth.

A further curious feature is the way in which this mesotrophism is confined to the surface rooting zone. Subsurface levels are much more in line with those of surrounding soils. It would appear, therefore, that the system is maintained by nutrient cycling and, since there is no evidence of any major input supply of bases

Table 15.7 Summary of the soil characteristics of mesotrophic *cerrado*

Means of two profiles	Surface (0–10 cm)	Subsurface (<10 cm)
Fractions of the total sample >2 mm (%)	n/a	n/a
Fractions of the fine earth sample (<2 mm) (%)		
sand (2–0.5 mm) (%)	n/a	n/a
silt (0.05–0.002 mm) (%)	n/a	n/a
clay (<0.002 mm) (%)	n/a	n/a
clay dispersible in water (%)	n/a	n/a
Degree of flocculation (%)	n/a	n/a
Acidity and nutrient status		
pH H_2O	6.3	5.4
pH KCl	5.3	4.5
Ca+Mg (cmol/kg)	19.20	0.11
Mg (cmol/kg)	2.58	0.08
K (<cmol/kg)	0.38	0.06
Sum (S) Ca, Mg, K, Na	22.16	0.25
Al (cmol/kg)	0.05	0.34
H	n/a	n/a
base saturation (%)	n/a	n/a
100Al/(S)+Al	57.6	42.0
Organic properties		
P (assimilable) ppm	2.4	0.6
C organic (%)	5.4	1.4
N total (%)	0.36	0.13
C/N	15.0	11.0
Oxides		
SiO_2 (%)	n/a	n/a
Al_2O_3 (%)	n/a	n/a
Fe_2O_3 (%)	n/a	n/a
TiO_2 (%)	n/a	n/a
Weathering indices	n/a	n/a
SiO_2/Al_2O_3 (ki)	n/a	n/a
SiO_2/R_2O_3 (kr)	n/a	n/a
Al_2O_3/Fe_2O_3	n/a	n/a

and indeed, a strong likelihood of a continuous "leakage" through eluviation, it would seem to be a system which is unlikely to persist. This raises the question of how it developed in the first place and whether it is, in fact, being maintained. Additional patches of similar vegetation are found not far away in the adjacent Reserve belonging to IBGE and in the Jardim Botânico de Brasília in similar topographic positions. One possibility is that the lateral seepage is bringing nutrients from some upslope supply. One rather unlikely possibility is that the patch represents soils which are residual from the weathering of a palaeo-outcrop of now weathered and eroded parent material. There are thin calcareous bands found throughout the region and a range of diverse parent materials which could act as sources for soil formation. A greater possibility is an anthropogenic origin, such as the collection of fertile soil by Indians to produce areas for rearing useful plants – a mechanism demonstrated by Anderson and Posey (1989). If this is the case, then position on the slope, lying above the wet galleries and grasslands, would have been likely to be a significant locational factor.

15.5 CONCLUSIONS

The soil landscape of Fazenda Água Limpa is, like that of central Brazil, predominantly dystrophic. There is little doubt that the most important edaphic factor affecting differentiation of vegetation is moisture, and that plant communities closely reflect the slope and drainage. Swampy gallery forest occurs along streams where the water-table is always high. Drier gallery forest occurs on somewhat better drained sites (such as the margins of wet galleries or headstream catchments). Open woody savanna and dense closed savanna woodland occur on well-drained soils over upper slopes, whilst wet grasslands are located where there is a wide seasonal fluctuation in groundwater but where saturation of the rooting zone is a frequent annual occurrence. The dry hill grasslands are found on the hyperdrained upper slopes especially on shallow soils. There seems to be little significant difference in the nutrient levels of the soils of these diverse vegetation types, although at micro-scale there may well be important controls exerted by soil depth and other physical factors such as texture.

The only soil discordant with the general dystrophic pattern is that of the mesotrophic facies *cerrado*, which has much higher pH and dramatically greater levels of calcium and magnesium in the surface horizons. The occurrence of characteristic marker species in this relatively restricted area demonstrates how sensitive and reliable they are as indicators of "better" soils. Furthermore, the fact that richer soil contained within shallow superficial horizons can bring about a strong floristic change in tree composition, shows that these species must gather their nutrients by surface feeding and maintain their supply by efficient cycling mechanisms. If these more fertile patches are of human origin, the fertility must still be maintained by surprisingly efficient recycling. It is evident that historical as well as environmental evidence needs to be considered before an accurate understanding of plant distribution can be obtained.

Observations over a number of years have demonstrated the occurrence of rapid vegetation succession. The availability of soil water is always a requirement, and soil chemical properties take a lesser role. In many areas protected from fire for a number of years, *cerrado* has been observed to thicken to become *cerradão*, whilst the dense woody savanna has begun to transform into forest. These changes involved the establishment of fire-sensitive species such as *Siphoneugena densiflora* and *Maprounea guianensis* in very large numbers. When a disastrous fire occurred in the *cerradão* reserve after about eight years of protection, enormous numbers of saplings of these *cerradão* and forest species were destroyed. Other indications of succession can be detected such as the presence of large relict *cerrado* trees well within the gallery forest at Corrego Capitinga, indicating a considerable upslope expansion of the forest area during the lifetime of these trees. Thus a 20-year span of observation has been sufficient to sense the dynamic flux of vegetation with forest migration over the lower, wetter savannas. The surface soils are rapidly affected whilst, at least for a period of time, the subsurface horizons persist in their previous state. These vegetation changes have been accredited to long-term climatic change acting through the water supply. In the toposequences characterizing the landscapes of central Brazil, slope and drainage determine these supplies and thereby shape the pattern of vegetation.

ACKNOWLEDGEMENTS

I am indebted to Dr J A Ratter, Royal Botanic Garden, Edinburgh for his advice based on long experience of *cerrado* botany and in particular, for the vegetation survey of the University of Brasília's ecological reserve. I should also like to pay tribute to numerous ecology students who helped with fieldwork or through discussion of related projects.

REFERENCES

Alvim, P. de T. and Araujo, W.A. 1952. Soil as an ecological factor in the development of vegetation in the central plateau of Brazil. *Turrialba*, 2, 153–160.

Anderson, A.B. and Posey, D.A. 1989. Management of a tropical scrub savanna by the Gorotiré Kayapó of Brazil. *Advances in Economic Botany*, 7, 59–173.

Diniz, M de Araujo Neto., Furley, P.A., Haridasan, M. and Johnson, C.E. 1986. The murundus of the cerrado region of central Brazil. *Journal of Tropical Ecology*, 2, 17–35.

Eiten, G. 1972. The cerrado vegetation of Brazil. *Botanical Review*, 38, 201–341.

EMBRAPA (Empresa Brasileira de Pesquisa Agropecuaria). 1978. *Levantamentos de Reconhecimento dos Solos do Distrito Federal*. SNLCS (Serviço National de Levantamento e Conservação de Solos) Boletim Técnico 53, Rio de Janeiro.

Felfily, J.M. and Silva Junior, M.C. da 1986. Inventário florestal de uma faixa de cerrado (152 ha) na Fazenda Água Limpa, DF. *5° Congresso Florestal Brasileiro*, Recife.

Felfily, J.M. and Silva Junior, M.C. da 1988. Distribuição dos diametros numa faixa de cerrado na Fazenda Água Limpa em Brasília. *Acta Botânico Brasileira*, 2, 85–104.

Felfily, J.M. and Silva Junior, M.C. da 1992. Floristic composition, phytosociology and comparison of cerrado and gallery forests at Fazenda Água Limpa, Federal District, Brazil. In P.A. Furley, J. Poctor and J.A. Ratter (eds), *Nature and Dynamics of Forest–Savanna Boundaries*. Chapman & Hall, London, pp. 393–416.

Furley, P.A. 1985. Notes on the soils and plant communities of Fazenda Água Limpa, Brasilia, DF. Occasional Publications No. 5, University of Edinburgh.

Furley, P.A. 1986. Classification and distribution of murundus in the cerrado of Central Brazil. *Journal of Biogeography*, 13, 265–268.

Furley, P.A. 1992. Edaphic changes across the forest-savanna boundary with particular reference to the neotropics. In P.A. Furley, J. Proctor and J.A. Ratter (eds), *The Nature and Dynamics of Forest–Savanna Boundaries*. Chapman & Hall, London, pp. 91–117.

Furley, P.A. and Ratter, J.A. 1988. Soil resources and plant communities of the central Brazilian cerrado and their development. In P.A. Furley, (ed.), *Biogeography and Development in the Humid Tropics*. Special Issue of *Journal of Biogeography*, 15, 97–108.

Furley, P.A., Haridasan, M. and Ratter, J.A. 1984. Uso de especies indicadoras de solo mesotrófico no cerrado. Resumos 35, *Congresso Nacional de Botanica*, Rio de Janeiro, p. 152.

Furley, P.A., Ratter, J.A. and Gifford, D.R. 1988. Observations on the vegetation of eastern Mato Grosso III. The woody vegetation and soils of the Morro de Fumaça, Torixoreu, Brazil. *Proceedings of the Royal Society of London*, B203, (191-)-208.

Goodland, R. and Pollard, R. 1973. The Brazilian cerrado vegetation; a fertility gradient. *Journal of Ecology*, 61, 2(19-)224.

Haridasan, M. 1982. Aluminium accumulation by some cerrado native species of central Brazil. *Plant & Soil*, **65**, 266–273.

Haridasan, M. 1985. Accumulation of nutrients by eucalyptus seedlings from acidic and calcareous soils of the cerrado region of central Brazil. *Plant & Soil*, **86** 35–45.

Lopes, A.S. and Cox, F.R. 1977a. Cerrado vegetation in Brazil: an edaphic gradient. *Agronomy Journal*, **69**, 828–831.

Lopes, A.S. and Cox, F.R.1977b. A survey of the fertility status of surface soils under cerrado vegetation in Brazil. *Soil Science Society of America Journal*, **41**, 742–747.

MME/SG (Ministerio das Minas e Energia/Secretaria Geral). 1982. Projeto Radambrasil. Folha SD 23, Brasília. *Levantamentos de Recursos Naturais 29*, Rio de Janeiro.

Oliveira-Filho, A.T. de and Furley, P.A. 1990. Monchão, cocuruto, murundu. *Ciência Hoje*, **11**, 30–37.

Oliveira-Filho, A.T. de and Ratter, J.A. 1995. A study of the origin of central Brazilian forests by the analysis of plant species distribution patterns. *Edinburgh Journal of Botany*, **52**, 141–(194.)

Oliveira-Filho, A.T. de., Shepherd, G.J., Martins, F.R. and Stubblebine W.H., 1989. Environmental factors affecting physiognomic and floristic variation in an area of cerrado in central Brazil. *Journal of Tropical Ecology*, **5**, 413–431.

Ponce, V.M. and da Cunha, C.N. 1993. Vegetated earthmounds in tropical savannas of central Brazil – a synthesis with special reference to the Pantanal do Mato Grosso. *Journal of Biogeography*, **20**(2), 2(19–)225.

Ratter, J.A. 1971. Some notes on two types of cerradão occurring in northeastern Mato Grosso. In M.G. Ferri (ed.), *3° Simpósio sobre o cerrado*. Universidade de São Paulo, pp. 100–102.

Ratter, J.A. 1980. *Notes on the Vegetation of Fazenda Água Limpa (Brasília, DF, Brazil)*. Department Publication no. 1, Royal Botanic Garden, Edinburgh.

Ratter, J.A. 1991. *Guia para a vegetaçao da Fazenda Água Limpa (Brasília DF)*. Editora Universidade de Brasília, Brasil.

Ratter, J.A. 1992. Transitions between cerrado and forest vegetation in Brazil. In P.A. Furley, J. Proctor and J.A. Ratter (eds), *Nature and Dynamics of Forest–Savanna Boundaries*. Chapman & Hall, London, pp. 417–430.

Ratter, J.A. and Dargie, T.C.D. 1992. Analysis of the floristic composition of 26 cerrado areas in Brazil. *Edinburgh Journal of Botany*, **49**(2), 235–250.

Ratter, J.A., Richards, P.W., Argent, G. and Gifford, D.R. 1973. Observations on the vegetation of the northern Mato Grosso I. The woody vegetation types of the Xavantina-Cachimbo Expedition area. *Philosophical Transactions of the Royal Society*, **B266**, 449–492.

Ratter, J.A., Askew, G.P., Montgomery, R.F. and Gifford, D.R. 1977. Observações adicionais sobre o cerradão de solos mesotróficos no Brasil central. In M.G. Ferri (ed.), *4° Simpósio sobre o Cerrado*. Universidade de São Paulo, pp. 303–316.

Ratter, J.A., Askew, G.P., Montgomery, R.F. and Gifford, D.R. 1978. Observations on forests of some mesotrophic soils in central Brazil. *Revista Brasileira Botânica*, **1**, 47–58.

Waibel, L. 1948. Vegetation and land use in the planalto central of Brazil. *Geographical Review*, **38**, 529–554.

16   Nitrous Oxide Emissions from a Range of Soil–Plant and Drainage Conditions in Belize

R. M. REES,[1] K. CASTLE,[1] J. R. M. ARAH[2] and P. A. FURLEY[3]
[1]*Soils Department, SAC, Edinburgh, UK*
[2]*Institute of Terrestrial Ecology, Penicuik, Midlothian, UK*
[3]*Department of Geography, University of Edinburgh, UK*

16.1   INTRODUCTION

The influence of low-angle slopes upon drainage and surface processes has rarely received attention. Low gradients are known to be able to shed water rapidly in intense or prolonged rainfall but generally tend to accumulate and retain moisture for longer than more steeply inclined slopes. The ability to drain water from field capacity results in a range of aeration and oxidation processes. In the humid tropics, reduction may predominate over oxidation for most of the year in the wettest regions, but gradually shifts towards an alternating oxidation and reduction regime in the seasonally humid tropics.

Belize lies at approximately 17°N and experiences persistently warm humid conditions over the south of the country, but is affected by a progressively stronger dry season in the north towards the Mexican border. Over and above these trends are the rainfall patterns from east to west, where the inland western districts experience higher temperatures and more clearly defined seasonality. The sites examined in this contribution extend from the centre to the northwest of the country and are consequently characterized by distinct wet and dry periods.

The argument presented here focuses upon local-scale differences within this national pattern. It aims to assess the effects of changes in vegetation, drainage and land use on nitrous oxide emissions and to examine the extent to which slope and drainage may influence such emissions.

16.2   ENVIRONMENTAL AND SOIL VARIATIONS IN BELIZE

To gauge the significance of varying N_2O-emissions, it is necessary to place the sampling locations in a regional framework (Figure 16.1). The principal topographic controls on drainage are the Maya Mountains, which occupy around one-third of the country, and consist of rolling uplands with deeply incised valleys and highly weathered ancient soils derived from Palaeozoic, metamorphic and volcanic rocks with later intruded granite (mostly freely drained Oxisols) (Figure 16.2). The broad

Advances in Hillslope Processes, Volume 1. Edited by M. G. Anderson and S. M. Brooks.

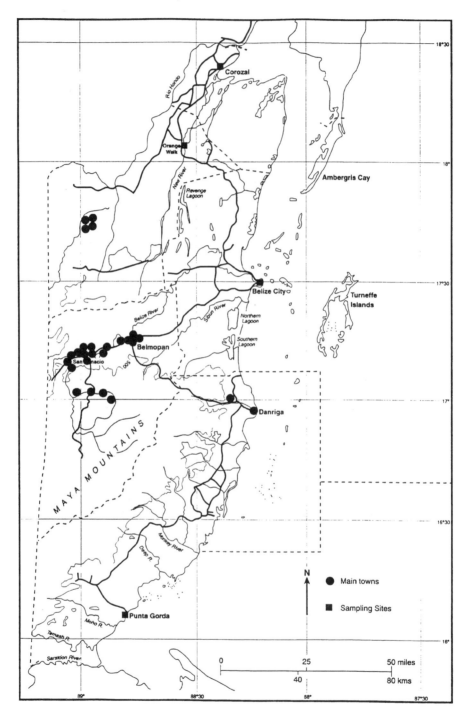

Figure 16.1 The distribution of sampling sites

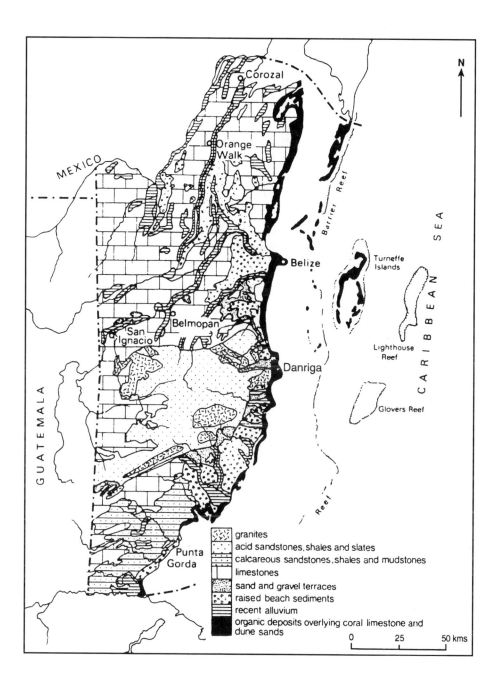

Figure 16.2 Soil parent materials in Belize

Table 16.1   The soil environments of Belize

---

1.   Maya Mountains (Oxisols, Utisols; sometimes plinthitic)
   (a) Mountain Pine Plateau: highly weathered igneous and metamorphic, coarse textured, acidic with low nutrient levels
   (b) Steeply incised slopes: shallow, stony, variable texture, acidic with low nutrient levels; typically Entisols and Inceptisols
2.   Clay soils of karstic limestone uplands and foothills (Mollisols, Inceptisols): shallow, reddish brown to black clays; rich in Mg and Ca
3.   Toledo mudstones and siltstones (Alfisols, Ultisols): shallow, well drained, moderately acid
4.   Pine Ridge (Ultisols, Entisols): coarse-textured surface overlying mottled clay subsoils; acidic, very low nutrient levels with problems of water availability
5.   Dark limestone soils of the Northern plain (Mollisols, Inceptisols): shallow, grey to black cracking clays; deep in depressions; neutral to alkaline
6.   Reddish limestone clays of the Northern plain (Mollisols and Alfisols): shallow and stony clays, slightly cracking, alkaline, low N and P
7.   Swamp soils (Entisols, Inceptisols, Histosols, Mollisols): varied soils, wet for part or all of the year, varied textures, often organic
8.   Young soils on river alluvium (Entisols, Inceptisols): floodplain, often wet subsoils, nutrient rich with more weathered older terraces
9.   Young coastal soils (Entisols, Inceptisols): poorly drained, weakly developed, varied provenance and properties

---

After Baillie *et al.* 1993.

influence of topography upon drainage and consequently upon soil properties is reasonably well understood (Furley 1992). Nine categories of soil have been identified in Belize, according to physiognomic groups (Table 16.1). Around the massif of the Maya Mountains lies an arc of normally well-drained, acidic and coarse-textured alluvial and colluvial soils derived from the mountains. To the south, mudstones and siltstones result in younger soils (Ultisols and Alfisols). To the north of the country over limestone parent materials (Mollisols/Inceptisols) there are varied alkaline soils, which are closely related to texture, depth and drainage. Along river valleys, in floodplains, the soils are also younger (Vertisols/Inceptisols). Finally, along the coastline, there are numerous wetlands, some highly brackish and covered by mangrove and its associates and others more alluvial in character affected by fresh water (Histosols). In these areas, reduction processes predominate.

16.3   NITROUS OXIDE EMISSIONS

The atmospheric concentration of nitrous oxide (N_2O) is currently increasing at a rate of about 0.3% per year (Prinn *et al.* 1990). This build-up has attracted widespread concern, as N_2O is not only a greenhouse gas which therefore contributes to global warming (Lashof and Ahuja 1990), but is also involved in the destruction of stratospheric ozone (Cicerone 1987). Tropical forest soils are a particularly important source of N_2O, with recent estimates suggesting that they may contribute approximately 50% of the global total (Keller *et al.* 1983, 1986). Although the N_2O flux from tropical forests tends to be relatively low, they occupy an area of $11.6 \times 10^6 \, km^2$ or 24.4% of the earth's land surface (Skole and Tucker 1993) and therefore contribute to a large total flux. The land use of many areas in the humid tropics is undergoing rapid change as a result of population pressures and deforestation. At present the situation in Belize is less severe, where the total forest loss between 1980 and 1990 was estimated at 5000 ha (0.2%), although the rate is accelerating and the figures available are of doubtful accuracy (WRI 1994). A number of studies have shown that N_2O emission is affected by such changes in vegetation and land use (particularly N fertilizer application) as well as by drainage. For example, Luizaõ *et al.* (1989) found that N_2O emissions from pasture soils in Brazil were more than five times those from adjacent forests, while Duxbury (1990) observed fluxes from perennial crops to be lower than those from annual crops in Florida. Nitrous oxide is released from soils as a result of nitrification and denitrification. Factors affecting these processes in the soil environment are numerous, with drainage and vegetation only indirectly affecting the direct influences (Figure 16.3). There is some uncertainty regarding the relative importance of the contributions of nitrification and denitrification to N_2O production in the tropics. Where soils are heavily fertilized, especially with nitrate-based fertilizers, it seems that denitrification is likely to predominate (Keller *et al.* 1988; Livingston *et al.* 1988; Bakwin *et al.* 1990). However, in undisturbed environments such as forests which cover much larger areas, it has been suggested that nitrification is the principal source of N_2O, particularly in soils subject to cyclical wetting and drying (Davidson *et al.* 1993). Within the type of climate outlined above this alteration of soil water regimes is particularly significant where associated with low gradients and footslope sites, which represent some of the more potentially useful agricultural soils in the country.

In this study, N_2O emissions were measured from a range of field sites. The sites were grouped into broad categories according to vegetation type and drainage in order to investigate the relationship between N_2O emissions and a number of environmental variables. Nitrous oxide measurements were repeated in samples returned to the laboratory, and an attempt was made to discriminate between the different N_2O source processes using C_2H_2. Short-term N dynamics were studied using freshly incubated soils, and ^{15}N labelled fertilizer.

16.4 METHODS

Forty study sites were chosen to represent a wide range of soil and vegetation types across Belize (Table 16.1 and Figure 16.1). Emission of N_2O was measured using three replicate metal cover boxes (100 mm high $\times$ 220 mm long $\times$ 230 mm wide). Cover boxes were sampled after approximately 45 min, after which time a 10 ml sample was transferred to an evacuated glass tube. Atmospheric samples were also collected at each site. All gas samples were returned to Edinburgh, and analysed for N_2O with a gas chromatograph fitted with an electron capture detector. Six replicate

352

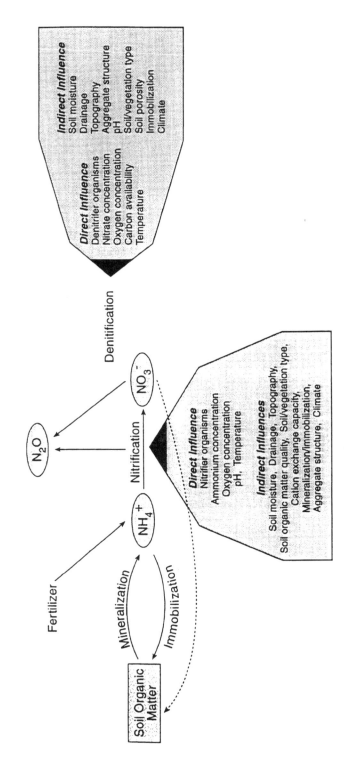

Figure 16.3 Factors influencing the release of nitrous oxide from terrestrial ecosystems

standards (10 ppmv N_2O were injected into sample tubes before travelling, taken to and from Belize, and analysed in the same way as the other samples on return.

Drainage status (assessed by visual examination of soil pits), soil and air temperatures were recorded at each site, and three 100 g soil samples were returned to the local laboratory at Central Farm, Belize, for analysis within 24 h (with the exception of samples 29 and 30, which were analysed after approximately 36 h). In the laboratory the soil pH of each sample was measured in 0.05M $CaCl_2$ using a pH meter, and nitrate-N was determined in filtered aqueous extracts using a portable nitrate meter (Nitrachek reflectometer). Soil contents were determined by oven-drying for 24 h at 100°C.

Nitrogen transformations at the different sites were studied in aerobic incubations. Nine 100 g samples of fresh soil collected from 0–200 mm at each site were weighed into polythene bags (100 mm × 140 mm). One millilitre of a solution containing $Ca(^{15}NO_3)_2$ (40 g N/1, with a ^{15}N enrichment of 11.3 atom %) was pipetted evenly across the surface of three of the samples; distilled water was added to the remaining six. The bags were left unsealed and placed outside, but protected from rain and direct sun. After 10 days the NO_3^- concentrations in three of the unamended soil samples were determined. The remaining soil samples amended and unamended were air dried to prevent further microbial activity. The dried soil samples were returned to the UK, and subdivided in order to carry out further analyses. Organic-C concentrations were determined on finely milled samples ($< 100 \mu m$) by wet oxidation. Available N was determined in 1 M KCl extracts using continuous flow analysis, and potentially available-N was measured using the method described by Waring and Bremner (1964). The ^{15}N enrichment of available N in soils amended with $^{15}NO_3^-$ was determined using the microdiffusion technique of Brooks et al. (1989).

In order to measure N_2O production from the soils returned to the UK, 8 ml of distilled water was added to duplicate 20 g samples (roughly simulating field moisture conditions after rain). The soils were sealed in plastic bags (catheter bags), equipped with three-way valves, and were allowed to equilibrate overnight (about 17 h) at room temperature (approximately 24°C); the bags were flushed with 200 ml of laboratory air and connected to a 32 port programmable injector (Arah et al. 1994). Samples from two injection loops were analysed for N_2O by a Pye Unicam PU4500 gas chromatograph fitted with an electron capture detector and for CO_2 and C_2H_2 by a Hewlett Packard HP 5890 Series II gas chromatograph fitted with a thermal conductivity detector. The PU4500 was equipped with a backflush pre-column to remove C_2H_2. Bags containing N_2O and CO_2/C_2H_2 were also analysed. N_2O and CO_2 production were corrected for gas dissolved in the liquid phase using temperature-adjustment partition coefficients (Hodgman et al. 1955).

Each bag was sampled once every 80 min. A sacrificial pre-sample was used to flush the system before each analysis. After the first analysis cycle, 5 ml of C_2H_2 was added to each bag, to inhibit nitrification and the terminal stage of denitrification ($N_2O \rightarrow N_2$). Where nitrification is the dominating N_2O producing process this should result in a decreased flux; where denitrification dominates then the flux is likely to rise. The concentration of C_2H_2 was used to calculate the volume of the incubation bag.

16.5 RESULTS

The environmental conditions encountered during this study were conducive to high rates of biological activity in general, and N_2O production in particular, as is often the case in the humid tropics. Soil temperatures at a depth of 100 mm ranged from 23.0 to 32.7°C with a mean of 27.6°C and gravimetric soil water contents in the surface 200 mm ranged from 0.02 to 0.35 g/g with a mean of 0.22 g/g. Annual rainfall in this region is in the range 1500–3000 mm, with a distinct dry season during the months of January–April.

Nitrous oxide emissions at the 40 study sites were found to be approximately log normally distributed, with a geometric mean of 3.98 ng N_2O m^2/s and a range of 0.4–527 ng/m^2/s (Figure 16.4). At two sites (Numbers 19 and 34, Table 16.2) the flux exceeded 200 ng/m^2/s, with high values in all three replicates at each site. These sites are responsible for the small secondary peak in the N_2O flux distribution (Figure 16.4). The analysis and presentation of log-normally distributed data can cause problems; arithmetic means and conventional statistical analyses are inappropriate as they give undue weight to very occasional large values. We employed a uniform minimum variance unbiased estimator, statistically the best choice for small samples drawn from what is believed to be a log-normal population (Parkin *et al.* 1988). However, neither approach revealed any statistical differences in N_2O emissions

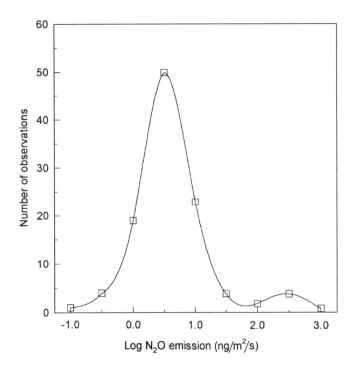

Figure 16.4 N_2O production at 40 sites in Belize during July and August 1991

Table 16.2 The vegetation and land use, soil texture and location of sites

Sample no.	Vegetation type	Drainage	Soil texture	Location
1	Perennial crop (citrus)	Good	Silt loam	Melinda Forest Station
2	Perennial crop (banana)	Poor	Clay	Central Farm
3	Perennial crop (citrus)	Poor	Clay	Central Farm
4	Perennial crop (coconut)	Poor	Clay	Central Farm
5	Perennial crop (citrus)	Imperfect	Sandy loam	San Ignacio
6	Perennial crop (coffee)	Good	Sandy loam	Mountain Pine Ridge
7	Perennial crop (citrus)	Good	Silt loam	Central Farm
8	Arable crop (maize)	Imperfect	Clay	Georgeville
9	Arable crop (peanut)	Good	Clay	Central Farm
10	Arable crop (recently ploughed)	Good	Clay	Central Farm
11	Arable crop (peanut)	Good	Clay	Central Farm
12	Arable crop (maize)	Good	Clay	Central Farm
13	Arable crop (maize)	Good	Clay	Spanish Lookout
14	Arable crop (dry rice)	Imperfect	Clay	Belmopan
15	Arable crop (soya)	Imperfect	Clay	Belmopan
16	Arable crop (maize)	Imperfect	Clay	Belmopan
17	Arable crop (peanut)	Imperfect	Clay	Belmopan
18	Arable crop (maize)	Imperfect	Loamy sand	San Ignacio
19	Arable crop (paddy rice)	Good	Silty clay	Central Farm
20	Arable crop (maize)	Good	Clay	
21	Arable crop (recently ploughed)	Good	Silt loam	Central Farm
22	Forest (Mangrove)	Good	Silt loam	Danriga
23	Forest	Imperfect	Clay	Norland
24	Forest	Good	Silt loam	Belmopan
25	Forest	Good	Clay	River Walk
26	Forest	Good	Clay	Mountain Pine Ridge
27	Forest	Poor	Sand	Mountain Pine Ridge
28	Forest	Good	Silt loam	Mountain Pine Ridge
29	Forest	Good	Clay	Rio Bravo
30	Forest	Good	Clay	Rio Bravo
31	Cleared forest	Good	Clay	Georgeville
32	Cleared forest	Imperfect	Clay	River Walk
33	Cleared forest	Poor	Clay	Belmopan
34	Cleared forest	Poor	Sandy clay	Iguana Creek
35	Cleared forest	Poor	Clay	Rio Bravo
36	Cleared forest	Good	Clay	Rio Bravo
37	Grass	Imperfect	Clay loam	Norland
38	Grass	Good	Clay	Norland
39	Grass	Good	—[a]	Floral Park
40	Grass	Good	Clay	Belmopan

[a]Not determined.

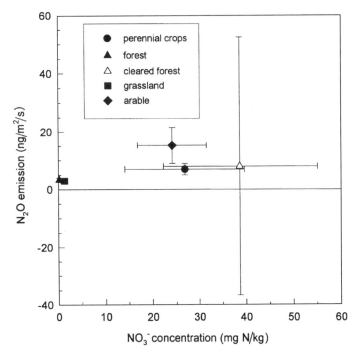

Figure 16.5 N_2O production (solid bars) abd NO_3^--N concentrations in the top 200 mm soil under different vegetation types

between different types of vegetation site (Table 16.3), although the cleared forest areas had significantly higher ($P<0.05$) concentrations of NO_3^- than those in undisturbed forest (33 and 0.2 mg/kg NO_3^- -N respectively; Figure 16.5).

Differences in N_2O emissions did occur in the different drainage categories. Using both arithmetic and geometric means, emissions of N_2O from poorly drained sites was significantly more ($P<0.05$) than that from good or imperfectly drained sites (Table 16.3). It was not possible to correlate N_2O production with soil water content, temperature, NO_3^- concentrations in the soil, or NO_3^- produced by aerobic incubations. When fresh soils were incubated for 10 days with K $^{15}NO_3$, it was found that $^{15}NO_3^-$ loss was generally greatest in soils from sites where large fluxes of N_2O had been measured in the field (Figure 16.6; $r^2=0.51$, $P<0.001$). The average $^{15}NO_3^-$ -N loss in all soils was 117 mg/kg. The losses of N from poorly drained sites (230 mg N/kg), were significantly higher ($P<0.05$) than those from sites that had good or imperfect drainage (104 and 58 mg N/kg respectively).

Little relationship was observed between N_2O production in the field, and N_2O production by the same soils after drying and then rewetting in the laboratory incubation (Figure 16.7). This was possibly a consequence of the uniform moisture contents of the incubated soils, providing near-optimal conditions for microbial activity. In the field, the low moisture contents at a number of sites would be likely to have reduced microbial activity. However, NO_3^- concentrations did not seem to

Table 16.3 Nitrous oxide production from soils under different types of vegetation in the field. All values are in ng/m²/s (standard errors in parentheses)

	Arithmetic mean	Geometric mean	UMVUE[a] mean
Vegetation type			
Perennial crops	7.31	5.26	6.96
	(4.12)	(1.54)	(1.91)
Arable	22.76	5.25	15.28
	(16.34)	(1.23)	(6.29)
Forest	3.35	2.44	3.32
	(1.15)	(0.72)	(0.70)
Cleared Forest	90.04	4.88	7.98
	(87.3)	(4.63)	(44.51)
Grass	2.91	2.08	3.01
	(1.44)	(1.27)	(1.03)
Drainage status			
Good	3.68	2.74	3.57
	(1.84)	(0.59)	(0.74)
Imperfect	4.39	2.96	4.47
	(2.23)	(0.84)	(1.47)
Poor	117.6	20.75	117.69
	(8.34)	(3.46)	(103.0)

[a]Uniform minimum variance unbiased estimator.

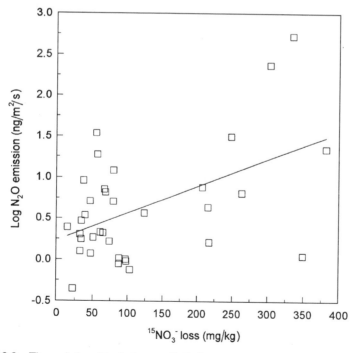

Figure 16.6 The relationship between N_2O flux in the field and $^{15}NO_3^-$ loss in soils incubated with 40 mg K $^{15}NO_3$-N for 10 days

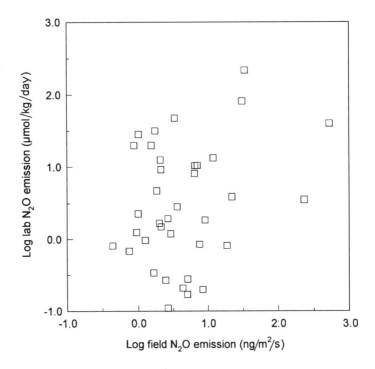

Figure 16.7 The relationship betweeen N_2O in the field and that from soil samples that were air dried, and then rewetted in a laboratory incubation

change much. The concentrations of nitrate measured in the fresh soils (as measured by the portable nitrate meter) were highly correlated ($r=0.91$; $P<0.001$) with those measured in the same soils after they had been dried, and returned to Scotland, although the amount of nitrate was increased by around 27%. Other variables that may have changed during the process of drying that are known to influence rates of N_2O emission include the nature of microbial populations, and the availability of carbon. These parameters were not examined in the present study. The addition of 10% C_2H_2 led to an increase of about 35% in N_2O production in most soils. When the N_2O ratio (N_2O production in the absence of C_2H_2 divided by N_2O production in its presence) was plotted against N_2O production without C_2H_2 (in the laboratory incubation), a positive correlation was observed (Figure 16.8), suggesting that nitrification was an important source.

The different drainage categories were distributed roughly evenly between the different classes of vegetation, but differences in fertility within the vegetation types were indicated by a number of the parameters measured. In addition to differences in N_2O emissions and NO_3^--N concentrations (Figure 16.5), there were significant differences in organic carbon concentrations, NH_4^+-N released by anaerobic incubation, and CO_2 production in the laboratory (Table 16.4). Despite these differences in NO_3^- concentrations and the amount of N released by aerobic

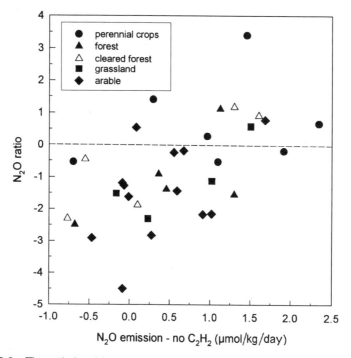

Figure 16.8 The relationship between the flux of N_2O from soils incubated in the laboratory, and the N_2O ratio (N_2O production in the absence of C_2H_2 divided by N_2O production in the presence of N_2O) of the same soil

Table 16.4 Indices of soil fertility within each class of vegetation (standard errors in parentheses)

Site classification	Carbon[a] (g C/kg)	CO₂ production[b] (mg C/kg/day)	NO₃⁻³ᶜ (mg N/kg)	NO₃⁻ production[d] (mg N/kg/day)	NH₄⁺ production[e] (mg N/kg/day)
Perennial crops	30	199	26.9	0.56	63
	(5.1)	(31)	(12.8)	(0.47)	(3.72)
Arable crops	21	188	24.2	0.81	5
	(2.9)	(8)	(7.4)	(0.53)	(2.7)
Forest	56	416	0.2	0.53	98
	(11.8)	(130)	(0.2)	(0.23)	(22.7)
Cleared forest	73	270	38.7	0.40	102
	(12.1)	(57)	(16.4)	(0.65)	(34.7)
Grass	26	121	1.3	0.25	62
	(9.7)	(26)	(0.7)	(0.22)	(5.8)

[a]Carbon concentration determined by wet oxidation.
[b]CO_2 produced in an aerobic laboratory incubation.
[c]Available soil-NO_3^- determined in fresh soils.
[d]NO_3^- production during a 10-day aerobic incubation of fresh soil.
[e]NH_4^+ production during a 10-day aerobic incubation of dry soil.
Further details of methods are given in the text.

incubations, no effect of site on the production of NO_3^- (as determined by aerobic incubation) was observed. The mean net production of NO_3^--N across all sites was 5.8 mg/kg during a 10-day incubation, which is lower than values quoted in comparable studies (Vitousek and Matson 1988).

16.6 DISCUSSION

16.6.1 Nitrous Oxide Production and Site Characteristics

The results of this study have shown that there was a significant correlation between drainage and N_2O flux measured in the field. It is interesting to note that other factors known to affect N_2O production (such as nitrate concentration and soil water content) were not correlated. It is possible that the soil's drainage status provides a better indicator of soil aeration than do measurements of water content, which are strongly influenced by recent climatic conditions (particularly rainfall). As indicated in Figure 16.3, drainage can effect N_2O emission from both nitrification and denitrification through its influence on oxygen concentration.

The high rates of loss of $^{15}NO_3^-$ measured, especially at the poorly drained sites, provides support for the hypothesis that nitrous oxide emission in this study results predominantly from denitrification. The addition of NO_3^- to soils is known to increase N_2O release both in tropical and temperate regions as a result of increased rates of denitrification particularly in wet and poorly drained soils. Livingston et al. (1988) and Keller et al. (1988) found that the addition of NO_3^- resulted in a very much larger production of N_2O than comparable additions of NH_4^+-based fertilizers, indicating that denitrification rather than nitrification was largely responsible. It is highly improbable that all of the $^{15}NO_3^-$ loss would have been emitted as N_2O as this would have resulted in an N_2O emission of around 230 μmol N_2O/kg. In the laboratory incubation the maximum N_2O emission was less than 3 μmol N_2O/kg. However, in a review of the available literature, Keller et al. (1988) found that N_2O production was always less than 3% of the N added as fertilizer. Other pathways of N loss also need to be considered; for example, Hutchinson and Brams (1992) found that losses of NO resulting from fertilizer-N additions were more than seven times greater than N_2O losses in the humid subtropics of southern Texas. An unknown fraction of the N_2O produced by denitrification is also further reduced to N_2 gas.

Another potential sink for the added $^{15}NO_3^-$ would be the soil microbial biomass and soil organic matter. Only the $^{15}NO_3^-$ remaining in the soil solution at the end of the incubation was measured, so NO_3^- immobilized during the incubation would not have been recovered by the extraction procedure. It has generally been observed that NO_3^- is not easily immobilized by microbial populations as the process is energetically unfavourable, especially where NH_4^+ is available (Robertson 1989). However, significant immobilization of NO_3^- has been observed where there is an adequate C supply. Recous et al. (1990) observed an immobilization rate of 27 mg NO_3-N/kg when NO_3^- was added in the presence of 500 mg/kg glucose.

Acetylene was used in the laboratory incubation to indicate the dominant source process. The relevant measured factor is the N_2O fraction (Arah 1990) defined as the ratio between N_2O production in the absence of C_2H_2, divided by N_2O production in the presence of C_2H_2, This was used to suggest the relative importance of nitrification and denitrification in the production of N_2O. An N_2O ratio of less than 1 indicates a soil in which denitrification predominates. The results presented in Figure 16.8 suggest that nitrification was responsible for the highest rates of N_2O production, with 71% of the N_2O produced having an N_2O ratio of greater than 1. These results appear to conflict with those of the ^{15}N study; however, the soils used in this experiment were air-dried prior to incubation (in the ^{15}N study fresh soils were used). Davidson et al. (1993) examined N_2O production in soils subject to cycles of wetting and drying, and also found that nitrification was the dominant source process. In the present study, the drying of soils prior to incubation probably altered the processes of N turnover. However, this is not altogether unrepresentative of field conditions; soils in this region are regularly exposed to wetting and drying cycles.

There was no significant effect of vegetation class on N_2O production. Of the five sites with the largest N_2O fluxes, three were under arable crops, one under perennial crops and one cleared forest. The probability that any particular site will have a large flux of N_2O is dependent on the coincidence of a number of controlling variables. Thus the highest flux of N_2O ($504\,ng/m^2/s$ occurred at a deforested site (34) with poor drainage, a higher than average rate of NO_3^- production ($1.5\,mg$ N/kg/day) and a higher than average gravimetric water content (33.8%). Although the mean NO_3^- concentration in deforested areas was significantly higher than that in other areas, this was not always accompanied by poor drainage (and other factors controlling N_2O production). The production of N_2O was therefore not consistently largest in these areas. The second highest flux of N_2O ($233\,ng/m^2/s$) occurred from a paddy rice field (site 19) which also had poor drainage, a plentiful supply of NO_3-N ($13.7\,mg/kg$) and a high gravimetric water content (32.1%). However, many of the N_2O fluxes from arable soils were low and were accompanied by good/imperfect drainage and low concentrations of NO_3^-, or both.

An underlying relationship may have been obscured by the addition of fertilizer to a number of the arable sites, or may have been masked by other soil variables. In particular, where N_2O production results from denitrification, the aeration state of the soil (Mosier et al. 1986), the nature of the soil's microbial populations (Weier and MacRae 1992) and the concentration and distribution of oxidizable organic matter are known to be important (Burford and Bremner 1975) as may have been the case in the present study. Furthermore the high degree of spatial variability associated with N_2O production can mean that even in soils with uniformly high concentrations of NO_3^-, significant reduction may take place in a limited number of "hotspots". The importance of variables other than NO_3^- concentration in determining N_2O production is illustrated when comparing the field and laboratory measurements. Despite a strong correlation between NO_3^- concentrations in the fresh and dried soils, no correlation was observed in N_2O production. This may have been the result of an alteration in the availability of carbon substrates, an alteration in the composition of the microbial populations, or, most probably, in differences in moisture content between the field and laboratory samples.

16.6.2 Regional N$_2$O Production

Although extrapolation from a small data set is open to numerous questions, particularly in relation to spatial and temporal variations, it can provide useful information particularly in regions where measurements are not commonly made. Given these limitations, it is possible to express the results on a regional scale, if land-use practices within that region are known. The advantage of this approach is that it relates the work more clearly to previous studies. Much of the experimental work took place in the Belize River Valley, an area occupying 3170 km^2, or 14% of the country. This region was surveyed by Jenkin *et al.* (1976), who quantified the various categories of land use. By far the largest category was moist tropical forest (occupying 79% of the area), with smaller areas of grassland, annual arable crops, perennial crops and a miscellaneous category, including built-up areas (occupying 19, 0.9, 0.3 and 0.8% respectively). Using the average flux measured in each area and multiplying by the appropriate land area, it is possible to estimate the regional annual flux (Table 16.5).

It is interesting to note that although the average N$_2$O flux from forested areas is relatively low (490 g N/ha per year), emissions from these areas make up 80% of the total, since they occupy a large proportion of the land surface. This has further implications for the importance of processes contributing to N$_2$O production. As discussed previously, the high fluxes are probably associated with denitrification, but the lower fluxes measured in the forested areas are more likely to be associated with nitrification. This low rate of nitrification in forested areas may be quantitatively more important than denitrification. In a review by Bouwman (1990), production of N$_2$O by different terrestrial ecosystems was found to be in the range 0.1–2.6 kg N$_2$O–N/ha/year), with production in tropical ecosystems lying at the upper end of this scale (0.4–2.6 kg N$_2$O–N/ha/year). The values presented in this study (0.4–1.1 kg N$_2$O–N/ha/year) fall within this range, and are broadly typical of other studies in tropical regions (e.g. Livingston *et al.* 1988; Davidson *et al.* 1993).

Table 16.5 Estimates of N$_2$O production in the Belize River valley, using measurements made in the present study

Land use	Area (km^2)	N$_2$O production			% from each area
		ng N$_2$O/m^2/s (geometric means)	g N$_2$O-N/ha/ year	kg N$_2$O-N/year	
Forest	2 493	2.44	490	12 216	80.4
Grass	614	2.08	417	2560	16.8
Annual crops	30	5.25	1 053	31	1.3
Perennial crops	9	5.26	1 056	10	0.7
Miscellaneous	24	ND	ND	ND	ND

16.7 CONCLUSIONS

This study has examined a sequence of low gradient sites which were classified according to drainage status, related to slope, and vegetation type. Nitrous oxide production was significantly higher in poorly drained sites than in those imperfectly or well drained. No relationship was observed between N_2O production and vegetation types, but this may have been a result of the wide range of drainage classes that were contained within each of the types of vegetation studied. Interpretation of the data suggests that patterns of nitrogen transformation do differ between vegetation type, and it seems likely that a larger number of samples chosen from areas of cleared forest (preferably with similar drainage characteristics), would have shown differences in N_2O production from forested areas.

The role of drainage is thus a critical factor in determining nitrogen transformations and it can be hypothesized that over a range of differing slopes, an increase in slope angle would lead to increased drainage, improved aeration and a consequent reduction in N_2O production. At very low angles there is a particularly sensitive balance between water input, retention in the soil plant system and drainage outflow. Thus relatively small changes in gradient may have a profound effect on sensitive soil indicators such as gaseous emissions.

This investigation represents only a snapshot of the processes contributing to N_2O production in Belize. High fluxes often occur only for short periods of time, and if an attempt is to be made to construct regional budgets for N_2O production both temporal and spatial patterns must be identified. Whilst denitrification processes may be responsible for short bursts of high N_2O production, the background production of N_2O resulting from nitrification is clearly important. Land-use change should also be related to timescales. Thus in the short term forest clearance may result in increased N_2O production. However, Sanhueza et al. (1990) found that where forests are replaced by savanna, production may decrease. It was suggested that this is a consequence of the lower water contents, and lower mineral-N contents of the latter.

ACKNOWLEDGEMENTS

The work in this chapter was carried out on a University of Edinburgh research expedition in Belize (ODA-OFI R4736). The authors would like to thank Sue Westoby and Hamer Dodds for assisting with field work, and Rab Howard for subsequent laboratory analyses. Thanks are also extended to the staff at Central Farm, Belize on whose premises much of the work was carried out, and to the Farmers Club and the AFRC/NERC Joint Initiative on Pollutant Transport in Soils and Rocks, for financial support.

REFERENCES

Arah, J.R.M. 1990. Diffusion-reaction models of denitrification in soil microsites. In N.P. Reusbech and J. Sorensen (eds), *Denitrification in Soil and Sediment*. Plenum Press, New York, pp. 245–258.

Arah, J.R.M., Crichton, I.J., Smith, K.A., Clayton, H. and Skiba, U.M. 1994. Automated gas chromatographic analysis system for micrometeorological measurement of trace gas fluxes. *Journal of Geophysical Research*, **99**, 16 593–16 598.

Baillie, I.C., Wright, A.C.S., Holder, M.A. and Fitzpatrick, E.A. 1993. Revised classification of the soils of Belize. Natural Resources Institute, NRI Bulletin 59, Chatham, UK.

Bakwin, P.S., Wofsy, S.C., Fan, S., Keller, M., Trumbore, S. and da Costa, J.M. 1990. Emission of nitric oxide (NO) from tropical forest soils and exchange of NO between the forest canopy and atmospheric boundary layers. *Journal of Geophysical Research*, **95**, 16 755–16 764.

Bouwman, A.F. 1990. Exchange of greenhouse gases between terrestrial ecosystems and the atmosphere. In A.F. Bouwman (ed.), *Soils and Greenhouse Effect*. Wiley, Chichester, pp. 61–128.

Brooks, P.D., Stark, J.M., McInteer, B.B. and Preston, T. 1989. Diffusion method to prepare soil extracts for automated nitrogen-15 analysis. *Soil Science Society of America Journal*, **53**, 1707–1711.

Burford, J.R. and Bremner, J.M. 1975. Relationships between the denitrification capacities of soils and total water-soluble and readily decomposable soil organic matter. *Soil Biology and Soil Biochemistry*, **7**, 389–394.

Cicerone, R.J. 1987. Changes in stratospheric ozone. *Science*, **237**, 35–42.

Davidson, E.A., Matson, P.A., Vitousek, P.M., Dundin, K., Garcia-Mendez, G. and Maass, J.M. 1993. Processes regulating soil emissions of NO and N_2O in a seasonally dry tropical forest. *Ecology*, **74**, 130–139.

Duxbury, J.M. 1990. Agriculture, nitrous oxide and our environment, *New York's Food and Life Sciences Quarterly*, **20**(3), 28–31.

Furley, P.A. 1992. Edaphic changes at the forest–savanna boundary with particular reference to the neotropics. In P.A. Furley, J. Proctor and J.A. Ratter (eds), *Nature and Dynamics of Forest–Savanna Boundaries*; Chapman & Hall, London, pp. 91–114.

Hodgman, C.D., Weast, R.C. and Selby, S.M. 1955. *Handbook of Chemistry and Physics*. Chemical Rubber, Cleveland.

Hutchinson, G.L. and Brams, E.A. 1992. NO versus N_2O emissions from an NH_4^+-amended Bermuda grass pasture. *Journal of Geophysical Research*, **97** (D9), 9889–9896.

Keller, M., Goreau, T.J., Wofsy, S.C., Kaplan, W.A. and McElroy, M.B. 1983. Production of nitrous oxide and consumption of methane by forest soils. *Journal of Geophysical Research*, **10**, 1156–1159.

Keller, M., Kaplan, W.A. and Wofsy, S.C. 1986. Emissions of N_2O, CH_4 and CO_2 from tropical forest soils. *Journal of Geophysical Research*, **91**, 11 791–11 082.

Keller, M., Kaplan, W.A., Wofsy, S.C. and Costa, J.M. 1988. Emissions of N_2O from tropical soils: response to fertilization with NH_4^+, NO_3^-, and PO_4^{3-}. *Journal of Geophysical Research*, **93**, 1600–1604.

Jenkin, R.N., Rose, R.R., Dunsmore, J.R., Walker, S.H. and Briggs, J.S. 1976. The agricultural development potential of the Belize Valley. Land Resource Study 24, Land Resources Division, Ministry of Overseas Development, London.

Lashof, D.A. and Ahuja, D.R. 1990. Relative contributions of greenhouse gas emissions to global warming. *Nature*, **344**, 529–531.

Livingston, G.P., Vitousek, P.M. and Matson, P.A. 1988. Nitrous oxide flux and nitrogen transformations across a landscape gradient in Amazonia. *Journal of Geophysical Research*, **93**, 1593–1599.

Luizão, F., Matson, P., Livingston, G., Luizão, R. and Vitousek, P. 1989. Nitrous oxide flux following tropical land clearing. *Global Biogeochemical Cycles*, **3**, 281–285.

Mosier, A.R., Guenzi, W.D. and Schweizer, E.E. 1986. Soil losses of dinitrogen and nitrous oxide from irrigated croplands in North-eastern Colorado. *Soil Science Society of America Journal*, **50**, 344–348.

Parkin, T.B., Meisinger, J.J., Chester, S.T., Starr, J.L. and Robinson, J.A. 1988. Evaluation of statistical estimation methods for lognormally distributed variables. *Soil Science Society of America Journal*, **52**, 323–329.

Prinn, R., Cunnold, D., Rasmussen, R., Simmonds, P., Alyea, F., Crawford, A., Fraser, P. and Rosen, R. 1990. Atmospheric emissions and trends of nitrous oxide deduced from 10 years of ALE-GAGE data. *Journal of Geophysical Research*, **95**, 18 369–18 385.

Recous, S., Mary, B. and Faurie, G. 1990. Microbial immobilization of ammonium and nitrate in cultivated soil. *Soil Biology and Biochemistry*, **22**, 913–922.

Robertson, G.P. 1989. Nitrification and denitrification in humid tropical ecosystems. In J. Proctor (ed.), *Mineral Nutrients in Tropical Forest and Savanna Ecosystems*. British Ecological Society Special Publication No. 9. Blackwell, Oxford, pp. 55–69.

Sanhueza, E., Hao, W.M., Scharffe, D., Donoso, L. and Crutzen, P.J. 1990. N_2O and NO emissions from soils of the northern part of the Guayana Shield, Venezuela. *Journal of Geophysical Research*, **95**, 22 481–22 488.

Skole, D. and Tucker, C. 1993. Tropical deforestation and habitat fragmentation in the Amazon: satellite data from 1978 to 1988. *Science*, **260**, 1905–1910.

Vitousek, P.M. and Matson, P.A. 1988. Nitrogen transformations in tropical forest soils. *Soil Biology and Biochemistry*, **20**, 361–367.

Waring, S.A. and Bremner, J.M. 1964. Ammonium production in soil under waterlogged conditions as an index of nitrogen availability. *Nature (London)*, **201**, 951–952.

Weier, K.L. and MacRae, I.C. 1992. Denitrifying bacteria in the profile of a brigalow clay soil beneath a permanent pasture, and cultivated crop. *Soil Biology and Biochemistry*, **24**, 919–923.

WRI (World Resources Institute) 1994. *World Resources 1994–95: A Guide to the Global Environment*. OUP, New York.

17 A Soil–Landscape Continuum on a Three-Dimensional Hillslope, Quantock Hills, Somerset

S. J. PARK, T. P. BURT and P. A. BULL
School of Geography, University of Oxford, UK

17.1 INTRODUCTION

Most geomorphologists assume that soils at different slope positions have different properties, because present and/or past physical and chemical processes on each part of the hillslope have been different. Consequently, the catena or toposequence is one of the most widely recognized concepts in both geomorphology and pedology, and a great number of investigations have been carried out to verify general soil–landform relationships. However, our knowledge of these relationships still remains qualitative and theoretical, and process-oriented pedogeomorphological studies designed to find out which mechanisms produce spatial soil variations have been quite limited. In his review of soil–landscape relationships in the British Isles, Gerrard (1990) cast serious doubt on widespread application of the catena concept in temperate climatic regions, even though he recognized that general relationships do exist between soils and landforms.

Many pedologists and geomorphologhists believe that the major limitation in understanding the soil–landscape continuum is the lack of understanding of the processes involved in their development (Hall and Olson 1991; McSweeney *et al.* 1994). Wilding and Drees (1983) argue that if enough observations are made, the error in predicting the landscape variability can be understood and explained. Gerrard (1990) notes that our failure to identify such relationships is due to a lack of proper understanding of the genetic heterogeneity of both landforms and soils, and the practical constraints involved with collecting and analysing sufficient numbers of soil samples. McSweeney *et al.* (1994) argue that a major challenge for geomorphologists is to characterize landscapes into domains where soil, hydrological and landform attributes can be considered products of common processes of formation. This kind of approach can provide useful data sets for interpreting landform evolution at different topographical positions (Trudgill 1986), ecological modelling (Christophersen *et al.* 1993), and earth surface management (Arnold and Wilding 1991; Moore *et al.* 1993), as well as allowing better understanding of the soil–landform relationships.

Advances in Hillslope Processes, Volume 1. Edited by M. G. Anderson and S. M. Brooks.
© 1996 John Wiley & Sons Ltd.

Our purposes in this chapter are (i) to describe the three-dimensional characteristics of soil profiles on a hillslope; (ii) to classify pedogeomorphological zones; and (iii) to identify the dominant processes in each zone. The basic assumption in this research is that soils are very good indicators of the environmental processes which have produced them, and the net result of these processes is to produce a particular soil profile. Even though the selection of meaningful soil properties still remains the most difficult part of such studies (Gerrard 1990), it is assumed that measurement of a wide range of morphological and chemical characteristics at a large number of sites will allow the dominant processes operating in soils at different slope positions to be identified.

17.2 STUDY AREA AND METHODS

The study area for this research is a small south-facing slope at Bicknoller Combe on the Quantock Hills in Somerset (Figure 17.1). In general, all slope profiles follow the sequence: flat interfluve–convex shoulder–steep backslope–weak concave footslope,

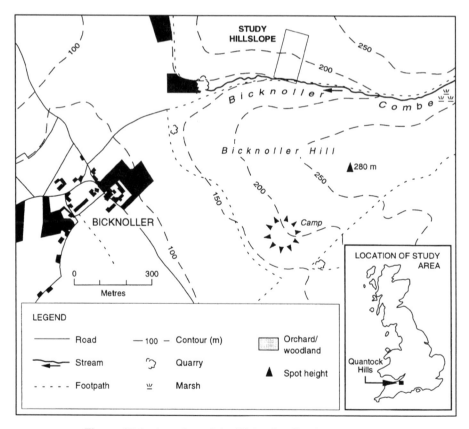

Figure 17.1 Location of the Bicknoller Combe study site

though the downstream spur lacks a concave footslope due to channel incision (Figure 17.2). There is a well-developed hollow starting from the upper convex shoulder slope, and extending toward the stream. A topographical survey was conducted using EDM and GPS methods (Fix and Burt 1996). The slope has been intensively studied in relation to subsurface flow generation (Anderson and Burt 1978; Burt 1978); solutional denudation (Crabtree and Burt 1983; Burt et al. 1984), and soil profile development in terms of Al and Fe redistribution (Fielder 1981).

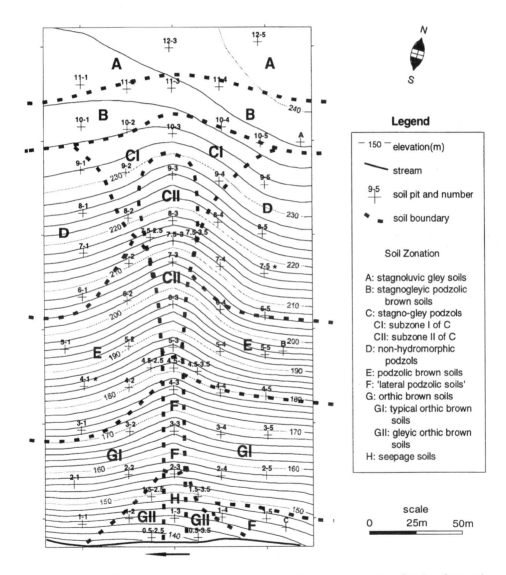

Figure 17.2   Soil zonation on a three-dimensional hillslope in Bicknoller Combe, Somerset (*extragrade: 4–1 luvic podzol, 7–5; ferric stagnopodzol)

The parent material belongs to the Hangman Grits (Trentishoe Grits) of the Middle (or Late) Devonian (Edmonds and Williams 1985). The soils which develop on these rocks have two important characteristics relevant to pedogenesis: high permeability and low base content. The regolith is therefore susceptible to podzolization under acidifying vegetation. The Soil Survey of England and Wales classified the soils on the study slope as Larkbarrow Associations (coarse loamy ferricpodzols and ferric stagnopodzols) and Rivington 2 Associations (coarse loamy brown earths and brown podzolic soils) (Findlay et al. 1984). Vegetation on the steep slopes is dominated by bracken and grass, becoming mixed heathland on the flat interfluve. The dominant vegetation within each soil zone is described below. The study slope is used as rough grazing for sheep. Two meteorological stations at different altitudes (96 m OD and 286 m OD) at Nettlecombe Court Field Centre (ST 057377), about 6 km southwest from Bicknoller Combe, indicate the general climate of this locality. Average rainfalls are 985 mm at the lower site and 1080 mm at the higher site. Mean air temperature at the two sites is 9.56°C and 8.93°C respectively; mean soil temperature at 30 cm depth is 11.41°C and 9.73°C (The Meteorological Office, 1969–1991).

Fifty-six soil pits were dug, based on a 25 m sampling grid, during the spring of 1994; and a further 10 soil pits were surveyed during February/March 1995; in total, 69 profiles including three pits for preliminary research were used for the soil classification exercise (Figure 17.2). Pits varied in depth from 50 cm to 110 cm depending on the stoniness of the subsoil. Soil description followed Hodgson (1976), and soils were classified at the subgroup level according to Avery (1990). Approximately 2 kg soil samples were collected for laboratory analysis at 10 cm sampling intervals. Even-depth sampling was selected for two reasons: (i) the main interest of this research is the vertical comparison of each soil profile; (ii) lateral continuity of soil horizonation is poor because of coarse soil textures and high slope angle. The 513 samples collected were dried at room temperature and gently ground by using a rubber mortar in order to pass a 2 mm sieve. Soil pH was measured by adding distilled water and 0.01M $CaCl_2$ solution in 1 : 2.5 ratio ($w : v$). Organic content was estimated by using low-temperature ashing at 375°C for 16 h (Hesse 1971). Clay content was measured by using a laser granulometer after removal of organic matter and oxides with H_2O_2 and sodium dithionite–citrate–bicarbonate (Kunze and Dixon 1986).

The oxides and hydroxides of Al, Fe and Mn in soils, particularly the poorly crystallized and microcrystalline forms, may be the most reactive components of acidic soils (Paterson et al. 1991). These oxides and hydroxides are also closely involved in many soil-forming processes such as weathering and brunification, podzolization, lessivage and gleying (van Schuylenborgh 1965; Blume and Schwertmann 1969). Consequently they are widely used as pedochemical indices to describe the type and intensity of pedogenic processes. Two well-known methods, acid ammonium oxalate method and sodium dithionite–citrate–bicarbonate (CBD) method were used in this chapter. These two methods have been widely applied to differentiate soil groups and identify major pedochemical processes, especially podzolization (e.g. McKeague and Day 1966; Blume and Schwertmann 1969; Loveland and Bullock 1976; Herbauts 1982). It is generally agreed that acid ammonium oxalate solution is most suitable for extracting organic Fe and poorly crystalline Fe oxides, this being most of the Fe involved in podzolization (Farmer *et al.* 1983; Loveland 1988; Schwertmann and Murad 1990). On the other hand, CBD mixed solution can extract most of the "free Fe" from soils (Jackson *et al.* 1986; Loveland 1988). The ratio Fe(o):Fe(d) ("active ratio") is commonly used as a relative measure of the degree of aging or crystallinity of free iron oxide caused by pedogenesis, particularly in humid temperate soils (Blume and Schwertmann 1969). While such clear differential extraction cannot be expected for Al and Mn oxides, oxalate and CBD extractable Al and Mn can be used for the estimation of relative depth functions of pedogenesis, and sometimes as more sensitive indicators for differentiation of important soil groups (Blume and Schwertmann 1969). The acid ammonium oxalate method followed the procedures of Ross and Wang (1993), and the modified CBD extraction procedures of Kunze and Dixon (1986) were used to extract Fe, Mn and Al after removal of organic matter by H_2O_2 treatment. All extracted solutions were analysed by atomic absorption spectrophotometer. Only Al(o) extracted by acid ammonium oxalate and Mn(d) extracted by CBD solution will be presented in this chapter as "free Al" and "free Mn", because these usually show higher concentrations than equivalent methods and are more pedochemically significant (Jackson *et al.* 1986). In their review of selective dissolution methods of oxides and hydroxides, Jackson *et al.* (1986) argue that CBD extraction may be used to extract free MnO_2 with any H_2O_2 pretreatment, and acid oxalate solution shows better selective extraction of non-crystalline pedogenic Al, including allophane and imogolite, than CBD extractable Al.

17.3 RESULTS

Notwithstanding spatial variations in the vertical distribution patterns of oxides and hydroxides, parent materials, and evidence of past pedogeomorphological processes at many soil profiles, eight major soil zones with four subzones can be recognized in terms of similarity of soil morphology, vertical functions of soil oxides, and consistence of spatial distribution (Figure 17.2). In making this classification, most emphasis was placed on the soil morphological features and their spatial distribution. The classification scheme and definition of diagnostic features followed Avery (1980, 1990) with some minor modifications. Table 17.1 summarizes the morphological criteria that were used to define the soil zones. Modifications are explained below the table and in the text. In this chapter, a soil zone is defined as a continuous area of the hillslope in which soils show genetically similar morphological and chemical features. Therefore profiles which do not fully satisfy the morphological criteria of a soil zone and spatially isolated soil profiles were classified as extragrades. Intergrades are transitional soils between soil zones which have very weak development of critical diagnostic features.

The results of soil zonation are given in Figure 17.2. Variations and dominant processes are described below for each soil zone. Full details of soil chemistry and morphological descriptions are contained in Park (1996).

Table 17.1 Morphological criteria for soil zonation in Bicknoller Combe, Somerset

Soil zone and soil subzone	Diagnostic morphological features (modified from Avery 1990)	Soil subgroups at each soil zone (no. of surveyed profiles)[a]
A. Stagnoluvic gley soils	1. A gleyed subsurface horizon[b] without podzolic B horizon 2. A slowly permeable subsurface horizon	Stagnoluvic gley soils (3)
B. Stagnogleyic podzolic brown soils	1. A podzolic B (Bh and/or Bs) horizon without an albic E horizon[d] and prominent Bh horizon[c] 2. Gleyic features over a slowly permeable subsurface horizon	Stagnogleyic podzolic brown soils (7) Intergrade to humoferric gley podzols (1)
CI. Subzone I of stagnogley podzols	1. An albic E horizon[d] over a podzolic B horizon 2. A podzolic B horizon that includes a prominent Bh horizon[c] 3. Gleyic features within or over podzolic B horizon	Luvic (or humoferric, gley-podzols (4) Humoferric gley-podzol intergrade to stagnogleyic podzolic brown soils (1) Luvic gley-podzols (?, 8)
CII. Subzone II of stagnogley podzols	1. Same morphological criteria with subzone I 2. A shallow subsurface eluviation features[e]	
D. Non-hydromorphic podzols	1. An albic E horizon[d] over a podzolic horizon and under darker colour A horizon 2. A podzolic B horizon that includes a prominent Bh horizon[c]	Luvic (or humoferric) podzols (9) Ferric stagnopodzols (1) Intergrades to (stagnogleyic) podzolic brown soils (2)
E. Podzolic brown soils	1. A podzolic B (Bs and/or Bh) horizon without intervening albic E horizon[d] and prominent Bh[c]	Typical podzolic brown soils (9) Stagnogleyic podzolic brown soils (2) Intergrades to podzols (2) Intergrade to orthic brown soils (1) Luvic podzol (1)
F. "Lateral podzolic soils"	1. Podzolic or brown topsoils 2. Loosely fabricated subsurface skeletal layer[f] with abundant fine granular aggregates	Podzolic brown soils (?, 5)
GI. Typical orthic brown soils	1. Brown soils without podzolic B horizon	Typical orthic brown soils (8) Typical luvic brown soils (1) Intergrade to podzolic brown soils (1) Gleyic orthic brown soils (4)
GII. Gleyic orthic brown soils	1. Brown soils without podzolic B horizon 2. Gleyic features without slowly permeable subsurface horizon	
H. Seepage soils	1. Soils in the seepage zone 2. Water saturation within 50 cm depth	Humic stagno-orthic gley soils (1) Gleyic podzolic brown soils (?, 1)

[a]Soil identifications were based on the surveyed depth.
[b]Depth limit of gleyed subsurface horizons extended from 40 cm to 50 cm according to Soil Survey Staff (1975).
[c]At least 3 cm thick and laterally continuous dark-coloured horizon formed by translocated humus in combination with sesquioxides.
[d]No colour criteria. Lighter colour than overlying A (or A/E) horizon and underlying podzolic (Bs or Bh) horizon.
[e]See text.
[f]See text and Soil Survey Staff (1975).

17.3.1 Soil Zones A and B: Stagnoluvic Gley Soils and Stagnogleyic Podzolic Brown Soils

These two soil zones will be considered together, because they show relatively similar pedogenic processes and are spatially intermixed with each other. These soils mainly occur on the interfluve and upper convex "shoulder" slope; however, stagnoluvic gley soils are much more dominant on the relatively flat interfluve, where surface ponding is sustained for a considerable time during the winter. Morphological criteria for differentiating these soils from adjacent stagnogley podzols and non-hydromorphic podzols on steeper downslopes are the absence of clear podzolic horizonation, and the existence of gleyic features at shallow depth (usually within 40–60 cm depth). The surface is usually fully covered with heathland vegetation that is dominated by bent–fescue (*Festuca–Agrostis*) grass with abundant small patches of gorse (*Ulex* spp.), heather (*Calluna vulgaris*), bell heather (*E. cinerea*) and bilberry (*Vaccinium myrtillus*). Bare patches also occur in places.

The general horizonation of stagnoluvic gley soils are (1) moderately developed brownish black colour A horizon under 4–10 cm thick heath or grass root mats and moder (or hydromoder) type organic layer; (2) massive or coarse blocky gleying horizon (Bg and/or Eg) with common to many coarse gleyic mottles, fine iron segregation, and/or bleached colour on upper parts of coats and peds; and (3) platy-massive reddish brown coloured indurated horizons (2Btx(g)). Sandstone fragments in the A and E horizons are usually highly weathered and easily broken.

Stagnogleyic podzolic brown soils show similar soil horizonation. Gleying features are far less intense; consequently there are quite well-developed Bs(g) or Bs horizons with many organic coats which indicate apparent podzolization. Indurated horizons are relatively deep and sometimes show less intense development than that of the stagnoluvic gley soil. However, gleyic features are still quite closely associated with the indurated horizons. In terms of spatial occurrence and general morphological features, stagnogleyic podzolic brown soils closely resemble the "podzols with gleying", which are described by Crampton (1963) in South Wales and reported from the adjacent Exmoor Forests (Curtis 1971), but without the prismatic thin iron pans. Avery (1990) categorized "podzols with gleying" in the podzolic brown soils, and described them as intergrades to stagnopodzols on upper slopes in perhumid upland localities.

In terms of present hydrological and pedochemical processes, the most significant morphological characteristic in these soils may be the occurrence of compacted and indurated horizons at shallow depth (usually at 30–60 cm). Their existence is easily recognized in the field, due to an abrupt textural change of soil matrix and colour, the high compactness and brittle soil consistence without significant cementation, an abrupt or disturbed boundary with the overlying horizon, and entire thick coatings of fine material (mainly silitan: Hodgson 1976) on upper parts of stones. The dominant soil colour of the indurated horizons is usually reddish brown or dark reddish brown (10R 4/4 to 2.5YR 3–4/4–6) with higher chroma coarse mottles or linings at the upper boundary. Soil structure is usually massive; however, most undisturbed soil faces show clear platy structure. These horizons were commonly found on steeper downslopes where the stagnogley podzols and non-hydromorphic podzols appear. The morphological appearance and spatial distribution of the indurated horizon closely resembles the "indurated layer" described in Glentworth (1944), "indurated horizon" (Fitzpatrick 1956), "fragipan" (Soil Survey Staff 1975) or "fragipan-like horizons" (Avery 1990). Payton (1992) mentions that there is still some confusion over the identification and definition of such horizons. Here, we use the more comprehensive morphological term, "indurated horizon" to describe the dense, closely packed subsurface horizon with general morphological characteristics described as above. Soil cubes collected from the indurated horizons are brittle when moist, and have a very firm or moderately strong soil consistence when dry. Most dry soil fragments slake in water, but some, especially when collected from well-developed podzols, fail to slake in water.

The presence of similar horizons on flat interfluves and upper slope elements in the Highlands, Southern Uplands and South Wales has been widely reported (Glentworth and Dion 1949; Crampton 1965; Pyatt 1970; Curtis 1971; Payton 1988). Indurated horizons are generally thought to have formed by frequent alternation in freezing and thawing under past periglacial environments (Fitzpatrick 1956; Crampton 1965; Payton 1988; Avery 1990; Payton 1992). When compared with the description of typical indurated horizons, however, there are some exceptional features at Bicknoller, like the absence of prismatic and/or vesicular structures, weak cementation and clay maxima usually coincident with these horizons. Because of the high stoniness of the study site and disturbance of original boundary along with the pedogenic modifications caused by podzolization and gleying processes, it is extemely difficult to identify the exact genesis by macromorphology and pedochemical data alone. However, it is quite clear that they are unlikely to be a product of any contemporary pedogenic processes, because the indurated horizons are not found in any fixed relationship to the surface or to other horizons in the profile. Furthermore, general morphological characteristics such as platy structure, disturbed boundary, and broken features downslope provide strong evidences of relict strong physical processes acting on these horizons.

In spite of the similarity of indurated horizons, morphological features in the overlying horizons are quite variable. The dominant pedological processes in these soil zones can be placed on a spectrum from strong surface-water gleying to incipient podzolization with weak gleying. The high compactness and low permeability of indurated horizons prevent vertical infiltration and promote lateral water movement. Any morphological differences are mainly due to the depth of the indurated horizon and its surface configuration. On some parts of the interfluve, where a low hydraulic gradient is combined with a well-established indurated horizon, there is strong gleying caused by a locally perched water-table. On the other parts of these slope segments, especially close to the relatively steep convex hillslope, there is relatively rapid lateral subsurface throughflow through the A horizon and root mats, and better aeration. This part of the soil zone shows very limited gleying features (such as bright colour mottles or lining on coats and peds), and weak podzolization, largely due to the base-deficient parent materials and better drainage.

The most common forms of gleying in many British upland areas are "surface-water gleying" and "superficial gleying" (Pyatt 1970). Surface-water gleying occurs in soils where vertical drainage is impeded at depth by a more compacted layer (Pyatt

1970; Duchaufour 1982; Avery 1990). On the other hand, superficial gleying is mainly caused by thick organic layers or root mats. The important role of the organic layer and root mats in holding water for considerable periods is widely believed to be an important gleying process on the flat interfluves, and shoulder slopes of the British uplands (Muir 1934; Crampton 1963; Pyatt and Smith 1983; Avery 1990). For these soil groups at Bicknoller, it seems more reasonable to assume that these two gleying processes work together. However, the relatively deep gleyic features usually associated with indurated horizons, the absence of any kinds of mottles inside the indurated horizons, and relatively thin organic and root mats indicate that surface-water gleying is the dominant process.

Vertical distribution patterns of oxides and hydroxides confirm the morphological characteristics of these soil zones. In a well-developed stagnoluvic gley soil (Figure 17.3(a)), Fe(d) levels are low in topsoil and Bg (or Eg) horizons, with a marked increase at the indurated horizon, indicating the likelihood that considerable amounts of oxides above the indurated horizon have been removed. Fe(o) and Al(o) also generally show similar patterns. Iron removal in horizons overlying the indurated horizon is mainly caused by gleying processes largely enhanced by impermeability of the indurated horizon and "superficial gleying". The most significant pedochemical feature in this soil zone is the high "active ratio" in the topsoil with marked decrease toward the indurated horizons. While such trends were not observed in the data reported by Blume and Schwertmann (1969), McKeague and Day (1966) and Blume (1988), the increase in the active ratio accompanying morphological evidence of gleying has been widely reported by several workers (Stonehouse and St Arnaud 1971; Moore 1973). Moore (1973) explains that the increase in the active ratio in gley soils results from the retarding effects of dehydration and crystallization of Fe oxides and hydroxides in the presence of waterlogged conditions.

The stagnogleyic podzolic brown soil (Figure 17.3(b)) shows quite different vertical distribution of Fe(d), Fe(o), Al(o) and active ratio. The major difference is the presence of a Bs horizon, which is usually associated with the maxima of most oxides and the active ratio compared to the underlying indurated horizon. Overall, the active ratio in stagnogleyic podzolic brown soil is lower than that in stagnoluvic gley soils. The relatively low active ratio probably indicates better drainage than stagnoluvic gley soils. The intervening illuvial horizons between topsoils and underlying indurated horizons reflect the fact that the convex shoulder slope occupies a transitional position between gleying on the flat interfluve and podzolization on steeper slopes.

In summary, the dominant hydrological process on the interfluve and less steep shoulder slopes appears to be shallow lateral subsurface flow, enhanced by anisotropy of infiltration rates between the organic layer and mineral soil horizon, and by the impermeable indurated horizon. The soil morphological patterns and oxide distribution patterns show that downward percolation beyond the indurated horizons is minimal on this part of the hillslope. Water circulation in these soils is mainly confined to the upper 50 cm. However, the depth of present pedogenesis and water circulation is slightly increased on more steep convex shoulder slopes, and these parts of the slope show better chemical leaching than the interfluve. The main

(a)

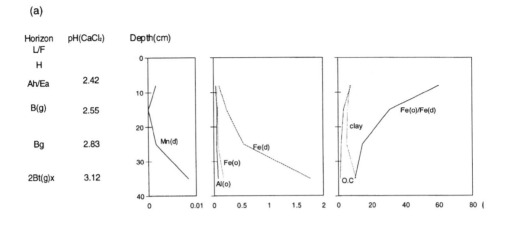

(b)

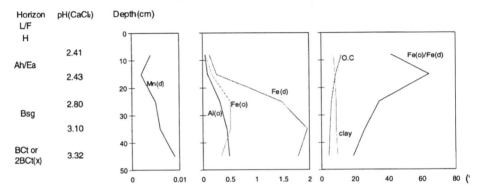

Figure 17.3   Vertical distributions of Fe, Al and Mn oxides, organic content (OC), and clay content in (a) stagnoluvic gley soil (profile 12-5) and (b) stagnogleyic podzolic brown soil (profile 11-4) in Bicknoller Combe, Somerset

pedochemical processes on the flat interfluve are acid hydrolysis and gleying under acidifying vegetation and seasonal anaerobic conditions, but this gradually gives way to incipient podzolization because of better aeration and more active leaching.

17.3.2   Soil Zone C: Stagnogley Podzols

This soil zone consists mainly of luvic gley podzols with some humoferric gley podzols. Identification of these two soil types depends on the presence of an indurated horizon at an appropriate depth. Two main morphological features, well-developed podzolic horizonation and gleyic features underlying or within podzolic B horizons, are used as general criteria for differentiating this soil group from others.

All these soils have a very clear podzolic horizon with few exceptions: albic Ea under darker-coloured Ah or EBh horizon, and prominent Bh or Bhs. Gleyic features are rather variable, but most of them occur under the Bh horizon with the underlying relatively impermeable indurated horizon notated as 2B(t)xg (or Bsg/2B(t)x), or weakly cemented Bs horizon. While stagnogleyic features, gleyed albic E or Bh, are also seen in a few profiles, most of them have more strongly developed gley mottles and/or ferruginous mottles under Bh or Bs horizons.

Thirteen soil profiles with one intergrade were identified in this group. They show fairly good spatial consistency, starting on the lower convex shoulder slope, and forming a "fan" shape centred on the hollow (Figure 17.2). One isolated ferric stagnopodzol was found on the middle spur (profile 7-5) inside the non-hydromorphic podzol zone. While some heath vegetation or dense bent–fescue grass was found at three sites on the upslope adjacent to the stagnogleyic podzolic brown soil zone, the dominant vegetation type is bracken (*Pteridium aquilinum*) with some grasses and mosses below. The boundary with stagnogleyic podzolic brown soils seems to correlate well with the upper limit of bracken growth. It is widely believed that bracken rhizomes are very sensitive to water saturation in the root zone (Pakeman and Marrs 1993).

This soil zone can be further subdivided into two subzones according to the morphology of the topsoil under the organic layer. The thickness of L/F layers is highly variable, from 2 cm to 13 cm. The H layers are weakly developed, less than 1 cm thick, and are not significantly mixed with the mineral fraction. Organic layers are followed by two different types of topsoil over an albic E horizon. One is the normal brownish black or dark brown coloured Ah (or Ah/Ea) horizon in which mineral fractions are mixed with humified organic matter in the presence of common bracken rhizomes (subzone CI, Figure 17.2). The other consists of two horizons with a lighter coloured eluviated horizon over a darker coloured and rather compacted EBh horizon (subzone CII, Figure 17.2).

The latter type of topsoil was observed in eight profiles out of thirteen stagnogley podzols; most of them occurring along the hollow. Similar horizons were also observed in two non-hydromorphic podzols (8-4 and 8-5). These two soil profiles occur at the junction of the convex shoulder slope and the straight backslope, which shows high profile curvature. Leached topsoils usually have 1–2 units higher colour value than the underlying EBh horizon, and about 15–20 cm thickness, within which bracken rhizomes are most common. Soil structures vary from a moderately developed fine and medium blocky structure to a skeletal structure mixed with bracken rhizomes, largely depending on the intensity of subsurface eluviation. The intensity of eluviation becomes gradually stronger in the middle hollow. As an extreme case, profile 6-3 has a 15 cm thick skeletal subsurface eluviated horizon under a partially disturbed L/F/H organic layer. The main constituents of this horizon are rounded medium and large stones with dense bracken rhizomes. The EBh horizon generally shows more compacted structures such as massive or moderately developed blocky ped. The darker colour might be caused by the vertical illuvation of humified organic substances from overlying horizons and from upslope.

The development of the surface Ea–EBh horizon sequence seems to be caused by shallow subsurface lateral flow which is largely enhanced by the steeper slope gradient and the presence of dense bracken rhizomes in the upper levels of soil profiles. The role of bracken rhizomes in enhancing lateral subsurface flow has been reported by Arnett (1976). During this survey, soil water flowing through and around coarse rhizomes was frequently observed. The spatial distribution of these shallow subsurface eluviation features clearly indicates that convergent subsurface flow from upslope is gathered into and flows down the hollow. Burt *et al.* (1984) mapped the relatively high hydraulic potential along the hollow at this point using tensiometer networks. These seem to be two simultaneous subsurface flow paths: (vertical) percolation resulting in a podzolic horizon sequence with EBh, and shallow subsurface lateral flow. The amount of percolating water may be quite limited because of the existence of the relatively impermeable indurated horizon and partially cemented podzolic B horizon, causing gleyic features.

Both topsoils are commonly underlain by an E or Ea(g) horizon. The thickness of the E horizon varies from 3 to 25 cm. Sometimes the colour of the E horizon does not fully satisfy the requirement of the albic E horizon defined in Avery (1980, 1990), but a more greyish, lighter colour can be easily recognized between the overlying Ah (or EBh) and underlying darker Bh horizons. The Bh horizon underlying the albic E horizon shows a clear black (2.5–5YR 2/1–2) colour and thick organic coating on the stones and mineral grains in the horizon with a sharp upper boundary. The lateral continuity of the Bh horizon is very poor, and mainly decided by the shape of the underlying Bs/2Btx (or Bs) horizon and stones. Underlying Bs/2Btx (or Bs) horizons show a marked decrease in permeability, sometimes with weak cementation. All the pedochemical data show high accumulation of sesquioxides, clay and organic matter at this horizon.

Gleying features are mainly associated with the Bh and underlying Bs/2Btx (or Bs) horizon, and show medium or coarse gley mottles and/or fine and medium ferruginous mottles. While it is very difficult to differentiate the spatial patterns of gleyic features, gley mottles are much more dominant in soil profiles of subzone CII, especially in lower slope positions. Most gley mottles show "pocket" shapes embraced by the highly compacted and/or partially disturbed Bs/2Btx (or Bs) and stones, mainly under the Bh horizon. When gleyic features occur in the Bh horizon, it is usually in the form of ferruginous mottles.

The presence and shape of the indurated horizon seems to play a vital role in the development of the Bh horizon and gleyic features in many of these soils. The morphology of this horizon is quite similar to that discussed earlier, but shows much stronger modification by present podzolization and disturbance. Three soil profiles in this zone had intact indurated horizons but with no lateral continuity, while many others had broken boundaries or other disturbance features like distortion, dentric shape fractures, or irregularity on the upper boundary. The bisequal development of podzols is quite common in the British Isles, especially in southern England, and many of them become podzolized following clay illuviation or fragipan formation (Avery 1990). In profile 8-2, which has a clear-cut indurated horizon, the development of a prominent Bh is very weak in the section lacking an underlying indurated horizon, and the main depth of gleying is also much deeper (about 80–90 cm depth) than in the part associated with the indurated horizon. In several other profiles with broken indurated horizons, in spite of entire organic coating on the mineral fraction, a clear Bh is sometimes not seen over fractures.

Figure 17.4(b) shows typical vertical sequences of Fe, Al and Mn oxides in a stagnogley podzol with an underlying indurated horizon and normal Ah topsoil. When compared with a non-hydromorphic podzol (Figure 17.5), the depth of maximum accumulation of the three oxides increases in the order Mn(d) > Al(o) > Fe(o) and Fe(d), a sequence widely reported for podzols (Duchaufour 1982). The clear podzolic morphological differentiation is mainly reflected in the distribution of Fe(o) and Fe(o):Fe(d), decreasing in the E horizon from the Ah and with a sharp peak in Bhs (or Bs) horizon. The effect of gleying, when compared with the non-hydromorphic podzols, is hardly detectable in the chemical data. On the contrary, a luvic gley podzol with subsurface eluviation features in the middle hollow (Figure 17.4(a)) shows a quite different vertical sequence of oxides: the maximum depth of oxide accumulation is much deeper and appears under the main gleyic horizons. The maximum Fe(o):Fe(d) occurs at the main gleying depth, possibly due to the retarding crystallization under reducing conditions (as explained in relation to the stagnoluvic

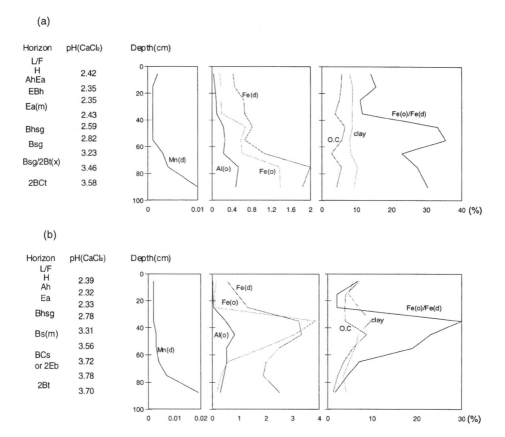

Figure 17.4   Vertical distributions of Fe, Al and Mn oxides, organic content (OC), and clay content in two two representative stagnogley podzols in Bicknoller Combe, Somerset. (a) luvic gley podzol (profile 7-3) and (b) humoferric gley podzol (profile 9-4)

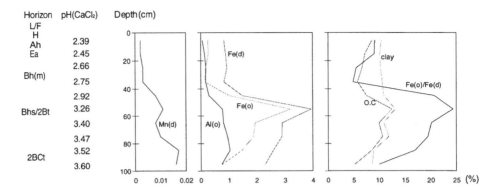

Figure 17.5 Vertical distributions of Fe, Al and Mn oxides, organic content (OC), and clay content in a representative non-hydromorphic podzol (profile 7-4) in Bicknoller Combe, Somerset

gley soils). Surface eluviation features are also visible in most distributions of extracted oxides and clay content. While most of stagnogley podzols in subzone CII show virtually the same patterns to those shown on Figure 17.4(b), few profiles found in the middle hollow (profiles 6-3 and 5-3) resemble those shown in Figure 17.4(a). This spatial pattern clearly indicates that large amounts of shallow subsurface water converge into the middle hollow. Profile 6-3 shows extremely complex gleying features between 50 and 70 cm depth, while profile 5-3 shows relatively weak gleying with the existence of a subsurface skeletal layer that will be described later (see subsection 3.5).

In summary, the general morphology of this soil zone provides evidence of hydrological and pedological significance. The surface or upper Ea (or Ah/Ea) horizon represents strong lateral eluviation by shallow subsurface lateral flow through the relatively permeable topsoils containing coarse bracken rhizomes. The spatial pattern clearly shows that shallow subsurface flow becomes stronger in the hollow than on the spurs, and is dominant in the middle hollow. The strong shallow subsurface impoverishment features suddenly disappear at the middle hollow, which forms a boundary between stagnogley podzols and "lateral podzolic soils" with the substitution of an impeding Bs/2Btx (or Bs) layers into the skeletal subsurface Bs horizon. Fine materials eroded from upslope may be deposited at the transitional soil zone around profile 4-3. Soil profile 4-3 is the only one where fine material was dominant among the surveyed profiles, and which showed an extremely low content of exchangeable bases. In spite of the strong shallow subsurface runoff, vertical infiltration of water results in a podzolic soil sequence with an albic E and prominent Bh horizon. The presence of indurated Bs/2Btx horizons and/or partially cemented podzolic horizon (Bs(m)) largely restrains free drainage and results in gleyic mottles. However, because of the strong disturbance of these horizons, one cannot assume there is a continuous perched water-table over the indurated and podzolic B horizons.

17.3.3   Soil Zone D: Non-hydromorphic Podzols

This soil zone mainly consists of luvic podzols and humoferric podzols, which are equivalents of luvic gley podzols and humoferric gley podzols without subsurface gleyic features. Their morphology and pedochemical features quite closely resemble most of the stagnogley podzols, especially those in the CI subzone (Figure 17.5). Indurated horizons with clay maxima are common; most of them have been severely disturbed. Cementation in Bs/2Btx (or Bs) seems to be relatively weaker than in the stagnogley podzols, and the soil is generally less compact. One outlier of this soil group was identified among the podzolic brown soil zone as a luvic podzol (profile 4-1) without apparent fragipan characteristics. Shallow subsurface eluviation features were also found in two soil profiles (profiles 8-4 and 8-5).

Ten non-hydromorphic podzols and two soils intergrading to podzolic brown soils were surveyed. The zone starts from the middle of the convex shoulder slopes on the spur and forms a triangle whose long axis extends downslope along the spurs (Figure 17.2). In its lower part, it merges into podzolic brown soil with some intergrades. The dominant vegetation is bracken with some other heathland vegetation, such a bilberry, ling, bell heather and gorse. The density of heathland vegetation gradually increases upslope; on lower slopes this soil zone is mainly covered by bracken.

When considering the spatial distribution of this soil zone and its general soil morphological pattern, the absence of gleyic features seems to be mainly caused by lower amounts of vertically infiltrated water reaching the underlying subsurface horizons rather than because of less well-developed subsurface horizons such as the indurated and cemented Bs horizon. As shown in Anderson and Burt (1978) and Burt *et al.* (1984) for this slope, the general topographic configuration influences the direction of subsurface flow. Given the general hillslope configuration, a large portion of rainwater in this zone flows down to the adjacent stagnogley podzol zone. Upslope drainage areas are small within this zone so there is much less influence of lateral flow than in the hollow where much lateral flow converges.

17.3.4   Soil Zone E: Podzolic Brown Soils

These soils occur as a band between podzols upslope and orthic brown soils below. Avery (1980, 1990) defines these soils as those which have a friable and (more or less) ochreous Bs horizon below an Ah or Ap horizon, with no intervening distinct E or Bh horizon. The presence of a Bs horizon and organic coating distinguishes podzolic brown soils from orthic brown soils, and the absence of a distinct albic E and a prominent dark Bh horizon differentiate podzolic brown soils from podzols. The podzolic soil zone seems to extend upslope on the flanks and merges into stagnopodzols and "lateral podzolic soils" in the middle hollow (Figure 17.2). Vegetation can be divided into two dominant types: bracken alone or grass-bracken with abundant bristle bent (*Agrostis setacea*). The latter vegetation mainly occurs on the eastern spur and flank, and is associated with fractured bedrock at about 40–50 cm depth.

The general horizonation of soils under bracken are as follows:

1. a mor or moder type organic layer under bracken remains;
2. a deep (about 20 cm) black or brownish black coloured Ah/Ea horizon with common bracken rhizomes;
3. a granular or weak fine blocky reddish brown Bs and/or Bhs horizon; and
4. a loose apedal BCs(t) horizon.

Even though a distinct Bh horizon is not seen, there are abundant black coloured organic coatings on peds and stones in the B and BC horizons, which reflect apparent podzolization. The development of indurated horizons, which are widespread upslope, is not significant in this soil zone. Even so, discontinuity and stratification of parent materials can be easily recognized. Such discontinuities sometimes impede vertical infiltration and result in weak gleyic features in the lower parts of soil profiles (6-4 and 6-5). Two soil profiles (5-4 and 4.5-3.5) under bracken with grass and bristle bent are an exception to this general horizonation; they show more advanced humification of organic matter and clear pellet-like aggregates between the fractured sandstone. Some of the "lateral podzolic soils", especially adjacent to stagnogley podzols in the middle hollows (4.5-3), show podzolic topsoils. However, the morphological and chemical characteristics of B and BC horizons are completely masked by strong lateral illuviation features (see sub-section 3.5).

In spite of clear morphological differences with podzols, many of the vertical distributions of pedochemical variables resemble those of podzols (Figure 17.6), differing only in (i) higher contents in topsoil and (ii) shallower depth and lower content of maxima of translocated oxides in podzolic brown soils. The lateral eluvial processes in the Ah/Ea horizons might also influence these depth functions of oxides. However, organic coating and crumb-like structure in the Bs horizon indicate relatively weaker translocation of sesquioxides in this soil group and less advanced podzolization (Herbauts 1982). The vertical distribution of the "active ratio" also confirms the illuviation of amorphous iron and the absence of an albic E horizon in podzolic brown soils. The common appearance of podzolic brown soils between

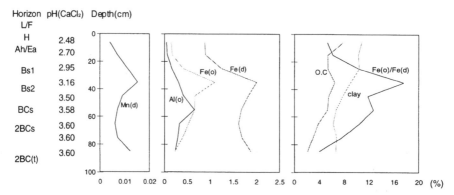

Figure 17.6 Vertical distributions of Fe, Al and Mn oxides, organic content (OC), and clay content in a representative podzolic brown soil (profile 5-2) in Bicknoller Combe, Somerset

podzols and brown soils as a catenary sequence has been widely reported in the British Uplands (Ball 1966; Avery *et al.* 1977). It is commonly regarded as resulting from an early stage of podzolization.

While there are no available permeability data, the absence of an impeding subsurface layer and general morphological patterns clearly indicate higher vertical percolation rates than for other soil groups upslope. However, shallow subsurface flow through the Ah/Ea horizon, which is usually related to the presence of bracken rhizomes, might enhance lateral subsurface flows. De Coninck (1980b) argues that during the development of the podzolic (spodic) horizon a rearrangement occurs in the distribution of the different pore sizes; the number of large pores decreases while fine and medium pores increase. Therefore, it might be expected that some decrease of infiltration rates would occur because of the decrease of pore size and the higher water-holding capacity of oxides and hydroxides in Bs horizon. Even so, this effect would be far less than for upslope soil groups.

17.3.5 Soil Zone F: "Lateral Podzolic Soil"

These soils are dominant in the middle hollow between the stagnogley podzol zone and seepage soil zone, and also occur as a patch on weakly concave footslopes in association with orthic brown soils (Figure 17.2). The surface vegetation is mainly bent–fescue grass with dense bracken. The only exception was a bracken monospecies near the boundary with the stagnogley podzol zone (profile 4.5-3). General horizonation shows a L/F/H–Ah–Bw–2BCst(g) sequence. The main morphological features are as follows:

1. The surface organic layer is usually of the mull or moder type with a clear thin H layer. Humification looks slightly less than that of orthic brown soils, but much higher than for podzolic soils.
2. There are moderately developed dark reddish-brown topsoils and extremely stony, loosely fabricated skeletal subsurface horizons (Soil Survey Staff 1975). The depth and development intensity of topsoil generally increases at higher slope positions. The dominant soil ped in the topsoil is fine granular as in the orthic brown soils.
3. There is strong subsurface deposition of sesquioxides and fine materials appear as coats and pellet-like fine aggregates in loosely fabricated subsurface skeletal layers.
4. Stones in lower parts (below 90 cm) are completely surrounded by fine materials (mainly clays) and show weak gleying features.

The clearly different morphological feature of this soil type, compared with other soil groups, is the presence of a loosely fabricated skeletal subsurface horizon (Soil Survey Staff 1975). Stones in this subsurface horizon are mainly large and very large sandstone fragments with gaps that are not completely filled by the fine soil matrix. Stones usually show more roundness and larger size than those in overlying horizon. The occurrence and upper position of this layer is spatially very consistent. It is first found in soil profile 5-3 as an intergrade from the luvic gley podzols at about 60 cm depth and extends downslope along the hollow. Its depth becomes shallower toward downslope, except profile 4-3, and eventually merges into the water-table in the seepage soils (profiles 2-3 and 1-3).

The easiest explanation for the formation of skeletal layers with discontinuous indurated horizons and frequent stratification of parent materials may be periglacial processes during cold periods. There is widespread evidence of strong periglacial processes in southwest England (Catt 1979; Edmonds and Williams 1985). Solifluction would have taken place under alternating conditions of freeze and thaw when frost-shattered debris lubricated by meltwater unable to penetrate the frozen subsoil was moved downslope by gravity. Seasonal meltwater from snow fields would concentrate in the hollow, and would wash out fine materials and even medium stones. The greater roundness of large stones in the skeletal layer than those of surface soil clearly indicates water action and the presence of significant amounts of flowing waters. The moderately developed surface soils are possibly a result of physical deposition by weak solifluction and surface wash processes during and after the reduction of snow cover.

Figure 17.7 shows vertical sequences of oxides and clay from two profiles in this group. In contrast to the orthic brown soils (Figure 17.8), all available pedochemical indices (including base saturation and organic matter content) show a marked increase with depth. Accumulation of sesquioxides appears as pellet-like dark aggregates of nodules on and between stones. It is believed that this high accumulation of fine soil materials is mainly caused by deep throughflow, for two reasons. First, vertical movement of oxides may be quite limited: vegetation, relatively high pH and base saturation in the surface soil clearly show that the dominant pedogenic process is brunification rather than podzolization (except for profile 4.5-3). Secondly, the skeletal subsurface layer may be working as a conduit for subsurface water gathered from areas further up the hollow and from the adjacent spurs. There is a high water content in the fine fractions stuck to stones, and weak gleying features, greyish colour and some ferruginous mottles, were found below 80 cm depth in three soil profiles (1-4, 3-3 and 4.5-3). The main deposition mechanism is possibly the high oxidation state in the skeletal layer caused by better aeration. High Eh environments rapidly oxidize the reduced Fe^{2+} and Mn^{2+}, which was mobilized upslope and moved into these soils in the lateral throughflow.

In terms of hillslope hydrological processes, this soil zone reflects a significant change in dominant flow paths from the stagnogley podzols, especially those found along the hollow. As discussed above, the flow paths in stagnogley podzols and non-hydromorphic podzols are mainly confined to shallow depths, because of the coarse bracken rhizomes in the A horizon and the presence of dense podzolic B and indurated horizons. The absence of such kinds of impeding layers and the occurrence of the skeletal layers enhance vertical infiltration and deep throughflow.

17.3.6 Soil Zone G: Orthic Brown Soils

These soils mainly occur on divergent, steep straight slopes on the lower spur and slopes flanking to hollow (Figure 17.2). They grade into podzolic brown soils as altitude increases, into seepage soils at the lower base of the hollow, and into "lateral

(a)

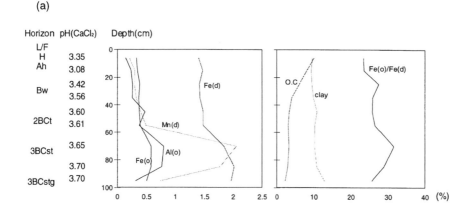

(b)

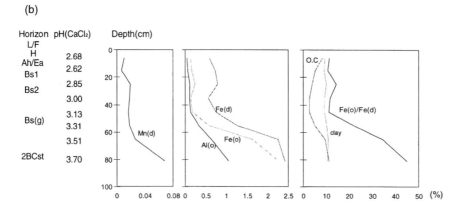

Figure 17.7 Vertical distributions of Fe, Al and Mn oxides, organic content (OC), and clay content in two representative "lateral podzolic soils" in Bicknoller Combe, Somerset: (a) profile 1-4, (b) profile 4.5-3

podzolic soils" on lower weak concave slopes and middle hollow. Fourteen soil profiles are identified as belonging to this soil group, and most of them can be subdivided into typical orthic brown soils and gleyic orthic brown soils (profiles 0.5-2.5, 0.5-3.5, 1.5-2.5 and 1-2). A minor variation is one luvic brown soil (profile 2-4).

General horizonation is L/F/H–Ah (or AhBw)–Bw–BCt (or BCt(g) in gleyic orthic brown soils) with few exceptions. The main morphological characteristics of this group are as follows:

1. A mull type organic layer with thin loose L/F layer (less than 2 cm). The H layer is usually very weakly developed and well mixed with mineral fractions.
2. Moderately or weakly developed fine granular peds in A and B horizon. This indicates active faunal activity and clay-oxides aggregates (Duchaufour 1982).

3. The soil profile generally shows an overall reddish brown colour inherited from bedrock, and has very weak horizonation, especially between A and B horizons. Horizonation from A to B horizons is normally distinguished only by a weak colour change, structure, or the decrease of bracken rhizomes.
4. Very prominent dark manganese coatings on stones and peds, especially in the A horizon. The intensity of manganese coating markedly decreases as depth and altitude increase.
5. Common stratification or discontinuity of parent material. Even though stones are mainly very small to medium ($<6\,cm$) subangular and angular sandstone fragments, discontinuities can be easily recognized by stone content and the compactness of stones.
6. There are common continuous or entire clay skins or coats on stones in layers designated BCt or BCtg, and coats frequently show a bleached (lighter) colour especially in dry soils. This feature is much stronger in gleyic orthic brown soils.

The occurrence of these orthic brown soils is strongly associated with the presence of grass vegetation. The catenary boundary between brown soils and podzolic brown soils is quite close to the point where bent–fescue grassland with bracken changes into the bracken-dominated area. The main processes forming brown soil (brunification) are widely believed to be as follows: iron oxides or hydroxides, freed from silicate minerals by weathering or inherited from parent materials, are bonded to silicate clays and organic matter and immobilized at or near the surface of soils (Duchaufour 1982; Avery 1990). Litter of bent–fescue grass yields easily decomposed organic matter; this promotes less acid, higher base saturation, and less surface organic matter accumulation of surface soil in acid environments (Miles 1985).

Figure 17.8 shows the vertical sequence of Fe, Al and Mn oxides of selected two soil profiles in this group. There are no significant vertical changes in "total free" iron Fe(d). The irregularity is mostly caused by discontinuities in parent material and the presence of an argillic horizon in profile 2-4. However, amorphous iron (Fe(o)) and active ratios of iron decrease with depth. Blume and Schwertmann (1969) mention that oxalate extractable Fe may be a better indicator of pedogenesis in soils where discontinuity of parent material is common. Deviations from this trend are shown by slightly lower values of Fe(o) and Fe(o)/Fe(d) in the Ah horizon (usually within 20 cm), and by the existence of variation of subsurface soil layers such as argillic horizons (profile 2-4) and skeletal subsurface horizons (profile 3-4). The decrease in amorphous or poorly ordered sesquioxidic materials in the A horizon is widely reported under semi-natural vegetation, and believed to be the result of strong leaching and weathering of silicates under the influence of acid organic matter rather than by translocation below (Avery 1990). Shallow subsurface lateral flows enhanced by the process of dense bracken rhizomes in the A horizon may be another important factor in the removal of amorphous oxides.

One of the significant pedochemical features in this soil zone is the high accumulation of secondary manganese extracted by both CBD and acid ammonium oxalate at or near surface horizons in footslope positions. This accumulation is easily recognized in the field as entire and continuous black colour coats on stones and peds can be seen. The value of Mn(o) and Mn(d) is well correlated with Fe(o)

(a)

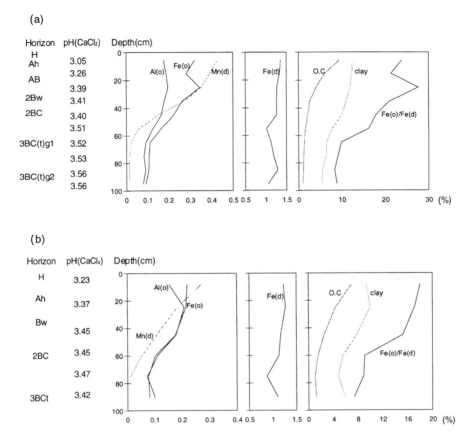

Figure 17.8 Vertical distributions of Fe, Al and Mn oxides, organic content (OC), and clay content in two representative orthic brown soils in Bicknoller Combe, Somerset: (a) gleyic orthic brown soil (profile 0.5–2.5); (b) typical orthic brown soil (profile 1-1)

vertically. However, one difference between Fe(o) and secondary manganese (Mn(d)) is that Mn(d) shows a clear lateral decrease as altitude increases. In orthic brown soils, the content of secondary Mn(d) is sometimes higher than that of Fe(o), but Mn in other soil groups at higher slope positions has concentration more than two orders of magnitude lower than that of Fe(o) (Figures 17.4 and 17.8). Within toposequences, several studies report the same decreasing patterns of secondary Mn (Yaalon *et al.* 1972; McDaniel and Buol 1991; McDaniel *et al.* 1992). Redistribution of Mn in soilscapes is closely tied to the oxidation–reduction dynamics of the system. Oxidation and reduction of Mn are thermodynamically favoured at relatively higher redox potentials than for Fe at a given pH. The maximum accumulation of Mn(d) in acid surface soils at lower slope positions occurs as follows: secondary Mn, once released by chemical weathering, moves in reduced form from upslope and accumulates in the A horizon at lower (or convergent) hillslope positions through formation of relatively stable Mn^{2+}-organic matter complexes under higher Eh conditions (Bloom 1983; McDaniel and Boul 1991; McDaniel *et al.* 1992).

The orthic brown soil zone can be classified into two subzones: typical orthic brown soil, and gleyic orthic brown soils which occur mainly around seepage soils. Criteria for these two soil groups provide evidence of water saturation in the lower part of soil profiles. There is no prominent gleyic feature even with a stagnant water-table at the lower parts of soil profiles. This may be caused by two factors: one is the high permeability of soils and the other is that the water-table (or "saturated wedges") varies considerably in height and is relatively rich in dissolved oxygen owing to the rapid rate of water renewal. In gleyic orthic brown soils, however, the seasonal rise of the water-table or high water content results into two clear pedogenic features: (1) a complete fine material coating around stones (not just on stones) and a brighter colour on coats; and (2) an increase in the percentage of base saturation as measured by the summation methods of $BaCl_2$–KCl mixed solution (Amacher *et al.* 1990). The formation of a seasonal saturation wedge on this part of the slope was shown by tensiometer measurements (Anderson and Burt 1978). The hydraulic potential (closely accordant with hillslope gradient) is quite well correlated with the distribution pattern of the gleyic orthic brown soil subzone. These morphological features are closely associated with an increase of base saturation in the water saturated parts that shows vertical "C" shapes, and strongly associated with decrease in exchangeable Al and increase of divalent exchangeable alkaline earths (Ca^{2+} and Mg^{2+}). The common features of iron removal in a reduced groundwater environment is not seen for the gleyic orthic brown soils.

In summary, pedogenic and hillslope hydrological processes in this soil zone can be readily inferred from the soil morphological and pedochemical patterns. Slope processes during recent periglacial periods produced deep, highly permeable parent materials which enhance the aeration of soils. Even with dense bracken rhizomes in the Ah horizon and a small reduction in permeability at the boundary of the A and B horizons, shallow lateral subsurface flows are less significant than for soils on higher slopes. Compared with those soils, most rainwater can percolate into the BC horizon until it meets heterogeneous subsurface layers such as more compacted subsurface layers or argillic horizons. This water may then flow as throughflow according to the hillslope configuration (Anderson and Burt 1978), to form a seepage zone and saturation wedge in the lower hollow resulting in gleyic orthic brown soils.

17.3.7   Soil Zone H: Seepage Soils

Only two soil profiles (2-3 and 1-3) are included in this soil zone. These are found in the lower part of the hollow which shows convergent plan and very weak concave profile configuration. Soils are humic stagno-orthic gley soils (profile 1-3) and gleyic (or stagnogleyic) podzolic brown soils (profile 2-3). The main morphological characteristics of these two profiles are that the skeletal structure starts from soil surface and the water-table is found at shallow depth (60 and 38 cm respectively). Abundant hydromorphic mor type organic matter is mixed with loosely fabricated large subrounded sandstone fragments. The upper part of profile 2-3 above the water-table resembles the subsurface skeletal layers in soil zone F in terms of depth function of oxides, so that this soil profile can be considered as an intergrade from a "lateral podzolic soil" zone to the seepage soil zone. Even though the soil matrix is fully saturated with water, the high porosity of the surface horizon does not allow any significant gleyic features in the soil matrix. The most significant pedochemical features in this soil group are the extremely high content of bases, high base saturation ($>95\%$) and pH, when compared with other soils.

Vegetation is also much more diverse than that of any other soil zone. Rushes (*Juncus* sp.), nettle (*Urtica urens*) and some other herbaceous species form a mosaic with bent–fescue grass and bracken. Three or four springs can be found in this part of hollow during the winter, but any return flow tends to infiltrate a short distance downslope through the stony soil surface. Springs mainly form because of abrupt textural changes formed by stratification of head deposits rather than because of any increase in water-table level. In general, the stony soil is sufficiently permeable to maintain *sub*surface flow even at high saturation (Anderson and Burt 1978).

17.4 SUMMARY AND DISCUSSION

There is a clear pattern of soil zonation on the study slope. This may be summarized as follows. Stagnoluvic gley soils have developed mainly on the flat interfluve. The dominant process in this soil zone is gleying under the presence of root mats and shallow subsurface indurated horizons. This soil zone is succeeded by stagnogleyic podzolic brown soils on the upper convex shoulder slopes in which gleying gradually gives way to podzolization because of better drainage and consequently more active chemical leaching. Stagnogley podzols on the upper and middle convergent slopes show well-developed podzolic horizons with gleying features under or inside the podzolic B horizon. They can be subdivided into two subsoil zones according to the intensity of shallow subsurface eluviated features. The stagnogley podzol zone extends downslope along the hollow with a general increase of shallow subsurface eluviation and gleyic features, and eventually merges into "lateral podzolic soils". Non-hydromorphic podzols on the lower convex shoulder and upper spur-flank slopes are equivalent of stagnogley podzols without subsurface gleyic features, and less intensive shallow impoverishment features. Podzolic brown soils appear as a catenary sequence between podzols and orthic brown soils on the middle spur and flank slopes. In terms of hillslope hydrological processes, this soil zone is also a transitional zone from shallow subsurface flow dominant upslope to vertical infiltration dominant downslope. Orthic brown soils have developed on the lower spur and flank slope, and they can be further divided into two subzones according to subsurface gleyic features caused by a seasonally fluctuating water-table. This soil zone shows much better vertical infiltration than any upslope soil zone. "Lateral podzolic soils" on the lower concave slope and middle hollow show high accumulation of sesquioxides and fine soil materials moved into relict skeletal subsurface layers by lateral subsurface flow. This soil zone merges into the seepage zone soils in the lower hollow.

The pattern of soil zonation can be easily explained in terms of hillslope hydrological processes, which are mainly controlled by slope configuration and relict geomorphological features, especially for the soils developed on the flat interfluve, convex shoulder slopes, and steep hollow. However, the processes causing soil differentiation on the lower convex slopes and steep backslopes along spurs and spur-flanks are not so clear. While common catenary sequences from podzols to brown soils are widely reported, it is hard to find any detailed explanation of the mechanisms which produced such variations. In the following paragraphs, we discuss the pedochemical and hydrological processes which may have caused this catena to have developed. A full description of soil chemical data (pH organic content, exchangeable cations, and oxides not discussed here) and the results of geostatistical analysis are contained in Park (1996).

In spite of the strong impact of the "inorganic" theory of podzolization proposed by Farmer and co-workers (Anderson *et al.* 1982; Farmer *et al.* 1980, 1983), many pedologists and soil chemists still believe in the traditional "organic" theory of podzolization in which the organo-metallic complexes play a dominant role in solubilizing Fe and Al in temperate acid environments (De Coninck 1980a; Duchaufour 1982; McKeague *et al.* 1983; Buurman and Van Reeuwijk 1984; Chesworth and Macias-Vasquez 1985; Ross 1989; Schwertmann and Murad 1990). In a recent review of Fe mobility, Schwertmann and Murad (1990) argue that the inorganic mechanism of Fe mobilization, involving the formation and migration of reasonably stable mixed Al–Fe–Si–hydroxy sols, may be particularly relevant for the movement of Al but less so for Fe.

In the traditional "inorganic theory" of podzolization, Duchaufour (1982) proposes a genetic series to show the stage of degradation of acid soils: acid brown soil–ochreous brown soil–podzolic brown soil–podzols. This evolutionary sequence is generally agreed by several workers (Duchaufour and Souchier 1978; Herbauts 1982; Payton 1988; Schwertmann and Murad 1990) and is relevant to the catenary sequence on this slope. Duchaufour (1976) argues that the distinction of sequences of podzolic soils is directly connected to the dynamics of organic matter: more intense humification results in less developed podzolic horizonation. Duchaufour and Souchier (1978) experimentally proved that contents of iron and clay in the acid parent materials control the type of humification of litter. Furthermore, iron and clay play more important roles in the humification processes than the amount of alkaline earths (Ca^{2+} and Mg^{2+}), soil pH, and the composition of litter. De Coninck (1980a) also proposed a similar mechanism: when mobile organic substances are formed during breakdown of plant remains, if at the top of the mineral soil enough polyvalent cations, especially Al and Fe either as free hydroxides or supplied by weathering of minerals, are available, the mobile organic substances formed are immobilized immediately and no migration occurs. In cases where insufficient amounts of Al and/or Fe are available to completely immobilize the mobile compounds, these cations are complexed by the mobile compounds and transported downward. A similar role for Mn oxides in the formation of mobile organo-metallic complexes is also proposed by Bartlett (1990) and De Coninck (1990).

On the study slope, we can assume a general similarity of parent material across the slope which affects the availability of polyvalent cations. Cleaned coarse sands from most of the samples commonly showed three or four different types of local sandstone or siltstone fragments, and there was no significant difference between

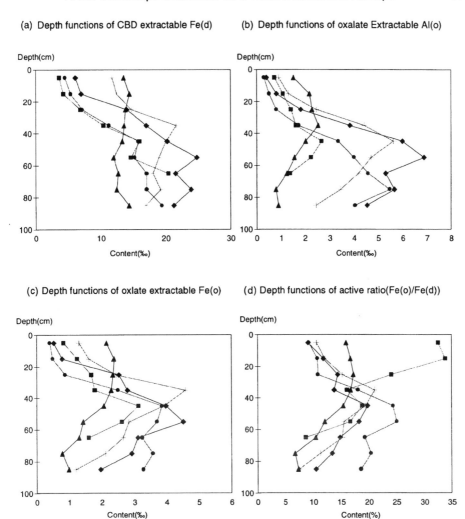

(a) Depth functions of CBD extractable Fe(d)

(b) Depth functions of oxalate Extractable Al(o)

(c) Depth functions of oxlate extractable Fe(o)

(d) Depth functions of active ratio(Fe(o)/Fe(d))

Figure 17.9 Vertical distributions of Fe(d), Al(o), Fe(o) and Fe(o)/Fe(d) of five soil groups in Bicknoller Combe. OBS (orthic brown soils, $n = 14$), PBS (podzolic brown soils, $n = 13$), NP (non-hydromorphic podzols, $n = 12$), SP (stagnogley podzols, $n = 13$), SLG (stagnoluvic gley soils and stagnogleyic podzolic brown soils, $n = 8$)

each soil profile, especially in the topsoil. Figure 17.9 shows the depth functions of average content of Fe(o), Fe(d), Fe(o)/Fe(d) and Al(o) in the five soil zones (stagnoluvic gley soils and stagnogleyic podzolic brown soils are considered together as stagnogley soils; the two extraordinary soil groups, seepage soils and "lateral podzolic soils" are omitted). Even though the figures have some limitations, being simple averages without consideration of different horizon depth and the heterogeneity of subsurface layers, it does show the general pattern of the depth functions of oxides for the different soil groups.

In stagnogleyic soils, while Fe(d) show the lowest content among the five soil groups, Fe(o), Al(o), Mn(d), clay content and soil pH have higher levels than in podzols (Figure 17.9). Furthermore, the "active ratio" has a much higher value than for any of the other soils. In the other four major soil zones, the maxima and minima of Fe(d), Fe(o) and Al(o) are more strongly expressed and move downward as the degree of podzolization increases. Similar patterns are reported in Herbauts (1982) and Schwertmann and Murad (1990). In the stagnogley podzols, however, the oxide maximum is lower and deeper than in non-hydromorphic podzols, possibly due to the gleying processes under increased amounts of subsurface water especially along the hollow. In the upper 20 cm, the contents of Fe(d) ("free Fe") and Al(o) ("free Al") increase as follows: stagnogley podzols < non-hydromorphic podzols < podzolic brown soils < orthic brown soils. While clay content also shows the same pattern, soil pH and exchangeable divalent cations (Ca^{2+} and Mg^{2+}) do not (Park 1996). Assuming that the contents of oxides and hydroxides in the surface soils are directly related to the availability of polyvalent cations in the formation and translocation of mobile organic substances as explained in De Coninck (1980), these soil chemical patterns in topsoils are exactly matched with the genetic sequence of podzolization and mobilizing processes of organo-metallic complex. However, as Duchaufour and Souchier (1978) mentioned, exchangeable base content is not well correlated with the genetic sequence. This indicates that the intensity of podzolization is basically decided by the amount of oxides and hydroxides and clay content available in the topsoils.

When considering the general distribution of these soil zones on a three-dimensional hillslope, we can conclude that the relative intensity of removal of oxides and hydroxides from topsoils results in catenary differentiation of podzolic soils. The dominant factor in this process is, without doubt, the contemporary hillslope hydrology as influenced by hillslope configuration, upslope area and the presence or absence of less permeable subsurface layers. This can explain the intensity of podzolization not only down, but also across the slope. The general upward extension pattern of orthic brown soils and podzolic brown soils on the flank slopes can be understood in the same way. However, it must be emphasized that the removal of oxides along the hollow is much stronger than on the spurs and flanks, because of convergent lateral subsurface flow; this results in downslope extension of stagnogley podzols and lesser amounts of oxides in subsurface horizons (see also Crabtree and Burt 1983; Burt 1986).

This general catenary trend might be enhanced or lessened by other environmental factors such as the presence or absence of less permeable horizons at depth, variations in parent material, vegetation and microclimate. However, it seems likely that these factors are of minor importance in the general soil–landform continuum on the study slope, compared to the distribution of percolation and lateral subsurface flow.

ACKNOWLEDGEMENTS

We are very grateful to Dr John Boardman and Dr P. H. T. Beckett for their helpful comments on an earlier draft of this chapter. We also wish to thank Ronald Fix and Oleh Hodun for their help with the field survey.

REFERENCES

Anderson, H.A., Berrow, M.L., Farmer, V.C., Hepburn, A., Russell J.D. and Walker A.D. 1982. A reassessment of podzol formation processes. *Journal of Soil Science*, **33**, 125–136.

Anderson M. G. and Burt T.P. 1978. The role of topography in controlling throughflow generation. *Earth Surface Processes*, **3**, 331-344.

Amacher, M.C., Henderson, R.E., Breithant, M.D., Seale C.L. and LaBauve J.M. 1990. Unbuffered and buffered salt methods for exchangeable cations and effective cation exchange capacity. *Journal of Soil Science Society of America*, **54**, 1036–1042.

Arnett R.R. 1976. Some pedological features affecting the permeability of hillslope soils in Caydale, Yorkshire. *Earth Surface Processes*, **1**, 3–16.

Arnold, R.W. and Wilding, L.P. 1991. The need to quantify spatial variability. In M.J. Mausbach and L.P. Wilding (eds), *Spatial Variabilities of Soils and Landforms*. SSSA Special Publication No. 28, Soil Science Society of America, Madison, WI, pp. 1–8.

Avery, B.W. 1980. *Soil Classification for England and Wales (Higher Categories)*. Technical Monograph No. 14, Harpenden.

Avery, B.W. 1990. *Soils of the British Isles*. CAB International, Wallingford.

Avery, B.W., Clayden, B. and Ragg, J.M. 1977. Identification of podzolic soils (spodosols) in upland Britain. *Soil Science*, **123**, 306–318.

Ball, D.F. 1966. Brown podzolic soils and their status in Britain. *Journal of Soil Science*, **17**, 148–158.

Bartlett, R.J. 1990. An A or an E: which will it be. In J.M. Kimble and R.D. Yeck (eds), *Proceedings of the Fifth International Soil Correlation Meeting (ISCOM IV) Characterization, Classification, and Utilization of Spodosols*. USDA, Soil Conservation Service, Lincoln, NE, pp. 7–18.

Bloom, P.R. 1983. Metal–organic matter interactions in soil. In R.H. Dowdy *et al.* (eds), *Chemistry in the Soil Environment*. ASA Special Publication No. 40, American Society of Agronomy—Soil Science Society of America, Madison, WI, pp. 129–149.

Blume, H.P. 1988. The fate of iron during soil formation in humid-temperate environments. In J.W. Stucki, B.A. Goodman and U. Schwertmann (eds), *Iron in Soils and Clay Minerals*. NATO ASI Series C, Mathematical and physical sciences, 217. D. Reidel Publishing, Dordrecht, pp. 749–777.

Blume, H.P. and Schwertmann, U. 1969. Genetic evaluation of profile distribution of aluminium, iron, and manganese oxides. *Proceedings of Soil Science Society of America*, **33**, 438–444.

Burt, T.P. 1978. Runoff processes in a small upland catchment with special reference to the role of hillslope hollows. Unpublished PhD Thesis, University of Bristol.

Burt, T.P. 1986. Runoff processes and solutional denudation rates on humid temperate hillslope. In S.T. Trudgill (ed.), *Solute Processes*. John Wiley & Son, Chichester, pp. 193–249.

Burt, T.P., Crabtree, R.W. and Fielder, N.A. 1984. Patterns of hillslope solutional denudation in relation to the spatial distribution of soil moisture and soil chemistry over a hillslope hollow and spur. In T.P. Burt and D.E. Walling (eds), *Catchment Experiments in Fluvial Geomorphology*. Geobooks, Norwich, pp. 431–446.

Buurman, P. and van Reeuwijk, L.P. 1984. Proto-imogolite and the process of podzol formation: a critical note. *Journal of Soil Science*, **35**, 447–452.

Catt, J.A. 1979. Soils and quaternary geology in Britain. *Journal of Soil Science*, **30**, 607–642.

Chesworth, W. and Macias-Vasquez, F. 1985. pe, pH and podzolization. *American Journal of Science*, **285**, 128–146.

Christopherson *et al.* 1993. Modelling the hydrochemistry of catchments: a challenge for the scientific method, *Journal of Hydrology*, **152**, 1–12.

Crabtree, R. W. and Burt, T.P. 1983. Spatial variation in solutional denudation and soil moisture over a hillslope hollow. *Earth Surface Processes and Landforms*, **8**, 151–160.

Crampton, C.B. 1963. The development and morphology of iron pan podzols in Mid and South Wales. *Journal of Soil Science*, **14**, 282–302.

Crampton, C.B. 1965. An indurated horizon in soils of South Wales. *Journal of Soil Science*, **16**, 230–241.

Curtis, L.F. 1971. *Soils of Exmoor Forest*. Soil Survey Special Survey No. 5, Harpenden.

De Coninck, F. 1980a. Major mechanisms in formation of spodic horizons. *Geoderma*, **24**, 101–128.

De Coninck, F. 1980b. The physical properties of spodosols. In B.K.G. Theng (ed.), *Soils Variable Change*, New Zealand Society of Soil Science, Lower Hutt, pp. 325–339.

De Coninck, F. 1990. Spodosols of Belgium. In J.M. Kimble and R.D. Yeck (eds), *Proceedings of the Fifth International Soil Correlation Meeting (ISCOM IV) Characterization, Classification, and Utilization of Spodosols*. USDA, Soil Conservation Service, Lincoln, NE, pp. 153–170.

Duchaufour, Ph. 1976. Dynamics of organic matter in soils of temperate regions: its action on pedogenesis. *Geoderma*, **15**, 31–40.

Duchaufour, Ph. 1982. *Pedology: Pedogenesis and Classification* (translated by T.R. Paton). George Allen and Unwin, London.

Duchaufour, Ph. and Souchier, B. 1978. Roles of iron and clay in genesis of acid soils under a humid, temperate climate. *Geoderma*, **20**, 15–26.

Edmonds, E.A. and Williams, B.J. 1985. *Geology of the Country around Taunton and the Quantock Hills*. Memoir for 1:50 000 geological sheet 295, New Series, British Geological Survey, London.

Farmer, V.C., Russell, J.D. and Berrow, M.L. 1980. Imogolite and protoimogolite allophane in spodic horizons: evidence for a mobile aluminium silicate complex in podzol formation. *Journal of Soil Science*, **31**, 673–684.

Farmer, V.C. Russell, J.D. and Smith B.F.L. 1983. Extraction of inorganic forms of translocated Al, Fe and Si from a podzol Bs horizon. *Journal of Soil Science*, **34**, 571–576.

Fielder, N.A. 1981. Distribution of iron and aluminium in some acidic soils. Unpublished PhD Thesis, University of Bristol.

Findlay, D.C., Colbourne, G.J.N., Cope, D.W., Harrod, T.R., Hogan, D.V. and Staines, S.J. 1984. *Soils and Their Use in South West England*. Soil Survey of England and Wales Bulletin No. 14, Harpenden.

Fitzpatrick, E.A. 1956. An indurated soil horizon formed by perma-frost. *Journal of Soil Science*, **7**, 248–254.

Fix, R.E. and Burt, T.P. 1996. The global positioning system: an effective way to map a small area or catchment. *Earth Surface Processes and Landform*, **20**, 817–827.

Gerrard, J. 1990. Soil variations on hillslopes in humid temperate climates. *Geomorphology*, **3**, 225–244.

Glentworth, R. 1944. Studies on the soils developed on basic igneous rocks in central Aberdeenshire. *Transactions of the Royal Society of Edinburgh*, **61**, 149–170.

Glentworth, R. and Dion, H.G. 1949. The association or hydrologic sequence in certain soils of the podzolic zone of north-east Scotland. *Journal of Soil Science*, **1**, 35–49.

Hall, G.F. and Olson, C.G. 1991. Predicting variability of soils from landscape models. In M.J. Mausbach and L.P. Wilding (eds), *Spatial Variabilities of Soils and Landforms*. SSSA Special Publication No. 28. Soil Science Society of America, Madison, WI, pp. 9–24.

Herbauts, J. 1982. Chemical and mineralogical properties of sandy and loamy-sandy ocherous earths in relation to incipient podzolization in a brown earth-podzol evolution sequence. *Journal of Soil Science*, **33**, 743–762.

Hesse, P.R. 1971. *A Textbook of Soil Chemical Analysis*. John Murray, London, pp. 209–231.

Hodgson, J.M. 1976. *Soil Survey Field Handbook*. Soil Survey Technical Monograph No. 5, Harpenden.

Jackson, M.J., Lim, C.H. and Zelazny, L.W. 1986. Oxides, hydroxides and aluminosilicates. In A. Klute (ed.), *Methods of Soil Analysis, Part 1. Physical and Mineralogical Methods*. Agronomy Monograph No. 9, American Society of Agronomy–Soil Science Society of America, Madison, WI, pp. 101–150.

Kunze, G.W. and Dixon, J.B. 1986. Pretreatment for mineralogical analysis. In A. Klute (ed.), *Methods of Soil Analysis, Part 1. Physical and Mineralogical Methods*. Agronomy Monograph No. 9, American Society of Agronomy–Soil Science Society of America, Madison, WI, pp. 91–100.

Loveland, P.J. 1988. The assay for iron in soils and clay minerals. In J.W. Stucki, B.A. Goodman and U. Schwertmann (eds), *Iron in Soils and Clay Minerals*. NATO ASI series, Series C, Mathematical and physical sciences, 217. D. Reidel Publishing, Dordrecht, pp. 99–140.

Loveland, P.J. and Bullock, P. 1976. Chemical and mineralogical properties of brown podzolic soils in comparison with soils of other group. *Journal of Soil Science*, **27**, 523–540.

McDaniel, P.A. and Buol, S.W. 1991. Manganese distributions in acid soils of the North Carolina Piedmont. *Journal of Soil Science Society of America*, **55**, 152–158.

McDaniel, P.A., Bathke, G.R., Buol, S.W., Cassel, D.K. and Falen, A.L. 1992. Secondary manganese/iron ratios as pedochemical indicators of field–scale throughflow water movement. *Journal of Soil Science Society of America*, **56**, 1211–1217.

McKeague, J.A. and Day, J.H. 1966. Dithionite- and oxalate-extractable Fe and Al as aids in differentiating various classes of soils. *Canadian Journal of Soil Science*, **46**, 13–22.

McKeague, J.A. De Coninck, F. and Franzmeier, D.P. 1983. Spodosols. In L.P. Wilding, N.E. Smeck and G.F. Hall (eds), *Pedogenesis and Soil Taxonomy. II. The Soil Orders*. Elsevier, Amsterdam, pp. 217–252.

McSweeney, K., Slater, B.K., Hammer, R.D., Bell, J.C., Gessler, P.E. and Pertersen, G.W. 1994. Toward a new framework for modeling the soil–landscape continuum. In R.G. Amundson, J.W. Harden and M.J. Singer (eds), *Factors of Soil Formation: A Fifties Anniversary Retrospective*. Special Publication No. 33, Soil Science Society of America, Madison, WI, pp. 127–145.

Miles, J. 1985. The pedogenic effects of different species and vegetation types and the implications of succession. *Journal of Soil Science*, **36**, 571–584.

The Meteorological Office 1966–1991. *Monthly Weather Report*, Vol. 83–108.

Moore, I.D., Turner, A.K., Wilson, J.P., Jensen, S.K. and Band, L.E. 1993. GIS and land-surface-subsurface process modeling. In M.F. Goodchild, B.O. Parks and L.T. Steyaert (eds), *Environmental Modeling with GIS*. Oxford University Press, New York and Oxford, pp. 196–234.

Moore, T.R. 1973. The distribution of iron, manganese, and aluminium in some soils from North-East Scotland. *Journal of Soil Science*, **24**, 162–171.

Muir, A. 1934. The soils of the Teindland State Forest. *Forestry*, **8**, 25–55.

Pakeman, R.J. and Marrs, R.H. 1993. Bracken. *Biologist*, **40**, 105–109.

Park, S.J. (1996), Modelling the soil–landscape continuum on a three-dimensional hillslope in Bicknoller Combe, Somerset, England (in prep.).

Paterson, E., Goodman, B.A. and Farmer, V.C. 1991. The chemistry of aluminium, iron and manganese oxides in acid soils. In B. Ulrich and M.E. Sunner (eds), *Soil Acidity*. Spring-Verlag, Berlin and Heidelberg, pp. 97–124.

Payton, R.W. 1988. Podzolic soils of the fell sandstone, Northumberland: their characteristics and genesis. *North of England Soils Discussion Group Proceedings*, **32**, 1–41.

Payton, R.W. 1992. Fragipan formation in argillic brown earths (Fragiudalfs) of the Midfield Plain, north-east England. I. Evidence for a periglacial stage of development. *Journal of Soil Science*, **43**, 621–644.

Pyatt, D.G. 1970. *Soil Groups of Upland Forests*. Forest Commission: Forest Record No. 71, HMSO, London.

Pyatt, D.G. and Smith, K.A. 1983. Water and oxygen regimes of four soil types at Newcastleton Forest, south Scotland. *Journal of Soil Science*, **34**, 465–482.

Ross, G.J. and Wang, C. 1993. Extractable Al, Fe, Mn, and Si. In M.R. Carter (ed.), *Soil Sampling and Methods of Analysis*. Canadian Society of Soil Science, pp. 239–246.

Ross, S. 1989. *Soil Processes*. Routledge, London.

Schwertmann, U. and Murad, E. 1990. Forms and translocation of iron in polzolized soils. In J.M. Kimble and R.D. Yeck (eds), *Proceedings of the Fifth International Soil Correlation Meeting (ISCOM IV) Characterization, Classification, and Utilization of Spodosols*. USDA, Soil Conservation Service, Lincoln, NE, pp. 319–341.

Soil Survey Staff 1975. *Soil Taxonomy: A Basic System of Soil Classification for Making and Interpreting Soil Surverys*. US Department of Agriculture, Handbook No. 436, Washington, DC.

Stonehouse, H.B. and St Arnaud, R.J. 1971. Distribution of iron, clay and extractable iron and aluminium in some Saskatchewan soils. *Canadian Journal of Soil Science*, **51**, 283–292.

Trudgill, S.T. 1986. Solute processes and landforms: an assessment. In S.T. Trudgill (ed.), *Solute Processes*. John Wiley & Sons, Chichester, pp. 497–503.

Van Schuylenborgh, J. 1965. The formation of sesquioxides in soils. In E.G. Hallsworth and D.V. Crawford (eds), *Experimental Pedology*. Butterworths, London, pp. 113–125.

Wilding, L.P. and Drees, L.R. 1983. Spatial variability and pedology. In L.P. Wilding, N.E. Smeck and G.F. Hall (eds), *Pedogenesis and Soil Taxonomy. I. Concepts and Interaction*. Elsevier, Amsterdam, pp. 83–116.

Yaalon, D.H. Jungreis, C. and Koyumdjisky, H. 1972. Distribution and reorganization of manganese in three catenas of Mediterranean soils. *Geoderma*, 71–78.

18 Modern Fluvial Processes on a Macroporous Drift-Covered Cavernous Limestone Hillslope, Castleton, Derbyshire, UK

PAUL HARDWICK and JOHN GUNN
Limestone Research Group, Department of Geographical and Environmental Sciences, University of Huddersfield, UK

18.1 INTRODUCTION

Karstified limestone aquifers are characterized by a low primary porosity and permeability, and a high secondary permeability due to the presence of fissures and caves enlarged over time by dissolution of the limestone during groundwater circulation. In bare karsts, where the soil is thin or absent, karst groundwaters may be rapidly and extensively contaminated by pollutants infiltrating with little modification via integrated fisssure networks in the vadose (unsaturated) zone. In contrast, in karsts extensively covered by soils or drift materials, the soil subsystem is important in modifying and buffering rainwater inputs to groundwater by controls on water storage, on hydrological pathways, and by its role in the dissolution of limestone bedrock via biogenic CO_2 production. Covered karst soils often form an important agricultural resource, and hence the soil zone is also important in buffering and modifying rainwater which may be contaminated with agrochemicals such as plant fertilizers, herbicides or pesticides. Most geomorphological, hydro-logical and pedological attention has focused on dolines, the typical landform on covered karst (e.g. Zambo 1986; Barany-Kevei 1988) and there have been few investigations of soils on those karst hillslopes where dolines are few or are absent. In addition, the soil medium, its hydrology, structure and relationship with underlying bedrock and caves receive little attention in most karst hydrological and geomorphological texts. This is considered detrimental to the effective management of karst groundwaters in agricultural catchments and to the conservation of limestone caves and their features.

It was hypothesized (Hardwick and Gunn 1990) that modern agricultural practices on land overlying caves may increase the rate of removal of soil materials by autogenic recharge due to

1. the presence of underlying joints and fissures which could facilitate the evacuation of soils from the surface;

Advances in Hillslope Processes, Volume 1. Edited by M. G. Anderson and S. M. Brooks.
© 1996 John Wiley & Sons Ltd.

2. irrigation techniques used to apply digested sewage sludge which might accelerate soil erosion via rapid subsurface runoff.

This chapter reports the results of a field and laboratory investigation of the role of soils in facilitating subsurface runoff and sediment transmission to the cavernous Dinantian Limestone of the Castleton karst, Derbyshire.

18.2 STUDY AREA

The investigations were conducted in Field 7559, a 12.5 ha enclosed field on a hillslope within the Castleton catchment. The north-facing site forms part of the northernmost exposure of the Carboniferous Limestone in Derbyshire and is underlain by most of the 2500 m long P8 cave (Figure 18.1).

18.2.1 Topography

The site is broadly linear in plan and concavo-convex in profile with a maximum slope angle of 6°. It ranges in elevation from around 340 m above OD at its northern end to around 370 m above OD at the southeastern end (Figure 18.1). The few closed depressions are confined to the southwest corner of the site adjacent to the B6061 Sparrowpit–Castleton road. The northeastern end of the site is bounded by a gully formed at the limestone–shale boundary which contains the P8 cave entrance. The western end of the gully grades into a shallow hollow (hollow A, Figure 18.1) which runs southward approximately following the line of the underlying cave. A second hollow (hollow B, Figure 18.1) is located to the west and runs from a further gully containing the P7 stream.

18.2.2 Hydrology

There are no perennial surface drainage channels on the study site. On some occasions during wet weather, hollow B focuses drainage from the site into the P7 gully. However, most of the site appears to drain via autogenic recharge to the underlying P8 Cave, and work on determining subcatchment boundaries for over 20 autogenic recharge sites in the cave is ongoing. An initial tracer experiment suggests that drainage from that part of the field siutated to the east of the cave may trend eastward into a dry valley between the site and Snels Low rather than entering the cave (Figure 18.1).

18.2.3 Land use

The field forms part of a typical upland farm in the Derbyshire Peak District, the principal land use being livestock grazing (mostly sheep with some cattle). Table 18.1 represents the best estimate of work on the field undertaken between 1984 and 1991. The field is regularly single-tine ploughed to a depth of 15–20 cm prior to the seeding of brassica fodder crops or rye/clover grass mixtures, and is also fertilized by chemical fertilizers, manure applications, and, since 1988, by digested sewage sludge (Table 18.2). The sludge is applied either by a rotary sprinkler or, when conditions are favourable, directly from a spreader valve on a road tanker which is driven round the site.

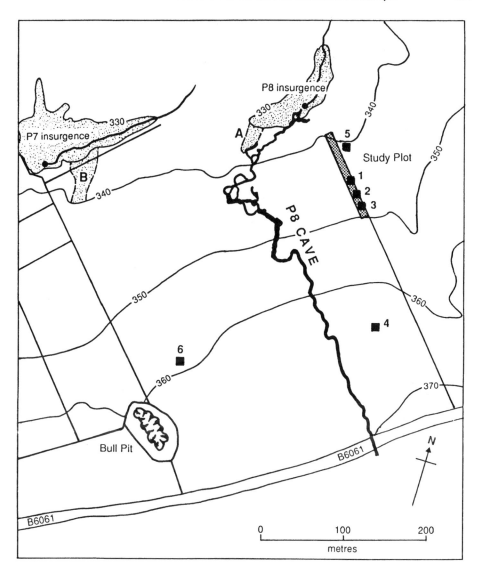

Figure 18.1 Field 7559 showing approximate position of P8 Cave, the P8 and P7 insurgence gullies and associated hollows A and B, and soil pits 1–6 (■)

18.3 METHODOLOGY

Six soil pits 1.5–2 m in width and length were excavated to bedrock (Table 18.3) Three were in a 780 m^2 fenced-off plot provided by the farmer to avoid interference with crops in the field (Figure 18.1, pits 1, 2 and 3). To avoid the possibility of results being due to very localized soil conditions, the farmer permitted the temporary

Table 18.1   Agricultural operations undertaken on Field 7559, 1984–1991 (source: Perryfoot Farm, December 1992)

Year	Operations
1984	Ploughed, crop of fodder rape NPK (20:10:10): 381 kg/ha
1985–1988	Re-seeded with Ryegrass/White Clover/Timothy Fertilizer: 90 kg N and 60 kg K annually Manure: *c.* 9.5 t/ha dry weight annually
1989	Ploughed, crop of fodder rape NPK (20:10:10): 381 kg/ha Manure: *c.* 9.5 t/ha dry wt
1990	Ploughed, re-seeded with ryegrass/white clover/timothy Fertilizer: 90 kg N and 60 kg K Manure: *c.* 9.5 t/ha dry wt Herbicide applied to control docks/weeds
1991	Lime: 5 t/ha Manure: *c.* 9.5 t/ha dry wt

Table 18.2   Digested sewage sludge disposal to Field 7559 (source: North West Water plc, October 1992)

Dates	Volume applied (m³)	Application rate (m³/ha)
7 June 1988–?	1384.2	110.48
23 May 1990–8 June 1990	954.5	76.18
14 Aug. 1990–31 Aug. 1990	177.3	14.15
28 May 1991–30 May 1991	463.6	37.00

excavation of two pits in soils subject to the same agricultural management some 250 m to the southeast and southwest (Figure 18.1, pits 4 and 6 respectively). A control pit was dug outside the site in land which had not been ploughed or worked for 20 years (Figure 18.1, pit 5) to determine any effect of management practices.

The soils of the study site were described using standard nomenclature (Munsell Colour Company 1953; Soil survey 1973; Gardiner and Dackombe 1983). Marked vertical, horizontal and lateral variations were observed in the soil characteristics of all the pits including the control (pit 5), and hence a three-dimensional sampling technique was developed to determine whether the variation was significant. A 1 m² monolith was isolated within the deeper soil of pit 3 and carefully excavated by removing horizontal slices with a sharpened spade, working to guide rails on a dexion framework. Each slice was 3.75 cm thick, the thickness being governed by the spacing between holes on the dexion frame. Each exposed horizontal profile was cleaned with a trowel, and visually contasting features (in terms of morphology or colour differences) such as macropores, and "matrix" (unaltered) material, were

Table 18.3   Soil profile descriptions: Field 7559 and environs

Al [1–6] 0–3/4 cm	Brownish black (7.5YR 3/2) sandy silt loam; stoneless; damp; intimate organic matter and many fine and coarse roots; pH 5.4; abrupt, wavy boundary to Ap
Ap [1–6] 3/4–25/28 cm	Brown (7.5YR 4/4) sandy silt loam; stoneless; damp; strong medium subangular blocky peds; very porous; no visible macropores; moderately weak soil strength; brittle; uncemented; slightly sticky; very plastic; pH 5.85; few distinct medium and coarse mottles of yellowish-red (5YR 5/8) from decomposing grit, mainly around 25 cm; intimate organic matter, some fine and coarse roots; clear deeply tonguing boundary (some tongues to 70 cm) to Eb1
Eb1 [1] 25/28–67/71 cm [2] 25/28–51/53 cm [3] 25/28–75/78 cm [4] 25/28–94/98 cm [5] 25/28–89/94 cm [6] 25/28–120/126 cm	Dull brown (7.5YR 5/4) sandy silt loam; occasional (1%) large angular/subangular sandstone clasts or small angular/subangular chert; damp; weak medium subangular blocky structure; moderately porous; very weak soil strength; brittle; uncemented; slightly sticky; moderately plastic; pH 4.9; few fine roots; 5–10% earthworm channels and irregular coarse macropores, some filled with brown (7.5YR 4/4) to brownish black (7.5YR 3/2) sandy loam; gradual boundary grades into Eb2
Eb2 [3] 75/78–85/88 cm	Bright brown (7.5YR 5/6) sandy silt loam; occasional (<1%) large angular/subangular sandstone clasts or small angular/subangular chert; damp; weak medium subangular blocky structure; moderately porous; very weak soil strength; brittle; uncemented; slightly sticky; moderately plastic; pH 4.8; few fine roots; 5–10% earthworm channels and irregular coarse macropores, some filled with brown (7.5YR 4/4) to brownish black (7.5YR 3/2) sandy loam; indistinct boundary grades into Bt
Bt [3] 85/88–107/111 cm [4] 94/98–119/122 cm [5] 89/94–114/118 cm [6] 120/126–148/153 cm	Bright brown (7.5YR 5/6) sandy clay loam; occasional (<1%) large angular/subangular sandstone clasts or small angular/subangular chert; damp; weak medium subangular blocky structure; moderately porous; firm soil strength; brittle; uncemented; sticky; very plastic; pH 4.45; 1–3% earthworm channels, some filled with brown (7.5YR 4/4) to brownish black (7.5YR 3/2) sandy loam, some lined small amounts ?clay coatings; few very fine fibrous roots; abrupt boundary to R
R [1] 67/71 cm+ [2] 51/53 cm+ [3] 107/111 cm+ [4] 119/122 cm+ [5] 114/118 cm+ [6] 148/153 cm+	Rock dominant material, jointed crystalline limestone; grey (10YR 6/1) and light grey (10YR 7/1); crevice infilling with bright brown (7.5YR 5/6) sandy clay loam and occasional brown (7.5YR 4/4) and brownish black (7.5YR 3/2) sandy loam. Occasional "contouring" of these materials over relief features of limestone

Numbers in square brackets refer to numbered soil pits (see Figure 18.1 for locations).

sampled using plastic 35 mm photographic film canisters under a stratified sampling strategy. A photographic record of each "slice" was also made. Measurements from the series of images were then used to build up three-dimensional diagrams of soil features (Figure 18.2). To determine whether the perceived differences were quantifiable, and to gain some insight into soil processes and particularly the function of soil macropores, the sediment samples were analysed for water and organic content and subject to standard mechanical analysis together with mineralogical analysis using X-ray diffractometry (XRD). ^{137}Cs analyses were also undertaken since the radionuclide is generally considered to be immobilized on the clay fraction of soils and hence studies of its spatial distribution have provided a valuable indicator of lateral soil erosion on hillslopes in a plethora of studies. However, few studies have used radionuclide tagging to study subsurface clay movement in suspension, a study by Scharpenseel and Kerpen (1967) being one exception.

Within the P8 cave, sediment collection tanks were installed in four contrasting recharge sites with the aim of collecting inwashed sediments to compare them with overlying soil and drift materials and to determine autogenic mechanical erosion rates. A 6 m^2 heavy gauge polythene sheet was suspended below the roof at two vadose seep sites (slow drippage from the cave roof) to catch the drippage and channel it into the collection tanks. A third vadose seep/wall flow was channelled into a tank using gutter pipe. At the fourth site, a small vadose stream issuing from a 20 cm wide fissure (the "Idiot's Inlet") was also channelled into a collection tank via a gutter pipe. Discharge was sufficient to allow continuous monitoring using a small quarter 90° V-notch weir cut in the wall of the tank together with a water-level recorder (Hardwick 1995). Collected sediments were subjected to mechanical and ^{137}Cs analysis (where sufficient material was obtained) and to XRD analysis.

18.4 RESULTS AND DISCUSSION

18.4.1 Soil Description

The sandy silt-loam soils of the study site (Table 18.3) are developed from aeolian loess deposits. They generally conform to the Malham Series of the Malham 2 Soil Association and are thus described as brown earths (Ragg et al. 1984). The soil profiles are characterized by an organic litter "Al" horizon some 3–4 cm in depth, grading into an agricultural "plough" or "Ap" horizon which extends to c. 25–28 cm depth. This has an abrupt or clear lower boundary, and represents material which has been mechanically cultivated. The Ap horizon is underlain by an eluviated "B" horizon ("Eb") of varying thickness, apparently unaltered by agricultural practices. The Eb horizon is underlain, in most cases, by an illuviated clay-rich "Bt" horizon. This in turn overlies Carboniferous Limestone (classified as "R" – hard or very hard bedrock, continuous except for cracks). The bedrock is a palaeokarst surface, pitted by smooth rounded runnels, and hollows c. 10–15 cm in diameter. The intervening positive relief features are also rounded.

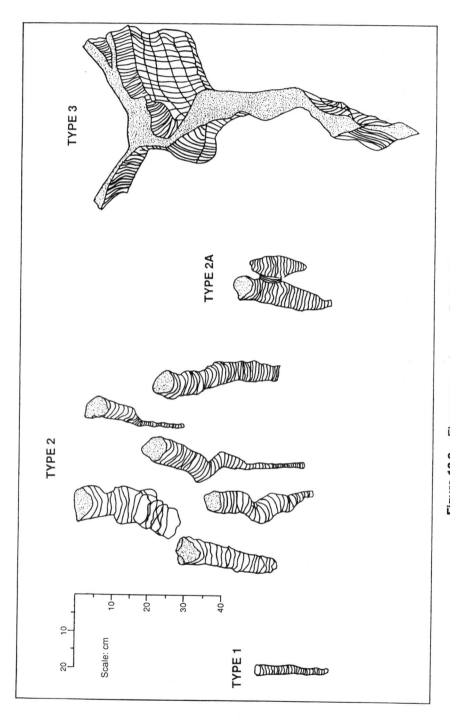

Figure 18.2   Eb macropore types, soil pit 3, Field 7559

The Bt horizon was absent from the shallow soils in pits 1 and 2. In pit 3, a marked colour change (reddening) was evident at the base of the Eb horizon. This phenomenon was not seen elsewhere but appears to represent a transition boundary between the Eb and Bt horizon and may be indicative of ferralitization due to more intense local leaching conditions. Elsewhere, these conditions have resulted in the palaeo-argillic brown earths of the Nordrach Series, also part of the Malham 2 Soil Association. However, Malham and Nordrach soils typically exhibit a stony subsoil of large subangular limestone clasts immediately above the limestone bedrock, and this was absent from all six profiles. Consequently, the shallower soils at this site are best described as a Malham variant, with the deeper soils such as those of pit 3 as a possible Nordrach variant.

Up to 10% of the surface area of each profile was occupied by zones of material of differing morphology, size, colour and apparent density to that of the Eb or Bt horizons. The zones included

1. large numbers of vertical tubular channels 0.25–1 cm in diameter ($1–2 \times 10^6$ channels/ha), some infilled with coarse-looking material, others completely open, and many occupied by earthworms (*Oligochaeta*);
2. large numbers of predominantly vertical infilled tubular features between 1 and 4 cm in diameter ($1–2 \times 10^3$ features/ha), much fewer *Oligochaeta*;
3. one irregularly shaped feature with no particular orientation, up to 20 cm wide and high, which appeared to be an infilled void.

All the profiles showed Ap-like material at depth, which in pits 1 and 2 appeared to have "spread" across the surface of the underlying bedrock. The "unaltered" Eb and Bt horizons are termed the "matrix", following Chamberlain (1972); and the various microgeomorphological features have been termed "macropores" following the Soil Classification of the Soil Survey of England and Wales (Avery 1973) and the review of macropore types and their morphology undertaken by Beven and Germann (1982). The macropores were classified on the basis of their characteristics, dimensions, and orientation in soil space (Figure 18.2 and Table 18.4). Two predominantly vertical structures (Type 1 and 2 macropores) dominate. One predominantly horizontal/subhorizontal structure (Type 3 macropore) was recorded together with several examples of a variant on the Type 2 macropores (Type 2A) due to a horizontal "offshoot" or "branch" from the main feature.

The macropore infill material appeared similar to that present in the upper Ap horizon of the soil, a phenomenon noted elsewhere (e.g. Gaiser 1952 *in* Gilman and Newson 1980) and to link the soil surface to the underlying bedrock via interconnecting networks of open conduits or voids and/or zones of higher permeability. In addition, following rainfall the macropores appeared wetter than the surrounding Eb or Bt matrix, suggesting that they may function as drainage conduits and that the flow through them may be turbulent, since some contained little clastic material and others appeared to contain more coarse materials than the surrounding matrix.

Table 18.4   Macropore classification: Soil Pit 3, Field 7559

Macropore description		Typical dimensions	
Type 1:	predominantly vertical; tubular; infilled or vacant; *Oligochaeta* occasionally present; diameter usually constant; occasional end in dish-shaped pit of same diameter	diameter length	1–3 cm 7–21 cm
Type 2:	predominantly vertical; tubular; infilled; *Oligochaeta* occasionally present; diameter reduces to Type 1 or ends by 38–63 cm depth, occasional >52 cm depth	diameter length	4–7 cm 15–38+ cm
Type 2A:	predominantly vertical; tubular; infilled; occasional lateral branching; diameter reduces to Type 1 or ends by 38–63 cm depth, occasional >52 cm depth	diameter (lateral) length	4–7 cm 9–17 cm 15–38+ cm
Type 3:	predominantly horizontal, infilled; non-linear; branched; continuous in horizontal plane	long axis width depth	85+ cm to 10 cm 60+ cm

18.4.2   Comparison of Soil Materials

Data on the physical properties of the various horizons and materials in pit 3 are shown in Tables 18.5–18.11. A statistical comparison of the data was undertaken using the non-parametric Mann–Whitney U test (Tables 18.12–18.14). The only difference between the brown coloured upper "Eb1" and bright brown coloured lower "Eb2" horizon was one of colour, probably due to ferralitization. There was no significant difference ($p = <0.01$) between the Ap horizon and Eb macropore materials, but there was a marked contrast between both these materials and the Eb horizon. The Ap horizon and macropores had a significantly lower dry bulk density, sand, silt and clay content than the Eb matrix; and a greater organic matter content. There was no significant difference in the H_2O content or >2 mm fraction. The Bt matrix contained significantly greater amounts of clay with correspondingly less sand and silt, and a greater dry bulk density than the Eb matrix. There was no significant difference in dry bulk density, organic matter or any of the inorganic fractions between the Eb macropores and Bt macropores. Contrary to the observations made during the initial field observations, the Ap horizon and Eb macropores were not significantly wetter than the Eb matrix ($p = <0.01$). Hence water content may not be the best determinant of hydrological function, and the tracing of soil waters to investigate drainage pathways was considered to be a more useful approach. The results of tracer experiments are considered further below.

18.4.3   Mineralogical Analysis of the Study Soils

X-ray diffractometry (XRD) was used to determine the main mineral constituents of the field soils (Table 18.15). The data were used to determine whether there was any

Table 18.5 Pit 3: Field 7559, Perryfoot, near Castleton, Derbyshire (SK 10875 81700). The Ap horizon 3/4–25/28 cm

Sample no.	Depth (cm)	Colour (wet)	Colour (dry)	Dry bulk density (g/cm³)	H₂O (% w/w)	Organic (% w/w)	Gravel >2 mm (% w/w)	Sand (% w/w)	Silt (% w/w)	Clay (% w/w)
1	6.25	–	brownish black 10YR 2/3	0.897	6.1	12.44	4.79	35.82	53.41	5.98
2	10.0	very dark brown 7.5YR 2/3	brownish black 10YR 2/3	0.761	30.4	25.44	0.39	37.85	57.81	3.95
3	13.75	very dark brown 7.5YR 2/3	dark brown 7.5YR 3/3	0.923	27.9	13.60	0.65	37.70	54.80	6.85
4	17.5	black 7.5YR 1.7/1	brownish black 10YR 2/3	0.869	28.3	16.33	0.59	29.46	65.38	4.55
5	17.5	brownish black 7.5YR 3/2	dull yellowish brown 7.5YR 4/3	1.112	28.9	15.85	0.87	34.90	70.20	4.05
6	21.25	very dark brown 7.5YR 2/3	brownish black 10YR 2/2	1.080	29.1	20.98	2.35	35.53	57.76	4.36
7	21.25	brownish black 10YR 2/3	dark brown 10YR 3/4	1.282	21.5	14.10	1.76	28.50	63.60	6.14
Mean				0.989	24.60	16.960	1.63	34.25	60.42	5.13
σ_{n-1}				0.177	8.65	4.65	1.57	3.77	6.12	1.17
Median				0.923	28.30	15.85	0.87	35.53	57.81	4.55

Table 18.6 Pit 3: Field 7559, Perryfoot, near Castleton, Derbyshire (SK 10875 81700). The Eb1 horizon c. 25–60 cm, matrix samples

Sample no.	Depth (cm)	Colour (wet)	Colour (dry)	Dry bulk density (g/cm³)	H_2O (% w/w)	Organic (% w/w)	Gravel (% w/w)	Sand (% w/w)	Silt (% w/w)	Clay (% w/w)
11	28.75	very dark brown 7.5YR 2/3	brown 10YR 4/4	1.426	9.3	12.44	1.50	39.63	52.36	6.51
14	32.50	very dark brown 7.5YR 2/3	brownish black 10YR 2/3	1.452	18.2	9.43	1.41	39.08	51.54	7.97
16	36.25	dark brown 7.5YR 2/3	dull brown 7.5YR 5/4	1.428	19.3	13.57	1.74	40.94	49.78	7.54
19	40.00	brownish black 10YR 2/3	dark brown 10YR 3/3	1.691	18.8	12.69	3.23	39.19	50.11	7.46
22	43.75	brown 7.5YR 4/6	brightish brown 7.5YR 5/6	1.395	20.5	6.60	1.10	37.7	52.37	8.82
24	47.5	brown 7.5YR 4/4	brown 7.5YR 4/4	1.417	18.1	7.76	3.56	39.55	50.23	6.75
31	50.25	brown 7.5YR 4/4	brown 7.5YR 4/4	1.295	N/A	5.54	0.86	42.57	47.54	9.03
33	54.0	brown 7.5YR 4/6	brightish brown 7.5YR 4/5	1.540	N/A	8.65	1.04	39.62	52.00	7.34
36	57.75	brown 7.5YR 4/4	brown 7.5YR 4/4	1.306	N/A	7.93	0.05	37.39	45.00	7.56
Mean				1.439	17.37	9.40	1.61	39.52	50.10	7.66
σ_{n-1}				0.119	4.05	2.86	1.12	1.56	2.47	0.84
Median				1.426	18.5	8.65	1.41	39.55	50.23	7.54

N/A: sampled at a later date and thus not used in H_2O analysis.

Table 18.7 Pit 3: Field 7559, Perryfoot, near Castleton, Derbyshire (SK 10875 81700). The Eb2 horizon c. 60–85 cm, matrix samples

Sample no.	Depth (cm)	Colour (wet)	Colour (dry)	Dry bulk density (g/cm³)	H_2O (% w/w)	Organic (% w/w)	Gravel (% w/w)	Sand (% w/w)	Silt (% w/w)	Clay (% w/w)
42	68.75	brown 7.5YR 4/4	brown 7.5YR 4/4	1.435	N/A	9.65	0.55	40.54	50.57	8.34
43	72.50	brightish brown 7.5YR 5/6	brown 7.5YR 4/6	1.263	N/A	5.21	1.01	34.35	58.07	6.57
45	76.25	brown 7.5YR 4/4	brown 7.5YR 4/6	1.429	N/A	5.43	2.35	41.38	46.92	9.35
47	80.00	brown 7.5YR 4/6	brown 7.5YR 4/4	1.856	N/A	12.70	4.03	39.35	50.16	6.46
49	83.75	brown 7.5YR 4/4	brown 7.5YR 4/4	1.475	N/A	9.89	0.78	38.75	52.47	8.00
Mean				1.492		8.52	1.74	38.87	51.64	7.74
σ_{n-1}				0.219		3.21	1.46	2.73	4.11	1.23
Median				1.435		9.65	1.01	39.35	50.57	8.00

N/A: sampled at a later date and thus not used in H_2O analysis.

Table 18.8 Pit 3: Field 7559, Perryfoot, near Castleton, Derbyshire (SK 10875 81700). The Bt horizon *c.* 85–107 cm, matrix samples

Sample no.	Depth (cm)	Colour (wet)	Colour (dry)	Dry bulk density (g/cm³)	H_2O (% w/w)	Organic (% w/w)	Gravel (% w/w)	Sand (% w/w)	Silt (% w/w)	Clay (% w/w)
51	88.75	brown 7.5YR 4/4	brown 7.5YR 4/4	1.507	N/A	2.63	0.05	1.56	35.56	62.83
52	92.50	brightish brown 7.5YR 4/4	brightish brown 7.5YR 5/6	1.634	N/A	1.87	0.00	3.78	26.85	69.40
54	96.25	brown 7.5YR 4/4	brown 7.5YR 4/4	1.702	N/A	0.58	0.00	1.00	30.84	68.16
56	100.0	brown 7.5YR 4/6	brightish brown 7.5YR 3/3	1.735	N/A	3.86	0.00	1.64	32.00	66.36
58	103.75	brightish brown 7.5YR 3/3	brightish brown 7.5YR 5/6	1.845	N/A	0.78	0.00	0.50	25.37	74.13
60	107	brown 7.5YR 4/6	brightish brown 7.5YR 5/6	1.670		1.57	0.2	1.90	36.58	61.24
Mean				1.682		1.88	0.04	1.73	29.83	67.02
σ_{n-1}				0.112		1.22	0.08	1.12	4.42	4.67
Median				1.686		1.72	0.00	1.60	28.85	67.26

N/A: sampled later than main sample set, thus H_2O data not applicable.

Table 18.9 Pit 3: Field 7559, Perryfoot, near Castleton, Derbyshire (SK 10875 81700). The Eb1 horizon c. 25–60 cm, macropore samples

Sample no.	Depth (cm)	Colour (wet)	Colour (dry)	Dry bulk density (g/cm³)	H₂O (% w/w)	Organic (% w/w)	Gravel (% w/w)	Sand (% w/w)	Silt (% w/w)	Clay (% w/w)
8	25.00	very dark brown 7.5YR 2/3	brownish black 10YR 2/2	0.656	35.1	–	–	–	–	–
10	28.75	dark brown 7.5YR 3/3	dark brown 7.5YR 4/4	0.987	27.9	17.51	1.87	36.29	56.14	5.71
12	32.50	very dark reddish brown 5YR 2/4	dark brown 7.5YR 3/3	0.731	27.1	–	–	–	–	–
13	32.50	brownish black 10YR 2/3	dark brown 10YR 3/4	1.130	23.8	9.70	1.16	37.15	55.73	5.95
15	32.50	very dark brown 7.5YR 2/3	brown 10YR 4/4	0.397	28.7	8.15	–	–	–	–
17	36.25	brown 7.5YR 4/4	dull brown 7.5YR 5/4	0.786	25.8	14.40	3.41	37.95	54.84	3.79
18	36.25	very dark brown 7.5YR 2/3	brown 7.5YR 4/3	0.824	30.0	19.82	0.00	36.93	58.35	4.72
20	40.00	very dark brown 7.5YR 2/3	dark brown 10YR 3/4	1.028	13.1	11.98	0.68	33.15	60.23	5.94
21	40.00	brownish black 10YR 2/3	brown 7.5YR 4/3	0.999	27.6	17.89	0.84	36.65	56.10	6.42
23	43.75	brownish black 10YR 2/3	brown 7.5YR 4/3	1.029	13.1	20.90	1.81	36.71	55.89	5.59
25	47.50	brownish black 7.5YR 2/2	dark brown 7.5YR 3/2	0.670	32.8	–	–	–	–	–
26	47.50	brown 7.5YR 4/4	brown 7.5YR 4/4	1.042	15.9	14.98	0.65	39.47	54.38	5.50
30	50.25	brown 7.5YR 4/3	brown 7.5YR 4/4	0.965	N/A	8.65	2.10	40.35	53.22	4.33
32	54.00	brown 7.5YR 4/4	brown 7.5YR 4/4	1.010	N/A	13.45	0.05	31.50	60.15	8.30
35	57.75	brownish black 10YR 2/3	brown 7.5YR 4/3	1.030	N/A	11.50	0.00	34.85	58.65	6.50
37	57.75	brown 7.5YR 4/4	brown 7.5YR 4/4	0.865	N/A	4.58	0.15	38.25	59.25	2.35
Mean				0.884	25.49	13.35	1.06	36.60	56.91	5.13
σ_{n-1}				0.194	6.69	4.88	1.06	2.49	2.34	1.17
Median				0.976	27.35	13.45	0.76	36.82	56.12	4.55

N/A: sampled later than main sample set and thus H₂O data not applicable.

Table 18.10 Pit 3: Field 7559, Perryfoot, near Castleton, Derbyshire (SK 10875 81700). The Eb2 horizon *c.* 60–85 cm, macropore samples

Sample no.	Depth (cm)	Colour (wet)	Colour (dry)	Dry bulk density (g/cm³)	H₂O (% w/w)	Organics (% w/w)	Gravel (% w/w)	Sand (% w/w)	Silt (% w/w)	Clay (% w/w)
41	68.75	brown 7.5YR 4/4	brown 7.5YR 4/4	0.765	N/A	10.85	0.05	30.65	64.94	4.36
44	72.50	brownish black 7.5YR 3/2	brownish black 7.5YR 3/2	0.995	N/A	14.12	0.00	38.00	55.18	6.82
46	76.25	very dark brown 7.5YR 2/3	brownish black 10YR 2/2	1.027	N/A	11.36	1.00	34.67	60.54	3.79
48	80.00	brown 7.5YR 4/6	brown 7.5YR 4/4	1.000	N/A	18.63	0.15	39.98	56.02	3.85
50	83.75	brownish black 7.5YR 3/2	brownish black 7.5YR 3/2	0.875	N/A	17.75	0.00	34.27	60.05	5.68
Mean				0.932		11.54	0.24	35.51	59.35	4.9
σ_{n-1}				0.110		2.07	0.43	3.61	3.73	1.32
Median				0.995		11.36	0.05	34.67	60.05	4.36

N/A: sampled later than main sample set, thus H₂O data not applicable.

Table 18.11 Pit 3: Field 7559, Perryfoot, near Castleton, Derbyshire (SK 10875 81700). The Bt horizon c. 85–107 cm, macropore samples

Sample no.	Depth (cm)	Colour (wet)	Colour (dry)	Dry bulk density (g/cm³)	H_2O (% w/w)	Organics (% w/w)	Gravel (% w/w)	Sand (% w/w)	Silt (% w/w)	Clay (% w/w)
53	88.75	brown 7.5YR 4/4	brown 7.5YR 4/4	1.025	N/A	9.65	0.55	29.35	64.96	5.14
55	92.50	brownish black 7.5YR 3/2	dark brown 10YR 3/4	1.057	N/A	11.38	0.02	34.50	58.36	7.12
57	96.25	brown 7.5YR 4/3	brown 7.5YR 4/4	0.967	N/A	17.36	0.00	36.75	59.57	3.68
59	100.00	brown 7.5YR 4/4	brown 7.5YR 4/4	0.935	N/A	12.35	0.00	30.00	66.00	4.00
61	103.75	very dark brown 7.5YR 2/3	brownish black 10YR 2/2	1.037	N/A	12.20	0.65	28.62	64.48	6.25
62	107	brown 7.5YR 4/6	brightish brown 7.5YR 5/6	0.837	N/A	19.89	0.20	38.60	55.10	6.10
Mean				0.976		13.81	0.237	32.97	56.41	5.38
σ_{n-1}				0.082		3.94	0.293	4.22	11.50	1.35
Median				0.996		12.27	0.11	32.25	58.97	5.62

N/A: sampled later than main sample set, thus H_2O data not applicable.

Table 18.12   Mann–Whitney U-test of Eb1 and Eb2 matrix materials, pit 3, Field 7559: Eb1 matrix versus Eb2 matrix

Eb1 matrix versus Eb2 matrix	Median Eb1	Median Eb2	Mean rank Eb1	Mean rank Eb2	Significance level	Result
Dry bulk density	1.43	1.44	6.9	8.6	0.505	Eb1=Eb2
Organic matter	8.65	9.65	7.8	7.0	0.790	Eb1=Eb2
Inorganic fraction: >2 mm	1.41	1.01	7.8	7.0	0.790	Eb1=Eb2
Inorganic fraction: sand	39.55	39.35	7.7	7.2	0.894	Eb1=Eb2
Inorganic fraction: silt	50.23	50.57	6.9	8.6	0.505	Eb1=Eb2
Inorganic fraction: clay	7.54	8.00	7.3	7.8	0.894	Eb1=Eb2

Results: $a > b$: the median of a is significantly greater than the median of b at the 0.01 significance level.
$a = b$: no significant difference between the medians of a and b at the 0.01 significance level.

qualitative difference in soil mineralogy between Ap, Eb/Bt and macropore materials, whether there were any mineralogical impacts of agricultural practices, and whether there was any qualitative difference between soils and sediments washed into the underlying cave by percolation waters.

Quartz and quartzite peaks dominate the diffractograms, and quartzite is found throughout the soil profile in all horizons and in macropores. The clay mineralogy is dominated by kaolinite or kaolinite/smectite, again found throughout. The distribution of chlorite, illite and illite/smectite was patchy; chlorite was present in the macropore samples and in one sample from the Ap horizon, whereas illite and illite/smectite was found in the Ap horizon but not in the macropores. Calcite was present in the matrix between -50 cm and -55 cm, and may derive from the limestone bedrock at 50–60 cm depth. However, calcite was also present in macropore material at -40 cm and -55 cm. The calcite in the shallower macropore may reflect backflushing of macropores with "return throughflow" during saturated conditions or recycling via biogenic activity (e.g. *Oligochaeta*). Residues of agricultural lime washed down from the soil surface are unlikely given the absence of calcite in any of the other matrix samples. Moreover, the field had not been limed for several years prior to soil sampling, although aeolian inputs of limestone dust from the Eldon Hill Quarry to the south are a possibility.

Chlorapatite [Francolite, $Ca_5 (PO_4)_3$ (F, Cl, OH)] was found in the Eb matrix at -50 cm and -55 cm and also in the macropores at -40 cm and -55 cm. This mineral is one product of the reaction of phosphate with calcite, and is indicative of an autogenic reaction between leached phosphates from chemical fertilizers or sewage sludge and the calcite found only in the same samples. Phosphate solutions are also known to react with calcite forming the highly soluble calcium hydrophosphate ($2CaHPO_4$) and are thus considered highly corrosive to calcite (Jakucs 1977).

The aforementioned minerals account for over 80% of peaks on the diffractograms, the balance comprising a large complex of much smaller peaks which were generally identified as mixtures of ferrihydrites or ferrisulphates. These

Table 18.13   Mann–Whitney U-test comparison, soil materials of Pit 3, Field 7559

a versus b	Median 1	Median 2	Mean rank 1	Mean rank 2	Signifi- cance level	Result
Dry bulk density						
Ap vs Eb matrix	0.92	1.43	4.1	14.4	0.0004	Ebmx > Ap
Ap vs Eb macropores	0.92	0.98	17.0	13.7	0.3670	Ap = Ebmac
Eb matrix vs Eb macropores	1.43	0.98	28.5	11.0	0.0000	Ebmx > Ebmac
H_2O content						
Ap vs Eb matrix (Eb1 data only)	28.9	18.5	10.0	3.5	0.0380	Ap = Ebmx
Ap vs Eb macropores	28.9	27.35	11.9	8.9	0.7350	Ap = Ebmac
Eb matrix vs Eb macropores	18.5	27.35	5.9	11.3	0.0549	Ebmac = Ebmx
Organic matter						
Ap vs Eb matrix	15.85	9.04	17.4	7.8	0.0008	Ap > Ebmx
Ap vs Eb macropores	15.85	12.37	17.6	11.2	0.0566	Ap = Ebmac
Eb matrix vs Eb macropores	9.04	12.37	11.8	20.1	0.0135	Ebmac = Ebmx
Inorganic fraction: > 2 mm						
Ap vs Eb matrix	0.87	1.26	10.2	11.4	0.7091	Ap = Ebmx
Ap vs Eb macropores	0.87	0.65	15.6	7.3	0.1721	Ap = Ebmac
Eb matrix vs Eb macropores	1.26	0.65	20.0	11.5	0.0276	Ebmx = Ebmac
Inorganic fraction: sand						
Ap vs Eb matrix	35.53	39.45	5.2	13.9	0.0028	Ebmx > Ap
Ap vs Eb macropores	35.53	36.71	9.7	13.6	0.2275	Ap = Ebmac
Eb matrix vs Eb macropores	39.45	36.71	21.4	11.5	0.0027	Ebmx > Ebmac
Inorganic fraction: silt						
Ap vs Eb matrix	57.81	50.40	17.4	7.8	0.0009	Ap > Ebmx
Ap vs Eb macropores	57.81	56.14	14.1	11.8	0.4848	Ap = Ebmac
Eb matrix vs Eb macropores	50.40	56.14	8.1	22.5	0.0000	Ebmac > Ebmx
Inorganic fraction: clay						
Ap vs Eb matrix	4.55	7.50	4.6	14.2	0.0009	Ebmx > Ap
Ap vs Eb macropores	4.55	5.59	12.8	12.4	0.9241	Ap = Ebmac
Eb matrix vs Eb macropores	7.55	5.59	23.4	9.9	0.0000	Ebmx > Ebmac

Results: *a* > *b*: the median of *a* is significantly greater than the median of *b* at the 0.01 significance level. *a* = *b*: no significant difference between the medians of *a* and *b* at the 0.01 significance level. mx = matrix; mac = macropores.

oxides and sulphates of iron impart the characteristic red, yellow or brown colouration of many soils. However, they are difficult to identify as they are usually only present in trace amounts, as mineral coatings on larger clasts. Ferrihydrites and sulphates were conspicuously absent from the macropores, perhaps indicating more intense leaching in these features.

18.4.4 [137]Cs Distribution in Soils

Table 18.16 shows the concentrations of ^{137}Cs measured in pit 3 soil materials. Ap horizon concentrations were significantly higher than in the Eb matrix or

Table 18.14   Mann–Whitney U-test comparison, Pit 3 soil materials, Field 7559

a versus b	Median 1	Median 2	Mean rank 1	Mean rank 2	Signifi-cance level	Result
Dry bulk density						
Eb matrix vs Bt matrix	1.43	1.69	8.2	15.8	0.0094	Btmx > Ebmx
Bt matrix vs Bt macropores	1.69	0.99	9.5	3.5	0.0051	Btmx > Btmac
Bt macropores vs Eb macropores	0.99	0.99	16.8	13.2	0.3359	Btmac=Ebmac
Organic matter						
Eb matrix vs Bt matrix	9.04	1.72	13.5	3.5	0.0006	Ebmx > Btmx
Bt matrix vs Bt macropores	1.72	12.28	3.5	9.5	0.0051	Btmac > Btmx
Bt macropores vs Eb macropores	12.28	12.37	13.5	12.2	0.7139	Btmac=Ebmac
Inorganic fraction: > 2 mm						
Eb matrix vs Bt matrix	1.26	0.00	13.4	3.8	0.0010	Ebmx > Btmx
Bt matrix vs Bt macropores	0.00	0.11	5.3	7.8	0.2673	Btmx=Btmac
Bt macropores vs Eb macropores	0.11	0.65	8.9	13.1	0.2076	Btmac=Ebmac
Inorganic fraction: sand						
Eb matrix vs Bt matrix	39.45	1.60	13.5	3.5	0.0006	Ebmx > Btmx
Bt matrix vs Bt macropores	1.60	32.25	3.5	9.5	0.0051	Btmac > Btmx
Bt macropores vs Eb macropores	32.25	36.70	8.0	11.1	0.0999	Btmac=Ebmac
Inorganic fraction: silt						
Eb matrix vs Bt matrix	50.40	28.85	13.5	3.5	0.0006	Ebmx > Btmx
Bt matrix vs Bt macropores	28.85	58.97	3.7	9.3	0.0082	Btmac > Btmx
Bt macropores vs Eb macropores	58.97	56.14	13.3	11.5	0.5995	Btmac=Ebmac
Inorganic fraction: clay						
Eb matrix vs Bt matrix	7.55	67.26	7.5	17.5	0.0006	Btmx > Ebmx
Bt matrix vs Bt macropores	67.26	5.62	9.5	3.5	0.0051	Btmx > Btmac
Bt macropores vs Eb macropores	5.62	5.59	12.5	11.8	0.6611	Btmac=Ebmac

Results: *a* > *b*: the median of *a* is significantly greater than the median of *b* at the 0.01 significance level.
        a=b: no significant difference between the medians of *a* and *b* at the 0.01 significance level.
mx=matrix; mac=macropores.

macropores, but there were significantly greater amounts in the Eb macropores than in the Eb matrix. Concentrations in the Ap horizon samples were significantly less than in samples from undisturbed surface soils, probably due to the mechanical intermixing by ploughing of caesium-rich and caesium-deficient materials (Ritchie and McHenry 1973; Wise 1980). ^{137}Cs concentrations also varied according to macropore type (Table 18.17). Type 2 macropores (Figure 18.2) contained the greatest concentrations, their lateral extensions (Type 2A macropores) and the one sample from a Type 3 macropore the least.

Table 18.15   Soil mineralogy, Pit 2: Field 7559

Mineral	\multicolumn{13}{c}{Depth below surface (cm $\pm$ 1.5 cm)}											
	05	10	15	20	25	30	35	40	40[a]	45	50	55	55[a]
Quartzite	✓	✓	✓	✓	✓	✓	✓	✓	✓	✓	✓	✓	✓
Quart	✓	✓	✓	✓	✓	✓	✓	✓	✓	✓	✓	✓	✓
Kaolinite	✓		✓	✓	✓	✓	✓	✓	✓	✓	✓	✓	✓
Kaolinite/smectite	✓	✓	✓	✓	✓	✓	✓	✓	✓	✓	✓	✓	✓
Chlorite	✓					✓		✓					✓
Illite		✓								✓	✓		
Illite/smectite					✓								
Chlorapatite									✓		✓	✓	✓
Calcite									✓		✓	✓	✓
Ferrihydrites/ sulphate-hydrates	✓	✓	✓	✓	✓	✓			✓		✓	✓	✓

✓=Presence of mineral peak on XRD diffraction trace.
[a] Soil macropores.

A fractionation and analysis of ^{137}Cs-rich Ap horizon material found that the radionuclide is concentrated in the fine silt and particularly the clay fraction (Table 18.18). However, the small size of the macropore samples precluded an investigation of the ^{137}Cs carrier fraction in those features. The distribution of ^{137}Cs suggests that macropores, particularly Type 2 macropores, facilitate the transmission of ^{137}Cs-labelled fine silts and clays to depth within the soil profile. However, transmission could occur via several mechanisms:

1. *Transmission in percolation waters to some depth before immobilization.* Gudzenko and Klimchouk (1994) reported concentrations of 9–15 mBq/l ^{137}Cs in percolation waters to the Marble Cave, Tchatyrdag massif, Crimea, following the Chernobyl nuclear accident. At the study site, such transmission would be facilitated by the low clay contents in the Eb matrix (*c.* 8% w/w), which elsewhere have resulted in ^{137}Cs remaining mobile within soils (Livens *et al.* 1991). However, matrix clay concentrations are higher than those of the macropores (*c.* 5%) or Ap horizon (*c.* 5%) where ^{137}Cs concentrations were greatest. Moreover, the clay mineral illite (the dominant sink for radiocaesium in soils; Livens *et al.* 1991) was only found in patches within the Eb horizon matrix where ^{137}Cs concentrations were effectively nil (Table 18.16). Consequently, the ^{137}Cs is probably adsorbed either onto the micaceous kaolinite and smectite minerals found throughout the soil profile (Table 18.15), or, more probably, onto organic matter–mineral complexes given the greater concentrations of organic matter in the Ap horizon (*c.* 17%) and macropores (*c.* 13%) than in the Eb/Bt horizon (*c.* 9%). There has been some dispute as to the effectiveness of near-surface immobilization in soils where organic matter is significant and may be expected to compete with clays as the caesium sink. In such situations, radiocaesium uptake by vegetation may be significant (Livens *et al.* 1991). However, in the undisturbed soils the effectiveness of the soil in immobilizing ^{137}Cs is shown by the relatively greater concentration in

Table 18.16   [137]Cs concentrations in Pit 3 materials, Field 7559

Spectra no.	Sample no.	Sample depth (cm)	[137]Cs at counting date (D2) (Bq/kg)	Error (%) (±)
Ap horizon				
5038	1	6.25	11.9	16.0
5033	2	10.00	16.67	10.0
5036	3	13.75	9.75	9.0
5040	4	17.50	10.13	9.8
5039	5	17.50	11.48	9.0
5035	6	21.25	8.91	5.0
5041	7	21.25	0.82	63.0
Eb/Bt horizon				
5034	11	28.75	1.15	88.0
5042	14	32.50	0.36	119.0[a]
5043	16	36.25	0.0	0.0
5044	19	40.00	0.17	216.0[a]
5045	22	43.75	1.16	143.0[a]
5046	24	47.50	0.41	102.0[a]
5037	26	51.00	0.57	113.0[a]
Macropores				
5011	8	25.00	7.56	16.0
5012	10	28.75	1.59	18.6
5014	12	32.50	0.36	25.5
5016	13	32.50	0.85	32.3
5015	15	32.50	7.42	19.1
5019	17	36.25	5.71	18.6
5010	18	36.25	3.98	7.5
5009	20	40.00	3.37	3.4
5017	21	40.00	5.47	18.9
5018	23	43.75	0.69	24.3
5047	25	47.50	7.50	15.0
5067	47	51.00	1.98	45.0
5066	57a	61.00	1.90	32.0
5068	57b	61.00	0.19	274.0[a]
5070	61	65.00	1.07	63.0

[a]Nil.

samples from 0–2 cm depth as against samples from 2–4 cm depth (Table 18.19). It was not possible to partition the caesium between organic and mineral fractions as the organic determination method removed the organic material but left the caesium, which would then have attached itself to the remaining mineral fraction.

2. *Biogenic redistribution.* Several of the macropores were found to contain *Oligochaeta*, which are known to ingest and redistribute soil materials (e.g. Canti 1992).

3. *Downwashing of ^{137}Cs-labelled clastic soil materials via turbulent macropore drainage.* Bevan and Germann (1982) suggested that macropores play an important role in the hydrology of some soils and that this has important

Table 18.17   [137]Cs concentrations in Pit 3, Field 7559: variation with macropore type

Description (type)	Sample depth (cm)	[137]Cs at counting date (D2) (Bq/kg)	Error % (±)
Dark brown tubular (2)	25.00	7.56	16.0
Grey-brown diiffuse (2)	28.75	1.59	18.6
Lateral (2A)	32.50	0.36	25.5
Lateral (2A)	32.50	0.85	32.3
Tubular (2)	32.50	7.42	19.1
Tubular (2)	36.25	5.71	18.6
Tubular "soft" (2)	36.25	3.98	7.5
Tubular (2)	40.00	3.37	3.4
Tubular (2)	40.00	5.47	18.9
Tubular "soft" (2)	43.75	0.69	24.3
Tubular (2)	47.50	7.50	15.0
Tubular (2)	51.00	1.98	45.0
Tubular (2)	61.00	1.90	32.0
Lateral (3)	61.00	0.19	274.0[a]
Tubular (2)	65.00	1.07	63.0

[a]Nil.

Table 18.18   Field 7559 Ap horizon soil fractionation: [137]Cs carrier fraction

Particle size (mm)	Description (BS 1377 (1967))	Fraction % (w/w)	[137]Cs at counting date (D2) (Bq/kg)	Error % (±)
>2	cobbles and gravel	2.05	0.00	297[a]
2–0.2	coarse and medium sand	9.49	0.15	156[a]
0.2–0.06	fine sand	19.36	0.45	105[a]
0.06–0.02	coarse silt	11.07	1.64	58.9
0.02–0.006	medium silt	23.95	2.87	12.1
0.006–0.002	fine silt	28.32	9.75	13.8
<0.002	clay	5.76	12.53	11.4

[a]Nil.

implications both for geochemical interactions and for the transmission of pollutants. Trudgill *et al.* (1983) found that stemflow from beech woodland overlying magnesian limestone was transmitted rapidly along root channels and macropores into the fissured bedrock, effectively bypassing the soil matrix. Trudgill *et al.* (1983) also suggested that soil water flow may be partitioned into two phases, based on the speed of transmission. First, rapid *preferential flow* through structural pathways such as inter-ped spaces and macropores, and second, slow flow or storage by uniform displacement. Solute losses via preferential flow have been shown to be significant and to have a rapid and detrimental impact on groundwater quality (Thomas and Phillips 1979; Coles and Trudgill 1985; Jury *et al.* 1986; Anderson 1988; Van der Molen and Van Ommen

Table 18.19   [137]Cs concentrations at 0–2 cm depth in "undisturbed" soils, Castleton Catchment, Derbyshire

Site	NGR (SK)	Elevation (m AOD)	[137]Cs at counting date (D2) (Bq/kg)	Error % ($\pm$)
2[a]	09875 82625	411	19.39	4.8
7[a]	10375 81825	330	42.83	5.0
6	10400 81625	330	55.04	9.7
P6/7	10400 81675	332	40.82	5.5
P6/7 (2–4 cm)	10400 81675	332	23.64	18.2
5	10600 81400	363	115.15	8.4
4	10700 81400	360	23.50	9.2
3	10775 81300	359	20.46	7.6
11	12600 82450	425	71.79	1.6
9	11250 81650	395	35.15	6.4
10	10075 81300	324	40.99	5.4
P8	10525 81700	338	24.28	6.5
P8 (2–4 cm)	10525 81700	338	12.45	9.6
Mean ^{137}Cs concentration (0–2 cm) (Bq/kg)			44.49	6.37

[a]On non-limestones.

1988; Schmaland 1992). Clay migration is recognized as being more rapid in soils with large pore spaces than in more consolidated soils (Hallsworth 1963; Scharpenseel and Kerpen 1967). For example, Simpson and Cunningham (1982) found that large interconnected macropores enhanced the rapidity of effluent-irrigation water flow through the soil, and that the runoff was erosive, thereby contributing to the further development of the macropore drainage conduits.

The results of the mechanical analysis, the ^{137}Cs analysis, and to some extent the presence of authigenic phosphates shown by the XRD investigation, suggest that the macropores of the study soil function as high-porosity drainage conduits, linking the Ap horizon to the underlying bedrock, largely bypassing the Eb and Bt horizons which comprise the bulk of the drift materials. This assertion is further supported by the results of dye-tracer investigations on pre-wetted soil monoliths from the site. Rhodamine WT dye was used to label infiltrating waters, and spectrofluorometric analysis of solvent extracts of soil materials from the treated monoliths found that dye concentrations were greater in macropore material than in the soil matrix (Tables 18.20 and 18.21). In addition, applied gypsum slurry was transmitted through structural macropores in Eb horizon monoliths at a rate of 150 cm/h from the surface to the base of a 50 cm long 25 cm diameter monolith. Several plaster casts of macropore voids were obtained.

It is clearly possible that labelled materials could have been washed through the soil and subcutaneous zone and into the underlying P8 cave. Hence, any materials washed into the P8 cave and subsequently trapped in the sediment collectors might contain measurable amounts of caesium. This possibility is considered further below.

Table 18.20   Rhodamine WT in various soil materials following a flooding infiltrometer experiment on an Eb horizon monolith from pit 4, Field 7559

Soil material	Depth in soil column	Soil extract fluorescence at excitation/emission wavelengths (λ)		
		530/575 nm	550/575 nm	Positive?
Macropore	top	33.0	56.1	✓
Macropore	middle	31.5	54.5	✓
Macropore	bottom	9.2	15.8	✓
Matrix	middle	1.6	1.9	✓
Matrix	bottom	7.6	8.3	✓

Table 18.21   Comparative Rhodamine WT concentrations in three 50 cm deep Eb horizon monoliths, Pits 3, 4 and 6, Field 7559

	Comparative sample extract fluorescence (530/575 nm)		
	Depth from top of monolith (cm)	Matrix samples	Macropore samples
Pit 3 Eb monolith (monolith top −28 cm below surface)	−2	13	36
	−5	0	45
	−15	0	65
	−17	18	161
Pit 4 Eb monolith (monolith top −28 cm below surface)	−2	6	141
	−4	0	139
	−6	0	84
Pit 6 Eb monolith (monolith top −28 cm below surface)	0	26	39
	−5	7	92
	−10	15	63
	−12	0	41
	−14	0	98
	−16	0	55
	−21	0	48

18.4.5   Macropore Genesis

Several authors have documented features similar to those observed at the study site in other loessic soils (Williams and Allman 1969; Burek 1977; Langohr and Sanders 1985; Reeve 1991; Canti 1992; Schmaland 1992). However, they were not recorded by Pigott (1962) in his work on 25 soil pits in the Peak District limestone outcrop, or by Carroll (1986) and Curtis *et al.* (1976) in their general reviews of the effects of podsolization, leaching, clay translocation and cryoturbation on soils formed on Carboniferous Limestone. However, Reeve (1991) described "macropores" and "very coarse vertical worm channels" in a Nordrach variant soil at Pikehall, Derbyshire. In addition, Canti (1992) provided an illustration of vertical features similar to those in Figure 18.2 in a profile of a sandy silt-loam developed from "solifluction deposits" at Lismore Fields, Buxton. Langohr and Sanders (1985, p. 364–5), working on soils of the Belgian loess belt, recorded similar features in soils both on plateau sites and on convex and straight valley side-slopes and described them as a "bleached eluvial horizon (A2gc) through a relict consolidated subsoil". Langohr and Sanders (1985) considered the macropores to be relict features formed by cryoturbation during the consolidation and de-calcification of the loess deposit in periglacial climatic conditions. They knew of no process active in a modern moist temperate forest environment which could develop the subsoil consolidation found. They attributed a Weichsalian age to the soils which corresponds to the late Devensian when there is abundant evidence of periglacial activity in Derbyshire (Burek 1977, 1978). Hence cryoturbation processes may have operated on the soils of the study site. Burek (1977, p. 99) cites an example of cryoturbation at Hopton (SK 263 545) where frost-heaving had created convolutions in the sandy clay loam to depths of 50–85 cm. In places, the sandy clay loam had been completely overturned forming "eyes" in the clay horizon. Burek (1977) considered the "silty drift" of Pigott (1962) to itself be evidence of cryoturbation, because of the inter-mixing of loess and underlying insoluble (chert) limestone residue. Some of the large non-tubular macropores evident in two-dimensional profiles at the study site are suggestive of cryoturbation (Burek, C., pers. comm. 1992). However, the "silty drift" with large limestone fragments (as defined by Pigott 1962) considered indicative of cryoturbation, is absent from the study site, although present elsewhere in the catchment. Nevertheless, it is possible that the type 2 macropores are indicative of cryoturbation in the underlying subsoil, the features having been enhanced during the Holocene by the infiltration of humic matter. The type 3 macropore (Figure 18.2) may represent part of a frost-crack feature, again subsequently modified during Holocene soil formation. Similar features are common in a variety of modern arctic soils and also in relict periglacigenic soils (e.g. Van Vleit-Lanoe 1985), including a variety of soils elsewhere in Britain (e.g. Rose et al. 1985).

A biogenic origin for the type 1 and 2 macropores is also an attractive proposition, given that large numbers of *Oligochaeta* are present in the soils of the study site, and also that biogenic reworking of soils has been shown to be significant elsewhere. For example, Canti (1992) attributes the "tonguing" of A horizon materials for up to 10 cm into an underlying B horizon to bioturbation via *Oligochaeta*, particularly the species *Allolobophora longa* and *Allolobophora nocturna*. These species are selective in the materials ingested, but are capable of ingesting clasts up to 2 mm in diameter (Kuhnelt 1961 *in* Canti 1992) and of bringing sizeable quantities of soil to the surface. Such upcasting and reworking is capable of burying larger clasts and items such as archaeological artifacts to some depth (Atkinson 1957 *in* Canti 1992). In the study soils, the dominant species in the surface horizons is *Aporrectodea rosea* with occasional specimens of the deep-burrowing *Lumbricus terrestris* (Thomas and Bottrell 1992). The soils are free-draining, with saturated hydraulic conductivity (K_{sat}) values of 324–612 mm/h for the Ap horizon and 36–324 mm/h for the Eb horizon. Thus, it is possible that earthworms may have been modifying and integrating surface and subsoil materials for several thousand years.

Gaiser (1952 *in* Gilman and Newson 1980) found that biogenic (tree-root) macropore densities approached 10 000 channels/ha, a similar density to that found for the type 2 macropores at the study site. During the Pleistocene and early Holocene much of the study site may have been covered by dwarf birch or similar tundra species, and the type 2 macropores may represent voids formed by the decomposition of tree roots which were subsequently infilled by clastic sediments. This would ascribe a remarkable longevity to the macropores, whereas research on loess soils in Germany indicates that biogenic (*Oligochaeta*) macropores are dynamic features largely dependent on soil moisture and that they may be absent from the soil during dry summers (Schmaland 1992).

It is considered that the more open type 1 macropores are recent and related to earthworm (*Oligochaeta*) activity and/or to crop root extension. The filled type 2 macropores are possibly the infilled voids of decayed root systems from an ancient tree cover. These, together with the one example of a (possible) remnant filled ice-wedge (type 3 macropore), are thus permanent features of this soil.

18.4.6 Autogenic Clastic Sediment Transmission to the P8 Cave

Table 18.22 provides data on the amounts and physical characteristics of sediments collected from four recharge sites in the cave between January 1989 and September

Table 18.22   Clastic sediments input via autogenic recharge to monitored sites in the P8 Cave: physical analysis

Date		Dry weight	>2 mm	Sand	Silt	Clay
From	To	(g)	(%)	(%)	(%)	(%)
Idiot's Inlet						
18/1/89	1/10/89	10.7	–	–	–	–
1/10/89	26/3/90	23.51	0.00	36.49	45.36	18.17
26/3/90	18/9/90	0.96	0.00	–	–	–
18/9/90	30/3/91	597.90	15.12	82.34[a]	2.2	0.34
1/4/91	19/9/91	3.65	0.00	–	–	–
19/9/91	20/9/92	14.20	0.00	–	–	–
Far Flats						
10/1/89	1/10/89	0.87	0.00	–	–	–
1/10/89	18/9/90	1.20	0.00	–	–	–
18/9/90	19/9/91	3.25	0.00	–	–	–
19/9/91	28/9/92	2.86	0.00	–	–	–
P8 (3)						
21/1/89	1/10/89	0.45	0.00	–	–	–
1/10/89	18/9/90	0.55	0.00	–	–	–
18/9/90	19/9/91	0.78	0.00	–	–	–
Stalactite Passage						
10/1/89	1/10/89	0.24	0.00	–	–	–
1/10/89	18/9/90	0.97	0.00	–	–	–
18/9/90	19/9/91	0.64	0.00	–	–	–

–, Not determinable.
[a] Comprising 44.0% coarse, 37.2% medium, 18.7% fine sand.

1992. Lesser than 1 g per annum dry weight of clastic sediment was output from vadose seeps (P8(3) and Stalacite Passage). Slightly greater amounts of material were output in vadose seepage from the Far Flats roof which is "contaminated" with clastic materials. One unidentified *Oligochaeta* 2 cm long and 2 mm in diameter and a 1.5 cm long millipede were also recovered from the sediment collector at the Far Flats roof site.

The vadose flow site (Idiot's Inlet) contributed the greatest amount of clastic material, most being collected in the sediment trap during October to March, consistent with an increased discharge during the winter months. The amount of material obtained between September 1990 and March 1991 was large in comparison to that obtained in previous and subsequent years. Most of the material appeared between 21 and 31 March 1991, suggesting that the sediment is supply-limited rather than transport-limited, the material not being present in the conduit network but, rather, being input to the network from an adjacent storage. The most likely explanation is that the material had slumped or collapsed into the conduit network from the overlying drift materials or soils and then been transported rapidly to Idiots Inlet.

The age of the inwashed sediment was estimated by measuring the ^{137}Cs concentrations of the materials. In general, the amounts of bulk sediment collected from the P8 cave sites were too small to count on the gamma spectrometer. However, two of the bulk sediment samples collected at the Idiot's Inlet site were analysed, and the results are shown in Table 18.23. The presence of ^{137}Cs-labelled sediments in P8 Cave at a point some 13 m below the cave entrance and around 17 m below the surface of the overlying gully provides clear evidence that clastic sediments have been transported from the surface of the overlying soils and vegetation to the cave over a period of less than 30 years. Migration velocities are thus in the order of 0.47 m/ year, an order of magnitude greater than reported by other workers (Denk and Felsmann 1987).

The presence of measurable ^{137}Cs in the collected sediment is surprising, given the coarse nature of the inwashed material (Table 18.22). The coarseness may in part be due to the poor design of the Idiot's Inlet sediment tank which may trap coarse fractions but allow fine fractions to escape. Hence the 15% gravel, 2.2% silt and 0.3% clay fraction of the material collected on 31 March 1991 (Table 18.22). Thus, actual ^{137}Cs concentrations could be much higher depending on the sediment delivery ratio of the percolation water stream.

Table 18.23 ^{137}Cs concentration in Idiot's Inlet sediments, P8 Cave

Sample	^{137}Cs at counting date (D2) (Bq/kg)	Error % ($\pm$)
Idiot's Inlet 2 (1 Oct. 1989–26 Mar. 1990)	6.68	26.7
Idiot's Inlet 3 (18 Sept. 1990–30 Mar. 1991)	30.53	12.5

18.4.7   Mineralogy of the Inwashed Cave Sediment

An investigation of sediment mineralogy was undertaken using X-ray diffractometry (XRD), to determine the main mineral constituents and to provide a comparison with the mineralogy of the field soils (Table 18.24). The mineralogy is broadly similar to that of the surface soil materials, although calcite features much more prominently in the cave sediments. Chlorapatite was found in the Idiot's Inlet cave sediments following winter flushing, and may be due to either of two mechanisms:

1. the mineral may be produced authigenically by the reaction of calcite and phosphate in the soil or subcutaneous zones, the sediment then acting as a vehicle for the transport of otherwise immobilized ions; or
2. the mineral may be produced by the reaction of phosphate in solution with clastic sediment in the conduit or with the limestone of the conduit walls.

The absence of chlorapatite in sediment collected during spring and summer is also more indicative of authigenic production in the overlying soil and subcutaneous zone, rather than in the conduit system, since the sediments collected during this time are more likely to have been transported a limited distance within the conduit by lower flows.

18.4.8   Subsurface Mechanical Erosion Rates, Idiots Inlet Catchment

The erosion rate in the Idiot's Inlet catchment was estimated from the 597.9 g of clastic material collected in the trap between September 1990 and September 1991 (Table 18.22). This amount was adjusted upward to 1476.7 g, according to the fine/coarse fraction ratio in the overlying soil, to take account of the fine fraction

Table 18.24   Mineralogy of sediments inwashed by percolation waters to the P8 Cave

Mineral	Cave autogenic sediments		
	Far Flats[a]	Idiot's Inlet 1[b]	Idiot's Inlet 2[c]
Quartzite	✓	✓	✓
Quartz	✓	✓	✓
Kaolinite	✓	✓	✓
Kaolinite/smectite	✓	✓	✓
Chlorite			
Illite		✓	
Illite/smectite		✓	
Chlorapatite		✓	
Calcite	✓	✓	✓
Ferrihydrite/sulphate-hydrates	✓	✓	✓

[a]1 Oct. 1989–18 Sept. 1990.
[b]1 Oct. 1989–26 Mar. 1990.
[c]26 Mar. 1990–18 Sept. 1990.

probably lost because of poor sediment trap efficiency (data on catchment area and bulk density from Hardwick 1995).

$$1476.7\,g \text{ of material from a } 1280\,m^2 \text{ catchment} = 1153.4\,kg/km^2$$
$$\text{mean soil bulk density} = 1.515\,g/cm^3$$
$$\therefore 1153.4\,kg = 0.76\,m^3$$
$$\therefore \text{ the erosion rate} = \underline{0.76\,m^3/km^2/year}$$

The autogenic clastic sediment erosion rate is around two orders of magnitude less than solutional erosion rates previously estimated for the outcrop (Christopher *et al.* 1977) and represents a comparatively negligible surface lowering of 0.8 mm/1000 years. This is also three orders of magnitude lower than the surface lowering rate of 0.8 mm/year due to lateral slope wash in Hungarian dolines recorded by Zambo (1986).

18.5 CONCLUSIONS

The soils developed on loess drift materials overlying cavernous Carboniferous Limestone are markedly anisotropic and contain macropores which appear to link the land surface and bedrock. Comparison of the various soil materials and features very strongly suggests that, although the origin of the macropores is unclear, they have a modern hydrological function, channelling percolation waters rapidly through the soil and into bedrock. The presence of ^{137}Cs in the macropores at depth but not in the surrounding soil matrix indicates that they also facilitate clast migration, probably by turbulent flow. ^{137}Cs was also found in sediments inwashed to the underlying cave by percolation waters. This indicates a direct connection between the cave and the soil surface via soil macropores. Hence, the macropores of the study site should be viewed as an extension of the karst drainage system into the soil zone.

Sampling of the soil features proved problematical primarily because of their size. For example, the type 1 macropores were very difficult to sample as they are often less than 1 cm in diameter and may consist largely of void! Problems of subjectivity in sampling could also arise where the boundaries between, for example, macropores and the surrounding soil matrix are unclear or gradual. It should be emphasized that sampling was undertaken on the basis of morphological and colour difference, which on the basis of a statistical analysis of small numbers of samples would appear to have been justified. Future work at the site will involve investigations of the hydrological connectivity of various macropores with the bedrock and underlying cave using *in situ* labelling of drainage pathways using staining dyes such as Rhodamine WT.

Although the presence of modern surface materials in the cave has been proved, these could have been inwashed irrespective of the level of agricultural management since macropores are present in the control pit 5 which had not been managed for *c.* 20 years. Thus a much wider study of autogenic sediment inputs to caves and of the nature and age of the sediments is recommended. More detailed investigations of

^{137}Cs distribution in soil features are also required and would be assisted by the use of more sensitive gamma spectrometric instrumentation capable of analysing small (<30 g) samples.

Geomorphologically insignificant amounts of clastic sediment were retained in the percolation water sediment traps installed in the cave and thus the estimated mechanical erosion rate is negligible. Thus autogenic recharge is not an important agent of mechanical erosion beneath the covered limestone hillslope of the site, despite the field being regularly ploughed and despite the presence of soil macropores. One caveat to the results is that the trap efficiency of the sediment traps may have been low since there was a marked reduction in the percentage of fine material retained in the sediment traps in comparison to that of the overlying soils.

The study has shown that surface sediments may be transported relatively rapidly to underlying caves via drainage through open and infilled macropores. However, the rate of erosion is low in comparison to rates of solutional erosion of the underlying bedrock. Nevertheless, erosion via macropore drainage may be more significant than via lateral surface runoff, since precipitation rates are rarely likely to exceed the typical saturated hydraulic conductivity values for the soils of 324–612 mm/h for the Ap horizon and 36–324 mm/h for the Eb horizon. However, the erosion rate of 0.76 m^3/km^2/year is similar to an estimated rate of soil formation of 0.75–1.5 m^3/km^2/year (based on solutional erosion rates from Pigott (1965) and a greater than 98.5% purity for the Carboniferous Limestone (Gatliff 1982)). Hence, perhaps the main significance of the inwashed modern sediments lies in any attached agrochemicals such as herbicides or pesticides. The investigation suggests that the presence of a thick cover of soil or drift deposits on cavernous limestone does not necessarily provide a buffer for the underlying cave or groundwater body against applied agrochemicals which are normally considered to be immobilized within the soil.

ACKNOWLEDGEMENTS

The authors would like to thank Dr L. Richards and Mr R. Wright of English Nature for their support and assistance during the Nature Conservancy Council Commissioned Research Project which led to this study; C. Allen, G. Dixey et al. (University of Huddersfield), H. Dawkes, J. Russell and M. Sapsford (Manchester Metropolitan University) who undertook undergraduate dissertations or project work on topics related to the study and Mr W. Ritter of Perryfoot Farm for provision of access to P8 cave and to the study site. The investigations formed part of doctoral work by the first author (Hardwick 1995).

REFERENCES

Anderson, J.M. 1988. Invertebrate mediated transport processes in soils. *Agriculture, Ecosystems and Environment*, **24**, 5–19.

Avery, B.W. 1973. Soil classification in the soil survey of England and Wales. *J. Soil Science*, **24**, 324–338.

Barany-Kevei, I. 1988. Man's impact on karstic areas in Hungary. *Proc. Int. Symp. Human Influence on Karst. Postojna 1988*, pp. 207–219.

Beven, K. and Germann, P. 1982. Macropores and water flow in soils. *Water Resources Research*, **18**(5), 1311–1325.

Burek, C. 1977. The Pleistocene Ice Age and after. In T.D. Ford (ed.), *Limestones and Caves of the Peak District*. Geobooks, Norwich.

Burek, C. 1978. Quaternary Deposits on the Carboniferous Limestone of North Derbyshire. *PhD Thesis*. Unviersity of Leciester.

Canti, M. 1992. *Soil Report from Lismore Fields, Buxton*. Report, Trent and Peak Archaeological Trust, University Park, Nottingham.

Carroll, D.M. 1986. Soils associated with Carboniferous Limestone in England and Wales. In M.M. Sweeting and K. Paterson (eds), *New Directions in Karst*. Geobooks, Norwich, pp. 141–149.

Chamberlain, T.W. 1972. Interflow in mountainous forest soils of coastal British Columbia. In H.O. Slaymaker and H.J. McPherson (eds), *Mountain Geomorphology*. Tantalus Research, Vancouver.

Christopher, N.S.J., Beck. J.S. and Mellors, P.T. 1977. Hydrology, water in the limestone. In T.D. Ford (ed.), *Limestones and Caves of the Peak District*. Geobooks, Norwich.

Coles, N. and Trudgill, S.T. 1985. The movement of nitrate fertiliser from the soil surface to drainage waters by preferential flow in weakly structured soils, Slapton, S. Devon. *Agriculture Ecosystems and Environment*, **13**, 241–259.

Curtis, L.F., Courtney, F.M. and Trudgill, S.T. (eds) 1976. *Soils of the British Isles*. Longman, London.

Denk, H.J. and Felsmann, M. 1987. Measurements and evaluation of a multicompartment model for estimating future activity profiles of radiocesium in undisturbed soil of pasture-land in North Rhine-Westphalia. In W. Feldt (ed.), *Proc. XVth Reg. Conf. IRPA. The Radioecology of Natural and Artificial Radionuclides, Gøtland, Sweden*, 1987, pp. 182–187.

Gardiner, V. and Dackombe, R.V. 1983. Geomorphological Field Manual. George Allen and Unwin, London.

Gatliff, R.W. 1982. *The Limestone and Dolomite Resources of the Country around Tideswell, Derbyshire: Resource sheet SK 17 and parts 18 and 27*. Mineral Assessment Report 98, Institute of Geological Sciences, HMSO, London.

Gilman, K. and Newson, M.D. 1980. *Soil Pipes and Pipeflow: A Hydrological Study in Upland Wales*. Res. Monogr. 1, British Geomorphological Research Group, Geobooks, Norwich.

Gudzenko, V.V. and Klimchouk, A.B. 1994. Chernobyl radiocaesium in a karst system. Marble Cave, Crimea. *Cave and Karst Science*, **21**(1), 13.

Hallsworth, E.G. 1963. An examination of some factors affecting the movement of clay in an artificial soil. *J. Soil Science*, **14**, 360–371.

Hardwick, P. 1995. The impact of agriculture on limestone caves with special reference to the Castleton karst. *PhD Thesis*, Manchester Metropolitan University.

Hardwick, P. and Gunn, J. 1990. Soil erosion in cavernous limestone catchments. In J. Boardman, I.D.L. Foster and J.A. Dearing (eds), *Soil Erosion on Agricultural Land*. John Wiley, Chichester, pp. 301–310.

Jakucs, L. 1977. *Morphogenetics of Karst Regions*. Adam Hilger, Bristol.

Jury, W.A., Elabd, H. and Resketo, M. 1986. Field study of napropamide movement through unsaturated soil. *Water Res.*, **22**, 749–755.

Langohr, R. and Sanders, J. 1985. The Belgian loess belt in the last 20,000 years: Evolution of soils and relief in the Zonien Forest. In J. Boardman (ed.), *Soils and Quaternary Landscape Evolution*. John Wiley, Chichester, pp. 359–371.

Livens, F.R., Horrill, A.D. and Singleton, D.L. 1991. Distribution of radiocesium in the soil plant systems of upland areas of Europe. *Health Physics*, **60**(4), 539–545.

Munsell Colour Company 1953. *Munsell Colour Charts*. Munsell Colour Company, Baltimore.

Pigott, C.D. 1962. Soil formation and development on the Carboniferous Limestone of Derbyshire. 1. Parent materials. *J. Ecology*, **50**, 145–156.

Pigott, C.D 1965. Limestone surfaces under superficial deposits in north Derbyshire. *Geographical J.*, **131**(1), 41–44.

Ragg, J.M., Beard, G.R., George, H., Heaven, F.W., Hollis, J.M., Jones, R.J.A., Palmer, R.C., Reeve, M.J., Robson, J.D. and Whitfield, W.A.D. 1984. *Soils and Their Use in Midland and Western England*. Soil Survey of England and Wales Bull. No. 12, Harpenden.

Reeve, M.J. 1991. *Properties and Distribution of the Soils of the White Peak with Special Reference to Dale-side Soils*. Commissioned Research Report for the Limestone Research Group, Manchester Polytechnic. Soil Survey and Land Research Centre, Shardlow Hall, Derby.

Ritchie, J.C. and McHenry, J.R. 1973. Vertical distribution of fallout cesium-137 in cultivated soils. *Radiation Data and Reports*, **12**, 727–728.

Rose, J., Allen, P., Kemp, R.A., Whiteman, C.A. and Owen, N. 1985. The Early Anglian Barham soil of eastern England. In J. Boardman (ed.), *Soils and Quaternary Landscape Evolution*. John Wiley, Chichester.

Scharpenseel, H.W. and Kerpen, W. 1967. Studies on tagged clay migration due to water movement. *Proc. Int. Atomic Energy Agency Symp, Instanbul*, pp. 287–290.

Schmaland, G. 1992. Macropores in agricultural soils on loess, a dynamic phenomenon. In H. Hotzl and A. Werner (eds), *Tracer Hydrology*. Balkema, Rotterdam, pp. 201–204.

Simpson, T.W. and Cunningham, R.L. 1982. The occurrence of flow channels in soils. *J. Environmental Quality*, **11**(1), 29–30.

Soil Survey 1973. *Soil Survey Field Handbook*. Soil Survey of England & Wales, Harpenden.

Thomas, G.W. and Phillips, R.E. 1979. Consequences of water movement in macropores. *J. Environmental Quality*, **8**(2), 149–152.

Thomas, R. and Bottrell, S. 1992. The role of Oligochaeta in the ecology of Speedwell Cavern, Derbyshire. *Cave Science, Trans. British Cave Research Assn.*, **19**(1), 21–23.

Trudgill, S.T., Crabtree, R.W., Pickles, A.M., Smettem, K.R.J. and Burt, T.P. 1983. Hydrology and solute uptake in hillslope soils on Magnesian Limestone: The Whitwell Wood Project. In T.P. Burt and D.E. Walling (eds), *Catchment Experiments in Fluvial Geomorphology*. Geobooks, Norwich, pp. 183–215.

Van der Molen, W. and Van Ommen, H. 1988. Transport of solutes in soils and aquifers. *J. Hydrol.*, **100**, 433–451.

Van Vliet-Lanoe, B. 1985. Frost effects in soils. In J. Boardman (ed.), *Soils and Quaternary Landscape Evolution*. John Wiley, Chichester.

Williams, R.E. and Allman, D.W. 1969. Factors affecting infiltration and recharge in a loess-covered basin. *J. Hydrology*, **8**, 265–281.

Wise, S.M. 1980. Caesium-137 and Lead-210: a review of the techniques and some applications in geomorphology. In R.A. Cullingford, D.A. Davidson and J. Lewin (eds), *Timescales in Geomorphology*. John Wiley, Chichester, pp. 109–127.

Zambo, L. 1986. *The impact of soil on karstic corrosion (an experimental study on the Aggtalek karst, Hungary)*. English Abstract C.Sc. Geogr. Thesis, Budapest Academy of Sciences, Hungary.

19 Examining the Factors Controlling the Spatial Scale of Variation in Soil Redistribution Processes from Southwest Niger

A. CHAPPELL,[1] M. A. OLIVER,[2] A. WARREN,[1] C. T. AGNEW[1] and M. CHARLTON[3]

[1]*Department of Geography, University College London, UK*
[2]*Department of Soil Science, University of Reading, UK*
[3]*Department of Physics and Astronomy, University College London, UK*

19.1 INTRODUCTION

There are many difficulties in measuring soil redistribution (Higgitt 1991), some of which arise because the measurement techniques do not encompass the spatial and temporal variation of the processes operating in the landscape. Hence, there is a dearth of information on soil flux (Lal 1988) and the factors that control it.

The radionuclide ^{137}Cs has considerable potential for measuring soil redistribution and its use overcomes many of the difficulties experienced with other methods. It provides accurate net soil flux measurements aggregated over the last 30 years (Ritchie and McHenry 1990; Sutherland and de Jong 1990; Walling and Quine 1991). However, its application to large areas has been limited because the measurement of ^{137}Cs is time-consuming and because many samples are needed to describe its spatial variation and thus soil redistribution (Sutherland 1994). One aim is to identify other properties that relate to soil ^{137}Cs concentration to minimize the number of samples that need to be analysed.

Topographic attributes from digital elevation models (DEMs) can be used as surrogate variables for modelling spatial processes that are not amenable to a truly physical approach (Moore *et al.* 1991). Many researchers have used slope gradient, surface curvature, catchment area and water ponding depth (Martz and de Jong 1991) to relate soil erosion rates to topography. Aeolian activity can be inferred from two simple topographic indices of wind sheltering–exposure, namely fetch and directional relief (Lapen and Martz 1993). Such relations are important for elucidating geomorphological processes, soil formation, its spatial distribution and for predicting soil properties precisely at unvisited locations (Odeh *et al.* 1991; Moore *et al.* 1993).

Linear regression models for relating soil (response) properties and topographic (environmental predictor) attributes are of limited value for large databases.

Advances in Hillslope Processes, Volume 1. Edited by M. G. Anderson and S. M. Brooks.

Ordination, such as principal components analysis (PCA), is one way of summarizing large amounts of data. By integrating regression and ordination, known as canonical ordination, relations between soil and topography can be effectively determined. The canonical form of PCA is known as redundancy analysis (RDA); axes are constrained to linear combinations of environmental variables to minimize the sum of squares (Odeh *et al.* 1991).

Point-survey data are generally interpolated to a grid to form a DEM. Implicit in methods of interpolation is the assumption that the data are spatially dependent, yet this is rarely determined. Thus, topographic attributes derived from these DEMs do not encompass the spatial variation in the processes they are acting as surrogates for and reduce the efficacy for predicting net soil flux (Martz and de Jong 1991; Moore *et al.* 1993).

The aim of this study was to elucidate the factors controlling soil redistribution processes by relating topographic attributes, soil properties and vegetation cover using redundancy analysis, to construct a predictive model of net soil flux. Variograms were computed for all properties and the leading principal components were analysed to determine the structure of the spatial variation. The parameters of the variogram models were then used to estimate all properties using block-kriging, a form of geostatistical estimation (Matheron 1971). A dynamic, non-hierarchical, multivariate soil classification of soil sample sites provided additional insight into the pattern of variation and helped to validate the relational models.

19.2   STUDY AREA

19.2.1   Climate

The study area, in southwest Niger, is approximately 60 km east of Niamey (Figure 19.1). It has a semi-arid climate typical of the southern Sahel, with a marked north to south precipitation gradient of about 1 mm/km (Lebel *et al.* 1992). The average rainfall at Niamey for the period 1905–1989 was about 560 mm. For the past 25 years the rainfall has been persistently below-average with a mean of about 495 mm.

During the summer the prevailing southwesterly winds can be overshadowed periodically by strong easterly winds which precede squalls with short-duration, high-intensity rainfall. The intensity of the rainfall may cause crusting, clay eluviation and rapid removal of unconsolidated material from the surface in runoff (Casenave and Valentin 1989). During the dry winter months the northeasterly winds of the Harmattan dominate and transport aeolian material primarily from ancient lake beds near to Bilma in Niger (McTainsh and Walker 1982). Drees *et al.* (1993) have reported recently that total dust infall in Niger during 1985–1989 was very seasonal. They suggested that frontal storms from the east were responsible for more of the dust infall (2 t/ha/year) than the Harmattan dust storms.

19.2.2   Topography and Soil

A schematic cross-section of a soil toposequence typical of the area is shown in Figure 19.2, extending from the ferricrete-capped plateau to the valley floor (Prince

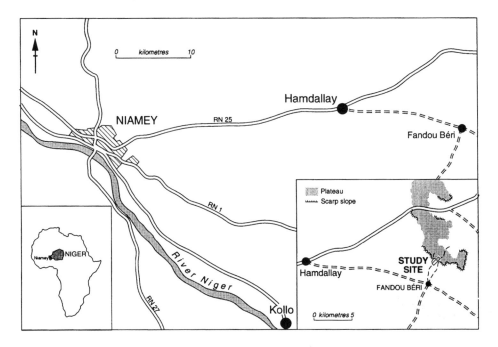

Figure 19.1   Location of study area in southwest Niger, West Africa

et al. 1995). The plateaux are remnants of peneplains with poorly developed soil from a highly weathered, humus-poor, sesquioxide-rich and mainly kaolinitic parent material. The topsoil is a gravelly loam over cemented ironstone gravels (Manu *et al.* 1991) with pH < 5, low nutrient status and small water storage capacity because of the soil's generally shallow depth and large content of coarse fragments. The surface of the plateaux has bare patches of soil interspersed with bands of vegetation of varying size, orientated with the long axis orthogonal to the slight slope towards the plateaux edge (Thiery *et al.* 1995). The vegetation appears to form stripes when viewed from the air and is appropriately named *brousse tigrée* or tiger bush. Many of the vegetation islands in the study area are visible on maps and aerial photographs from 1950, 1975 and 1992. The soil within these islands of vegetation is different from that dominating the plateaux; it is humus-rich with a pH of 5–6 and is approximately 20 cm thick above ferricrete cobbles.

Between the plateau and the valley there is a transition zone called the plateau edge with three components: talus slope, sand skirt and shelf (Figure 19.2). The base of the talus slope is marked by a decrease in surface gradient, but the gradient gradually increases again downslope in the sand skirt. Here the terrain is hummocky, suggesting spatial difference in soil loss. Vegetation-capped pedestals or mounds of acid soil (pH < 5) are typically 30–40 cm high and from 10 to 30 m in width. Some are sharp ridges at right angles to the prevailing wind direction, while others are rounded. The surface gradient also decreases at the shelf between the sand skirt and

432

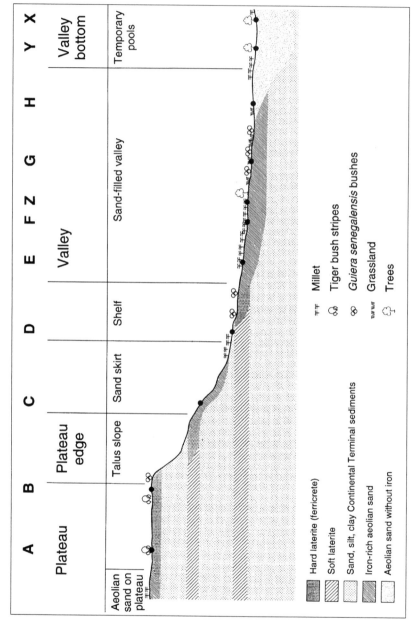

Figure 19.2 Schematic cross-section of a toposequence typical of the region (adapted from Prince *et al.* 1995)

the valley. These distinct changes in surface gradient are controlled by the proximity of the underlying ferricrete to the surface. In these locations the vegetation is sparse and the soil surface is extremely compacted. The exposure of clay-enriched, indurated B horizons at the surface suggests that topsoil may have been removed.

On the plateau there is evidence of localized rilling and surface wash which feed gullies on the plateau edge that cut into the soil and provide pathways for soil movement onto the valley (Figure 19.2).

19.2.3 Indigenous Land Use

The plateau is used for grazing herds of goats and cattle and also as a source of fuelwood and of plants for medicinal purposes (Manu *et al.* 1991). The valley is dominated by fields of cultivated millet which are grazed after harvesting. Increasing population pressure, prolonged periods of recurrent drought and successive crop failure in the region have reduced the fallow periods to between three and five years (Manu *et al.* 1991). There are few areas of undisturbed land within the study area except some small vegetated areas at the base of the toposequence.

19.3 METHODOLOGY AND ANALYSIS

19.3.1 Sampling Technique and Soil Property Measurements

This study is part of a wider investigation of the spatial variation of net soil flux (Chappell 1995). Since we had little information about the spatial scale of variation of the soil properties, the aim was to sample at a range of distances to ensure that we would identify the main source of the variation at the scale of the investigation. The sampling strategy was based on a nested grid with intersections at 5 m, 20 m and 100 m. In designing the sampling scheme, account was taken of the main physiographic units which were the plateau and valley (Figure 19.3). At the grid nodes, soil samples were collected between July and October 1992 and *in situ* measurements were also made. In general, soil properties (Table 19.1) were selected because they were easy to determine or because of their known correlation with ^{137}Cs fixation in soil (McHenry and Ritchie 1977). In addition, soil surface magnetic susceptibility was measured *in situ* at 458 sites on a 1 m^2 grid using a Bartington Instruments survey loop connected to a portable meter. This property was calibrated in the laboratory by measuring low and high frequency (Bartington Instruments 0.47 kH and 4.7 kH, respectively) mass specific susceptibility, χ_{Lf} and χ_{Hf}, respectively. These parameters are related to the concentration of ferrimagnetic minerals and were used to derive χ_{Fd} which provides information on the grain-size assemblage within a sample. Shear vane soil measurements near to the surface (5 cm, 10 cm and 15 cm) were recorded to indicate the cohesiveness of surface layers. Topsoil was sampled using a bulk density ring and the bulk density was measured in the laboratory. Total vegetation cover was expressed approximately as a percentage of a 1 m^2 quadrat at all sites.

Soil samples for laboratory analyses were obtained in one of two ways depending on the site and the depth of the ^{137}Cs profile. On the plateau, a 20 cm deep pit

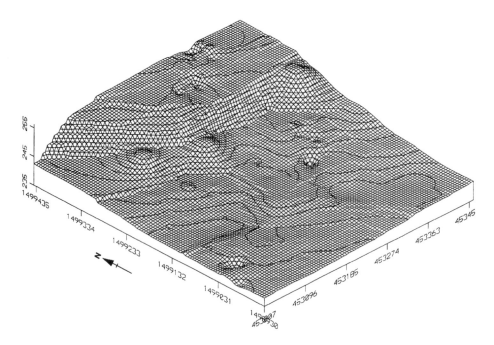

Figure 19.3 An orthographic projection using UTM coordinates and block-kriging (18.6 m) of elevation in the study area relative to an arbitrary datum

(600 cm² area) was dug and a complete soil and ^{137}Cs profile was taken. Soil was sampled to a depth of 51 cm in the valley using a Dutch auger (6.25 cm² internal area) to obtain the complete ^{137}Cs profile. At each site the soil from the profile was bulked and homogenized to provide a representative sample of the soil profile, and

Table 19.1 Soil properties measured, their RDA labels, types of data and total number of sites for which measurable data were available

Soil property	Label	Sites	Units	Type
Organic matter	loi	213	%	Numeric
Bulk density	bd	158	kg/m³	Numeric
^{137}Cs	Bq/kg	81	Bq/kg	Numeric
^{137}Cs	Bq/m²	81	Bq/m²	Numeric
Soil strength	csv1	158	k/Pa	Numeric
Relative elevation		177	m	Numeric
χ_{Hf} and χ_{Lf}	chilf	458	$10^9 \chi$/m³/kg	Numeric
χ_{Fd}		458	%	Numeric
Soil colour	hue	208	Munsell	Ordinal
pH	pH	208	pH	Numeric
Particle size	clay, silt	210	%	Numeric
Vegetation cover	veg	217	%	Numeric

subsampled by hand. The following properties were then determined using air-dried soil samples.

Soil colour was recorded from the Munsell charts. pH was measured with a glass electrode in suspensions of soil in water at a ratio of 1:5. Soil organic matter content was obtained by standardizing soil moisture at 105°C prior to weight loss at 375°C. Particle-size distribution was determined by a Malvern 2600 laser sizer on subsamples with the organic matter removed prior to disaggregation using ultrasonic action. The threshold values for percentage clay, silt and sand content used in subsequent analyses were $2\,\mu m$, $60\,\mu m$ and $1128\,\mu m$. The ^{137}Cs activity was measured by γ-ray energy spectrometry. The soil samples were lightly ground with mortar and pestle to pass a 1 mm sieve. The samples were placed in a standard Marinelli (re-entrant) beaker which was then located on a horizontally oriented 20% relative efficiency hyper-pure germanium (HpGe) γ-ray detector. The detector was coupled to spectroscopy grade amplifiers and a PC-based data collection system. Calibration samples were used to derive the absolute ^{137}Cs activities in each sample.

19.3.2 Calculation of Net Soil Flux from [137]Cs

The ^{137}Cs concentration of material in both profiles and trapped material is ten times greater in soil samples from the plateau than from the valley. The probable preferential accumulation of ^{137}Cs-enriched dust and surface wash beneath vegetation makes it difficult to identify a "stable" [137]Cs reference site.

At an unvegetated site (Z) in the valley where there has been no cultivation or wood cutting for at least 30 years, limited erosion only is likely to have taken place. A mass-specific exponential model was fitted to the ^{137}Cs profile at Z ($r^2 = 0.97$). The model is:

$$Log_{10}C_i = 0.933 - 0.034\,D_i \tag{19.1}$$

where C is the mass ^{137}Cs concentration (Bq/kg), D is the soil depth (cm), and i is the sample increment. The ^{137}Cs inventory predicted by the model for site Z ($2066 \pm 125\,\mathrm{Bq/m^2}$) is the best approximation of the reference ^{137}Cs inventory for an "undisturbed" site in the study area (Chappell 1995).

To calculate the net soil flux, a relation between the movement of ^{137}Cs and the redistribution of soil must be established (Walling and Quine 1990). In this rangeland area only the gravimetric model (Zhang et al. 1990) was applicable. It uses the exponential coefficient of the ^{137}Cs depth-distribution. A value for this coefficient of 0.026 was used for the plateau region and one of 0.034 (equation (19.1)) for the valley. By assuming that surface lowering has reduced the total ^{137}Cs activity, soil loss at a site can be estimated from measurements of the ^{137}Cs remaining in the profile relative to the ^{137}Cs reference inventory.

To interpret net soil gain, Chappell (1995) showed that the proportional model (Sutherland and de Jong 1990) could be modified and applied to depositional sites associated with vegetation. The total residual ^{137}Cs activity as a percentage of the reference level was reduced by the ratio of the average ^{137}Cs concentration of material on the plateau to the average ^{137}Cs concentration of material in the valley, thus accounting for the preferential accumulation of ^{137}Cs on the plateau.

19.3.3 Rationale for Analysis

The data were analysed using several multivariate statistical and geostatistical methods. The sites were classified non-hierarchically (Oliver and Webster 1987a) at the outset to see whether there was any clustering in the data, and if so, whether it related to the physiography. The data were standardized first to zero mean and unit variance and the samples were then subdivided arbitrarily to 10 groups. The criterion chosen for optimization was SS_w, the sums of squares of deviations from group means, which measures the dispersion within classes. The test criterion was calculated and individuals were moved from group to group, with the test criterion being recalculated after each reallocation. If the criterion was reduced the change was retained, otherwise it was not. The procedure continued iteratively until no further improvement was possible.

By transforming the data to principal components the relations between sites in terms of their properties alone can be determined. The leading principal axes usually explain a large proportion of the total variation. The latent vector loadings show which variables load heavily on the axes and often enable an interpretation of them that has some environmental meaning.

Linear RDA was used to relate net soil flux, the soil properties, vegetation cover and several topographic attributes. Missing values present a problem in RDA, therefore values were estimated by kriging.

Geostatistical methods based on regionalized variable theory (Matheron 1971) were used for analysing the data spatially. The central tool is the variogram which measures the degree of spatial dependence or autocorrelation in the data arising from the underlying spatial structure in the variation (Oliver and Webster 1987b). The variogram model can then be used with the data to estimate properties from sample data (Oliver *et al.* 1989a) by kriging. Essentially, kriging is a method of local weighted averaging where the weights are derived from the variogram. Hence, the estimates are based on the way that the property varies spatially. Kriging has been found to be one of the most reliable two-dimensional spatial estimators (Laslett *et al.* 1987), but it is not suitable where there are abrupt spatial changes in properties, i.e. boundaries. Voltz and Webster (1990) have shown how soil classification can be combined with kriging to use the advantages of both approaches.

The statistical distribution of the data was examined for normality and any properties with a skewness more than one were transformed to ensure that the variances were stable. To test for directional variation or anisotropy, directional variograms were computed for northeast, east and southeast using only magnetic susceptibility (χ_{Lf}) in the valley region because these variograms require more data than were available for other properties. Since variation was isotropic for this property it was assumed that it was representative of all other properties. Thus, omni-directional variograms were computed for all variables.

Mathematical models were fitted to the experimental semivariance to gain insight into the structure of spatial variation and also for kriging. A weighted least-squares procedure was used to fit a variety of models to the variograms. The parameters of the variogram model can be used for local estimation by kriging.

19.3.4 Landform Analysis

Relative elevation was determined by ground-based survey. Block-kriging was then used to estimate values on a regular grid with a spacing of 18.6 m to produce a DEM (Figure 19.4) so that primary and compound topographic attributes (Moore *et al.* 1991) could be calculated. Although slope-form topographic attributes are less strongly associated with geomorphological processes and pedological properties than slope position variables, both types were calculated. Slope-form attributes included aspect (ASPECT), slope gradient (GRDNT), profile curvature (PROFCV) and contour curvature (CONTCV).

Slope position topographic attributes that provide indirect measures of aeolian processes (fetch and directional relief) were derived from simple topographic indices

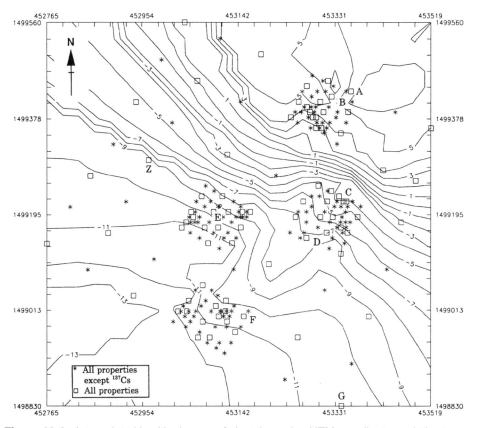

Figure 19.4 Interpolated isarithmic map of elevation using UTM coordinates, relative to an arbitrary datum, and showing sample sites where all properties were measured and all properties except ^{137}Cs were measured

of wind sheltering–exposure (Lapen and Martz 1993). A range of values for the obstacle height increment (0.012, 0.025 and 0.05 m/m) was used to calculate fetch values in an easterly (F*E) and northeasterly (F*NE) direction (where * is equivalent to 1, 2 or 3 referring to the respective values used). These directions were chosen to coincide with the easterly squalls and northeasterly Harmattan, both of which have a significant effect on aeolian processes in this region (Drees *et al.* 1993). Directional relief was calculated for these same directions (R1E and R1NE). These indices reflect topographic shelter and exposure zones but exclude the effect of vegetation cover which may be particularly important in providing shelter zones. A simple method of incorporating the effect of vegetation cover is to increase the elevation of the DEM by the average height of vegetation where it occurs. Fetch (VF*E and VF*NE) and directional relief (VR1E and VR1NE) were calculated using the modified DEM in the same directions and with the same obstacle height increments as those above.

Indirect measures of surface wash were also provided by slope position topographic attributes (Boundy and Martz 1988) and included FILL, SINK, local catchment area (LCAT) and global catchment area (GCAT) at every element of the DEM (Martz and de Jong 1988). Compound topographic variables (Moore *et al.* 1991) based on slope form and slope position variables were used for modelling spatial processes that cannot be analysed directly using a truly physical approach. Three hydrologically based compound indices that can predict the spatial distribution of soil properties are the stream power index (POW), the wetness index (SWC) and a sediment transport capacity (LS) index (Moore *et al.* 1993). These indices have been used in conjunction with local (LCAT) and global (GCAT) catchment area to produce six compound topographic attributes (LCATPOW, GCATPOW, LCATSWC, GCATSWC, LCATLS and GCATLS measured in metres per degree).

19.4 RESULTS AND DISCUSSION

19.4.1 Multivariate Analysis

The non-hierarchical classification was performed with all of the soil variables and at all sites where these were measured; six classes were found to be optimal. The locations of the sampling sites labelled according to the class they belong to are shown in Figure 19.5. The separation of valley (classes 1 and 5) from plateau (classes 2, 3, 4 and 6) is evident. This distribution suggests that there is less spatial variation in the valley than on the plateau partly because the latter includes a transition zone. The class averages for each soil property (Table 19.2) aid the interpretation of these classes. Class 5 dominates the valley; it has intermediate ^{137}Cs concentrations, low χ_{Fd}, organic matter and silt content and high sand content and bulk density. Class 1 is related mainly to the transitional sloping land between the plateau and the valley and seems to coincide with gullies. Its silt and organic matter content, bulk density, χ_{Fd}, vegetation cover and ^{137}Cs are intermediate. Classes 2, 4 and 6 dominate the plateau. All contain large amounts of organic matter, ^{137}Cs and silt, and are associated with the most vegetation cover. Class 3 is very small (four members) and seems to contain aberrant sites on both the plateau and valley. It is associated with bare ground and with soil of low bulk density, ^{137}Cs and high χ_{Fd}.

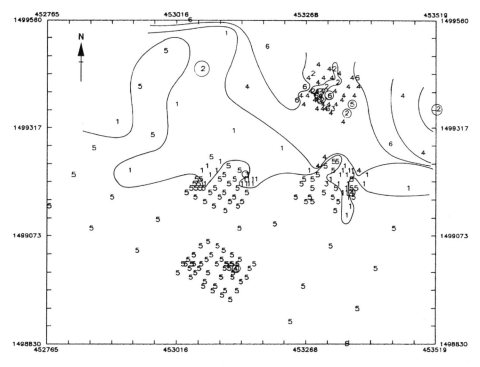

Figure 19.5 Spatial location of sites classified using a non-hierarchical procedure, separated by five isarithmic lines

Soil on the plateau with a large silt content probably results from the accumulation of ^{137}Cs-enriched dust beneath vegetation. Reworking of this material by surface wash and wind redistributes it to some nearby unvegetated sites resulting in large silt and ^{137}Cs contents. Thus, the interpretation of the classification is that the plateau is separated into areas of dust accumulation (Classes 2 and 6) and others that are eroding (Class 4).

Table 19.2 Summary of the soil within the study area using the average soil property for each group

Soil properties	Classes					
	1	2	3	4	5	6
Organic matter (%)	0.50	1.51	0.75	1.25	0.38	1.25
Bulk density (kg/m^3 × 10^3)	1.54	–	1.53	1.57	1.59	–
^{137}Cs (Bq/m^2 × 10^3)	0.46	1.55	0.04	0.22	1.24	2.91
χ_{Fd}(%)	10.13	9.98	11.04	10.10	9.76	10.03
% Silt content	15.30	20.00	25.10	27.00	9.30	28.90
% Vegetation	7.13	30.00	0.00	34.34	23.00	43.00

– , Identifies properties that were not available.

The vegetation cover is less and the soil depth greater in the valley than on the plateau. Thus, illuviation will probably take place in valley soil exposed to intense rainfall, causing eluviation of silt in the valley soil profiles. Hence, sites in the valley with little vegetation cover are exposed to more erosion and have more silt and less ^{137}Cs in the compact subsoil (Class 1) than sites with greater vegetation cover and less erosion (Class 5). Class 3 sites are in unvegetated areas, yet most are not associated with surface lowering. It is likely that reworking and preferential selection of redistributed soil has produced this disparate group.

The data were transformed to a new set of axes orthogonal to one another by principal components analysis. The data comprised 15 soil properties at the principal 217 sites. Since the measurements were on several different kinds of scale the analysis was performed on data transformed to between 0 and 1 (Odeh *et al.* 1991) to avoid bias against the ordinal data.

The first two principal axes account for over 57% of the variation (Table 19.3), suggesting strong correlation among some properties. The eigenvector loadings (Table 19.3) above a threshold of 0.3 (emphasized in italic) were used to interpret them. Note that two of the three particle-size measurements only were included in the analysis. Principal component one appears to be an axis representing variation in particle size, pH and organic matter content. The pH is larger and the organic matter content greater where the soil is finer. Component two apparently accounts for variation in bulk density, soil strength, magnetic susceptibility and ^{137}Cs content. Where bulk density is large, soil strength and low-frequency magnetic susceptibility (χ_{Lf}) have high values. The concentration of ^{137}Cs is negatively correlated with these properties.

Table 19.3 Principal component eigenvalues and loadings of soil properties from sites within regions of the study area

	Principal component loadings	
	1	2
Eigenvalues	5.48	3.14
Variance %	36.52	20.97
Soil property		
Organic matter	−0.356	0.058
Bulk density	0.293	0.355
^{137}Cs (Bq/kg)	−0.291	−0.379
^{137}Cs (Bq/m^2)	−0.207	−0.391
Strength (5 cm)	−0.200	0.332
Strength (10 cm)	−0.123	0.282
Strength (15 cm)	−0.305	−0.254
χ_{Lf}	−0.222	0.313
χ_{Fd}	0.231	−0.314
Hue	−0.118	0.224
Value	0.117	0.086
Chroma	0.242	−0.042
pH	−0.376	0.147
% Silt content	−0.376	0.147
% Sand content	0.374	−0.150

19.4.2 Geostatistical Analysis

Experimental variograms were computed using the principal components scores for sites throughout the study area (global region, Figure 19.6). These variograms effectively summarize the spatial structure of the properties that load heavily on the leading components (Oliver *et al.* 1989b). In common with other studies (Oliver and Webster 1987b), the first principal component in the present study explains most of the spatially dependent variation (Figure 19.6). Generally, higher-order components for the global region show decreasing variation and a reduction in the spatially dependent variation. These results suggest that properties that load heavily on principal component one (particle size, pH and organic matter content) are those exhibiting strong spatial control of the variation, whereas those properties associated with principal component two (bulk density, soil strength, magnetic susceptibility and ^{137}Cs content) will be only weakly spatially dependent.

The function that fitted the experimental semivariances of the first principal component scores for the global region was a periodic one with a power function. The variogram increases without bound suggesting that with increasing lags more sources of variation are incorporated. The full periodic model calculates the semivariances of the average lag distance $\gamma(h)$ from:

$$\gamma(h) = c_0 + c_1 \cos(2\gamma h/r) + c_2 \sin(2\gamma h/r) + c_3 h^\alpha \qquad (19.2)$$

where r is the wavelength; c_0 is the nugget variance; c_1 is the coefficient of the sine term; c_2 is the coefficient of the cosine term; and c_3 is the coefficient of the power

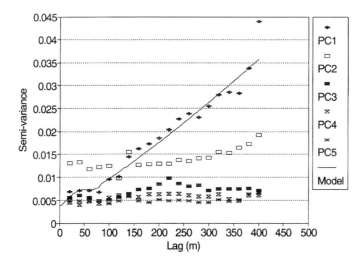

Figure 19.6 Variograms and models of all principal component scores from soil properties for the global region

term. The above model is authorized for kriging in only one dimension (Oliver *et al.* 1989b) but for kriging in two dimensions a damped periodic model can be used:

$$\gamma(h) = c_0 + c_1 \cos(2\gamma h/r)/h + c_2 \sin(2\gamma h/r)/h + c_3 h^\alpha \qquad (19.3)$$

This function suggests repetition in the variation and that spatial dependence does not decrease continuously with distance, as observations spaced far apart may be more similar than those which are more closely spaced. The distances at which the minima of cyclical variograms occur correspond to the wavelengths of pronounced repetitive elements (Olea 1977). Although few examples of periodicity are evident in the literature, Oliver *et al.* (1989b) fitted periodic models to variograms of slope angles of valley sides in Dartmoor where there were terraces separated by more steeply sloping sections.

The classification results suggested that there were two major regions. Since geostatistics is based on a model that assumes the spatial variation is continuous within a region, it is sensible to stratify the data for spatial analysis. The variogram of the first principal component for the plateau (Figure 19.7) suggests that this region is more variable spatially than the valley because the semivariances increase more rapidly at short lag distances and are also much larger than for the valley. This kind of variation was attributed by McBratney *et al.* (1991) to an abrupt change in properties associated with a distinct boundary. Hence, it seems that these variograms support the interpretation of the numerical classification that there is a distinct landform boundary between the plateau and valley. It is possible that the periodicity in the variograms of the valley reflects the ferricrete terraces in a similar way to that suggested by Oliver *et al.* (1989b). The wavelength of this variogram is much larger than that of the first principal component for the plateau (Figure 19.7), suggesting

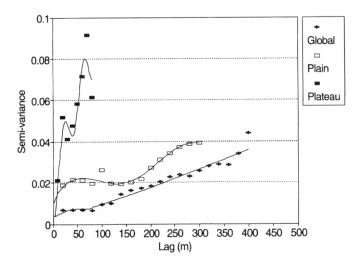

Figure 19.7 Variograms and models of the first principal component scores from soil properties for global, plateau and valley regions

that processes or controlling factors with different spatial scales are, or have been, in operation in each region.

The experimental variograms of relative elevation for the whole area and the two regions showed evidence of drift or trend, suggesting that there is some systematic variation with distance. Although the plateau region appeared to be approximately level, these results suggest that the plateau slopes continuously towards the valley, which accords with Thiery *et al.* (1995). In the presence of global drift or trend, several assumptions of the regionalized variable theory do not hold and the trend must be removed and the variogram estimated a second time from the residuals of a fitted polynomial (Olea 1977). The variograms of the residuals for relative elevation in the valley and plateau regions are shown in Figure 19.8. These are both cyclical but the best-fitting model for the valley was a damped periodic model with an exponential function, whilst for the plateau it was a periodic model with a power function. The differences in the wavelengths of these two variograms are considerable (Table 19.4).

Periodic functions provided the best fit to many of the variograms of the properties listed in Table 19.4. Those for which the wavelengths are given are spatially dependent, whilst the other variograms were pure nugget, i.e. there was no spatial dependence at the scale of the sampling. The wavelengths of the valley variograms are consistently larger than those of the plateau.

Since geostatistics is based on a model of continuous variation it is not sensible to disregard marked boundaries, especially for kriging. Ideally our area should be subdivided between each of the soil classes and these areas kriged separately. However, the regional variograms were likely to be unreliable because there were too few samples. We overcame this by computing a pooled within-class variogram (Voltz and Webster 1990) so that the entire area could be kriged using a single model.

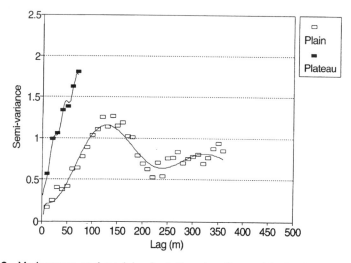

Figure 19.8 Variograms and models of relative elevation residuals after removal of global drift on plateau and valley regions

Table 19.4 Wavelengths of variograms computed for all properties in each region

Property	Global	Valley	Plateau	PWC
Component 1	*94*	*257*	*19*	–
Organic mater	266	*267*	*44*	411
Bulk density	165	*178*	–	–
^{137}Cs (Bq/kg)	*255*	149	42	380
Soil strength (5 cm)	266	*338*	–	–
Soil strength (10 cm)	157	192	–	–
Soil strength (15 cm)	178	156	–	–
Residual elevation	254	*190*	*19*	–
χ_{Lf}	56	*97*	*39*	–
Hue	*398*	*170*	32	*337*
pH	*414*	*107*	91	*414*
Clay content	221	281	*30*	210
Silt content	227	275	26	386
Sand content	181	273	*42*	–
Vegetation cover	*244*	340	33	*117*

– , Identifies properties that were not available or were not calculated.

Of the pooled within-class variograms, only pH and vegetation cover had any significant spatially dependent variation remaining within the classes. Thus, the PWC variograms for pH and vegetation were used for kriging. These results showed how effective the classification was in accounting for the spatial variation in all other soil properties. However, spatial dependence was present in the regional variograms for organic matter, ^{137}Cs, residual relative elevation, χ_{Lf} and sand content suggesting that continuous variation still remained within the regions. Thus, block-kriging (18.6 m) for each region separately was conducted, with the exception of ^{137}Cs which was not kriged in this way. The global variogram was used for block-kriging ^{137}Cs (18.6 m) because the sparsity of data in each region caused instability in the block-kriging equations. Furthermore, the estimates from block-kriging were deemed to be more accurate to predict values at unsampled sites than using the class mean values of the properties in the presence of continuous variation. The block-kriged estimates provided continuous data for RDA.

19.4.3 Redundancy Analysis (RDA)

The program CANOCO (ter Braak 1988) was used for redundancy analysis (RDA). The data were transformed to between 0 and 1 as before and tested for the appropriate type of RDA (Odeh *et al.* 1991). Ordinary (centred) RDA explained more variance in the transformed soil properties than standard RDA and the former was used for subsequent analyses.

Based on the results described earlier, RDA was carried out for the valley and plateau regions separately. To predict net soil flux, the analysis related net soil flux to topographic attributes alone, soil properties alone and then to topographic attributes and the soil properties combined. To select the most significant predictive variables the multicollinearity between these variables was reduced by a forward selection procedure which uses a Monte Carlo permutation test at each selection-step to test the statistical significance of the variance explained by each additional variable (ter Braak 1988). Table 19.5(a) shows that the topographic attributes explain more of the net soil flux variance than the soil properties, but that a combination of both topographic attributes and soil properties explains more net soil flux variance than either separately. This result accords well with that of others (Moore *et al.* 1993).

The intra-set correlations for variables ranked according to their explanation of variance during forward selection for the plateau and the valley are given in Table 19.5(b). The results confirm quantitatively the intuitive impression of soil deposition. Slope gradient explains most of the net soil erosion variation on the plateau, confirming the importance of surface wash in soil redistribution. Thus, as the slope gradient increases towards the edge of the plateau, net soil erosion increases. Soil bulk density and surface strength also reflect soil erosion here; the former increases and the latter decreases as net soil erosion increases. Among the other most important variables on the plateau is small-scale vegetation relief and small-scale

Table 19.5 (a) Ordinary RDA with forward selection removal of multicollinear variables to explain the net soil flux variance using topographic attributes, soil properties and a combination of both on data from plateau and valley regions. (b) Intra-set correlations and regression coefficients of variables predictive of net soil flux after removal of multicollinearity (ranked according to explanation of variance during forward selection)

(a)

	Net soil flux variance explained (%)			
	Topographic attributes	Soil properties	Combined	Collinear removed
Plateau RDA	34.6	25.6	56.3	52.8
Valley RDA	24.6	19.6	31.2	–

(b)

Plateau			Valley		
Environmental variables	Correlation	Coefficient	Environmental variables	Correlation	Coefficient
GRADNT	−0.4697	−0.37	VR1E	−0.4084	−0.717
bd	−0.3520	−0.27	csv1	0.2983	0.355
csv1	0.2379	0.38	bd	−0.1581	−0.449
clay	0.1929	0.48	GCATSWC	0.0923	0.230
VF1NE	−0.2668	−0.07	veg	0.1311	0.346
pH	0.2227	−0.68	F3E	−0.0460	−0.315
VR1NE	−0.3847	−1.00	PROFCV	−0.0386	−0.211
F3E	0.0533	0.42			
hue	0.0262	−0.28			
PROFCV	−0.0195	−0.27			

perturbation fetch with vegetation, both in a northeasterly direction (Table 19.5(b)). As vegetation fetch and vegetation relief increase, both in a northeasterly direction, net soil erosion increases. The inference is that decreased vegetation cover exposes soil and increased height of vegetation increases turbulence, increasing wind erosion on the plateau. Conversely, the presence of vegetation reduces the fetch and the decreased height of vegetation reduces the turbulence and forms a shelter against aeolian processes from the northeast, resulting in the deposition of aeolian material amongst the vegetation islands. The net soil flux patterns on the plateau do not appear to be associated with fetch and relief from an easterly direction, suggesting that the aeolian events preceding summer squalls are not important on the plateau. Instead, those events associated with the winter Harmattan from the northeast are dominant.

Vegetation relief in an easterly direction in the valley explains more of the variance in net soil flux than any soil property or topographic attribute (Table 19.5(b)). This suggests that aeolian activity is more important in the valley than on the plateau and that summer squalls are largely responsible. This also implies that vegetation cover is the controlling factor for exposure to aeolian activity in the valley rather than position in the landscape. The results suggest that net soil erosion increases as vegetation relief increases and that the converse is also true. Furthermore, as bulk density increases and soil strength decreases, net soil erosion increases. These changes in the surface structure of the soil must be due mainly to aeolian activity. Furthermore, the inclusion of soil-water content as an important property in the prediction of net soil flux and its propensity to decrease as soil erosion increases suggests that soil moisture limits soil erosion by wind. Profile curvature in the valley is also included as a significant property in the prediction of net soil flux. As it increases, net soil erosion increases, suggesting that slope form probably controls soil erosion by surface wash in the valley.

19.5 CONCLUSION

An heuristic–geomorphological model would separate the study area into plateau and valley. The numerical classification of the soil properties at each sampling site separated the study area similarly, indicating a strong relationship between *in situ* soil-forming and soil-redistribution processes. The only subdivision of the initial heuristic–geomorphology model would separate the *brousse tigrée* bands from the bare lanes in-between these bands on the plateau. The soil classification produced several distinct groups here which are probably closely related to the pattern of vegetation and may provide a more subtle distinction between the redistributed dust and *in situ* soil, than the intuitive model. The classification distinguished the less from the more badly eroded areas in the valley; the former a transition zone onto the plateau.

The variogram analyses confirmed this landscape classification and they also identified periodicity in the spatial variation of the soil properties which was not evident from other analyses. The differences between the variograms of the residuals for relative elevation for the two regions suggested that the relief has different origins. The wavelength of the residual relative elevation variogram on the plateau was similar to the average separation between the narrow vegetation islands and the wider bare surfaces, i.e. 40 m. Hence, it is probable that the periodicity reflects the accumulation of soil beneath vegetation and the loss of material where the surface is exposed to surface wash and aeolian activity. The periodicity in the valley appears to be related closely to the proximity of the underlying ferricrete layer to the soil surface, indicated by compact red exposed subsoil. The wavelength of the residual relative elevation variogram in the valley is close to the average distance between these areas where the change in gradient has probably increased wind and water erosion, i.e 150 m.

The results of RDA for the plateau supported the interpretation of the variograms. Net soil flux on the plateau is strongly correlated with the surface gradient, soil bulk density and strength, vegetation and northeasterly Harmattan winds and in the valley with easterly squalls, vegetation, soil bulk density and strength. This suggests that for the study area, aeolian processes active on the plateau are different from those in the valley, suggesting that the nature of the aeolian activity is related to topographic position. This is largely attributable to differences in vegetation cover. Vegetation cover decreases downslope with proximity to the village, suggesting that wind erosion and soil structure modification is linked to land use. The dominance of slope gradient in the RDA results for the slightly sloping plateau and its absence from the results for the steeper slopes of the valley suggests that surface wash and the consequent modification of soil structure is important on the plateau but not in the valley. This difference is probably a consequence of large infiltration rates in the valley and extensive soil crusting and low infiltration rates on the plateau, showing that topographic attributes alone are inadequate for predicting soil redistribution processes in this region.

ACKNOWLEDGEMENTS

The research for this chapter formed part of a study conducted whilst A.C. was in receipt of a NERC framework award (GT4/91/AAPS/35). A.C. would like to thank P. Kabat for enabling the research to be conducted under the auspices of HAPEX-Sahel, and R. Sutherland for encouragement to undertake the work. A.C. is also grateful to R. Webster for guidance and provision of software for the geostatistical analysis, and to L. Martz for providing the programs for calculating the topographic attributes. The GPS equipment (No. 396/0393) was generously loaned to A.C. by the NERC Geophysical Equipment Pool. The assistance from D. Walling and J. Grapes with the gamma-ray calibration is also appreciated. We are grateful for the technical support provided by the staff of the Geography Department Drawing Office at UCL.

REFERENCES

Boundy S.L. and Martz, L.W. 1988. *Measuring topographic variables from digital terrain models: a discussion and listing of four FORTRAN programs*, University of Saskatchewan, Department of Geography, Saskatchewan.

Casenave, A. and Valentin, C. 1989. *Les Etats De Surface De La Zone Sahelienne: Influence sur l'infiltration*. Editions de l'ORSTOM, Paris.

Chappell, A. 1995. Geostatistical mapping and ordination analysis of ^{137}Cs-derived net soil flux in south-west Niger. *PhD Thesis*, University of London.

Drees, L.R., Manu, A. and Wilding, L.P. 1993. Characteristics of aeolian dusts in Niger, West Africa. *Geoderma*, **59**, 213–233.

Higgitt, D.L. 1991. Soil erosion and soil problems. *Prog. in Phys. Geog.*, **15**(1), 91–100.

Lal, R. 1988. Soil erosion by wind and water: problems and prospects. In R. Lal (ed.), *Soil Erosion Research Methods*. Soil and Water Conservation Society, Ankeny, Iowa, USA, pp. 1–6.

Lapen, D.R. and Martz, L.W. 1993. The measurement of two simple topographic indices of wind sheltering-exposure from raster digital elevation models. *Computers and Geosciences*, **19**(6), 769–779.

Laslett, G.M., McBratney, A.B., Pahl, P.J. and Hutchinson, M.F. 1987. Comparison of several spatial prediction methods for soil pH. *J. Soil Sci.*, **38**, 325–341.

Lebel, T., Sauvegeot, H., Koepffner, M., Desbois, M., Guillot, B. and Hubert, P. 1992. Rainfall estimation in the Sahel: the EPSAT-Niger experiment. *Hydrol. Sci. J.*, **37**, 201–215.

Manu, A., Geiger, S.C., Pfordresher, A., Taylor-Powell, E., Mahamane, S., Ouattara, M., Isaaka, M., Salou, M., Juo, A.S.R., Puentes, R. and Wilding, L.P. 1991. Integrated management of agricultural watersheds (IMAW): characterisation of a research site near Hamdallaye, Niger. *TropSoils Bulletin* No. 91–103. Soil Management CRSP North Carolina State University, USA/S and CSD Texas A and M University, USA/INRAN Niamey, Niger/USAID Niamey, Niger.

Martz, L.W. and de Jong, E. 1988. CATCH: a Fortran program for measuring catchment area from digital elevation models. *Computers and Geosciences*, **14**, 627–640.

Martz, L.W. and de Jong, E. 1991. Using caesium-137 and landform classification to develop a net soil erosion budget for a small Canadian Prairie watershed. *Catena*, **18**, 289–308.

Matheron, M.A. 1971. *The Theory of Regionalized Variables and its Applications*. Cahiers du Centre de Morphologie Mathématique de Fountainebleau no. 5.

McBratney, A.B., Hart, G.A. and McGarry, D. 1991. The use of region partitioning to improve the representation of geostatistically mapped soil attributes. *J. Soil Sci.*, **42**, 513–532.

McHenry, J.R. and Ritchie, J.C. 1977. Physical chemical parameters affecting transport of caesium-137 in arid watersheds. *Wat. Res. Res.*, **13**(6), 923–927.

McTainsh, G.H. and Walker, P.H. 1982. Nature and distribution of Harmattan dust. *Z. Geomorph. NF*, **26**, 417–435.

Moore, I.D., Gessler, P.E., Nielsen, G.A. and Peterson, G.A. 1993. Soil attribute prediction using terrain analysis. *Soil Sci. Soc. Am. J.*, **57**, 443–452.

Moore, I.D., Grayson, R.B. and Ladson, A.R. 1991. Digital terrain modelling: a review of hydrological, geomorphological and biological applications. *Hydrological Processes*, **5**, 3–30.

Odeh, I.O.A., Chittleborough, D.J. and McBratney, A.B. 1991. Elucidation of soil–landform interrelationships by canonical ordination analysis. *Geoderma*, **49**, 1–32.

Olea, R.A. 1977. *Measuring Spatial Dependence with Semi-Variograms*. Series on Spatial Analysis no. 3. Kansas Geological Survey, Lawrence.

Oliver, M.A. and Webster, R. 1987a. The elucidation of soil pattern in the Wye Forest of the West Midlands, England. I Multivariate distribution. *J. Soil. Sci.*, **38**, 279–291.

Oliver, M.A. and Webster, R. 1987b. The elucidation of soil pattern in the Wye Forest of the West Midlands, England. II Spatial distribution. *J. Soil. Sci.*, **38**, 293–307.

Oliver, M., Webster, R. and Gerrard, J. 1989a. Geostatistics in physical geography. Part I: theory. *Trans. Inst. Br. Geogr.*, **14**, 259–269.

Oliver, M., Webster, R. and Gerrard, J. 1989b. Geostatistics in physical geography. Part II: applications. *Trans. Inst. Br. Geogr.*, **14**, 270–286.

Prince, S., Kerr, Y.H., Goutorbe, J.-P., Lebel, T., Tinga, A., Bessemoulin, P., Brouwer, J., Dolman, A.J., Engman, E.T., Gash, J.H.C., Hoepffner, M., Kabat, P., Monteny, B., Said, F., Sellers, P. and Wallace, J.S. 1995. Geographical, biological and remote sensing aspects of the Hydrologic Atmospheric Pilot Experiment in the Sahel (HAPEX-Sahel). *Remote Sens. Environ.*, **51**, 215–234.

Ritchie, J.C. and McHenry, J.R. 1990. Application of radioactive fallout cesium-137 for measuring soil erosion and sediment accumulation rates and patterns: A review. *J. Environ. Qual.*, **19**, 215–233.

Sutherland, R.A. 1994. Spatial variability of ^{137}Cs and the influence of sampling on estimates of sediment redistribution. *Catena*, **21**, 57–71.

Sutherland, R.A. and de Jong, E. 1990. Estimation of sediment redistribution within agricultural fields using caesium-137, Crystal Springs, Saskatchewan, Canada. *Applied Geography*, **10**, 205–221.

Ter Braak, C.J.F. 1988. *CANOCO: A FORTRAN Program for Canonical Community Ordination by [partial] [detrended] [canonical] Correspondence Analysis and Redundancy Analysis (version 2.1).* Agricultural Mathematics Group, Wageningen, The Netherlands.

Thiéry, J., d'Herbès, J.M. and Valentin, C. 1995. A model simulating the genesis of banded vegetation pattern in Niger. *J. Ecology*, **83**, 497–507.

Voltz, M. and Webster, R. 1990. A comparison of kriging, cubic splines and classification for predicting soil properties from sample information. *J. Soil Sci.*, **41**, 473–490.

Walling, D.E. and Quine, T.A. 1990. Calibration of caesium-137 measurements to provide quantitative erosion rate data. *Land Degradation and Rehabilitation*, **2**, 161–175.

Walling, D.E. and Quine, T.A. 1991. The use of caesium-137 to investigate soil erosion on arable fields in the UK – potential applications and limitations. *J. of Soil Sci.*, **42**, 146–165.

Webster, R. and Oliver, M.A. 1992. Sample adequately to estimate variograms of soil properties. *J. Soil Sci.*, **43**, 177–192.

Zhang, X., Higgitt, D.L. and Walling, D.E. 1990. A preliminary assessment of the potential for using caesium-137 to estimate rates of soil erosion in the Loess Plateau of China. *Hydrological Sciences Journal*, **35**, 267–276.

20 Variations in Tree Communities and Soils with Slope in Gallery Forest, Federal District, Brazil

MANOEL C. DA SILVA JÚNIOR,[1] PETER A. FURLEY[2] and
JAMES A. RATTER[3]
[1] Departamento de Engenharia Florestal, Universidade de Brasília, Brazil
[2] Department of Geography, University of Edinburgh, UK
[3] Royal Botanic Garden, Edinburgh, UK

20.1 INTRODUCTION

20.1.1 Context

While the role of topography in determining the character of soil properties at a landscape or hillslope scale is well known, the microscale is less understood. Equally, although the general principles of soil–plant relationships have been well established, the reciprocal effects between soils and plant communities and their dual relationship with slope and drainage have attracted little attention, at least in the humid tropics. Detailed phytosociological research is now revealing subtle relationships between plant associations and their edaphic environment. Until recently, gallery forests have been considered as relatively homogeneous forest communities, lying in valley footslope sites within savanna regions and consistent with permanently damp soils. Floristic variations within the galleries were not studied in detail until the late 1980s and very little work has examined the environmental processes responsible.

Contemporary research on Brazilian gallery forests indicates much greater floristic and phytosociological variety than was previously realized and poses questions concerning the origin, development and controlling factors of such communities. Although slope has been suggested as a determining factor in the distribution of gallery forest vegetation world-wide (Newberry and Proctor 1984; Baillie *et al.* 1987; Burke *et al.* 1989; Oliveira-Filho *et al.* 1990, 1994a,b; Furley 1992; Schiavini 1992; Felfili 1993; Bendix 1994; Ramos 1995; Silva Júnior 1995), the precise relationship between the pattern of tree species and their environment has not been explored. The differing soil moisture regimes over slopes appear to act selectively in species establishment and success, resulting in identifiable communities. These communities interact with and transform the soil environment, producing a mutually dependent equilibrium where cause and effect become difficult to distinguish.

Advances in Hillslope Processes, Volume 1. Edited by M. G. Anderson and S. M. Brooks.
© 1996 John Wiley & Sons Ltd.

The present investigation is a detailed, comparative study of communities in the Pitoco gallery forest of the IBGE Ecological Reserve (RECOR) near Brasília, close to the centre of the savanna (*cerrado*) formation in Brazil (Figure 20.1). The main reason for the selection of this site was its distinct floristic composition and also its protection from major disturbance for at least 20 years.

The approach in this study is conceived as a sequence of complementary analyses (classification and ordination) to investigate the association of particular floristic communities with differing environmental conditions. In particular, it aims to assess the extent to which slope and drainage may account for the spatial patterns of the gallery communities.

Figure 20.1 Aerial photograph of the RECOR-IBGE ecological reserve showing the Pitoco gallery forest (1) and related galleries

20.1.2 Gallery Forests of the *Cerrado*

The *cerrado* biome of central Brazil covers a landscape of 2 million km², dominated by tree and shrub savanna but with gallery forests distributed in a dendritic pattern along watercourses. Such galleries occur where surface and subsurface water supply is sufficiently constant to allow maintenance of forest vegetation. They tend to be limited upslope by seasonally drier conditions, where they are normally replaced by a band of grassland (usually rich in sedges) which is saturated in the wet season but can resist the fall in the water-table that takes place during the dry season. The *cerrado* vegetation requires soils that are freely drained throughout the year and is therefore characteristic of slopes lying above the gallery-fringing grassland.

It is likely that galleries have been the sites (refugia) maintaining forest species during the drier and colder periods of the glacial maxima in the Pleistocene. During this time most of the present areas of tropical forest are presumed to have resembled today's savanna landscape (Whitmore and Prance 1987; Meave and Kellman 1994; Oliveira-Filho and Ratter 1995). The gallery forests of the *cerrado* biome are of diverse floristic origin, with relationship to Amazonian and Atlantic forests (*sensu lato*) and, to some extent, *cerradão* (savanna woodland) and central Brazilian semi-deciduous forest, thus producing an outstanding richness of tree species (Pires 1984; Prance 1987; Oliveira-Filho and Ratter 1995).

The relationship of some gallery species and communities to distinct subhabitats has been suggested, but quantitative analysis of this has been attempted in very few studies (Schiavini 1992; Felfili 1993; Oliveira-Filho *et al.* 1994a and b; Silva Júnior 1995). Detecting the constituent plant communities of the galleries, their associated environments and the dynamic processes which affect them, should provide a better understanding of the spatial and temporal patterns of this very complex vegetation formation and also generate information vital to their management.

20.2 SITE, MATERIAL AND METHODS

20.2.1 The Pitoco Valley

The Pitoco gallery lies in the northeastern section of the RECOR-IBGE ecological reserve, with most of the area occupied by well-drained Cambisols (mostly Inceptisols, US Soil Taxonomy). It is wider at the stream head (160 m) and narrows downstream (to 120 m or less), where the forest is concentrated in a narrow zone at the foot of the steep valley. It then joins a second gallery (the Monjolo), uniting to form the Roncador Stream, and picks up a further major tributary gallery (the Taquara) further downstream, eventually flowing into Lago Paranoá (Figure 20.1).

20.2.2 Floristic Composition and Phytosociology

The Point-Centred-Quarter (PCQ) method (Cottam and Curtis 1956) was used to survey trees with DBH $\geq$ 5 cm. Grids of 250 sampling points, spaced at 10 m intervals from the riverbank to the forest edge along sampling lines 10 m apart were sited in each study area. Figure 20.2 shows the disposition of sampling along the Pitoco Stream.

454

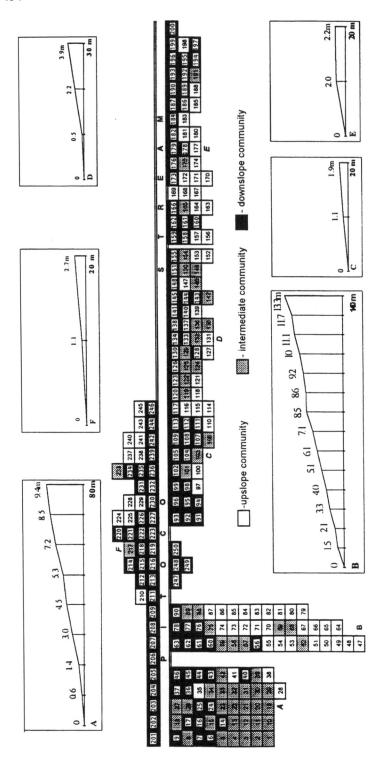

Figure 20.2 Tree communities and elevation profiles for the Pitoco gallery forest

Species vouchers were deposited in the herbaria of the IBGE Ecological Reserve (RECOR) and the Royal Botanic Garden, Edinburgh (E).

Field data were processed by means of phytosociological analysis (Mueller-Dombois and Ellenberg 1974). Diversity was assessed using the Shannon and Wiener Index (H').

20.2.3 Hierarchical Classification

A hierarchical classification (TWINSPAN, Hill 1979), based on species presence and absence, was applied to seek patterns in the vegetation data that could be meaningfully related to the environmental features of the gallery. A phytosociological analysis of the categories defined by this classification was then carried out to characterize their floristic composition, species density and species dominance (basal area). Species showing increases within a community of more than 50% of that scored in the total sampling are indicated as "related" to it.

20.2.4 Soil Properties

Mixed soil samples (0–10 cm) were collected for each of the sites and grouped according to the classification of vegetation communities (shown in Figure 20.2).

Chemical and physical analyses were carried out at the soil laboratory of EMBRAPA/CPAC following the procedures specified by EMBRAPA (1979). Analyses were made of $pH(H_2O)$, Al, H + Al, Ca + Mg, Ca, P, K, Mn, Cu, Zn, Fe, organic matter and the percentages of sand, silt and clay. From these results, Al saturation (%), cation exchange capacity (CEC), total base saturation (TEB%) and the percentage of base saturation (V) were calculated.

Soil pH was measured with a potentiometer in a 1:2.5 soil-water suspension. Available P, K, Fe, Mn, Zn and Cu were determined in soil extracts of a di-acid mixture (0.05 N HCl + 0.025 N H_2SO_4). Exchangeable Ca + Mg, and exchangeable Al were determined in soil extracts of 1 N KCl solution; Al was titrated with 1 N NaOH solution while Ca + Mg and Ca were separately titrated with 0.025 N EDTA. Total acidity (H + Al) was obtained using a 1 N calcium acetate extracting solution at pH 7.0. Organic carbon was determined by the Tuirin method and the percentage of organic matter was calculated using the Bremelem constant (× 1.724). Texture was obtained by the Bouyoucos hydrometer method.

20.3 THE VEGETATION AND ENVIRONMENTAL RELATIONSHIPS

The computer program CANOCO, version 3.1, was used to run a detrended correspondence canonical analysis (DCCA) of the data sets. This ordination incorporates correlation and regression within the ordination analysis itself, thus summarizing their variability (Kent and Cocker 1992). The detrended version of the analyses is intended to remove the "arch effect", which represents the second axis of the ordination as a quadratic distortion of the first (Gauch 1982).

20.3.1 Species Data Set

The basal area per groups of sampling points (upslope dry, intermediate and downslope wet (Figure 20.2) was calculated for each of 33 species with $\geqslant 10$ individuals in the overall gallery forest survey. Species with less than 10 individuals were eliminated as they have little or no influence on the results (Gauch 1982; Causton 1988).

20.3.2 Environmental Data Set

Elevation from stream to the forest–savanna transition (Figure 20.2) was measured giving an indirect assessment of the level of the water-table, whilst soil water status was recorded descriptively at profile and sampling sites.

Three nominal variables were included in the analysis: the "upslope dry", "intermediate" and "downslope wet" communities. Their inclusion provided a meaningful grouping of species and samples, once they were categorized in accordance with TWINSPAN community classification.

Soil data samples were logarithmically transformed for normalization purpose, and standardized (zero mean and unit variance), thereby giving equal weights to all of them (Noy-Meir *et al.* 1975; Newbery and Proctor 1984; James and McCulloch 1990). The environmental variables that were found to be highly correlated with each other were eliminated from the analysis as indicated by a high variance inflation factor (VIF) (Ter Braak 1988). The CANOCO's Monte Carlo test of significance was run to test whether or not the canonical axes were significantly related to the environmental variables (Ter Braak 1988).

In the diagrams yielded from the DCCA, species and samples were represented as points and environmental factors as arrows indicating the direction of maximum change. The arrow lengths are proportional to the correlation between the variables and that axis (Ter Braak and Prentice 1988). To clarify the results, the ordination of species, samples and environmental variables are presented separately. TWINSPAN community classification is labelled in the species and sample diagrams to show community trends along the gradients.

20.4 RESULTS

20.4.1 Floristic Composition and Phytosociology

The total number of species recorded is 99 in an estimated area of 5.2 ha, with an estimated density of 1914 trees/ha and basal area of 37.7 m²/ha. The curve for species versus number of sampling points (Mueller-Dombois and Ellenberg 1974) suggests that 250 sampling points are sufficient to provide a representative floristic survey and also give adequate data for estimating density and basal area of the more important species (Silva Júnior 1995).

The Shannon & Wiener diversity index is calculated as 3.86. Generally the values of this index are higher for gallery forests than for *cerrado* (*sensu stricto*) (Felfili and Silva Júnior 1993) or *cerradão* (*cerrado* woodland) (Felfili and Silva Júnior 1992), the neighbouring communities in the *cerrado* biome. The present value is comparable with those for Amazonian and Atlantic rain forest which range from 3.7 to 4.3 (Silva and Shepherd 1986).

20.4.2 The Gallery Forest Communities

A broad band of forest up to 170 m across occurs close to the stream head, where the bed is 0.5 to 1.0 m deep and the inclination of the slope is 9.5 to 11.8%. As one goes downstream, the riverbed becomes shallower but the adjacent slopes become steeper (11 to 13.5%) and the gallery much narrower (c.40 m). As the topography of the valley becomes progressively steeper, the beginning of the "dry upslope" community comes as close as 30 m to the stream margin.

TWINSPANs first division of sites revealed a major pattern related to elevation. There is a clear separation into a wet downslope community occurring close to the stream bank and a dry community further upslope. The downslope and upslope communities show a very dissimilar floristic composition, and therefore a further TWINSPAN division was run to try to discover other vegetation groups that could be associated with site features. This division failed to reveal any meaningful subgroups in the downslope communities, but yielded two usable classes in the dry community. These subcommunities are related to slope and distance from the stream bank and are designated as intermediate and dry upslope.

20.4.3 The Dry Upslope Community and Constituent Species

Table 20.1 presents density and basal area values for the most important species of the dry upslope, intermediate and wet downslope communities. Strong habitat preferences are shown and only one species, *Tapirira guianensis* (a well known and geographically very widespread "generalist"), occurs in all three communities. *Callisthene major*, which holds the first position in the Importance Value Index for the whole sampling (Silva Júnior 1995), is confined to the upslope and intermediate habitats, showing only 6.6% of the density and 14% of the basal area in the intermediate as compared to the upslope community. *Copaifera langsdorffii*, the species with the highest basal area in the total sampling, is confined to the intermediate community, while *Protium almecega* and *Pseudolmedia guaranitica* are exclusive to the downslope community and show the highest species densities there. Table 20.1 demonstrates that all species show strong community preferences.

20.4.4 The Wet Downslope Community and Constituent Species

The most striking finding is the predominance of *Protium almecega*, which is ranked as the third most important species in the total sampling. However, in the downslope community, its density and basal area scores are 2.5 times higher than in the total, with no observations at all recorded for the dry community. Scores for other species related to the community are illustrated in Table 20.1.

The differences in species density and basal area within the Pitoco gallery are considered as indicative of significant environmental preferences, characterizing distinct plant communities. Topography and consequent soil moisture differences

Table 20.1 Variation in density and basal area values for species of the communities upslope, intermediate and downslope. Scores of the total sampling are also displayed for comparisons

Species/communities	Density (n/ha)				Basal areas (m²/ha)			
	Upslope	Intermediate	Downslope	Total	Upslope	Intermediate	Downslope	Total
Callisthene major Mart. (Call Majo)	362	24	–	122	10.1	1.42	–	3.54
Lamanonia ternata Vell. (Lama tem)	82	–	–	31	4.62	–	–	2.14
Pera glabrata Poepp. ex. Baill. (Pera glab)	68	–	–	31	0.74	–	–	0.32
Platypodium elegans Vog. (Plat eleg)	61	12	–	24	0.61	0.15	–	0.28
Guettarda viburnioides Cham. & Schecht. (Guet vibu)	55	24	–	24	0.53	0.28	–	0.19
Eriotheca pubescens Schott. & Endl. (Erio pube)	75	–	–	22	–	–	–	–
Faramea cyanea Muell. Arg. (Fara cyan)	7	289	–	91	–	–	–	–
Copaifera langsdorffii Desf. (Copa lang)	–	241	–	55	–	15.9	–	3.71
Jacaranda puberula Cham. (Jaca pube)	14	253	–	47	0.02	1.94	–	0.65
Inga alba Willd. (Inga alba)	–	96	–	22	–	–	–	–
Matayba guianensis Aublet. (Mata guia)	27	48	–	28	0.2	0.37	–	0.22
Bauhinia rufa Steud (Bauh rufa)	41	48	–	22	0.35	0.56	–	0.21
Protium almecega Marchand. (Prot alme)	–	0	266	106	–	–	4.99	1.92
Tapirira guianensis Aublet. (Tapi guia)	61	60	255	126	–	2.66	4.03	2.14
Pseudolmedia guaranitica Hassl. (Pseu guar)	–	0	266	100	–	–	2.7	1
Emmotum nitens Miers. (Emmo nite)	7	0	90	45	–	–	4.14	1.87
Ocotea aciphylla Mez. (Ocot acip)	–	12	80	31	–	0.53	1.18	0.45
Licania apetala (E. Mey.) Fritsch. (Lica apet)	–	24	53	28	–	–	–	–
Virola sebifera Aublet (Viro sebi)	20	0	58	35	–	–	0.49	0.3

seem to be the most important factors in accounting for the floristic and structural variation. This is explored further in the later sections of this chapter.

20.4.5 Soil Variation

The average values, standard deviations, coefficients of variation and t-tests, from the soils analyses are presented in Table 20.2.

Table 20.2 Average values (Av), standard deviations (sd), coefficients of variance (CV) and t-tests for soil samples (0–10 cm) of the Pitoco gallery forest communities. CS=coarse sand and FS=fine sand

	pH (H2O) / Zn ppm	OM (%) / Mn	Al / CEC	H+Al / TEB cmol/Kg	Ca / V	Mg / Al sat	Ca+Mg / Clay	K / Silt	P / CS	Fe / FS	Cu / Texture
						cmol/Kg		%	%		
Intermediate community											
Av	4.7	10.05	2.50	19.06	0.32	0.66	0.98	0.27	2.90	110.32	0.77
	2.30	17.73	20.31	1.25	15.8	328.09	59.90	20.80	4.80	14.50	
sd	0.3	2.96	0.99	4.18	0.31	0.49	0.78	0.15	0.84	39.98	0.16
	0.81	16.90	4.38	0.81	4.83	123.81	6.38	3.26	1.48	3.54	
CV (%)	5.9	29.5	39.7	21.9	94.4	74.2	79.0	54.8	28.9	36.2	20.4
	35.4	95.3	21.60	64.80	30.5	37.7	10.7	15.7	30.7	24.4	
Downslope wet community											
Av	4.7	12.66	2.96	21.70	0.19	0.37	0.57	0.25	4.17	137.97	0.96
	2.93	5.43	22.51	0.82	12.5	397.76	49.39	19.00	9.87	21.74	
sd	0.2	4.75	1.05	5.88	0.08	0.13	0.14	0.10	1.77	87.45	0.34
	1.01	3.23	5.99	0.17	3.22	133.55	15.67	3.95	3.95	9.27	
CV (%)	4.7	37.5	35.5	27.1	39.8	35.7	25.1	41.7	42.5	63.4	34.9
	34.5	59.4	26.60	20.70	25.7	33.6	31.7	20.8	40.1	42.6	
Upslope dry community											
Av	5.0	10.38	1.93	18.03	0.91	0.98	1.89	0.29	2.54	89.77	0.80
	2.87	37.39	20.21	2.18	21.3	246.87	57.48	18.96	6.96	16.60	
sd	0.3	3.25	1.34	5.37	1.22	0.74	1.89	0.12	1.45	43.19	0.36
	1.50	30.80	5.04	1.91	9.26	178.11	10.61	3.37	4.12	6.66	
CV(%)	6.7	31.3	69.1	29.8	134.0	74.9	100.1	41.2	56.9	48.1	45.4
	52.4	82.4	24.80	87.60	43.4	72.1	18.5	17.8	59.2	40.1	
Totals for all gallery communities											
Av	4.8	11.23	2.44	19.66	0.52	0.69	1.21	0.27	3.25	112.43	0.86
	2.79	21.33	21.14	5.37	16.5	320.71	54.69	19.29	7.74	18.28	
sd	0.3	3.99	1.25	5.58	0.87	0.59	1.41	0.12	1.67	66.93	0.33
	1.23	25.81	5.37	1.43	11.1	165.70	12.97	3.60	6.37	7.90	
CV (%)	6.6	35.5	51.1	28.4	166.3	86.4	116.7	43.6	51.4	59.5	38.6
	43.9	121.0	25.40	26.60	33.2	51.7	23.7	18.7	82.3	43.2	

T-tests
I/U. 1%: pH; 5%: Zn, Mn, V, Al saturation, silt, coarse sand.
I/D. 1%: Mg, Mn, clay, coarse sand, fine sand; 5%: organic matter, Ca, P, Cu, Zn, V.
U/D. 1%: pH, Al, Ca, Mg, Ca+Mg, P, Mn, TEB, V, Al saturation; 5%: organic matter, H+Al, Fe, clay, fine sand.
I=Intermediate community, U=upslope dry community, D=downslope wet community.

The pH results range from 4.3 to 5.7 and average 4.8. The majority of the samples (67.2%) are classified as very acid (<5.0). The standard deviation for the averages give coefficients of variation lower than 9% for all the soil pHs. The dry community soils always display significantly ($P<0.01$) higher pHs when compared with the downslope wet community soils, showing a clear distinction between the Pitoco constituent communities.

All the soil samples are considered to have high levels of organic matter, ranging from 5.79 to 23.41% ($\chi=11.23$). The OM standard deviations give coefficients of variation around 30% of the average. The downslope wet community predictably shows significantly ($P<0.05$) higher levels than the drier upslope communities.

Only 3.4% of the samples show low levels of exchangeable Al, which has an average of 2.43 cmol/kg. Aluminium saturation is high in 98.3% of the samples, of which 96.5% show an allic character ($>50\%$ of Al saturation). Total acidity is classified as high for all soil samples. The coefficients of variation are high, suggesting care in the interpretation of the differences between average values. The soils of the downslope wet gallery community consistently have Al levels higher than those of the dry communities. The dry communities always show significantly ($P<0.01$) lower total acidity and Al saturation levels. The average standard deviation results in coefficients of variation around 30% for H + Al and Al saturation.

Six Al-accumulator species are ranked amongst the 10 species of highest IVI. They comprise *Callisthene major*, *Protium almecega*, *Copaifera langsdorfii*, *Pseudolmedia guaranitica*, *Faramea cyanea* and *Emmotum nitens*. This suggests a direct relationship between soil Al levels and the occurrence of Al-accumulator species, as also indicated by Silva (1991).

Among the samples, 89.6% have very low levels of Ca + Mg (<3.0 cmol/kg) with an average of 1.2. Calcium represents, in general, 43% of the Ca + Mg scores but is consistently found at very low levels ($\chi=0.52$). Standard deviation of the Ca and Mg averages indicate high variability ($>80\%$) among the individual samples. Dry community soils have significantly higher ($P<0.01$) Ca levels when compared with the downslope wet community. Magnesium levels in the communities follow the same patterns found in the Ca analysis.

The Ca and Mg levels in each community may be interpreted as a gradient, ranging from the downslope community with the lowest levels, to the upslope community with the highest levels. Although it was not pursued further during the present study, this provides a field experiment where tree species could have their performance evaluated.

Results for K range from 0.11 to 0.61 cmol/kg. Individual samples are variable, resulting in high coefficients of variance (43.6%). As a result, the different communities identified in the forest show no statistical differences in their K content.

Phosphorus levels range from 1.3 to 7.8 ppm ($\chi=3.25$). Variability between individual samples results in high standard deviations (51.4%). Downslope wet community soils show significantly higher P levels than those of the upslope communities. The range of values found in the communities do not suggest that P is a limiting factor in the distribution of vegetation. However, P is one of the main constraints to crop growth in the *cerrado* region and presumably also affects natural vegetation.

Copper results for the gallery forest soils range from 0.3 to 1.6 ppm ($\chi = 0.86$). No significant differences are found for Cu levels in the comparisons between the soils of the constituent gallery communities.

The results for Fe demonstrate a wide range of levels from 19.2 to 418.1 ppm ($\chi = 112.4$). In this case the downslope wet community soils have significantly higher averages than the upslope dry communities.

The results indicate generally high levels of Mn ranging from 1.58 to 110.5 ppm ($\chi = 21.3$). The dry community soils always have significantly higher Mn levels than the wet communities.

The Zn values range from 1.19 to 7.95 ppm ($\chi = 2.79$). All the results are above the proposed critical level of 1 ppm (Lopes and Cox 1977). Soils of the dry upslope community have significantly higher Zn levels than soils of the downslope community.

Results for cation exchange capacity (CEC) show a variation of 13.8–36.7 cmol/kg with an average of 23.6 cmol/kg. Standard deviations give coefficients of variation between 20 and 30% for the soils of the forest community as a whole. Tests of significance do not show differences between most of the soil samples. This reflects the high H and Al levels found in the majority of the soil samples which are classified as of high CEC (>13.3 cmol/kg). These results indicate that the soils are better provided for CEC than the average values for the *cerrado*.

The results for total exchangeable bases (TEB) indicate a range from 1.6 to 11.9 with an average of 3.9 cmol/kg. Most of the samples (70.7%) have medium levels (2.62–6.3). Standard deviation for the average values give 27% of coefficient of variance. The upslope dry community soils has significantly higher TEB levels than those for the downslope wet communities. The base saturation (V) scores of up to 50% denote that all soils are dystrophic.

Most of the soil samples (90%) are classified as clay-rich (≥35%). Individual levels vary considerably, giving high standard deviations for the average values. Silt contents range from 11 to 26%. Standard deviation results in coefficients of variation lower than 20% taking all the soil samples into account. The coarse sand contents were significantly higher in soils of the downslope community. The content of fine sand for most of the soils is of similar level.

To summarize, soil properties associated with each constituent community of the gallery forest are found to be significantly different for most of the soil variables examined.

The wet downslope community soils shows the highest OM, Al, H + Al, P, Fe, Cu, Zn, CEC, Al saturation and fine and coarse sand contents, intermediate levels for pH and silt, and the lowest Ca, Mg, K, Mn, TEB, V and clay contents. Ross (1989) indicated that the much lower CO_2 solubility in moist soils results in the formation of acid compounds, thus reducing pH. Temperature is also reduced because of the cooling effect of the water. Consequently microbial activity is also reduced and OM accumulation might be predicted. Under wetter, more acid and OM-richer soil conditions, the solubility of Al and micronutrients is expected to be higher (Brady 1990). All of the soils are hydromorphic and frequently gleyed or highly reduced.

The dry upslope community soils show the highest pH, Ca, Mg, K, Mn, TEB and V, intermediate levels for OM, Cu, Zn, CEC, clay, coarse and fine sand, and the lowest Al, H + Al, P, Fe, Al saturation, and silt. Most profiles described in these soils (Silva Júnior 1995) are regarded as Cambisols in which weatherable material is still present in B horizons and higher base saturation is expected because of the relative youth of the soils which are constantly renewed due to slope. However, it is worth mentioning that most of these soils in the Federal District are derived from metamorphic bedrock which produces acid soils of low nutrient status with time (Adamoli *et al.* 1985; Haridasan 1991).

Analysis of the soils of the intermediate community indicates levels which lie in between the upper and lower slope sites for 11 out of the 21 properties analysed, confirming their position on a progressive gradient of soil properties from the stream margins to the forest–*cerrado* border. Its soils have the highest clay and silt contents, intermediate Al, H + Al, Ca, Mg, K, P, Fe, Mn, TEB, V and Al saturation, and the lowest pH, OM, Cu, Zn, CEC, coarse and fine sand levels.

20.4.6   Ordination of Plant Species and Environmental Variables (DCCA)

The variable loadings on the ordination axes are shown in Table 20.3. The analysis resulted in eigenvalues for the axes of 0.472, 0.178, 0.098 and 0.052 respectively. The cumulative percentage variance accounted for by the axes are respectively 10.8, 14.8, 17.1 and 18.3% for species data, and 33.6, 51.0, 58.9 and 62.3% for the species–environment relations. The sum of all canonical eigenvalues is 1.408. Axes III and IV explain only a little of the variance and are not discussed further here. These results indicate that the environmental variables considered in the analysis are apparently sufficient to explain most of the species' basal area variation within the gallery forest samples. These values of cumulative percentage of variance normally indicate that a variety of other environmental features not included in the analysis may be involved in determining the species distribution patterns.

The Monte Carlo test for the first two axes shows that species basal area distributions are significantly ($p > 0.01$) related to the environmental variables included in the analysis, indicating differences in the species basal area among samples.

20.4.7   The Environmental Gradients and Plant Distribution

Figure 20.3 illustrates the environmental variables selected by showing the highest correlations with axes I and II. Table 20.3 shows the weighted correlations between the environmental variables, the species and environmental axes.

The first ordination axis is highly and positively correlated with the nominal variable "stream" effect, indicating a high water-table influence, and also with the highest P, OM, Al, H + Al and Fe levels. The most prominent negative correlations are found with the higher elevation and forest–*cerrado* border effect, both indicating soils less affected by high water-table, and also with higher Mn, Mg and clay contents.

Axis I is thus interpreted as representing a topographic–moisture gradient along which a strongly marked soil fertility and textural gradient is also inferred, ranging

Table 20.3 Detrended canonical correspondence analysis (DCCA) for the Pitoco gallery forest: matrix of weighted correlations between the species axis (SP) and environmental axis (EN) and the environmental variables

	SP AX1	SP AX2	SP AX3	SP AX4	EN AX1	EN AX2	EN AX3	EN AX4
SP AX1	1							
SP AX2	−0.0299	1						
SP AX3	0.0266	−0.009	1					
SP AX4	0.0738	0.1164	−0.1315	1				
EN AX1	0.9429	−0.0192	−0.0142	0.0576	1			
EN AX2	−0.0236	0.7671	−0.0653	0.0708	−0.025	1		
EN AX3	−0.019	−0.0708	0.7079	−0.1378	−0.0201	−0.0923	1	
EN AX4	0.0909	0.0909	−0.1634	0.5972	0.0964	0.1185	−0.2308	1
OM	0.5289	0.2936	−0.1608	−0.0298	0.5609	0.3827	−0.2272	−0.0498
Al	0.5909	0.0098	0.1697	−0.072	0.6267	0.0128	0.2398	−0.1205
H+Al	0.5435	0.1925	−0.1791	−0.1078	0.5764	0.251	−0.2531	−0.1805
Mg	−0.4994	0.0753	−0.0426	−0.2197	−0.5296	0.0982	−0.0601	−0.368
P	0.595	−0.0811	−0.0532	−0.2269	0.631	−0.1057	−0.0752	−0.38
Fe	0.5612	0.2835	−0.2299	0.0751	0.5952	0.3695	−0.3248	0.1257
Mn	−0.6189	0.1648	−0.0667	−0.1778	−0.6563	0.2148	−0.0942	−0.2978
Clay	−0.5159	−0.091	−0.0014	0.2311	−0.5471	−0.1187	−0.002	0.387
Elevation	−0.756	−0.054	−0.087	−0.0643	−0.8017	−0.0704	−0.123	−0.1077
Border Zone	−0.6808	0.3904	0.1586	−0.1951	−0.7221	0.5089	0.2241	−0.3267
Stream Zone	0.8397	−0.0812	−0.0005	−0.0067	0.8906	−0.1059	−0.0008	−0.0113

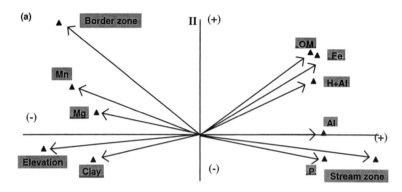

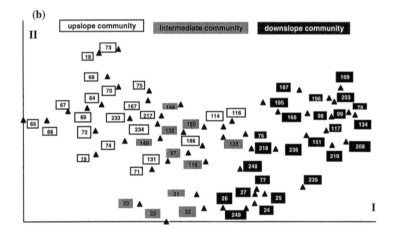

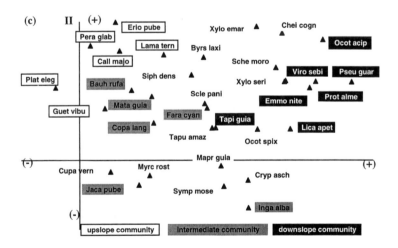

from the stream margins to the forest–*cerrado* border. Correlation values with the secondary environmental gradient are lower, indicating weak relationships.

Differences within the total population of soil samples have been picked out earlier. Soil variables detected by the DCCA analysis to distinguish species and samples, are significantly different between the wet downslope and dry upslope communities but are not significantly different between the intermediate and dry upslope communities (Table 20.2).

In Figure 20.3, samples are ordinated as a continuum along the axis, indicating that the sampling design covers the whole range of the environmental variation. It can also be clearly observed that the samples from upslope and downslope communities are separate, forming distinguishable groups at either end of the environmental gradient provided by axis I. The intermediate community samples are situated between the extreme groups.

In Figure 20.3 there is a clear definition of groups of species related to the upslope and downslope communities, which are ordinated at the end points of axis I. At the dry end, species such as *Callisthene major*, *Eriotheca pubescens*, *Guettarda viburnioides*, *Lamanonia ternata*, *Pera glabrata* and *Platypodium elegans* seem to avoid soil saturation, at least in the superficial soil layer, and exhibit their largest basal areas in the drier, Mg-, Mn- and clay-rich soils. Species of this group seem to show another common characteristic in demanding higher light intensity (Gandolfi 1991; Felfili 1993; Oliveira-Filho *et al.* 1994a; Silva Júnior 1995), which is reported as an important aspect of the forest upslope communities (Kellman and Tackaberry 1993). *Eriotheca pubescens*, *Pera glabrata* and *Platypodium elegans* can also be found colonizing the more open *cerrado* communities, while *Callisthene major* and

---

Figure 20.3 (*opposite*) Ordination of trees with basal areas $\geqslant 5$ cm DBH of the 33 species represented by $\geqslant 10$ individuals, using detrended correspondence canonical analysis. (a) Axes I and II which represent linear combinations of the environmental variables selected in the analysis, by showing the highest correlations which are displayed in Table 20.3. These environmental variables maximize the dispersion of the species scores represented in diagram (c). Each arrow can be interpreted as an axis that represents the variation of each variable. The arrows point in the direction of maximum correlation. The longest arrows represent the groups of variables interpreted as "border", elevation and "stream" effects. They are the ones most strongly correlated with the ordination axes, and therefore most closely related to the pattern of variation in samples and species composition, shown in diagrams (b) and (c). Thus, axes I and II in the diagram define an ordination space representing a topographic–moisture and chemical–textural gradient as a consequence of the correlations between the environmental variables and the axis.
(b) Sample ordination which arranges site points in a continuum, where points which are closer correspond to sites that are similar in species composition, and points which are far apart correspond to sites which are dissimilar. The projection of the sample points on to the arrows of diagram (a) represents the main relationship between samples and each of the environmental variables. Samples are labelled with their respective community classification provided by TWINSPAN to allow visual interpretation of the relationships.
(c) Species ordination showing position along the environmental axes. The position of the species if superimposed on diagram (a) would represent the main relationships between species and each of the environmental variables. Species are toned to show their community classification provided by TWINSPAN.

Lamanonia ternata are frequently recorded as tall and large trees with full crowns at the upslope sites.

The species typical of the downslope community, *Emmotum nitens*, *Licania apetala*, *Ocotea aciphylla*, *Pseudolmedia guaranitica* and *Virola sebifera* seem to perform best at the wettest, most acid and P-, OM- and Al-rich soils as indicated by the producting their greatest basal areas. They can also be regarded as species able to survive seasonal soil saturation, since *Licania apetala* and *Pseudolmedia guarantitica* are well-known moist sites colonizers. On the other hand, *Emmotum nitens* and *Virola sebifera*, also commonly occur in *cerradão* (dense, tall *cerrado* woodland), where drier soil conditions and higher light availability obtain. This paradox of co-occurrence of species of both wet and dry soils may be partially explained by the site characteristics. The Pitoco valley comprises steep topography which provides a spatial mosaic where these species are presumably finding "subsites" appropriate for their needs.

The species of the intermediate community show two tendencies. A group of species including *Bauhinia rufa*, *Copaifera langsdorffii*, *Faramea cyanea*, *Jacaranda puberula* and *Matayba guianensis* are ordinated closer to the dry upslope community, while *Inga alba*, is closer to the wet downslope ground of species.

A number of species are not considered to be related to any community (see Table 20.1). They are ordinated towards the middle of the axis, showing their greatest basal area at intermediate conditions on the environmental gradient. However, trends are still observable, for instance, species such as *Byrsonima laxiflora*, *Cupania vernalis*, *Myrcia rostrata* and *Siphoneugena densiflora* are ordinated closer to the "upslope" community. In contrast, *Cheiloclinium cognatum*, *Cryptocarya aschersoniana*, *Ocotea spixiana*, *Schefflera morototoni*, *Xylopia emarginata* and *X. sericea* tend to be more closely associated with the downslope community. Finally, species recorded as widely dispersed among the communities, such as *Maprounea guianensis*, *Sclerolobium paniculatum* var. *rubiginosum*, *Symplocos mosenii* and *Tapura amazonica*, seem to be best adapted to intermediate positions of the axis I gradient.

A topographical–moisture gradient from the stream margin to the forest–*cerrado* border is found in the analysis. It strongly suggests that species related to the downslope community occur preferentially in the very acid, Al- Fe- and OM-richer soils of the stream margins. On the other hand, species related to the upslope communities are closely associated with the soils richer in Mn, Mg and clay at the forest–*cerrado* boundary. These differences in soil chemical and textural variations along the topographical gradient are strongly correlated with the floristic variations in the tree community. *The forest–savanna transition is sharply defined and the apparent upslope migration of forest appears to have been sufficiently dynamic to eliminate any edge effect.*

20.5 CONCLUSIONS

The woody vegetation of the Pitoco gallery is characterized and classified to provide constituent communities which closely associate with segments of the slope from the stream margin to the forest–*cerrado* border.

Topography, which is directly related to water-table levels, is the main determining factor of the forest boundaries, structure, floristic composition, richness and density, as also reported in many other studies (Camargo *et al.* 1971; Ratter 1980; Metzler and Damman 1985; Furley 1985, 1992; Powell 1984; Dunhan 1989; Oliveira-Filho 1989; Ribeiro 1991, Felfili 1993; Oliveira-Filho *et al.* 1994a,b; Ramos 1995).

Variations of soil physical and chemical properties within the forest are correlated with position on the slope, and the distribution of tree species is considerably influenced by these factors. Other variables are undoubtedly important in explaining community and species patterns but their influence is disguised by the strength of the topographical effects.

Processes affecting soil moisture status change constantly. Dynamic changes in the stream courses, changes in climate or seasonal climatic patterns can create new situations of drought or excessive water supply, changing the soil characteristics and consequently affecting the boundaries between the vegetation communities. The forest–savanna boundary has been shown to be migrating upslope with gallery forest expansion (Ratter 1992). In the present study, evidence of the pace of change is provided by the presence of *cerrado* species recorded within the gallery, such as *Blepharocalyx salicifolia* and *Pouteria ramiflora*. A permanent series of markers has been established within the gallery and this will make it possible to monitor the dynamic relationships between the vegetation and environment.

Natural vegetation has profound effects on soil chemical and physical properties which, in turn, affect the predominance of some species over others. Cause and effect relationships are difficult to define and can only be established through experimentation. What has been established is that the distinct communities of the gallery forest are closely associated with changes in environmental factors such as slope, drainage and soil.

REFERENCES

Adamoli, J., Macedo, J., Azevedo, L.G. and Madeira Neto, J. 1985. Caracterização da região dos cerrados. In W.J. Goedert (ed.), *Solos dos Cerrados*. EMBRAPA/Nobel, Brasília, pp. 33–73.

Baillie, I.C., Ashton, P.S., Court, M.N., Anderson, J.A.R., Fitzpatrick, E.A. and Tinsley, J. 1987. Site characteristics and the distribution of tree species in Mixed Dipterocarp Forest on tertiary sediments in central Sarawak, Malaysia. *Journal of Tropical Ecology*, **3**, 201–220.

Bendix, J. 1994. Among-site variation in riparian vegetation of the Southern California Transverse ranges. *The American Midland Naturalist*, **132**(1), 136–151.

Brady, N.C. 1990. *The Nature and Properties of Soils*, 10th edn. Macmillan/Collier Macmillan, New York.

Burke, I.C., Reiners, W.A. and Olson, R.K. 1989. Topographic control of the vegetation in a mountain big sagebrush steppe. *Vegetatio*, **84**, 77–86.

Camargo, J.C.G., Cesar, A.L., Gentil, J.P., Pinto, S.A.F. and Troppmair, H. 1971. *Estudo fitogeográfico da vegetação ciliar do rio Corumbataí (SP)*. *Série Biogeográfica 3*, Instituto de Geografia, Universidade de São Paulo.

Causton, D.R. 1988. *An Introduction to Vegetation Analysis*. Unwin Hyman, London.

Dunham, K.M. 1989. Vegetation–environment relations of a middle Zambezi floodplain. *Vegetatio*, **82**, 13–24.

Felfili, J.M. 1993. Structure and dynamics of a gallery forest in Central Brazil. *DPhil. thesis*, University of Oxford.

Felfili, J.M. and Silva Júnior, M.C. 1992. Floristic composition, phytosociology and comparison of cerrado and gallery forests at Fazenda Água Limpa, Federal District, Brazil. In P.A. Furley, J. Proctor and J.A. Ratter (eds), *Nature and Dynamics of Forest–Savanna Boundaries*. Chapman & Hall, London, pp. 393–415.

Felfili, J.M. and Silva Júnior, M.C. 1993. A comparative study of cerrado (sensu stricto) vegetation in central Brazil. *Journal of Tropical Ecology*, **9**(3), 227–289.

Furley, P.A. 1985. *Notes on the Soils and Plant Communities of Fazenda Água Limpa (Brasilia, D.F. Brasil)*. Occasional Publication No. 5, Department of Geography, University of Edinburgh.

Furley, P.A. 1992. Edaphic changes at the forest–savanna boundary with particular reference to the neotropics. In P.A. Furley, J. Proctor and J.A. Ratter (eds), *Nature and Dynamics of Forest–Savanna Boundarie*. Chapman & Hall, London, pp. 91–117.

Gandolfi, S. 1991. Estudo florístico e fitossociológico de uma floresta residual na área do aeroporto internacional de São Paulo, Município de Guarulhos, SP. *Master's thesis*, Universidade Estadual de Campinas, Campinas, SP.

Gauch, H.G. 1982. *Multivariate Analysis in Community Ecology*. Cambridge University Press, Cambridge.

Haridasan, M. 1991. Solos do Distrito Federal. In M.N. Pinto (ed.), *Cerrado, Caracterização, Occupação e Perspectivas*. Editora Universidade de Brasília, Brasília, pp. 309–330.

Hill, M.O. 1979. *TWINSPAN – A FORTRAN Program for Arranging Multivariate Data in an Ordered Two-way Table by Classification of the Individuals and Attributes*. Cornell University, Ithaca, New York.

James, F.C. and McCulloch, C.E. 1990. Multivariate analysis in ecology and systematics: Panacea or Pandora's box. *Annual. Rev. Ecol. Sys.*, **21**, 129–166.

Kellman, M. and Tackaberry, R. 1993. Disturbance and tree species coexistence in tropical riparian fragments. *Global Ecol. Biogeog. Lett.*, **3**, 1–9.

Kent, M. and Coker, P. 1992. *Vegetation Description and Analysis – A Practical Approach*. Belhaven Press, London, 363 pp.

Lopes, A.S. and Cox, F.R. 1977. A survey of the fertility status of surface soils under cerrado vegetation in Brazil. *Soil Science Society of America Journal*, **41**, 742–747.

Meave, J. and Kellman, M. 1994. Maintenance of rain forest diversity in riparian forests of tropical savannas: implications for species conservation during Pleistocene drought. *Journal of Biogeography*, **21**, 121–135.

Metzler, K.J. and Damman, W.H. 1985. Vegetation patterns in the Connecticut river flood plain in relation to frequency and duration of flooding. *Le Naturaliste Canadien*, **112**, 535–547.

Mueller-Dombois, D. and Ellenberg, H. 1974. *Aims and Methods of Vegetation Ecology*. J. Wiley & Sons, New York.

Newbery, D. McC. and Proctor, J. 1984. Ecological studies in four contrasting lowland rain forests in Gunugu Mulu National Park, Sarawak. IV. Association between tree distribution and soil factors. *Journal of Ecology*, **72**, 475–493.

Noy-Meir, I., Walker, D. and Willians, W.T. 1975. Data transformation in eco-logical ordination. II. On the meaning of data standardisation. *Journal of Ecology*, **63**, 779–800.

Oliveira-Filho, A.T. 1989. Composição florística e estrutura comunitária da floresta de galeria do Córrego da Paciência, Cuiabá (MT). *Acta Botanica Brasilica*, **3**, 91–112.

Oliveira-Filho, A.T. and Ratter, J.A. 1995. A study of the origin of Central Brazilian forests by the analysis of plant species distribution patterns. *Edinburgh Journal of Botany*, **52**, 141–194.

Oliveira-Filho, A.T., Ratter, J.A. and Shepherd, G.J. 1990. Floristic composition and community structure of a central Brazilian gallery forest. *Flora*, **184**, 103–117.

Oliveira-Filho, A.T., Vilela, E.A., Gavilanes, M.L. and Carvalho, D.A. 1994a. Effect of flooding regime and understorey bamboos in the physiognomy and tree species composition of a tropical semideciduous forest in Southeastern Brazil. *Vegetatio*, 113, 99–124.

Oliveira-Filho, A.T., Vilela, E.A., Gavilanes, M.L. and Carvalho, D.A. 1994b. Comparison of the woody flora and soils of six areas of Montane semideciduous forest in southern Minas Gerais, Brazil. *Edinburgh Journal of Botany*, 51(1), 355–389.

Pires, J.M. 1984. The Amazonian forest. In H. Sioli (ed.), *The Amazon–Limnology and Landscape Ecology of a Mighty Tropical River and its Basin*. Junk Pub., Dordrecht.

Powell, G.R. 1984. Forest cover on two watersheds of the nashwaak experimental watershed project in west-central Brunswick. *Naturalist can. (Rev. Écol. Syst.)*, 111, 31–44.

Prance, G.T. 1987. Biogeography of neotropical plants. In T.V. Whitmore and G.T. Prance (eds), *Biogeography and Quaternary History of Tropical America*. Clarendon Press, Oxford, pp. 46–65.

Ramos, P.C.M. 1995. Vegetation communities and soils in the National Park of Brasília. *PhD thesis*, University of Edinburgh.

Ratter, J.A. 1980. *Notes on the Vegetation of Fazenda Agua Limpa (Brasilia, DF, Brazil), Including a Key to the Woody Genera of the Dicotyledons of the Cerrado*. Royal Botanic Garden, Edinburgh.

Ratter, J.A. 1992. Transitions between cerrado and forest vegetation in Brazil. In P.A. Furley, J. Proctor and J.A. Ratter (eds), *Nature and Dynamics of Forest–Savanna Boundaries*. Chapman & Hall, London, pp. 417–429.

Ribeiro, J.F. 1991. Environmental heterogeneity in space and time and plant life history traits on zonation of five riparian woody species in the California Central valley. *PhD thesis*, University of California, Davis.

Ross, S.M. 1989. *Soil Processes. A Systematic Approach*. Routledge, London.

Schiavini, I. 1992. Estrutura das comunidades arbóreas de mata de galeria da Estação Ecológica do Panga (Uberlândia, MG). *PhD thesis*, Universidade Estadual de Campinas, Campinas, SP.

Silva, A.F. and Shepherd, G.J. 1986. Comparacões floristicas entre algumas matas brasileiras utilizando analises de agrupamento. *Revta. bras Bot.*, 9(1), 81–86.

Silva, P.E.N. 1991. Estado nutricional de communidades arbóreas em quatro matas de galeria na região dos cerrados do Brasil central. *Master's thesis*, Universidade de Brasília, Brasília, DF.

Silva Júnior, M.C. 1995. Tree communities of the gallery forests of the IBGE Ecological Station, Federal District, Brazil. *PhD thesis*, University of Edinburgh.

Ter Braak, C.J.F. 1988. *CANOCO – A Fortran Program for Canonical Community Ordination by Partial Detrended Canonical Correspondence Analysis and Redundancy Analysis (Version 2.1)*. Technical report LWA-88-02. Netherlands.

Ter Braak, C.J.F. and Prentice, I.C. 1988. A theory of gradient analysis. *Adv. Ecol. Res.*, 18, 271–317.

Whitmore, T. and Prance, G.T. (eds) 1987. *Biogeography and Quaternary History of Tropical America*. Clarendon Press, Oxford.

21 The Significance of Soil Profile Differentiation to Hydrological Response and Slope Instability: A Modelling Approach

S. M. BROOKS[1] **and A. J. C. COLLISON**[2]
[1]*Department of Geography, University of Bristol, UK*
[2]*Department of Geography, King's College London, UK*

21.1 INTRODUCTION

Physically based modelling, particularly involving coupling of soil hydrology with slope stability assessment, has recently been introduced into geomorphological research for interpretation of shallow translational hillslope failure mechanisms. In various applications combined soil hydrology–slope stability modelling has elucidated some of the controls on slope failure more fully than previously possible, specifically with reference to shallow translational failures in podsols developing during the Holocene (e.g. Brooks *et al.* 1993a). In certain circumstances, physically based models have been able to make a potentially significant contribution to furthering understanding of processes which have traditionally been assessed through correlation of field evidence, such as dating slope failures followed by their association with periods of significant climatic deterioration (Starkel 1966; Grove 1972; Brazier and Ballantyne 1989), or anthropogenic vegetation modification (Innes 1983). The introduction of physically based models has opened up possibilities for exploring such linkages through direct consideration of failure mechanisms, for time periods unavailable to direct observation, and for situations in which there are several possible explanations for the occurrence of slope failure. Such models can suggest the likely relative significance of different explanations, by providing evidence which favours one explanation over another.

This chapter presents a new angle from which to assess the contribution of modelling, by considering the relative roles of soil profile properties, vegetation and climate on shallow translational hillslope failure. Previous modelling approaches (e.g. Kirkby 1976, 1989) have attempted an integration of soils, vegetation and climate with long-term slope development, and have achieved some success in elucidating the effect of regolith formation, weathering and removal on slope formation. However, in this research it is difficult to overcome the effects of scale. Vegetation and soil processes operate on much shorter timescales than does slope development, making outputs from vegetation/soil models incompatible with inputs

Advances in Hillslope Processes, Volume 1. Edited by M. G. Anderson and S. M. Brooks.
© 1996 John Wiley & Sons Ltd.

to slope development models. Thus a compromise needs to be achieved by ignoring short-term fluctuations in soil/vegetation processes and considering changes averaged over longer (around 10 000 year) periods. In some respects this can restrict elucidation of the detailed processes and their links between the different systems. Considering long-term pedogenesis and shallow translational slope failures, the effect of temporal development of *one* soil type (freely draining humus–iron podsol) on slope stability has been investigated in detail (Brooks *et al.* 1993a). That paper also included treatment of the role of climatic variation in the Holocene. Climatic change in Scotland in the Holocene is thought to occur through shifts in the position of the westerly air stream, bringing storms of different character to the region (Lamb 1977, 1982). The detailed effect of these storm variations on the stability of a *mature podsol* has also been evaluated (Brooks and Richards 1994). A recent extension of the combined soil hydrology–slope stability model includes routines treating the effect of vegetation characteristics (Collison 1993) and this has enabled investigation of the relative effect of vegetation and climate on the stability of podsols at different stages of development (Brooks *et al.* 1995). Considering soil–climate–vegetation interaction through modelling suggests that simple associations between climatic deterioration and slope instability, or vegetation removal and slope failure, are difficult to establish, which has helped to broaden the debate on the causes and mechanisms which underlie enhanced slope instability for different periods in the Holocene.

Since pedogenesis during stable phases creates conditions under which climate or vegetation change can become initiators of subsequent instability, it is important to consider temporal alterations in soil properties and the rate at which these occur. Combining chronosequence studies (Bockheim 1980; Protz *et al.* 1984; Ellis and Richards 1985) with physically based soil hydrology–slope stability model (Brooks *et al.* 1993a) has led to some success in elucidating the interaction between pedogenesis and slope stability, indicating that soil formation over a period of about 8000 years is sufficient to promote failure in freely draining podsols on relatively steep slopes. However, when taking into account uncertainty in model parametrization, such interpretation of model outputs is less straightforward and clear-cut. By including a range in parameter values reflecting spatial variability and measurement uncertainty, results show that anything from between 10 000 and 6000 years is associated with the onset of mass movements, and that the predicted gradients on which mass movement occurs, as well as the location of the failure surface within the soil profile, are difficult to predict. This issue related to the effect of parameter uncertainty on model outputs and process interpretation is a key area for further research.

An important extension of modelling the link between soil development and mass movement is to evaluate other soil types since stability criteria are likely to vary from one type to the next, just as they alter with the stage of development in podsols. Horizon differentiation is an important pedogenic control on slope stability. With different efficacy of leaching and translocation, soils develop different horizon properties. Podsols ("Spodosols" is the terminology used by the United States Department of Agriculture 7th Approximation 1975) develop where there is considerable chemical leaching potential under humid climates, with freely draining parent material and acidic humus. The process primarily involves movement of iron, aluminium and organic material to lower depths, although the precise mechanism whereby this occurs is disputed (Farmer and Fraser 1982; Buurman 1985). The result is sharply defined horizons, with a bleached (eluvial) horizon overlying a red-brown illuvial horizon enriched with iron and organic matter. With slightly less available water, less permeable parent material and less acidic vegetation, the typical soil type is that of a brown podsolic or brown earth ("Mollisol" is the terminology used by the United States Department of Agriculture 7th Approximation 1975) (Bridges 1975). Finally, where there is a more continental climate strong clay movement occurs resulting in a distinctive textural B horizon within the profile ("Alfisols" is the terminology used by the United States Department of Agriculture 7th Approximation 1975). Such soil profiles have been termed "sol lessivès" by Duchaufour (1982).

The general trend in these soil types is towards the development of a textural B horizon, bringing a change in the hydrological characteristics significant to slope stability and potentially involving an association of different soil types with different angles of limiting stability. Thus different soil types might occupy different landscape facets as a result. Geomorphological research has shown qualitatively how different soil profiles occupy different positions in the landscape, which results in varying susceptibility to different erosional processes (Dalrymple *et al.* 1968; Arnett and Conacher 1973). While this generalized scheme applies to a wide range of slope gradients and processes, we might begin to apply physically based models to investigate and explain some of the trends observed for high-angled ($<25°$) slopes undergoing mass movement. The success of such an approach relies on having appropriate data for model parametrization and validation, only applying the model within the bounds for which it is validated, and being confident that process representation is adequate for the application under study. Such issues will be discussed in the light of the various scenarios considered below. This chapter therefore seeks to establish whether the modelling approach adopted in previous research can be extended to examine the hydrological and slope stability consequences of changes in climate and vegetation over a range of soil types, and to try to elucidate further some of the observed associations between soils and landscapes utilizing an explanation based on the probability of slope failure occurrence.

21.2 MODEL STRUCTURE AND PARAMETRIZATION

The model employed here has a variety of features which make it particularly suitable for this application. It consists of a simple vertical array of cells which can represent thin horizons of differing depths. Thus soil profiles can be defined in some detail, including in particular vertical variations in soil hydrological and geotechnical properties dependent on varying degrees of weathering and translocation. Pore-water pressures are calculated for each horizon assuming moisture redistribution occurs according the Richards Equation. Thus the model has the advantage of being able to predict vertical and temporal variations in pore-water pressures for soil profiles which differ in terms of detailed hydrological behaviour within and between horizons. However, the model is not suitable where flow takes place predominantly via macropores or laterally. The model also allows separate specification of geotechnical properties, which are then combined with the pore-water pressure data to calculate the factor of safety against failure assuming different slope angles. The detailed structure of the model is described by Anderson and Howes (1985) in one of its first applications, and also by Anderson *et al.* (1988), although its use in geomorphological investigations is somewhat restricted. Recent modifications to the model enable inclusion of a vegetation layer which incorporates interception, enhanced infiltration and root reinforcement (Collison 1993), which potentially broadens the use of this model in geomorphological applications.

21.2.1 Parameter Values for Vegetation and Climate

Model parametrization for vegetation and climate has been described fully elsewhere (Brooks and Richards 1994; Brooks *et al.* 1995), so a brief discussion only will be presented here. Simple vegetation effects are considered by comparing bare soils with those having pine forest, although more complex vegetation scenarios can be included where data are available for parametrization. Addition of forest involves modification of rainfall character (loss of water volume and lag in its receipt at the ground surface) as well as enhanced permeability and material shear resistance in the root zone (Collison 1993). Below the root zone no direct effects of vegetation are included, but the underlying horizons are affected by vegetation modification to moisture receipt at the ground surface.

Rainstorms are simulated using hourly rainfall totals, which permit inclusion of varying total rainfall, storm duration and associated intensity. Additionally, rainfall distribution within any given storm can be included (Brooks and Richards 1994). The storms used in the simulations below adopt rainfall totals ranging from 10 to 200 mm, and a constant storm duration of ten hours, which results in rainfall intensity variation ranging from 1 to 20 mm/h.

21.2.2 Model Requirements for Differentiated Soil Profiles

Modelling the effect of differentiated soil profiles on hydrology and stability has been achieved previously using a finite-difference column of ten cells, which can be divided into hydrologically distinct layers. For each soil horizon, model inputs include a value for the saturated hydraulic conductivity, the soil moisture characteristic curve, the saturated moisture content, bulk density, cohesion and the angle of internal friction. The method of establishing parameter values for these inputs, discussed below, is an important process in the modelling, especially where small changes in input values can lead to large differences in outputs. Several methods are available based on a large amount of recent research into the hydrology of different soil types (Hendrickx 1990).

21.2.3 Methods for Estimating Soil Hydrological Parameters

The soil profile characteristics for the soil types considered in the model simulations are outlined in Table 21.1. The humus–iron podsol is a relatively coarse-grained soil, with a slightly finer-grained eluvial than illuvial horizon. This particle differentiation is commonly reported from chronosequence studies of podsols (Protz *et al.* 1984; Ellis and Richards 1985), with the surface layer progressively increasing in silt through time (Brooks *et al.* 1993a; Brooks and Richards 1993). The brown podsolic and brown earth soils are finer-grained than the humus–iron podsol, but again have a finer-grained surface horizon than the underlying B horizon. However, strong mechanical translocation results in soil profiles showing two significant differences from the podsol, brown podsolic and brown earth soils. First, the particle size distribution is much finer overall, and secondly, the underlying B horizon is considerably finer than the surface horizon (Bridges 1975).

Soil hydrology is now a well researched area, providing a large database for considering the behaviour of different soil types. Of particular significance to this chapter is the work of Brakensiek *et al.* (1981), Carsel and Parrish (1988), Rawls *et al.* (1982), Saxton *et al.* (1986), Van Genuchten (1980), and Wosten and Van Genuchten (1988). These papers provide well-tested methods for establishing the soil hydraulic parameters required by the model from readily available profile data, such as particle size, organic matter and bulk density. Combining these methods with information from soil profile descriptions permits consideration of hydrological variation within differentiated soil profiles, and using a physically based coupled soil hydrology–slope stability model, consequences for shallow translational failure might be determined.

Many of the studies which relate hydraulic parameters to textural data do so using multiple regression analysis. For example, the soil moisture content at specified tensions may be found using a relationship of the following form:

$$\theta_p = a + b(\% \text{ sand}) + c(\% \text{ silt}) + d(\% \text{ clay}) + e(\% \text{ organic matter}) + f(\text{bulk density})$$

$$(21.1)$$

θ_p is the soil moisture content for a given tension, and a–f are regression coefficients. A soil moisture characteristic curve can thereby be derived for different soil textures, but the method is limited because the few points derived cannot define a continuous curve. The derivation of a continuous curve is necessary for modelling, especially when simulations involve a small range of suction data for which detailed measurements are not available.

Continuous curves can be derived by fitting mathematical functions through empirical data, and these have the advantage of permitting comparison of soil types over the full suction range, as well as being computationally more efficient. One of the earliest and frequently used methods is that determined by Brooks and Corey (1966), who suggested that the following function describes the soil moisture characteristic curve adequately for most purposes;

$$\frac{\theta - \theta_r}{\theta_s - \theta_r} = (h/h_b)^{-\beta}$$

$$(21.2)$$

where θ_r is the residual water content, θ_s is the saturated moisture content, h_b is the air entry suction and β is a parameter which varies with soil type. It has since been shown to work poorly at the low suctions which prevail when the soil is close to saturation. This is precisely the range which needs to be considered when evaluating slope stability criteria, as most soils are stable at high tensions. The method of Van Genuchten (1980) appears to work better in the low suction range, and this adopts the following function:

$$\frac{\theta - \theta_r}{\theta_s - \theta_r} = [1/\{1 + (\alpha h)^n\}]^m \qquad (21.3)$$

where α, m and n are parameters which vary with soil type. Equations (21.2) and (21.3) produce similar results at high tensions, but in the range appropriate to the current modelling the latter should be employed. Furthermore, derivation of the parameters of the Van Genuchten model (α, n and m) have formed the basis for subsequent research which has enabled them to be predicted from particle size, bulk density and organic matter, thereby defining the whole curve (Wosten and Van Genuchten 1988).

Of greater relevance in the present context is the way in which these parameters have been derived for individual soil textural classes (Carsel and Parrish 1988). Thus the information readily available for textural differentiation between the horizons of the different soil types can be used to evaluate the soil moisture characteristic curve. The parameter values and textural classes for the soils employed in the modelling simulations are given in Table 21.1, and Figure 21.1 depicts the curves derived using the values of Carsel and Parrish (1988). One major effect is that textural differentiation in horizons of the podsol and brown earth is insufficient to produce any variation in soil moisture characteristic curves between horizons. This could, in part, reflect insensitivity in the method used to convert the particle size distribution to a soil moisture characteristic curve. However, comparison of the two soil types shows distinct curves. The most heavily mechanically leached soil profile does show clear profile differentiation, with the B horizon curve indicating greater moisture retention for a given suction. Both horizon differentiation and changes between soil types have implications for stability which can be assessed using the model simulations.

The saturated hydraulic conductivity can also be predicted using textural data, and a number of equations are available. While this is simpler than for the soil moisture characteristic curve since the saturated hydraulic conductivity takes a single value, there are still problems when relating matrix texture to gross soil permeability

Table 21.1 Van Genuchten parameter values used to predict the soil moisture characteristic curves (from Saxton et al., 1986)

Soil type	Horizon	Textural class	α	m	n	θ_s	θ_r
Podsol	Surface	sandy silt	0.145	0.627	2.68	0.43	0.045
(Spodosol)	B	medium sand	0.145	0.627	2.68	0.43	0.045
Brown earth	Surface	v. fine sandy loam	0.075	0.479	1.89	0.41	0.065
(Mollisol)	B	fine sandy loam	0.075	0.479	1.89	0.41	0.065
Sol lessivè	Surface	silt loam	0.02	0.291	1.41	0.45	0.067
(Alfisol)	B	silty clay loam	0.01	0.187	1.23	0.43	0.089

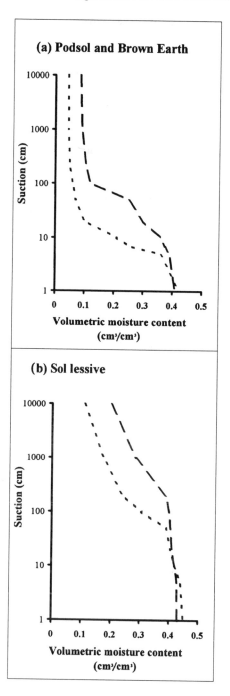

Figure 21.1 Soil moisture characteristic curves predicted using the Van Genuchten (1980) model: (a) podsol (— — —) and brown earth (·······); (b) sol lessivè: A horizon (·····) and B horizon (— — —)

in the presence of macropores. This becomes especially serious with clay soils, where shrinkage cracks frequently exist, but in the soil types considered here, the texture is largely coarser than fine silt, precluding the development of macropores. Carsel and Parrish (1988) have derived a relationship between saturated hydraulic conductivity and textural class. Values found from this are employed in the model simulations for consistency with soil moisture characteristic curve derivation, and for their wide applicability. The values appropriate to each of the horizons of the soil profiles are depicted in Figure 21.2.

In addition to the hydrological parameters, the model requires values for material shear resistance. Again this varies according to the texture of the individual horizons. Generally, as texture becomes finer, angles of internal friction decrease (Kirkby 1973), although porosity, particle size mixture and mineralogy are also significant (Statham 1974). Typical values for different materials of the three soil profiles are given in Lambe and Whitman (1969), and those employed in the modelling for the soil profiles are shown in Table 21.2.

Given the soil hydrology, soil shearing resistance, climate and vegetation parameter variation described above, there are several scenarios which can be

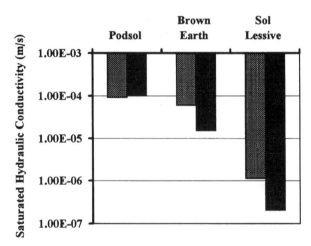

Figure 21.2 Saturated hydraulic conductivities for the soil profiles (A horizon: light grey; B horizon: dark grey)

Table 21.2 Angles of internal friction for the soils and horizons

Soil type	Horizon	Angle of internal friction
Podsol	Surface	36°
(Spodosol)	B	38°
Brown earth	Surface	31°
(Mollisol)	B	30°
Leached soil	Surface	30°
(Alfisol)	B	28°

Data from Lambe and Whitman (1969).

evaluated using the model to assess the relative significance of different soil–vegetational–climatic combinations on slope stability.

21.3 MODEL SIMULATION RESULTS FOR DIFFERENT SOIL PROFILES

For each of the soil types, two basic scenarios were simulated to assess the impact of vegetation. The first involved an unvegetated surface and the second involved forest cover. To consider climate, six storms were included, involving a total rainfall of 10 mm to 200 mm. The soil profiles were initially drained from saturation for 24 h to produce appropriate initial conditions. In the first set of simulations a slope angle of 30° was employed and each soil profile was subjected to a 10 mm rainstorm of 10 h duration. The minimum factor of safety developed at different depths is shown in Figure 21.3 for three different soil types. The effects of root reinforcement of the surface layers is apparent, resulting in great stability in vegetated profiles. A second effect relates to enhanced permeability in vegetated surface horizons, whereby water is delivered at a faster rate to the underlying B horizon, resulting in reduced stability just below the root zone. This is most apparent and propagated to the greatest depth in the sol lessivè, where the permeability in the B horizon is lowest. In this soil there is the greatest contrast in saturated hydraulic conductivity between the horizons. In all soils, however, the deepest cell is the one of minimum stability. At this depth the effect of vegetation is negligible in all but the sol lessivè where vegetation has a small *destabilizing* role. On the whole, in these soils both beneficial and adverse effects of vegetation are suggested, but both relate to depths well above the potential shear surface.

Of greater significance is the apparent decline in stability from the humus–iron podsol, through the brown earth to the sol lessivè, evident from Figure 21.3. To explore this further, a range of slope angles was employed in a series of simulations. Figure 21.4 illustrates changes in limiting stable slope angles for the different soil profiles of almost 10°, from 39° for the podsol to 31° for the sol lessivè, with the brown earth having intermediate angles. An important potential consequence is that different soils might well occupy different positions in the landscape, depending on the maximum stable slope angle on which they can develop. Podsols appear to occupy the steepest slopes while sols lessivè seem to be stable only on slopes of lower gradient. Figure 21.4 also clearly illustrates the comparatively negligible effect of vegetation on angles of limiting stability. Arguments supporting vegetation change as a major factor in initiating mass movement activity seem to have little support from these modelling results. However, climatic factors also need to be considered, and their interaction with soils and vegetation make interpretation less straightforward.

That slope failures are more prevalent during high-magnitude rainstorms is not disputed. However, storms can be of high magnitude in terms of duration, intensity and total rainfall, each parameter having a different outcome for slope stability depending on soils and vegetation. To investigate this, simulations were carried out for each soil type involving different storm totals but constant duration (10 h), which produced variations in rainfall intensity. For each soil type the relationship between

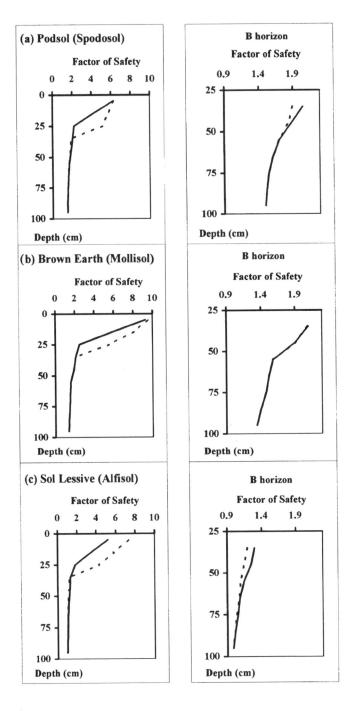

Figure 21.3 Comparison of stability for different soils under vegetated and non-vegetated conditions (vegetated profiles: dotted line; non-vegetated profiles: solid line)

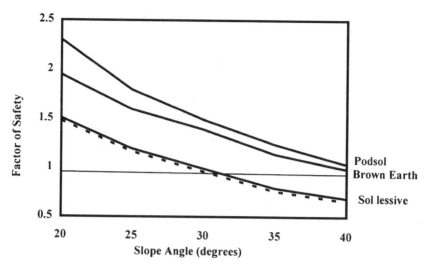

Figure 21.4 Angles of limiting stability for the soil profiles (non-vegetated: solid line; vegetated: dashed line)

rainfall total and minimum factor of safety (Figure 21.5) shows the varying sensitivity of the soil types to changing rainfall total. For the humus–iron podsol even the highest totals do not promote failure on 30° slopes. However, brown earths are vulnerable at rainfall totals of 125 mm ($i = 12.5$ mm/h) and sols lessivè at totals of 50 mm ($i = 5$ mm/h). Of particular interest is the implication that periods of climatic deterioration might have more severe consequences for soil profiles having undergone mechanical translocation, although this depends on the associated slope angles on which different soil profiles actually develop.

Thus far, modelling results suggest that for the scenarios considered, the effect of vegetation on slope stability is of minor significance compared with soil character and climate, but these latter factors interact to produce various interpretations. Clearly, diminishing permeability in subsurface horizons is of major significance to slope stability. Further simulations reveal the potential complexity of this process. By decreasing gradually the permeability in the B horizon in a manner likely to result from progressive increases in translocation, the effect on the factor of safety was ascertained. The results, plotted in Figure 21.6 depict a decline in the factor of safety as B horizon permeability decreases from 1.5×10^{-5} to 3×10^{-6} m/s, for a 30° slope and a 100 mm rainstorm of 10 h duration ($i = 1$ mm/h). Under such conditions the soil profiles considered here fail, but the depth of the shear surface varies according to the ratio of surface: B horizon permeability. As well as providing a greater propensity for failure, mechanical translocation results in different depths to the potential shear surface.

In summary, these modelling results suggest complexities in slope failure initiation which cannot be considered using field evidence alone. In particular, for periods where direct observation is impossible, such a modelling approach is of great value. This chapter has indicated that there are complex interactions between climate and

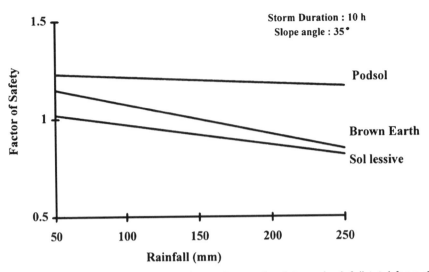

Figure 21.5 Relationship between minimum factor of safety and rainfall total for a 10 h storm on a 35° slope

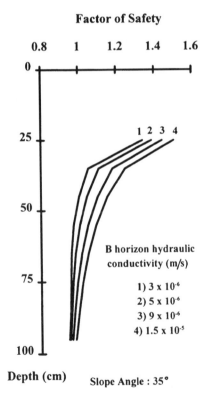

Figure 21.6 Decline in factor of safety as B horizon permeability is reduced from 1.5×10^{-5} to 3×10^{-6} m/s

soil properties which need to be considered in greater detail when discussing periodicity in mass movement activity. Suggestions based on field evidence for mass-movement initiation in the Holocene have frequently focused on climatic factors in isolation, failing to consider parameter interaction. It also appears that vegetation is of relatively minor significance, although there are issues relating to model formulation and parametrization, discussed below, which need to be assessed. Soil hydrology is undoubtedly highly significant to shallow translational failure. One major manifestation of this is the varying angles of limiting stability which appear to exist for different soil profiles. Such findings provide an interesting new perspective from which to consider the relationship between soil type, slope position and geomorphological processes of soil removal, hitherto discussed solely from a qualitative viewpoint.

21.4 DISCUSSION

The above results have indicated that different responses to rainfall characterize different soils developed under varying degrees of translocation. Soils which are subjected to the effects of podsolization (chemical movement) appear to be more stable than those which have undergone substantial physical translocation involving redeposition of silt- and clay-sized particles into the B horizon. Profile forms in-between exhibit intermediate degrees of stability, depending on the intensity of translocation experienced. Thus a highly significant pedogenic effect on hydrology and slope stability is the resulting change in B horizon permeability and moisture retention resulting from mechanical translocation. This is potentially of major significance to interpretations of long-term slope stability, and to explanations of existing soil–landscape associations, and should be explored in greater detail.

It has long been recognized that soils develop different characteristic profiles depending on the climatic regime, parent material and topography (Jenny 1941; Kubiena 1953). More recently much research has been conducted into the way in which soil properties control profile hydrology (Saxton et al. 1986; Carsel and Parrish 1988; Wosten and Van Genuchten 1988; Hendrickx 1990). Furthermore, the development of physically based coupled hydrology–stability modelling is continuing (Anderson and Howes 1985; Anderson et al. 1988). This chapter has attempted to draw together these three areas to provide a challenging new scheme for geomorphologists to interpret the complex relationship between soils and slopes, and their change over time.

Over timescales such as the Holocene, soil-forming factors are unlikely to remain constant. The modelling results help to consider the implications of this for soil and slope evolution. It has been demonstrated that the different soils each have a varying sensitivity to climatic change. Evidence suggests that during phases of climatic deterioration it is the soil profiles having greater mechanical translocation which are most likely to undergo failure, rather than there being generalized, widespread slope failure throughout a particular region. However, the slope angle range actually occupied by each soil type needs to be taken into account. Vegetation change can also affect slope stability, but the above modelling has indicated that for relatively freely-draining soils, where root reinforcement does not extend to the potential shear surface, vegetation change is unlikely to be a significant underlying cause of slope failure. In fact, the presence of vegetation may even promote failure as it speeds up the delivery rate of water to deeper regions in the profile, where a shear surface is more likely to develop. With the soil profiles considered here, this effect seems to be of low significance, although marginally more obvious on the finer-grained, lower permeability soils.

Most early geomorphological schemes of long-term slope development considered slope profiles to develop independently of soil cover, although evolving soil "metastability" (Penck 1924) represents an interesting connection between soil and slope development. The neglected significance of soil profile variations was further stressed by Chorley (1959) and later interpretations introduced the idea that regolith (as opposed to soil) properties controlled the angle of shearing resistance and permeability, thereby exerting some influence on slope stability (Carson and Petley 1979). However, there is a need to consider more fully the interaction between soils and slopes. Physically based modelling offers one possibility for this. Modelling results presented in this chapter indicate that detailed variations in the soil profile hydrology might be fundamental to slope stability (Brooks et al. 1993b). Hence as soils and slopes co-evolve different soil profiles have the potential to develop depending on the prevailing slope angle. Sols lessivè are unlikely to be stable on slopes over 35° unless they remain relatively shallow, whereas podsols could be stable. Thus the landscape position occupied by different soil profiles depends on their properties, a view previously expressed in a qualitative manner through the nine-unit landsurface model. However, this qualitative model assumes that soil profiles are largely the outcome of the various hydrological and erosional processes operating on the different facets rather than playing an interactive role in process operation. One of the main conclusions of the work presented here is that soil profile character can determine the processes operating, suggesting a more interactive model is required.

Finally, the above simulations have drawn upon several areas of research to enable generation of new ideas which might merit further investigation. However, there are issues related to the use of physically based models which need to be addressed. Well-formulated models can elucidate process mechanisms, and permit close investigation of the combined effects of different controlling parameters. The above application has indicated their potential for addressing aspects of geomorphology previously approached through field or laboratory assessment. However, model limitations exist which may restrict the applicability of the conclusions. The model used above considers flow to be wholly Darcian without including the effect of larger pores providing preferential flow pathways (Germann 1990). For the simple soil profiles used here as a starting point, macropore development is unlikely as soils are relatively coarse-grained. The model operates as a one-dimensional column and is therefore restricted to situations in which widespread saturation and lateral flow do not occur. Again, in the case of the above simulations this is a reasonable assumption as the soil remains unsaturated and one-dimensional flow is a reasonable approximation of true conditions. An extension for saturated throughflow would require a two-dimensional model, and the next challenge for developing soil–slope models is to provide a realistic representation of slope profile geometry combined with comparatively thin soil layers which underpin hydrological response. Furthermore, detailed data for model parametrization and validation are required, necessitating detailed quantification of the relationship between soils and slopes, downslope variations in soil hydrological properties and consequences for pore-water pressure distribution. For slope stability assessment, detail is required on the relationship between geotechnical properties and slope position. Thus although physically based models offer considerable future potential for process elucidation within geomorphology, particularly for time periods where direct measurement is difficult, the complexity and variety of landscape processes means that we must address issues of model parametrization and validation for new schemes to consider the relationship between soils and slopes over the long term.

REFERENCES

Anderson, M.G. and Howes, S. 1985. Development and application of a combined soil water–slope stability model. *Quarterly J. Engineering Geology*, **18**, 225–236.

Anderson, M.G., Kemp, M. and Lloyd, D.M. 1988. Applications of soil water finite difference models to slope stability problems. *5th International Landslide Symposium*, Lausanne, pp. 525–530.

Arnett R.R. and Conacher, A.T. 1973. Drainage basin expansion and the nine unit landsurface model. *Australian Geographer*, **12**, 237–249.

Bockheim, J.G. 1980. Solution and use of chronofunctions in studying soil development. *Geoderma*, **24**, 71–85.

Brakensiek, D.L., Engleman, R.L. and Rawls, W.J. 1981. Variation within textural classes of soil water parameters. *Transactions ASAE*, **24**, 335–339.

Brazier, V. and Ballantyne, C.K. 1989. Late Holocene debris cone evolution in Glen Feshie, western Cairngorm Mountains, Scotland. *Transactions Royal Society of Edinburgh: Earth Sciences*, **80**, 17–24.

Bridges, E.M. 1975. *World Soils*. Cambridge University Press.

Brooks, R.H. and Corey, A.T. 1966. Properties of porous media affecting fluid flow. *Jl. Irrig. Drain Div., Am. Soc. Civil Eng.*, **92**(IR2), 61–88.

Brooks, S.M., and Richards, K.S. 1993. Establishing the role of pedogenesis in changing soil hydraulic properties. *Earth Surface Processes and Landforms*, **18**, 573–578

Brooks, S.M. and Richards, K.S. 1994. The significance of rainstorm variations to shallow translational hillslope failure. *Earth Surface Processes and Landforms*, **19**, 85–94.

Brooks, S.M., Richards, K.S. and Anderson, M.G. 1993a. Shallow failure mechanisms during the Holocene: utilisation of a coupled slope hydrology–slope stability model. In D.S.G. Thomas and R.J. Allison (eds), *Landscape Sensitivity*. Wiley, Chichester.

Brooks, S.M., Richards, K.S. and Anderson, M.G. 1993b. Approaches to the study of hillslope development due to mass movement. *Progress in Physical Geography*, **17**, 3–49.

Brooks, S.M., Collison, A.J.C. and Anderson, M.G. 1995. Modelling the role of climate, vegetation and pedogenesis in shallow translational hillslope failure. *Earth Surface Processes and Landforms*, **20**, 231–242.

Buurman, P. 1985. Carbon/sesquioxide ratios in organic complexes and the transition to albic-spodic horizon. *Journal of Soil Science*, **36**, 255–260.

Carsel, R.F. and Parrish, R.S. 1988. Developing joint probability distributions of soil water retention characteristics. *Water Resources Research*, **24**, 755–769.

Carson, M.A. and Petley, D.J. 1970. The existence of threshold slopes in the denudation of the landscape. *Transactions Institute of British Geographers*, **49**, 71–95.

Chorley, R.J. 1959. The geomorphic significance of some Oxford soils. *American Journal of Science*, **57**, 503–515.

Collison, A.J.C. 1993. Assessing the influence of vegetation on slope stability in the tropics. *PhD thesis*, University of Bristol.

Dalrymple, J.B., Blong, R.J. and Conacher, A.J. 1968. A hypothetical nine-unit landsurface model. *Zeitschrift für Geomorphologie*, **1**, 60–76.

Duchaufour, P. 1982. *Pedology, Pedogenesis and Classification* (translated by TR Paton). George Allen and Unwin, London.

Ellis, S. and Richards, K.S. 1985. Pedogenic and geotechnical aspects of Late Flandrain slope instability in Ulvadalen, west central Norway. In K.S. Richards, R.R. Arnett and S. Ellis (eds), *Geomorphology and Soils*. George Allen and Unwin, London.

Farmer, V.C. and Fraser, A.R. 1982. Chemical and colloidal stability of sols in the Al_2O_3–Fe_2O_3–SiO_2–H_2O system: their role in podsolisation. *J. Soil Science*, **33**, 737–742.

Germann, P. 1990. Macropores and hydrological hillslope processes. In M.G. Anderson and T.P. Burt (eds), *Process Studies in Hillslope Hydrology*. Wiley, Chichester.

Grove, J.M. 1972. The incidence of landslides, avalanches and floods in western Norway during the Little Ice Age. *Arctic and Alpine Research*, **4**, 131–138.

Hendrickx, J.M.H. 1990. Determination of hydraulic soil properties. In M.G. Anderson and T.P. Burt (eds), *Process Studies in Hillslope Hydrology*. Wiley, Chichester.

Innes, J.L. 1983. Landuse changes in the Scottish Highlands in the 19th century: the role of pasture degeneration. *Scottish Geographical Magazine*, **99**, 141–149.

Jenny, H. 1941. *Factors of Soil Formation*. McGraw-Hill, New York.

Kirkby, M.J. 1973. Landslides and weathering rates. *Geologia Applicata e Idrogeologia*, Bari, **8**, 171–183.

Kirkby, M.J. 1976. Deterministic continuous slope models. *Zeitschrift für Geomorphologie*, NF SB25, 1–19.

Kirkby, M.J. 1989. A model to estimate the impact of climatic change on hillslope and regolith form. *Catena*, **16**, 321–334.

Kubiena, W.L. 1953. *The Soils of Europe*. Murby, London.

Lamb, H.H. 1977. *Climate: Past, Present and Future*, vol. 2. Methuen, London.

Lamb, H.H. 1982. *Climate, History and the Modern World*, Methuen, London.

Lambe, T.W. and Whitman, R.V. 1969. *Soil Mechanics*, SI version. Wiley, New York.

Penck, W. 1953. *The Morphological Analysis of Landforms*. Translated by H. Czech and K.C. Boswell from *Die Morphologie Analyse*, Stuttgart, 1924. London.

Protz, R., Ross, G.J., Martini, R.P. and Terasmae, J. 1984. Rate of podsolic soil formation near Hudson Bay, Ontario. *Canadian J. Soil Science*, **64**, 31–49.

Rawls, W.J., Brakensiek, D.L. and Saxton, K.E. 1982. Estimation of soil water properties. *Trans. ASAE*, **25**, 1316–1320, 1328.

Saxton, K.E., Rawls, W.J., Romberger, J.S. and Papendick, R.I. 1986. Estimating generalised soil water characteristics from texture. *Soil Sci. Soc. Am. J.*, **50**, 1031–1036.

Starkel, L. 1966. Post glacial climate and the modelling of European relief. In *Proceedings of the International Symposium on World Climate, 8000–0BC*. Royal Meteorological Society, London, pp. 15–32.

Statham, I. 1974. The relationship of porosity and angle of repose to mixture proportions in assemblages of different sized materials. *Sedimentology*, **21**, 149–162.

Van Genuchten, M.Th. 1980. A closed-form equation for predicting the hydraulic conductivity of unsaturated soils. *Soil Sci. Soc. Am. J.*, **44**, 892–898.

Wosten, J.H.M. and Van Genuchten, M.Th. 1988. Using texture and other properties to predict the unsaturated hydraulic functions. *Soil Sci. Soc. Am. J.*, **52**, 1762–1770.

Section 5

SOIL EROSION ON HILLSLOPES

22 Soil Erosion by Water: Problems and Prospects for Research

J. BOARDMAN

Environmental Change Unit and School of Geography, University of Oxford, UK

22.1 INTRODUCTION

This short review of research needs is based of necessity on personal experience and viewed from a northwest European perspective. In order to deal succinctly with a broad subject, arbitrary divisions are imposed: experimentation, modelling and monitoring. These accord with the approach of the GCTE (Global Change and Terrestrial Ecosystems) Soil Erosion Working Group (Ingram 1994).

The aim of erosion research is to understand erosional processes in order to enhance understanding of landform development. Equally important in the short and medium term is the provision of advice to policy-makers, in particular in the agricultural planning area, on the risk of erosion, the significance of erosion (including the long-term view) and on how to stop it, if its significance is such that this is considered to be worthwhile. Off-site impacts of erosion have also to be assessed since in temperate western Europe and North America they are of greater immediate economic importance than erosion *per se*. Within this broad framework of aims an understanding of geomorphological processes of erosion, transport and deposition is vital. The complementary approaches of experimentation, monitoring and modelling contribute to this understanding.

22.2 EXPERIMENTATION

Experimenters have traditionally attempted the application of reductionist approaches holding constant certain factors in order to isolate and quantify others. Usually, some form of hypothesis testing is adopted. Much of our understanding of erosional processes has been achieved by the use of experimental plots, simulated rainfall and the laboratory flume. Precision and control are gained but the transfer of results to the field is hampered by problems of scale and simplification; how, for example, do results obtained on single soil aggregates relate to the complex effects of rainsplash and overland flow on the field surface?

Soil erosion studies have been inextricably linked to the adoption in the USA in the 1920s of the experimental plot, eventually resulting in the Universal Soil Loss Equation (USLE) (Wischmeier and Smith 1978). Thus experimentation has been

Advances in Hillslope Processes, Volume 1. Edited by M. G. Anderson and S. M. Brooks.
© 1996 John Wiley & Sons Ltd.

central to US erosion studies and has dominated approaches to prediction of erosion in the field. The USLE was developed using plot data but with the aim of predicting long-term erosion rates at the field scale. Much experimental effort has continued to be directed at adjustments and refinements to USLE (Yoder and Lown 1995). Although Wischmeier recognized limitations and warned of the misuse of USLE, there was little discussion of whether it actually worked (Wischmeier 1976). Were the predictions it gave reliable (Risse *et al.* 1993)?

Unfortunately long-term rates of erosion on agricultural fields have not been measured and therefore the application of USLE to field-sized areas is untestable. Data from plots do exist and therefore a very limited test can be carried out. Burwell and Kramer (1983) compare observed and USLE predicted rates from plots (Figure 22.1). It can be seen that USLE overpredicts for rates below about 10 t/ha and probably underpredicts for higher rates. The average annual predicted soil loss was 1.9 times the observed loss for conventionally tilled corn over 24 years and 1.6 times for conservation-tilled corn. However, to take the average disguises the variability (Figure 22.1) and that median values are more appropriate for typically skewed erosion data (Table 22.1). It must be emphasized that this poor performance is achieved under ideal conditions – experimental plots; this is not a test of how well USLE performs at the field scale.

Govers (1991) reports a poor correlation between mean rill erosion rate on fields in central Belgium and erosion amounts predicted by USLE. This is for a three-year study measuring only rill erosion. The mean value predicted by USLE is about double the measured rate; the median and shape of the distributions are strikingly different (Figure 22.2).

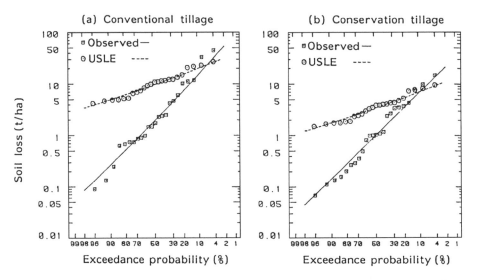

Figure 22.1 Comparison of observed and USLE-predicted annual soil loss exceedance probability for corn plots under conventional and conservational tillage. The lines are Log–Pearson distribution fits to the data (from Burwell and Kramer 1983, reproduced by permission of the Soil and Water Conservation Society, Ankeny, Iowa)

Table 22.1 A comparison of observed and predicted soil loss values (t/ha) as plotted in Figure 22.1

	Mean	Median
Conventionally tilled		
observed	5.8	1.5
predicted	11.1	10.0
Conservation tilled		
observed	2.48	1.0
predicted	3.96	4.0

Discussion of the work by Burwell and Kramer (1983) and that by Govers (1991) clarifies the problem of erosion prediction based on experimentation. There are two potential sources of error that tend to go unrecognized. First, the assumption that experimental plots can represent the agricultural landscape and processes therein (this is an analogous problem to using a rain-gauge to estimate the amount of rain falling on a catchment); secondly, that a model (USLE) derived from plot experiments can predict rates of erosion on fields when it clearly has limited success even on plots!

Evans recognized these problems in the 1970s. They were some of the reasons for the selection of a monitoring scheme rather than plot experiments to assess erosion in England and Wales between crop years 1981/82 and 1985/86. As Evans notes in his report (1988) to the Soil Survey, with regard to the selection of a technique to monitor erosion,

> In March 1980 a meeting was called by the Soil Management Committee of ADAS (Agricultural Development and Advisory Service) to discuss a possible project to monitor erosion and its effects over a period of five years. Two approaches were considered. The one chosen was to assess the numbers of eroding fields in selected strips of country, their rates of erosion and the effects of erosion and deposition on the environment; the other approach was to monitor plot experiments.

Experimental results from standard USDA or Wischmeier plots are limited in their applicability and should be used with care. No distinction is made between rill and interrill processes; indeed rills may not always form on slope lengths of 22 m – the length of Wischmeier plots. Lateral concentration of runoff in topographic depressions does not occur on plane plots; also crusting and compaction are not explicitly dealt with since the USLE K factor is usually held constant.

The simplification of topographic variability inherent in plot studies is a major problem. The initiation and development of rills is frequently associated with convexities in the landscape (Evans 1980), and with various forms of flow concentration along linear elements such as vehicle wheel tracks, field boundaries, headland furrows, depressions and dry valley floors (De Ploey 1989). While useful experimentation can occur at the laboratory scale, for example on the initiation of rills, much will have to be at the field or watershed scale to understand the development of topographically related erosional features.

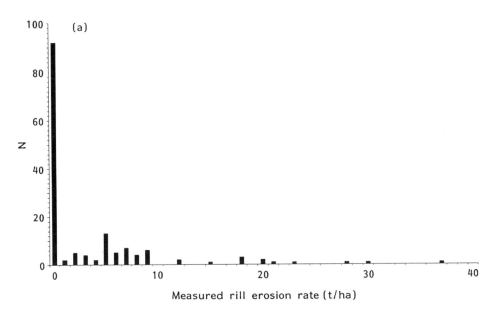

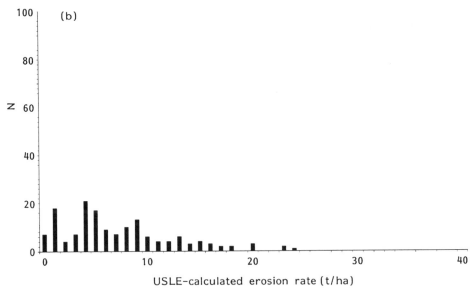

Figure 22.2 (a) Measured and (b) USLE-calculated erosion rate for 86 fields in central Belgium (data from Govers 1991; replotted by Favis-Mortlock 1994)

A further problem with plot-based experiments is that the storage of runoff and sediments is ignored. Assessment of erosion at the field and watershed scale has to take storage into account. Working at the watershed scale, fundamental work by Meade (1982) and Trimble (1983) in the USA shows that high rates of erosion may result in massive storage of sediment on tributary floodplains and gradual release of sediment to major rivers.

The sediment yield of rivers is frequently equated uncritically to "erosion rate". The interpretation of sediment yield results is not without its pitfalls (Walling and Webb 1981; Stocking 1987). The link between hillslope and stream has been provided by the "delivery ratio" concept which is generally agreed to be a crude and unsatisfactory tool (Wolman 1977; Walling 1983; Novotny and Chesters 1989). Sediment yields are therefore unlikely to provide reliable estimates of erosion rates on fields (e.g. Boardman and Favis-Mortlock 1992; Evans 1993b). Current approaches to the problem of using the record of streams or lake sedimentation to predict erosion rates are focusing on sediment fingerprinting using mineral magnetic or radioactive signatures (e.g. Foster et al. 1990; Walling 1990). Recent empirical approaches have attempted to measure delivery ratio at the scale of soil loss from individual fields (Evans 1990a; Quine and Walling 1991); both emphasize the control exerted by slope and soil texture. Because of time lags in the transfer process, sediment stored on hillslopes may only occasionally be moved to floodplains – as in the example described by Slattery et al. (1994). The connectivity of hillslope-stream systems in arable landscapes is a neglected topic.

How can experimentation advance our understanding in the face of these real difficulties? First it is necessary to distinguish research problems which are amenable to plot experimentation: gullying clearly is not! Govers and Poesen (1988) using a plot of $7500 \, m^2$ show that the relationship between rill and interrill processes is complex and varies in time and space. Reviews of other studies also show that the relative contribution of rilling varies greatly but is most commonly 70–80% of total erosion.

Secondly, small-scale processes such as crusting should be investigated in the laboratory and on experimental plots; equally important, however, is to relate such experiments to cultivation practices and natural rainfall in order to predict sites and periods of risk. Papy and Boiffin (1989) point out that on loamy soils in northern France around 450 mm of rain are required to decrease the roughness grade of fields from that typical of "recently ploughed" to a "strongly crusted state". On the South Downs in southern England mean rainfall for October to February is about 10 mm below this threshold; in wetter winters it is exceeded. However, the picture is complicated by the stoniness of the soil, the lack of residue left on the surface and earlier ploughing and drilling dates in southern England (Boardman 1991). Assessment of the risk of erosion in this case must be studied using both experimentation and field observation.

Thirdly, the results of some experiments emphasize the complexities of the field situation. For example, Slattery and Bryan's (1992) carefully controlled flume experiment examined various hydraulic parameters as a means of separating rilled from unrilled flows. Rather than threshold values they found zones of transition within which both types of flow occurred. Govers et al. (1990) also show the extreme sensitivity of runoff erosion to be related to variations in initial moisture content.

Finally, experimental work must recognize the problems of scale. For example, Poesen and Ingelmo-Sanchez (1992) discuss the effects of stones on the surface of soils. At a small scale $(1 \, m^2)$ they may generate runoff, whereas at the scale of a large plot runoff generated from stony areas may cause erosion of surface crusts downslope leading to an increase in infiltration in these zones.

22.3 MONITORING

Boardman (1994) defines erosion monitoring as field-based measurement of erosional and/or depositional forms over a significant area (e.g. $> 10\,km^2$) and for a period of > 2 years. In its purest form an inductive approach is adopted although the choice of what to measure influences the conclusions. However, in contrast to an experimental approach, monitoring is not hypothesis driven. Indeed, in terms of global change studies, Burt (1994, p. 493) argues that this is one of its strengths:

> It is not possible to anticipate the exact questions which may be asked in several decades time. All that can be done is to design and maintain a network of sites where measurements are made which seem most likely to provide indications of change in the future.

Monitoring is necessary for several reasons. Evans (1993a) claims that, "to assess the problem of water erosion, its extent, frequency and rates need to be known, and how these parameters vary between different landscapes". Quantitative assessment of the role of processes such as gullying which, because of their scale or infrequency are difficult to study experimentally, is only possible through monitoring. Monitoring schemes provide data for and the opportunity to field test models.

Few countries have the data to be able to make sensible decisions regarding erosion risk. In some cases it may be available in the form of maps based on the application of USLE (e.g. Eilers *et al.* 1989), or other forms of hazard mapping (e.g. Prickett 1984). In England and Wales the National Soil Map indicates the extent of erosion-prone soil associations (Soil Survey of England and Wales 1983). Recent updating of this information attempts to rank soils in terms of their susceptibility to erosion under recent land use and climate conditions (Evans 1990b). Both the National Soil Map and Evans' updating are based on monitoring schemes in selected localities.

Monitoring erosion has been neglected mainly because of a perception that it is "not scientific". The experimental approach involving plots has, despite major problems (Evans 1995), been preferred. There is also the assumption that monitoring schemes are costly. Evans' (1992) figures for the National Soil Erosion Monitoring Scheme for England and Wales (1982–86) are however modest: on average 5.1 working days to interpret the aerial photographs and 23.3 days per year to carry out fieldwork; plus a further 212 days to analyse the data. To this should be added time and costs involved in air photography, and report and paper writing.

The GCTE Soil Erosion Network has collected information on field-based monitoring schemes (Boardman 1994; Poesen *et al.* 1996). Many have now ceased to operate, usually for funding reasons (Table 22.2). Despite this, valuable information has been assembled: Evans (1993a) estimates median erosion rates on eroded fields in 17 localities in England and Wales of between 0.5 and $4.0\,m^3/ha/year$; a 10-year study of erosion in the South Downs gave ranges of $0.5–5.0\,m^3/ha/year$ for

Table 22.2 Field-based soil erosion monitoring schemes

Country	Area	Period	Source
Belgium (Central)	86 fields	1981–85	Govers (1991)
Belgium (Central)	c. 50 km^2	1982–93	Poesen (1993)
Belgium (Central)	3 catchments (20–200 ha)	1989–92	Vandaele (1993) Vandaele and Poesen (1995)
France (North)	20 catchments (3.7–100 ha)	1988–90	Auzet et al. (1993)
Italy (Central)	230 km^2	1990–94	Busoni et al. (1995)
Norway	9 catchments (50–680 ha)	1992–96	Ludvigsen
Sweden	90 km^2	1986–88	Alstrom and Bergman-Akerman (1992)
England and Wales	17 localities (c. 826 km)	1982–86	Evans (1992, 1993a)
England and Wales	13 localities (10–179 ha)	1989–93	Chambers et al. (1992)
England (South Downs)	c. 36 km^2	1982–91	Boardman (1990)
Scotland (Northeast)		1984–86	Watson and Evans (1991)
USA (Oklahoma)	several 1.6 ha	1977–94	Williams and Nicks (1988)
Scotland (East)	20 km^2	1993–	Wade (unpublished)
England and Wales		1993–	Evans and McLaren (1994)

catchment areas (Boardman and Favis-Mortlock 1993). Such rates are lower than those from many plot studies (Evans 1995). On the South Downs long-term monitoring showed that 71% of the total soil eroded in the decade 1981–92 was lost in three years. The role of high-magnitude, low-frequency events is difficult to study experimentally; monitoring on the Downs included the autumn of 1987 when erosion rates exceeded 20 m^3/ha on 15 fields and 200 m^3/ha on one (Boardman 1988a). Unfortunately the National Monitoring Scheme had finished in 1986 and therefore did not include the results of the particularly wet autumn in 1987 (Evans 1993a). Monitoring on the South Downs suggested that thresholds at which rilling was initiated on winter cereals fields were around 30 mm of rain in two days (Boardman 1992), a finding partly based on plot studies nested into the broad-scale monitoring approach (Robinson and Boardman 1988).

Examples of mapped erosion data based on monitoring schemes are shown in Figure 22.3. Monitoring includes broad spatial coverage which allow site characteristics to be related to topography, rainfall, soil and land use across an area (Boardman 1988a). Figure 22.3(a) identifies sites with erosion rates greater than 10 m^3/ha/year. Within such a scheme detailed studies of particular sites may be made (Figure 22.3(b)). Mapping in central Belgium has allowed Poesen (1993) to distinguish bank gullies from ephemeral gullies and suggest different origins and rates of development (Figure 22.3(c)).

(a)

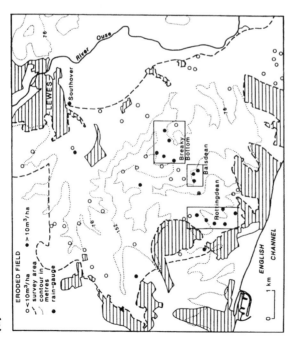

ERODED FIELD
o <10m³/ha   ● >10m³/ha
-- survey area
contour in metres
● rain-gauge

(b)

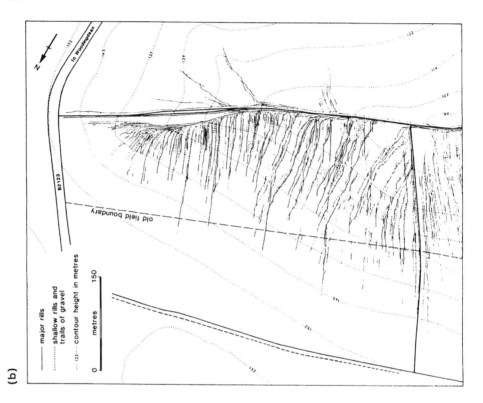

— major rills
⋯⋯ shallow rills and trails of gravel
⋯122⋯ contour height in metres

metres   150
0

(c)

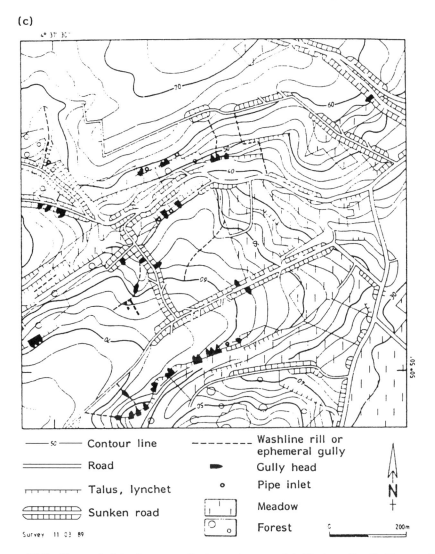

Contour line

Road

Talus, lynchet

Sunken road

Survey 11 03 89

Washline rill or ephemeral gully

Gully head

Pipe inlet

Meadow

Forest

N

200 m

Figure 22.3 Mapped data from monitoring schemes: (a) Eastern South Downs, UK, showing boundary of monitored area and eroded fields 1987–88 (Boardman 1988a, reproduced by permission of Catena Verlag); (b) the rill system at Bevendean, South Downs, autumn 1982 (Boardman and Robinson 1985, reproduced by permission of Elsevier Science Ltd); (c) typical locations of bank gullies and ephemeral gullies in central Belgium. All gully heads located on a bank are bank gullies (Poesen 1993, reproduced by permission of Elsevier Science–NL)

One of the benefits of monitoring schemes has been the realization in western Europe of the great importance of ephemeral gullies, valley bottom rills or thalweg gullies in erosional systems (e.g. Evans and Cook 1986; De Ploey 1989; Poesen 1989; Auzet *et al.* 1993; Vandaele 1993). Decades of experimental work at the laboratory and plot scale had failed to focus attention on these features. Allied to this has been the recognition that in many areas of the European loess belt (and elsewhere) the main short-term impact of erosion is in off-site damage due to flooding (e.g. Evans 1988; De Ploey 1989; Auzet *et al.* 1990; Boardman *et al.* 1994).

Why are monitoring schemes needed in the future? The threat of climate change poses obvious problems. Changed climates may increase the risk of erosion; monitoring along present-day climate gradients or across areas of particular sensitivity can investigate these issues. Data from monitoring schemes under conditions of climatic *variability* can be used to model the effects of future climate change.

Land-use change, driven either by climate, or more likely by economic and political forces, is the greater threat. In the 1970s and 1980s in Britain a massive switch to the growing of winter cereals explains much of the increase in erosion (Evans and Cook 1986; Boardman 1990). In the 1990s the introduction of "Set-Aside" promises to reduce rates of erosion if it is sensibly targetted (Boardman 1988b). Monitoring of a catchment with flood-prone housing in Sussex suggests that Set-Aside can be used to reduce risk but that long-term and carefully targetted reversions to grass are required (Boardman and Evans 1991). Solving economic and political problems is a greater challenge than the physical one of reducing runoff from arable catchments!

Benefits derived from Set-Aside may be offset by the introduction of new arable crops such as maize and linseed vulnerable to summer thunderstorm activity (Boardman *et al.* 1996). Increased intensification on remaining arable land may also be a problem. Other land-use trends are difficult to foresee because they relate to global trading patterns, subsidies, weather conditions and food shortages.

Investigation of the impact of climate and land-use change in the future can, to a certain extent, be carried out by experimental and modelling approaches. However, in view of the problems of spatial scale, the impact of high-magnitude events and the need for data and validation, it is clear that monitoring schemes should also play a central role in soil erosion research.

22.4 MODELLING

For the past 30 years erosion prediction has been dominated by application of the USLE or variations on its basic form such as SLEMSA (Stocking 1987). In the 1970s and 1980s more complex computer-based erosion models have been developed; many of these are based on USLE wholly or in part. Table 22.3 lists a selection of these models but excludes two recent process-based models that are yet to become operational: WEPP and EUROSEM. These developments have taken place despite an increasing perception of the problems regarding application of USLE. It could be argued that all problems relate to *misuse* of USLE since application should only be to hillslope elements that are homogenous as regards topography, soils, land use, climate and farming practices, thus resembling experimental plots.

Clearly the USLE has been applied with some disregard for these principles. Jager (1994), for example, calculates USLE-based erosion rates for the state of Baden–

Table 22.3 A comparison of some sediment yield models

Model	Field or watershed?	Time series output?	USLE based	Handles deposition?	Handles streambank erosion?
GLEAMS[a]	field	yes	(no)[2]	yes	no
EPIC[b]	field	yes	yes[4]	(yes)[5]	no
GAMES[c]	watershed (cell)	no	yes	(yes)[6]	no
SWRRB[d]	watershed (cell)	yes	yes[7]	yes	no
ANSWERS[e]	watershed (grid)	yes	no[2]	yes	no
AGNPS[f]	watershed (grid)	yes	yes[8]	yes	no

Model	Crop growth component?	Tillage component?	Chemistry component?	Complexity of data requirement[1]
GLEAMS[a]	yes[3]	no	yes	medium
EPIC[b]	yes	yes	yes	high
GAMES[c]	no	no	no	low
SWRRB[d]	yes[3]	yes[3]	no	high
ANSWERS[e]	no	no	no	high
AGNPS[f]	no	no	yes	medium

Notes:
[1] Necessarily subjective
[2] However, some USLE components used
[3] Fairly simple
[4] USLE, MUSLE, or Onstad-Foster may be used
[5] Implicitly only, when MUSLE or Onstad-Foster are used
[6] Implicitly only, in the sense that delivery ratios may be less than 1 for individual cells
[7] Uses MUSLE
[8] Uses slightly modified USLE
Sources:
[a] Leonard *et al.* (1987); Davis *et al.* (1990)
[b] Sharpley and Williams (1990); Williams *et al.* (1990)
[c] Dickinson and Rudra (1990)
[d] Arnold *et al.* (1989)
[e] Beasley *et al.* (1980); De Roo *et al.* (1989)
[f] Young *et al.* (1989); Panuska *et al.* (1991)

Wurttemberg, southwestern Germany. USLE factors are calculated for $2 \times 2\,km^2$ cells. Results are validated by comparison with published experimental plot data. Average rates of erosion in some areas are predicted to be $> 30\,t/ha/year$. Such rates seem unlikely in view of recently published field measurements from the same area (Baade *et al.* 1993).

Validation of models has been neglected: Burwell and Kramer's (1983) simple plot-based test of USLE and that of Govers (1991) have already been referred to. In some cases the shaky basis for assertions regarding erosion rates is not made clear. In an explanation of the CORINE erosion-risk mapping project, Giordano *et al.* (1995) write:

> In order to have an idea about the magnitude of the soil erosion in the southern countries of the EU, the mean soil erosion rate of some small watersheds are given below: . . . Subcatchments of the Segura river near Murcia (Spain) 40 t/ha/yr (Romero Diaz *et al.* 1992).

In fact, the erosion rate from the Segura is estimated by application of USLE to the catchments (Romero Diaz et al. 1992). Thus, erosion rates obtained in CORINE by a very crude USLE-like approach are given some validity by references to a USLE estimate.

Even the most sophisticated process-based models such as WEPP also assume homogeneity of the cell or field and require large amounts of data for quite small sites. Detailed topographic information may be available in the future via technologies such as GPS and DTM but the same may not be true of information on land use and particularly soil properties. Acquisition of good-quality data with which to develop and test models is, and is likely to remain, a major problem. In a recent GCTE model-comparison exercise, common plot data were used from sites in Canada, USA and Portugal, but there were inadequacies in almost all data sets, such as lack of detailed soils data and rainfall data of inadequate temporal resolution (Boardman and Favis-Mortlock 1996). These were the best data sets that the GCTE erosion network could acquire. If it is difficult for such a group to acquire data for such comparative exercises, how much more difficult will it be to run the models in the field?

In the past models have poorly represented or ignored some important processes in erosion. These include crusting, stoniness, rolling and gullying. Some of these deficiencies are now being addressed. In addition, crucial processes such as the partition of precipitation into runoff and infiltration have been very crudely modelled using the SCS curve method (e.g. Boughton 1989). Boiffin et al. (1988) point out that moisture status and microtopography of the soil surface is not explicitly dealt with in the USLE, which puts "perhaps exaggerated stress" on crop cover. Indeed, the instruction for the setting up of bare fallow Wischmeier-type erosion plots in 1961 contained the injunction to cultivate "at other times when necessary to eliminate serious crust formation" (Mutchler et al. 1988). Bollinne (1985) shows that the C factor for winter wheat in the USLE should be 176% of that for cultivated bare plots because of wheel tracks on the former. Bollinne's suggested adjustments for western European conditions are a useful advance but they are still based on measurements on 22 m long plots and therefore unlikely to bear much relation to the field situation.

Several of the new generation of soil-erosion models make use of rainfall generator models to simulate rainfall from known parameters of a region's rainfall. However, there is evidence that some generators at least fail to fully represent high-magnitude, low-frequency events in the distributions produced: this may lead to underprediction of erosion by as much as 40% (Favis-Mortlock 1995).

Modelling has become easier and quicker in part due to the availability of large data sets: for example, the EPIC and WEPP can access large data sets of soils and weather for many parts of the USA but not elsewhere. Similarly, technical advances allow a sophisticated presentation of results from those models which are applicable to large areas using GIS. However, there is a continuing danger that in some cases gross inadequacy of data or the model is disguised by the cosmetic of sophisticated presentation (e.g. Flacke et al. 1990). Similarly, the ANSWERS model has been linked to GIS but validation in the field is lacking (De Roo et al. 1989; De Roo and Walling 1994).

Empirical models, principally USLE, have severe limitations, have been widely misused and are largely untestable. The development of physically based models (e.g. WEPP and EUROSEM) has been proceeding for several years but it is still too early to assess their effectiveness and value in anything other than a tentative way. A major challenge will be the provision of high-quality data for anything other than very small areas or experimental plots. A recent review criticized the development of needlessly complex models: "it would not be far wrong to assume that some 99% of the published material consists of theoretical exercises which have little, if any, bearing on the planning and management of environmental systems" (Biswas 1990, p. 5).

Alternative approaches to conventional modelling have been explored. For example, Harris and Boardman (1990) used an Expert System to induce a series of rules from a set of examples contained in an extensive database of erosion events for the South Downs. The approach yields good results but requires a large data set based on monitoring. Dependence on measurement can be reduced or eliminated by use of a "soil erosion expert" to construct the rules. Unmeasured or non-quantified parameters can be added to the data set, e.g. date of drilling in relation to erosion-generating rainfall (as a surrogate for crop cover).

22.5 CONCLUSIONS

Too often the connections between experimenters, fieldworkers and modellers have been tenuous. The aim for the future must be to incorporate the best experimental and monitored results into models but also to ensure that the testing of models is an opportunity to identify parameters which are poorly understood and require further investigation. The GCTE Soil Erosion Network is the first attempt to bring together these three groups of researchers and as such is an important development. Finally, these collaborative research efforts should not lose sight of the aim of providing predictive tools that are usable by the people and in the areas where they are most needed.

ACKNOWLEDGEMENTS

I thank Dr David Favis-Mortlock, Dr Michael Slattery and Professor John Catt for comments on earlier versions of this chapter.

REFERENCES

Alstrom, K. and Bergman-Akerman, A. 1992. Contemporary soil erosion rates on arable land in southern Sweden. *Geografiska Annaler*, **74**A(2–3), 101–108.

Arnold J.G., Williams, J.R., Nicks, A.D. and Sammons, N.B. 1989. *SWRRB (A Basin Scale Simulation Model for Soil and Water Resources Management) User's Manual*, Texas A&M University Press, USA.

Auzet, A.V., Boiffin, J., Papy, F., Maucorps, J. and Ouvry, J.F. 1990. An approach to the assessment of erosion forms and erosion risk on agricultural land in the northern Paris basin, France. In J. Boardman, I.D.L. Foster and J.A. Dearing (eds), *Soil Erosion on Agricultural Land*. Wiley, Chichester, pp. 383–400.

Auzet, A.V., Boiffin, J., Papy, F., Ludwig, B. and Maucorps, J. 1993. Rill erosion as a function of the characteristics of cultivated catchments in the North of France. *Catena*, **20**, 41–62.

Baade, J., Barsch, D., Mausbacher, R. and Schukraft, G. 1993. Sediment yield and sediment retention in a small loess-covered catchment in SW-Germany. *Zeitscrift fur Geomorphologie*, **92**, 217–230.

Beasley, D.B., Huggins, C.F. and Monke, E.J. 1980. ANSWERS: a model for watershed planning. *Transactions of the American Society for Agricultural Engineers*, **23**, 939–944.

Biswas, A.K. 1990. Environmental modelling for developing countries: problems and prospects. In A.S. Biswas, T.N. Khoshoo and A. Khosla (eds), *Environmental Modelling for Developing Countries*, Tycooly Publishing, London, pp. 1–12.

Boardman, J. 1988a. Severe erosion on agricultural land in East Sussex, UK October 1987. *Soil Technology*, **1**, 333–348.

Boardman, J. 1988b. Public policy and soil erosion in Britain. In J.M. Hooke (ed.), *Geomorphology in Environmental Planning*. Wiley, Chichester, pp. 33–50.

Boardman, J. 1990. Soil erosion on the South Downs: a review. In J. Boardman, I.D.L. Foster and J.A. Dearing (eds), *Soil Erosion on Agricultural Land*. Wiley, Chichester, pp. 87–105.

Boardman, J. 1991. Land use, rainfall and erosion risk on the South Downs. *Soil Use and Management*, **7**(1), 34–38.

Boardman, J. 1992. The sensitivity of Downland arable land to erosion by water. In D.S.G. Thomas and R.J. Allison (eds), *Landscape Sensitivity*. Wiley, Chichester, pp. 211–228.

Boardman, J. 1994. Erosion monitoring programmes. Paper to GCTE Workshop, *Soil Erosion under Global Change*, Paris, 29–31 March 1994.

Boardman, J. and Evans, R. 1991. Flooding at Steepdown. Report to Adur District Council.

Boardman, J. and Favis-Mortlock, D.T. 1992. Soil erosion and sediment loading of watercourses. *SEESOIL*, **7**, 5–29.

Boardman, J. and Favis-Mortlock, D. 1993. Simple methods of characterizing erosive rainfall with reference to the South Downs, southern England. In S. Wicherek (ed.), *Farm Land Erosion: In Temperate Plains Environment and Hills*. Elsevier, pp. 17–29.

Boardman, J. and Favis-Mortlock, D.T. (eds) 1996. *Modelling Soil Erosion by Water*. NATO ASI Series, Springer Verlag, in press.

Boardman, J. and Robinson, D.A. 1985. Soil erosion, climatic vagary and agricultural change on the Downs around Lewes and Brighton, autumn 1982. *Applied Geography*, **5**, 243–258.

Boardman, J., Ligneau, L., de Roo, A. and Vandaele, K. 1994. Flooding of property by runoff from agricultural land in northwestern Europe. *Geomorphology*, **10**, 183–196.

Boardman, J., Burt, T.P., Evans, R., Slattery, M.C. and Shuttleworth, H. 1996. Soil erosion and flooding as a result of a summer thunderstorm in Oxfordshire and Berkshire, May 1993. *Applied Geography*, **16**(1), 21–34.

Boiffin, J., Papy, F. and Eimberck, M. 1988. Influence des systèms de culture sur les risques d'érosion par ruissellement concentré. 1. Analyse des conditions de déclenchement de l'érosion. *Agronomie*, **8**, 663–673.

Bollinne, A. 1985. Adjusting the Universal Soil Loss Equation for use in western Europe. In S.A. El-Swaify, W.C. Moldenhauer and A. Lo (eds), *Soil Erosion and Conservation*. Soil Conservation Society of America, Ankeny, Iowa, pp. 206–213.

Boughton, W.C. 1989. A review of the USDA SCS curve number method. *Australian Journal of Soil Research*, **27**, 511–523.

Burt, T. 1994. Long-term study of the natural environment – perceptive science or mindless monitoring? *Progress in Physical Geography*, **18**(4), 475–496.

Burwell, R.E. and Kramer, L.A. 1983. Long-term annual runoff and soil loss from conventional and conservation tillage of corn. *Journal of Soil and Water Conservation*, **38**(3), 315–319.

Busoni, E., Salvador Sanchis, P., Calzolari, C. and Romagnoli, A. 1995. Mass movement and erosion hazard patterns by multivariate analysis of landscape integrated data: the Upper Orcia Valley (Sienna, Italy) case. *Catena*, **25**, 169–185.

Chambers, B.J., Davies, D.B. and Holmes, S. 1992. Monitoring water erosion on arable farms in England and Wales, 1989–90. *Soil Use and Management*, **8**(4), 163–170.

Davis, F.M., Leonard, R.A. and Kaisel, W.G. 1990. *GLEAMS User Manual*, US Department of Agriculture, Agricultural Research Service Southeast Watershed Research Laboratory, Tifton, Georgia, USA.

Dickinson W.T. and Rudra, R.P. 1990. *GAMES (Guelph model for evaluating the effects of Agricultural Management Systems on Erosion and Sedimentation)*, User's Manual v3.01. Technical Report 126-86, School of Engineering, University of Guelph, Guelph, Ontario, Canada.

De Ploey, J. 1989. Erosional systems and perspectives for erosion control in European loess areas. *Soil Technology*, **1**, 93–102.

De Roo, A.P.J., Hazelhoff, L. and Burrough, P.A. 1989. Soil erosion modelling using "ANSWERS" and Geographical Information Systems. *Earth Surface Processes and Landforms*, **14**, 517–532.

De Roo, A.P.J. and Walling, D.E. 1994. Validating the "ANSWERS" soil erosion model using [137]Cs. In R.J. Rickson (ed.), *Conserving Soil Resources: European Perspectives*. CAB International, Wallingford, UK, pp. 246–263.

Eilers, R.G., Langman, M.N. and Coote, D.R. 1989. *Water Erosion Risk: Manitoba*. Land Resource Research Centre, Agriculture Canada, Contribution Number 87-12.

Evans, R. 1980. Characteristics of water-eroded fields in lowland England. In M. De Boodt and D. Gabriels (eds), *Assessment of Erosion*. Wiley, Chichester, pp. 77–87.

Evans, R. 1988. Water erosion in England and Wales 1982–1984. Report to Soil Survey and Land Research Centre, Silsoe, UK.

Evans, R. 1990a. Water erosion in British farmers' fields – some causes, impacts, predictions. *Progress in Physical Geography*, **14**(2), 199–219.

Evans, R. 1990b. Soils at risk of accelerated erosion in England and Wales. *Soil Use and Management*, **6**(3), 125–131.

Evans, R. 1992. Assessing erosion in England and Wales. *Proceedings 7th ISCO Conference*, Sydney, Australia, **1**, 82–91.

Evans, R. 1993a. Extent, frequency and rates of rilling of arable land in localities in England and Wales. In S. Wicherek (ed.), *Farm Land Erosion: In Temperate Plains Environment and Hills*. Elsevier, pp. 177–190.

Evans, R. 1993b. On assessing accelerated erosion of arable land by water. *Soils and Fertilizers*, **56**(11), 1285–1293.

Evans, R. 1995. Some methods of directly assessing water erosion of cultivated land – a comparison of measurements made on plots and in fields. *Progress in Physical Geography*, **19**(1), 115–129.

Evans, R. and Cook, S. 1986. Soil erosion in Britain. *SEESOIL*, **3**, 28–58.

Evans, R. and McLaren, D. 1994. *Monitoring Water Erosion of Arable Land*. Friends of the Earth, London.

Favis-Mortlock, D.T. 1994. Use and abuse of soil erosion models in southern England. *PhD thesis*, University of Brighton.

Favis-Mortlock, D.T. 1995. The use of synthetic weather for soil erosion modelling. In D.F.M. McGregor and D.A. Thompson (eds), *Geomorphology and Land Management in a Changing Environment*. Wiley, Chichester, pp. 265–282.

Flacke, W., Auerswald, K. and Neufang, L. 1990. Combining a modified Universal Soil Loss Equation with a Digital Terrain Model for computing high resolution maps of soil loss resulting from rain wash. *Catena*, **17**, 383–397.

Foster, I.D.L., Grew, R. and Dearing, J.A. 1990. Magnitude and frequency of sediment transport in agricultural catchments: a paired lake-catchment study in Midland England. In J. Boardman, I.D.L. Foster and J.A. Dearing (eds), *Soil Erosion on Agricultural Land*. Wiley, Chichester, UK, pp. 153–171.

Giordano, A., Peter, D. and Maes, J. 1995. Erosion risk in southern Europe. In D. King, R.J.A. Jones and A.J. Thomasson (eds), *European Land Information Systems for Agro-Environmental Monitoring*. Institute for Remote Sensing Applications, Joint Research Centre, European Community, EUR 16232 EN.

Govers, G. 1991. Rill erosion on arable land in central Belgium: rates, controls and predictability. *Catena*, **18**, 133–155.

Govers, G. and Poesen, J. 1988. Assessment of the interrill and rill contributions to total soil loss from an upland field plot. *Geomorphology*, **1**, 343–354.

Govers, G., Everaert, W., Poesen, J., Rauws, G., De Ploey, J. and Lautridou, J.P. 1990. A long flume study of the dynamic factors affecting the resistance of a loamy soil to concentrated flow erosion. *Earth Surface Processes and Landforms*, **15**, 313–328.

Harris, T.M. and Boardman, J. 1990. A rule-based Expert System approach to predicting waterborne soil erosion. In J. Boardman, I.D.L. Foster and J.A. Dearing (eds), *Soil Erosion on Agricultural Land*. Wiley, Chichester, pp. 401–412.

Ingram, J. (ed.) 1994. Report of the GCTE Workshop "Soil Erosion under Global Change", Paris, 29–31 March 1994. GCTE Focus 3 Association Office, Oxford, UK.

Jager, S. 1994. Modelling regional soil erosion susceptibility using the Universal Soil Loss Equation and GIS. In R.J. Rickson (ed.), *Conserving Soil Resources: European Perspectives*. CAB International, Wallingford, UK, pp. 161–177.

Leonard, R.A., Kaisel, W.G. and Still, D.A. 1987. GLEAMS: Groundwater Loading Effects of Agricultural Management Systems. *Translations of the American Society of Agricultural Engineering*, **30**(5), 1403–1418.

Ludvigsen, G.H. 1995. Soil monitoring programme in Norway 1992–1996. *European Society for Soil Conservation Newsletter*, **12**, 25–27.

Meade, R.H. 1982. Sources, sinks and storage of river sediment in the Atlantic drainage of the United States. *Journal of Geology*, **90**(3), 235–252.

Mutchler, C.K., Murphree, C.E. and McGregor, K.C. 1988. Laboratory and field plots for soil erosion studies. In R. Lal (ed.), *Soil Erosion Research Methods*. Soil and Water Conservation Society, Ankeny, Iowa, pp. 9–36.

Novotny, V. and Chesters, G. 1989. Delivery of sediment and pollutants from nonpoint sources: a water quality perspective. *Journal of Soil and Water Conservation*, **44**(6), 568–576.

Panuska, J.C., Moore, I.D. and Kramer, L.A. 1991. Terrain analysis: integration into the Agricultural Nonpoint Source (AGNPS) pollution model. *Journal of Soil and Water Conservation*, **46**(1), 59–64.

Papy, F. and Boiffin, J. 1989. The use of farming systems for the control of runoff and erosion. *Soil Technology*, **1**, 29–38.

Poesen, J. 1989. Conditions for gully formation in the loam belt of Belgium and some ways to control them. *Soil Technology*, **1**, 39–52.

Poesen, J. 1993. Gully typology and gully control measures in the European Loess Belt. In S. Wicherek (ed.), *Farm Land Erosion in Temperate Plains Environment and Hills*, Elsevier, pp. 221–239.

Poesen, J. and Ingelmo-Sanchez, F. 1992. Runoff and sediment yield from topsoils with different porosity as affected by rock fragment cover and position. *Catena*, **19**, 451–474.

Poesen, J., Boardman, J., Wilcox, B. and Valentin, C. 1996. Soil erosion monitoring and experimentation for global change studies. *Journal of Soil and Water Conservation*, in press.

Prickett, R.C. 1984. *Sheet 23 Oamaru Erosion Map of New Zealand*. National Water and Soil Organisation, Wellington, New Zealand.

Quine, T.A. and Walling, D.E. 1991. Rates of soil erosion on arable fields in Britain: quantitative data from caesium-137 measurements. *Soil Use and Management*, **2**(4), 169–176.

Risse, L.M., Nearing, M.A., Nicks, A.D. and Laflen, J.M. 1993. Error assessment in the Universal Soil Loss Equation. *Soil Science Society of America Journal*, **57**(3), 825–833.

Robinson, D.A. and Boardman, J. 1988. Cultivation practice, sowing season and soil erosion on the South Downs, England: a preliminary study. *Journal of Agricultural Science, Cambridge*, **110**, 169–177.

Romero Diaz, M.A., Cabezas, F. and Lopez Bermudez, F. 1992. Erosion and fluvial sedimentation in the Segura Basin (Spain). *Catena*, **19**, 379–392.

Sharpley, A.N. and Williams, J.R. 1990 (eds). *EPIC – Erosion/Productivity Impact Calculator: 1. Model Documentation*, US Department of Agriculture, Technical Bulletin 1768.

Slattery, M.C. and Bryan, R.B. 1992. Hydraulic conditions for rill incision under simulated rainfall: a laboratory experiment. *Earth Surface Processes and Landforms*, **17**(2), 127–146.

Slattery, M.C., Burt, T.P. and Boardman, J. 1994. Rill erosion along the thalweg of a hillslope hollow: a case study from the Cotswold Hills, central England. *Earth Surface Processes and Landforms*, **19**, 377–385.

Soil Survey of England and Wales 1983. *Soil Map of England and Wales. Scale 1:250 000*. Soil Survey of England and Wales, Harpenden.

Stocking, M. 1987. Measuring land degradation. In P. Blaikie and H. Brookfield (eds), *Land Degradation and Society*. Methuen, London.

Trimble, S.W. 1983. A sediment budget for Coon Creek basin in the Driftless area, Wisconsin, 1853–1977. *American Journal of Science*, **283**, 454–474.

Vandaele, K. 1993. Assessment of factors affecting ephemeral gully erosion in cultivated catchments of the Belgian Loam belt. In S. Wicherek (ed.), *Farm Land Erosion in Temperate Plains Environment and Hills*. Elsevier, pp. 125–136.

Vandaele, K. and Poesen, J. 1995. Spatial and temporal patterns of soil erosion rates in an agricultural catchment, central Belgium. *Catena*, **25**, 213–226.

Walling, D.E. 1983. The sediment delivery problem. *Journal of Hydrology*, **11**, 129–144.

Walling, D.E. 1990. Linking the field to the river: sediment delivery from agricultural land. In J. Boardman, I.D.L. Foster and J.A. Dearing (eds), *Soil Erosion on Agricultural Land*. Wiley, Chichester, UK, pp. 129–152.

Walling, D.E. and Webb, B.W. 1981. The reliability of suspended load data. In *Erosion and Transport Sediment Measurement*. International Association of Hydrological Sciences Publication no. 133, Wallingford, UK, pp. 177–194.

Watson, A. and Evans, R. 1991. A comparison of estimates of soil erosion made in the field and from photographs. *Soil and Tillage Research*, **19**, 17–27.

Williams, R.D. and Nicks, A.D. 1988. Using CREAMS to evaluate filter strip effectiveness in erosion control. *Journal of Soil and Water Conservation*, **34**(1), 108–112.

Williams, J.R., Jones, C.A. and Dyke, P.T. 1990. The EPIC model. In Sharpley, A.N. and Williams, J.R. (eds), *EPIC – Erosion/Productivity Impact Calculator. 1. Model Documentation*, US Department of Agriculture Technical Bulletin No. 1768, pp. 3–92.

Wischmeier, W.H. 1976. Use and misuse of the universal soil loss equation. *Journal of Soil and Water Conservation*, **31**(1), 5–9.

Wischmeier, W.H. and Smith, D.D. 1978. Predicting rainfall erosion losses – a guide to conservation planning. *US Department of Agriculture, Agriculture Handbook No. 537*.

Wolman, M.G. 1977. Changing needs and opportunities in the sediment field. *Water Resource Research*, **13**, 50–59.

Yoder, D. and Lown, J. 1995. The future of RUSLE: inside the new Revised Universal Soil Loss Equation. *Journal Soil and Water Conservation*, **50**(5), 484–489.

Young, R.A., Römkens, M.J.M. and McCool, D.K. 1990. Temporal variations in soil erodibility. In Bryan R.B. (ed), *Soil Erosion – Experiments and Models*, Catena Supplement 17, Catena Verlag, Cremlingen, Germany pp. 41–53.

23 The Particle-Size Selectivity of Sediment Mobilization from Devon Hillslopes

P. M. STONE and D. E. WALLING
Department of Geography, University of Exeter, UK

23.1 INTRODUCTION

Information on the particle-size characteristics of sediment eroded from hillslopes is necessary from the geomorphological perspective of understanding sediment mobilization and redistribution, and also from the agricultural perspective, where there is a need to understand how agricultural practices and associated erosion alter soil characteristics (e.g. Parsons *et al.* 1991). Information concerning the particle-size composition of eroded sediment is also important from a pollution perspective, since the transport of contaminants is frequently preferentially associated with the clay fraction. However, the available literature on the particle-size characteristics of eroded sediment and the relationship of these to those of source material is limited. Documented studies have generally involved field- or laboratory-based rainfall simulation experiments undertaken on bare or cropped plots (e.g. Young 1980; Parsons *et al.* 1991; Burney and Edwards 1992; Meyer *et al.* 1992; Durnford and King 1993), although one notable exception is the recent work of Slattery and Burt (1993), which investigated the size characteristics of eroded sediment in the field under natural conditions. Existing research has tended to concentrate on two aspects: first, the existence, size and stability characteristics of aggregates in the eroded sediment, and secondly, the size-selective nature of the erosion process itself.

For most soil types, it is recognized that rainsplash erosion processes detach particles both in clusters or aggregates and as primary particles (e.g. Young 1980; DePloey 1981; Loch and Donnollan 1982; Meyer *et al.* 1992; Slattery and Burt 1993). It is therefore important to distinguish what Ongley *et al.* (1981) have termed the *in situ* or effective size distribution, which includes composite particles, from the absolute size distribution provided by conventional laboratory particle-size analysis which relates to the individual primary mineral particles. It has been suggested by Young (1980) that erosion of cohesive soils generally produces both primary particles and aggregate or composite particles, whilst non-cohesive soils tend to erode mainly as primary particles (*cf.* Parsons *et al.* 1991). The precise behaviour of the clay particles within a particular soil is, however, open to some uncertainty. DePloey (1981), Loch and Donnollan (1982), and Young (1980) indicate that most

Advances in Hillslope Processes, Volume 1. Edited by M. G. Anderson and S. M. Brooks.
© 1996 John Wiley & Sons Ltd.

of the clay is eroded as aggregates and that little clay tends to be eroded as primary particles. In contrast, Meyer *et al.* (1992) found that clay is eroded both as aggregates and also as primary particles, whilst Slattery and Burt (1993) found that most of the clay eroded as primary particles. The breaking down of eroded aggregates during transport is suggested by Loch and Donnollan (1982) to generate primary clay particles. Walling (1990) has emphasized that, although it is known that much of the material eroded from cultivated fields in the UK is in the form of aggregates, there is little quantitative information on the size characteristics or composition of these aggregates. He also suggested that this information is an essential prerequisite for modelling and understanding the enrichment mechanisms associated with sediment delivery.

By comparing the particle-size distribution of the detached sediment with that of the matrix soil, it is possible to assess the degree to which there has been selective removal or enrichment of certain classes in the detached sediment. LeRoux and Roos (1983), for example, compared the particle-size distribution of sediment deposited in a reservoir 50 km south of Bloemfontein with that of soil samples from the catchment area and found that the sediment in the reservoir was significantly finer than the soils. Parsons *et al.* (1991), Poesen and Savat (1980) and Rhoton *et al.* (1979), amongst others, have also reported clear evidence of selectivity in the detachment process, but again there is some uncertainty as to its exact nature. Rhoton *et al.* (1979) suggested that the clay fraction was preferentially eroded, whereas Young (1980), in a study of silty soils, found that silt-size particles were erdoed more readily than clay or sand particles. In contrast, Slattery and Burt (1993) found that for larger rills, with higher velocity flows, there was essentially no difference between the size characteristics of the chemically dispersed soil and the sediment mobilized in runoff and that erosion was therefore not size selective. Selectivity was, however, evident in smaller rills, where discharge and velocity were lower. Particle-size selectivity may also vary through time, with changes in the relative proportions of different size classes eroded occurring as a rainfall event proceeds. Durnford and King (1993) and Miller and Baharuddin (1987) both reported changes in the relative proportions of the different size classes during a storm event, although Miller and Baharuddin (1987) attributed this selectivity to the transport limitations of runoff. Loch and Donnollan (1982), however, did not find any change in the size composition of material in runoff over time, but they did not compare the size characteristics of the eroded material with those of the matrix soil.

It is against this background of uncertainty concerning the precise relationship between the particle-size composition of eroded sediment and that of the source material that this study was undertaken. Sediment tracing studies of the sources of suspended sediment in Devon rivers have already suggested that a substantial proportion of such sediment is derived from pasture land (Walling *et al.* 1993) and in the majority of cases this is likely to be mobilized by low-intensity, longer-duration rainfall events. Very few data are available on the particle-size characteristics of sediment eroded under such conditions. The aim of this investigation was therefore to quantify these particle-size characteristics and to assess the extent to which selective mobilization of material from the hillslopes occurs. The size distribution of the material mobilized from the hillslopes has also been compared with that in adjacent streams, by documenting suspended-sediment particle-size distributions.

23.2 FIELD SITE AND METHODS

The basin of the River Dart selected for the study is $46 \, km^2$ in size and its land use comprises 10.8% cultivated, 82.2% grassland and 7% woodland (MAFF 1993). The catchment is underlain by Upper Carboniferous Culm Measures, which comprise sandstones interbedded with shales, mudstones and siltstones. The basin has well-developed interfluves and incised valleys with an average slope angle of 11°. Permanent pasture is important on the steeper slopes. Suspended-sediment tracing studies suggest that over the longer term c. 48% of the sediment yield is derived from grassland, c. 31% from cultivated land and c. 21% from channel banks (Walling *et al.* 1993). Mean annual precipitation is almost 900 mm, with 60% falling in the winter season. A comparison of rainfall intensity/duration data with peak suspended-sediment discharges in the river suggests that storms mobilizing significant amounts of sediment during the winter tend to be of longer duration (3–10 h) and of lower maximum 1 h intensity (2–8 mm/h) than summer storms which are typically of shorter duration (< 3 h) and have higher maximum 1 h intensities (8–12 mm/h) with short bursts of very high intensity rainfall (up to 50 mm/h). Rainfall intensity/duration data assembled for the River Dart indicate that a storm with an intensity of 7 mm/h lasting for approximately 3 h has a 1 year return period and is representative of important sediment-generating events.

In order to control the variables involved in the soil erosion process, a field portable rainfall simulator was used to mobilize sediment from natural plots under constant rainfall characteristics. The rainfall simulator employed was based on the design described by Bowyer-Bower and Burt (1989). This provided simulated rainfall over a plot $1 \times 0.5 \, m$ with a median drop diameter of 2.4 mm, a fall height of 1.50 m and a range of intensities from 3 mm/h to 50 mm/h. The plots used were unbounded, since it was felt that the presence of a boundary could interfere with surface runoff and alter the particle-size distribution of the mobilized sediment. Runoff and sediment were collected using a trough at the base of the plot. The trough was 0.5 m long and collected runoff across the full width of the plot. This meant that the majority of the runoff generated on the plot was collected in the trough. Slight losses undoubtedly occurred from the sides of the plot, since a boundary was not used. The trough was tightly sealed to the ground surface so that no water could seep beneath it. The 3-h 7 mm/h design storm was judged to be representative of both summer and winter storms and was used for the simulation experiments. In an attempt to represent the heavy bursts of rainfall associated with summer rainfall, 30 mm/h rainfall was applied for 20 mm, if little or no runoff was generated after 140 min of 7 mm/h simulated rainfall.

Simulation experiments were undertaken on pasture areas, since these represent the main suspended sediment source. In order to investigate the effect of seasonal changes in ground conditions on sediment production, a site was chosen with permanent pasture and a constant slope angle of 11° and simulation experiments were undertaken on this site at monthly intervals during the period October 1994 to June 1995. Replicate measurements were obtained whenever possible. The representativeness of these results, in relation to the whole basin, was then tested by selecting three further sites where single experiments were performed. The land use and underlying geology of these sites was the same as for the main site, but the slope varied. A simple measure of antecedent soil moisture was calculated for each monthly visit to the single plot and for each of the experiments on the additional plots, to facilitate comparisons.

The total volume of runoff and associated sediment generated during each experimental run was collected every 20 min from the trough at the base of the plot. One of the samples collected during peak runoff was quickly returned to the laboratory for analysis of its effective particle-size distribution using a Coulter LS130 laser diffraction device. Runoff volume was recorded for all samples and the sediment recovered from the samples was subsequently freeze-dried to determine its mass. The sediment was then treated with hydrogen peroxide to remove the organic fraction, dispersed in sodium hexametaphosphate and disaggregated using ultrasonics. The dispersed particle-size distribution was again measured using the Coulter LS130 laser diffraction equipment. Soil samples, taken adjacent to the plot before the simulation run, were subjected to similar pretreatment and their absolute (dispersed) particle-size distributions were measured. Measurements of both effective and absolute particle-size composition were used to characterize the sediment mobilized from the plot experiments, since the former reflects the nature of the mobilized particles whereas the latter reflects the primary grain-size composition of the mobilized sediment. Comparison of the absolute particle-size composition of mobilized sediment with that of the parent soil provides a definitive measure of the particle-size selectivity of sediment mobilization, whereas any attempt to investigate the detachment- and transport mechanisms involved should be based on the effective size distributions. Suspended sediment samples for comparison with the mobilized sediment were collected from the river at the downstream gauging station during a range of storm events, and these were characterized by measurements of both effective and absolute particle-size distributions.

23.3 RUNOFF GENERATION AND SEDIMENT MOBILIZATION

The runoff amounts generated during the simulation runs undertaken each month exhibit two general trends related to variation during individual simulated storms and seasonal variation (Figure 23.1). In all cases no runoff was generated within the first 20 min of the simulation. With the exception of the October and February simulation runs, runoff volume during the autumn/winter (October to March) period subsequently increased steadily up to 80–100 min, when peak runoff occurred. Volumes then decreased slightly towards the end of the simulation. This decrease in runoff towards the end of the simulation run probably reflects minor changes in the area contributing to the trough at the base of the unbounded plot. The spring simulation runs (April to June) did not produce such clear patterns of runoff variation during simulated storms. Peak runoff for the April simulation occurred

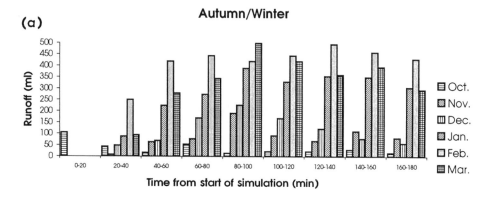

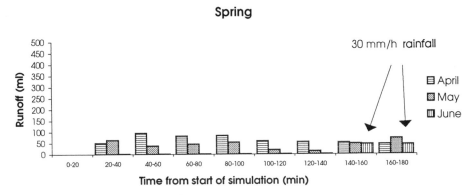

Figure 23.1 Runoff volumes measured during monthly simulation runs for (a) autumn/winter and (b) spring

after 40–60 min and steadily declined to the end of the simulation, whilst runoff from the May simulation displayed no clear trend. The June simulation demonstrated the importance of rainfall intensity in generating runoff, since only a trace was collected at each sampling time up to 140 min, after which 30 mm/h rainfall was applied and significant runoff was collected.

A clear seasonal variation in the volume of runoff generated can be identified, with the greatest volumes being generated towards the end of the winter season (February and March). The smallest volumes of runoff were generated during the April to June simulations, with virtually no runoff generated from 7 mm/h rainfall during the June simulation. Since the monthly rainfall simulation experiments involved a site with a constant slope and essentially constant ground cover of permanent pasture, the key variable influencing variations in runoff amount is clearly the antecedent moisture status of the soil. Considerably more natural rainfall had fallen on the plot over the winter months and the average infiltration rate changed from 180 mm/h in October to 2 mm/h in March. The infiltration rate increased after the winter, with little natural rainfall occurring during May or June. The runoff associated with the June simulation experiment was only generated when higher intensity rainfall, which exceeded the infiltration rate, was applied. These findings are consistent with those of Istok and Boersma (1986) who found that in regions of low-intensity rainfall, runoff and soil loss are closely related to antecedent soil moisture conditions, since there is no high intensity rainfall causing infiltration excess overland flow and erosion. They found that low-magnitude events can generate runoff where there has been sufficient antecedent rainfall and they suggest that runoff is not necessarily linked to high-intensity rainfall. Kwaad (1991) also noted the difference between saturation overland flow in winter, as opposed to infiltration-excess overland flow in the summer in his study of soil erosion on cultivated loess soils in the Netherlands.

The amounts of sediment in the runoff collected from the monthly simulation runs evidence similar patterns to runoff, both during individual runs and seasonally (Figure 23.2). The greatest amounts of sediment were collected from the March simulation which also produced the greatest runoff. The smallest amounts of sediment were produced during the spring simulations. The amount of sediment mobilized by the higher-intensity rainfall during the June simulation run was small in relation to that produced by the low-intensity rainfall during the winter simulation runs. Sediment concentrations did not vary substantially from month to month, except for the February and March runs where the concentrations were considerably greater than those for the other months. This would suggest that antecedent moisture is also important in controlling the amount of sediment in runoff during the erosion process. Kwaad (1991) found that soil losses were high in summer due to higher-intensity rainfall, but also found that soil loss increased with soil wetness during winter, when rainfall intensities are lower.

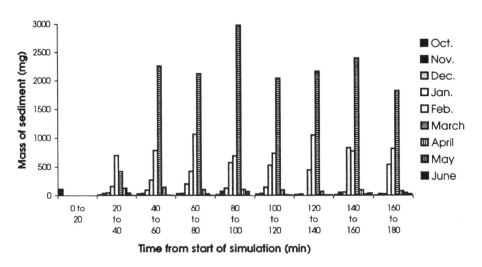

Figure 23.2 The total amount of sediment associated with runoff samples collected at 20-min intervals during the monthly simulation runs

23.4   SIZE-SELECTIVE MOBILIZATION OF SEDIMENT

At its simplest level, the particle-size selectivity of erosion from hillslopes can be measured by comparing the absolute (chemically dispersed) size characteristics of sediment in runoff with the equivalent absolute particle-size distribution of the source material or soil being eroded. A soil sample was collected each month immediately adjacent to the site of the simulation run before each of the monthly experiments and its absolute particle-size distribution was measured. This was compared with the absolute particle-size distribution of the sediment collected in the runoff samples. The soil samples showed slight variations in the volume percentage of clay ($<2\,\mu$m), silt (2–63 μm) and sand ($>63\,\mu$m) between the individual months, with clay percentages ranging from 3.35% (November) to 7.47% (March), silt from 31.12% (May) to 39.93% (November) and sand from 56.32% (January) to 61.84% (May). However, these variations are likely to reflect the natural spatial variability of soil properties rather than any significant seasonal trend. The sediment in runoff displayed variations in absolute particle-size composition both between months and to a lesser extent during individual simulation runs. For the purposes of determining whether the erosion process results in selective mobilization of sediment, or whether material was detached *en masse* from the soil, the absolute size class data for the soil and collected sediment were each averaged over the study period to permit a simple comparison (Figure 23.3). Figure 23.3 clearly shows that the particle-size distribution of the sediment in runoff is not the same as that for soil samples. There is more clay and silt and less sand in the sediment compared to the soil sample. Relative to the soil, the runoff sample is enriched 3.72 times in clay and 2.11 times in silt, whereas the sand is depleted 16.39 times. There is thus clear evidence that clay and silt are preferentially mobilized from the soil, whilst the sand-sized material is retained. Analysis of the absolute particle-size distributions of the sediment collected in all the individual 20-min runoff samples indicates that the maximum particle size observed was 229.7 μm from the 080 to 100 min sample collected during the March simulation run. This sample contained 4.6% of sand-sized material in comparison to the average of 3.6% in all samples and thus, although Figure 23.3 is based on

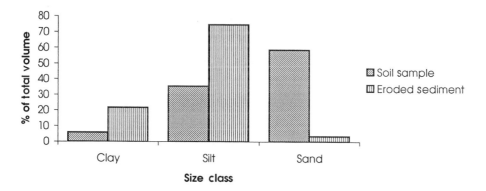

Figure 23.3   A comparison of the mean absolute particle-size distributions of source soil and sediment mobilized in runoff

average particle size data, the selective erosion identified would appear to be generally representative.

23.5 SIZE CHARACTERISTICS OF MOBILIZED MATERIAL

As indicated previously, it is widely recognized that much of the material mobilized from hillslopes consists of both primary and composite particles, or aggregates, which can be represented by the effective particle-size distribution. Comparison of measurements of the effective size distribution of samples collected from the monthly simulation experiments, obtained by quickly returning the samples to the laboratory, with equivalent measurement of the same samples after chemical dispersion, confirmed that the sediment collected comprised both aggregates and primary particles. One sample from each simulation run was analysed in this way, except in the case of the October simulation run when no measurements were made. The cumulative effective particle-size distribution associated with each of these effective size distribution measurements is shown in Figure 23.4. There is a clear difference between the effective size distributions measured during the November to March period and those measured between April and June. The overall size distributions in the latter group evidence larger particles than those obtained for samples collected between November and March. The median particle size (D_{50}) of the November to March group ranges from 9.8 μm in January to 14.0 μm in November, whilst the D_{50} of the April to June group varies from 22.4 μm in June to 25.8 μm in April. Meyer *et al.* (1980), amongst others, also report seasonal differences in the size distributions of

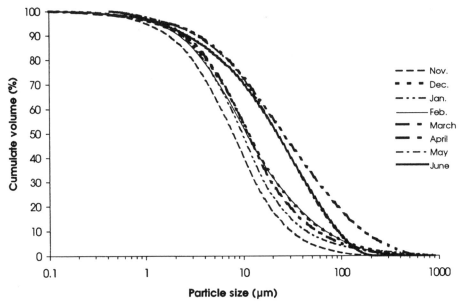

Figure 23.4 Effective particle-size distributions of sediment collected from the monthly simulation runs

eroded sediment. It would appear that the seasonal differences in runoff volume and amount of sediment mobilized identified previously are also reflected by the effective particle-size characteristics of the mobilized sediment. More particularly there would appear to be an inverse relationship between the amount of sediment mobilized and the D_{50} of this sediment.

The contrast between the effective and absolute particle-size distributions of sediment recovered from the runoff samples is illustrated by Figures 23.5(a) and (b) which present results for the January and April samples. The January sample shows a 3.14-fold increase of clay-sized ($<2\,\mu m$) material in the dispersed sample, but it is also evident that 7.5% of the volume of sediment represented by the effective size distribution was detached as primary clay particles. There is essentially no difference in the magnitude of the fine silt (2–$20\,\mu m$) fraction associated with the effective and ultimate size distributions, but there is a 2.87-fold increase in the coarse silt fraction

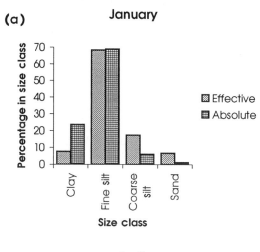

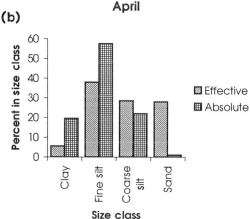

Figure 23.5 A comparison of the effective and absolute particle-size characteristics for samples collected from monthly simulation runs during (a) January and (b) April

(20–63 μm) and a 5.34-fold increase in the sand fraction ($> 63 \mu$m) associated with the effective size distribution, as compared to the absolute size distribution. The April sample shows a 3.45-fold increase in the proportion of clay in the ultimate size distribution as compared to the effective distribution, and that 5.63% of the effective size distribution is composed of primary clay. However, the magnitude of the fine silt fraction associated with the chemically dispersed sample is 1.52 times greater than that associated with the effective distribution, rather than being closely similar as in the case of the January sample. As with the January sample, the magnitudes of the coarse silt and sand fractions associated with the effective size distribution are increased. In this case, however, the values are 1.29 and 26.16 times respectively. The larger effective size fractions associated with the samples collected between April and June are composed mainly of sand-sized composite particles, since only 1.07% of the chemically dispersed April sample is composed of primary sand-sized material. The smaller increase in coarse-silt sized particles between the absolute and effective size distributions for the April sample confirms that the aggregates in this sample are generally sand-sized. This is in contrast to the January sample for which the effective size data indicate that significant amounts of both coarse silt- and sand-sized aggregates occur. This in turn would explain why the January effective particle-size distribution is finer than that of the April sample.

The mobilization of clay-sized material shows a slight variation between the January and April samples, with respectively 7.54% and 5.63% of each sample being composed of clay-sized material in the effective particle-size distributions. Similar increases in the clay content of the absolute size distribution relative to the effective distribution for the two samples of 3.14 times and 3.45 times respectively suggest that the additional clay associated with the absolute size distribution can be accounted for in terms of clay particles eroding as components of larger aggregates.

This seasonal variation in aggregate particle size is consistent with available information on seasonal variations in soil aggregate stability. For example, Bullock et al. (1988) in an analysis of three western US soils found that cooling temperatures in the autumn increased water content, and subsequent freezing were capable of destroying aggregate bonds so as to create less stable "microaggregates". These bonds then reformed as the non-freezing season progressed, so that by the end of the summer season the aggregates were at their most stable. Mulla et al. (1992) measured wet aggregate stability in the Palouse region of eastern Washington where the majority of rainfall for the year fell in the November to March period and very little fell from July to October. The coldest temperatures were also found in the November to March period. They found that the aggregate stability ratio decreased significantly from October through to March and increased significantly between March and June. Similarly, Caron et al. (1992) found a negative linear relationship between wet aggregate stability and gravimetric water content, suggesting that this was due to an increase of slaking and mellowing processes during wetting. Imeson and Kwaad (1990) cite the results of a study of a loess soil in Dutch South-Limbourg by Van Eijsden (1986) which looked at variations in aggregate stability during the growing season, and suggest that aggregate stability is lowest during the wetter May to July period and higher in the drier August to September period. Studies of aggregate stability in Ontario undertaken by Perfect et al. (1990) also demonstrated that stability decreased with increasing soil moisture. The results from the monthly rainfall simulations which evidence the mobilization of larger aggregates during the months with drier ground conditions and of smaller aggregate particles during wetter winter months are therefore consistent with seasonal contrasts in soil aggregate stability associated with changing soil moisture conditions. The more stable aggregates associated with drier ground conditions, are mobilized as larger aggregate particles.

Using the chemically dispersed or absolute particle-size data from the monthly simulation runs, the variation in the primary grain-size composition of the sediment mobilized during the simulation runs was characterized according to the percentage of clay, silt and sand. These results are presented in Figure 23.6. No clear pattern can be identified for the 20 min time increments during the simulations for any of the size classes. The monthly mean percentages for each of the three size classes (Table 23.1) suggest that there may be a seasonal variation in the mobilization of clay-sized material, but seasonal variations in silt- or sand-sized material are less clear. In order to compare the results from each of the monthly simulation runs statistically, a Mann–Whitney test with 95% confidence level was applied to the 20 min data from each monthly simulation run for each of the clay, silt and sand fractions. The results suggest that there was a distinct seasonal variation between the October to March (autumn/winter) and the April to June (spring) simulation runs. When applied to the clay fraction, the test shows that there is no significant difference between the proportion of clay in the November to February samples or in the April to June samples. The October and March samples were found to be statistically similar but were both different from the other autumn/winter samples. The November to February samples were also significantly different to those from April to June. This suggests that there is a difference in the proportion of clay in mobilized sediment between the autumn/winter and spring seasons.

The silt content of mobilized sediment was found not to vary significantly within the April to June simulations. Similarly, no significant pattern could be identified within the October to March simulation runs. However, a significant difference was identified between the November to March and April to May simulations, suggesting that the seasonal contrasts in clay content identified above may also exist for silt.

The tests for significant differences between the simulation results for sand suggest that there is no significant difference within the defined seasons or between seasons. The one notable exception to this is the proportion of sand associated with the February simulation which is significantly different from that in all other runs, except for January.

The mean values for absolute particle size associated with each of the monthly simulator runs, presented in Table 23.1, indicate that the clay content of mobilized sediment is generally greatest in the November to February period, whilst sand percentages are generally largest in April to June. The variations in the percentages of silt between seasons have been shown to be statistically significant, but the differences involved are relatively small. By further breaking down the silt class into fine (2–20 μm) and coarse (20–63 μm) silt, monthly and seasonal differences can be identified (Figure 23.7). Monthly percentages suggest that there is an increase in the fine silt content of mobilized sediment from October to February and then a decrease

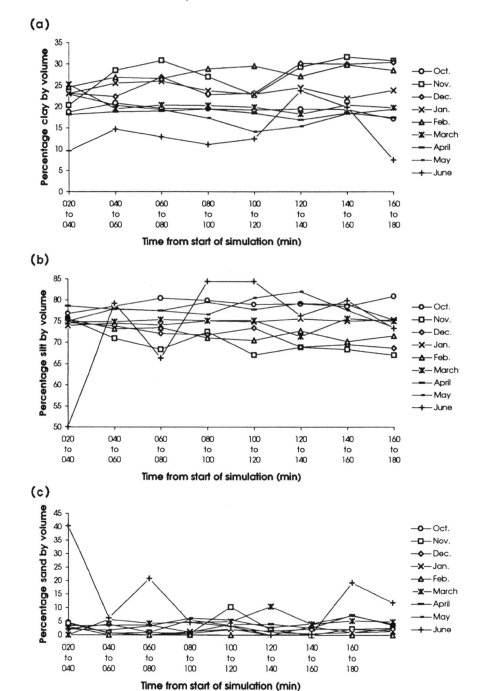

Figure 23.6 Temporal variation in the absolute particle-size composition of mobilized sediment from monthly simulation runs as demonstrated by the magnitude of the (a) clay, (b) silt and (c) sand size classes

Table 23.1 The mean volume percentage particle-size composition of chemically dispersed sediment collected from the monthly simulation runs

	Oct.	Nov.	Dec.	Jan.	Feb.	March	April	May	June
Clay (%)	19.3	27.6	26.1	23.9	27.7	20.5	18.3	18.4	14.0
SE	0.3	1.4	1.3	0.5	0.6	0.7	0.3	1.0	1.9
Silt (%)	79.2	69.8	71.6	74.8	72.2	74.6	78.1	77.5	74.2
SE	0.4	1.0	0.8	0.2	0.6	0.5	0.5	1.0	4.0
Sand (%)	1.6	2.6	2.3	1.3	0.0	4.9	3.6	4.1	11.8
SE	0.4	1.2	0.6	0.5	0.0	1.0	0.6	0.7	5.0

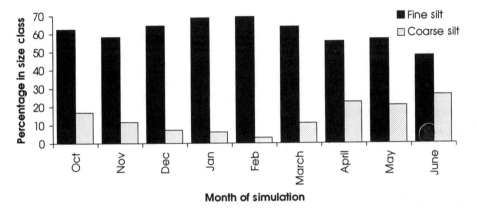

Figure 23.7 The proportions of fine and coarse silt associated with the absolute particle-size distributions of sediment collected from monthly simulation runs

from February to June. The coarse silt fraction decreases from October to February and then increases from February to June. The fine silt fraction collected from the February simulation is 22.39 times greater than the coarse silt fraction, whilst it is only 1.8 times greater for the June simulation. These results indicate that preferential removal of fine silt occurs during the autumn/winter months, whilst greater amounts of coarse silt are removed during the spring months.

23.6 EXTENSION OF THE MONTHLY SIMULATION RUN RESULTS TO THE BASIN SCALE

In an attempt to apply the results of the monthly simulation runs in developing an improved understanding of the particle-size characteristics of sediment mobilized from the hillslopes of the River Dart basin as a whole, three additional comparative simulation experiments were undertaken during the period April to May 1995 at randomly chosen locations within the basin. The three sites were located at Uppincote, Yate and Thongsleigh and one simulation run was undertaken at each. Both the land use and the characteristics of the rainfall applied were the same as for the regular monthly simulation runs. The slope angle of the three sites varied from 9° to 12°. The measured infiltration rate for each of the plots was similar to that of the April monthly simulation run and data from this run were therefore chosen for comparison. The size distribution of the natural and dispersed sediment samples from all four simulation runs are shown in Figure 23.8. The effective size distributions of all the samples contain similar proportions of each size fraction. The D_{50} of the samples range from 21.6 μm at Uppincote to 30.8 μm at Yate. The D_{50} of the effective distribution for the April run was 25.8 μm. The median grain size and size proportions characteristics of these effective particle-size distributions are similar to those obtained from the monthly simulation runs for the April to June group and reflect the similarities between ground conditions at that time and those of the comparative plots. The chemically dispersed samples also display similar size characteristics between sites for all of the classes. A Mann–Whitney significance test (95% level) of a comparison of the 20 min effective particle-size data from the April monthly simulation run with that from the Thongsleigh simulation shows that there is a significant difference between the percentages of clay in the samples, but that there is no significant difference between the magnitude of the fine silt, coarse silt and sand fraction. Whilst the size class data from Yate and Uppincote are not identical to

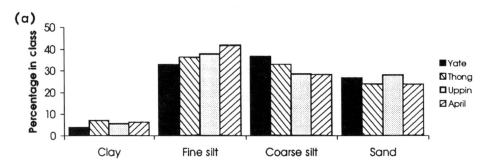

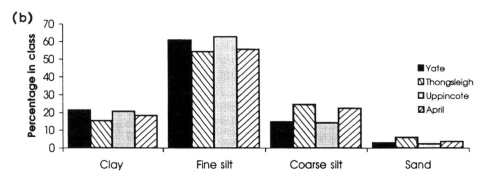

Figure 23.8 A comparison of (a) effective and (b) absolute particle-size class data obtained for simulation experiments carried out at other sites within the River Dart basin. Equivalent data for the April simulation run are also presented

those of the April or Thongsleigh results, the particle-size characteristics are generally similar to those associated with the samples from the monthly experiments undertaken when the ground conditions were drier. The monthly simulations suggested that when the infiltration rate is high and the ground cover more dense, the aggregates eroded tend to be mainly sand-sized and larger than those mobilized during periods with lower infiltration rates and less dense cover. Whilst not exactly the same as the monthly simulation run results, the results from comparative simulations displayed the characteristic particle-size composition associated with the monthly simulation runs with similar ground and antecedent conditions. In this respect they can be regarded as similar, and from this evidence it is suggested that the results from the monthly simulation runs are broadly representative of the particle-size characteristics of material mobilized throughout the basin under similar conditions.

23.7 COMPARISON BETWEEN SEDIMENT MOBILIZED FROM THE HILLSLOPE AND SUSPENDED SEDIMENT COLLECTED FROM THE RIVER

Direct comparisons between the grain-size composition of suspended sediment collected from the basin outlet and the results of the simulation experiments are difficult, because the simulation runs are by their very nature controlled experiments. Natural rainfall conditions are characterized by substantial variations in intensity and duration. These variations may cause subtle differences in the size distribution of eroded material and thus render direct comparisons difficult. Furthermore, it is also necessary to take account of the transformation of the particle-size composition of mobilized sediment as it is transported downslope and through the channel system. Unstable aggregates may be broken up in overland flow, or transport limitations may exist such that only the finest material reaches the base of the hillslope. There is little evidence available to indicate whether the aggregates found in suspended sediment represent surviving soil aggregates or the development of new aggregates within the in-stream system. In addition, a proportion of the suspended sediment may be derived from other sources such as channel banks. However, the trends identified from the monthly simulation runs, involving the presence of coarser aggregates during the months with drier ground conditions and seasonal variations in the absolute grain-size distribution of mobilized sediment might be expected to be paralleled by suspended sediment transport by the stream.

Suspended sediment samples collected from individual storm events will vary in their particle-size composition through an event, but for the purposes of this comparison a representative selection of samples was used so that several storms could be considered. Five samples collected from individual floods occurring at different times of the year are used here in an attempt to provide samples representative of a range of ground moisture and cover density conditions. The effective and absolute grain-size composition of these samples is shown in Figure 23.9. The results from the monthly simulation runs presented previously showed that the effective particle-size distributions of the autumn/winter samples were finer than

(a)

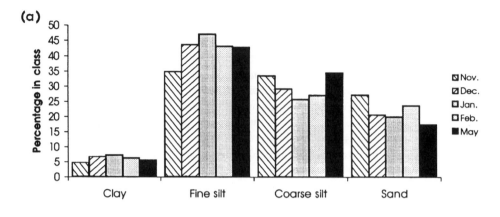

(b)

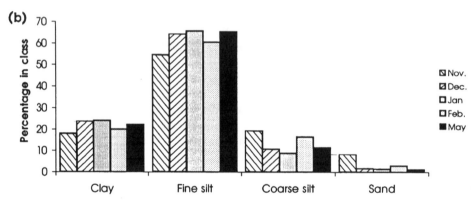

Figure 23.9 (a) Effective and (b) absolute particle-size composition of suspended sediment samples representative of individual months

those measured in spring. The effective size distribution data for the suspended sediment samples presented in Table 23.2 show that there is a slight variation in the proportion of clay-sized material, with values ranging from 5.5% in May to 7.3% in January and a similar pattern is evident for the fine silt-sized fraction with values ranging from 42.8% in May to 47.1% in January. The proportion of coarse silt-sized material is least in January (25.7%) and greatest in May (34.4%). Variations in the

Table 23.2 The volume percentage effective particle-size composition of suspended sediment samples collected from the outlet of the study catchment

	Nov.	Dec.	Jan.	Feb.	May
Clay (%)	4.65	6.67	7.29	6.26	5.49
Fine silt (%)	34.82	43.63	47.09	43.09	42.75
Coarse silt (%)	33.44	29.12	25.73	27	34.38
Sand (%)	27.09	20.58	19.89	23.65	17.38

percentages of sand-sized material associated with the effective grain-size distribution does not suggest any clear pattern, since the percentage of sand-sized material encountered in January is similar to that in May (19.9% and 17.4% respectively). The differences in clay- and silt-sized material suggest that the effective size distribution of suspended sediment is slightly coarser for the May sample than that collected in January. However, a Mann–Whitney significance test (95% confidence) applied to the effective size distributions of the suspended sediment shows that there is no significant difference between the samples. This situation contrasts with that for the effective size data obtained from the simulation runs which showed that the January and May samples were statistically different. The seasonal differences identified in the monthly simulation runs are therefore less clear in the effective particle-size data for suspended sediment.

The absolute grain-size distributions for suspended sediment (Table 23.3) exhibit similar trends between months to those of the effective grain-size distributions, but the variations are again not statistically significant and as with the effective particle-size distributions, the seasonal variations in absolute particle size identified from the monthly simulation runs are less clear for the suspended sediment.

There is, however, clear evidence of increases in the proportion of clay- and fine-silt sized material in the chemically dispersed samples, as compared to the effective distribution, and similarly there is an increase in the proportion of coarse silt- and sand-sized material in the effective size distributions. Comparison of the effective and absolute particle-size distributions of suspended sediment therefore suggests that the aggregate particles are predominately sand sized, but with a substantial fraction in the coarse silt size. The aggregates collected from the simulation experiments showed a seasonal variation. The effective particle-size measurements for suspended sediment do not exhibit this clear seasonal distinction.

A comparison of a rainfall simulator runoff sample and a suspended sediment sample for January and May is presented in Table 23.4. In the case of the January samples, the ultimate particle-size distributions are statistically similar, but the effective size distribution of the suspended sediment is associated with an increase in sand and coarse silt-sized material and a reduction in fine silt, as compared to the eroded sediment. The suspended sediment aggregates are coarser than those associated with the eroded sediment. The effective particle-size distributions of the two May samples are statistically similar, but when the absolute distributions are compared, the clay and sand fractions are seen to be similar, whilst there is an increase of fine silt and a reduction of coarse silt in the suspended sediment sample.

Table 23.3 The volume percentage absolute particle-size composition of suspended sediment samples collected from the outlet of the study catchment

	Oct.	Nov.	Dec.	Jan.	Feb.	March	April	May
Clay (%)	20.86	18.08	23.55	24.13	19.93	21.99	22.93	22.15
Fine silt (%)	61.9	54.52	64.12	65.59	60.49	66.87	65.71	65.25
Coarse silt (%)	15.25	19.24	10.73	8.91	16.57	9.26	9.95	11.46
Sand (%)	1.99	8.16	1.6	1.37	3.01	1.88	1.41	1.14

Table 23.4 A comparison of the effective and absolute particle-size composition of suspended sediment and runoff samples collected in January and May

	January				May			
	Suspended sediment		Runoff		Suspended sediment		Runoff	
	Absolute	Effective	Absolute	Effective	Absolute	Effective	Absolute	Effective
Clay (%)	24.1	6.3	23.9	7.5	22.2	5.5	18.4	5.7
Fine silt (%)	65.6	43.1	68.7	68.4	65.3	42.8	56.8	40.7
Coarse silt (%)	8.9	27.0	6.1	17.4	11.5	34.4	20.8	35.3
Sand (%)	1.4	23.7	1.3	6.7	1.1	17.4	4.1	18.3

The absolute particle-size distribution of suspended sediment is therefore finer than that of the sample of mobilized sediment. The effective particle-size distribution for the January suspended sediment is statistically similar to that for both the May mobilized sediment and suspended sediment. The seasonal differences in the grain-size composition of sediment mobilized from the hillslopes that was demonstrated by the monthly simulation runs is therefore not evident for the monthly suspended sediment data.

23.8 DISCUSSION

The results of the monthly simulation runs show clear evidence of particle-size selectivity when the grain-size distributions of sediment collected in runoff and the original soil are compared. The mobilized sediment consists of a combination of aggregates and primary particles which were found to vary in size, composition and relative contribution between the defined seasons. Analysis of the chemically dispersed or ultimate particle-size distribution of mobilized sediment indicated that grain-size selectivity varied on a seasonal basis, such that, in general, more clay- and fine silt-sized material was eroded during the autumn/winter months than during the spring months. During the spring months relatively more coarse silt- and sand-sized material is eroded. Since the simulation runs were undertaken at a site with constant slope and land use, using a constant rainfall intensity and duration, these seasonal differences in particle-size characteristics must reflect natural changes in plot condition between months. From October to March the average infiltration rate at the site decreased from 180 mm/h to 2 mm/h and the ground conditions almost reached saturation. At the same time, the density of the grass cover decreased, so that in February and March significantly more bare soil was exposed. From April to June there was very little natural rainfall on the site and the average infiltration rate increased to over 250 mm/h. The density of the grass cover also increased at this time so that very little bare soil was exposed. Within the simulation runs, an increase in soil moisture and a decrease in ground cover caused increased volumes of runoff, increased amounts of sediment in runoff and an overall fining of the absolute particle size of eroded sediment, due to increasing proportions of clay and silt. As the soil moisture decreased and ground cover increased, volumes of runoff decreased, the amount of sediment in runoff decreased and the absolute particle size of eroded sediment coarsened. Effective particle-size distributions also vary with these differences in ground conditions since larger aggregates are produced from the plots when the ground is drier and more densely covered. The relative importance of primary particles and composite particles or aggregates also varies with the changing ground conditions.

Parsons *et al.* (1991) suggest that the particle-size selectivity commonly apparent when comparing mobilized sediment and the original soil is not a result of rainsplash detachment, but rather a reflection of the competence of the overland flow to transport the material away. The selectivity is thus in their view transport- rather than supply-controlled. They found that splash erosion droplets typically consisted of 86% sand, 14% silt and 0% clay, whilst the average composition of transported sediment was 50.6% sand, 30.3% silt and 19.1% clay. This transport control has also been proposed by workers such as Walker *et al.* (1978), Miller and Baharuddin (1987) and Laird and Howard (1982). Laird and Howard (1982) suggest that there is a preferential aggregate settling and clay enrichment in low flows. The results from the monthly simulation runs undertaken in this study show that the coarser sand-sized material is not removed in runoff and the results are therefore consistent with the transport limiting theory. However, the *in situ* or effective particle-size distribution of the sediment showed that the overall size distribution of mobilized sediment collected from the April to June simulations was coarser than that for the November to March simulation runs. This is not entirely consistent with the transport limiting theory, since runoff volumes measured were lower for the April to June group. The coarser distribution of the April to June simulation can be related to the different size of the detached aggregates. The aggregates eroded during the November to March period are in general smaller than those associated with the April to June simulation runs and vary in their constituent particle composition. Transport limiting theory would suggest that the larger sized material should be deposited in the lower flows, but, although the physical size may be larger, the density of the aggregate may be lower than that of an equivalent primary particle. Very few data are available on aggregate densities. One exception is the work of Rhoton *et al.* (1983) who found that the densities of eroded wet aggregates ranged between 1.98 and 2.22 Mg/m^3, in comparison with the assumed primary particle density of 2.65 Mg/m^3. Aggregates play an important role in particle-size selectivity, not only in the supply of material to runoff, but also in the transport of that material. The implication is that ground conditions have an effect on the nature of the eroded material and consequently the characteristics of transported material. Therefore the mobilization of eroded sediment could be viewed as being supply controlled.

23.9 CONCLUSIONS

The absolute particle-size distribution of sediment mobilized by simulated rainfall and collected in runoff was found to differ from the absolute particle size characteristics of the surface soil. Selective mobilization meant that clay- and silt-sized material was removed preferentially and that the majority of the sand-sized material was not mobilized. The results from the monthly simulation runs show that seasonal variations exist in the volume of runoff produced and in both the amounts of sediment in runoff and in the effective and absolute particle-size characteristics of the mobilized sediment. Measurements of the *in situ* or effective particle-size characteristics of eroded material and comparison of these with the absolute particle-size characteristics showed that sediment was eroded as a combination of aggregate particles and primary particles. Seasonal variations in effective particle size showed that, in general, larger aggregate particles were eroded during the spring months than during the autumn/winter months. When soil moisture levels are high, the density of ground cover is less and infiltration rates are low, the aggregate particles tend to be smaller. The absolute particle-size distributions also showed a seasonal variation with sediment collected in the winter season being generally finer, and containing greater proportions of clay-sized material.

A comparison of the particle-size characteristics of suspended sediment with those of sediment mobilized in runoff, indicates that suspended sediment showed little evidence of the seasonal trends in effective or absolute particle-size characteristics that had been demonstrated by the results of the monthly simulation runs, and no clear link could be established between the two.

ACKNOWLEDGEMENTS

The support of the UK Natural Environment Research Council in providing a Research Studentship (P.M.S.) is gratefully acknowledged. The co-operation of local farmers in allowing access to their land is appreciated.

REFERENCES

Bowyer-Bower, T.A.S. and Burt, T.P. 1989. Rainfall simulators for investigating soil response to rainfall. *Soil Technology*, **2**, 1–16.

Bullock, M.S., Kemper W.D. and Nelson, S.D. 1988. Soil cohesion as affected by freezing, water content, time and tillage. *Soil Science Society of America Journal*, **52**, 770–776.

Burney, J.R. and Edwards, L.M. 1992. Size distribution of sediment in rill and runoff in response to variations in ground cover, freezing, slope and compaction of a fine sandy loam. *Journal Agricultural Engineering Research*, **56**, 99–109.

Caron, J., Kay, B.D., Stone, J.A. and Kachanoski, R.G. 1992. Modeling temporal changes in structural stability of a clay loam soil. *Soil Science Society of America Journal*, **56**,1597–1604.

DePloey, J. 1981. Crusting time-dependent rainwash mechanisms on a loamy soil. In R.P.C. Morgan (ed.), *Soil Conservation: Problems and Prospects*. Wiley, pp. 139–145.

Durnford, D. and King, P. 1993. Experimental study of processes and particle-size distributions of eroded soil. *Journal Irrigation and Drainage Engineering*, **119**(2), March/April, 383–398.

Imeson, A.C. and Kwaad, F.J.P.M. 1990. The response of tilled soils to wetting by rainfall and the dynamic character of soil erodibility. In J. Boardman, I.D.L. Foster and J.A. Dearing (eds), *Soil Erosion on Agricultural Land*. Wiley, Chichester, pp. 129–152.

Istok, J.D. and Boersma, L. 1986. The effect of antecedent rainfall on runoff during low-intensity rainfall. *Journal of Hydrology*, **88**, 329–342.

Kwaad, F.J.P.M. 1991. Summer and winter regimes of runoff generation and soil erosion on cultivated loess soils (the Netherlands). *Earth Surface Processes and Landforms*, **16**, 653–662.

Laird, D. and Howard, M.E. 1982. Relationship of the nature of suspended clay minerals to hydrologic conditions. *Journal Environmental Quality*, **11**, 433–436.

LeRoux, J.S. and Roos, Z.N. 1983. The relationship between top soil particle sizes in the catchment of Wuras Dam and the particle sizes of the accumulated sediment in the reservoir. *Zeitschrift für Geomorphologie*, **27**(2), June, pp. 161–170.

Loch, R.J. and Donnollan, T.E. 1982. Field rainfall simulator studies on two clay soils of the Darling Downs, Queensland. II Aggregate breakdown, sediment properties and soil erodibility. *Australian Journal Soil Research*, **21**, 47–48.

MAFF 1993. Unpublished data.

Meyer, L.D., Harmon, W.C. and McDowell, L.L. 1980. Sediment sizes from crop row sideslopes. *Transactions of the American Society of Agricultural Engineers*, **23**(4), 891–898.

Meyer, L.D., Line, D.E. and Harmon, W.C. 1992. Size characteristics of sediment from agricultural soils. *Journal of Soil and Water Conservation*, **47**(1), 107–111.

Miller, W.P. and Baharuddin, M.K. 1987. Particle-size of interill-eroded sediments from highly weathered soils. *Soil Science Society of America Journal*, **51**, 1610–1615.

Mulla, D.J., Huyck, L.M. and Reganold, J.P. 1992. Temporal variation in aggregate stability on conventional and alternative farms. *Soil Science Society of America Journal*, **56**, 1620–1624.

Ongley, E.D., Bynoe, M.C. and Percival, J.B. 1981. Physical and geochemical characteristics of suspended solids, Wilton Creek, Ontario. *Canadian Journal of Earth Science*, **18**, 1365–1379.

Parsons, A.J., Abrahams, A.D. and Shiu-Hungluk Luk 1991. Size characteristics of sediment in interrill overland flow on a semi-arid hillslope, S. Arizona. *Earth Surface Processes and Landforms*, **16**, 143–152.

Perfect, E., Kay, B.D., van Loon, W.K.P., Sheard, R.W. and Pojasok, T. 1990. Factors influencing soil structure stability within a growing season. *Soil Science Society America Journal*, **54**, 173–179.

Poesen, J. and Savat, J. 1980. Particle-size separation during erosion by splash and runoff. In M. DeBoodt and D. Gabriels (eds), *Assessment of Erosion*. Wiley-Interscience. John Wiley and Sons, pp. 427–439.

Rhoton F.E., Smeck N.E. and Wilding L.P. 1979. Preferential clay mineral erosion from watersheds in the Maumee River Basin. *Journal of Environmental Quality*, **8**(4), 547–550.

Rhoton, F.E., Meyer, L.D. and Whisler, F.D. 1983. Densities of wet aggregated sediment from different textured soils. *Soil Science Society of America Journal*, **47**, 576–578.

Slattery, M.C. and Burt, T.P. 1993. Size characteristics of sediment eroded from agricultural soil: dispersed versus non-dispersed, ultimate versus effective. In E.J. Hickin (ed.), *River Geomorphology*. Wiley and Sons.

Van Eijsden, G.G. 1986. *Bodemerosie op landbouwpercelen in een loessebied*. Thesis, Laboratory of Physical Geography and Soil Science, University of Amsterdam.

Walker, P.H., Kinnell, P.I.A. and Green, P. 1978. Transport of a noncohesive sandy mixture in rainfall and runoff experiments. *Soil Science Society of America Journal*, **42**, 793–801.

Walling, D.E. 1990. Linking the field to the river. Sediment delivery from agricultural land. In J. Boardman, I.D.L. Foster and J.A. Dearing (eds), *Soil Erosion on Agricultural Land*. Wiley, Chichester, pp. 129–152.

Walling, D.E., Woodward, J.C. and Nicholas, A.P. 1993. A multi-parameter approach to fingerprinting suspended-sediment sources. In *Tracers in Hydrology (Proceedings of the Yokohama Symposium, July 1993)*, IAHS Publication Number 215.

Young, R.A. 1980. Characteristics of eroded sediment. *Transactions of the American Society of Agricultural Engineers*, **23**, 1139–1142, 1146.

24 Shifts in Rates and Spatial Distributions of Soil Erosion and Deposition under Climate Change

D. T. FAVIS-MORTLOCK[1] and M. R. SAVABI[2]
[1] *Environmental Change Unit, University of Oxford, UK*
[2] *National Soil Erosion Research Laboratory, US Department of Agriculture–Agricultural Research Service, West Lafayette, USA*

24.1 INTRODUCTION

There is a good deal of current interest in such notions as "environmental geomorphology" (McGregor and Thompson 1995) and "geomorphology at work" (Thorpe 1995). Whether known by these labels, or the more prosaic "applied geomorphology", the potential for geomorphological understanding to inform the environmental decisions of policy-makers cannot be doubted.

An example is the estimation of the impacts of future climate change upon soil erosion by water. However, between the kinds of responses which policy-makers require and the answers which scientists are able to give there is commonly a mismatch (Henderson-Sellers 1991). Simulation models are frequently the scientist's tool of choice when prediction is required: yet such models may well be far from ideal at elucidating the broad yet dependable trends needed by policy-makers. From a potential users point of view, models in general tend to fall into one of two categories. The first represents the system in question in a very detailed way: this however renders it ill-suited to generalization. The second is well suited to producing widely applicable results, but necessarily omits systemic richness of response in the interest of generality.

Soil erosion models are no exception. In an effort to capture the great richness of response of the erosional system, complex physically based models such as EPIC (e.g. Williams *et al.* 1990) and WEPP (e.g. Flanagan and Nearing 1995) have been constructed.

24.1.1 Physically Based Models

Conceptually, physically based soil erosion models have much in common with GCMs (note: a list of acronyms is given at the end of the chapter). Both attempt to capture, mathematically, observed responses to physical laws: in the one case, those which are understood to operate within the climate–vegetation–soil system; in the other, the global atmosphere–ocean system. Their physical basis makes GCMs the

Advances in Hillslope Processes, Volume 1. Edited by M. G. Anderson and S. M. Brooks.
© 1996 John Wiley & Sons Ltd.

preferred tools for studies of climate under conditions of cautious extrapolation, such as those of an atmosphere with an increased content of greenhouse gases (Viner 1994). Similarly, under novel conditions physically based soil-erosion models may be expected to be more reliable than empirically based models, which make use only of relationships between main inputs and outputs (Wilmott and Gaile 1992). No existing erosion models, though, are completely process-based: all incorporate some empirical elements.

GCMs and physically based erosion models also share other characteristics: complexity, and a resultant appetite for data. Since GCMs must, by definition, be global in their coverage, their complexity means that they can be run only on the largest supercomputers. The implication for physically based erosion models is rather different. Their stringent data requirements confine them to use on well-documented – and hence usually relatively small – areas: a clear disadvantage in supplying the broad trends required for policy decisions (cf. Brignall et al. 1994). Almost all model-based studies of the effects of climate change upon erosion have been confined to a description of the effects upon a small area (Favis-Mortlock and Boardman 1995; Favis-Mortlock et al. in press).

24.1.2 Climate Change and Erosion

Anthropogenically produced climate change is driven by increased levels of greenhouse gases, including CO_2: this will give rise to both direct and indirect effects upon erosion. Any increase in rainfall amount or intensity has the potential to increase erosion: this constitutes the most direct effect (cf. Boardman et al. 1990). However, CO_2-induced climate change can affect rates of erosion in at least two other, more indirect, ways (Favis-Mortlock et al. 1991; Boardman and Favis-Mortlock 1993; Favis-Mortlock and Boardman 1995). The first of these is the influence of changed temperature, rainfall and solar radiation upon rates of vegetation growth; the second results from the ability of increased levels of atmospheric CO_2 to enhance plant growth.

Since 1804 it has been known that increased concentrations of atmospheric CO_2 can enhance plant growth (Kimball et al. 1993). This is because in many plants increased concentrations of CO_2 cause a higher exchange resistance to gases such as CO_2 and water vapour between leaf and ambient air, which reduces water losses by transpiration and increases water use efficiency (e.g. Kimball and Idso 1983; Kimball et al. 1993; Wolf 1993). However, there is a great range of responses between plant species. Plants with the C3 photosynthetic pathway (such as wheat) are generally more responsive to elevated CO_2 concentrations than C4 plants (such as maize). Interactions between temperature and enhanced CO_2 are, however, not yet fully understood, and crop responses to increased CO_2 in the (relatively) competitive ecosystem of a field may be very different from those noted during the isolation of a pot experiment (Tegart et al. 1990; Kimball et al. 1993). In particular, implications for the early stages of growth – the most critical period for soil erosion, as the young plant covers the ground – are unclear (J. I. L. Morison, pers. comm., 1994). In the field, the emerging plant will also be affected by such factors as the extent to which the soil is crusted (R. Evans, pers. comm., 1995).

Interactions between direct and indirect effects complicate erosion's response to climate change. In a study using the EPIC model, Favis-Mortlock and Boardman (1995) noted that the interplay between climate change-induced shifts in the timing and amount of rainfall and the rate of crop growth in a changed and CO_2-enriched climate gave rise to complex non-linear responses (cf. Woo *et al.* 1992). In particular, simulated erosion rates for wet years were seen to increase much more than those for dry years. To what extent, though, are these responses model-specific?

It is clear that any model which is used to study climate change and erosion must consider both direct and indirect effects. Until recently, EPIC was the only available erosion model which incorporated the effects of CO_2 concentration on plant growth. The development of a CO_2-sensitive version of WEPP (Savabi *et al.* in press), however, meant that this model might now also be used as a tool for climate change studies. This appears to offer some advantages. The EPIC model, while possessing a strong physically based hydrological component in particular (Williams *et al.* 1990), none the less still retains several empirical elements. These include the use of the SCS Curve Number approach for infiltration estimation (USDA-ARS 1972) and the use of variants of the Universal Soil Loss Equation (Wischmeier and Smith 1978): thus rill and interrill erosion processes are not considered separately, and soil crusting is not explicitly represented. Additionally, EPIC considers only plane slopes, and can only estimate an erosion rate for a point on a hillslope (i.e. deposition is not explicitly considered) (Williams *et al.* 1990). By contrast, WEPP is more strongly based on fundamental relationships for infiltration, surface runoff, soil consolidation and erosion mechanics (Lane *et al.* 1992). A version of the Green and Ampt equation, modified to consider ponding, is used for infiltration. Soil detachment, transport and deposition processes are represented using a steady-state continuity equation which considers rill and interrill processes, under both sediment- and transport-limited conditions (Lane *et al.* 1992): consequently, WEPP has the ability to estimate both erosion and deposition on complex slope forms (Flanagan and Nearing 1995). Soil crusting is considered in WEPP by a relationship which assumes an exponential decrease of hydraulic conductivity with total rainfall since tillage (Alberts *et al.* 1995). However, the plant growth component of WEPP is very similar to that of EPIC (Arnold *et al.* 1995), while the water balance submodel is derived from that of SWRRB (Williams *et al.* 1985) which is itself similar to that of EPIC. A detailed account of the EPIC model is given by Williams *et al.* (1990); Lane *et al.* (1992) present a good overview of WEPP, while a more detailed description is given in Flanagan and Nearing (1995).

The availability of a CO_2-sensitive version of WEPP ("WEPP-CO2") therefore offered the opportunity both to verify the earlier findings of Favis-Mortlock and Boardman (1995), and also to evaluate how climate change may affect the balance of erosion and deposition on a hillslope. The simulations described in this chapter thus have three main aims:

- to evaluate the CO_2-sensitive version of WEPP for the study of climate change and erosion on UK agricultural hillslopes;
- to compare WEPP results with those from earlier work using the EPIC model;

• to investigate the effect of climate change upon spatial patterns of erosion and deposition.

24.2 EVALUATION OF THE MODEL

A first step was to evaluate the ability of the WEPP hillslope model to simulate erosion under present conditions, and crop growth under present and increased-CO_2 conditions.

24.2.1 The site

WEPP was run with data for a 7.8 ha hillslope site (Figure 24.1) at Woodingdean, East Sussex (TQ360069), on the UK South Downs about 6 km southwest of Lewes. In this area, current annual rainfall varies between 750 and 1000 mm with a peak in autumn. Mean annual temperature is about 9.8 °C, the January mean being about 3.9 °C and the July mean about 16.3 °C (Potts and Browne 1983). Autumn-planted

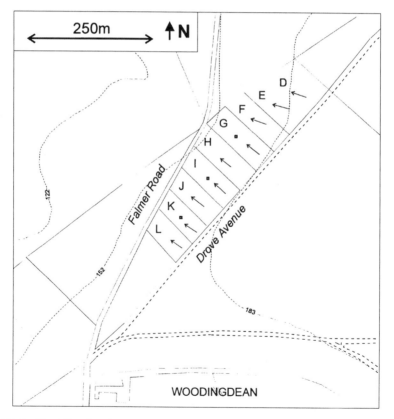

Figure 24.1 The Woodingdean site. Contour heights are in metres; the small arrows point down the line of greatest slope. Soil samples were taken at the sites marked by squares

cereals are common (Boardman 1993). Soil at the Woodingdean site is typical of the South Downs: a silty and moderately stony rendzina of the Andover series (Jarvis *et al.* 1984), less than 30 cm in thickness upon chalk. The site is described in detail by Favis-Mortlock *et al.* (1991) and Favis-Mortlock (1994).

The field was notionally divided into nine strips (D to L on Figure 24.1) with boundaries approximately down the line of greatest slope. These have a range of slope lengths and angles (Table 24.1). It was assumed that there was no significant movement of runoff between the strips; however, there is a slight convexity toward the middle of the field (strips F to J). Profiles of each strip are shown in Figure 24.2

Measured rates of erosion are available annually for the whole field for 1982–1988: in 1985–1986 measurements were also taken on individual strips. Meteorological data are available from the National Rivers Authority station at Ditchling Road, Brighton, which is about 4 km west-southwest of the field, supplemented with wind data from Gatwick Airport. Soils data were obtained from literature, and supplemented by the analysis of soil samples taken within the field (Favis-Mortlock 1994). Winter wheat was simulated: this used tillage operations typical of the South Downs (Favis-Mortlock 1994) with the crop being drilled on 28 September and harvested on 29 July.

24.2.2 Erosion

Favis-Mortlock (in press) used the WEPP model with data from this site for part of the GCTE evaluation of erosion models under present conditions of climate and land use. However, the GCTE study used the 95.7 version of WEPP; whereas the CO_2-sensitive version of the model used in this study is based on a slightly earlier release (95.001). Results from both these versions were compared with measured values.

When either version of the model was used without calibration, estimated erosion rates were much too high. This appears to be related to the stoniness (about 38% at the surface) of the Andover soil: the empirical relationships recommended for the estimation of values for baseline rill and interrill erodibility (Alberts *et al.* 1995, p. 7.1) and baseline effective hydraulic conductivity (Alberts *et*

Table 24.1   Strips of the Woodingdean hillslope

Strip	Area (ha)	Slope length (m)	Mean slope (%)
D	0.78	135	14.0
E	0.85	150	12.0
F	0.93	165	15.0
G	1.00	180	17.0
H	1.00	180	20.0
I	0.93	165	19.0
J	0.85	150	19.0
K	0.80	140	18.0
L	0.60	125	18.0

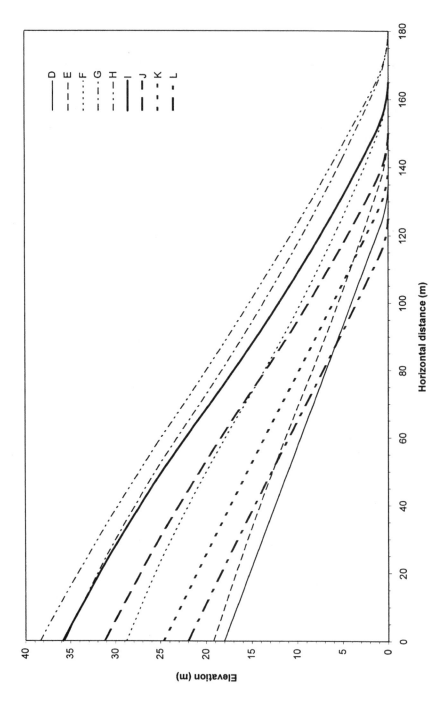

Figure 24.2 Slope profiles for strips D to L at Woodingdean. Vertical exaggeration is 2.7 : 1

al. 1995, p. 7.9) for input to WEPP do not consider stoniness. Stone content, however, can affect both infiltrability and (particularly) erodibility, which can be both positively and negatively affected (e.g. Poesen 1992). Apparent erodibility decreases with the presence of a stone cover (Evans, 1996), due to the stones' shielding effect (G. Govers, pers. comm., 1995). Boardman (1992) noted a long-term decrease in erodibility of soils on the South Downs due to increasing stoniness.

Values for the three erodibility parameters of WEPP-CO2 were therefore subjectively calibrated against measured values of soil loss. These needed to be reduced to about 50% of the values calculated using the relationships given by Alberts *et al.* (1995). The value used for the baseline effective hydraulic conductivity parameter, however, needed to be increased from that calculated, by about 40%. In a sensitivity analysis of WEPP, Nearing *et al.* (1989) found this to be the most sensitive parameter tested; the parameter for rill erodibility was also very sensitive.

Both versions of the model were then run using the same data sets. Results for simulated runoff under present conditions are shown in Table 24.2, and for simulated erosions in Table 24.3. WEPP-CO2 estimates rather lower values for

Table 24.2 Simulated runoff (mm) for strips D to L of the Woodingdean field from two versions of the WEPP model. Measured runoff was not available

Model version	Weather data	D	E	F	G	H	I	J	K	L	Mean	Rainfall	% of total rainfall
WEPP 95.001 CO2	generated 30 years	63.9	62.3	62.9	62.7	63.2	64.3	65.5	65.1	66.7	64.1	801.2	8.0
WEPP 95.7	generated 30 years	70.6	68.6	70.0	70.0	70.5	72.2	73.2	72.5	74.2	71.3	801.2	8.9
WEPP 95.7	measured 1975–88	89.7	87.6	89.4	90.2	91.2	93.6	95.6	93.4	96.1	91.9	812.1	11.3

Table 24.3 Simulated soil loss (m³/ha) from two versions of the WEPP model. Measured soil loss 1982–88 for the whole field is 2.8 m³/ha. Further details of the WEPP 95.7 validation are given in Favis-Mortlock (in press)

Model version	Weather	D	E	F	G	H	I	J	K	L	Mean
WEPP 95.001 CO2	generated 30 years	1.8	1.7	4.7	7.5	9.5	9.0	7.5	4.4	3.8	5.6
WEPP 95.7	generated 30 years	3.4	3.1	8.6	13.1	16.4	15.6	13.3	8.1	7.0	9.8
WEPP 95.7	measured 1975–88	4.5	4.2	11.3	17.1	22.0	21.0	18.1	10.9	9.6	13.2

runoff and clearly lower values (about 50% lower) for soil loss. Note that, for WEPP 95.7, the use of measured (rather than generated) weather data resulted in higher values of both runoff and erosion rate. Favis-Mortlock (1995) found that the use of the rainfall sequences produced by the EPIC weather generator instead of measured weather resulted in lower rates of erosion; this was due to an insufficient proportion of large rainfall events. However, although the CLIGEN weather generator used with WEPP is a development of the EPIC weather generator (Nicks et al. 1995), it appears much better at reproducing rainfall frequency distributions (Favis-Mortlock, in press).

24.2.3 Crop Growth

As a preliminary evaluation of WEPP's ability to simulate crop growth under current conditions, simulated winter wheat yields from both WEPP-CO2 and EPIC (version 3090) were compared. The data sets used for both models were similar in terms of management and crop parameters. WEPP does not model fertilizer-limited growth constraints in crops (Arnold et al. 1995). EPIC was therefore run without soil fertility constraint.

For current climate and CO_2 levels (350 ppmv), WEPP estimated a slightly higher yield (6.2 t/ha) than EPIC (5.8 t/ha). Measured yields for the South Downs are usually moderately low, in the range 5–6 t/ha (Favis-Mortlock et al. 1991), due to limitations of water availability (Evans and Catt 1987; Evans 1990; Burnham and Mutter 1993): thus results from both models are realistic for the South Downs.

WEPP-CO2 incorporates relationships for the effects of carbon dioxide on crop growth which were originally developed by Stockle et al. (1992a) for the EPIC model. This CO_2-sensitive version of EPIC was subsequently validated by Stockle et al. (1992b), Easterling et al. (1992a), and Rosenberg et al. (1992) and then used for a study of climate change impacts on the productivity of the US Great Plains (Easterling et al. 1992b,c). Subsequently, it was used for studies of climate-change impacts on soil erosion by Favis-Mortlock et al. (1991), Boardman and Favis-Mortlock (1993), Favis-Mortlock and Boardman (1995), and Lee et al. (in press). An initial validation of WEPP-CO2's responses to increased CO_2 was accomplished by comparing simulated yields from both models at various levels of CO_2 with values from both experimental and modelling studies in several countries (Figure 24.3).

Calibration was again necessary. Once calibrated, WEPP has a generally similar sensitivity to EPIC in response to changing CO_2 levels, although the sensitivity of WEPP decreases slightly compared to that of EPIC at very high levels of CO_2 (beyond those used in the climate change scenarios of this study). Both models, though, appear to be at the higher end of the sensitivity envelope. Results from experimental studies reported by Wolf (1993) and plotted on Figure 24.3 indicate that wheat in water-limited situations responds more strongly to increased CO_2 than wheat with no water shortage (although yields in water-limited situations are of course lower). The responses to CO_2 forcing simulated by both WEPP and EPIC appear to correspond more with those found in water-limited situations.

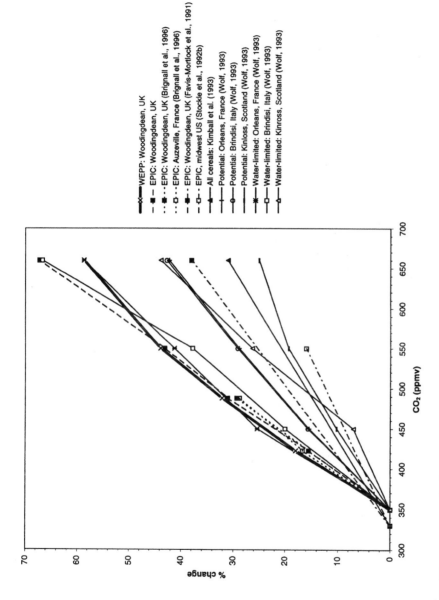

Figure 24.3 The response of winter wheat to increased CO₂. The WEPP results are shown by a thick line; for all others, measured values are shown by solid lines, simulated by broken lines. Y-axis values are percentage change in yield, except those of Wolf (1993) which are change in dry matter. For those of Favis-Mortlock *et al.* (1991), temperature was also increased (by 1.5°C)

24.3 CLIMATE-CHANGE SIMULATIONS

24.3.1 Climate Scenarios

The physical mechanisms behind the greenhouse effect are now quite well understood, and GCMs are beginning to be able to replicate large-scale atmospheric features and patterns of change (Viner 1994). However, there is still a good deal of uncertainty as to how regional climate may change in the future. Wilby (1995) gives a useful overview of the limitations of GCMs with regard to deriving projections from them for change at the regional scale; none the less, methodologies are beginning to emerge which enable the downscaling of GCM-estimated change (Viner 1994). One such is used here.

As in Brignall *et al.* (1996), projections of monthly temperature and precipitation change from MAGICC (Viner and Hulme 1992, 1993) were used to construct climate change scenarios for 2025 and 2050 for the area of the southern UK including the South Downs (Table 23.4). These projections assumed IPCC emissions scenario IS92a (Houghton *et al.* 1992), a global climate sensitivity of 2.5 °C to CO_2 doubling, and made use of the spatial change fields from the UK Meteorological Office High-Resolution GCM (Mitchell *et al.* 1990). Effects of sulphate aerosols were not considered in the GCM run. These projections suggest a global climate forcing of 0.6 °C by 2025, and of 1.4 °C by 2050.

The scenarios for the South Downs in Table 24.4 project marked increases in winter rainfall (up to 21%) by 2050. Changes in monthly rainfall are distributed in a fairly complex way; this contrasts with the simpler distribution used in earlier work (e.g. Favis-Mortlock *et al.* 1991). The CO_2 levels used here are lower than those in previous work. In all other respects the scenarios do not differ greatly from those used in earlier studies such as Favis-Mortlock and Boardman (1995).

These scenarios were then used (as in Favis-Mortlock and Boardman 1995) to perturb a 30-year sequence of present-day daily data. This was generated by CLIGEN (Nicks *et al.* 1995). Also as with previous studies, several parameters were held unchanged from present values (for lack of information regarding future change). These include the variability of both temperature and rainfall, the relative frequency of wet and dry days, and storm duration and peak intensity (Nicks *et al.* 1995). Note though that the variability of rainfall in particular is unlikely to remain unchanged in a changed climate (Waggoner 1990): rainfall events in the UK may become both more intense and less frequent under a greenhouse climate (Rowntree *et al.* 1993). WEPP-CO2 was then run with these climate scenarios.

24.3.2 Results

24.3.2.1 Long-Term Mean Rates

Changes in mean erosion rate generated by WEPP-CO2 for each 30-year climate sequence are shown in Figure 24.4. Note that these rates are for areas of net detachment (i.e. no deposition), not for the whole strip. The changes are very similar to those projected by EPIC for this site in earlier studies (Boardman and Favis-Mortlock 1993). However, the models differed in their estimates of absolute rates:

Table 24.4 Current climate details and climate change scenarios

Year	Parameter	J	F	M	A	M	J	J	A	S	O	N	D
1990	T_{max} (°C)	6.27	6.11	8.33	11.44	14.62	17.68	19.69	19.55	17.23	13.90	10.04	7.99
	T_{min} (°C)	1.32	1.09	2.70	4.42	7.47	10.65	12.69	12.76	10.94	8.23	4.87	3.41
	Rain (mm)	85.6	47.3	78.8	48.9	56.3	52.7	53.5	60.6	69.6	107.7	83.1	88.8
	Wind speed (m/s)	2.79	2.87	2.87	2.62	2.67	2.29	2.29	2.20	2.32	2.37	2.73	2.80
	CO_2 (ppmv)							350					
2025	Δ_{Tmax} (°C)	1.1	1.1	0.8	0.7	0.7	0.6	0.7	0.9	0.9	0.7	0.6	0.7
	Δ_{Tmin} (°C)	1.1	1.1	0.9	0.7	0.7	0.6	0.6	0.7	0.7	0.7	0.5	0.7
	Δ_{rain} (%)	9.4	6.7	7.9	5.5	0.5	1.5	-1.2	1.1	-0.5	4.4	0.9	6.1
	$\Delta_{wind\ speed}$ (%)	3.3	2.6	2.9	1.1	-0.5	2.0	0.9	1.7	-0.7	0.9	-0.6	1.6
	CO_2 (ppmv)							423					
2050	Δ_{Tmax} (°C)	2.3	2.3	1.8	1.5	1.7	1.4	1.5	1.9	1.9	1.5	1.4	1.7
	Δ_{Tmin} (°C)	2.5	2.3	2.1	1.7	1.5	1.4	1.4	1.7	1.7	1.5	1.1	1.5
	Δ_{rain} (%)	21.0	14.9	17.5	12.3	1.1	3.3	-2.8	2.5	-1.1	9.8	2.1	13.7
	$\Delta_{wind\ speed}$ (%)	7.3	5.8	6.5	2.3	-1.1	4.6	2.1	3.7	-1.7	1.9	-1.2	3.5
	CO_2 (ppmv)							488					

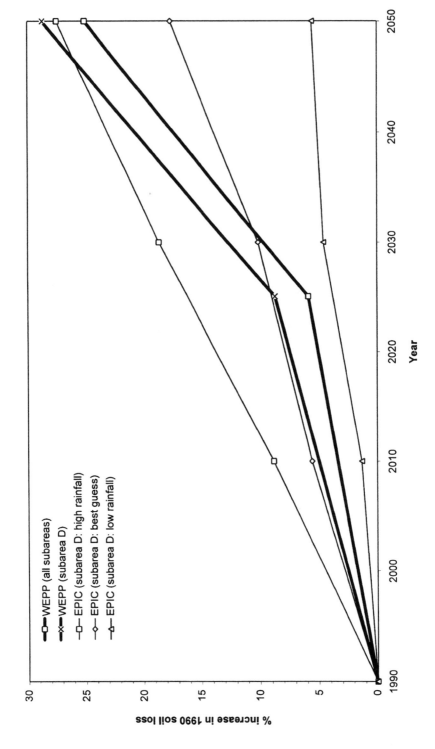

Figure 24.4 Change in mean erosion rate simulated by WEPP and EPIC (EPIC results redrawn from data in Boardman and Favis-Mortlock 1993)

the value of soil loss rate from EPIC for strip D only in 1990 was $3.2\,m^3/ha$, while the rate from WEPP was $1.8\,m^3/ha$ (Table 24.3).

The earlier EPIC results for strip D form an envelope bounded by results from the low and high rainfall scenarios. The WEPP results for strip D in 2025 fall near the centre of this envelope, very close to those of the EPIC "best guess" rainfall scenario. The WEPP results for 2050 are higher however – just outside the EPIC envelope. If the WEPP results for the whole field (strips D to L) are averaged, then these remain within the EPIC envelope.

Estimates of changes in long-term mean rate by WEPP are not distributed equally over the whole field (top line of Table 24.5). Absolute rates of change increase toward the centre of the field, although percentage changes are reasonably constant for all strips.

24.3.2.2 Rates for Individual Years

If (as in Favis-Mortlock and Boardman 1995) simulation years are divided into "wet" (annual rainfall above average for the simulation) and "dry" (annual rainfall below average), larger increases in erosion rate are estimated in wet years than in dry years (middle lines of Table 24.5). In wet years, absolute rates of change appear to increase toward the centre strips, with percentage rates fairly constant. In dry years, by contrast, changes in erosion rate appear roughly constant over the whole field in absolute terms.

However, some caution is needed in inferring spatial patterns of change in wet years and dry years since some of these responses are artefacts which result from the very low rates simulated (Table 24.6). In simulation years 5, 6 and 21 for example, change in rate between the 1990 and 2050 scenarios is very small in absolute terms, yet sizeable when expressed as a percentage. None the less, it can be seen from Figure 24.5 that while the same years are years of high or low erosion rate for both strips D and H, there are differences in the relative magnitudes of their responses for individual years. The centre areas in general show a greater range of response between wet and dry years than do the edges (bottom line of Table 24.5).

24.3.2.3 Rates for Individual Erosion Events

Figure 24.6 shows the relationship between 1990 rainfall and crop cover in 1990 and 2050 on strip H for two contrasting years: simulation year 2 (when erosion decreased

Table 24.5 Change in WEPP-simulated long-term mean erosion rates 1990–2050: all years, wet years and dry years. Differences Δ are in m^3/ha

Years	D		E		F		G		H		I		J		K		L		Mean	
	Δ	%	Δ	%	Δ	%	Δ	%	Δ	%	Δ	%	Δ	%	Δ	%	Δ	%	Δ	%
All	0.5	28.7	0.5	28.8	1.3	26.7	1.8	23.8	2.4	24.8	2.2	24.0	1.8	24.0	1.1	26.0	1.0	28.8	1.4	25.1
Wet	1.0	31.4	0.9	29.2	2.7	33.1	3.8	30.5	4.7	29.5	4.5	30.2	3.7	29.6	2.4	31.3	2.1	31.5	2.9	30.7
Dry	0.2	23.8	0.2	26.4	0.3	13.2	0.4	10.6	0.7	12.6	0.6	12.6	0.5	12.6	0.3	14.8	0.3	15.8	0.4	15.8
Wet −dry	0.8	7.6	0.7	2.8	2.4	19.9	3.4	19.9	4.0	16.9	3.9	17.6	3.2	17.0	2.1	16.5	1.8	15.7	2.5	14.9

Table 24.6 Change in WEPP-simulated annual mean erosion rates 1990–2050 for selected simulation years, strips D and H. Differences Δ are in m³/ha

Year of simulation	D		H	
	Δ	%	Δ	%
5	<0.1	33.3	0.5	35.6
6	<0.1	12.5	0.4	225.0
9	1.8	133.3	7.7	41.9
12	<0.1	50.0	<0.1	33.3
15	−0.6	−62.8	−1.6	−29.5
16	−0.8	−55.7	−1.2	−11.6
17	1.2	47.2	5.5	22.0
18	4.1	36.8	6.8	18.9
20	<0.1	22.2	<0.1	36.4
21	<0.1	100.0	<0.1	100.0
22	−0.1	−7.9	3.3	50.3
28	<0.1	7.1	−0.8	−31.6
29	1.0	109.8	4.7	45.8

under the 2050 scenario) and year 9 (when it increased). The timing of rainfall is critical. In year 9, the large rainfall event of 64.2 mm on 9 October occurred at a time when crop cover was near zero, as did the subsequent smaller events on 13, 15 and 17 October. Increased rainfall on these dates in the 2050 scenario (not shown in Figure 24.6) was therefore able to produce more erosion (Table 24.7). However, crop cover had increased by the time of the 25.8 mm rainfall on 18 November; and the greater increase in cover for 2050 compared with 1990 resulted in a lower rate of erosion in 2050. (In passing, note that the occurrence of any erosion at all on 23 December in the 2050 scenario is surprising, considering the high crop cover on this date.) In year 2, rainfall events during the critical period of low crop cover were all relatively small (maximum 21.5 mm on 24 October) and produced almost no erosion. However, the similarly-sized events of 26, 27 and 29 November (19.6, 18.0 and 23.6 mm respectively) did produce some erosion in the 1990 scenario (crop cover being less than 30%). Greater crop cover in the 2050 scenario decreased the size of these events.

Why should the later rainfall events in year 2 have produced more erosion than similarly sized ones earlier? Within WEPP, the crusting routine decreases surface random roughness over time from the value produced by the last tillage operation; together with cumulative rainfall energy since the operation, this operates to decrease effective hydraulic conductivity (Alberts *et al.* 1995) (Figure 24.7: compare Figure 13.2 in Boardman 1993). This effect thus works in the contrary direction to the increasing protection of the growing crop, so that the late November rainfalls of year 2 produced more erosion than those of October in the same year. Note the "spike" in conductivity on 9 October of year 9 which coincides with the large rainfall event.

24.3.2.4 Downslope Changes in Erosion and Deposition

Figure 24.8 shows the distributions of long-term soil loss and deposition along the profiles of strips D to L. In all cases, erosion increases gradually downslop – the

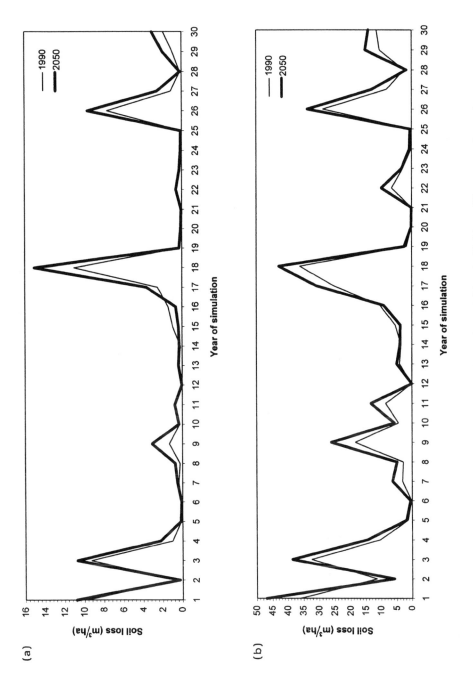

Figure 24.5 Simulated annual erosion rates for (a) strip D and (b) strip H. Note different y-axis scales

544

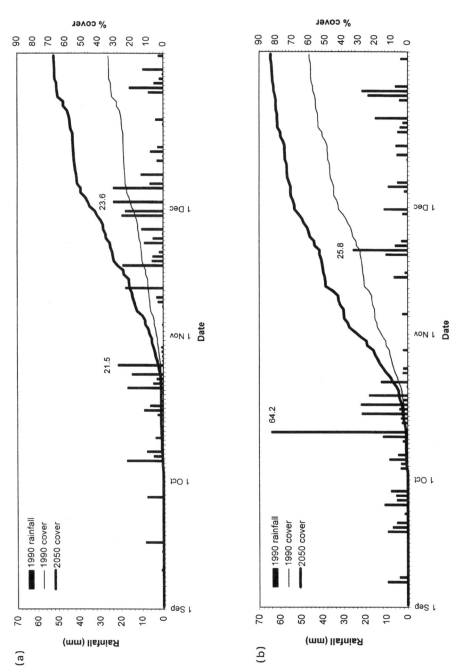

Figure 24.6 Rainfall and crop cover for simulation (a) year 2 and (b) year 9, strip H

Table 24.7 Erosion events (September to December only) in simulation years 2 and 9

Year of simulation	Date	Soil loss in 1990 scenario (m³/ha)	Soil loss in 2050 scenario (m³/ha)	Change (m³/ha)
2	3 Oct.	0.0	0.0	0.0
	22 Oct.	0.0	0.0	0.0
	24 Oct.	0.0	0.0	0.0
	10 Nov.	0.0	0.0	0.0
	15 Nov.	0.0	0.0	0.0
	23 Nov.	0.0	0.0	0.0
	26 Nov.	0.9	0.6	−0.2
	27 Nov.	1.3	1.2	−0.1
	29 Nov.	4.0	3.2	−0.8
	2 Dec.	0.0	0.0	0.0
	3 Dec.	0.0	0.0	0.0
	10 Dec.	0.0	0.0	0.0
	28 Dec.	0.0	0.0	0.0
9	9 Oct.	7.9	12.2	4.4
	13 Oct.	0.5	1.2	0.7
	15 Oct.	3.7	5.5	1.9
	17 Oct.	2.9	5.0	2.1
	20 Oct.	0.0	0.0	0.0
	18 Nov.	1.3	0.1	−1.2
	17 Dec.	0.0	0.0	0.0
	22 Dec.	0.0	0.0	0.0
	23 Dec.	0.0	1.2	1.2
	24 Dec.	0.0	0.0	0.0

influence of the convexity on strips F to J can clearly be seen in the shape of this increase – and decreases more rapidly as the slope flattens at its foot. Deposition at the footslope is always confined to a small area but can be substantial (more than 30 m³/ha on strip H). In all cases, there is a greater increase in erosion from 2025 to 2050 compared to 1990 to 2025 (compare Figure 24.4).

Differences between the 1990 and 2050 distributions are shown in Figure 24.9. The largest increases in erosion occur toward the middle of the field (compare top line of Table 24.5), as do the largest increases in deposition: over 11 m³/ha on strip H.

24.4 DISCUSSION

As with previous studies of future climate change, the above results rely upon many assumptions. Several of these have been previously discussed in Favis-Mortlock *et al.* (1991) and Favis-Mortlock and Boardman (1995). The discussion below covers other points.

• Probably the most fundamental limitation concerns future changes in land use. Holding land use (including the timings of tillage operations) unchanged under

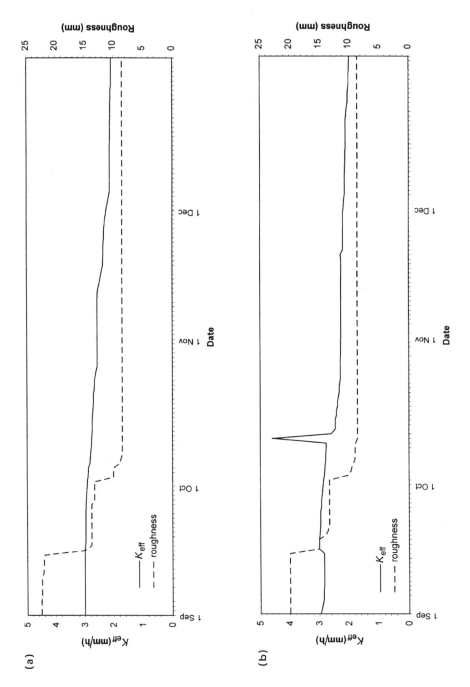

Figure 24.7 Hydraulic conductivity (K_{eff}) and surface roughness for (a) simulation year 2 and (b) simulation year 9, strip H, 1990 scenario

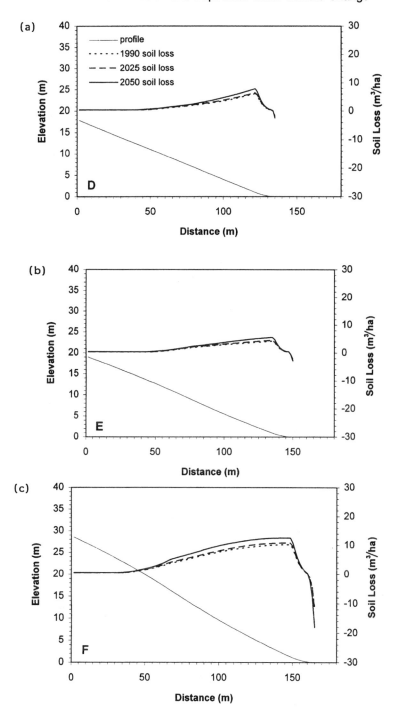

Figure 24.8 Simulated average annual soil loss (+ve) and deposition (−ve) along the profiles of strips D to L. See strip D for key

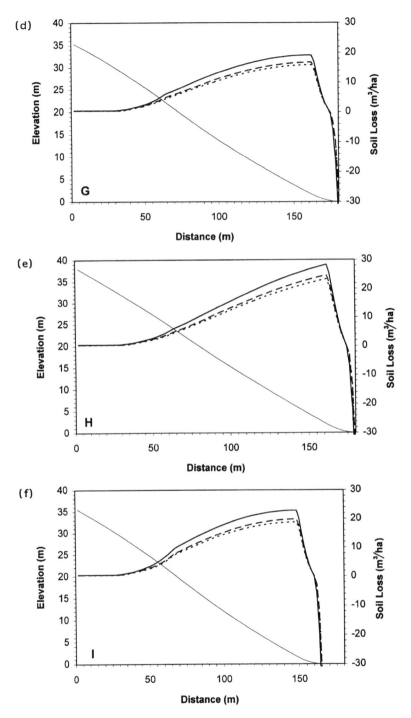

Figure 24.8   *(Continued)*

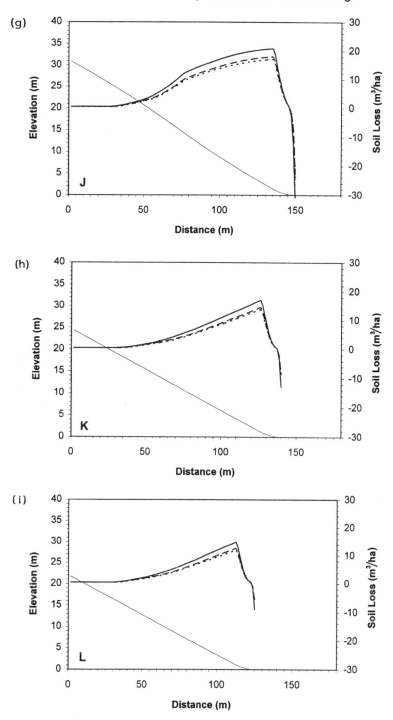

Figure 24.8 *(Continued)*

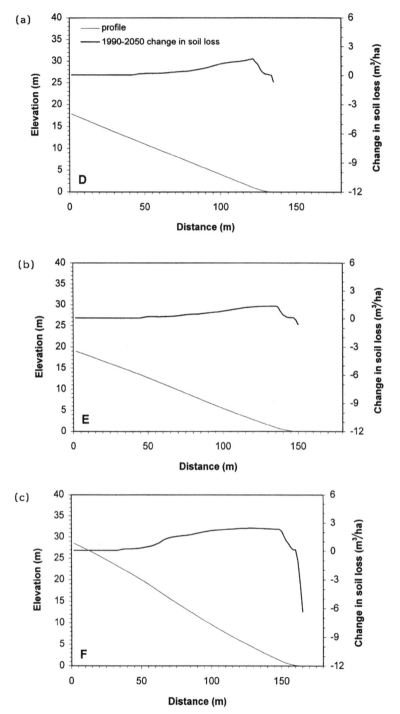

Figure 24.9   The difference between simulated average annual soil loss in 1990 and 2050 along the profiles of strips D to L. See strip D for key

(d)

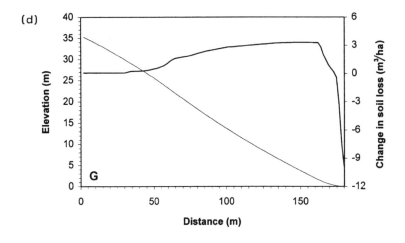

(e)

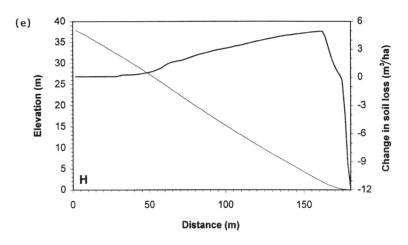

(f)

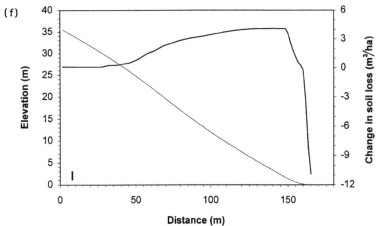

Figure 24.9   *(Continued)*

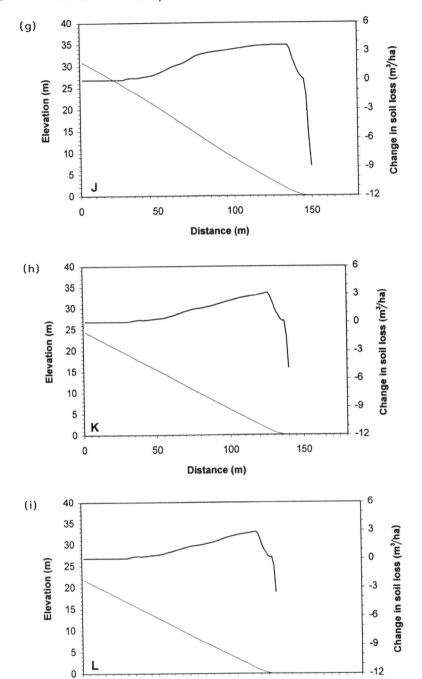

Figure 24.9   *(Continued)*

future climate scenarios is a commonly used simplification (the "dumb farmer" approach of Easterling *et al.* 1992b) which is in the spirit of scientific reductionism. But do studies which take this approach teach us anything useful? In a study of past erosion, Favis-Mortlock *et al.* (in press) used the EPIC model to simulate soil loss from 7000 BP to the present. During this period, both land use and climate varied. The influence of land-use change on simulated erosion rates was far more noticeable than the influence of climate. Future land-use changes, therefore, may well influence rates of soil loss more than future climate change. None the less, studies such as that described here have a value which is more than merely theoretical, if only in highlighting those present land uses which, if continued unchanged (though this is unlikely), would result in excessive erosion under a future climate.

- A second potentially major weakness is the necessity for calibration, both to achieve reasonable values for present-day erosion and also (but less importantly, in this case) to match results for crop yield. In effect, this type of calibration is a substitute for inadequate process description within the model, with the calibration substituting for a representation of the "missing" processes. While accepting that no model can be completely physically based, the principal virtue of the process-based approach is a strong foundation on (presumably universally applicable) physical laws; the principal vice of the empirical approach is a reliance on unknown linkages between input and output. Any calibration thus diminishes virtue and increases vice; once applied to data other than that on which it was calibrated, the model may continue to give results which are acceptable, as long as the data is within the range for which the importance of (or the linkage to) the "missing" processes is not greatly different from the calibration data. But since the "missing" processes are missing just because they are unknown, there is no way of knowing exactly what this range is. If a calibrated model is to be used in novel conditions, it is not possible to know if results remain reliable.

  For the present study, calibration of conductivity and erodibility parameters appeared to be necessary because of the stoniness of the soil at the Woodingdean site; WEPP has produced good results elsewhere without calibration (M. Nearing, pers. comm., 1995). Use of the calibrated model in this study therefore assumes that stoniness will still be influencing conductivity and erodibility under 2050 climatic conditions. It must be emphasized though that this calibration will probably *not* significantly affect the changes in rates produced by input of changed climate data. In this connection, it is worth noting that Barfield *et al.* (1991), speaking of an earlier generation of erosion models, suggested that only relative – rather than absolute – results could be relied upon. It is not clear to what extent this is still true of current models.

- Only "equilibrium" (i.e. with no trend) rather than "transient" (with trend) climate scenarios were used. The main implication of this approach, discussed by Favis-Mortlock and Boardman (1995), is that it may result in inappropriate soil data being used for the future climate simulations. For example, South Downs soil profiles in 2050 will have experienced a further 50 years of erosion compared to 1990s profiles, and will as a result be stonier than at present. Transient EPIC simulations undertaken by Favis-Mortlock and Boardman (1995) suggested a decrease in erodibility because of this. Thus the rates and changes of rate estimated by WEPP in this study (for 2050 in particular) may be a little on the high side.

- Results from this study are applicable only to hillslope sites: they cannot be extended to the wider landscape, which will also include valley-bottom ("talweg") erosion (cf. Evans 1993).

- Finally, the indirect effects of future climate and CO_2, while clearly important in changing crop cover after the crop is established, may also affect the speed with which it becomes established. This would be of crucial importance for erosion (cf. Boardman 1993). While the EPIC and WEPP crop models are presumably adequate in the first respect, too little research has been done to evaluate them for the second.

24.5 IMPLICATIONS

The introductory section of this chapter questioned the usefulness of physically based models in providing policy information. Within the limits imposed by the study, this section attempts to draw out some practical implications of the findings.

First, the WEPP results for change in long-term mean erosion rate reinforce those obtained in earlier studies by Boardman and Favis-Mortlock using EPIC. Since the two models take different approaches to describing the processes which produce erosion, this similarity of results – while of course confirming neither – must none the less go some way toward improving confidence in the models. Note too that a different methodology was used for the construction of the climate-change scenarios in the present study. It seems increasingly likely, therefore, that long-term erosion rates on hillslopes on the South Downs would increase under a future changed climate, although increased stoniness would slow this increase to some extent.

The study also confirmed the result of Favis-Mortlock and Boardman (1995), that over the long term, erosion rate would increase more in wet years than dry years, with this increase however occurring in a narrower "window of opportunity" (e.g. Figure 13.3 in Boardman 1993) before increased vegetation cover reduces the risk of erosion. The WEPP simulations here additionally indicate that erosion on the longer and steeper slopes toward the centre of the field increases more in absolute terms, though there is no proportional increase. All rates were shown to increase more between 2025 and 2050 than between 1990 and 2025.

The picture which emerges is of a strongly non-linear spatial and temporal response of soil erosion on the South Downs to climate change. Spatially, this increase is patchy, with those locations which are current hillslope "hotspots" for erosion (e.g. long, steep slopes below convexities) seeing the greatest increase. In temporal terms, the prediction is for more erosion to occur in a shorter period following drilling, but with this increase also patchily distributed, occurring more in wetter-than-average years; and the rate of increase will itself increase over time.

This rather gloomy scenario of course assumes no change in current farming methods. Would such increases in erosion by themselves move the farming community to taking preventative action? Evans (1992) suggests that UK farmers are only concerned with erosion when more than about 5% of the area of a field is affected (this includes both rilling and deposition). Using the results from the present study, two methods are available to make a crude estimate of the area of the Woodingdean field which might be affected on a long-term basis in the future. The first approach uses the spatial distributions of erosion and deposition estimated by WEPP (Figure 24.8). From these, the area in which deposition occurs can be obtained directly. An estimate of the area affected by rill erosion can be made by assuming that rilling begins at the point on the profile where the erosion rate in Figure 24.8 rises above the interrill "background" rate, and also assuming that, in the long term, rills will occur only over half the width of the strip, but on this they are on average spaced 3 m apart and are 15 cm wide. Results are in Table 24.8 (the upper three lines of the 1990 and 2050 scenarios).

The second method uses an empirical relationship derived by Evans (1992) between the volume of soil eroded and the area of field affected by both rills and deposition. This was calculated from a survey of 97 fields on sandy and clayey soils. This expression can be reworked as:

$$A = 0.29 \ V - 0.08$$

where A is the area affected (%) and V is the volume of soil loss (m^3/ha). This was used to calculate a second estimate ("Evans' method") in Table 24.8. There is a rough agreement: however, the important point is that both sets of results are below 5%, even for the centre strips under the 2050 scenario. This implies that future erosion on the South Downs as estimated by WEPP will not, in the long term, be perceived as a problem by farmers.

Table 24.8 Estimates of the area of the field affected by erosion (see text for explanation)

		D	E	F	G	H	I	J	K	L	Mean
1990	% affected by deposition	0.02	0.04	0.09	0.17	0.12	0.18	0.11	0.05	0.05	0.09
	% affected by rills	1.52	1.60	1.84	1.97	1.99	1.96	1.88	1.77	2.05	1.84
	% total	1.54	1.64	1.94	2.14	2.11	2.15	1.99	1.83	2.11	1.94
	soil loss (m^3/ha)	1.77	1.68	4.73	7.54	9.54	8.99	7.55	4.42	3.78	5.56
	% affected by rills and deposition (Evans' method)	0.43	0.41	1.29	2.11	2.69	2.53	2.11	1.20	1.02	1.53
2050	% affected by deposition	0.02	0.04	0.09	0.17	0.19	0.18	0.11	0.05	0.05	0.10
	% affected by rills	1.59	1.64	1.89	1.99	2.04	1.99	1.93	1.82	2.11	1.89
	% total	1.61	1.69	1.98	2.16	2.23	2.17	2.04	1.87	2.16	1.99
	soil loss (m^3/ha)	2.28	2.17	5.99	9.33	11.90	11.15	9.35	5.56	4.80	6.95
	% affected by rills and deposition (Evans' method)	0.58	0.55	1.66	2.63	3.37	3.15	2.63	1.53	1.31	1.93

24.6 CONCLUSIONS

- WEPP-CO2 is a useful tool for climate change studies, although calibration is still needed for stony soils.
- The results of earlier studies were confirmed: erosion under a future changed climate at hillslope sites on the South Downs is likely to (assuming no change in land use) increase. This increase is estimated to occur more in wet years than dry years. The interplay between change in rainfall and increasing crop cover resulting both from climate and CO_2 change is critical, resulting in a narrower "window of opportunity" for erosion after drilling during which erosion is, however, likely to be more severe.
- These increases will probably not occur linearly: rates may rise more steeply in the second quarter of the next century.
- Spatially, erosion may increase most of those hillslope areas on which it is already most severe.
- None the less, the areas affected by erosion and deposition will in the long term probably still be too limited for erosion to be perceived as a problem by farmers.

ACRONYMS

CLIGEN	Climate Generator model
EPIC	Erosion-Productivity Impact Calculator
GCTE	Global Change and Terrestrial Ecosystems
GCM	Global Circulation Model
MAGICC	Model for the Assessment of Greenhouse gas Induced Climate Change
SWRRB	Simulator for Water Resources in Rural Basins
WEPP	Water Erosion Prediction Project

ACKNOWLEDGEMENTS

We wish to thank John Boardman (Oxford) and Bob Evans (Anglia) for comments on an earlier draft of this chapter; also James Morison (Essex) and Gerard Govers (Leuven) for discussion regarding CO_2 responses and soil stoniness respectively. DFM would also like to thank Mark Nearing (USDA) for assistance with WEPP, and Arlin Nicks (USDA) for assistance with CLIGEN. Thanks also to Steve Taylor (National Rivers Authority) for supplying climate data. This chapter is a contribution to the Soil Erosion Network of the GCTE, which is a Core Research Project of the International Geosphere–Biosphere Project.

REFERENCES

Alberts, E.E., Nearing, M.A., Weltz, M.A., Risse, L.M., Pierson, F.B., Xhang, X.C., Laflen, J.M. and Simanton, J.R. 1995. Soil component. In D. C. Flanagan and M. A.

Nearing (eds), *USDA – Water Erosion Prediction Project Hillslope Profile and Watershed Model Documentation*. NSERL Report No. 10, USDA–ARS, West Lafayette, Indiana, pp. 7.1–7.20.

Arnold, J.R., Weltz, M.A., Alberts, E.E. and Flanagan, D.C. 1995. Plant growth component. In D.C. Flanagan and M.A. Nearing (eds), *USDA – Water Erosion Prediction Project Hillslope Profile and Watershed Model Documentation*. NSERL Report No. 10, USDA–ARS, West Lafayette, Indiana, pp. 8.1–8.41.

Barfield, B.J., Haan, C.T. and Storm, D.E. 1991. Why model? In D.B. Beasley, W.G. Knisel and A.P. Rice (eds), *Proceedings of the CREAMS/GLEAMS Symposium*. Agricultural Engineering Dept, University of Georgia, Athens, Georgia, pp. 3–8.

Boardman, J. 1992. Current erosion on the South Downs: implications for the past. In M. Bell and J. Boardman (eds), *Past and Present Soil Erosion*. Oxbow Monograph 22, Oxbow Books, Oxford, UK, pp. 9–19.

Boardman, J. 1993. The sensitivity of Downland arable land to erosion by water. In D.S.G. and R.J. Allison (eds), *Landscape Sensitivity*. Wiley, Chichester, pp. 211–228.

Boardman, J. and Favis-Mortlock, D.T. 1993. Climate change and soil erosion in Britain. *Geographical Journal*, **159**(2), 179–183.

Boardman, J., Evans, R., Favis-Mortlock, D.T. and Harris, T.M. 1990. Climate change and soil erosion on agricultural land in England and Wales. *Land Degradation and Rehabilitation*, **2**(2), 95–106.

Brignall, A.P., Hossell, J.E., Favis-Mortlock, D.T. and Rounsevell, M.D.A. 1994. Climate change and crop potential in England and Wales. *Journal of the Royal Agricultural Society of England*, **155**, 140–161.

Brignall, A.P., Downing, T.E., Favis-Mortlock, D.T., Harrison, P.A. and Orr, J. 1996. Climate change and agricultural drought in Europe: site, regional and national effects. In T.E. Downing and A.A. Oolsthorn (eds), *Climate change and extreme events*. Institute for Environmental Studies, Vrije Universiteit, Amsterdam, The Netherlands, 51–96.

Burnham, C.P. and Mutter, G.M. 1993. The depth and productivity of chalky soils. *Soil Use and Management*, **9**(1), 1–8.

Easterling III, W.E., Rosenberg, N.J., McKenney, M.S., Jones, C.A., Dyke, P.T. and Williams, J.R. 1992a. Preparing the erosion productivity impact calculator (EPIC) model to simulate crop response to climate change and the direct effects of CO_2. *Agricultural and Forest Meteorology*, **59**, 17–34.

Easterling III, W.E., McKenney, M.S., Rosenberg, N.J. and Lemon, K.M. 1992b. Simulations of crop response to climate change: effects with present technology and no adjustments (the 'dumb farmer' scenario). *Agricultural and Forest Meteorology*, **59**, 53–73.

Easterling III, W.E., Rosenberg, N.J., Lemon, K.M. and McKenney, M.S. 1992c. Simulations of crop responses to climate change: effects with present technology and currently available adjustments (the 'smart farmer' scenario). *Agricultural and Forest Meteorology*, **59**, 75–102.

Evans, R. 1990. Crop patterns recorded on aerial photographs of England and Wales: their type, extent and agricultural implications. *Journal of Agricultural Science, Cambridge*, **115**, 369–382.

Evans, R. 1992. Rill erosion in contrasting landscapes. *Soil Use and Management*, **8**(4), 170–175.

Evans, R. 1993. Extent, frequency and rates of rilling of arable land in localities in England and Wales. In S. Wicherek (ed.), *Farmland Erosion in Temperate Plains, Environments and Hills*. Elsevier, Amsterdam, The Netherlands, pp. 177–190.

Evans, R. 1996. Some soil factors influencing accelerated water erosion on arable land. *Progress in Physical Geography*, **20**(2), 205–215.

Evans, R. and Catt, J.A. 1987. Causes of crop patterns in eastern England. *Journal of Soil Science*, **38**, 309–324.

Favis-Mortlock, D.T. 1994. Use and abuse of soil erosion models in southern England. PhD Thesis, University of Brighton, UK.

Favis-Mortlock, D.T. 1995. The use of synthetic weather for soil erosion modelling. In D.F.M. McGregor and D.A. Thompson (eds), *Geomorphology and Land Management in a Changing Environment*. Wiley, Chichester, UK, pp. 265–282.

Favis-Mortlock, D.T. in press. Evaluation of field-scale erosion models on the UK South Downs. In J. Boardman and D.T. Favis-Mortlock (eds), *Modelling Soil Erosion by Water*. NATO-ASI Series, Springer-Verlag, Heidelberg, Germany.

Favis-Mortlock, D.T. and Boardman, J. 1995. Nonlinear responses of soil erosion to climate change: a modelling study on the UK South Downs. *Catena*, 25(1–4), 365–387.

Favis-Mortlock, D.T., Evans, R., Boardman, J. and Harris, T.M. 1991. Climate change, winter wheat yield and soil erosion on the English South Downs. *Agricultural Systems*, 37(4), 415–433.

Favis-Mortlock, D.T., Boardman, J. and Bell, M. in press. Modelling long-term anthropogenic erosion of a loess cover. South Downs, UK, The Holocene.

Favis-Mortlock, D.T., Quinton, J.N. and Dickinson, W.T. in press. The GCTE validation of soil erosion models for global change studies. *Journal of Soil and Water Conservation*, 51(5).

Flanagan, D.C. and Nearing, M.A. 1995. *USDA Water Erosion Prediction Project: Hillslope Profile and Watershed Model Documentation*, NSERL Report No. 10, USDA–ARS National Soil Erosion Research Laboratory, West Lafayette, Indiana, USA.

Henderson-Sellers, A. 1991. Policy advice on greenhouse-induced climatic change: the scientist's dilemma. *Progress in Physical Geography*, 15(1), 53–70.

Houghton, J.T., Callander, B.A. and Varney, S.K. (eds) 1992. *Climate Change 1992. The Supplementary Report to the IPCC Scientific Assessment*. Cambridge University Press, Cambridge, UK.

Jarvis, M.G., Allen, R.H., Fordham, S.J., Hazelden, J., Moffat, A.J. and Sturdy, R.G. 1984. *Soils and Their Use in South-East England*. Soil Survey of England and Wales Bulletin 15, Harpenden, UK.

Kimball, B.A. and Idso, S.B. 1983. Increasing atmospheric CO_2: effects on crop yield, water use and climate. *Agricultural Water Management*, 7, 55–72.

Kimball, B.A., Mauney, J.R., Nakayama, F.S. and Idso, S.B. 1993. Effects of elevated CO_2 and climate variables on plants. *Journal of Soil and Water Conservation*, 48(1), 9–14.

Lane, L.J., Nearing, M.A., Laflen, J.M., Foster, G.R. and Nichols, M.H. 1992. Description of the US Department of Agriculture water erosion prediction project (WEPP) model. In A.J. Parsons and A.D. Abrahams (eds), *Overland Flow*. UCL Press, London, pp. 377–391.

Lee, J.J., Phillips, D.L. and Dodson, R.F. in press. Sensitivity of the US corn belt to climate change and elevated CO_2: II. Soil erosion and organic carbon. *Agricultural Systems*.

McGregor, D.F.M. and Thompson, D.A. 1995. Geomorphology and land management in a changing environment. In D.F.M. McGregor and D.A. Thompson (eds), *Geomorphology and Land Management in a Changing Environment*. Wiley, Chichester, UK, pp. 1–10.

Mitchell, J.F.B., Manabe, S., Meleshko, V. and Tokioka, T. 1990. Equilibrium climate change – and its implications for the future. In J.T. Houghton, G.J. Jenkins and J.J. Ephraums (eds), *Climate Change: the IPCC Scientific Assessment*, Cambridge University Press, Cambridge, UK, pp. 131–172.

Nearing, M.A. Ascough, L.D. and Chaves, H.M.L. 1989. WEPP model sensitivity analysis. In L.J. Lane and M.A. Nearing (eds), *USDA – Water Erosion Prediction Project: Hillslope Profile Version*. USDA–ARS National Soil Erosion Research Laboratory Research Report No. 2, West Lafayette, Indiana, pp. 14.1–14.33.

Nicks, A.D., Lane, L.J. and Gander, G.A. 1995. Weather generator. In D.C. Flanagan and M.A. Nearing (eds), *USDA–Water Erosion Prediction Project: Hillslope Profile and Watershed Model Documentation*. USDA–ARS National Soil Erosion Research Laboratory Report No. 10, West Lafayette, Indiana, pp. 2.1–2.22.

Poesen, J.W.A. 1992. Mechanisms of overland-flow generation and sediment production on loamy and sandy soils with and without rock fragments. In A.J. Parsons and A.D. Abrahams (eds), *Overland Flow*. UCL Press, London, pp. 275–305.

Potts, A.S. and Browne, T.E. 1983. The climate of Sussex. In Geographical Editorial Committee (eds), *Sussex: Environment, Landscape and Society*. Alan Sutton, Gloucester, pp. 88–108.

Rosenberg, N.J., McKenney, M.S., Easterling, W.S. and Lemon, K.M. 1992. Validation of EPIC model simulations of crop responses to current climate and CO_2 conditions: comparisons with census, expert judgement and experimental plot data. *Agricultural and Forest Meteorology*, **59**, 35–51.

Rowntree, P.R., Murphy, J.M. and Mitchell, J.F.B. 1993. Climatic change and future rainfall predictions. *Journal of the Institution of Water and Environmental Management*, **7**, 464–470.

Savabi, M.R., Smith, D., Micklin, P.P. and Shamilievna, O. in press. Possible effect of climate change on Aral Sea-level fluctuation. Proceedings of International Geographical Union Conference on Global Changes and Geography, Moscow, Russia, 14–18 August 1995.

Stockle, C.O., Williams, J.R., Rosenberg, N. and Jones, C.A. 1992a. A method for estimating the direct and climatic effects of rising atmospheric carbon dioxide on growth and yield of crops: Part I – modifications of the EPIC model for climate change analysis. *Agricultural Systems*, **38**, 225–238.

Stockle, C.O., Dyke, P.T., Williams, J.R., Jones, C.A. and Rosenberg, N.J. 1992b. A method for estimating the direct and climatic effects of rising atmospheric carbon dioxide on growth and yield of crops: Part II – sensitivity analysis at three sites in the midwestern USA. *Agricultural Systems*, **38**, 239–256.

Tegart, W.J.McG., Sheldon, G.W. and Griffiths D.C. (eds) 1990. *Climate Change: the IPCC Impacts Assessment*. Australian Government Publishing Service, Canberra, Australia.

Thorne, C.R. 1995. Editorial – geomorphology at work. *Earth Surface Processes and Landforms*, **20**(7), 583–584.

USDA–Agricultural Soil Conservation Service 1972. *National Engineering Handbook. Section 4, Hydrology*.

Viner, D. 1994. Climate change modelling and climate change scenario construction methods for impacts assessment. In *Climate Impacts LINK Symposium Proceedings*. Climatic Research Unit, University of East Anglia, Norwich, UK.

Viner, D. and Hulme, M. 1992. *Climate Change Scenarios for Impact Studies in the UK*. Report for UK Department of the Environment. Climatic Research Unit, University of East Anglia, Norwich, UK.

Viner, D. and Hulme, M. 1993. *Construction of Climate Change Scenarios by linking GCM and STUGE Output*. Technical Note for UK Department of the Environment. Climatic Research Unit, University of East Anglia, Norwich, UK.

Waggoner, P.E. 1990. Anticipating the frequency distribution of precipitation if climate change alters its mean. *Agricultural and Forest Meteorology*, **47**, 321–337.

Wilby, R.L. 1995. Greenhouse hydrology. *Progress in Physical Geography*, **19**(3), 351–369.

Williams, J.R., Nicks, A.D. and Arnold, J.G. 1985. Simulator for water resources in rural basins. *Journal of Hydrological Engineering*, **111**(6), 970–986.

Williams, J.R., Jones, C.A. and Dyke, P.T. 1990. The EPIC model. In A.N. Sharpley and J.R. Williams (eds), *EPIC – Erosion/Productivity Impact Calculator. 1. Model Documentation*, US Department of Agricultural Technical Bulletin No. 1768, pp. 3–92.

Wilmott, C.J. and Gaile, G.L. 1992. Modeling. In R.F. Abler, M.G. Marcus and J.M. Olson (eds), *Geography's Inner Worlds*. Rutgers University Press, New Jersey, pp. 163–186.

Wischmeier, W.H. and Smith, D.D. 1978. *Predicting Rainfall Erosion Losses*. US Department of Agriculture, Agricultural Research Service Handbook 537.

Wolf, J. 1993. Effects of climate change on wheat and maize production potential in the EC. In G.J. Kenny, P.A. Harrison and M.L. Parry (eds), *The Effects of Climate Change on*

Agricultural and Horticultural Potential in Europe. Environmental Change Unit, University of Oxford, pp. 93–119.

Woo, M.-K., Lewkowicz, A.G. and Rouse, W.R. 1992. Response of the Canadian permafrost environment to climatic change. *Physical Geography*, **13**(4), 287–317.

25 Simulation of Radiocaesium Redistribution on Cultivated Hillslopes using a Mass-Balance Model: An Aid to Process Interpretation and Erosion Rate Estimation

T. A. QUINE,[1] D. E. WALLING[1] and G. GOVERS[2]
[1]*Department of Geography, University of Exeter, UK*
[2]*Laboratory for Experimental Geomorphology, Catholic University of Leuven, Belgium*

25.1 INTRODUCTION

The caesium-137 (^{137}Cs) technique for soil erosion assessment is becoming established as a valuable tool for erosion research (cf. Ritchie and McHenry 1990). However, uncertainty concerning the derivation of quantitative erosion rate estimates from radiocaesium data currently hampers the widespread acceptance of the results from the technique (Walling and Quine 1990, 1992; Quine 1995). This chapter first addresses some of the causes of the uncertainty surrounding these estimates and then explores a new approach to erosion rate estimation which recognizes the recent advances in understanding the geomorphic impact of soil tillage. The application of the approach to the interpretation of ^{137}Cs data from Dalicott Farm in Shropshire demonstrates the improved simulation of ^{137}Cs redistribution and the access to process-specific erosion rate data provided by the approach.

25.2 CALIBRATION OF CAESIUM-137 DATA – UNCERTAINTY AND EXPLANATION

The derivation of quantitative erosion rate estimates from ^{137}Cs data (often termed calibration) is essential to maximize the benefit obtained from application of the technique. However, some uncertainty surrounds such calibration, and this detracts from the results of the approach. This section examines the two principal sources of uncertainty and considers possible explanations. The first source of uncertainty is reflected in the range of calibration relationships (relating ^{137}Cs loss and gain to soil

Advances in Hillslope Processes, Volume 1. Edited by M. G. Anderson and S. M. Brooks.
© 1996 John Wiley & Sons Ltd.

loss and gain) which have been proposed and the varying estimates of erosion which some of these approaches have yielded (Walling and Quine 1990; Loughran *et al.* 1990). The second source of uncertainty lies in the evident differences in magnitude between ^{137}Cs-derived erosion rates and those estimated from field observation of soil erosion by water (cf. Quine and Walling 1993).

25.2.1 Variation in Calibration Relationships

Detailed discussion of the range of calibration relationships that have been used would be superfluous because these have been comprehensively reviewed elsewhere (Walling and Quine 1990). However, it is important to examine briefly the approaches employed as a context for the remainder of the chapter. Two broad approaches have been widely applied to the establishment of calibration relationships: use of mass-balance models to simulate ^{137}Cs loss in association with eroded soil (and ^{137}Cs gain with deposition) and of empirically derived relationships relating ^{137}Cs loss to soil loss.

Much of the uncertainty surrounding calibration procedures has concerned the empirical relationships. This uncertainty reflects the variation between individual empirical relationships and their apparent divergence from relationships derived using mass-balance models. The variation between empirical relationships is to be expected, because each relationship must, by definition, be site-specific and represents only the conditions at the collection site for the data used in establishing the relationship (Walling and Quine 1990). In this respect two points bear reiteration. First, because of the continuous nature of the processes of radioactive decay and ^{137}Cs loss with eroded soil, the relationships are time-specific. Secondly, because of the mixing of ^{137}Cs through the plough layer, the relationships may only be applied at sites with the same plough depth and bulk density of the plough zone. These considerations explain some of the differences observed between empirical relationships, such as those proposed by Ritchie and McHenry (1975) and Wilkin and Hebel (1982), and also the divergence from contemporary relationships based on mass-balance models. However, failure to recognize the constraints on the use of empirical relationships has led to their erroneous application for inappropriate sites or time periods. The scarcity of suitable erosion rate data for establishing empirical calibration relationships further highlights the limitation imposed by the constraints outlined above. A further example of the limitation of the empirical approach is provided by the very wide divergence of an empirical relationship derived by Loughran *et al.* (1990) from most other calibration relationships. Recent evidence has indicated that the erosion-plot data, on which this relationship was based, may have been seriously flawed (R. J. Loughran, pers. comm.). It has become apparent that problems relating to the collection and measurement of the eroded sediment led to significant underestimation of erosion rates and the calibration relationship therefore also underestimates erosion rates for a given magnitude of ^{137}Cs loss.

As a result of the problems associated with the empirical approach, the use of mass-balance models as a means of establishing calibration relationships has attracted considerable interest. Such mass-balance models simulate ^{137}Cs loss/gain associated with specific levels of soil loss/gain from a single soil profile and most conform to the following basic form:

$$C_l = S_l \cdot f(C_p, C_e, C_s) \tag{25.1}$$

where C_l is ^{137}Cs loss; S_l is soil loss; C_p is ^{137}Cs content of plough layer (assumed mixed); C_e is particle size enrichment of the ^{137}Cs in the eroded soil; and C_s is enrichment of ^{137}Cs at the soil surface.

Various parameters are used to describe the profile to obtain a simulation that is representative of specific site conditions. A series of simulations is conducted to establish the impact of a range of erosion/deposition rates on the ^{137}Cs inventory of the profile. The resultant predictions are then used to establish a calibration relationship relating ^{137}Cs loss to soil loss, and ^{137}Cs gain to soil gain. The estimated soil redistribution rate for each point under investigation is then established by "reading-off" the estimated erosion/deposition rate which corresponds to the measured level of ^{137}Cs loss or gain.

The models vary according to the parameters included, especially C_e and C_s. However, there is general agreement over the basic structure and this is reflected in soil redistribution rate predictions for a particular magnitude of ^{137}Cs loss/gain which may vary by a factor of 2–5, but not by an order of magnitude. The most widely used has been the simplest model ("directly-proportional" model) which takes no account of C_e and C_s and assumes that soil loss is directly proportional to ^{137}Cs loss (cf. Mitchell et al. 1980; de Jong et al. 1983; Martz and de Jong 1987; Kachanoski 1987; Fredericks and Perrens 1988). This method will usually provide upper estimates of erosion and, in general, models which account for the sources of potential caesium enrichment in eroded sediment may be expected to provide more realistic estimates (cf. Kachanoski and de Jong 1984; Fredericks and Perrens 1988; Quine 1989, 1995). The accuracy of these estimates will, clearly, depend on the quality of the simulation of ^{137}Cs redistribution and especially mechanisms of ^{137}Cs enrichment. These are, therefore, areas of considerable importance in the development of reliable and realistic approaches to erosion rate estimation.

This section has sought to highlight three points. First, the theoretical and practical limitations associated with the use of empirically derived calibration relationships. Secondly, the broad agreement that exists concerning the use of mass-balance models in calibration and the extent of the divergence in estimates provided by existing models. Thirdly, the evident need for further development of models of soil-associated ^{137}Cs redistribution to provide a reliable basis for erosion rate estimation.

25.2.2 Comparison of ^{137}Cs-derived Erosion Rates and Field Observations

The greatest uncertainty concerning the validity of ^{137}Cs-derived erosion rate estimates has resulted from comparison with the results of field observations of water erosion. This is clearly demonstrated by a comparison of a number of ^{137}Cs-derived rates of gross erosion (not accounting for in-field deposition) obtained for UK fields (Walling and Quine 1991; Quine and Walling 1991) with estimates of the rates of water erosion for the same fields based on measurement of rill dimensions (Evans

Table 25.1 Comparison of ^{137}Cs-derived gross erosion rates and estimates of water erosion based on rill volumes

Site	^{137}Cs-derived gross erosion rate (t/ha/year)	Rill-based water erosion rate (t/ha/year)	Period of rill monitoring (years)
Dalicott Farm	10.2	0.3[a]	9
Lewes	4.3	c. 1.0[b]	8

[a]Evans (1992).
[b]Boardman (pers. comm.).

1988; Boardman 1990). Table 25.1 summarizes the data for two fields for which direct comparison is most appropriate; the first at Dalicott Farm (52°33′N 2°20′W) in Shropshire and the second near Lewes in Sussex (location withheld at farmer's request). If it is assumed that soil redistribution on agricultural land is dominated by rill erosion, the very evident differences in the ^{137}Cs-derived estimates and the rill-based estimates give serious cause for concern. However, three lines of evidence should be taken into account when evaluating the reliability of the ^{137}Cs-derived rates:

1. the internal consistency of the rate estimates;
2. agreement with comparable independent data;
3. the redistribution of soil by tillage.

The internal consistency of the soil redistribution rate estimates may be examined by consideration of depositional areas within fields. Such locations are characterized by two properties in relation to ^{137}Cs: first, they have elevated ^{137}Cs inventories (activity per unit area) which are used to estimate deposition rates from calibration relationships; secondly, they are characterized by "over-deepened" ^{137}Cs depth distributions (cf. Walling and Quine 1990, 1992). The extent of this "over-deepening" may be used to provide an alternative estimate of the rate of deposition, thereby providing a check for internal consistency. If it is established that the estimation of deposition based on the calibration relationship is reliable, then the total amount of deposition estimated for the field using the calibration relationship may be used as a minimum estimate of total erosion. This procedure was carried out for the field at Dalicott Farm referred to in Table 25.1 and a high degree of internal consistency was established (Quine and Walling 1993). The ^{137}Cs-derived gross aggradation rate of 3.7 t/ha per year, therefore, provides a *minimum* estimate of erosion which, although only 36% of the gross erosion rate, is an order of magnitude higher than the estimate of rill erosion.

Comparison of ^{137}Cs-derived redistribution rates with independent evidence for total soil redistribution is necessarily hampered by the scarcity of such data. However, on one of the few occasions where this has been possible, Kachanoski (1987) compared ^{137}Cs and soil loss from a runoff plot at the University of Guelph over the period from 1965 to 1976, and found a close agreement between measured soil loss and that predicted from the ^{137}Cs measurements using the "directly proportional" calibration method. Furthermore, independent evidence for soil redistribution on a second field at Dalicott Farm has been presented elsewhere (Quine and Walling 1993) and the data are examined in detail below.

The final line of evidence in evaluating ^{137}Cs-derived erosion rate estimates is the growing body of research demonstrating the significant geomorphic effect of tillage operations. Experimental studies have shown that the downslope flux associated with tillage is proportional to the slope angle and that on slopes of only 10% a single tillage operation using a mouldboard plough may produce a downslope flux of 20–30 kg/m (Lindstrom et al. 1990, 1992; Revel et al. 1993; Govers et al. 1994). Therefore, an area with a slope curvature of only 0.5%/m would be subject to a soil redistribution rate of 1–1.5 kg/m^2 per year. It is, therefore, evident that on rolling agricultural land in temperate environments, soil redistribution associated with tillage may be of equal or greater magnitude than that associated with water erosion. Significant divergence between ^{137}Cs-derived rates of erosion and field-based measurements of water erosion may, therefore, be expected as the ^{137}Cs-derived rates relate to all processes of soil redistribution.

This section has focused on one of the main objections to the ^{137}Cs technique, namely the disagreement between ^{137}Cs-derived rates and field observations of water erosion. Evidence has been presented to support the ^{137}Cs-derived rates and also to explain the observed disagreement. The validity of these explanations may be seen in a number of recent studies which have demonstrated the close agreement between patterns and rates of soil erosion and deposition derived using ^{137}Cs and those observed or expected as a result of both tillage and water erosion (Govers et al. 1993, 1996; Quine et al. 1994). However, the evidence for soil redistribution by tillage has important implications for erosion rate estimation from ^{137}Cs measurements. These implications are considered below after a brief outline of recent advances in our understanding of tillage.

25.3 SOIL REDISTRIBUTION BY TILLAGE

In view of the importance of soil redistribution by tillage in the interpretation of ^{137}Cs redistribution, it is useful to summarize some of the attributes of the process that have been revealed by a number of experimental studies including those by Lindstrom et al. (1990, 1992), Revel et al. (1993) and Govers et al. (1994). The most important result of the experimental studies has been the demonstration of the role of topography in determining rates and patterns of soil redistribution by tillage. This role is most fundamentally described by the relationship between slope angle and the mean displacement distance of the plough layer as a result of a single tillage operation:

$$D = aS + b \tag{25.2}$$

where D is the displacement distance in direction of tillage (m); S is the sine of the downslope angle (i.e. negative upslope); a is a slope constant (m); and b is the intercept (m).

Table 25.2 Published values of the coefficients a and b obtained in tillage experiments and published values of T_k (equivalent to Ma in equation (25.4))

Author	Implement/direction	a (m)	b (m)	T_k (kg/m)
Govers et al. (1994)	mouldboard (up and down)	0.28	0.62	234
	chisel (up and down)	0.15	0.55	111
Lindstrom et al. (1992)	mouldboard (up and down)	0.24	1.02	363
	mouldboard (across)	0.24	1.12	330

After Govers et al. (1994).
Bulk density, 1350 kg/m³; plough depth, 0.28 m.

Table 25.2 lists the values of a and b from published experimental studies. As Govers et al. (1994) have demonstrated, equation (25.2) may be used to calculate the sediment flux as a result of a single tillage operation and the annual flux as a result of tillage in opposing directions (i.e. up one year, down the next):

$$Q_t = M(aS + b) \tag{25.3}$$

$$Q_{to} = MaS \tag{25.4}$$

where Q_t is sediment flux due to a single tillage pass (kg/m); Q_{to} is the average sediment flux due to tillage in opposing directions (kg/m); and M is the mass of the plough layer (kg/m²).

Because the rate of soil redistribution due to tillage is equivalent to the rate of change of the flux, these relationships may be used to establish the rate of net soil redistribution due to tillage:

$$R_{td} = Ma(S_{i-1} - S_i)/d \tag{25.5}$$

$$R_{tu} = Ma(S_i - S_{i+1})/d \tag{25.6}$$

$$R_{to} = 0.5 \times Ma(S_{i-1} - S_{i+1})/d \tag{25.7}$$

where R_{td} is the redistribution rate due to downslope tillage at point i (kg/m²); R_{tu} is the redistribution rate due to upslope tillage at point i (kg/m²); R_{to} is the redistribution rate due to tillage in opposing directions at point i (kg/m²); S_i is the sine of slope at point i (i+1 is downslope, i−1 upslope); d is the distance between points (m).

In relation to geomorphic change the most important relationships are defined by equations (25.5)–(25.7) which highlight the potential for tillage to contribute to erosion and deposition. Furthermore, these equations demonstrate that the soil redistribution rate due to tillage is unrelated to tillage direction for a point on a slope of continuous curvature (a significant result in relation to the development of soil conservation strategies). However, in the context of tracer studies the absolute movement of the plough layer (equation (25.5)) is of equal significance to the net redistribution rate because this lateral movement may be expected to be associated with lateral mixing. This is examined further in the following section.

Table 25.3 Differences between tillage and water erosion of greatest significance for the simulation of ^{137}Cs redistribution

Property	Tillage	Water Erosion
Size selectivity	aselective	*splash/interrill*: preferential removal of fine sediment
		rill: aselective
Source of soil subject to redistribution	whole plough layer	splash/interrill: preferential removal from surface
		rill: potential incision throughout/below the plough layer
Spatial Distribution		
upslope boundary	erosion	no change
slope convexity (low slope length)	erosion	low erosion
linear steep slope	no change	high erosion
slope concavity	deposition	high erosion
downslope boundary	deposition	deposition

25.4 SOIL REDISTRIBUTION BY TILLAGE: IMPLICATIONS FOR CAESIUM-137 REDISTRIBUTION

The evidence for soil redistribution by tillage has three key implications for the interpretation of ^{137}Cs distributions and the simulation of ^{137}Cs redistribution:

• the need to account for ^{137}Cs redistribution by both water and tillage,
• the need to account for plough layer displacement,
• the reduction of spatial variability in inventories.

25.4.1 ^{137}Cs Redistribution by Water and Tillage

The derivation of calibration relationships using mass-balance models (cf. equation (25.1)) is evidently dependent on the efficacy of the simulation of ^{137}Cs redistribution. The need to take account of both tillage and water erosion processes in this simulation may be demonstrated by considering three properties which differ markedly for the processes and which are important for ^{137}Cs redistribution. These differences are summarized in Table 25.3 and their significance is discussed below.

As indicated in equation (25.1), simulation of ^{137}Cs redistribution requires estimation of the ^{137}Cs content of eroded soil. The first two properties in Table 25.3 are important in deriving such estimates. The first property, the size-selectivity of the redistribution processes, is important because of the well-documented preferential adsorption of ^{137}Cs by the fine particle fraction. It is clear that contrasts in the size-selectivity of the processes involved will lead to differing relationships between ^{137}Cs loss and soil loss for tillage and water erosion and for rill and interrill erosion.

The second property in Table 25.3, which reflects differences in the source of soil subject to erosion, will be important if the sources are characterized by differing

[137]Cs contents. Although, under most conditions, the mixing of the plough layer will lead to equal ^{137}Cs contents in the soil subject to water erosion and that subject to tillage redistribution, there are two important exceptions. The first exception, which must be taken into account in all circumstances where water erosion operates, relates to the period of significant atmospheric fallout. During this period ^{137}Cs would have had an uneven depth distribution because the fallout deposited between tillage events would have been concentrated at the soil surface. Therefore, soil eroded by water from the surface during this period would have been associated with higher ^{137}Cs contents than soil redistributed by tillage, irrespective of any size-selectivity. The second exception relates to depths of rill incision in excess of the plough depth. Where this occurs, the ^{137}Cs content of soil eroded by water will be dependent on the ^{137}Cs content of both the plough and subplough zones. The proportion of eroded sediment derived from the subplough layer may vary from zero to several tens of per cent. In contrast, the ^{137}Cs content of soil redistributed by tillage will reflect the ^{137}Cs content of the current plough layer (note that where net soil loss occurs, each ploughing episode incorporates into the plough zone some material which was formerly in the subplough zone; however, in any one tillage episode, the contribution of former subplough soil to the mass redistributed by tillage will typically be less than 1%).

The third property in Table 25.3, the spatial distribution of the erosion processes, is important because application of the basic mass-balance model of ^{137}Cs redistribution summarized in equation (25.1) is based on prediction of ^{137}Cs loss for specified erosion rates. The model, therefore, implicitly assumes that sample sites may be exclusively divided into eroding and depositional locations. The broad descriptions of the spatial distributions of tillage and water erosion shown in Table 25.3 demonstrate that when both processes are taken into account this simple subdivision is no longer valid. Indeed, it is quite probable that areas subject to the highest rates of water erosion will coincide with areas of high tillage deposition (cf. Quine *et al.* 1994).

On the basis of this discussion it can be seen that optimum simulation of ^{137}Cs redistribution must incorporate independent parallel simulation of ^{137}Cs redistribution by both water erosion and tillage processes.

25.4.2 Plough Layer Displacement

Use of calibration curves to estimate point erosion rates from point measurements of ^{137}Cs is based on the assumption that the ^{137}Cs measurement represents the impact of erosion processes at the point sampled. This apparently obvious statement requires careful evaluation in the light of the descriptions of tillage given in equations (25.2) to (25.7). If it is assumed that the displacement of the plough layer described by these equations is accompanied by mixing, a number of impacts may be predicted from these relationships:

1. significant lateral movement and mixing of the plough layer will take place even on level surfaces subject to tillage in opposing directions (equation (25.2));

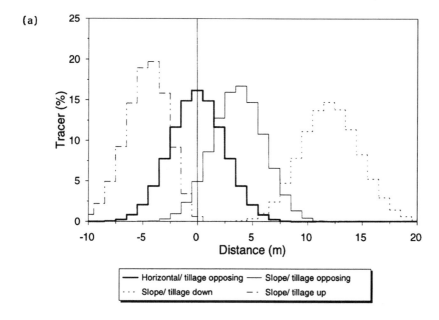

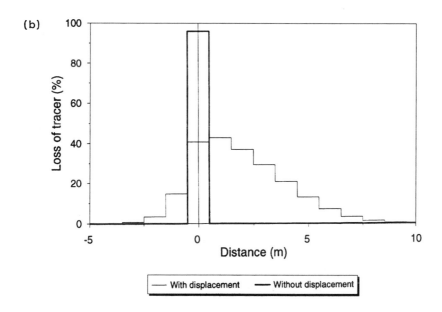

Figure 25.1 (a) Distribution of tracer applied at point 0 after 30 years of redistribution by tillage: the influence of slope (horizontal v 20%) and tillage direction. (b) Pattern of tracer loss after 30 years of 10% annual depletion at point 0: the influence of displacement due to tillage on a 20% slope. Displacement distances were calculated using a relationship derived by Govers *et al.* (1994)

2. sites subject to tillage in a single direction will be subject to significant lateral displacement of the plough layer (equations (25.2) and (25.3));
3. preferential downslope movement of the plough layer will take place on sloping sites subject to tillage in opposing directions (equation (25.4)).

Simple simulations based on published parameter values allow these impacts to be illustrated (Figure 25.1(a)). The following conditions were employed:

^{137}Cs distribution:	100 at origin in year 0; 0 at all other sites; no subsequent additions of ^{137}Cs
Equation (25.2) parameters	$a = 0.62$, $b = 0.28$ m (Govers et al. 1994)
Cell size:	1 m^2
Assumption:	complete mixing within cells
Slope:	20%, except the horizontal site
Time period:	30 years

These simulations indicate that even at a horizontal site, only c. 16% of the tracer is predicted to remain at its original location within the field after 30 years of tillage. The simulations may also be used to predict the source of the tracer present at the origin by reversing the x-axis scale; i.e. for tillage in opposing directions, c. 17% of the tracer present at the origin after 30 years is predicted to have been present 4 m upslope in year 0. This is only of academic interest until a location is subject to erosion, when this lateral displacement becomes of significance. Figure 25.1(b) illustrates this for a simulation under the following conditions:

^{137}Cs distribution:	100 at all points in year 0; no subsequent additions of ^{137}Cs
Annual erosion	
– at point 0:	10% of labelled soil
– at other points:	zero
Equation 2 parameters:	$a = 0.62$, $b = 0.28$ m (Govers et al. 1994)
Cell size:	1 m^2
Assumption:	complete mixing within cells
Slope:	20%
	no correction for erosion
Time period:	30 years

If the point residuals (% loss/gain of ^{137}Cs) shown in Figure 25.1(b) were used in isolation to calculate point erosion rates, the erosion rate at the origin would be greatly underestimated and the rate downslope would be greatly overestimated. Although this simulation is clearly artificial because of the point distribution of erosion and the lack of slope correction, it does illustrate the potential significance of downslope dilution in the interpretation of tracer distributions. Optimum simulation of ^{137}Cs redistribution should, therefore, take account of this potential.

25.4.3 Reduction in Variation in Point Inventories

As has been indicated, most ^{137}Cs studies are based on a series of point measurements of ^{137}Cs inventories. It is, therefore, of vital importance that the point measurements may be assumed to be representative of the *inventory* of the surrounding area (even if that inventory has a complex derivation). This assumption has come under scrutiny because of the recent focus of interest in the spatial variability of ^{137}Cs inventories at undisturbed, uncultivated reference sites (cf. Sutherland 1991). Although it is not yet clear whether the observed variability at reference sites is a product of variation in fallout inputs or variation in the receptor surface (presence of tussocks, macro-pores, etc.), it is clearly important to examine the significance of this potential variation for interpretation of ^{137}Cs inventories from cultivated sites.

It is suggested that lower variation in inventories will be found at cultivated sites because of the mixing effect of tillage in both homogenizing the ^{137}Cs distribution and in reducing local variability in soil properties (a potential source of variation in the fallout receptor surface). The latter is difficult to test but the relationships discussed above may be used to examine the role of tillage mixing in reducing potential spatial variation in inventories. In order to accomplish this a simple simulation was undertaken:

Initial ^{137}Cs distribution:	random number generator (Quattro Pro); range, 90–110; mean, 99; initial standard deviation, 6%
Equation (25.2) parameters:	$a = 0.62$, $b = 0.28$ m (Govers *et al.* 1994)
Cell size:	1 m^2
Assumption:	complete homogenization of soil within cells

The simulated reduction in local variability as a result of tillage is shown in Figure 25.2. A rapid reduction in the standard deviation of the population occurred (20% after 1 year, 48% after 5 years, 71% after 30 years) and the maximum deviation between adjacent cells was reduced from 17% in the initial distribution to 1.5% after 30 years. These results indicate that considerable confidence may be placed in the representativeness of point inventories from cultivated sites, notwithstanding potential initial micro-variation in fallout.

25.5 IMPROVED SIMULATION OF CAESIUM-137 REDISTRIBUTION

The foregoing discussion allows the identification of elements which should be included in improved simulation of ^{137}Cs redistribution:

1. separation of tillage and water erosion,
2. particle-size selectivity – separation of rill and interrill erosion,
3. surface enrichment of ^{137}Cs,
4. ^{137}Cs content of plough and subplough layers,
5. potential for water erosion and tillage deposition at the same site,
6. lateral displacement and mixing of the plough layer.

(a)

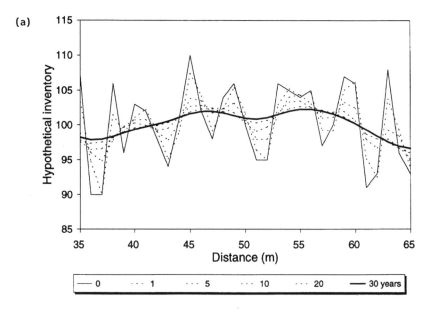

(b)

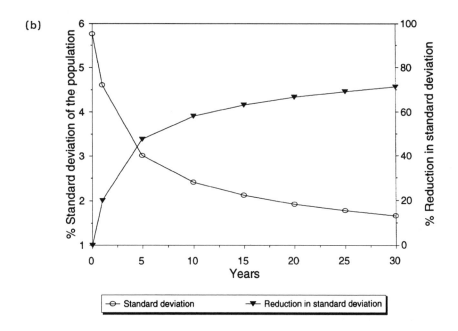

Figure 25.2 Reduction in hypothetical local variability in ^{137}Cs inventories: (a) change in pattern of inventories; (b) change in standard deviation

As has been indicated, past approaches to erosion rate estimation from ^{137}Cs data have been based on simulation of a single soil profile with the results of separate runs used to provide points on a single calibration relationship for the site under study. Although this soil profile approach can be adapted to account, at least in part, for the first five of the criteria listed above (cf. Walling and Quine 1990; Quine 1995), inclusion of the sixth criteria demands a radical change from the single profile approach to the simulation of ^{137}Cs redistribution over space as well as through time. Furthermore, a one- or more dimensional approach has two significant advantages over the point soil profile model, in addition to allowing simulation of lateral displacement of the plough layer. First, simulation of the ^{137}Cs content of sediment deposited by water is greatly improved, because the source area of the sediment is simulated in parallel. Secondly, estimation of erosion rates from the ^{137}Cs data is based on comparison of spatial distributions of ^{137}Cs inventories, rather than examination of point values in isolation. The remainder of the chapter provides a brief outline of a model developed by the authors to facilitate this new approach to erosion rate estimation, an example of its application and a discussion of its limitations and potential.

25.6 A SLOPE TRANSECT MODEL

The model developed by the authors simulates ^{137}Cs redistribution by both tillage displacement and water erosion and deposition along an entire slope transect from the upper to the lower margin of the field. Rates of soil redistribution by both water and tillage are established for each point by repeating the simulation until there is good agreement between the simulated distribution of ^{137}Cs inventories and the measured distribution. This involves a three-stage process which is outlined below. Detailed description of the model lies beyond the scope of this chapter. However, the stores employed in the model are listed in Table 25.4, the transfers of soil and ^{137}Cs between stores are illustrated diagrammatically in Figure 25.3 and the most important components of the transfers shown in Figure 25.4.

Table 25.4 Stores of soil and [137]Cs

Soil	[137]Cs
For each segment:	
Plough mass: described by depth and bulk density of the plough layer	New [137]Cs: deposited during tillage cycle and concentrated at the surface
	Old [137]Cs: mixed through the plough layer
Sub-plough mass: described by depth and bulk density of the ^{137}Cs-bearing sub-plough layer	Sub-plough [137]Cs
For the whole transect:	
Water-eroded soil	Water-eroded [137]Cs

(a)

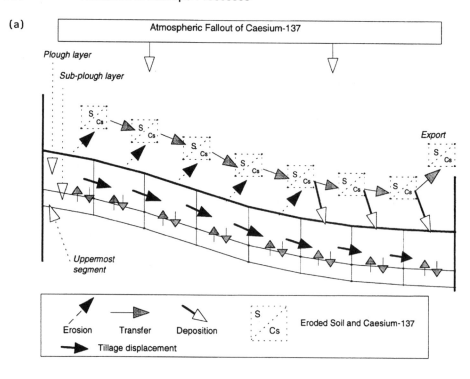

(b)

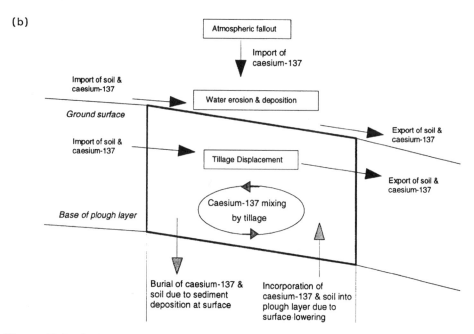

Figure 25.3 A diagrammatic representation of the model of ^{137}Cs redistribution: (a) arrangement of stores; (b) transfers at each slope segment

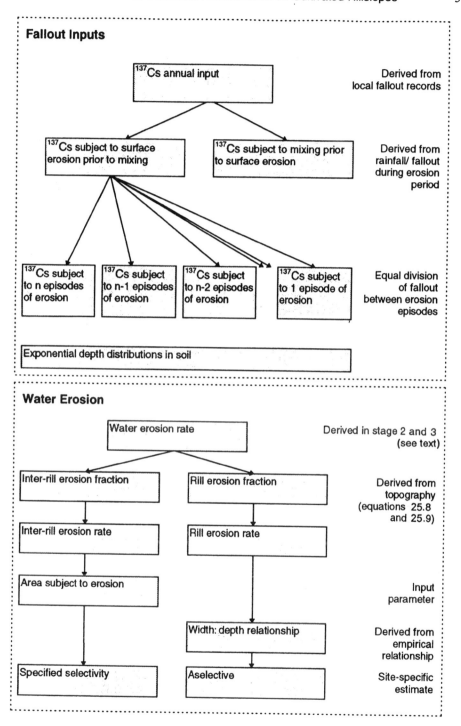

Figure 25.4 The major transfers and parameters in the slope transect model

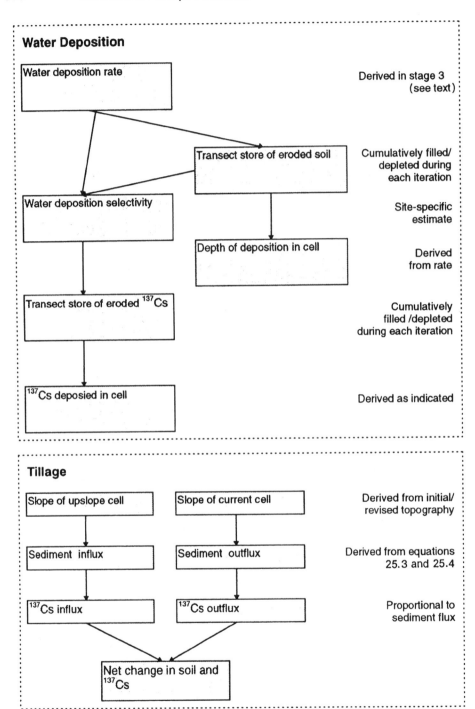

Figure 25.4 *(Continued)*

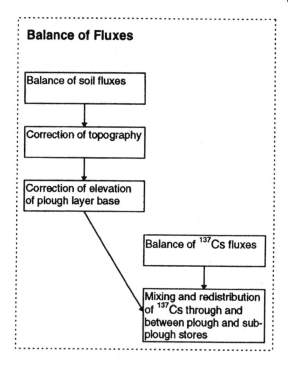

Figure 25.4   *(Continued)*

25.6.1   Derivation of Erosion Rate Estimates

Derivation of erosion rate estimates involves the three-stage process which is described in this section. At the beginning of each simulation the following data are input for each cell: position of mid-point, height of mid-point, water erosion value and interrill fraction (the derivations of the last two are explained below). A single simulation is run iteratively, with each iteration representing the period between tillage events. The number of iterations employed in the simulation is determined by the time elapsed between the date of initiation of atmospheric fallout and the date of sampling, and by the number of tillage events per year.

25.6.1.1   First Stage – Tillage Only

The first group of simulations assumes zero water erosion and examines tillage alone. Tillage fluxes are calculated using either equation (25.3) or equation (25.4), depending on the tillage directions, and the values of parameters a and b are based on published data (Table 25.2). The transfer of ^{137}Cs between cells is assumed to be proportional to the transfer of soil.

The initial topographic data are used to calculate tillage fluxes in the first iteration of the simulation. At the end of each iteration the height of each cell is re-calculated to account for the simulated soil transfers and the new topographic data are used to calculate tillage fluxes in the following cycle.

At the end of the simulation, the simulated and measured ^{137}Cs distributions are compared. If there is significant divergence between the simulated and measured ^{137}Cs distributions, the simulation is repeated to account for soil and ^{137}Cs redistribution by water erosion.

25.6.1.2   Second Stage – Tillage and "Unlimited" Water Erosion

The second group of simulations use the same iterative basis and topographic correction as the first, but include simulation of the impact of specified rates of water erosion along the transect (see Figures 25.3 and 25.4). The *pattern* of rill and interrill erosion values over the transect are estimated from the initial topographic data using elements of the model developed by Govers *et al.* (1993):

$$E_r = n.r.S^p.L^t \tag{25.8}$$

$$E_{ir} = u.r.S^w \tag{25.9}$$

where E_r is the rill erosion value (kg/m^2), E_{ir} is the interrill erosion value (kg/m^2), L is the horizontal distance from the upper boundary or interfluve (m), S is the sine of the slope, r is the bulk density of the plough layer (kg/m^3), and n, p, t, u and w are coefficients.

The initial water erosion *value* for each cell is equal to the sum of the rill and interrill components (assuming no transport limitation). These water erosion values and the interrill fraction (IRF, equation (25.11)) are read for each cell and are considered to be valid for the duration of the simulation. The magnitude of the erosion *rate* estimates for each simulation is determined by a water erosion factor (WF, equation (25.12)). This allows rapid alteration of the magnitude of the rates without alteration of the pattern.

$$E_c = E_r + E_{ir} \tag{25.10}$$

$$\mathrm{IRF} = E_{ir}/E_c \tag{25.11}$$

$$E_w = E_c \cdot \mathrm{WF} \tag{25.12}$$

where E_c is the water erosion value for the cell (kg/m^2), E_w is the water erosion rate for the cell (kg/m^2), IRF is the interrill fraction for the cell, and WF is the water erosion factor for the simulation.

A series of simulations are undertaken with varying values of WF until agreement is found between the simulated and measured ^{137}Cs inventories over the upper part of the slope, where transport limitation is most likely to be insignificant.

25.6.1.3   Third Stage – Tillage and Water Erosion to "Fit"

The third group of simulations is used to derive agreement between the predicted and measured ^{137}Cs inventories over the portion of the slope where poor agreement was found in the second stage. This is achieved by altering the specified water erosion values (E_c) for the cells over the relevant part of the slope and re-running the

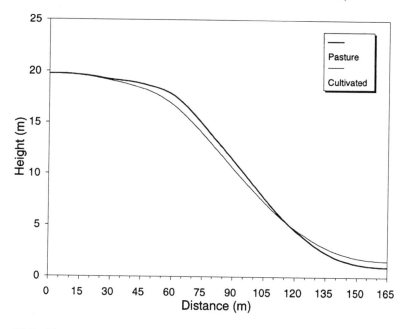

Figure 25.5 The topography of the sampled cultivated field and adjacent pasture at Dalicott Farm

simulation (all other parameters as in the second stage). This procedure is repeated until agreement is obtained between the predicted and measured ^{137}Cs inventories.

25.6.2   Application of the Model – Dalicott Farm, Shropshire

The application, limitations and potential of the slope transect approach may be illustrated by the study of a field at Dalicott Farm in Shropshire. A series of samples were collected for ^{137}Cs analysis along a transect parallel to a field boundary which was perpendicular to the contour. Both the transect topography and that of an adjacent pasture field were recorded (Figure 25.5).

25.6.2.1   Stage 1

The results of the first group of simulations are shown in Figure 25.6(a). Water erosion was assumed to be zero and tillage was simulated in opposing directions in alternate years (up and down the slope). Values of a were used which reflect the range of published measurements of diffusive constants (T_k, Table 25.2; plough depth $= 0.25$ m; bulk density $= 1350$ kg/m^3). Relatively little variation was found in the predicted soil redistribution rates and ^{137}Cs residuals with values of T_k from 250 to 300 kg/m (Figure 25.6(a)). The model may, therefore, be considered to be robust regarding this parameter for this simulation. However, the divergence between the measured ^{137}Cs residuals and the simulated residuals is very evident. It is, therefore, necessary to account for water erosion to explain the observed distribution of ^{137}Cs.

(a)

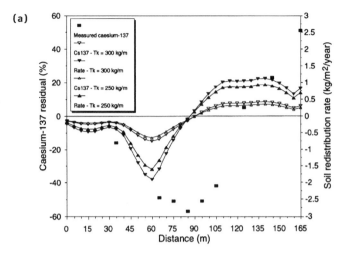

(b)

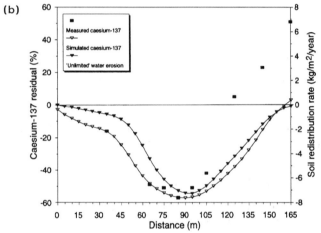

(c)

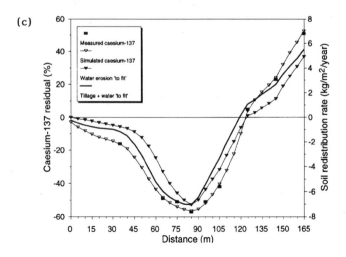

25.6.2.2 Stage 2

A remarkably high level of agreement was obtained between the measured and simulated ^{137}Cs distributions over the upper 85 m of the slope when tillage ($a = 0.89$; $T_k = 300$ kg/m) was combined with water erosion using a value of WF of 2.8 (coefficients for equations (25.8) and (25.9) as suggested by Govers *et al.* (1993): $n = 3 \times 10^{-4}$; $p = 1.45$; $t = 0.75$; $u = 3.68n$; $w = 0.8$). The simulated water erosion rates and the simulated and measured ^{137}Cs residuals are shown in Figure 25.6(b). However, despite the high level of agreement over the upper part of the slope, there is clear divergence between simulated and measured ^{137}Cs over the remainder of the slope. This suggests that sediment transport is limited over the lower portion of the slope.

25.6.2.3 Stage 3

In the third stage, the specified water erosion rates over the lower 80 m of the slope were adjusted and the simulation repeated until agreement was obtained between predicted and measured ^{137}Cs residuals over this part of the slope. Figure 25.6(c) shows the rates of water erosion and deposition rates required to obtain agreement between the measured and predicted ^{137}Cs residuals.

25.6.3 Limitations

The model appears to provide a powerful method for interpretation of ^{137}Cs data in terms of quantitative rates of different erosion processes. However, before examining the potential of the approach, it is important to recognize three potential limitations of the model.

The first potential limitation of the approach is the requirement for the *initial* topography of the transect. However, in *most* cases in temperate environments (long-slope lengths and moderate relief) the changes to the initial form of the transect, over the period of 30–40 years which is simulated, will be insufficient to cause substantial variation in the simulated rates of tillage redistribution and the predicted pattern of water erosion. Therefore, in the absence of other data, it will usually be appropriate to employ the slope form at the time of sampling as an approximation of the initial form. Nevertheless, it should be noted that on short terraces tillage redistribution can lead to substantial changes in slope form over relatively short periods of time and some attempt must be made to account for this in determining the initial slope form (Quine *et al.* 1993).

The second potential limitation is that the deviation of measured inventories from simulated inventories is attributed to variation in water erosion processes. The rationale for this assumption lies in the high level of consistency in experimental studies of soil redistribution by tillage, the simplicity of the assumptions involved in

---

Figure 25.6 *(Opposite)* Application of the slope transect model to the Dalicott Farm slope transect: redistribution rates and predicted ^{137}Cs residuals: (a) stage 1 – simulation of tillage; (b) stage 2 – simulation of tillage (T_k=300 kg/m) and "unlimited" water erosion; (c) stage 3 – simulation of tillage (T_k=300 kg/m) and water erosion rates required to obtain a fit between

simulating tillage and the robustness of the model to variation in the tillage parameters (cf. Figure 25.6(a)). While it is important to recognize that this assumption is made, it is unlikely that it will lead to large errors in the estimation of water erosion rates.

The third potential limitation lies in the simulation of the impact of water erosion processes on ^{137}Cs redistribution (clearly this is a limitation of all calibration procedures). The approach employed attempts to optimize this simulation by using parameters for which independent data are available and accounting for both rill and interrill erosion (cf. Quine 1995). Nevertheless, it cannot be claimed that the estimated rates are without uncertainty. Of the factors which exercise the greatest control over the relationship between ^{137}Cs loss and soil erosion by water, the initial depth distribution of ^{137}Cs after fallout is subject to the greatest uncertainty. The approach employed assumes that the initial depth distribution has the following form:

$$C_d = C_f \cdot k \cdot e^{-kd} \qquad (25.13)$$

$$k = \frac{\ln(10)}{NL} \qquad (25.14)$$

where C_d is caesium-137 activity at depth d (Bq/m^2/cm), C_f is caesium-137 fallout input (Bq/m^2), d is depth (cm), k is a constant, and NL is the depth above which 90% of fallout input is found (cm).

The simulated surface concentration and, therefore, the amount of ^{137}Cs lost per unit mass of soil, during the period of fallout, is determined by the value used for parameter NL. The larger the value, the lower the surface concentration and, therefore, the higher the erosion rate required to obtain the same level of ^{137}Cs-depletion. Experimental evidence for the value of NL, for levels of ^{137}Cs deposition equivalent to those observed in fallout, is difficult to obtain because of the low concentrations involved and because of the problems of sampling soil at millimetre levels of precision. Nevertheless, in experiments undertaken by He (1993) and Owens (1994), a value of 0.8 cm was obtained for NL on gleyed brown earths of the Rixdale series in Devon (15% clay; 42% silt; 43% sand). The soils at Dalicott were much coarser and therefore, although total adsorption of atmospheric input would still occur, a larger value of NL could be expected because of greater infiltration and a lower concentration of adsorption sites per unit mass. However, for the simulations discussed above, a slightly *smaller* value (0.5 cm) was used. It is intended that this should account for synchronous fallout deposition and water erosion, which could be expected to depress the value of NL. Evidently, in the present state of knowledge, this value represents an informed estimate rather than a definitive value. It is, therefore, important to give brief consideration to the impact of variation in the value of NL. Table 25.5 shows the impact of variation in the value of NL on the deviation between measured and simulated ^{137}Cs residuals and on the value of WF required to obtain an optimum fit. These data confirm the importance of the value of NL, but also provide some evidence for the robustness of the model in regard to this parameter. Even if the value of NL is reduced to 0.2 cm (25% of the experimentally derived value), the water erosion rates required are 75% of those required with a

Table 25.5 Influence of variation in NL on the deviation between measured and simulated ^{137}Cs residuals (water erosion alone), and on the water erosion rates used to obtain an optimum fit

Value of NL (cm)	Deviation at:	Deviation (%)					Water erosion cf. NL=0.5 (%)
		35 m	65 m	75 m	85 m	Mean	
0.5		−2.0	−4.9	1.9	0.4	2.3	100
0.2		1.1	−3.8	1.7	−0.6	1.8	75
0.1		4.5	−2.0	−2.3	−0.6	2.3	57
0.05		7.7	−1.1	1.8	−1.7	3.1	39

value of NL of 0.5 cm. While further reduction of the value of NL yields lower water erosion estimates, this reduction is paralleled by a reduction in the goodness of fit (indicated by the per cent deviations in Table 25.5), especially on the upper part of the slope, where the impact of tillage will lead to further deviation between the measured and simulated pattern. Therefore, while it is important to recognize the influence of the variable NL on the results of the simulation, it is clear that values which offer a satisfactory combination of "goodness-of-fit" and comparability with experimental data provide similar erosion rate estimates. These rate estimates will be given further consideration below.

25.6.4 Validity

It is clearly desirable to assess the *validity* of the results derived from this improved approach to estimating erosion rates from ^{137}Cs data. The agreement obtained in stage 2 between measured and predicted ^{137}Cs residuals over the upper 85 m of the slope is itself very encouraging. However, it would be desirable to compare the output with independent data. In this respect the field at Dalicott Farm offers suitable evidence. As was shown in Figure 25.5, the cultivated field subject to investigation lies adjacent to a pasture field and the boundary between the two is marked by a significant lateral step. If it is accepted that negligible soil redistribution has occurred in the pasture field during the lifetime of the boundary, then the step between the fields may be attributed to soil redistribution within the cultivated field over the same time period. The magnitude of the step may, therefore, be used to indicate the pattern of long-term total soil redistribution and to provide independent data for comparison with the model output. The total soil redistribution rates were calculated as follows:

$$M_i = (CE_i - PE_i)r \qquad (25.15)$$

where M_i is measured total soil redistribution at point i (kg/m^2), CE_i is elevation of cultivated field at point i (m), PE_i is elevation of pasture field at point i (m), and r is soil bulk density (kg/m^3).

In order to compare the pattern of soil redistribution indicated by the topography of the field boundary with that derived from the model discussed above, a parallel procedure was adopted:

1. The soil redistribution rate (T) at each point as a result of a single tillage operation was simulated ($T_k = 200$ kg/m).
2. The "unlimited" water erosion value (UW) was predicted for each point.
3. The data from 1 and 2 for the slope section from 35–100 m were regressed against (M) the measured total soil redistribution ($r = 1350$ kg/m³).
4. The regression equation ($M = 564T + 616UW$; $R^2 = 0.97$) was used to estimate the period of cultivation of the field (564 years) and the value of WF for the period (616/564 = 1.1). Values of T_k of 100 and 300 kg/m result in periods of 1124 years and 374 years and WF values of 0.5 and 1.6, respectively.
5. Water erosion rate estimates "to fit" were obtained for the lower part of the slope by subtracting the predicted tillage redistribution from the total measured redistribution for the slope section from 100–165 m.

Figure 25.7(a) shows the rate of total soil redistribution, estimated from the topographic data, and the rates of soil redistribution by tillage and water, derived using the procedure outlined above ($T_k = 200$ kg/m). The rates are based on the following assumptions: (i) the period of significant water erosion is coincident with the period of cultivation; (ii) the pattern of tillage redistribution over time will have been relatively constant, but the rates may be expected to have been lower than at present (a value of 200 kg/m was used for T_k in this simulation instead of 300 kg/m used in simulation of ^{137}Cs data). Both the assumptions appear to be appropriate and, therefore, although the rates must be regarded as tentative they provide a reasonable basis for comparison with the ^{137}Cs-derived data.

The rates of soil redistribution by water derived from the topographic data ($T_k = 100$–200 kg/m) and slope transect model (NL = 0.2–0.5 cm) are compared in Figure 25.7(b). The similarity in the patterns of water erosion over the upper part of the slope is to be expected, because the same procedure was adopted to simulate "unlimited" water erosion over this section. However, the similarity of the patterns over the lower slope is significant because of the differing methods used in derivation of these patterns. The "topographic" rates were derived from the difference between the tillage simulation rates and the total soil redistribution rates. The ^{137}Cs-derived rates were obtained using a similar procedure but based on matching measured ^{137}Cs inventories and those predicted using the slope transect model. The similarity in the patterns, therefore, provides strong support for the contention that the ^{137}Cs data reflect total soil redistribution and for the reliability of the slope transect model.

25.6.5 Potential

If the validity of the model output is accepted then the considerable potential of the approach becomes apparent. Where the ^{137}Cs data alone are available the approach offers the opportunity to examine:

Figure 25.7 *(Opposite)* Soil redistribution rates derived from topographic and ^{137}Cs data for the slope transect at Dalicott Farm: (a) rates of soil redistribution derived from the topographic data (T_k=200 kg/m); (b) comparison of ^{137}Cs-derived and topographically derived water erosion rates; (c) comparison of ^{137}Cs-derived (NL=0.5 cm; T_k=300 kg/m) and topographically derived (T_k=200 kg/m) rates of change in elevation

(a)

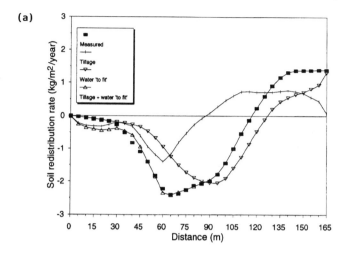

(b)

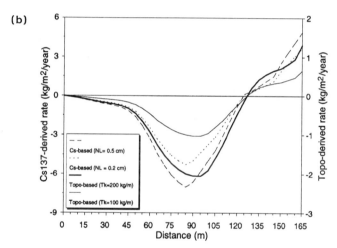

(c)

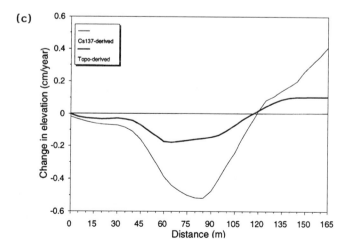

1. rates of water erosion and tillage redistribution and their relative importance (cf. Figures 25.6(a) and (c));
2. the pattern of landscape evolution expressed in total soil redistribution rates (cf. Figure 25.6(c));
3. controls on sediment transport indicated by the deviation in the water erosion rates from the "unlimited" estimates (cf. Figures 25.6(b) and (c)).

Where other data are available, the approach offers potential to obtain a rare insight into changes in rates of erosion and patterns of landscape evolution over time. This is well illustrated by the Dalicott example. As Figure 25.7(b) shows, the water erosion rate over the last c. 40 years appears to be three to four times the water erosion rate of the last c. 6–11 centuries. This result is in line with much recent work, indicating an acceleration of water erosion in response to the intensification of agriculture in the post-war period, but provides a rare opportunity to quantify the impact. This increase in water erosion also has significance for slope evolution (Figure 25.7(c)). Over the last c. six centuries the impacts of water and tillage have been similar (export of sediment from 0–85 m section: 57% water; 43% tillage) leading to maximum denudation at 60 m downslope and almost parallel retreat over the 60–90 m section. Whereas, *on this slope*, over the last 40 years the balance has shifted towards water erosion (export of sediment from 0–85 m section: 76% water; 24% tillage), leading to progressive increase in denudation to a maximum rate at 85 m downslope.

Finally, it is interesting to note that even when tillage erosion has been taken into account, both the topographic data and the ^{137}Cs data indicate rates of water erosion (0.25–0.5 and 1.1–1.5 kg/m^2 per year) which are very significantly higher than the rill-based estimate for the field (0.1 kg/m^2 per year) obtained by Evans (1992). This may be due to the fact that the rates discussed in this chapter correspond to a small part of the field, characterized by the highest slope angle and length (and possibly concentration of wheelings in the proximity of the field edge), whereas the rill-based estimates represent erosion over the whole field area. However, if this is considered to be an insufficient explanation then there are two other obvious potential explanations. First, both the topographically derived and ^{137}Cs-derived rates may be overestimates. Secondly, the rill-based estimate may be an underestimate. The preceding discussion has explicitly stated the procedures involved in the derivation of the ^{137}Cs-derived and topographically derived estimates and it is difficult to identify a source for such a large systematic error. Furthermore, reduction of the topographically derived erosion rates would only be possible if the field boundary was very significantly older – reduction in the topographically-derived water erosion rate by the factor necessary to obtain agreement with the rill-based estimates would require the continuous presence of the hedge, and continuous contrasting land use either side of it for over 2500 years! It is, therefore, appropriate to consider the possibility that the rill-based rates may be underestimates. As Figure 25.7(c) shows, the maximum rates of ^{137}Cs-derived denudation and accretion on the slope are 0.4–0.5 cm/year and 0.3–0.4 cm/year, respectively. Even if these rates were concentrated on only 40% of the area (e.g. in furrows) this would represent only 1.0–1.3 cm/year maximum incision and 0.7–0.9 cm/year maximum deposition. Surface impacts of this magnitude would not be readily identifiable by field observation.

25.7 CONCLUSIONS

Recent advances in the understanding of soil redistribution by tillage demand a re-evaluation of erosion rate estimates derived from ^{137}Cs data and of methods used in their derivation. It is suggested that a new approach to erosion rate estimation from ^{137}Cs data is required in which attention focuses on simulation of ^{137}Cs distributions, rather than focusing on individual point values. A slope transect model of soil-associated ^{137}Cs redistribution has been developed and this greatly increases the potential value of the ^{137}Cs technique. The model facilitates the derivation of process-specific erosion rate data and, therefore, both allows evaluation of the relative importance of processes of soil redistribution and provides evidence of value in understanding the operation of water erosion processes. Finally, it is suggested that the approach offers a sound basis for the derivation of net rates of soil redistribution from ^{137}Cs data.

ACKNOWLEDGEMENTS

The financial support for the work at Dalicott provided by NERC and the assistance with fieldwork given by Dr Qingping He are gratefully acknowledged.

REFERENCES

Boardman, J. 1990. Soil erosion on the South Downs: a review. In J. Boardman, I.D.L. Foster and J.A. Dearing (eds), *Soil Erosion on Agricultural Land*. John Wiley, Chichester, pp. 87–105.

De Jong, E., Begg, C.B.M. and Kachanoski, R.G. 1983. Estimates of soil erosion and deposition for some Saskawatchewan soils. *Canadian Journal of Soil Science*, **63**, 617–607.

Evans, R. 1988. *Water Erosion in England and Wales, 1982–1984*. Report for Soil Survey and Land Research Centre, Silsoe.

Evans, R. 1992. Erosion at Dalicott Farm, Shropshire – extent, frequency and rates. European Society for Soil Conservation, First International Congress, Post-Congress Tour Guide, pp. 56–62.

Fredericks, D.J. and Perrens, S.J. 1988. Estimating erosion using caesium-137: II. Estimating rates of soil loss. In M.P. Bordas and D.E. Walling (eds), *Sediment Budgets*. Proc. Porto Alegre Symposium, IAHS Publication 174, pp. 233–240.

Govers, G., Quine, T.A. and Walling, D.E. 1993. The effect of water erosion and tillage movement on hillslope profile development: a comparison of field observations and model results. In S. Wicherek (ed.), *Farmland Erosion in Temperate Plains Environment and Hills*. Elsevier, pp. 285–300.

Govers, G., Vandaele, K., Desmet, P.J.J., Poesen, J. and Bunte, K. 1994. The role of tillage in soil redistribution on hillslopes. *European Journal of Soil Science*, **45**, 469–478.

Govers, G., Quine, T.A., Desmet, P.J.J. and Walling, D.E. 1996. The relative contribution of soil tillage and overland flow erosion to soil redistribution on agricultural land. *Earth Surface Processes and Landforms*, in press.

He, Q. 1993. Interpretation of fallout radionuclide profiles in sediments from lake and floodplain environments. PhD thesis, University of Exeter.

Kachanoski, R.G. 1987. Comparison of measured soil cesium-137 losses and erosion rates. *Canadian Journal of Soil Science*, **67**, 199–203.

Kachanoski, R.G. and de Jong, E. 1984. Predicting the temporal relationship between soil cesium-137 and erosion rate. *Journal of Environmental Quality*, **13**, 301–304.

Lindstrom, M.J., Nelson, W.W., Schumacher, T.E. and Lemme, G.D. 1990. Soil movement by tillage as affected by slope. *Soil and Tillage Research*, **17**, 255–264.

Lindstrom, M.J., Nelson, W.W. and Schumacher, T.E. 1992. Quantifying tillage erosion rates due to moldboard plowing. *Soil and Tillage Research*, **24**, 243–255.

Loughran, R.J., Campbell, B.L. and Elliott, G.L. 1990. The calculation of net soil loss using caesium-137. In J. Boardman, I.D.L. Foster and J.A. Dearing (eds), *Soil Erosion on Agricultural Land*. John Wiley, Chichester, pp. 119–126.

Martz, L.W. and de Jong, E. 1987. Using cesium-137 to assess the variability of net soil erosion and its association with topography in a Canadian Prairie landscape. *Catena*, **14**, 439–451.

Mitchell, J.K., Bubenzer, G.D., McHenry, J.R. and Ritchie, J.C. 1980. Soil loss estimation from fallout cesium-137 measurements. In M. DeBoodt and D. Gabriels (eds), *Assessment of Erosion*. John Wiley, Chichester, pp. 393–401.

Owens, P. 1994. Toward improved interpretation of caesium-137 measurements in soil erosion studies. *PhD thesis*, University of Exeter.

Quine, T.A. 1989. Use of a simple model to estimate rates of soil erosion from caesium-137 data. *Journal of Water Resources*, **8**, 54–81.

Quine, T.A. 1995. Estimation of erosion rates from caesium-137 data: the calibration question. In I.D.L. Foster, A.M. Gurnell and B.W. Webb (eds), *Sediment and Water Quality in River Catchments*. John Wiley, Chichester, pp. 307–329.

Quine, T.A. and Walling, D.E. 1991. Rates of soil erosion on arable fields in Britain: quantitative data from caesium-137 measurements. *Soil Use & Management*, **7**, 169–176.

Quine, T.A. and Walling, D.E. 1993. Assessing recent rates of soil loss from areas of arable cultivation in the UK. In S. Wicherek (ed.), *Land Erosion in Temperate Plains and Hills*. Elsevier, Amsterdam, pp. 357–371.

Quine, T.A., Walling, D.E. and Zhang, X. 1993. The role of tillage in soil redistribution within fields on the Loess Plateau, China: an investigation using cesium-137. In K. Banasik and A. Zbikowski (eds), *Runoff and Sediment Yield Modelling*. Warsaw Agricultural University Press, Warsaw, pp. 149–155.

Quine, T.A., Desmet, P.J.J., Govers, G., Vandaele, K. and Walling, D.E. 1994. A comparison of the roles of tillage and water erosion in landform development and sediment export on agricultural land near Leuven, Belgium. In L. Olive, R.J. Loughran and J. Kesby (eds), *Variability in Stream Erosion and Sediment Transport*. Proc. Canberra Symposium, December 1994. IAHS Publication 224, pp. 77–86.

Revel, J.C., Guiresse, M., Coste, N., Cavalie, J. and Costes J.L. 1993. Erosion hydrique et entraînement hydrique des terres par les outils dans les côteaux dus sud-ouest de la France. In S. Wicherek (ed.), *Farm Land Erosion in Temperate Plains Environment and Hills*. Elsevier, Amsterdam, pp. 551–562.

Ritchie, J.C. and McHenry, J.R. 1975. Fallout Cs-137: a tool in conservation research. *Journal of Soil and Water Conservation*, **30**, 283–286.

Ritchie, J.C. and McHenry, J.R. 1990. Application of radioactive fallout cesium-137 for measuring soil erosion and sediment accumulation rates and patterns: a review. *Journal of Environmental Quality*, **19**, 215–233.

Sutherland, R.A. 1991. Examination of caesium-137 areal activities in control (uneroded) locations. *Soil Technology*, **4**, 33–50.

Walling, D.E. and Quine, T.A. 1990. Calibration of caesium-137 measurements to provide quantitative erosion rate data. *Land Degradation and Rehabilitation*, **2**, 161–175.

Walling, D.E. and Quine, T.A. 1991. The use of caesium-137 measurements to investigate soil erosion on arable fields in the UK: potential applications and limitations. *Journal of Soil Science*, **42**, 147–165.

Walling, D.E. and Quine, T.A. 1992. The use of caesium-137 measurements in soil erosion surveys. In J. Bogen, D.E. Walling and T. Day (eds), *Erosion and Sediment Transport Monitoring Programmes in River Basins*. Proc. Oslo Symposium, August 1992. IAHS Publication 210, pp. 143–152.

Wilkin, D.C. and Hebel, S.J. 1982. Erosion, redeposition, and delivery of sediment to midwestern streams. *Water Resources Research*, **18**, 1278–1282.

26 Erosional Response to Variations in Interstorm Weathering Conditions

R. BRYAN
Soil Erosion Laboratory, University of Toronto (Scarborough), Ontario, Canada

26.1 INTRODUCTION

The importance of climate as a dominant influence on soil erosion has long been understood and recognized in soil erosion models. Particular significance has been attached to rainfall aggressivity (Fournier 1960), erosivity (Wischmeier *et al.* 1958) and kinetic energy (Hudson 1965), although De Ploey (1972) has shown that severe erosion can occur in areas of low rainfall erosivity. Soil resistance has usually been regarded as of secondary significance and in the Universal Soil Loss Equation was treated as a constant parameter, independent of the climatic factor (Wischmeier and Smith 1978). In fact, soil erodibility is dynamic, reflecting changes in soil physical, chemical and biological properties, particularly as these affect soil aggregation. These changes respond directly or indirectly to climatic conditions, as shown by much experimental research (e.g. Bisal and Nielsen 1967; Bryan 1971; Reid and Goss 1982; Tisdall and Oades 1982; Utomo and Dexter 1982; Kemper *et al.* 1987; Shiel *et al.* 1988; Mostaghimi *et al.* 1988; Truman *et al.* 1990). As a result, systematic seasonal variations in soil erodibility can often be recognized, particularly where seasonal contrasts are strong (Dickinson *et al.* 1982; Mutchler and Carter 1983; Young *et al.* 1983; Kirby and Mehuys 1987), and some attempts have been made to develop seasonally adjusted soil erodibility values (El-Swaify and Dangler 1977; Zanchi 1983; Young *et al.* 1990).

Any major change in climatic conditions such as annual rainfall or the frequency of freeze–thaw activity would certainly affect soil erosion rates and perhaps also dominant erosion processes. It is likely, however, that the initial stages of climatic change would be gradual with rather subtle changes in climatic conditions, such as small changes in evaporation rates or the interval between storms. Little is known about the effect of such small perturbations on soil erosion processes or rates. One important effect is on soil moisture. The influence of antecedent moisture conditions on processes such as soil slaking and surface sealing are well-documented, but exactly how these effects are integrated at plot or hillslope scale is not clearly understood. Several field studies (e.g. Emmett 1970; Oostwoud Wijdenes 1992; Bryan 1994) have reported initial "flushes" of sediment loss during simulated rainfall tests, even after very brief interstorm intervals. Such flushes have been attributed to "weathering preparation" (e.g. Carson and Kirkby 1973) but the precise processes

Advances in Hillslope Processes, Volume 1. Edited by M. G. Anderson and S. M. Brooks.
© 1996 John Wiley & Sons Ltd.

involved, the effect of interval duration, or how these vary on different soils, have not been established.

Laboratory experiments were carried out to provide initial information on the effect of interstorm conditions on soil erodibility and dominant soil erosion processes. This information is essential for prediction of the effect of minor climatic change on soil erosion rates.

26.2 EXPERIMENTAL DESIGN

Experiments were carried out in a large experimental flume under simulated rainfall. *A priori* reasoning suggests that the effects of interstorm conditions will be most pronounced on cohesive, well-aggregated soils, and will be significantly affected by clay mineralogy. Two Canadian soils, Gobles silt loam grey-brown luvisol, developed on mixed moraine and till near Belmont, Ontario, and Swinton silt loam black chernozem, developed on loess near Swinton, Saskatchewan, were selected for the study. Major soil properties are shown in Table 26.1. Both soils are cohesive and well-aggregated but despite textural similarities, differ in clay mineralogy, Swinton being dominated by smectite, and Gobles by vermiculite, kaolinite and micaceous clays, with some smectite.

The experimental flume used is 0.8 m wide and 22.5 m long, and consists of impermeable perspex sections set, in this case, as a uniform rectilinear 0.061 (3.5°) slope. The flume base was covered in sequence by an aluminium mesh screen, permeable erosionet cloth, a 0.1 m deep layer of homogeneous coarse sand, more erosionet, and finally a 0.1 m deep layer of the test sample. Bulk soil samples were air-dried, sieved through an 8 mm sieve, and placed in the flume in two layers of equal depth. The first layer was manually compacted to a bulk density of 1.25 Mg/m³, then raked before addition of the second layer, which was also compacted, raked, and lightly brushed to remove linear marks. The ideal experimental design would have included numerous different storm sequences with adequate replication, but this was logistically impracticable. The full 22.5 m flume length was necessary for

Table 26.1 Basic properties of soil samples

Sample properties	Gobles silt loam	Swinton silt loam
Sand (%)	22.87	38.36
Silt (%)	59.76	41.96
Clay (%)	17.37	19.67
Water stable aggregation >2.8 mm diameter	40.4	31.2
Ex K^+ (cmol/kg)	0.47	0.35
Ex Ca^{++} (cmol/kg)	5.62	10.60
Ex Mg^{++} (cmol/kg)	1.56	3.15
Ex Na^+ (cmol/kg)	0.07	0.02
CEC (cmol/kg)	12.38	17.93
Organic C (%)	1.07	1.62
pH	6.81	5.73

realistic simulation of hillslope-scale processes. Each experiment required approximately 2.25 tonnes of soil, and because soil character changed during storms, each sample could be used only for one test sequence.

Tests involved simulated rainfall from Lechler and Spraco nozzles at 5 m height above the bed along the full flume. Storms lasted 1 or 2 h with mean intensity 55.6 mm/h (range 46.4–69.7 mm/h), and kinetic energy approximately 80% of similar natural rainfall. Three storm sequences were tested on each soil, each consisting of two or three 1–2 h storms with varying interstorm intervals (Table 26.2). During each experiment cumulative water (Q) and sediment discharge (Q_s) were monitored at a terminal weir, and suspended sediment load, bed load, standard hydraulic measurements and observations of erosion patterns and processes were taken at four or five hydraulic stations along the flume. The methodology has been described in detail by Bryan (1990). Detailed hydraulic observations will be discussed in another paper.

Moisture patterns within the experimental bed were monitored by eight micro TDR probes linked to a datalogger. At intervals of 2–4 days between storms, surface moisture content (gravimetric), bulk density, penetrometer resistance and shear strength with a standard vane shear apparatus and a highly sensitive vane apparatus designed for very wet soils were measured at four or five locations along the flume. Sampling in the flume was limited to reduce disturbance before subsequent tests, but aggregation data were collected from parallel weathering experiments with smaller samples. These were subjected to rainfall for one hour, then allowed to dry naturally with regular testing of aggregate stability by ultrasonic dispersion (Imeson and Vis 1984).

26.3 RESULTS

Summary results for all flume experiments are shown in Table 26.2, and are discussed separately below for each soil type.

26.3.1 Gobles Silt Loam

26.3.1.1 Simulated Rain Storms

Cumulative water (Q), sediment discharge (Q_s), instantaneous water discharge (Q_{is}) and sediment concentration (O_s) patterns at the weir are shown for each of three storm sequences in Figures 26.1 and 26.2 respectively. Initial storms in each sequence produced consistent results, with differences due to variations in rainfall intensity (R_i), matched by lag times to runoff. Once runoff started, hydraulic conditions rapidly reached threshold values usually associated with rill initiation (e.g. Govers 1985; Bryan 1990; Slattery and Bryan 1992) except at the top of the flume. However, no rill incision occurred during the first 60 min and entrainment was extremely localized. Limited scouring led to a pronounced stepped bed morphology which precisely matched results in a preliminary comparative field experiment near Belmont, western Ontario. O_s values during the first 60 min were similar on 11 June and 22 September but, apart from an initial peak, were lower on 26 November. The

Table 26.2 Water and sediment discharge during experimental storm sequences

Date	Duration[a]	Precipitation[b]	Q_{60}[c]	Q_{s60}[d]	Rate[e]	Q_{s60} Peak[d]	R_{r60}[f]	Q_{120}[h]	Q_{s120}[i]	Rate[e]	Q_{s120} Peak[i]	R_{r120}[j]
Gobles silt loam												
26 Nov.	2	66.9	0.896	3.8	0.057	10.3	0.66	2.059	20.8	0.155	27.0	0.76
10 Dec.	1	69.7	0.714	5.6	0.081	8.7	0.51					
21 Jan.	2	62.8	0.570	10.2	0.163	25.3	0.45	1.517	44.2	0.352	47.0	0.60
11 June	2	46.4	0.372	4.3	0.092	49.0	0.44	0.972	11.1	0.119	13.3	0.57
23 July	2	51.5	0.540	14.6	0.283	33.0	0.58	1.309	34.4	0.334	33.0	0.69
6 Aug.	2	48.4	0.700	12.2	0.251	19.7	0.79	1.445	24.0	0.248	19.7	0.82
22 Sept.	1	51.6	0.517	5.0	0.096	16.8	0.55					
27 Oct.	2	55.1	0.382	5.5	0.100	20.7	0.38	1.018	28.6	0.259	20.7	0.51
Swinton silt loam												
10 Dec.	2	53.7	0.363	2.3	0.044	8.6	0.37	1.120	23.7	0.221	8.6	0.57
17 Jan.	2	53.0	0.300	5.9	0.111	25.3	0.31	1.051	27.8	0.262	25.3	0.54
23 Mar.	1	57.7	0.315	2.7	0.047	12.6	0.32					
6 Apr.	2	55.2	0.838	10.0	0.181	18.5	0.83	1.719	28.7	0.259	25.5	0.54
9 Dec	2	54.0	0.141	0.8	0.016	6.5	0.20	0.742	11.2	0.104	24.4	0.86
11 Jan.	2	56.7	0.476	23.2	0.409	56.6	0.46	1.282	61.8	0.545	56.6	0.62
25 Jan.	1	51.4	0.710	28.9	0.710	49.2	0.76					

[a]Rainstorm duration (h).
[b]Rainfall intensity (mm/h).
[c]Water discharge, 1st 60 min (m^3).
[d]Sediment discharge, 1st 60 min (kg).
[e]Sediment loss:unit ppt. (kg/mm).
[f]Runoff ratio, 1st 60 min.
[g]Peak sediment concentration (g/cm^3).
[h]Water discharge, complete storm (m^3).
[i]Sediment discharge, complete storm (kg).
[j]Runoff ratio, complete storm.

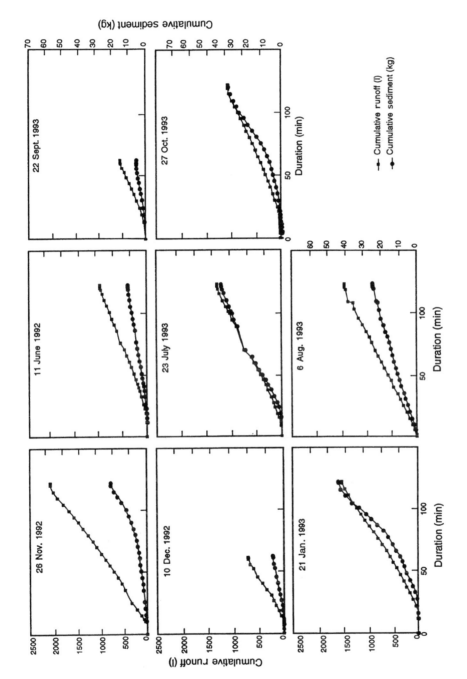

Figure 26.1 Weir cumulative runoff and sediment discharge (Gobles silt loam)

594

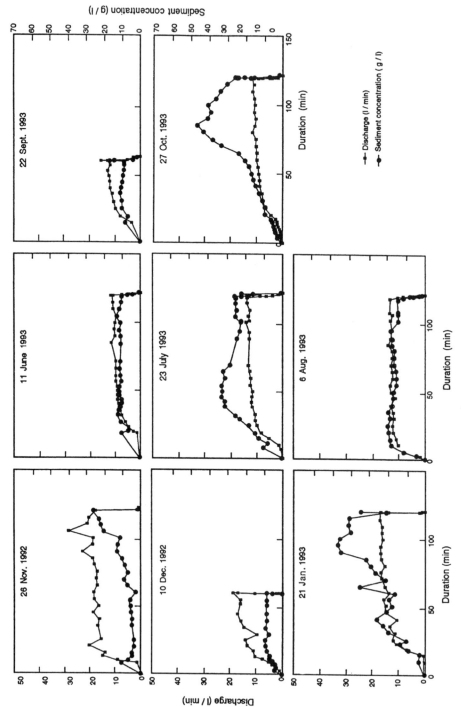

Figure 26.2   Weir instantaneous discharge and sediment concentration (Gobles silt loam)

26 November results reflect more effective aggregate breakdown and surface sealing with higher R_i and drier antecedent moisture conditions (Table 26.2). The result was lower Q_{s60}, and a sediment yield rate of 0.057 kg/mm, compared with 0.092 and 0.096 kg/mm for 11 June and 22 September. O_s values remained essentially constant for the second hour on 11 June, but increased rapidly on 26 November after 80 min due to major rill incision 1.5 m above the weir. Increased bed shear stress in strongly concentrated flow incising into encrusted subsoil produced peak O_s of 27 g/l near the end of the storm.

Despite existing rill incision, high antecedent moisture contents (Table 26.2) and higher R_i, erosion rates on 10 December were much lower than in the initial storm, with Q_{s60} of 5.65 kg, peak O_s of only 8.7 g/l, and Q_{is} values consistently below those of 26 November. Soil moisture levels remained high between the storms (Figure 26.3) with a few, small desiccation cracks, so the lower discharge must be attributed primarily to rill channel infiltration. The second storm in the June–August sequence (23 July) was at lower R_i, after a longer interstorm interval but due to air humidity, antecedent moisture levels remained high. Runoff lag-time was shorter, but otherwise discharge patterns on 11 June and 23 July were almost identical (Figure 26.1(b)). Q_{s60} on 23 July was, however, significantly higher (14.56:4.28 kg), due to rill incision on 23 July after 20 min. This started as a discontinuous rill, interrupted by sheetwash sediment deposition zones. Rills extended but did not join into a continuous channel during the storm. Near the weir a saturated wedge developed, with a sheetwash zone, sedimentation, and replacement of the original 3.5° slope by one at 1°. This increased deposition and led to declining weir O_s late in the storm (Figure 26.2).

The pattern during the second storm in the September–October sequence was quite different. In the initial 60 min storm 22 September erosion was limited to localized scour steps in mid-flume. Interstorm weather was continually hot and dry, giving low antecedent moisture content (Figure 26.3) and extensive desiccation cracking before 27 October. Crack infiltration loss was initially high giving a 60 min R_c of only 38%. Most cracks sealed, but some channelled runoff and developed into rills, with channel geometry strongly influenced by crack patterns (Figure 26.6). Rill incision caused rapid increase in O_s in the second 60 min to 46.25 g/l.

Third storms in the November–January and June–August sequences produced divergent results. The 21 January storm took place after a very dry 42-day interval, so the bed dried thoroughly to an average moisture content of 9.85%. The flume was covered by a dense network of wide (> 1.5 cm) desiccation cracks which penetrated to the upper erosionet. Nevertheless, runoff started swiftly on dense surfaces, cracks filled and overtopped, and flow was continuous after 10 min. In the lower flume cracks closed by swelling, but in the upper flume cracks widened by sloughing and developed into prominent transverse steps. These retreated headwards and developed into linear incisions interspersed with colluvial deposition, in a discontinuous rill system (Bryan and Oostwoud Wijdenes 1992). The incisions grew and joined (at about 85 min) with the original major incision to form a continuous channel. Channel linkage caused rapid colluvial scour and a marked, short-lived sediment pulse at the weir (peak O_s : 47 g/l).

The 6 August test followed a hot, humid interstorm interval. Average antecedent moisture was much higher (24.5%) than on 21 January and humid conditions caused

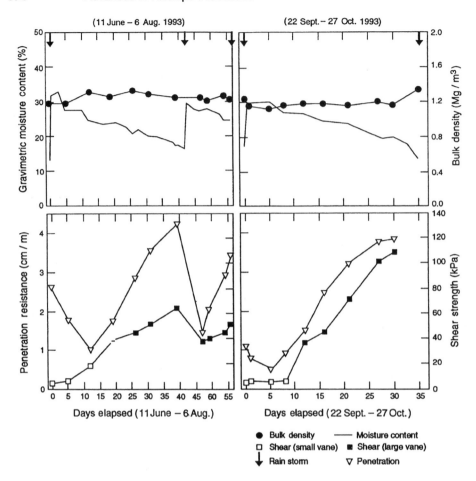

Figure 26.3 Change in soil moisture content, bulk density, penetration resistance and shear strength (Gobles silt loam)

moss and algal growth, particularly on deposition zones. High antecedent moisture, combined with algal hydrophobicity, resulted in almost instantaneous runoff, and an ultimate R^c of 82.9%. Despite high discharge, sediment transport was much lower than on 23 July or 21 January (Q_s 24.2, 34.4 and 44.2 kg respectively) with peak O_s of 19.7 g/l. This was partly due to protection of vulnerable colluvial areas by biological crusts, but these did not colonize the rill channels formed on 23 July. Despite high discharge, these expanded little, and did not join into a single channel as on 21 January.

26.3.1.2   Interstorm Conditions

Results of interstorm monitoring of the second and third storm sequences are shown in Figure 26.3. Because of instrument problems, only moisture data are available for the first sequence. Data are averages for five or six measuring points (eight for gravimetric moisture), and obviously obscure considerable spatial variability. The main spatial effect is that interstorm soil drying was much slower near the weir. Generally higher antecedent moisture conditions in this zone contributed to rapid saturated wedge development in subsequent storms. Apart from moisture content, all measurements started immediately after the initial storm.

Prevailing ambient interstorm conditions largely determined patterns of change in soil moisture content. Amongst the measures of soil physical properties chosen, bulk density was least satisfactory, showing no clear pattern and statistically insignificant differences. The core sampling method used was apparently insufficiently sensitive to discern subtle near-surface changes, particularly on very wet soils. The penetrometer was also rather insensitive, but did show interesting patterns, with clear decline in resistance for a few days after the initial test in both sequences. During this period, soil shear strength remained very low. Once soil moisture content dropped below about 27%, both penetration resistance and soil strength increased rapidly. Peak shear strength reached before the 27 October storm was much higher than the 23 July storm, with much lower antecedent moisture, but peak penetration resistance was very similar. Although soil shear strength declined significantly during the 23 July storm, it did not drop to the very low values measured at similar moisture contents on 11 June.

Soil aggregation was not measured directly in the flume but parallel tests of aggregate stability during drying over a 28 day period (Figure 26.4) showed interesting patterns. Aggregate stability showed no significant change until soil moisture content fell below about 15%, after which it dropped swiftly to close to zero. Comparison with flume soil moisture contents shows that the critical threshold was not reached in the June–August test sequence, was just reached before the 27 October storm, and in the first sequence was reached well before the 21 January storm (Figure 26.5(a)).

26.3.2   Swinton Silt Loam

26.3.2.1   Simulated Rain Storms

There were three test sequences on Swinton silt loam. Patterns of Q, Q_s, O_s and Q_{is} are shown in Figures 26.7 and 26.8 respectively. Because of intense, deep rill incision which reached the erosionet, each of the first two sequences included only two storms. The major difference between initial storms was the runoff lag time. On 10 December and 23 March, slight differences in R_i and significant differences in antecedent moisture content apparently cancelled out, and surface sealing occurred quickly in both cases with nearly identical lag times. Patterns of Q_s and Q in the first 60 min were almost identical, with runoff coefficients of 37.1% and 30.5% and Q_s: 2.34 and 2.73 kg respectively. No rilling occurred in the first 60 min on 10 December, but on 23 March a short rill developed 10 m above the weir, and a marked transverse step at 1 m (Figure 26.9(a)). On 10 December, rapid rill incision started at 6–10 m after 70 min, with a steep rise in O_s at the weir (Figure 26.8) 5 min later. Scouring was so rapid that runoff became disordered with much deposition causing a temporary drop in O_s at the weir, which then rose again as rills were re-established. Subsequent

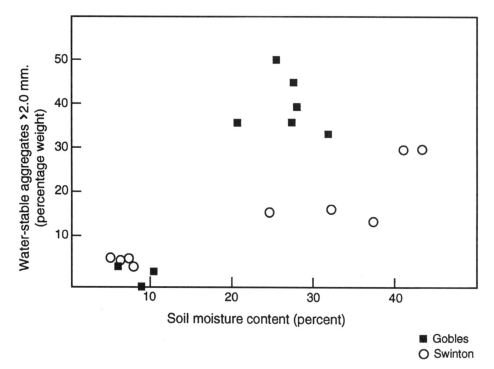

Figure 26.4 Relationship between soil moisture content and aggregate stability (●, Gobles; ○, Swinton)

rill evolution was extremely rapid with deep incision, frequent bank collapse, headcut retreat rates of several centrimetres per minute, and peak O_s of 58.8 g/l. Sediment yield during the second 60 min was 21.4 kg and by 120 min a deep central rill extended between 5 and 14 m above the weir.

The second storm in the first sequence on 17 January followed prolonged drying to an initial moisture content of only 3.5%. Intense desiccation cracking occurred (>2 cm wide, 10 cm deep) which reached the erosionet. Local runoff developed quickly on the dense, sealed soil, but which was stored in cracks, or routed beneath the test bed, as interflow at the erosionet. This routing was not apparent in discharge patterns at the weir. Rill flow was very limited until the second part of the storm, and even then was much less active than in the initial storm, with little bank collapse or headcut retreat. Micropiping along longitudinal cracks in the unrilled section above 14 m was, however, conspicuous. Hydraulic conditions measured in pipes were close to those in rills, causing rapid development, eventual collapse and extension of the rill system (Figure 26.10). Micropipes also developed along transverse cracks, producing a trellised rill system.

The response on 6 April was very different. Humid interstorm conditions gave an initial moisture content of 28.3%, and very limited, superficial cracking. Runoff began instantly and Q quickly reached equilibrium. Peak Q was significantly lower

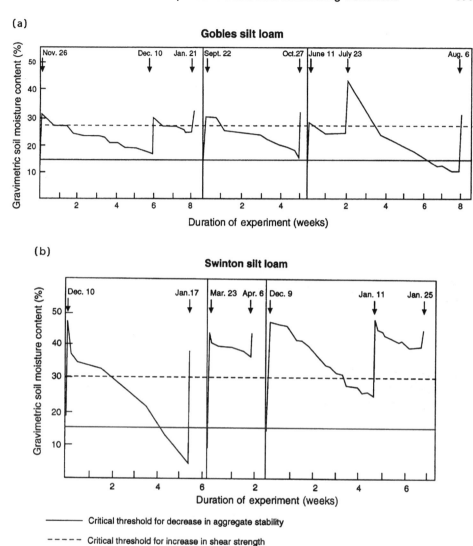

Figure 26.5 (a) Changes in soil moisture content in relation to critical thresholds for soil shear strength and soil aggregate stability during (a) three test sequences with Gobles silt loam and (b) three test sequences with Swinton

than on 17 January, despite similar R_i, but the eventual runoff coefficient was > 85%. Sediment yield and peak O_s were almost identical to 17 January, but the erosion pattern was very different. There was no micropiping in the upper flume, but the rill headcut at 10 m expanded vigorously, with much sediment deposited in the rill channel immediately below (Figure 26.11). Incipient headcuts in the lower flume did not develop into rills, but changed to transverse steps as a saturated wedge progressed upflume, eventually reaching 8 m above the weir.

Figure 26.6 Control of rill incision and orientation by desiccation cracks (Gobles silt loam: 27 October)

Results from the third sequence of storms (9 December–11 January) differed from both previous sequences. Lag time in the first storm was prolonged (35 min), resulting in a low R_c (37.7%), limited localized scouring, and low Q_s (11.2 kg). Once runoff started, Q_{is} and O_s patterns closely resembled initial storms in earlier

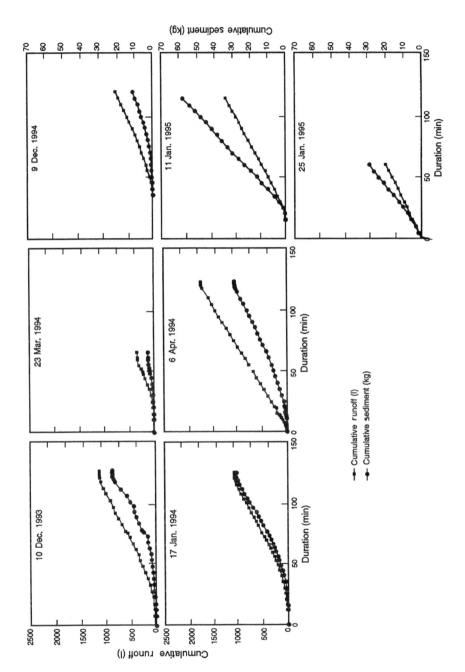

Figure 26.7 Weir cumulative runoff and sediment discharge (Swinton silt loam)

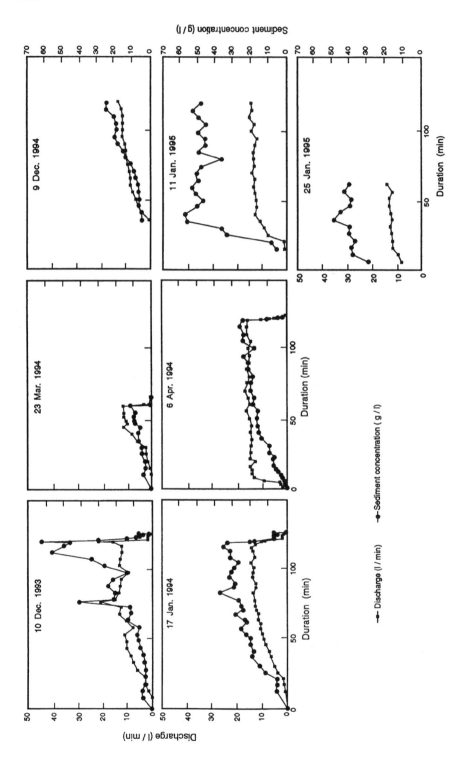

Figure 26.8 Weir instantaneous discharge and sediment concentration (Swinton silt loam)

Figure 26.9 Microstep scour related to development of saturated wedge near weir (Swinton silt loam: (a) 23 March; (b) 6 April)

experiments. This prolonged lag cannot be explained by differences in bed preparation, rainfall or antecedent moisture conditions. The only obvious influential factor is soil ageing in storage for over one year, which apparently significantly reduced surface sealing susceptibility. Initial runoff response on 11 January was

Figure 26.10 Micropiping developed along desiccation cracks in upper flume (Swinton: 17 January)

slightly delayed by crack storage, but after 15 min Q rose steeply, with vigorous rill incision in mid-flume (8–12 m), extremely high O_s values, and, ultimately, the highest Q_s for any experiment. Unlike other tests, significant scouring occurred at 1 m above the weir. O_s values in the final storm (25 January), were significantly lower, despite very high antecedent moisture (30%). Desiccation cracking was limited and no micropiping occurred during this storm sequence, though a trellised deeply-incised tributary system did develop perpendicular to the main rill channel.

26.3.2.2 Interstorm conditions

Changes in soil moisture content and physical properties for the three Swinton silt loam storm sequences are shown in Figure 26.12. Again, soil moisture contents

Figure 26.11   Deposition of sediment in rill below active knickpoint (Swinton, 6 April)

reflect prevailing weather conditions during interstorm periods, but both saturation moisture content and moisture retention are considerably higher than for the Gobles soil. Again, bulk density data showed no clear patterns. For the second two-storm sequence, penetration resistance showed the same post-storm drop, but only for 1–2 days, after which resistance increased swiftly with declining moisture. Soil shear strength dropped to very low levels during storms, and remained low for a considerable time, at no point approaching the peak levels shown by the Gobles soil in very dry conditions. The critical moisture content for significant increase in shear strength of about 30% was reached only during the first and last storm sequences (Figure 26.5(b)).

Aggregate stability was significantly lower than the Gobles soil at high moisture contents, but there was no significant difference in dry soils (Figure 26.4). As in the Gobles soil, aggregate stability declined dramatically below a threshold moisture

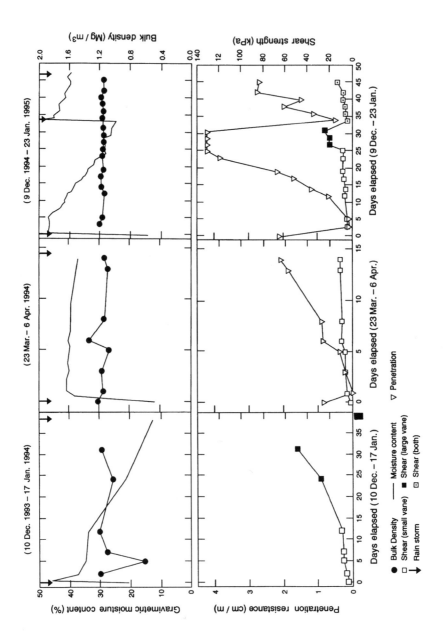

Figure 26.12 Change in soil moisture content, bulk density, penetration resistance and shear strength (Swinton silt loam)

content of about 15%. This value was reached only in the first storm sequence, immediately before the 17 January storm (Figure 26.5(b)).

26.4 DISCUSSION

The object of the study was examination of the effect of interstorm conditions on erosional response and identification of indicators of the effect of minor climatic change on hillslope erosion. The experimental design adopted provides information on two storm circumstances. The initial storm in each sequence occurred on newly prepared soils which had no "memory" of prior conditions. In these conditions erosional response was dominated by storm rainfall and by soil aggregation. All subsequent storms in each sequence were strongly influenced by interstorm conditions and, to some extent, by erosional artefacts from the initial storm. In the latter case, the duration and precise erosional effects of the initial storm became important.

There were strong similarities in Q and Q_s response patterns during initial 60 min storm periods. On the Gobles soil, variations reflected rainfall conditions. Response was similar on 11 June and 22 September, with R_i of 46.4 and 51.6 mm/h respectively, but on 26 November, a R_i of 66.9 mm/h caused greater aggregate disintegration and surface crusting, giving higher Q and lower Q_s. On the Swinton silt loam response was simpler with R_i of 53.7, 57.7 and 54.0 mm/h respectively. Response on 10 December and 23 March was identical, while on 9 December the runoff lag time was much longer, probably due to soil ageing, discussed below. Variations in antecedent moisture do not appear to have significantly affected response in these initial storms, but possibly enhanced aggregate disruption on 26 November.

Four of the six initial storms lasted for 2 h and sediment yield was considerably higher during the second half of the storm, particularly on 26 November (Gobles) and 10 December (Swinton), when rill incision occurred. In each case this caused marked increase in Q_s, and the rills formed affected response in subsequent storms. The dominant factors determining response in subsequent storms were changed in soil moisture in the interstorm interval and moisture level before the storm. These strongly influenced two important properties related to soil erodibility: shear strength and aggregate stability. They also determined the intensity of desiccation cracking. Soil moisture changes in interstorm intervals are plotted with critical thresholds for shear strength and aggregate stability in Figures 26.5. In the first Gobles sequence, drying between 26 November and 10 December was sufficient to significantly increase soil strength, but not to reduce aggregate stability. As a result, Q_s during the 10 December storm was very low, despite the highest R_i of 69.7 mm/h. Increased soil strength also affected the inherited rill channel, which showed no headcut retreat or further development. Prolonged drying before the 21 January storm reduced soil moisture below the critical level for aggregate stability, and also caused marked desiccation cracking. The combined result was greatly enhanced rill erosion over most of the flume with very high Q_s.

The same controls are apparent for the other storm sequences (Figure 26.5(a)). Between 11 June and 23 July, soil moisture closely approached the critical level, producing high Q_s on 23 July, but between 23 July and 6 August soil strength increased, so that Q_s in the third storm was significantly lower, despite a well-developed rill. Intense drying between 22 September and 27 October again reduced soil moisture below the critical level, resulting in rill incision and high Q_s in the second storm.

Response patterns for the Swinton silt loam experiments were more complex (Figure 26.5(b)). In the latter part of the initial 10 December storm, rill incision was intense with high Q_s. Interstorm drying before 17 January produced extreme desiccation cracking on this smectite-rich soil. Although a well-incised rill was inherited from the first storm, Q on 17 January was reduced by crack infiltration, particularly along the major rill, which showed little headcut recession or further development. Significant Q_s did not occur until the second half of the storm when micropiping developed in the upper flume. In the 23 March–6 April storm sequence drying was not sufficient to significantly increase shear strength, and rill incision was intense from the start of the second storm. However, much of the sediment mobilized was deposited in the rill (Figure 26.12) and did not contribute to Q_s at the weir.

The third storm sequence was apparently influenced by soil ageing and it is not clear that the same threshold moisture levels apply. Soil ageing effects have not been specifically examined on either of the soils tested, but age-hardening of agricultural soils has been reported by Utomo and Dexter (1982). There was no evidence of time-dependent change in properties or response for the Gobles soil during the 11-month experimental period, but the smectite-rich Swinton soil appeared to be more sensitive during a 14-month period. Although drying between 9 December and 11 January should have increased soil strength, erosion was intense on 11 January resulting in the highest Q_s recorded. Moisture remained high from 11 to 25 January, and erosion continued at the same high rate in the third storm.

As noted in the experimental design, only a limited range of storm sequences could be tested, and replication of experiments was certainly suboptimal. Nevertheless, the results discussed above clearly indicate the dominant influence of interstorm weathering on erosional response in sequential storms, at least on cohesive soils in which shear strength and aggregate stability are significantly affected by soil moisture contents. The occurrence, pattern and intensity of desiccation cracking has also been shown to be highly significant, affecting both initiation and morphology of the rill network. After prolonged drying on the Swinton soil, cracking resulted in interflow and a change in dominant process from rilling to micropiping. As in rainfall simulation experiments in the smectite-rich Alberta badlands (Bryan *et al.* 1978), this change was not reflected in Q or Q_s patterns at the weir.

Although the effect of interstorm drying on erosion during subsequent storms is clearly shown by the results, interpretation in terms of soil erodibility is complicated. Decline in aggregate stability with soil moisture content is clear, and aggregate stability can be a good soil erodibility index (e.g. Bryan 1968). However, aggregate stability should only control erosion resistance in storms on initially disturbed soil, where entrainment is resisted by the mass of individual particles. Once the soil crusts or compacts into a coherent body, soil shear strength should be a more effective index of erodibility. This means that shear strength should be the most effective index of erodibility for most of the tests. Soil shear strength increases with drying, and one would therefore expect lower erosion rates on drier soils. This apparently contradicts the experimental results. The reason appears to be desiccation cracking which tends to concentrate flow, enhancing rill incision, and exposing less coherent subsoils where aggregate stability is still influential.

The results suggest several significant conclusions related to study and modelling of soil erosion. First, rainstorm characteristics, and particularly intensity or erosivity, are of dominant importance when storms occur on newly disturbed or tilled surfaces, but become much less diagnostic on surfaces which have experienced several storms. The overwhelming importance attached to rainfall erosivity in models such as the Universal Soil Loss Equation probably reflects its origins in experiments on disturbed agricultural soils. The results show that recognition of interstorm weathering and particularly soil moisture change is at least as important in soil erosion experiments and models for predicting hillslope evolution or the effects of climatic change.

The second conclusion concerns the complexity of erosional patterns on experimental slopes. The discussion has been based largely on Q and Q_s patterns at the terminal weir. These data cannot be adequately interpreted without detailed knowledge of erosion and deposition patterns on the complete slope. The absence of such information is a critically important limitation in many erosion studies, notably those from standard 22 m USLE erosion plots. This is particularly important early in storm sequences before development of continuous rill systems. The early stages invariably involve localized scouring and deposition, leading in many cases to cyclic rill incision and colluvial deposition, and eventually, depending on storm duration, to continuous rilling. Weir data will obviously reflect juxtaposition to zones of incision or deposition. Information on the parameters which control cyclic rilling is scant and further study is a clear priority. The threshold storm duration necessary before integration of the rill system into a continuous channel occurs is particularly significant, as this will usually result in considerably increased weir Q_s. A relatively minor increase in average storm duration, which crosses this threshold, could greatly increase soil erosion rates, and markedly affect hillslope development.

One further consideration should be noted in interpreting weir data. Rill initiation on these soils usually involves incision of a narrow, arcuate, knickpoint or headcut (e.g. Savat 1976; Bryan and Poesen 1989; Slattery and Bryan 1992). This may reflect critical hydraulic conditions or, as shown by recent studies with automated microstandpipes (Bryan and Rockwell 1996), loss of soil strength associated with perched water-table development. The resulting incision is narrow and elongated. On 23 March (Swinton) significant scouring only occurred near the weir where a saturated wedge developed. In this case, the knickpoint was a shallow, broad, step-like feature (Figure 26.9) similar to those described by Bryan and Oostwoud Wijdenes (1992). On 6 April, the saturated wedge extended progressively upslope, and dominant rill morphology changed from narrow incisions to broad steps. This change in morphology generally changes patterns of incision and deposition, causing increased sediment storage in the rill system. Again the inference is that interpretation of weir data must be adjusted to dynamic changes in the erosional system, in this case related to appearance and development of perched water-tables and saturated wedges.

26.5 CONCLUSIONS

The experiments described demonstrate that erosional response at hillslope scale in sequential storms can be strongly influenced by moderate changes in interstorm conditions. During initial 60 min storms on disturbed soils, approximating those on newly tilled agricultural fields, erosional response was dominated by rainfall intensity. The only exception was for the last test sequence on the Swinton silt loam, which was affected by soil ageing in storage.

During some initial storms, particularly those which lasted 2 h, significant scouring and rill incision occurred, accompanied by deposition further downslope in a cyclic pattern. Surface morphology was therefore quite varied, reflecting precise patterns of erosion and deposition.

Soil moisture changes in interstorm intervals varied with duration and ambient atmospheric conditions, and were reflected in soil physical conditions. Bulk density showed no clear patterns, but soil shear strength and penetration resistance increased, and soil aggregate stability declined with progressive drying. Significant changes were also noted in desiccation crack size and density, particularly on the Swinton soil.

Subsequent erosional response was strongly influenced by inherited surface morphology, by crack patterns, and by antecedent moisture. Soil resistance showed little change with limited drying and erosion rates rapidly returned to the level of the initial storm. More prolonged drying, which reduced soil moisture below a threshold for significantly increased soil strength (Kettle Creek: 27%; Swinton: 30%) reduced erosion levels in subsequent storms. More prolonged drying, sufficient to reduce soil moisture below the threshold for aggregate stability decline and to cause intense cracking (15% for both soils), resulted in a considerable increase in erosion, accompanied by intense rill development or micropiping.

Identification of critical soil moisture thresholds for changes in strength, aggregation and cracking provides a simple index of the effect of defined changes in interstorm conditions. Precision of the threshold levels can be improved by further research, and the precise levels would be expected to vary with soil type. Evidence of the effect of shear strength and aggregate stability is somewhat contradictory, but it appears that shear strength dominates response until cracking is significant, and thereafter the influence of aggregation and cracking is dominant.

The results indicate that very minor changes in interstorm conditions close to either critical threshold can produce major changes in erosional response in subsequent storms, without significant change in rainfall characteristics. Rainfall characteristics are clearly important for erosion prediction on disturbed, agricultural soils, but attempts to predict erosional response on hillslopes in sequential storms should include measures of antecedent moisture and soil physical properties. In the latter case, vegetation and biological influences such as algal crust development must also be considered.

Apart from the importance of monitoring interstorm conditions, three methodological issues emerge from the study. First is the occurrence of cyclic rilling and deposition. It is clearly impossible to interpret sediment yield data at a terminal weir accurately unless information is available on erosional and depositional patterns along the slope, and particularly their juxtaposition to the collecting trough. Second, scour morphology, which influences entrainment and deposition, is strongly affected by subsurface moisture conditions. Third, influence of ageing effects on the Swinton soil indicates a potential problem for all replicate testing of cohesive, aggregated soils. It is clearly important to establish ageing patterns for samples stored before testing, and the threshold period after which these changes become significant.

ACKNOWLEDGEMENTS

This study was supported by the Natural Sciences and Engineering Research Council, Canada. Many people assisted with laboratory experiments, particularly R. Hawke, A. Farenhorst, G. Yap, J. Ramisch, T. Van Seters, M. Crow, D. Rockwell, N. Kuhn, E. Brun, A. Brunton, R. Bezner-Kerr and A. Yair. G. Wall of the University of Guelph and A. Mermut of the University of Saskatchewan arranged for collection of bulk soil samples. All this assistance is gratefully acknowledged.

REFERENCES

Bisal, F.M. and Nielsen, K.F. 1967. Effect of frost action on the size of soil aggregates. *Soil Science*, **104**, 268–272.

Bryan, R.B. 1968. The development, use and efficiency of indices of soil erodibility. *Geoderma*, **2**, 5–26.

Bryan, R.B. 1971. The influence of frost action on soil aggregate stability. *Transactions of the Institute of British Geographers*, **54**, 71–88.

Bryan, R.B. 1990. Knickpoint evolution in rillwash. *Catena*, Supp. Bd., **17**, 111–132.

Bryan, R.B. 1994. Microcatchment hydrological response and sediment transfer under simulated rainfall on semi-arid hillslopes. *Advances in Geoecology*, **27**, 71–97.

Bryan, R.B. and Poesen, J.W.A. 1989. Laboratory experiments on the influence of slope length on runoff, percolation and rill development. *Earth Surface Processes and Landforms*, **14**, 211–231.

Bryan, R.B., Yair, A. and Hodges, W.K. 1978. Factors influencing the initiation of runoff and piping in Dinosaur Provincial Park badlands, Alberta, Canada. *Zeitchrift für Geomorphologie*, Suppl. Bd., **29**, 151–168.

Bryan, R.B. and Oostwoud Wijdenes, D. 1992. Field and laboratory experiments on evolution of microsteps and scour channels on low-angle slopes. *Catena*, Suppl. Bd., **23**, 1–29.

Bryan, R.B. and Rockwell, D.L. 1996. Water table control on rill initiation and implications for erosional response. *Geomorphology*, in press.

Carson, M.A. and Kirkby, M.J. 1973. *Hillslope Form and Process*. Cambridge University Press, Cambridge.

De Ploey, J. 1972. A quantitative comparison between rainfall erosion capacity in a tropical and mid-latitude region. *Geografica Polonica*, **23**, 141–150.

Dickinson, W.T., Pall, R. and Wall, G.J. 1982. Seasonal variations in soil erodibility. *American Society of Agricultural Engineers Paper 82-2573*. ASAE Winter Meeting, Chicago, 14–17 December.

El-Swaify, S.A. and Dangler, E.W. 1977. Erodibilities of selected tropical soils in relation to structural and hydrologic parameters. In *Soil Erosion: Prediction and Control*. Soil Science Society of America, Ankeny, pp. 105–114.

Emmett, W.W. 1970. *Climat et Erosion: La Relation Entre l'Erosion du Sol par l'Eau et les Précipitations Atmospheriques*. Resses Universitaire de France, Paris.

Govers, G. 1985. Selectivity and transport capacity of thin flows in relation to rill erosion. *Catena*, **12**, 35–50.

Hudson, N.W. 1965. The influence of rainfall on the mechanics of soil erosion. *MSc thesis*, University of Capetown.

Imeson, A.C. and Vis, R. 1984. Assessing soil aggregate stability by water-drop impact and ultrasonic dispersion. *Geoderma*, **34**, 185–200.

Kemper, W.D., Rosenau, R.C. and Dexter, A.R. 1987. Cohesion development in disrupted soils as affected by clay and organic matter content and temperatures. *Soil Science Society of America Journal*, **51**, 860–867.

Kirby, P.C. and Mehuys, G.R. 1987. Seasonal variation of soil erodibilities in southwestern Quebec. *Journal of Soil and Water Conservation*, **42**, 211–215.

Mostaghimi, S., Young, R.A., Wilts, A.R. and Kenimer, A.L. 1988. Effects of frost action on soil aggregate stability. *Transactions of the American Society of Agricultural Engineers*, **32**(2), 435–439.

Mutchler, C.K. and Carter, C.E. 1983. Soil erodibility during the year. *Transactions of the American Society of Agricultural Engineers*, **4**, 1102–1104, 1108.

Oostwoud Wijdenes, D. 1992. The dynamics of gully-head on a semi-arid piedmont plain. *PhD thesis*, University of Toronto.

Reid, J.B. and Goss, M.J. 1982. Effect of living roots of different plant species on the aggregate stability of two arable soils. *Journal of Soil Science*, **32**, 521–541.

Savat, J. 1976. Discharge velocities and total erosion of a calcareous loess: a comparison between pluvial and terminal runoff. *Revue de Géomorphologie Dynamique*, **24**, 113–122.

Shiel, R.S., Adey, M.A. and Ladder, M. 1988. The effect of successive wet/dry cycles on aggregate size distribution in a clay texture soil. *Journal of Soil Science*, **39**, 71–80.

Slattery, M.C. and Bryan, R.B. 1992. Hydraulic variations for rill incision under simulated rainfall: a laboratory experiment. *Earth Surface Processes and Landforms*, **17**, 127–146.

Tisdall, J.M. and Oades, J.M. 1982. Organic matter and water-stable aggregation in soils. *Journal of Soil Science*, **33**, 141–163.

Truman, C.C., Bradford, J.M. and Ferris, J.E. 1990. Antecedent water content and rainfall energy influence on soil aggregate breakdown. *Soil Science Society of America*, **54**, 1385–1392.

Utomo, W.H. and Dexter, A.R. 1981. Age-hardening of agricultural top soils. *Journal of Soil Science*, **32**, 335–350.

Utomo, W.H. and Dexter, A.R. 1982. Changes in soil aggregate water stability and drying cycles in non-saturated soil. *Journal of Soil Science*, **33**, 623–637.

Wischmeier, W.H. and Smith, D.D. 1978. Predicting rainfall erosion losses. *Agricultural Handbook 537*, USDA, Washington, DC.

Wischmeier, W.H., Smith, D.D. and Uhland, R.E. 1958. Evaluation of factors in the soil-loss equation. *Agricultural Engineering*, **39**(8), 458–462, 474.

Young, R.A., Onstad, C.A., McCool, D.K. and Benoit, G.R. 1983. Temporal changes in soil erodibility. *ARS-30 Proceedings of the Natural Resources Modelling Symposium*, Colorado, 16–21 October.

Young, R.A., Romkens, M.J.M. and McCool, D.K. 1990. Temporal variations in soil erodibility. *Catena*, **17**, 41–53.

Zanchi, C. 1983. Influenza dell'azione battente della pioggia e del ruscellamento nel processo erosivo e variazioni dell'erodibilitta del suolo nei diversi periodi stagionali. *Annali Istituto Sperimentale Studie e Difesa Suolo*, **14**, 347–358.

27 Field Studies of Runoff Processes on Restored Land in South Wales and the Design of Channels for Erosion Control

R. A. HODGKINSON[1] and A. C. ARMSTRONG[2]
[1]*ADAS Land Research Centre, Cambridge, UK*
[2]*ADAS Land Research Centre, Mansfield, UK*

27.1 INTRODUCTION

Land disturbed by opencast mining activities in the UK is required by statute to be restored post-working to a similar physical condition to that which existed prior to working. In practical terms, especially on sites in areas where the original land quality was low and parts of the site suffer from industrial dereliction, this may just mean restoration to a suitable condition for the intended after-use.

The problems involved in the restoration of land after opencast coal working in South Wales are very different to those found in either lowland or upland England. The sites in Wales are typically located on the sides of steep valleys and restoration is normally carried out in such a way that the restored contours integrate with the existing landform. The soils found on these sites prior to working are frequently strongly gleyed with peaty or humose topsoil such as the Wilcocks series (Clayden and Hollis 1984). These soils have proved to be a material with which it is difficult to achieve a successful restoration. Common practice has therefore been to use mudstones and shales from depth as soil-making material to construct a new soil profile in place of the original soil. The soil-making material, mudstone and shale, weathers rapidly to produce material on which vegetation can be established, but which, in contrast to a natural soil, is unstable and is particularly vulnerable to erosion by water. If surface flows are not controlled adequately this combination of steep slopes, erodible soils and high rainfall gives a high risk of erosion. Under these conditions successful land restoration after opencast mining requires the installation of a drainage network to control surface runoff and ensure that neither surface runoff nor flow within the channel erodes the restored soil.

Sites in South Wales are often returned to common ownership once restoration has been completed and are invariably managed as low-input grazing. Conventional underdrainage would not have been installed on these sites prior to opencast working, nor is it normally installed during restoration. Control of surface water is achieved by a series of channels (grips) following the contours of the land. These

Advances in Hillslope Processes, Volume 1. Edited by M. G. Anderson and S. M. Brooks.
© 1996 John Wiley & Sons Ltd.

grips terminate in a carrier ditch running down the main slope which transports the water off site. Depending upon the slope and the expected rate at which water will be discharged, armouring may be required in carrier ditches to prevent bed erosion.

The design of surface drainage channels for erosion control on land restored after opencast coal mining has, for many years, been undertaken using rule of thumb criteria developed from experience. A lack of firm design criteria has resulted in systems being amended to take account of either financial or field layout criteria rather than being based on the need to control the risk of erosion for the flows generated by a specific return period of rainfall. Systems installed therefore have the potential to be either overdesigned with the consequential impact on the landscape of large numbers of channels, or underdesigned, leading to serious erosion problems. Serious erosion problems on some sites, following failure of the grips, have required major remedial works. British Coal commissioned ADAS to produce an improved method for the design of surface drainage systems for erosion control. Consequently, hydrological data on the rate of runoff from this land were required for use in the design process and two recently restored opencast coal sites in South Wales were instrumented. The approach selected for the design procedure was to develop a computer-based system to predict the amount of surface runoff expected, its erosive effect on the slope and the erosive potential of the flow once collected into a channel.

This chapter presents the results of the monitoring exercise, the subsequent modelling of the observations and the development of a practical design tool.

27.2 METHODS

27.2.1 Sites

Two sites were chosen for detailed monitoring. These were the Bryn Pica and Maes y Marchog restored opencast coal sites (Figure 27.1), on the northern side of the South Wales coalfield area. These sites offered a contrast in age of restoration, but otherwise were restored from very similar material (Table 27.1), and both were restored to grassland, as are all such sites in the area. The dominant particle size produced by weathering, once the underlying shales used as soil-making material are exposed, falls in the fine silt range. There is also a proportion of larger particles, but these are generally made up of unweathered shale fragments which also break down to silt size fairly rapidly (within 3–4 years) in the surface horizons. At both sites, the restoration process had included the installation of channels across the slope, intercepting water moving down the slope. The channels were designed to provide no obstacle to vehicular access, and so were shallow and broad, but sometimes had a more incised section at the centre where the majority of flow took place (Figure 27.2). Material excavated to form a channel is normally placed on the downslope side of the channel, creating a ridge which results in an asymmetrically shaped channel, the downslope bank being the steeper.

The Bryn Pica site (grid reference SO 005055) is located at an altitude of 260 m on the side of a valley above Aberdare. The site was restored in 1985. Surface channels had been installed across the 9% slope at spacings of 40–90 m. The Maes y Marchog site (grid reference SN875087), at an altitude of 275 m, was restored in 1988 with

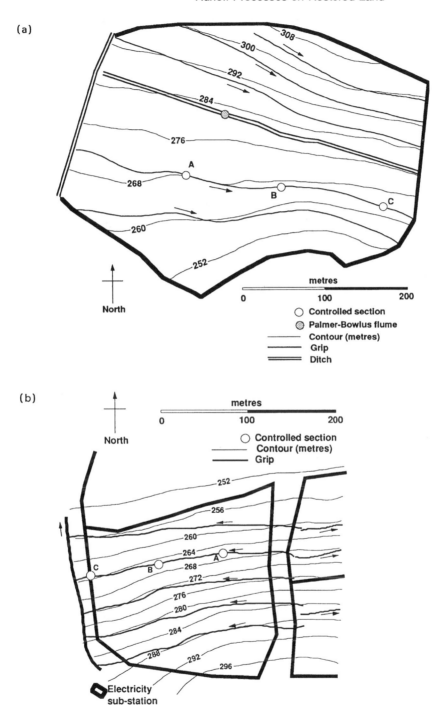

Figure 27.1   Site plans and location diagram: (a) Bryn Pica OCCS; (b) Maes y Marchog OCCS

Table 27.1 Particle-size distribution for the restored material at both sites

Size	Maes y Marchog	Bryn Pica
>600 μm	15.4	9.0
212–600 μm	6.2	8.7
63–212 μm	4.6	13.3
20–63 μm	9.6	15.8
2–20 μm	39.7	30.9
<2 μm	24.5	22.3

slopes up to 23%. The climate of the area is relatively extreme, reflecting their elevation. Mean annual rainfall for the agroclimatic area in which both sites fall is 1729 mm (Smith and Trafford 1976).

27.2.2 Hydrological Instrumentation

On each site, channels were instrumented by installing shallow V-shaped control sections (flumes) 150 mm deep and 600 mm wide at the top, made of pre-formed steel, and pre-calibrated in the laboratory. Water level was recorded both autographically and by an electronic transducer that was interrogated electronically every five minutes. The slope segments that were monitored on each site were defined by an upper ditch. At Bryn Pica a single grip, with a catchment area of 2.6 ha, was monitored at several points along its length. Two adjacent grips were monitored at Maes Y Marchog, each having a catchment area of 0.75 ha. The grips at Maes y Marchog were primarily monitored at the outfall, but supplementary points were established on the grip to determine the rate of transmission of peak discharge along the grip. The parameters describing the channels monitored are presented in Table 27.2. Rainfall was monitored on each site using a tipping-bucket rain-gauge recorded electronically every five minutes. Flows were monitored at both sites during the winters of 1990/91 and 1991/92; this provided a more than adequate sample of rainfall events for modelling.

Soil water status on the slopes above the monitored sections was investigated using conventional mercury tensiometers installed at nominal depths of 15, 30, 45 and 60 cm below the surface. These were read manually for two periods sufficient to characterize the hydrological response of the slopes. Long-term monitoring of soil moisture conditions was not possible because of the constraints of access to the land. However, as the aim was to establish the mechanisms, and not to monitor all the events, and these two periods gave clear information, this restricted monitoring exercise was considered sufficient to meet the objectives.

27.2.3 Soil Hydrological Characterization

Critical to the restoration is the nature of the material making up the restored slope. At both sites, the material was made up of weathered shale fragments, previously excavated, and replaced by mechanical means. The restoration process on these sites

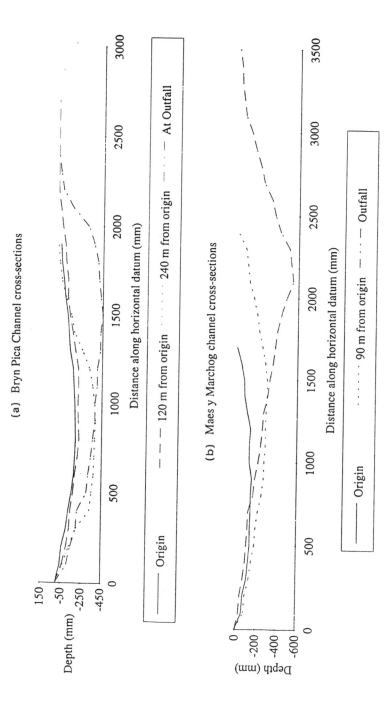

Figure 27.2 Mean ditch cross-sections: (a) Bryn Pica; (b) Maes y Marchog

Table 27.2 Characteristics of monitored subcatchments

	Bryn Pica	Maes y Marchog
Slope length	50–90 m	30 m
Mean slope steepness	1 in 6	1 in 3.9
Distance along ditch	385 m	250 m

would either have been to tip material from a dump truck and level it off with a bulldozer/grader or to spread it with a motor scraper. Either approach, particularly the second, will tend to produce a soil profile with distinct horizontal laminations. This will be enhanced by the laminar shape of fragments of shale or mudstone which will tend to align themselves with their long axis parallel to the direction of spreading. The soil profiles produced on these sites thus exhibit horizontal layering with the net result that there are very few paths for vertically downward movement of the water.

The hydrological consequences of this horizontal zonation were expected to be a marked difference in the vertical and horizontal hydraulic conductivity of the soil. This was investigated by the use of gypsum-encased monoliths (Bouma and Dekker 1981). Monoliths of soil 30 cm in each dimension were excavated, encased in gypsum *in situ* and removed from their location. The base of the monolith was then supported clear of the ground to give a free outfall, and water was added to the top of the monolith. The rate of transmission was then recorded, to give an estimate of the vertical hydraulic conductivity.

The equivalent horizontal conductivity was recorded by encasing the surface and bottom faces of the monolith, and freeing two opposing side faces. The block was then placed on its side so that these two free faces became top and bottom, and the rate of transmission again recorded. Although it is difficult to establish the exact physical meaning of the numerical values so obtained, because of the difficulty of establishing the exact hydraulic gradients involved, and equally difficult to relate the values so obtained to conditions in the field, the contrasting values for vertical and horizontal measurements are nevertheless a good indication of the relative permeabilities in these two directions.

These measurements of the hydraulic conductivity, in both the vertical and horizontal directions (Table 27.3), showed that the possibilities for vertical movement of water were very restricted indeed. By contrast, the horizontal hydraulic conductivity is high – at least an order of magnitude higher than the vertical. The physical condition of the soil, that of having very restricted permeability in the

Table 27.3 Hydraulic conductivity (mm/h) measured at Maes y Marchog using encased gypsum blocks

Block	Vertical Ksat	Horizontal Ksat
1	4	225
2	4	100

vertical direction, is established unequivocally by these results. Under these conditions, with a strong slope, lateral movement of water is to be anticipated.

It also proved possible to use the same blocks to estimate the drainable porosity of the soil. At the conclusion of the saturated hydraulic conductivity tests, the soil was saturated to the surface by placing it on a gravel bed in a tank full of water to the level of the soil surface. The surrounding water was then removed, the monolith allowed to drain for 48 h, and the amount of water draining collected. This method gave an estimate of a drainable porosity of 5%. Equivalent measurements made on subsamples taken from the block, equilibrated for 48 h on a tension table at 50 cm tension, gave a higher value of 12%.

It would thus appear that infiltration into the soil can be ignored as a negligible process on these sites for two reasons. First, the vertical hydraulic conductivity of these soils is very low. The second factor is the low value observed for drainable porosity; this results in the soil having a very low available storage capacity at the start of most storms, unless it has been dried below field capacity by evapotranspiration. On this type of site rainfall events are frequent, so there is no time for the soil to drain significantly between storms. Consequently, the soil remains close to saturation throughout the winter period.

27.3 RESULTS

Figures 27.3 and 27.4 show the flows from the two sites for two typical periods. Data from Maes y Marchog are presented for a period in April 1991 and those from Bryn Pica are for a period at the end of March 1992. These show that the runoff response from the two sites was very rapid indeed. Consideration of the volumes involved showed that nearly all of the rainfall appeared as stream flow. Water balances are presented for a number of events at Maes Y Marchog (Table 27.4). At Bryn Pica, in the following year the average runoff percentage was 65% and the maximum was 96% and the data indicated that runoff was proportional to rainfall with the proportionality constant tending towards 100% for larger storms. This suggests that at peak rainfall intensity in the design storm it is reasonable to assume 100% runoff. During this time, visual observations of field conditions indicated that the soil was saturated throughout the period for which flow was taking place. There was thus no possibility for the rainfall to infiltrate in the soil, and the water reaching the channel was flowing over the surface as simple saturation excess overland flow. The tensiometer studies at Bryn Pica, undertaken in January 1991, identified quantitatively that the soil profile was completely saturated whilst surface flow was occurring. Measurements several days after a rainfall event indicated that little drainage had taken place as soil moisture tensions remained low and that the soil was still close to saturation. Given the soil characterization data already presented, and the general similarity of the soil-making material used to restore sites in the South Wales coalfield, it is not unreasonable to extrapolate these findings to other sites.

The consequence of this observation is that the controlling factor for nearly all winter storm flows is the incident rainfall. Equally, there was virtually no flow between storm events, indicating that there was no slow drainage of the soil. These

620

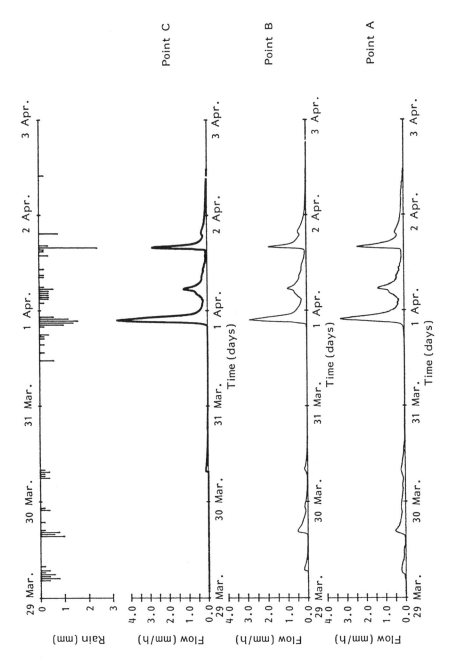

Figure 27.3 Bryn Pica: rainfall and runoff, 29 March 1992 to 2 April 1992

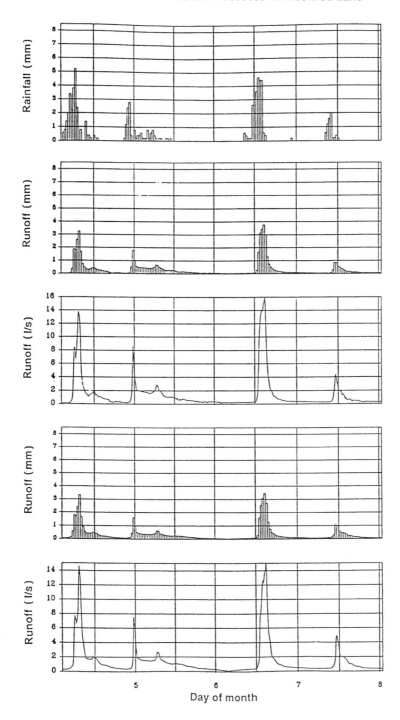

Figure 27.4   Maes y Marchog: rainfall and runoff, 02:25, 4 April 1991 to 00:55, 8 April 1991

Table 27.4   Waterbalances for selected events at Maes y Marchog

| Date | Percentage rainfall recorded as flow | | Rainfall (mm) |
	Channel 1	Channel 2	
16 Mar. 1991	missing data	116	12
18 Mar. 1991	missing data	95	84
4 Apr. 1991	68	73	26
4 Apr. 1991	125	134	12
6 Apr. 1991	90	92	21
7 Apr. 1991	93	127	5
25 Apr. 1991	107	86	7
29 Apr. 1991	103	94	38

results are comparable with those of Carling *et al.* (1993), who reported that on an upland site undergoing afforestation, storage of runoff on the hillslope was negligible during a discrete storm, with close to 100% of incident rainfall appearing as runoff within a few hours.

27.4 MODELLING OF FLOWS

Because of the need to provide predictive models suitable for design purposes, the flows from the site were modelled using a simple kinematic wave model for overland flow assuming all the incident rainfall became runoff. The kinematic wave model (Lighthill and Whitam 1955) is by now well described (e.g. by Miller 1984) and has been widely used for hydrological modelling, and is a component of several erosion packages, such as CREAMS (Knisel 1980) and EPIC (Williams *et al.* 1989).

The implementation of the kinematic wave model requires a value for the surface roughness coefficient. We used the Chezy representation of the flow equation:

$$Q = C\sqrt{sh}$$

in which s is the slope, h is the depth, and C is the roughness coefficient which was given a value of 5.0, derived from the data in the table of Thornes (1980). The model assumed that all incident rainfall becomes surface runoff, as infiltration could be assumed to be negligible. Hourly rainfall increments were used, although internal steps of the model were shorter.

The results from the model (Figures 27.5 and 27.6) showed excellent agreement between the observations and the predictions. Because the object of the model was to predict the peak flow rates, particular effort was made to check this component of the model. Predictions of peak flow rates are compared to the observations in Figures 27.7 and 27.8. These showed excellent agreement and it was thus concluded that the kinematic wave model gave acceptably accurate predictions of peak flow rates from these sites.

The results from this study have shown that winter flows from restored opencast sites in South Wales are generated almost entirely by overland flows. The dominant

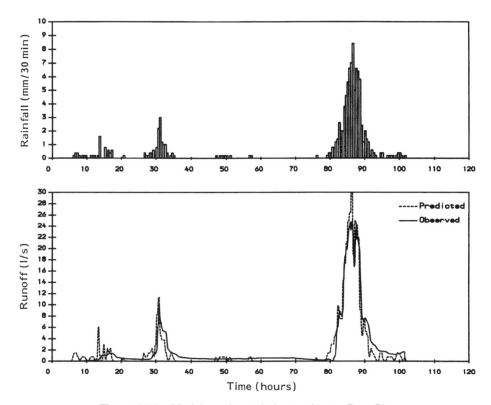

Figure 27.5   Model results and observations – Bryn Pica

hydrological process is thus that of saturation-excess overland flow of incident rainfall which cannot infiltrate because of the nature of the soil.

27.5 DEVELOPMENT OF A DESIGN TOOL

The hydrological studies undertaken in this project were the preparatory stage for the development of a tool to evaluate the design of these channels.

A three-stage computer package has been developed, as an aid to the design of such channels. The three stages are

1. estimation of design rainfall,
2. prediction of overland flows, and
3. estimation of erosion risk.

The aim is to present the user with the tool to evaluate the consequences of design options – in terms of the risk and degree of erosion. To this end the results are presented for all the tabulated return periods, and thus the designer can identify the

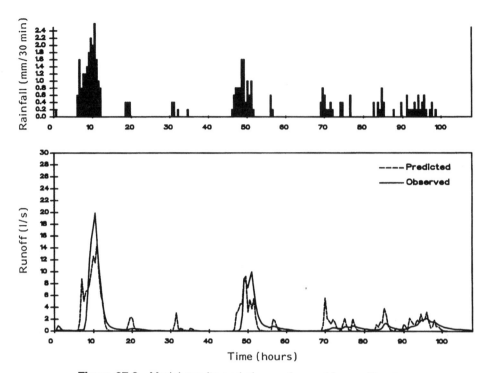

Figure 27.6 Model results and observations – Maes y Marchog

degree of risk associated with any single design decision. This philosophy is carried right through the design package.

27.5.1 Estimation of Design Rainfall

This is based on a rainfall intensity database obtained from the UK Meteorological Office. Data are available at 3.33 km grid resolution for six return periods and seven durations. This database was produced from the network of gauging stations used for the Flood Studies Report (NERC 1975), by use of interpolation techniques and growth factors. A sample output from this database is presented in Table 27.5.

These data give the best estimate available of the maximum rainfall intensity with a given duration and frequency. The design package retrieves the nearest data point and uses results for all periods of less than 24 h, and carries the information for all frequencies forward into subsequent stages of the package.

27.5.2 Prediction of Overland Flow

From the tabulated rainfall data, a design rainfall hyetograph is constructed which lasts 24 h and is symmetrical around 12 h. The values are generated to match all the

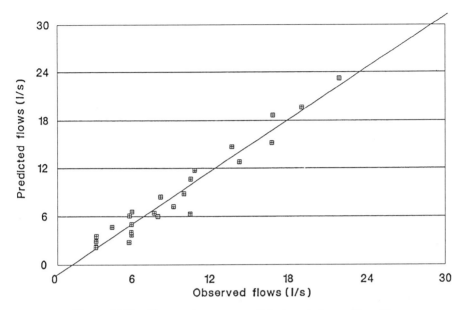

Figure 27.7 Observed versus predicted peak flow – Bryn Pica

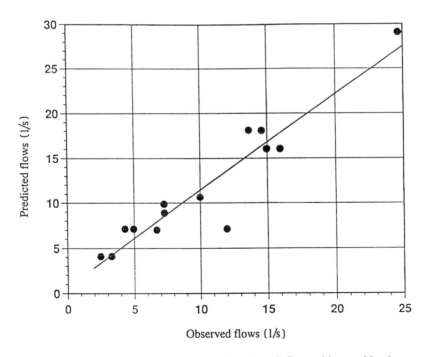

Figure 27.8 Observed versus predicted peak flow – Maes y Marchog

Table 27.5 Sample meteorological data for location: SO005055

| Duration (h) | Return period (years) | | | | | |
	1	2	5	10	25	100
1	11.7	14.6	18.9	22.7	27.6	37.0
2	16.6	20.3	25.8	30.7	37.5	50.0
3	20.2	24.4	30.6	36.2	44.1	58.6
6	28.0	33.3	40.8	47.5	57.3	75.5
12	39.0	45.5	54.4	62.3	73.8	94.9
24	54.1	62.2	72.7	81.9	95.0	118.2

intensities between 24 h and 1 h, which was achieved by fitting a log–log relationship between rainfall intensity and duration and interpolating values for intermediate points. The resultant hyetograph is more strongly peaked than those normally observed because high-intensity rainfall is normally short-lived. This approach was adopted to ensure that the design procedure focuses on those events that represent the greatest erosion risk.

The design hyetograph is input into the kinematic wave model to predict the flow across an inclined surface. The model provides estimates of the depth and velocity of flow as a function of distance downslope from the start of the catchment, which are then used to estimate the increases in shear force that take place as slope length increases.

The estimation of the erosive potential of these flows requires an estimate of the sediment transport capacity of the flow. The kinematic wave model gives values for flow, depth and velocity, and these can be coupled to the slope to estimate the erosive potential. This is estimated using the Yalin (1963) equation, which is identical to that adopted by the CREAMS model (Knisel 1980), after the review by Foster (1982). The Yalin equation has the advantage that it includes the important concept of critical shear stress, below which a flow is incapable of moving material from its bed. This critical stress is defined by the Shield's diagram, which was digitized and approximated by a polynomial. The sediment transport capacity was thus calculated from the peak flow of the design hydrograph and the resultant values for the hydraulic variables and the erosive potential plotted as a function of the distance downslope (Figure 27.9).

The values of shear stress are examined in relation to a number of values for the initiation of erosion, derived from the tabulated data of Chow (1959). For sample sediments virtually identical results were obtained using the critical tractive shear stress as a function of particle diameter as defined by Raudkivi (1967). Four critical stress values were defined (Table 27.6).

- Condition A represents the slopes at their most vulnerable, immediately after restoration when they are unvegetated.
- Condition B represents the same vegetative state, but after removal of the fine particles and relates to the "armoured" state of the channels.

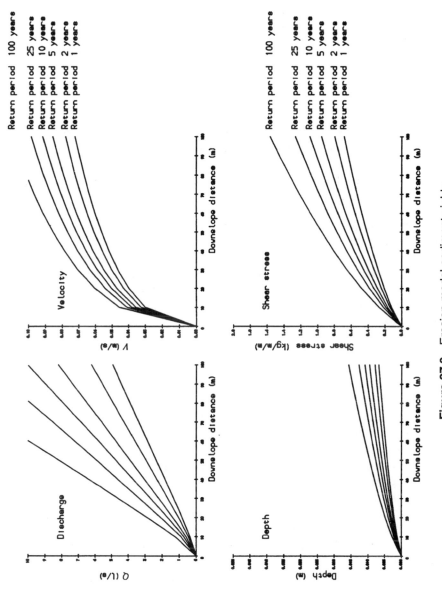

Figure 27.9 Erosion model sediment yields

Table 27.6 Critical stress values

State	Critical shear stress values (kg/m²)	Description
A	0.235	Unvegetated non-colloidal silts
B	0.367	Unvegetated fine gravel
C	0.961	Partially vegetated (gravel) soils
D	1.466	Fully vegetated (gravel) soils

- Condition C is recommended as the design value for the slopes and channels of this area, as it makes allowance for the rather patchy development of vegetation in the early stages.
- Condition D is probably the safe long-term value for the slopes and channels after the establishment of vegetation.

These four conditions are plotted on the shear stress graph (Figure 27.10), in order to enable the designer to balance the slope length against erosion potential to provide a design appropriate to his need. It is, however, noted that the use of condition A, the unvegetated state, would lead to very closely spaced channels. This demonstrates the urgent need to re-establish protective vegetative cover on these sites.

Once the designer has chosen a slope length using the information from the kinematic wave prediction (Figure 27.10), it is then possible to consider the processes along the channel.

27.5.3 Prediction of Channel Erosive Capacity

The observed downslope hydrograph per unit segment of slope width at the chosen slope length is multiplied by the catchment area to estimate the growth of discharge along the channel. This approach ignores the routing of the flows down the channel, which our observations suggest is an acceptable approximation since propagation of peak flow along the channel was found to be rapid, typically less than 5 min between stations 300 m apart. Thus we can predict successively larger catchments by just increasing the slope width rather than needing to use the sophistication of the kinematic wave to route flows down the channel. However, this does limit the application of the package to sites where the channel lengths are relatively short. As most opencast coal sites in upland areas are relatively small, or are broken up into small areas to match the topography this does not constitute a major limitation in practice.

As with the slope, the model predicts depth and velocity along the channel. Critical to the calculation of these values is the choice of channel slope and cross-sectional form. The designer is required to give the channel slope as a gradient for upslope and downslope sides (following the normal design procedures).

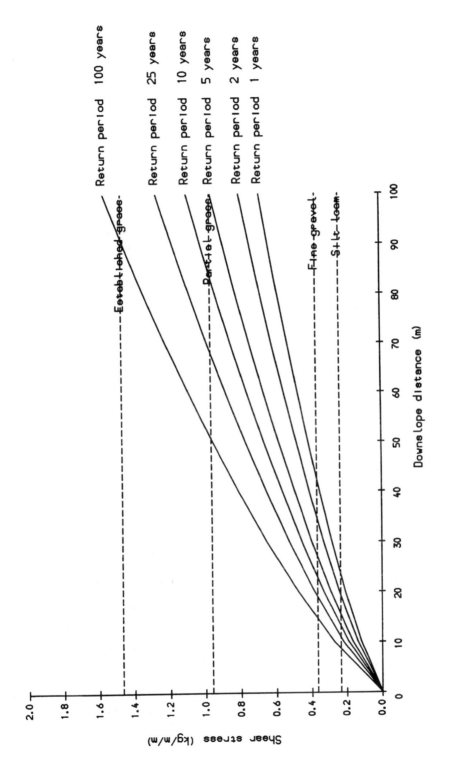

Figure 27.10 Slope erosion model – predicted shear stress

The flow conditions along the channel, flow depth, velocity and bed shear stress, are defined from the discharges. The unit rate of discharge from the slopes is multiplied by channel length, and this is converted into flow characteristics using the Manning flow equation for the channel geometry. The user is thus required to define the value of Manning's n to define the channel roughness. A value of $n=0.25$ is recommended based on the values tabulated by Chow (1959). The Yalin equation is used to define bed shear stress in exactly the same way as for the flows over the slopes. The package thus plots the channel shear stress as a function of distance down the channel (Figure 27.11).

Again the user is presented with a choice of channel conditions and critical shear stress values, the same as was presented for the slopes. The user is thus offered a choice of a design length that gives an acceptable erosion risk. More important, however, is the option to redefine the channel parameters. Frequently the channel length may be constrained by the site layout, in which case the user can cycle through this sector of the analysis to define the channel slope parameter that gives an acceptable erosion risk for a predefined length.

An analysis of the erosion potential was undertaken for the existing channels at both study sites. At Bryn Pica, where the channel was 425 m long, the predicted discharge at the outfall was 150 l/s at the 1 in 1 year return period, which would result in a depth of flow of around 150 mm. This is well within the carrying capacity of the channel, which is 450 mm deep and would not be overtopped at even the 1 in 100 year storm. However, consideration of the shear stress exerted on the channel bed indicated that the critical shear stress for partially vegetated channels will be exceeded even at the 1 in 1 year return period for most of its length. Field observations of the channel indicated that some bed erosion was occurring while the sides remained stable. Since the sides of the channel were fully vegetated whereas the base was only sparsely so, this observation is in general agreement with the predictions of the model.

At Maes y Marchog, the channels were shorter, only 250 m long, and were sufficiently deep to discharge flows at up to the 1 in 100 year return period. In terms of shear stress the model indicated that the channels were stable against flows with a return period of around 1 in 2 years for partial vegetation and 1 in 100 years for fully vegetated. This is in general agreement with the field observations of the channels which indicated that no bed erosion was taking place at this site.

27.6 DISCUSSION

This chapter has identified the runoff processes operating, and discussed the modelling of them, on restored opencast coal sites in the South Wales area. The dominant process in runoff generation was found to be surface flow. This process dominated because of the impermeable nature of the soil which, coupled with the high rainfall climate, causes the soil profile to be saturated for much of the year. Due to the simplicity of the processes a kinematic wave was found to be adequate to provide a runoff model to describe these sites. A sediment transport model was

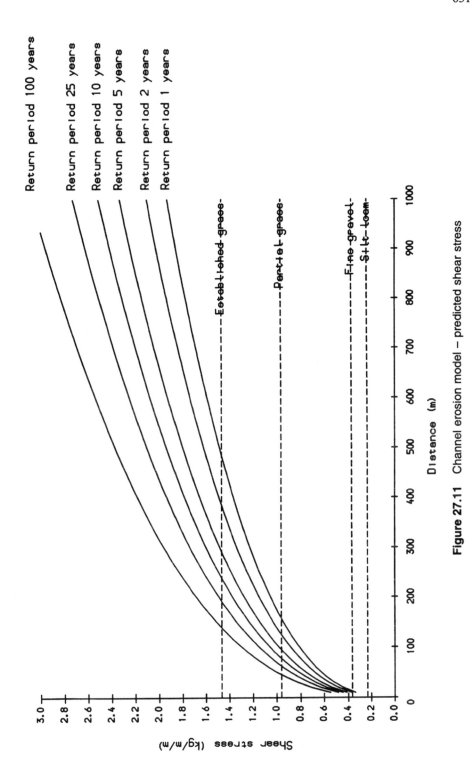

Figure 27.11 Channel erosion model – predicted shear stress

added to the runoff model by use of the Yalin (1963) equation. A similar approach was adopted to model the passage of the flows along the channel.

The studies have led to the implementation of a computer modelling package that allows the design of measures for surface runoff control to be based on an assessment of the acceptable risk rather than by empirical rules of thumb. While on the two sites studied use of the modelling approach would have brought about no changes in design, it is worth noting that on some sites in the locality grip spacing has had to be substantially reduced following serious erosion. Infilling with additional grips following a failure is bad restoration practice as soil material may have been lost from the site and vegetation damaged. Given that the statutory aftercare of such sites only lasts five years, such an occurrence reduces the likelihood of achieving a successful restoration by the end of this period.

This chapter shows how an understanding of the processes operating can be translated into a practical tool, which is of use to the engineer concerned with the design of channels installed on this land. The use of such a tool must result in a higher standard of restoration and reduced impact on the downstream water environment as well as less visual intrusion.

ACKNOWLEDGEMENTS

We are grateful to Andrew Baird for providing the code that became the centre of our kinematic wave model. This work was funded by British Coal Open Cast executive.

REFERENCES

Bouma, J. and Dekker, L.W. 1981. A method for measuring the vertical and horizontal Ksat of clay soils with macropores. *Soil Science Society of America Journal*, **45**, 662–663.

Carling, P.A., Glaister, M.S. and Flintham, T.P. 1993. Soil erosion and conservation on land cultivated and drained for afforestation. *Hydrological Processes*, **7**, 317–333.

Chow, Ven te 1959. *Open-Channel Hydraulics*. McGraw-Hill, New York.

Clayden, B. and Hollis, J.M. 1984. *Criteria for Differentiating Soils Series*, Technical Monograph No. 17. Soil Survey of England and Wales, Harpenden, Herts.

Foster, G.R. 1982. Modelling the erosion process. In C.T. Hann, H.P. Johnson and D.L. Brakensiek (eds), *Hydrological Modeling of Small Watersheds*. ASAE Monograph 5, ASAE St Joseph, Michigan, pp. 297–380.

Knisels, W.F. (ed.) 1980. *CREAMS: A Field Scale Model for Chemicals, Runoff and Erosion from Agricultural Management Systems*. US Department of Agriculture, Conservation Research Report No. 26, Washington, DC.

Lighthill, M.J. and Whitam, G.B. 1955. On kinematic waves. I. Flood movement in long rivers. *Proceedings of the Royal Society, Series A*, **229**, 281–316.

Miller, J.E. 1984. *Basic Concepts of Kinematic Wave Models*. United States Geological Survey Professional Paper 1302, Washington.

Natural Environment Research Council 1975. Flood Studies Report, NERC, London.

Raudkivi, A.J. 1967. *Loose Boundary Hydraulics*. Pergamon Press, Oxford.

Smith, L.P. and Trafford, B.D. 1976. *Climate and Drainage*. HMSO, London.

Thornes, J.B. 1980. Erosional processes of running water and their spatial and temporal controls: a theoretical view point. In M.J. Kirkby and R.P.C. Morgan (eds), *Soil Erosion*. John Wiley, Chichester, pp. 129–182.

Williams, J.R., Jones, C.A., Kiniry, J.R. and Spanel, D.A. 1989. The EPIC crop growth model. *Transactions of the American Society of Agricultural Engineers*, **32**, 497–511.

Yalin, Y.S. 1963. An expression for bed-load transport. *Journal of the Hydraulics Division, Proc. of the ASCE*, **89**(HY3), 221–250

28 On the Complexity of Sediment Delivery in Fluvial Systems: Results from a Small Agricultural Catchment, North Oxfordshire, UK

M. C. SLATTERY[1] **and T. P. BURT**[2]

[1] *Department of Geography, East Carolina University, USA*
[2] *School of Geography, University of Oxford, UK*

28.1 INTRODUCTION

One of the most perplexing problems in investigations of erosion and sedimentation is the satisfactory linkage of on-site rates of erosion on upland areas and sediment yields as measured at the drainage basin outlet (Hadley 1986). This relationship between erosion and sediment yield, referred to as *sediment delivery*, has received attention from hydrologists and geomorphologists for several decades (e.g. Glymph 1954; Maner 1958; Roehl 1962; Renfro 1975; Walling 1990) yet still remains one of the least understood components of the basin sediment system. Walling (1983), for example, in a review of the sediment delivery process, emphasized that "The linking of on-site rates of erosion and soil loss within a drainage basin to the sediment yield at the basin outlet, and improved knowledge and representation of the associated processes of sediment delivery, represent a major research need within the field of erosion and sedimentation." Likewise, Ebisemiju (1990), in a recent study of sediment delivery in Nigeria, stated that "Research efforts in the sediment delivery field should focus firstly on detailed empirical investigations and modelling of sediment erosion, transport, deposition and conveyence on catchment slopes to low-order channels and progress through the low-order to the more complex higher-order channels and drainage systems."

A general problem in studying sediment delivery is the lack of a satisfactory method with which to elucidate the processes involved. It is now increasingly recognized that only a proportion of the sediment mobilized by erosion within a basin, and perhaps a rather small proportion, will find its way to the outlet (Walling 1990; Burt 1993). Lumped methods, such as the sediment delivery ratio (SDR), were developed initially to provide a linkage between on-site erosion estimates and downstream sediment measurements. The sediment delivery ratio for a particular basin is defined as the ratio of sediment delivered at the basin outlet to the gross erosion occurring within the basin, and is written simply as:

Advances in Hillslope Processes, Volume 1. Edited by M. G. Anderson and S. M. Brooks.

$$SDR = \frac{SY}{EROS} \qquad (28.1)$$

where SDR is the sediment delivery ratio (%), SY is basin sediment yield (t/km^2/year), and EROS is catchment gross erosion (t/km^2/year). Recent work (e.g. Walling 1988) has emphasized the many problems associated with the concept of a simple relationship between sediment yield and gross erosion. These problems relate, in particular, to the spatial and temporal lumping inherent in the concept and to its black box nature. In addition, considerable uncertainty surrounds the values of gross erosion used in equation (28.1) which are generally *estimated* rather than measured, and may therefore be very much in error. Hence, studies using this type of lumped framework, while useful in providing a broad regional picture of sediment yield, shed little light on the nature of the sediment delivery process; neither the source of sediment is known nor the path and timing involved in the transport of sediment through the basin.

An alternative approach to elucidating and quantifying the linkages interposed between the erosion processes operating within a drainage basin and the downstream suspended sediment yield is the establishment of a sediment budget (e.g. Dietrich and Dunne 1978; Trimble 1981; Sutherland and Bryan 1991). With this, an attempt is made to quantify the various sediment sources and sinks, in order to provide an improved understanding of the linkages involved. However, as Walling and Webb (1992) note, whilst the sediment budget approach is important in focusing attention on the processes governing the suspended sediment yield at the basin outlet, it has remained an essentially conceptual approach, since it is difficult to assemble precise information on the rates and fluxes involved for anything but a relatively small drainage basin.

A third and related approach to studying sediment delivery involves the construction and analysis of storm-period sediment concentration/discharge relationships. Walling and Webb (1982) note that since the hysteresis form of such relationships reflects the overall pattern of erosion and sediment delivery operating in the upstream area, they provide a useful framework for isolating and interpreting the salient features of the basin sediment response. In effect, the hysteresis approach works backwards from the stream channel to the slopes: the storm event behaviour of sediment within the channel is first examined, this then forms the basis for differentiating which factors are possibly involved in the delivery process (e.g. Rieger *et al.* 1988). The major problem with this approach is that it too, like the sediment delivery ratio, is lumped. For example, positive hysteresis can be interpreted in terms of sediment derived from either within-channel or surficial sources (cf. Doty and Carter 1965; Bryan and Campbell 1986; Arnborg *et al.* 1967; Bogen 1980; Klein 1984). In the absence of additional evidence, it is not possible to use the occurrence of clockwise hysteretic rating loops to infer sediment origin and to elucidate the sediment delivery mechanism.

The key point that emerges from the preceding discussion is that detailed comprehension of the sediment delivery process cannot be derived from aggregated or lumped data. Both the sediment delivery ratio approach and the hysteretic loop approach are necessarily restricted by the lumped nature of the results obtained and by their failure to elucidate the relative importance of individual sediment sources and the sediment generation and conveyance processes operating within the basin. There is clearly a need for a more refined approach in order to improve the representation and parametrization of the processes involved in the delivery of sediment from its source to the basin outlet. Whilst the development of sophisticated predictive models remains an ultimate goal, there still remains a need for field investigations of the delivery system involved, in order to provide a basis for an improved understanding of the processes that will need to be modelled and to provide data which could be used to parametrize and test such models (Walling 1990).

This chapter reports the results of a 12-month field investigation conducted in a small agricultural catchment in north Oxfordshire, UK. The overall objective of the study was to elucidate the sediment delivery systems operating within the basin, in particular, the linkage between on-site erosion of hillslopes and stream channel and the downstream sediment at the basin outlet. These linkages are assessed by adopting a spatially distributed approach to the basin sediment system; one in which sediment is monitored at every stage from its entrainment and transport on slopes to its eventual delivery at the basin outlet.

28.2 RESEARCH DESIGN

The study basin is located in the Cotswold Hills in south central England (Figure 28.1). It has a drainage area of $6.2 \, \mathrm{km}^2$ and its altitude ranges from 202 m on the northern divide to 126 m at the basin outlet (UK grid reference: SP 356 362). Slopes are gentle on the interfluves (around 1°) but with steeper slopes in the central part of the basin ($> 5°$). The stream network is moderately incised into the valley floors, but bank heights are commonly less than 1 m. In the central portion of the basin, Northamptonshire Sandstone is overlain by *brown calcareous earths* of the Aberford Series; these are moderately stony, well-drained fine loam soils. On the southern and northern peripheries, Great Oolite Limestone is overlain by *ferritic brown earths* of the Banbury Series; these are stony, well-drained fine loamy soils. Land use is mixed arable farming with extensive autumn sowing of wheat and barley.

A 90° V-notch weir was installed at the basin outlet in July 1992 (Figure 28.1). A Campbell data logger was used to provide a continuous record of stage with an Ott stage recorder as back-up. Two rain-gauges were installed in the field: a tipping-bucket gauge was positioned close to the weir and attached to the logger, measuring each 0.2 mm rainfall; a Casella autographic natural siphon gauge was installed at Lower Nill Farm (Figure 28.1). Stream suspended sediment was sampled every 15 min during storm events and every 4–6 h during low flows with a Rock and Taylor automatic pump water sampler, located just upstream from the gauging station. We also installed five piezometers along the slope at the gauging station (see Figures 28.1 and 28.2). The piezometers were located about 6 m apart from each other, with the total length of monitored slope being 25 m. Pressure transducers (Druck PDCR 830) were placed in the piezometers to measure water-table elevation, and linked to the data logger.

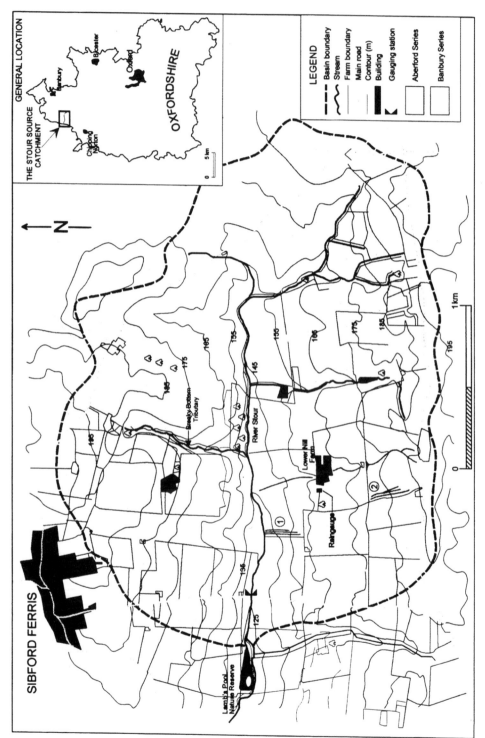

Figure 28.1 Map of the study catchment showing location of the gauging station and distribution of soils. Areas 1 and 2 show the location of the rilling referred to in the text

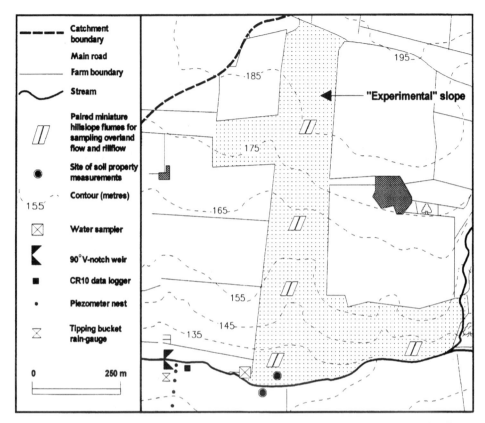

Figure 28.2 The experimental hillslope showing instrumentation layout in detail

Catchment hillslopes were monitored for surface runoff and erosion between August 1992 and July 1993. A single slope was "nested" within the catchment for detailed monitoring of the sediment delivery process (see Figure 28.2). The site was visited weekly in order to take several soil physical property measurements (see Burt and Slattery, Chapter 4, this volume), as well as during each major storm event. Portable hillslope flumes were constructed to sample interrill overland flow during storms. These flumes were modified versions of an original design by Parsons and Abrahams (1989). A major advantage of this type of flume over the traditional sediment trap is that it offered minimal disturbance to the natural pattern of overland flow so that further measurements could be made at other sites downslope.

Runoff and sediment was sampled at several locations on the nested slope during storms, as well as at other locations in the basin. For rill flow, samples were collected by hand by simply placing 0.5-l plastic bottles into the flow. Concentrated runoff along wheelings was sampled by routing flow across the portable hillslope flumes. These samples, along with the stream suspended samples, were returned to the laboratory immediately following an event to determine sediment concentration, sediment discharge and the particle-size distribution of the transported sediment (see

Slattery and Burt 1995). After several storms, slopes were surveyed using a Kern theodolite and electronic distance measurer and details of the erosion process mapped.

28.3 SEDIMENTOLOGICAL RESPONSE OF THE CATCHMENT

The processes by which storm runoff was generated in the study catchment have been described by Burt and Slattery (Chapter 4, this volume). The generation of storm runoff was found to be complex. The majority of stormflow was generated via subsurface flow, or throughflow, but there were significant contributions from surface runoff during larger storms, both in the form of Hortonian overland flow along roads, wheelings and some crusted surfaces, and saturation overland flow near to the stream channel. In the following discussion, the *link* between the generation and routing of runoff through the basin and the coincident entrainment and conveyance of sediment within that runoff, is examined. We focus our attention initially on the channel sediment response; this is then followed by an investigation of sediment production on slopes and its routeing to the stream channel.

28.3.1 Channel Sediment Dynamics

The sediment discharge data calculated for each month during the period of record, as well as the annual data, are given in Table 28.1. Two points must be noted in relation to these data before further discussion. The first is that daily sediment discharge rates were calculated using daily mean discharge and daily mean suspended sediment concentration during low flows, but during storm events, sediment discharge was computed using the 15-min suspended sediment samples. Whilst this stratified approach may appear to be the most logical for calculating sediment discharge rates, many investigations continue to use daily mean data

Table 28.1   Sediment yield data during the period of record: August 1992–July 1993

Month	Total sediment discharge (t)	Sediment yield (t/km²)	Sediment yield (t/ha)	Proportion of annual yield (%)
Aug. 92	3.34	0.54	0.005	2.50
Sept. 92	27.58	4.45	0.044	20.65
Oct. 92	4.72	0.76	0.008	3.53
Nov. 92	15.44	2.49	0.025	11.56
Dec. 92	37.30	6.02	0.060	27.92
Jan. 93	24.69	3.98	0.040	18.48
Feb. 93	2.49	0.40	0.004	1.86
Mar. 93	1.88	0.30	0.003	1.41
Apr. 93	12.67	2.04	0.020	9.49
May 93	2.56	0.41	0.004	1.92
June 93	0.62	0.10	0.001	0.46
July 93	0.29	0.05	0.001	0.22
Year	133.60	21.60	0.220	100.00

irrespective of flow conditions. There were significant differences between calculated sediment discharge values for storm days when daily mean data were used instead of the 15-min samples. The second point is that total sediment discharge had to be estimated during two major storm events, on 25 September 1992 and 13 January 1993. For the January storm, this proved to be less critical as there was some indication of the maximum suspended sediment concentration during the hydrograph peak through hand-held sampling in the stream. For the September storm, however, much of the recording equipment was damaged, including the automatic pump sampler, and thus potential suspended sediment concentrations had to be interpolated based on maximum suspended sediment concentrations during other large storms, specifically the three large events in December 1992. In spite of these limitations, it seemed better to try and obtain an approximate value for total sediment discharge during the 25 September 1992 storm rather than omit the event from the calculations altogether, which would have seriously masked the importance of the autumn period in terms of sediment yield.

The total sediment yield from the study basin between August 1992 and July 1993 was calculated at 133.6 tonnes. This translates into a rate of $21.6\,t/km^2/year$ or $0.22\,t/ha/year$. During the autumn months (i.e. September–November), 35.7% of the annual sediment load was exported from the basin, whilst 48.3% was exported during the winter months (December–February). Individually, the most important months were September, December and January, which, when combined, accounted for just over 67% of the total annual sediment load. As expected, the sediment yield during the summer months was low, with just over 3% of the annual load exported from the basin.

The next point to emerge from analysis of the sediment yield data was that most of the sediment exported from the basin was transported during large storms which occurred relatively infrequently. For example, Table 28.2 shows total sediment load during nine storms which together accounted for just over 59% of the total basin sediment yield. For much of the time the stream transported very little suspended

Table 28.2 Sediment discharge data for major storms during the study period

Date	Total rain (mm)	Peak discharge (l/s)	Total storm discharge (m³)	Total sediment discharge (kg)	Proportion of monthly total (%)	Proportion of annual total (%)
25 Sept. 92	31.2	?	?	19 008.0	68.92	14.23
3 Oct. 92	13.2	126.1	1 721.4	1 051.4	22.29	0.88
27–28 Nov. 92	9.6	127.9	1 103.7	1 761.7	11.41	1.32
30 Nov. 92	14.5	187.3	3 186.9	7 213.9	46.72	5.40
2 Dec. 92	23.5	244.3	7 116.6	9 272.5	24.86	6.94
6 Dec. 92	19.5	346.2	7 925.5	7 782.0	20.86	5.83
18 Dec. 92	18.7	353.6	6 796.2	10 713.0	28.72	8.02
13 Jan. 93	17.9	344.5	7 816.7	15 644.0	63.35	11.71
9 Apr. 93	20.7	142.3	3 582.0	6 468.7	51.04	4.84

sediment and the water was essentially clear (the minimum sediment concentration measurable using weighing and drying was 20 mg/l). This finding led the authors to seek a more generalized statement of process magnitude in the study basin and, in particular, to address the question of how much work (i.e. erosion/transport) do flows of a given size perform and when? Magnitude and frequency of work statements were prepared in tabular form for specific frequencies, as shown in Table 28.3. In generalizing the data in Table 28.3, it can be stated that the highest flood flows (i.e. 200 l/s or greater) occurring 0.9% of time (i.e. the wettest three days in any year, on average) transport just over 35% of the total suspended load in 6% total runoff, i.e. high flows are extremely efficient at transporting fine sediment through the drainage basin. However, flows between 150 and 200 l/s, which occur on average 7 days per year, carry 12.3% of the suspended load in 7% runoff, whilst flows between 100 l/s and 150 l/s (i.e. 20 days) carry 14.4% of the suspended load in 12.7% of runoff. Finally, the lowest flows between 0 and 50 l/s, which occur 79% of the time, only carry about 18% of the total suspended load. These data indicate that it is the highest flows which individually carry the greatest sediment loads in absolute terms and which do the most work in total, despite the fact they occur so infrequently (cf. Webb and Walling 1984). A second point of interest emerges from Table 28.3: the time taken to transport half (50%) the suspended load. For this particular catchment, half the suspended load is transported in flows of 135 l/s or greater which occur 2.5% of the time (or on 9 days a year). However, one important point worth emphasizing in relation to this analysis is that the magnitude and

Table 28.3 Summary of magnitude/frequency data (a) for various time intervals; (b) for various discharge thresholds

(a)

% TFEE	Qdm (l/s)	% S	% R
0.1	230.0	15.0	3.0
1	195.0	36.5	8.0
5	102.0	60.0	25.0
10	72.5	73.0	36.0
50	23.8	92.0	78.5
95	9.2	99.9	99.1

(b)

Q class (l/s)	% TFEE	% S	% R
>200	0.9	35.2	6.0
>150	2.0	47.5	13.0
>100	5.7	61.9	25.7
>50	21.1	82.1	52.0

% TFEE=% time flows equalled or exceeded.
Qdm=daily mean discharge.
% S=suspended load.
% R=runoff.

frequency statements are based on only a 12-month data record. Most magnitude/ frequency calculations assume a long hydrograph record and thus the results presented above cannot safely be applied to other basins in the area or be used safely for longer-term calculations.

In terms of individual storm sediment response, all storms, except those on 31 March 1993 and 26 May 1993, showed positive or clockwise hysteresis in the suspended sediment/discharge relationships (see Figure 28.3). However, as discussed earlier, although hysteresis is generally interpreted in terms of variations in sediment supply, these cannot necessarily be related to any specific sediment source. The work of Bryan and Campbell (1986) showed that production of sediment from surface erosion declines through time from an early peak, thereby producing clockwise hysteresis in storms where the sediment was derived from surficial sources. Others (e.g. Carling 1983; Klein 1984) have ascribed clockwise hysteresis to the depletion of channel sources of sediment, either to the entrainment of sediment deposited on the bed of the stream during the previous recession flow, or to the exhaustion of channel bank sediment during the rising limb. Identification of specific sediment sources was only possible by sediment "fingerprinting" (see Slattery *et al.* 1995). These data were consistent with the hysteresis loops suggesting that, for small storms, where overland flow was not observed, almost all of the suspended sediment originates from within the channel, either as freshly eroded sediment from the channel banks or from the re-entrainment of sediment deposited along the channel bed. During large storms, where much of the sediment is still derived from channel sources, additional contributions are received from surficial sources, most predominantly as a result of the erosion of floodplain soils (see discussion below).

Whilst fingerprinting provided detailed information on the relative importance of various sediment sources within the basin, the hysteresis loops still yielded some interesting information on the dynamics of sediment supply both within and between individual storm events. For example, the hysteresis loops for the storms on 2–3 October 1992 and later on 19–20 October 1992 (Figure 28.3(b)) show that the latter storm, despite generating a smaller peak discharge, is characterized by considerably higher suspended sediment concentrations. This is due to the lag between the two storms, in this case just over two weeks, during which time sediment is made "available", either on the slopes or within the channel, for re-entrainment during the second event. This process of sediment being made available between storms is also evident for the storms on 6 December 1992 and later on 18 December 1992 (Figure 28.3(g)), although here the effect is apparently not as pronounced as during the October storms due to the greater temporal resolution in suspended sediment sampling during the 18 December 1992 event. However, for storms which followed in close succession, the effect on sediment availability was essentially reversed in that an *exhaustion* of sediment between storms was observed. This is clearly evident for the two storms shown in Figure 28.3(c) where sediment concentrations during the first storm (i.e.14–15 November 1992) were almost double those of the second storm (16–17 November 1992) despite significantly higher discharges during the latter. A similar pattern of inter-event sediment depletion is evident during the three storms between 29 November 1992 and 6 December 1992 (Figure 28.3(e) and (f)) as well as during the two storms on 10 January 1993 and 11 January 1993 (Figure 28.3(i)). One

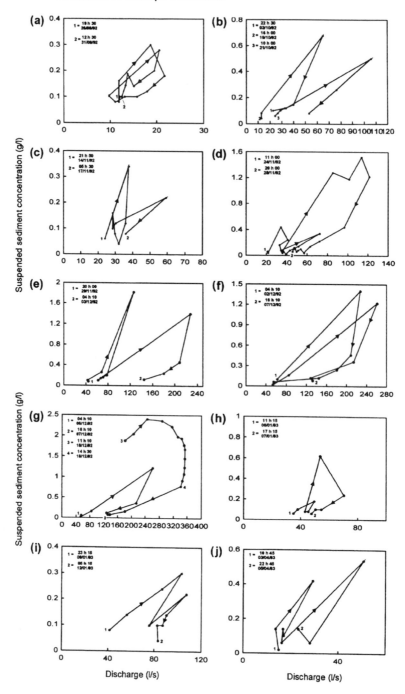

Figure 28.3 Suspended sediment/discharge relationships for selected storms during the study period

sequence of storms worth noting are the three events between 24 and 28 November 1992 (Figure 28.3(d)). An exhaustion of sediment between the first two storms is evident despite peak discharge during the second storm (72 l/s at 2300 h, 25 November 1992) being almost double that of the earlier storm (42 l/s at 2300 h, 24 November 1992). However, during the third storm, peak discharge increases to in excess of 120 l/s with suspended sediment concentrations almost four-fold greater than during the second event. Thus, it seems that despite an exhaustion of sediment between the first two storms, once some "threshold" discharge is exceeded, sediment concentrations increase again suggesting possibly the mobilization of sediment from a new source area.

The sequence of storms presented in Figure 28.3 also show evidence of intra-storm exhaustion of sediment, as indicated by the "openness" of the hysteresis loops. This is perhaps best illustrated by the storms on 2 and 7 December 1992 (Figure 28.3(f)), where suspended sediment concentration declines from an initial value of 1.88 g/l at 1110 h (at a corresponding discharge of 182.2 l/s) to a value of 0.16 g/l at 1830 h (at a corresponding discharge of 181.51 l/s).

28.3.2 Slope Sediment Delivery

The discussion on storm runoff generation presented by Burt and Slattery (Chapter 4, this volume) noted the predominance of subsurface stormflow as the major runoff mechanism during the late summer and early autumn. Surface runoff was only observed along paved roads and some dirt tracks, and thus sediment delivery from the slopes was essentially negligible during this period. The only erosion of any significance during the early autumn occurred along the valley floor during the 25 September 1992 storm as a result of overbank flooding. Large amounts of soil were removed from the recently cultivated floodplain, particularly where streamflow had incised into the soil along plough lines (Figure 28.4). Nevertheless, it was still extremely difficult to obtain an estimate of the total volume of soil eroded as much of the floodplain erosion occurred as "sheet" erosion (i.e. the removal of a thin layer of soil across extended areas) whilst other areas were subject to deposition, particularly in the vicinity of the gauging station.

The first notable erosion from the cultivated slopes occurred during the storm of 3 October 1992 when runoff down wheelings in the eastern section of the basin resulted in modest soil losses. However, very little of this sediment eroded along wheelings actually reached the stream channel, most being deposited in small alluvial fans along the edge of fields. Many of the subsequent storms during October and November generated runoff along wheelings with small amounts of sediment entrained in the runoff. Measurements were still extremely difficult to make, however, due to the fact that almost all these storms occurred during the early morning hours.

The first really complete set of slope measurements were obtained during the 18 December 1992 storm. The authors initially stopped at site 2 (shown in Figure 28.1) during the early part of the storm. Here, runoff down wheelings, which had been observed during the late November/early December storms, had begun to erode a series of rills (Figure 28.5) which ultimately developed into a complex braided rill network. Much of the sediment eroded from these rills was deposited in a fan along

Figure 28.4 Photograph showing erosion along the valley floor near the gauging station during the 25 October 1992 storm. The stream channel runs along the hedgerow behind the figure

the lower hedgerow, although significant amounts of runoff and sediment were routed beyond the field boundary and along the road towards Lower Nill Farm.

The 18 December 1992 storm was also the first one in which runoff was observed on the experimental hillslope (see Figure 28.2). Infiltration capacities were at a minimum during this period, never exceeding 2.5 mm/h (Burt and Slattery, Chapter 4, this volume). Measurements taken during the latter part of the storm also showed significant amounts of sediment being moved downslope, predominantly along wheelings, but, again, much of this was deposited in small fans along the valley floor. A series of shallow rills was also monitored at site 1 near the water sampler (Figure 28.1). This was one of the few sites where sediment was being discharged directly into the stream channel.

A significant outcome of the 18 December 1992 storm was the discovery the following day of the thalweg rill network across three fields in the southern section of the basin (Figure 28.6). In the uppermost section, developed within subcatchment A on what we termed Field 1A, closest to the basin divide, the rill extended from almost the basin divide to the field outlet, terminating in an extensive alluvial fan along the lower hedgerow. The thalweg rill in this upper field was supplied with runoff and sediment by a series of "feeder rills" developed along the steeper valley-side slopes (~5°). These were essentially linear features that had incised into the soil surface along vehicle wheelings. The rill system extended into the adjacent field downslope (i.e Field 1B), although here the system contained only a single channel in the valley floor with a smaller fan deposit immediately above Lower Nill Farm. It was only during the storm of 13 January 1993 that the rill network became extended

Figure 28.5 Braided rill network at Site 2 (see Figure 28.1) during the 18 December 1992 storm

below the farm. Flow from the rill system upslope diverged and broke through the farmyard but became preferentially concentrated along a major wheeling in the field immediately below the farm (Field 1C), scouring out the wheeling to form a single rill channel with a fan deposit directly adjacent to the stream. The entire rill system, once fully developed, formed a continuous and efficient link for the delivery of water and sediment between the upper fields and the valley floor (Figure 28.7).

The dynamics of the thalweg rill system have been discussed in detail elsewhere (Slattery *et al.* 1994) and only summary information is presented here. Total volumetric soil loss from the main rill on Field 1A was calculated at 25.46 m^3. Soil loss from the feeder rills totalled 6.82 m^3 giving a total volume of soil eroded from Field 1A of 32.28 m^3. This translates into a rate of 2.2 m^3/ha for the entire field. However, if the rate is calculated for the catchment area of the rill system (i.e. subcatchment A, Figure 28.6), which appears to be geomorphologically more meaningful, the rate increases to 3.01 m^3/ha. These values are only slightly less than the median values of soil loss during "severe erosion" on the South Downs reported by Boardman (1988). Several soil bulk density measurements were made on Field 1A in order to calculate soil loss on a weight basis. Mean bulk density was 1.3 Mg/m^3, which relates to 42 tonnes of total soil loss. The corresponding rates were 2.8 t/ha for the entire field and 3.91 t/ha for the contributing catchment. These values are very similar to the mean rill erosion rate of 3.6 t/ha calculated for 86 fields with loessic soil in Belgium (Govers 1991).

The process of rill development, and the simultaneous transfer of water and sediment from valley-side sources to rill channels and thence to the stream is, in

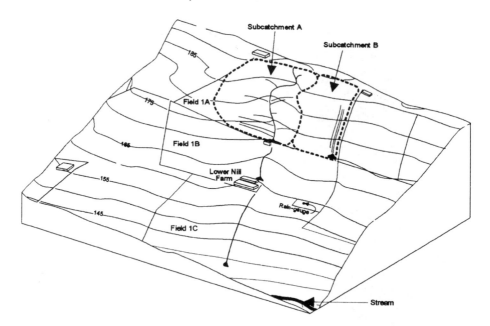

Figure 28.6 Block diagram showing the location of the thalweg rill system across three fields in the southern section of the catchment. The rills in subcatchment B are those shown in Figure 28.5

reality, not as simple as the previous discussion suggests. This is because of the storage of sediment along the conveyance route. We calculated the volume of sediment in the fan on Field 1A to be $16.19\,m^3$, or 50.2% of the total eroded volume (see Figure 28.6). This indicates that 49.8% of the material eroded by the rill system was transported beyond the field boundary (very little sediment was stored elsewhere *within* the field). This value effectively corresponds to the sediment delivery ratio (SDR) for this field for all storms prior to 13 January 1993. There are many problems, however, in calculating delivery ratios for individual fields: not only are volumes of stored material difficult to quantify but there is also the complication of spatial and temporal discontinuity within the transfer process. These problems have been fully reviewed by Walling (1983) and Novotny and Chesters (1989).

It is worth emphasizing here that the discussion of soil loss presented above relates only to rill erosion. Although we did observe some sheet wash, specifically on the valley-side slopes, amounts of soil moved by interrill processes appeared insignificant in comparison with the effect of rill development. Other researchers (e.g. Govers and Poesen 1988) have found that the contributions of interrill erosion to the total eroded volume at the field scale ranges between 10 and 20%. Evans (1990b) states that, in general, wash on most soils under arable crops will transport $<0.3\,m^3/ha/$ year, and that this probably applies to most sloping arable fields in Britain.

The fact that both February and March were dry months was disappointing as we had hoped to conduct several more measurements in this rill system, in particular, to determine the spatial variation in discharge across the entire system through

Figure 28.7 Photograph of the thalweg rill in subcatchment A at maximum flow (*c.* 120 l/s) during the 13 January 1993 storm

simultaneous monitoring on all three fields. However, by early April, the crop cover in the fields had become fully established and whilst we were actually in the field during the storm on 11 April 1993, no further surface runoff was observed, either in the rill system, or on any other fields, although paved roads continued to be a source of storm runoff production. These observations provided further confirmation that subsurface stormflow was the dominant storm runoff mechanism during the spring and summer months.

28.4 DISCUSSION

It is important to set the sediment data obtained in the present study within a wider regional context, so that a more general statement can be made regarding the sediment delivery system within the study basin. Figure 28.8, based on the work of Walling (1990), provides a map of Britain on which most of the available values of annual suspended sediment yield derived from river-monitoring programmes and some of the data available from reservoir surveys have been superimposed. This information indicates that the suspended sediment yields of British rivers are low by world standards and lie typically in the range 50–100 t/km^2/year. The highest values are generally located in upland regions, whereas over most of lowland Britain, the area where arable cultivation represents a significant proportion of the land use, yields rarely exceed 50 t/km^2/year (0.5 t/ha^2/year). The annual suspended sediment

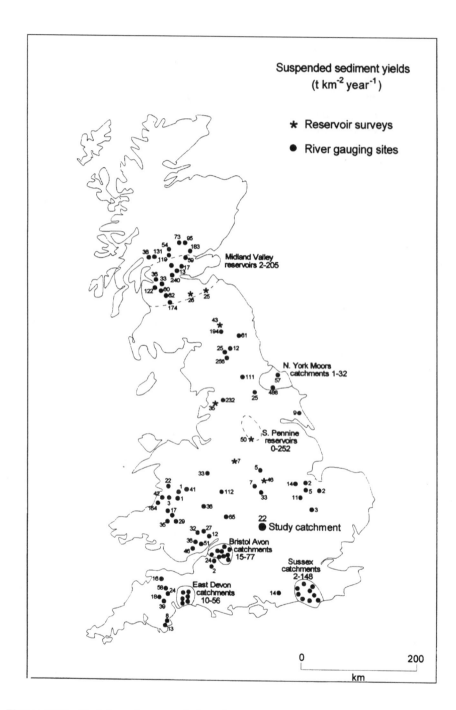

Figure 28.8 Variation of suspended sediment yields in Britain, including data from the present study (redrawn after Walling 1990)

yield from the Stour catchment, also plotted on Figure 28.8, is entirely consistent with these data from lowland Britain.

The low levels of suspended sediment yield associated with rivers draining the agricultural land of lowland Britain are generally seen as evidence of low rates of soil erosion, although, as Walling (1990) succinctly points out, it must be recognized that the values of sediment yield involved represent areal averages for drainage basins within which areas experiencing appreciable erosion may be juxtaposed with areas experiencing low or negligible rates of soil loss. This point is well illustrated by the data from the present study, where the on-site rate of erosion in a single field during the 12-month period was calculated at 2.8 t/ha, an order of magnitude greater than the 0.2 t/ha obtained from the suspended sediment yield data. This is the result of substantial transmission losses associated with the movement of sediment from the field to the channel system. In fact, almost none of the sediment eroded from this field reached the river channel, which supports statements made earlier by Walling (1990) and Burt (1993) that only a proportion, and perhaps a rather small proportion, of the soil eroded within a drainage basin will find its way to the basin outlet.

In terms of sediment export from the basin, most was evacuated during the autumn and winter, the wettest time of the year. Although the *effectiveness* of an event of a certain magnitude has proved difficult to define (see, for example, Carling and Beven 1989), the data from this study indicated that the highest flows did the most work in total. As noted earlier, the highest flows carried the greatest sediment loads in absolute terms despite their infrequent occurrence. This conclusion is somewhat contradictory to the early concepts of effectiveness in the geomorphological literature noted by Wolman and Miller (1960): "Analysis of the transport of sediment by various media indicate that a major portion of the work is performed by events of moderate magnitude which recur relatively frequently rather than by rare events of unusual magnitude" (p. 70). As Walling and Webb (1992) point out, the true significance of extreme events, for which records are rarely available, still remains somewhat uncertain. A much longer period of study would be needed to see if very large, rare flows transport more sediment in total than slightly smaller, more common flood events.

Vehicle wheelings were the most active runoff-generating components on cultivated slopes within the catchment and were also important zones of sediment production. As noted by Burt and Slattery (Chapter 4, this volume), vehicle wheelings were characterized by increased bulk density and shear strength and decreased infiltration capacity. Most storms during the autumn and winter produced runoff along wheelings, although soil losses tended to be low, except where runoff incised rills, with much of the sediment transported along wheelings deposited along the edge of slopes in small alluvial fans. In many ways, November and December were *too* wet (despite the fact that the success of the project was dependent upon significant rainfall and erosion) because many of the fields in the catchment and, in particular, the experimental hillslope, became impassable to farm traffic. The net result was that only one set of wheelings was imparted on the slope during rolling on 15 October 1992, which significantly reduced the likelihood of runoff concentration and rilling. Rilling was nevertheless the dominant erosion process in the catchment as a whole, resulting mainly from initial runoff concentration along wheelings. The thalweg rill was entirely topographically defined, resulting from convergent surface runoff along a dry valley bottom. The results presented here further emphasize the need for a spatially distributed approach to the study of runoff and erosion at the catchment scale in order to fully understand the processes involved in rill generation.

The development of rill systems increases, often by orders of magnitude, the amount of erosion compared to unrilled slopes. However, the erosion rates reported in this study were not exceptional by British standards (see, for example, Colborne and Staines 1985; Morgan 1985; Evans and Cook 1986; Reed 1986), and field-scale losses of over $100\,m^3/ha$ have been reported in several areas, including the South Downs (Boardman and Robinson 1985; Boardman 1988), Norfolk (Evans and Nortcliff 1978), West Sussex (Boardman 1983), the West Midlands (Reed 1979) and, recently, in Oxfordshire itself (Boardman et al. 1996 in press). In fact, Boardman (1988) reported extensive rill erosion at Rottingdean on the South Downs, where soil losses reached $200\,m^3/ha$ for one 12 ha field drilled with wheat. Defining on average rate of erosion is difficult, but 2 t/ha/year has been suggested as characteristic of sandy and chalky soils in southern England (Morgan 1980). Blackman (1990) suggests that these values may give a distorted view of overall rates of erosion within any particular locality since they are the exceptional results of extreme events. Notwithstanding the shortness of the soil erosion and stream load records presented here, it is worth pointing out that, in both cases, it is the largest events which have transported most sediment. The soil erosion rates may not be "exceptional" therefore, merely indicative of the fact that soil erosion is more severe in larger events. The problem, of course, is that the events which produce most soil erosion may not necessarily be the same events which produce the largest sediment export from the basin and even if they are, there is no guarantee of conveyance of eroded soil to the stream; the dominant source of *stream* sediment need not be the slopes, but, as is the case here, floodplain and channel. As noted above, a long-term study would be needed to establish which size of events achieve most in terms of slope erosion, sediment delivery and channel transport. The complexity of the sediment delivery process may mean that there is, in any case, no simple answer to such a question.

The results presented in this chapter also support the view that erosion in Britain occurs predominantly in the autumn and winter, largely as a result of autumn sowing of crops, specifically winter cereals, which are a high-risk crop in terms of erosion due to the prevalence of large areas of bare, smooth ground during the autumn and early winter, the wettest time of the year. A recent review by Boardman (1990) suggests that there has been a very real increase in the incidence of erosion over the past two decades, largely as a result of the growing of autumn planted cereals, and Evans (1990a) estimates that 36% of arable land in England and Wales is now at moderate to very high risk of erosion.

Serious erosion has traditionally been associated with the semi-arid and semi-humid areas of the world, rather than the temperate zone (Kohnke and Bertrand 1959; Hudson 1967, 1971); the temperate, maritime climate of the UK was not thought to be conducive to soil water erosion, due to the frequent but predominant low-intensity rainfall events. Hudson (1967), for example, suggested that rainfall intensities of 25 mm/h were required to initiate erosion, and argued therefore that only some 5% of British rainfall was erosive. However, more recent studies propose lower thresholds at which rainfall becomes erosive. Evans and Nortcliff (1978), in a study of erosion in Norfolk, use 7.5 mm as a measure of daily rainfall necessary to produce erosion. Field studies on sandy soils in Bedfordshire led Morgan (1980) to suggest that low-intensity rainfall still possessed enough kinetic energy to cause erosion and he suggested a threshold of 10 mm/h. More recently, Fullen and Reed (1986) demonstrated that in the West Midlands, intensities as low as 1.5–2 mm/h could cause runoff and erosion on compacted and capped soils, provided that 5–10 mm of rain were provided in the rainfall event. Boardman and Favis-Mortlock (1992) report that, on silty soils on the South Downs, soils usually return to field capacity during October; thereafter 30 mm of rain in two days is often sufficient to initiate rilling on winter cereal fields with less than 30% crop cover.

Thus, it is clear that water erosion does occur in Britain and that large rainfall amounts are not the primary cause, or even a fundamental prerequisite. Indeed, daily rainfall totals as low as 6 mm were sufficient to cause runoff and erosion on several fields in the study catchment during the winter; even lower totals produced runoff along tracks and wheelings. Most of the erosion in cultivated areas of Britain must therefore be attributed to management practices, in particular, the adoption of autumn sowing of crops, specifically winter cereals. Autumn-sown cereals grow relatively slowly over the winter period and this exposes the smooth, relatively bare seedbeds to the intense rain of autumn and winter. This frequently leads to surface sealing which is crucial in determining runoff and erosion. The development of seals significantly reduces the infiltration rate, thus leading to more frequent runoff and, ultimately, a greater potential for erosion. The erosion risk is further increased by other management practices, such as enlarging fields by the removal of field boundaries such as hedges, which increases catchment areas and slope lengths, thereby increasing the depth and speed of any runoff generated, and rolling, which results in a more compacted surface. This is particularly dangerous on fields sown to winter cereals, as these are generally bare during the wettest time of the year, as discussed above.

Part of the justification of conducting this research in this particular area was the fact that so little information presently exists on the extent, frequency and rate of erosion and sediment delivery on Cotswold soils. The data presented in this study suggests that erosion is perhaps more widespread and significant on Cotswold soils than previously thought – a view that was supported by more general surveys conducted in the area during the study period. The question that is then raised is are these soils, namely the Aberford and Banbury, particularly erodible, or was this just an unusually wet winter with above-average erosion? The first part of this question can be addressed with reference to Evans' (1990a) work on soil-erosion risk. The author classified the 296 soil associations of the National Soil Map of England and Wales into five categories of erosion risk, based on land use, landform and soil properties, from 1 = very small risk, to 5 = very high risk. The Aberford soil series was classified as being at small risk of erosion (category 2) and the Banbury soil series as being at moderate risk of erosion (category 3) although, as Evans points out, an individual field or group of fields may erode more severely than indicated by the risk category due to localized factors, such as different rainfall rates and slope angles. Nevertheless, the risk categories to which the Aberford and Banbury associations have been assigned seem entirely reasonable given the data presented here, although it is suggested that the Aberford association be annotated to show that erosion may be locally more common than usual for the association due to overbank flooding.

Whether the 1992–1993 field season was "average" or not in terms of erosion is, however, more difficult to assess. Certainly, the autumn of 1992 was wet with September, October and November rainfall totals all above the long-term mean. At Oxford, it was the twelfth wettest autumn on record and the wettest since 1974 (Burt and Shahgedanova 1993). However, winter as a whole was relatively dry with rainfall just a little below average, although this was masked by February's unusually low total. Despite the fact that rainfall was *in general* above average, no individual storm was truly exceptional, with only that on 25 September 1992 having a return period greater than one year. There is no reason to believe that the erosion rates reported in this study are excessive or even "above average". Indeed, observations made at the time of writing confirmed that erosion was still occurring in the catchment although (unfortunately?) the thalweg rill field had been placed in set aside for the 1993/94 winter, and no further erosion had taken place.

In concluding this discussion on runoff and sediment delivery mechanisms in this basin, attention must be drawn once more to the fundamental question of scale and scale transference. As noted above, rilling was found to be the most significant process affecting soil loss on the cultivated slopes of the catchment. However, information about rill erosion rates and mechanisms continues to be collected from plot studies and from laboratory flumes, and this level of investigation imposes several limitations. Conventional plot-based research methods singularly fail to reproduce topographically defined flow paths and exclude the potentially critical effects of water running on to the plot. Thus, potentially important erosional processes, such as thalweg rill development, are not being considered in such work. Based on the observations presented here, it is concluded that there is an urgent need to study erosion at the catchment scale in order to develop a more complete and integrated view of erosional systems within catchments. The identification of sites of surface flow convergence seems to be a particularly important task in this respect.

REFERENCES

Arnborg, L., Walker, H.J. and Pieppo, J. 1967. Suspended load in the Colville River, Alaska, 1962. *Geografiska Annaler*, **49A**, 131–144.

Blackman, J.D. 1990. Variation and Change in the Aggregate Stability and Erodibility of Downland Soils. PhD Thesis, University of Sussex.

Boardman, J. 1983. Soil erosion at Albourne, West Sussex. *Applied Geography*, **3**, 317–329.

Boardman, J. 1988. Severe erosion on agricultural land in East Sussex, UK, October 1987. *Soil Technology*, **1**, 333–348.

Boardman, J. 1990. Soil erosion in Britain: costs, attitudes and policies. *Education Network for Environment and Development*, Social Audit Paper 1, University of Sussex.

Boardman, J. and Favis-Mortlock, D. 1992. Soil erosion and sediment loading of watercourses. *SEESOIL*, **7**, 5–29.

Boardman, J. and Robinson, D.A. 1985. Soil erosion, climatic vagary and agricultural change on the Downs around Lewes and Brighton, autumn 1982. *Applied Geography*, **5**, 243–258.

Boardman, J., Burt, T.P., Evans, R., Slattery, M.C. and Shuttleworth, H. in press. Soil erosion and flooding as a result of a summer thunderstorm in Oxfordshire and Berkshire, May 1993. *Applied Geography*.

Bogen, J. 1980. The hysteresis effect of sediment transport systems. *Norsk Geografisk Tidsskrift*, **34**, 45–54.

Bryan, R.B. and Campbell, I.A. 1986. Runoff and sediment discharge in a semi-arid ephemeral drainage basin. *Zeischrift für Geomorphologie, Supplement Band*, **8**, 121–143.

Burt, T.P. 1993. From Westminster to Windrush: Public policy in the drainage basin. *Geography*, **78**, 389–400.

Burt, T.P. and Shahgedanova, M. 1993. The weather at Oxford during 1992. *Journal of Meteorology*, **18**, 86–90.

Carling, P.A. 1983. Particulate dynamics, dissolved and total load, in two small basins, northern Pennines. *Hydrological Sciences Journal*, **28**, 355–375.

Carling, P.A. and Beven, K. 1989. The hydrology, sedimentology and geomorphological implications of floods: an overview. In K. Beven and P.A. Carling (eds), *Floods: Hydrological, Sedimentological and Geomorphological Implications*. John Wiley and Sons, Chichester, pp. 1–9.

Colborne, G.J.N. and Staines, S.J. 1985. Soil erosion in south Somerset. *Journal of Agricultural Science, Cambridge*, **104**, 107–112.

Dietrich, W.E. and Dunne, T. 1978. Sediment budget for a small catchment in mountainous terrain. *Zeitschrift für Geomorphologie*, **29**, 191–206.

Doty, C.W. and Carter, C.E. 1965. Rates and particle-size distributions of soil erosion from unit source areas. *Transactions of the American Society of Agricultural Engineers*, **8**, 309–311.

Ebisemiju, F.S. 1990. Sediment delivery ratio prediction equations for short catchment slopes in a humid tropical environment. *Journal of Hydrology*, **114**, 191–208.

Evans, R. 1990a. Assessment of soil erosion risk in England and Wales. *Soil Use and Management*, **1**, 127–131.

Evans, R. 1990b. Water erosion in British farmers' fields – some causes, impacts, predictions. *Progress in Physical Geography*, **14**, 199–219.

Evans, R. and Cook, S. 1986. Soil erosion in Britain. *SEESOIL*, **3**, 28–58.

Evans, R. and Nortcliff, S. 1978. Soil erosion in north Norfolk. *Journal of Agricultural Science, Cambridge*, **90**, 185–192.

Fullen, M.A. and Reed, A.H. 1986. Rainfall, runoff and erosion on bare arable soils in East Shropshire, England. *Earth Surface Processes and Landforms*, **11**, 413–425.

Glymph, L.M. 1954. Studies of sediment yields from watersheds. *International Association of Hydrological Sciences*, Publication 36, pp. 173–191.

Govers, G. 1991. Rill erosion on arable land in Central Belgium: rates, controls and predictability. *Catena*, **18**, 133–155.

Govers, G. and Poesen, J. 1988. Assessment of the interrill and rill contribution to total soil loss from an upland field plot. *Geomorphology*, **1**, 343–354.

Hadley, R.F. (ed.) 1986. *Drainage Basin Sediment Delivery*. Proceedings of the Albuquerque Symposium, International Association of Hydrological Sciences, Publication 159.

Hudson, N.W. 1967. Why don't we have soil erosion in England? In J.A.C. Gibb (ed.), *Proceedings of the Agricultural Engineering Symposium*. Institute of Agricultural Engineers Paper 5/b/42.

Hudson, N.W. 1971. *Soil Conservation*. Batsford, London.

Klein, M. 1984. Anti-clockwise hysteresis in suspended sediment concentration during individual storms. *CATENA*, **11**, 251–257.

Kohnke, H. and Bertrand, A.R. 1959. *Soil Conservation*. McGraw-Hill, New York.

Maner, S.B. 1958. Factors influencing sediment delivery rates in the Red Hills physiographic area. *Transactions of the American Geophysical Union*, **39**, 669–675.

Morgan, R.P.C. 1980. Soil erosion and conservation in Britain. *Progress in Physical Geography*, **4**, 24–27.

Morgan, R.P.C. 1985. Assessment of soil erosion risk in England and Wales. *Soil Use and Management*, **1**, 127–131.

Novotny, V. and Chesters, G. 1989. Delivery of sediments and pollutants from nonpoint sources: a water quality perspective. *Journal of Soil and Water Conservation*, **44**, 568–576.

Parsons, A.J. and Abrahams, A.D. 1989. A miniature flume for sampling interrill overland flow. *Physical Geography*, **10**, 85–94.

Reed, A.H. 1979. Accelerated erosion of arable soils in the UK by rainfall and runoff. *Outlook on Agriculture*, **10**, 41–48.

Reed, A.H. 1986. Soil loss from tractor wheelings. *Soil and Water*, **14**, 12–14.

Renfro, G.W. 1975. Use of erosion equations and sediment-delivery ratios for predicting sediment yield. In *Present and Prospective Technology for Predicting Sediment Yields and Sources*. US Department of Agriculture, Publication ARS-S-40, pp. 33–45.

Rieger, W.A., Olive, L.J. and Gippel, C.J. 1988. Channel sediment behaviour as a basis for modelling delivery processes. In *Sediment Budgets*. Proceedings of the Porto Alegre Symposium, International Association of Hydrological Sciences, Publication 174, pp. 41–48.

Roehl, J.E. 1962. Sediment source areas, delivery ratios and influencing morphological factors. *International Association of Hydrological Sciences*, Publication 59, pp. 202–213.

Slattery, M.C. and Burt, T.P. 1995. Size characteristics of sediment eroded from agricultural soil: dispersed versus non-dispersed, ultimate versus effective. In E.J. Hickin (ed.), *River Geomorphology*. John Wiley and Sons, Chichester, pp. 1–17.

Slattery, M.C., Burt, T.P. and Boardman, J. 1994. Rill erosion along the thalweg of a hillslope hollow: a case study from the Cotswold Hills, central England. *Earth Surface Processes and Landforms*, **19**, 377–385.

Slattery, M.C., Burt, T.P. and Walden, J. 1995. The application of mineral magnetic measurements to quantify within-storm variations in suspended sediment source. *International Association of Hydrological Sciences*, Publication No. 229, pp. 143–151.

Sutherland, R.A. and Bryan, R.B. 1991. Sediment budgeting: A review and case study in the Katiorin drainage basin, Kenya. *Earth Surface Processes and Landforms*, **16**, 383–398.

Trimble, S.W. 1981. Changes in sediment storage in Coon Creek Basin, Driftless Area, Wisconsin, 1853–1975. *Science*, **214**, 181–183.

Walling, D.E. 1983. The sediment delivery problem. *Journal of Hydrology*, **69**, 209–237.

Walling, D.E. 1988. Measuring sediment yield from river basins. In R. Lal (ed.), *Soil Erosion Research Methods*. Soil and Water Conservation Society, Ankeny, Iowa, pp. 39–73.

Walling, D.E. 1990. Linking the field to the river: sediment delivery from agricultural land. In J. Boardman, I.D.L. Foster and J.A. Dearing (eds), *Soil Erosion on Agricultural Land*. John Wiley and Sons, Chichester, pp. 129–152.

Walling, D.E. and Webb, B.W. 1982. Sediment availability and the prediction of storm-period sediment yields. *International Association of Hydrological Sciences*, Publication 137, pp. 327–337.

Walling, D.E. and Webb, B.W. 1992. Water quality I. Physical characteristics. In P. Calow and G.E. Petts (eds), *The Rivers Handbook, Volume 1*. Blackwell Scientific Publications, Oxford, pp. 48–72.

Webb, B.W. and Walling, D.E. 1984. Magnitude and frequency characteristics of suspended sediment transport in Devon rivers. In T.P. Burt and D.E. Walling (eds), *Catchment Experiments in Fluvial Geomorphology*. Geobooks, Norwich, pp. 399–415.

Wolman, M.G. and Miller, J.C. 1960. Magnitude and frequency of forces in geomorphic processes. *Journal of Geology*, **68**, 54–74.

29 Verification of the European Soil Erosion Model (EUROSEM) for Varying Slope and Vegetation Conditions

R. P. C. MORGAN
Silsoe College, Cranfield University, UK

29.1 INTRODUCTION

The European Soil Erosion Model (EUROSEM) is a process-based dynamic and distributed model designed to simulate the erosion, transport and deposition of sediment by water over the land surface in an individual storm (Morgan *et al.* 1994; Morgan *et al.* in prep.). It has been developed by a group of European scientists with financial support from the European Commission to provide a means of assessing the risk of soil erosion from fields and small catchments under existing land-use conditions and of evaluating the likely impact of future changes in the environment and land use, including the implementation of erosion control measures. The model is currently undergoing a programme of validation and testing against measured data from field sites in both northern and Mediterranean Europe.

Another aspect of validation, however, is to assess whether the model provides realistic simulations of the way different factors control erosion when outputs are compared with trends reported in the literature. This chapter describes how EUROSEM (Version 3.3) simulates the effects of changes in slope steepness and vegetation cover on erosion from small hillslope plots. The results are compared with observations reported in previously published work.

29.2 BASIS OF EUROSEM

The structure of EUROSEM is illustrated in Figure 29.1. EUROSEM uses break-point rainfall data for a storm and computes, in turn, the interception of rainfall by the plant cover, the generation of runoff as infiltration excess, the detachment of soil particles by raindrop impact and runoff, the transport capacity of the runoff and the deposition of sediment. Soil loss is computed as a sediment discharge using a numerical solution of the dynamic mass balance equation for sediment. EUROSEM uses the runoff generator and the routeing routines for water and sediment contained

Advances in Hillslope Processes, Volume 1. Edited by M. G. Anderson and S. M. Brooks.
© 1996 John Wiley & Sons Ltd.

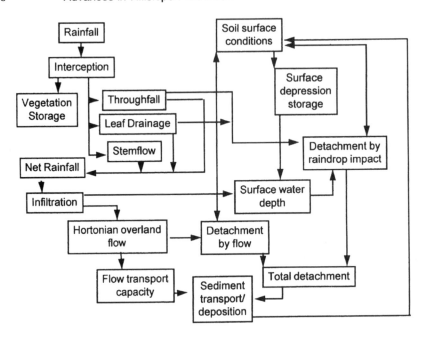

Figure 29.1   Flow chart for EUROSEM

in the KINEROS model (Woolhiser *et al.* 1990). A detailed description of the model with documentation of the operating equations is given in Morgan *et al.* (in prep.).

In contrast to other physically based soil-erosion models, EUROSEM provides for explicit simulation of rill and interrill flow, the effects of rock fragments and vegetation cover. Where rills are present and the interrill flow length exceeds 1 m, sediment transport and deposition is accounted for on the interrill areas and these processes control the delivery of sediment to the rills. Where rock fragments occur on the soil surface, EUROSEM models their effects on saturated hydraulic conductivity, the relative volume of soil not acting as a porous medium and the proportion of the surface soil exposed to raindrop impact. A physically based approach is adopted to describe the effect of vegetation which, through its properties of height, percentage cover, leaf shape, basal area, roughness and rooting system, influences the volume and energy of the rain reaching the ground surface, the infiltration of water into the soil, the velocity of flow and the strength of the soil.

The model can be run for a single slope plane or element, for predicting erosion from small fields within which slope, soil and land cover conditions are reasonably uniform; or for a series of multiple planes arranged in a cascading sequence, for predicting erosion from small catchments. Outputs of the model for a storm include the total runoff and soil loss, the hydrograph and the sediment graph. When operated for a small catchment, it is also possible to simulate the location of erosion (sediment sources) and deposition (sediment sinks). In this chapter, the model is operated for a single plane and only the output for total soil loss is examined.

29.3 SIMULATION OF A BASE CONDITION

Before investigating how EUROSEM output responds to changes in slope and plant cover, the model was applied to a base condition to test whether its response was realistic. The condition chosen was an 11° mid-slope segment of the field site near Silsoe, Bedfordshire, for which data on runoff and erosion were collected between 2 May 1973 and 16 August 1979 (Morgan *et al.* 1986). The slope has a sandy soil (Typic Quartzipsamment) derived from underlying Lower Greensand rock. The soil was kept free of vegetation using rotovation. Measurements were made using two 0.5 m wide Gerlach troughs to give a plot 10 m long and 1.0 m wide. Measured runoff and soil loss were respectively 15.6 mm and 18.9 t/ha for a storm of 53.2 mm on 7 August 1979. Erosion on the plot on that day was by a combination of rill and interrill erosion. Table 29.1 gives the cumulative rainfall over time for the storm as recorded at Silsoe, some 2 km from the plot. Table 29.2 gives the input data used to describe the slope segment. Table 29.3 gives the predictions obtained without model calibration. These are close to the observed values and indicate that the model gives a realistic simulation, particularly when applied by a user (the author) with a good knowledge of both the model and the field site.

29.4 SIMULATION OF SLOPE EFFECT

29.4.1 Effects of Slope Steepness as an Independent Factor

The first simulation was made for overland flow only, i.e. without rills, using the base conditions described in Table 29.1 and changing only the slope steepness. The resulting relationship between simulated soil loss and slope is shown in Figure 29.2. Soil loss increases slowly at first, as slope rises from 1% to 5%, then rapidly to a slope of 100% after which it levels off and becomes insensitive to further increases in steepness. The trend matches very closely that produced by Schmidt (1992) in his EROSION 3D model. Unfortunately no experimental work exactly supports this

Table 29.1 Data for the rainstorm used in the simulations

Time from start of storm (min)	Cumulative rainfall (mm)
0	0.0
10	2.2
20	8.1
30	19.4
40	30.7
50	36.3
60	41.9
70	47.6
80	51.0
90	53.2
200	53.2

Table 29.2 Base input data for erosion plot at Silsoe, Bedfordshire

Variable	Input value	Remarks
Plot length	10.0 m	Measured plot length
Plot width	1.0 m	Measured plot length
Slope	0.19	Measured in field
Interrill Manning's n	0.2	Estimated value for overland flow on the plot (Morgan 1980)
Rill Manning's n	0.02	Typical value for bare soil for flow in a channel
Saturated hydraulic conductivity	26.0 mm/h	Typical value for the sandy loam soil
Net capillary drive	240.0 mm	Value from User Guide (Morgan *et al.* 1993)
Porosity	0.45	Value from User Guide (Morgan *et al.* 1993)
Initial volumetric soil moisture content	0.40	Estimated value
Maximum volumetric soil moisture content	0.42	Value from User Guide (Morgan *et al.* 1993)
Proportion of the soil occupied by stones	0.0	Observed value
Infiltration recession factor	10.0 mm	Estimated from field measurements of surface roughness
Depth to non-erodible layer	3.0 m	Large value chosen because of no impediment to erosion
Number of rills	1	Observed value
Rill width	0.25 m	Typical field value
Rill depth	0.20 m	Typical field value
Rill side slope	1:1	Typical field value
Surface roughness	20 cm/m	Value from User Guide (Morgan *et al.* 1993) for rotovated soil
Interrill slope	0.19	Assumed the same as the plot slope
Median particle size of the soil	250 μm	Measured value
Detachability of the soil	2.6 g/J	Value from User Guide (Morgan *et al.* 1993)
Soil cohesion	7.0 kPa	Value from User Guide (Morgan *et al.* 1993); close to measured values for the soil

Table 29.3 Comparison of base simulations with observed data

	Runoff (mm)	Soil loss (t/ha)
Observed	15.58	18.90
Simulated	16.13	16.37

trend. As shown by Bryan (1979), the relationship between soil loss and slope can be very varied in form, depending upon the characteristics of the soil.

Nevertheless, some of the relationships presented by Bryan (1979), particularly for sandy and silty soils, display rapid increases in erosion with slope up to about 20–30%, levelling off thereafter, as do the results presented by Foster and Martin (1969) and Lattanzi *et al.* (1974).

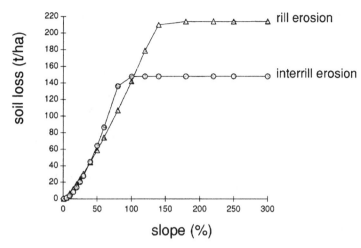

Figure 29.2   Simulated soil loss as affected by changes in slope steepness

For slopes between 1% and 80%, the relationship between soil loss (*A*) and slope steepness (*S*), as simulated by EUROSEM, can be expressed by the equation:

$$A = 0.02\ S^{2.08} \tag{29.1}$$

The value of the exponent lies beyond the range of 1.35–1.5 widely reported in the literature (Zingg 1940; Musgrave 1947; Kirkby 1969; Bryan 1979) but is in line with values closer to 2.0 obtained for high rainfall conditions (Hudson and Jackson 1959). The value obtained seems reasonable considering that, in the local area, the simulated storm would have a return period of about 100 years.

A second simulation was made with a single rill channel down the centre of the plot but otherwise maintaining the conditions of the first simulation. The rill was given the same dimensions as that used in the base simulation (Table 29.2). The result was a similar shaped curve (Figure 29.2) but peaking at a higher slope (140%) and at a higher magnitude of soil loss. For slopes between 1% and 140% the relationship between soil loss and slope steepness is:

$$A = 0.0126\ S^{2.16} \tag{29.2}$$

The higher exponent for rilled compared with unrilled flow is expected (Kirkby 1969), as is the higher magnitude of soil loss on the steeper slopes.

29.4.2   Effects of Slope Steepness with Interactions of Other Factors

A further set of simulations was made in which other inputs were allowed to change with increasing slope, following the findings of Poesen (1986) for soils prone to sealing. He showed that sealing intensity was greatest on low-angled slopes and decreased as slope became steeper; as a result, infiltration rates increased and soil strength decreased with increasing slope. In these simulations the saturated hydraulic conductivity and soil cohesion were varied from the base conditions given in Table

29.1. The degree of variation was kept within the guide limits for input values listed in the EUROSEM User Guide (Morgan *et al.* 1993). Thus, saturated hydraulic conductivity was increased from 7 mm/h on a 1% slope to 55 mm/h on a 55% slope, whilst cohesion was decreased from 10 kPa to 3 kPa (Table 29.4). Figure 29.3 shows the resulting trends with erosion increasing with slope steepness on slopes up to 30% and decreasing thereafter. Again, as expected, the soil loss from rill erosion was greater than that from overland (unchannelled) flow. Such trends have been reported by Heusch (1970) on natural slopes in Morocco and by Odemerho (1986) on cut slopes on road banks in Nigeria. The results also match the trend proposed theoretically by Rowlison and Martin (1971).

For slopes between 5% and 20%, the following relationships were obtained between soil loss and slope steepness:

Table 29.4   Input data used in simulation of effect of slope steepness

Percentage slope	Saturated hydraulic conductivity (mm/h)	Soil cohesion (kPa)
1	7.0	10
5	10.0	9.3
10	15.0	8.5
15	20.0	7.8
20	25.0	7
25	30.0	6
30	35.0	5
40	45.0	4
50	55.0	3

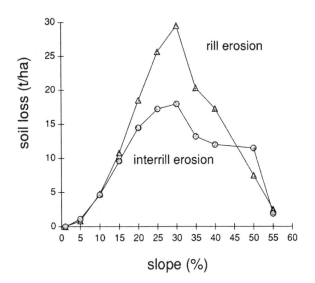

Figure 29.3   Simulated soil loss as affected by slope steepness with associated variations in saturated hydraulic conductivity and soil cohesion

$$A = 0.06\, S^{1.85} \qquad \text{for overland flow (no rills)} \tag{29.3}$$

$$A = 0.025\, S^{2.23} \qquad \text{with rills} \tag{29.4}$$

The exponents in these equations conform well to those cited earlier.
For slopes in the 30–55% range, the relationships were

$$A = 272\,402\, S^{-2.78} \qquad \text{without rills} \tag{29.5}$$

$$A = 11\,028\,844\, S^{-3.71} \qquad \text{with rills} \tag{29.6}$$

The exponents compare well with the value of -3.8 obtained by Heusch (1970) for slopes between 11% and 65% but are higher than the value of -1.4 obtained by Odemerho (1986) for slopes between 19% and 50%.

Overall, EUROSEM Version 3.3 predicts plausible trends in erosion with slope steepness, matching closely those proposed theoretically or found experimentally by other researchers. The exact shape of the curves relating soil loss to slope steepness will, of course, depend on the values used to describe the soil, vegetation and surface microtopography of a given slope element. Thus it should be possible to calibrate the model to simulate the trends observed for a given set of conditions.

29.5 SIMULATION OF VEGETATION EFFECTS

29.5.1 Effect of a Low-Growing Cover (Grass)

Simulations were made for increments of 10% cover between 0 and 100%. The height, interception capacity and percentage basal area of the grass, Manning's n and soil cohesion were varied with increasing percentage cover (Table 29.5) but otherwise the conditions were those of the base simulation. The results (Figure 29.4) showed that soil loss decreased exponentially with increasing cover and that no erosion was simulated when cover exceeded 60%. The simulation matches closely the results observed by many workers for low-growing vegetation or any other form of contact ground cover (e.g. mulches) (Elwell and Stocking 1976; Lang and McCaffrey 1984). The following relationships were obtained between soil loss (A) and percentage cover (C):

$$A = 1.59\, e^{-0.0596C} \qquad \text{for overland flow (no rills)} \tag{29.7}$$

$$A = 1.75\, e^{-0.0569C} \qquad \text{with rills} \tag{29.8}$$

The exponents are remarkably close to the value of -0.06 proposed by Elwell (1981) and within the range of -0.01 to -0.10 obtained in many studies (Brown *et al.* 1989).

29.5.2 Effect of a Canopy Cover (Maize)

Simulations were again made for increments of 10% cover between 0 and 100% and height, interception capacity, percentage basal area, Manning's n and soil cohesion were varied with increasing percentage cover. In this case, however, the cover applied

Table 29.5 Input data used in the simulation with grass cover

Percentage cover	Interception capacity (mm)	Percentage basal area	Plant height (m)	Manning's n	Soil cohesion (kPa)
0	0.0	0.0	0.00	0.20	7
10	0.5	0.2	0.02	0.30	7
20	1.0	0.3	0.03	0.40	7.5
30	1.5	0.4	0.04	0.45	8
40	2.0	0.5	0.05	0.50	8.5
50	2.3	0.55	0.06	0.60	9
60	2.3	0.6	0.06	0.65	10
70	2.5	0.7	0.06	0.70	11
80	2.5	0.8	0.06	0.75	12
90	2.5	0.8	0.06	0.80	13
100	2.5	0.8	0.06	0.80	14

Note: plant shape factor=1; plant angle=60°.

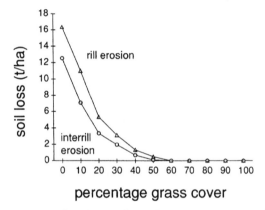

Figure 29.4 Simulated effect of a grass cover on soil loss

to the canopy and not to the ground surface which was assumed to be bare (Table 29.6). Thus, in these simulations, leaf drainage was predicted and soil detachment by raindrop impact occurred as a result of the action of direct throughfall and leaf drip combined. An exponential decline in soil loss with increasing cover was predicted (Figure 29.5) but with much lower exponent values, indicating a less sensitive relationship as cover increases from 0 to 50% compared with covers in contact with the ground. The relationships obtained were:

$$A = 1.029\, e^{-0.00057C} \quad \text{for overland flow (no rills)} \quad (29.9)$$

$$A = 1.217\, e^{-0.00044C} \quad \text{with rills} \quad (29.10)$$

Since virtually all previous work reported in the literature has been concerned with erosion under either very low (grasses, mulches) or very high canopies (trees), there is no direct experimental support for such a relationship. Such low exponent values, however, would suggest an almost linear relationship and, indeed, Wischmeier (1975) proposed a series of linear decay relationships to describe the effect of canopy cover on erosion in which, as the height of the canopy increased, the slope of the line decreased. This effect can be seen in Figure 29.4 by dividing each of the curves shown into two segments. The resulting linear functions are

$$A = 11.95 - 0.17\,C \qquad \text{no rills, 0–40\% cover} \qquad (29.11)$$
$$A = 16.56 - 0.14\,C \qquad \text{rills, 0–50\% cover} \qquad (29.12)$$
$$A = 6.92 - 0.04\,C \qquad \text{no rills, 40–100\% cover} \qquad (29.13)$$
$$A = 12.49 - 0.06\,C \qquad \text{rills, 50–100\% cover} \qquad (29.14)$$

Table 29.6 Input data for the simulations with maize cover

Percentage cover	Interception capacity (mm)	Percentage basal area	Plant height (m)	Manning's n	Soil cohesion (kPa)
0	0.0	0.00	0.0	0.2	7
10	0.3	0.05	0.3	0.3	7
20	0.4	0.10	0.4	0.4	7
30	0.5	0.12	0.6	0.5	7
40	0.6	0.14	0.8	0.6	7.5
50	0.7	0.16	0.9	0.7	7.5
60	0.75	0.18	1.0	0.8	7.5
70	0.8	0.20	1.25	0.9	8
80	0.8	0.20	1.5	0.95	8
90	0.8	0.20	1.75	1.0	8
100	0.8	0.20	2.0	1.0	8

Note: plant shape factor=2; plant angle=60°.

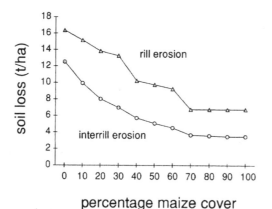

Figure 29.5 Simulated effect of a maize cover on soil loss

Thus, when the maize is in its early stage of growth and the canopy is close to the ground surface, the slope of the line is relatively steep. At higher canopy covers, a linear relationship with a lower slope is obtained. For the interrill erosion, the relationship up to 30% cover could still be reasonably well described by an exponential decay with an exponent equal to -0.01 in value.

Although the rather slow decline in soil loss with increasing cover above 50% can be explained by the increasing importance of soil detachment by the impact of large leaf drips (Morgan 1985), EUROSEM's simulations do not truly reflect this. If soil particle detachment by raindrop impact is calculated from EUROSEM's predictions of the combined kinetic energy of direct throughfall and leaf drainage, it shows an exponential decay with increasing cover rather than the increase observed in the field by Morgan (1985) and proposed experimentally by Styczen and Høgh-Schmidt (1988).

29.5.3 Effect of Tall Canopies (Trees)

A further simulation was attempted for a 20 m high tree canopy and no ground cover. The results showed that even with only 10% cover, erosion was reduced to zero. This was largely due to the high value (0.99) chosen for the basal area ratio of the trees. The basal area ratio is the total area of the base of the plant stems, expressed as a proportion (between 0 and 1) of the total area of the slope plane. Following the work of Holtan (1961), it is used in EUROSEM to increase the saturated hydraulic conductivity of the soil when a vegetation or crop cover is present. The value used increased the saturated hydraulic sufficiently to prevent runoff being simulated. In order to analyse better the effect of the canopy, a simulation was therefore run with an artificially low basal area ratio of 0.5 in order to simulate enough runoff to transport any detached soil particles. Soil loss was then found to reduce linearly and very gradually with increasing cover. Such a trend fits reasonably well the relationships proposed by Wischmeier (1975), even though his work does not consider canopies taller than 4 m. However, soil particle detachment by raindrop impact still fails to increase with percentage canopy, as would be expected from many studies under tree canopies (Styczen and Morgan 1995); instead it shows a very gradual linear decline.

29.5.4 Summary of the Simulations with Vegetation

Overall, the simulations for vegetative covers match well with the relationships expected from the theoretical and experimental work of others. EUROSEM, however, does not model soil detachment under canopy covers taller than 0.5 m in a way that accords many recent studies.

29.6 CONCLUSIONS

EUROSEM is a process-based model developed for simulating soil erosion by water on hillslopes and in small catchments. When operated for a single slope plot, it simulates the relationships between soil loss and slope steepness and soil loss and vegetation cover in accordance with most previous theoretical and experimental studies. Since the exact shape of the curves depends on the values used to describe the soil, vegetation and surface microtopography of a given slope element, it should be possible to calibrate the model to simulate the trends observed under a given set of conditions. Consideration should be given to improving the way the model simulates soil particle detachment by leaf drainage since this appears to be underestimated at present for canopy covers above 0.5 m in height.

ACKNOWLEDGEMENTS

The work on EUROSEM was funded by Directorate General XII of the Commission of European Communities under their Research and Development Programme in the Field of the Environment (1986–1990) and the Science and Technology for Environmental Protection (STEP) Programme (1989–1992).

REFERENCES

Brown, L.C., Foster, G.R. and Beasley, D.B. 1989. Rill erosion as affected by incorporated crop residue and seasonal consolidation. *Transactions of the American Society of Agricultural Engineers*, **32**, 1967–1978.

Bryan, R.B. 1979. The influence of slope angle on soil entrainment by sheetwash and rainsplash. *Earth Surface Processes*, **4**, 43–58.

Elwell, H.A. 1981. A soil loss estimation technique for southern Africa. In R.P.C. Morgan (ed.), *Soil Conservation: Problems and Prospects*. Wiley, Chichester, pp. 281–292.

Elwell, H.A. and Stocking, M.A. 1976. Vegetal cover to estimate soil erosion hazard in Rhodesia. *Geoderma*, **15**, 61–70.

Foster, R.L. and Martin, G.L. 1969. Effect of unit weight and slope on erosion. *Journal of the Irrigation and Drainage Division ASCE*, **95**, 551–561.

Heusch, B. 1970. L'érosion du Pré-Rif. Une étude quantitative de l'érosion hydraulique dans les collines marneuses du Pré-Rif occidental. *Annales de Recherches Forestières de Maroc*, **12**, 9–176.

Holtan, H.N. 1961. *A Concept for Infiltration Estimates in Watershed Engineering*. USDA Agricultural Research Service Publication ARS-41-51.

Hudson, N.W. and Jackson, D.C. 1959. Results achieved in the measurement of erosion and runoff in Southern Rhodesia. In *Proceedings, Third Inter-African Soils Conference*, Dalaba, pp. 575–583.

Kirkby, M.J. 1969. Erosion by water on hillslopes. In R.J. Chorley (ed.), *Water, Earth and Man*. Methuen, London, pp. 229–238.

Lang, R.D. and McCaffrey, L.A.H. 1984. Ground cover: its effects on soil loss from grazed runoff plots, Gunnedah. *Journal of the Soil Conservation Service of New South Wales*, **40**, 56–61.

Lattanzi, A.R., Meyer, L.D. and Baumgardner, M.F. 1974. Influence of mulch rate and slope steepness on interrill erosion. *Soil Science Society of America Proceedings*, **38**, 946–950.

Morgan, R.P.C. 1980. Field studies of sediment transport by overland flow. *Earth Surface Processes*, **3**, 307–316.

Morgan, R.P.C. 1985. Effect of corn and soybean canopy on soil detachment by rainfall. *Transactions of the American Society of Agricultural Engineers*, **28**, 1135–1140.

Morgan, R.P.C., Martin, L. and Noble, C.A. 1986. *Soil Erosion in the United Kingdom: A Case Study from mid-Bedfordshire*. Silsoe College, Occasional Paper No. 14.

Morgan, R.P.C., Quinton, J.N. and Rickson, R.J. 1993. *EUROSEM: A user guide*, Version 2. Silsoe College, Cranfield University.

Morgan, R.P.C., Quinton, J.N. and Rickson, R.J. 1994. Modelling methodology for soil erosion assessment and soil conservation design. *Outlook on Agriculture*, **23**, 5–9.

Morgan, R.P.C., Quinton, J.N., Smith, R.E., Govers, G., Poesen, J.W.A, Auerswald, K., Chisci, G., Torri, D. and Styczen, M.E. in preparation. The European Soil Erosion Model (EUROSEM): a process-based approach for prediction of soil loss from fields and small catchments. *Earth Surface Processes and Landforms*.

Musgrave, G.W. 1947. The quantitative evaluation of factors in water erosion: a first approximation. *Journal of Soil and Water Conservation*, **2**, 133–138.

Odemerho, F.O. 1986. Variation in erosion–slope relationship on cut slopes along a tropical highway. *Singapore Journal of Tropical Geography*, **7**, 98–107.

Poesen, J. 1986. Surface sealing as influenced by slope angle and position of simulated stones in the top layer of loose sediments. *Earth Surface Processes and Landforms*, **11**, 1–10.

Rowlison, D.L. and Martin, G.L. 1971. Rational model describing soil erosion. *Journal of the Irrigation and Drainage Division ASCE*, **97**, 39–50.

Schmidt, J. 1992. Modelling long-term soil loss and landform change. In A.J. Parsons and A.D. Abrahams (eds), *Overland Flow: Hydraulics and Erosion Mechanisms*. UCL Press, London, pp. 409–433.

Styczen, M. and Høgh-Schmidt, K. 1988. A new description of splash erosion in relation to raindrop size and vegetation. In R.P.C. Morgan and R.J. Rickson (eds), *Erosion Assessment and Modelling*. Commission of the European Communities Report No. EUR 10860 EN, pp. 147–198.

Styczen, M.E. and Morgan, R.P.C. 1995. Engineering properties of vegetation. In R.P.C. Morgan and R.J. Rickson (eds), *Slope Stabilization and Erosion Control: A Bioengineering Approach*. E & FN Spon, London, pp. 5–58.

Wischmeier, W.H. 1975. Estimating the soil loss equations cover and management factor for undisturbed areas. In *Present and Prospective Technology for Predicting Sediment Yields and Sources*. USDA Agricultural Research Service Publication ARS-S-40, pp. 118–124.

Woolhiser, D.A., Smith, R.E. and Goodrich, D.C. 1990. *KINEROS: A Kinematic Runoff and Erosion Model: Documentation and User Manual*. USDA Agricultural Research Service Publication ARS-77.

Zingg, A.W. 1940. Degree and length of land slope as it affects soil loss in runoff. *Agricultural Engineering*, **21**, 59–64.

30 Validation Problems of Hydrologic and Soil-Erosion Catchment Models: Examples from a Dutch Erosion Project

A. P. J. DE ROO
*Department of Physical Geography, Utrecht University,
The Netherlands*

30.1 INTRODUCTION

Our ability to make large complex models of the real world has grown, assisted by the growing computing facilities and our growing theoretical knowledge of processes. Recently, many complex, physically based, hydrologic and soil-erosion models have been developed, such as SHE (Abbott *et al.* 1986), SHETRAN (Wicks and Bathurst 1996), KINEROS (Smith 1981; Woolhiser *et al.* 1990), ANSWERS (Beasley *et al.* 1980; De Roo 1993), WEPP (Lane *et al.* 1992), EUROSEM (Morgan 1994) and LISEM (De Roo 1996a). However, it seems that our ability to properly validate these models is decreasing. Physically based models that simulate hydrologic and soil-erosion processes in catchments contain many parameters and variables, which are not always easy to measure or to determine otherwise. On the other hand, in order to validate the model, one needs measurements of discharge, soil moisture content, sediment load and on-site erosion or deposition rates. For example, for the validation of LISEM, extensive field and laboratory measurements in three research catchments and six subcatchments during two and a half years were available. Furthermore, five extra years of rainfall/runoff data are available for two catchments. Also, at 12 sites in the catchments along four slope profiles, continuous soil suction measurements have been undertaken to assess both vertical and lateral transport of water in the soil. However, even with this relatively large database, the validation and/or calibration of the LISEM and ANSWERS model proved difficult. In many cases, these detailed data are available for only one or two points in the catchment, mainly at the catchment outlet. In those cases, proper validation of catchment models is not possible. This chapter gives some examples of the problems related to model validation. Therefore, there seems to be a slight retreat from fully distributed models, because they offer little or no additional forecasting ability and are costly in their data requirements (Anderson and Burt 1990).

Advances in Hillslope Processes, Volume 1. Edited by M. G. Anderson and S. M. Brooks.
© 1996 John Wiley & Sons Ltd.

30.2 PROBLEMS WITH MODEL CALIBRATION

One of the first questions that has to be answered before we discuss validation problems, is whether or not it is permitted to calibrate physically based models. According to Beven *et al.* (1988), physically based models are "not well suited to calibration of model parameters because many different parameter sets will give equally acceptable fits to observed catchment behaviour". Also, calibration prevents identification of errors in process description and catchment representation (De Roo 1993). Furthermore, if a physically based model is calibrated, its physically based nature is compromised or is lost entirely. However, in almost every physically based modelling study, calibration has been applied. Very often, however, calibration is also handicapped by short data series (James 1982). In an attempt to validate the SHE model, Bathurst (1986) concluded that "a degree of calibration or optimisation of parameter values is likely to be needed to minimize the differences between measured and simulated hydrographs". Thus, we are confronted with the fact that calibration is not really desirable, but probably necessary to obtain reasonable fits.

Mostly, calibration is performed on measured hydrographs at the catchment outlet. It has been shown by Bathurst (1986), for example, that this operation does not provide sufficient information on model performance. In his example of the 10 September 1976 event in the Wye catchment he showed that the simulation of the flood hydrograph at the catchment outlet was good, but that the simulations for its tributaries were significantly in error. The errors cancelled each other. Therefore, it is frequently difficult to interpret the model validation results that many authors present, because only catchment outlet results are shown. For example, Calver (1988) presents results from the IHDM4 catchment hydrology model. After constructing an optimized parameter set for winter storms, she validated the model on three storms and obtained peak flow predictions within 11% of the observed values. This seems to be a good result, but as noted above does not prove that the model is good. Thus, it can be concluded that validation at more than one measured location is always desirable to obtain a clear assessment of the validity of the model.

Occasionally, difficulties with calibration exist, as can be shown by an example for the LISEM model. LISEM (LImburg Soil Erosion Model) is a physically based hydrological and soil-erosion model, which can be used for planning and conservation purposes (De Roo *et al.* 1996a). Vertical movement of water in the soil is simulated using the Richard's equation. The detachment and transport equations are similar to the EUROSEM model. For the distributed flow routing, a four-point finite-difference solution of the kinematic wave is used together with Manning's equation. For the Limburg data, LISEM did not need to be calibrated for the summer rainfall events. Although measured and simulated discharge and soil loss are different in many cases, in general the model did not seem to under or overpredict. Several test simulations (calibrated runs) indicated that calibration of single events yields better results, but using such a calibration data set on other events yields worse results. However, for the winter events, the situation is somewhat different. A number of simulations have been carried out using several calibrated values of initial pressure head in the upper soil layers for individual storms (Table 30.1). The choice of pressure head for calibration is based on the sensitivity analysis

Table 30.1 Simulated peak discharge in the Ransdaal drainage basin (Limburg, the Netherlands) with varying initial pressure head

Date	31 Dec. 1993	2 Jan. 1994	3 Jan. 1994
Measured peak discharge (l/s)	700	134	164
Simulated peak discharge (l/s) using:			
Soil saturated from 0 to 120 cm	1008	271	110
Soil saturated from 0 to 100 cm	300	146	63
Soil saturated from 0 to 70 cm	270	32	6
Soil saturated from 0 to 50 cm	212	13	2
Soil saturated from 0 to 30 cm	112	6	1
Simulated peak discharge (l/s) using field estimated pressure heads	91	0	0
Calibrated depth of saturation (cm)	110	90	>120

and the fact that pressure head can easily change in the field, e.g. changes from -5 to -100 or from -100 to -1000 cm occur frequently. From Table 30.1 it is clear that the measured peak discharge can be simulated correctly using a certain initial pressure head. For the 31 December storm, the calibrated pressure head is 0 cm from 0 down to 110 cm of the soil profile. For 2 January the calibrated pressure head is 0 down to 90 cm, and for 3 January the pressure head is 0 down to 120 cm. However, one would expect the soil profile to become wetter as a consequence of continuing rainfall. The calibration of the 2 January storm shows that there are difficulties with this assumption.

In conclusion, from a theoretical viewpoint, calibration of a physically based model should be avoided. However, in practice it is often necessary to obtain reasonable fits between measured and observed data. In some cases finding a general applicable calibrated parameter set is difficult.

30.3 VALIDATION: PREDICTED AND OBSERVED DISCHARGE AND SEDIMENT

Many examples exist of validation of physically based models, which are based on a split record test: calibrating the model on one set, and validating the model on another independent data set (Calver 1988; De Roo and Walling 1994). Often, reasonably good results are obtained. However, taking the above discussion into account, calibration is not desirable. Examples of the validation of uncalibrated models are hard to find in the literature. This seems strange, since many of these models should be applicable to predicting discharge and soil loss in areas for which no measured data exist!

Whenever such an uncalibrated model is used, the model results will often look like Quinton's results with EUROSEM (Quinton 1994). He observed a discrepancy between observed and predicted values of runoff. When plotting the cumulative distribution of the differences between predicted and observed runoff, he showed that 50% of the simulations were within 0.8 mm of the observed runoff. Some 20%

of the simulations differ more than 1.5 mm from measured runoff. Also, several studies have shown many problems with the sediment load predictions of physically based soil-erosion models. In the same EUROSEM model exercise, Quinton (1994) showed a discrepancy between measured and predicted soil loss, particularly when the measured soil loss was low (< 0.7 ton/ha/event).

In a study using the ANSWERS model in three catchments, De Roo (1993) showed that the validation results obtained using non-calibrated data sets indicated that the hydrographs cannot be simulated closely in many cases. The main reasons for the bad fits are difficulties in simulating soil crusting properly, the importance of subsurface lateral flow in one catchment (Yendacott), and the spatial variability of rainfall in catchments larger than 50 ha (Etzenrade). It was demonstrated that the predictions of total discharge and peak discharge for one catchment (Etzenrade) obtained using simple multiple regression models were better. Nevertheless, the observed overland flow paths and erosion and deposition locations in the catchment did match the simulated patterns. When a crusting submodel was used, the hydrographs of large storms in one of the catchments (Catsop) could be simulated with reasonable success. The sediment load associated with the discharge remained difficult to simulate correctly. The observed overland flow paths and erosion and deposition locations in the catchments matched the simulated patterns. The predictions obtained using ANSWERS in the Catsop catchment are better than the predictions of total discharge and peak discharge using simple regression equations. The better results for the Catsop catchment compared to the Etzenrade catchment might be explained by the fact that the Catsop catchment is much smaller (45 ha, compared to 225 ha), so problems with spatial variability of rainfall are smaller.

In a study with the LISEM model, which was developed to overcome the theoretical problems with ANSWERS, De Roo et al. (1996a,b) demonstrated similar discrepancies between measured and simulated discharge and soil loss as with ANSWERS (Figures 30.1 and 30.2). They concluded that the difference between measured and observed discharges can be explained fully by variations in hydraulic conductivity and initial pressure head, and thus the soil moisture content at the start of a rainstorm. Initial pressure head values for an entire catchment cannot be obtained. Thus, the model input map of initial pressure head is often based on a limited number of point measurements. Also, there are often gaps in the "continuous" data set so that for a number of periods no data are available. So, spatial and temporal interpolation of the pressure head values is necessary and causes uncertainties in data input, which has its repercussions on the model results. Figures 30.1 and 30.2 show that although there are a number of reasonable simulations, about 40% of the simulated hydrographs have significantly different peak discharges or other deviations from the observed discharges. Based on a sensitivity analysis and field observations, the main reason for these differences seems to be the spatial and temporal variability of the soil hydraulic conductivity, which is especially high in the tilled soils of the research catchment. Furthermore, spatial and temporal variability of the initial pressure head at the basin scale is an important factor. It proved difficult to translate three or six individual points per basin to an entire drainage basin.

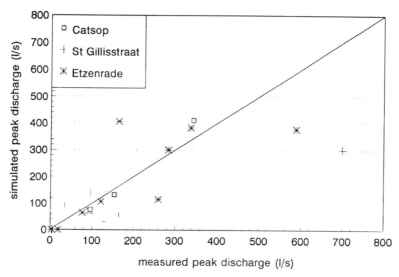

Figure 30.1 Measured and LISEM simulated peak discharge for all events in the three experimental basins

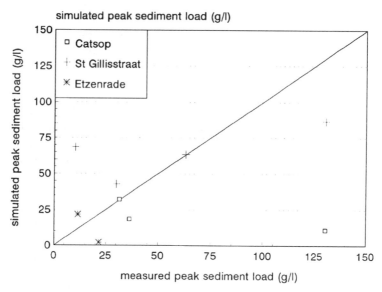

Figure 30.2 Measured and LISEM simulated peak sediment concentration for all events in the three experimental basins

In some winter events, field measurements and the observed hydrograph seem to suggest the importance of subsurface flow, which is at present not simulated by LISEM. This explains a part of the differences between measured and simulated hydrographs. Besides spatial and temporal variability of sensitive input variables, another reason for the differences between measured and simulated results are the deficiencies in our theoretical understanding of hydrological and soil-erosion processes. It was clear from the Limburg data that summer and winter response to rainfall are quite different, which might partly be explained by subsurface flow. But even within the main seasons there are significant different responses to rainfall due to tillage operations and biological activity, such as worms.

30.4 SPATIAL VALIDATION OF EROSION MODELS USING ^{137}CS DATA

Ideally, a spatial distributed hydrological and soil-erosion model should be validated in a spatial context, and not only by a few (sub-)outlet comparisons. This was undertaken in studies by De Roo (1993) and De Roo and Walling (1994) who compared ANSWERS spatial erosion estimates with soil loss rates calculated from ^{137}Cs data (De Roo 1991). Furthermore, because some authors (Anderson and Howes 1986; DeCoursey 1988) claim that simple models may perform as well (or as badly) as complex physically based models, three other soil-erosion models which can be used at a catchment scale, ranging from simple to complex physically based, were compared: the USLE (Wischmeier and Smith 1978; Jetten *et al.* 1988), the Morgan/Morgan/Finney model (Morgan *et al.* 1984; De Jong 1994) and KINEROS. One of the reasons for the fact that simple models might perform as well as complex models is that "complex models generally require more parameters, each of which has a range of equally likely values, that can collectively produce greater error than the fewer parameters in simple models" (DeCoursey 1988). De Roo (1993), for example, demonstrated that simple regression equations can predict total discharge, peak discharge and total soil loss from a catchment as well as, or even better than, the complex distributed ANSWERS model. Gregory and Walling (1973) also demonstrated the predictive value of multiple regression equations to estimate catchment discharge and suspended sediment yield. Although those equations may have a highly explanatory, and, therefore, predictive value in a statistical sense, they do not necessarily provide a good conceptual explanation (Morgan 1995).

To perform the model comparison and spatial validation, three major rainfall events in the Etzenrade catchment (Limburg, The Netherlands) (De Roo 1993) were chosen for the simulations. The USLE and MMF were run in the "original" way to predict annual soil loss. For the single storm models KINEROS and ANSWERS, the summation of simulated soil loss of these three events (Figure 30.3) has been used for comparisons with the (long-term) ^{137}Cs data and the ^{137}Cs-estimated soil loss (Figure 30.4) for validation. Details of the working methods can be found in De Roo (1993). Table 30.2 shows that the correlations of the simulated soil-erosion rates for the four models with the measured ^{137}Cs in the soil and the ^{137}Cs-estimated soil loss in the Etzenrade catchment are similar.

All correlations are low, and the differences are not very large. The correlations between Cs and the USLE, MMF and KINEROS are negative, because a small ^{137}Cs content corresponds with a high soil-erosion rate, which is indicated by positive values in these models. In the ANSWERS model, deposition is indicated with positive values, and soil erosion with negative values, so correlations with Cs are

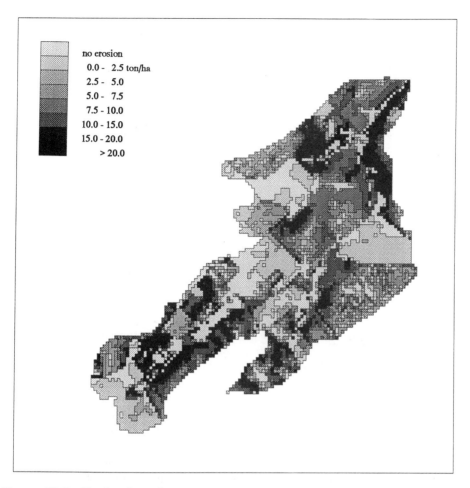

no erosion
0.0 - 2.5 ton/ha
2.5 - 5.0
5.0 - 7.5
7.5 - 10.0
10.0 - 15.0
15.0 - 20.0
> 20.0

Figure 30.3 Simulated total soil loss in the Etzenrade catchment, Limburg, the Netherlands, using the ANSWERS model, 1987–1990. Results of four years of individual storms are summed

expected to be positive. Recent results with the LISEM model (De Roo *et al.* 1996a) show a similar low correlation of -0.20 (Figure 30.5). The correlations between the USLE, MMF and ANSWERS are high. The fact that ANSWERS and MMF use similar methods such as the C and the P-factor of the USLE is the most likely cause of this correlation. Furthermore, for all three models slope gradient is very important. The correlations of the KINEROS model with the other models are poorer, probably due to the lumping that occurs because the number of elements is reduced from 5621 elements in the USLE, MMF, ANSWERS and LISEM, to 56 in KINEROS. The fact that intra-model correlations are all better than the correlations between the model results and the ^{137}Cs data might suggest that (1) the latter are considerably in error, or (2) that all models are equally poor.

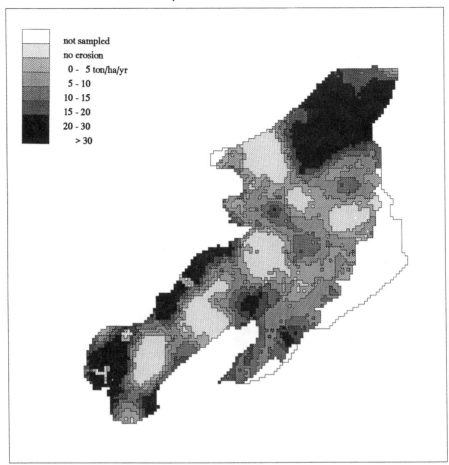

Figure 30.4 ¹³⁷Cs-estimated soil loss in the Etzenrade catchment, Limburg, the Netherlands (143 point measurements interpolated using block kriging)

Table 30.2 Spearman rank correlations (r) of the simulated soil erosion rates for the three storms in the Etzenrade catchment using four models (comparison of the individual ¹³⁷Cs sampling points and the Cs-derived soil erosion estimates)

3 storms, n=134	USLE	MMF	KINEROS	ANSWERS (nc)	ANSWERS (cr)
USLE	–				
MMF	0.927**	–			
KINERSOS	0.524**	0.581**	–		
ANSWERS (original)	−0.669**	−0.747**	−0.377**		–
ANSWERS (crust-model)	−0.692**	−0.756**	−0.376**	0.963**	–
¹³⁷Cs	−0.250**	−0.226**	−0.158*	0.173**	0.205**
¹³⁷Cs – estimated soil loss	−0.181*	−0.138	−0.109	0.113	0.141

**Significant at 95%.
*Significant at 90%.

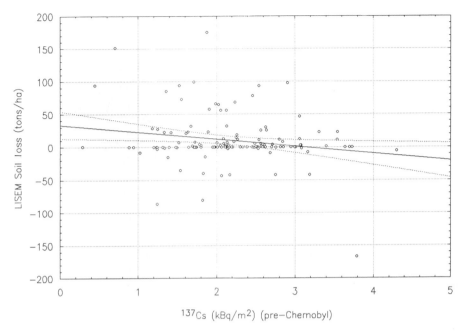

Figure 30.5 Comparison of ^{137}Cs in the soil and soil loss calculated with the LISEM model. As an approximation, ^{137}Cs values below 2.0 kBq/m² indicate erosion, whilst values above 3.0 kBq/m² indicate deposition

The results show that simple soil-erosion models are indeed little worse than more complex physically based soil-erosion models. The correlations with the measured ^{137}Cs data are similar for the four models being examined. This might be caused both by the quality of the model and/or by the quality of the ^{137}Cs data. Considering the soil-loss estimates from Cs data, it has been shown by Govers *et al.* (1993) and Quine *et al.* (1994) that tillage processes have a significant influence in soil redistribution. This might be one reason for the overall low correlations. The truly distributed character of the ANSWERS and LISEM model, with upslope and downslope influences, seems not to improve the results of the model significantly. The KINEROS model demonstrates the influence of both lumping in space, resulting in different soil-erosion estimates due to the averaging out of slope gradient, and lumping in time, resulting in averaging out short periods of large rainfall intensities. Despite the theoretical advantages of physically based distributed models like ANSWERS, KINEROS and LISEM, such as generating hydrographs and soil-erosion maps, assessing the effects of conservation planning strategies, testing of scientific knowledge, and indicating future research needs, the quantitative results of those models are not significantly better than the results of simple soil-erosion models, such as the USLE and MMF.

It seems that the level of prediction achieved by the models has remained the same, although the potential use and possibilities of models such as ANSWERS and LISEM for application have grown; for example, simulating extreme events, calculating dimensions of storage basins and discharges of small drainage ditches and canals, calculating the spatial effects of the introduction of erosion control measures, etc.

30.5   WHY VALIDATION MIGHT NOT BE POSSIBLE AT ALL

30.5.1   Uncertainties and Errors

Many users struggle to validate their catchment models. However, keeping in mind the issues mentioned above, are we able to perform validation successfully? A number of reasons exist which suggest that it might not be possible to properly validate complex distributed models. There is concern that the magnitude of error involved in distributed models prevents the realization of reliable forecasts (Anderson and Howes 1986). The specific problem of soil-erosion models is that erosion processes are superimposed, non-linearly, on hydrologic variability. Because describing and understanding the hydrologic variability is difficult, resulting in a poorly modelled system, the erosion variability, which is superimposed on the hydrology, is even more difficult to describe and understand.

Four main sources of uncertainty in distributed models can be identified. First, there can be *uncertainties in the theoretical structure of the model*: there are gaps in our understanding of hydrological and soil-erosion processes. In hydrology, no completely theoretical operational models exist: all hydrologic models contain empirical relationships (Haan 1988). For example, the ANSWERS model lacks several important subprocesses (throughflow, saturation overland flow, preferential flow through macro pores, gully erosion and bank erosion), and simplifies other processes (interception, infiltration and detachment of soil particles); LISEM lacks throughflow and gully erosion.

Second, *implementation error* may also be introduced by the choice of *mathematical solution* and the assumptions made to perform it. An example of error with a mathematical solution is the sensitivity of the ANSWERS model to the orientation of the outlet of the catchment with respect to the grid, as was shown by De Roo (1993).

Third, *a calibration error* exists. Every so-called "physically based" model has been calibrated. However, physically based models are not well suited to calibration of model parameters because they have so many parameters (Beven *et al.* 1988). Physically interacting parameters will inevitably result in many different parameter sets giving equally acceptable fits to observed catchment behaviour. Also, calibration is frequently handicapped by short data series (James 1982). Therefore, using the parameters obtained by calibration for the validation of a model using an independent data set always produces uncertainties.

Fourth, *uncertainties in the data* exist. These uncertainties include obvious sources of error (age of data, areal coverage, map scale, density of observations), uncertainties resulting from natural variation or from original measurements error (positional error, accuracy of content, measurement errors, laboratory errors, spatial variability), errors arising through processing (numerical errors in computers), digitizing errors, errors resulting from rasterizing a vector map, uncertainties in estimating parameter values (e.g. pedo-transfer functions) and generalization error (e.g. representing complex boundaries) (Burrough 1986).

30.5.2 Spatial and Temporal Variability

The major source of data error in distributed models arises from spatial and temporal variability of input variables that is not, or is insufficiently, taken into account. In distributed hydrological and soil-erosion models, the four most important variables which cause large uncertainties are rainfall, infiltration, initial soil moisture content and soil erodibility.

The effects of spatial and temporal variability of *rainfall* on the output of hydrological and soil-erosion models have been discussed by other workers, e.g. Wilson *et al.* (1979), James (1982) and Sivapalan and Wood (1987). In the past, it has been argued that uncertainties in the rainfall input are damped when routed through a catchment. However, Wilson *et al.* (1979) concluded that the spatial distribution of rain and the accuracy of the precipitation input have a marked influence on the outflow hydrograph from a small catchment. They analysed the influence of the spatial distribution of the rainfall input on the discharge by using one rain-gauge or 20 rain-gauges to record storms. They found that when the spatial pattern of rainfall is not appropriately preserved, serious uncertainties may occur in the total volume, peak and time-to-peak of the estimated runoff hydrograph. When storms are not of the frontal type, but intense, short in duration, and more localized in space (e.g. thunderstorms), the application of rainfall–runoff models without an appropriate description of the spatial character of the input may lead to unacceptable uncertainties. Thus, it is important to have a sufficient number of rain-gauges. Another problem that needs to be tackled is the transformation of these rainfall measurements at discrete points into areal rainfall. Most hydrologic models, including the ANSWERS model and LISEM, divide the catchment into subcatchments over which the rainfall input is assumed to be lumped. Sivapalan and Wood (1987) developed a method to describe storm movement over a catchment. Optimally, it should be possible to define the temporal variations in rainfall intensities for every single pixel! Finally, measurement errors of rainfall can occur due to the size of the tipping bucket (0.2 or 0.1 mm or smaller) and the minimum time interval allowed to store the data (1 min, or every tip, stored with minute and second!). Using too coarse buckets and time interval leads to an averaging out of rainfall intensity extremes, which are often very important.

The effects of spatial variability of *infiltration* on the output of hydrological and soil-erosion models have been discussed in De Roo *et al.* (1992). The results of the Monte Carlo experiments showed that large variations may occur due to the variability and error in the saturated infiltration rate: coefficients of variation up to 14% occur during the simulations of small storms. Ignoring this spatial variability and/or error of the saturated infiltration rate can lead to false judgements of the performance of model. If only a single simulation is carried out and the results are compared to the measured data, no reliable judgement on model performance can be made. Errors up to 40% can occur. However, not only spatial variability but also the temporal variability of infiltration due to crusting, tillage, drying and wetting, and the activity of soil microfauna is an important source of error if it is not taken into account.

The third important variable leading to large uncertainties in hydrological and soil-erosion modelling is the *soil moisture content* just before the start of the rainfall event. The soil moisture content varies both in space and time. At the catchment scale it is not feasible to measure soil moisture content continuously. Often, only a few measurement sites are investigated.

Furthermore, the temporal variability of *soil erodibility* is an important source of error in soil-erosion models if it is ignored. The USLE K-factor, used in ANSWERS, is assumed to be constant during the year. However, temporal variability of soil erodibility is often observed in the field (Bryan *et al.* 1989; Imeson and Kwaad 1990).

Finally, uncertainties in *soil and channel roughness* data may lead to uncertainties in sediment predictions. Temporary obstructions in channels, such as tree branches and leaves, cause the sediment to be trapped in the channels. Often these very local obstructions are not incorporated in the channel roughness data, causing an overestimation of sediment transport by the model.

30.5.3 Uncertainties in Representing the Data in the Model

Several uncertainties arise from representing the available data in the model. The most obvious errors are the errors produced when *digitizing* a map (Burroughs 1986) and when a vector map is *rasterized* to the grid size required by the model. Rasterizing areas with linear boundaries (agricultural fields, roads, lynchets) will lead to errors if they are oriented at angles not equal to zero. The errors are at maximum when they are oriented at 45° to the grid. Therefore, Beasley and Huggins (1982) advised that the orientation of the map should be chosen carefully, taking into account the dominant directions of the linear elements in a catchment. However, in large catchments there always will be different orientations, and consequently error. Areas with irregular boundaries will always lead to error. Errors may also arise from digitizing hyetographs (rainfall), sedigraphs (sediment) and hydrographs (discharge) or other graphs stored on paper.

Also, the *grid size* influences the model results. A smaller grid reduces the rasterizing errors. Small important topographical features, such as lynchets and incised ("hollow") roads in South Limburg, can only be represented by small grids. Especially in soil-erosion models, a simplification of reality by choosing too coarse a grid leads to serious errors. Large changes of slope gradients are averaged out in a large grid. The hydrological part of a model is sensitive to the grid size in the vertical direction (Amerman and Monke 1977). Special features that are too small to be incorporated in a grid can be treated in a special way, such as the incorporation of roads smaller than the pixel size in ANSWERS and LISEM.

Finally, if a GIS is used in combination with a hydrological and soil-erosion model, errors arise from *estimating slope gradient values from an altitude matrix*. In this study, field measurements were carried out to select the best method and to validate the calculated slope values.

30.5.4 Uncertainties in the Measured Data used for Validation

Recognizing all sources of error discussed above, and possibly using error propagation techniques to assess the consequences of these uncertainties (Heuvelink *et al.* 1989), the model is run and validated using data measured in the field. However, these measured data used for validation, such as discharge, sediment load, and spatial soil erosion and deposition estimates, also contain error.

Laboratory errors during the determination of the *sediment concentration* in the water samples can be considered very small, although there may be error in some of the turbidity measurements due to organic matter, which colours the water.

Secondly, the measurement of *discharge* contains errors. In general, the relative error is largest at small discharges. Uncertainties occur with the water-depth measurement itself, using, for example, ultrasonic measurements or vertical float recorders. Furthermore, uncertainties occur due to sediment deposits in the bottom of the flume. The flumes have to be cleaned regularly. Finally, uncertainties may occur due to the time interval used for storing the discharge measurements.

The *soil-erosion estimates using ^{137}Cs* are useful to validate distributed soil-version models, but do contain uncertainties (De Roo 1993). Also, tillage effects should be taken into account (Govers *et al.* 1993).

30.6 CONCLUSIONS

Many sources of error and uncertainty affect the results of distributed hydrological and soil-erosion models. A knowledge of error in physically based models is important to make a correct judgement of a model. Without a perfect knowledge of initial and boundary conditions it is impossible to validate, accept or reject such models. There can be uncertainties in the theoretical structure of the model, errors in the mathematical solution methods in the model, calibration errors, and errors and uncertainties in the data.

Given the fact that in general only limited data are available for validation, e.g. only outlet measurements, a meaningful assessment of physically based models is very difficult. On the other hand, it was shown here that complex model predictions are not necessarily better than predictions using simple models, although complex models provide more additional useful information.

Many scientists concentrate on the development of the theoretical structure of the model and try to fill the gaps in our understanding of hydrological and soil-erosion processes, which is very important work. However, the error in the data may have a larger effect on the simulation results obtained with the model. Furthermore, ignoring the uncertainties in the data may lead to the erroneous conclusion that the theoretical structure of a model is not correct. If obtaining good simulation results at the catchment scale is our main aim, maybe we are approaching the point at which improvement of data-collection methods becomes more important than model development.

REFERENCES

Abbott, M.B., Bathurst, J.C., Cunge, J.A., O'Connell, P.E. and Rasmussen, J. 1986. An introduction to the European Hydrological System – Système Hydrologique Européen, "SHE", 2: Structure of a physically-based distributed modelling system. *Journal of Hydrology*, **87**, 61–77.

Amerman, C.R. and Monke, E.J. 1977. Soil water modelling II: On sensitivity to finite difference grid spacing. *Transactions of the ASAE*, **20**, 478–484.

Anderson, M.G. and Burt, T.P. 1990. Process studies in hillslope hydrology: an overview. In M.G. Anderson and T.P. Burt (eds), *Process Studies in Hillslope Hydrology*. John Wiley, Chichester,, pp. 1–8.

Anderson, M.G. and Howes, S. 1986. Hillslope hydrology models for forecasting in ungauged watersheds. In A.D. Abrahams (ED.), *Hillslope Processes*. The Binghampton Symposia in Geomorphology, International Series. Allen & Unwin, Boston, pp. 161–186.

Bathurst, J.C. 1986. Physically-based distributed modelling of an upland catchment using the Système Hydrologique Européen. *Journal of Hydrology*, **87**, 79–102.

Beasley, D.B. and Huggins, L.F. 1982. ANSWERS Users manual. US-EPA, Region V, Chicago, Illinois, Purdue University, West Lafayette, IN., 54 pp.

Beasley, D.B., Huggins, L.F. and Monke, E.J. 1980. ANSWERS: A Model for Watershed Planning. *Transactions of the ASAE*, **23–4**, 938–944.

Beven, K.J., Wood, E.F. and Sivapalan, M. 1988. On hydrological heterogeneity – catchment morphology and catchment response. *Journal of Hydrology*, **100**, 353–375.

Bryan, R.B., Govers, G. and Poesen, J. 1989. The concept of soil erodibility and some problems of assessment and application. *Catena*, **16**, 393–412.

Burrough, P.A. 1986. *Principles of Geographical Information Systems for Land Resources Assessment*. Monographs on soil and resources survey no. 12, Clarendon Press, Oxford.

Calver, A. 1988. Calibration, sensitivity and validation of a physically-based rainfall runoff model. *Journal of Hydrology*, **103**, 103–115.

DeCoursey, D.G. 1988. A critical assessment of hydrologic modeling. In *Modeling Agricultural, Forest, and Rangeland Hydrology*. Proceedings of the 1988 International Symposium. ASAE Publication 07-88, St Joseph, Michigan, pp. 478–493.

De Jong, S.M. 1994. Applications of reflective remote sensing for landdegradation studies in a Mediterranean environment. *Netherlands Geographical Studies*, **177**, 237 pp.

De Roo, A.P.J. 1991. The use of ^{137}Cs as a tracer in an erosion study in South-Limburg (The Netherlands) and the influence of Chernobyl fallout. *Hydrological Processes*, **5**, 215–227.

De Roo, A.P.J. 1993. Modelling surface runoff and soil erosion in catchments using Geographical Information Systems: validity and applicability of the "ANSWERS" model in two catchments in the loess area of South Limburg (The Netherlands) and one in Devon (UK). *Netherlands Geographical Studies*, **157**, Utrecht.

De Roo, A.P.J. and Walling, D.E. 1994. Validating the "ANSWERS" soil erosion model using 137Cs. In R.J. Dickson (ed.), *Conserving Soil Resources: European Perspectives*. CAB International, Cambridge, pp. 246–263.

De Roo, A.P.J., Hazelhoff, L. and Heuvelink, G.B.M. 1992. The use of Monte Carlo simulations to estimate the effects of spatial variability of infiltration on the output of a distributed hydrological and erosion model. *Hydrological Processes*, **6**, 127–143.

De Roo, A.P.J., Wesseling, C.G. and Ritsema, C.J. 1996a. LISEM: a single event physically-based hydrologic and soil erosion model for drainage basins. I: Theory, input and output. *Hydrological Processes*, in press.

De Roo, A.P.J., Offermans, R.J.E. and Cremers, N.H.D.T. 1996b. LISEM: a single event physically-based hydrologic and soil erosion model for drainage basins. II: Sensitivity analysis, validation and application. *Hydrological Processes*, in press.

Govers, G., Quine, T.A. and Walling, D.E. 1993. The effect of water erosion and tillage movement on hillslope profile development: a comparison of field observations and model results. In S. Wicherek (ed.), *Farm Land Erosion in Temperate Plains and Hills*. Elsevier, Amsterdam, pp. 285–300.

Gregory, K.J. and Walling, D.E. 1973. *Drainage Basin Form and Process: A Geomorphological Approach*. Edward Arnold.

Haan, C.T. 1988. Parametric uncertainty in hydrologic modeling. In: Modeling agricultural, forest, and rangeland hydrology. Proceedings of the 1988 International Symposium. *ASAE Publication*, **07–88**, 330–346. St. Joseph, Michigan.

Heuvelink, G.B.M., Burrough, P.A. and Stein, A. 1989. Propagation of errors in spatial modelling with GIS. *International Journal of Geographical Information Systems*, **3**, 303–322.

Imeson, A.C. and Kwaad, F.J.P.M. 1990. The response of tilled soils to wetting by rainfall and the dynamic character of soil erodibility. In J. Boardman, I.D.L. Foster and J.A. Dearing (eds), *Soil Erosion on Agricultural Land*. John Wiley, pp. 3–14.

James, L.D. 1982. Precipitation–runoff modeling: future directions. In Singh (ed.), *Applied Modeling of Catchment Hydrology*. Water Resources Publications, pp. 291–312.

Jetten, V.G., Henkens, E.J. and De Jong, S.M. 1988. *The Universal Soil Loss Equation. Version 1.0, release 1.0, distributed*. Department of Physical Geography, Utrecht University.

Lane, L.J., Nearing, M.A., Laflen, J.M., Foster, G.R. and Nichols, M.H. 1992. Description of the US Department of Agriculture Water Erosion Prediction Project (WEPP) Model. In A.J. Parsons and A.D. Abrahams (eds), *Overland Flow: Hydraulics and Erosion Mechanics*. UCL Press Limited, London, pp. 377–391.

Morgan, R.P.C., Morgan, D.D.V. and Finney, H.J. 1984. A Predictive Model for the Assessment of Soil Erosion Risk. *Journal of Agricultural Engineering Research*, **30**, 245–253.

Morgan, R.P.C. 1994. The European Soil Erosion Model: an update on its structure and research base. In R.J. Rickson (ed.), *Conserving Soil Reserves: European Perspectives*. CAB International, Cambridge, pp. 286–299.

Morgan, R.P.C. 1995. *Soil Erosion and Conservation*, 2nd edn. Longman, London.

Quine, T.A., Desmet, P.J.J., Govers, G., Vandaele, K. and Walling, D.E. 1994. A comparison of the roles of tillage and water erosion in landform development and sediment export on agricultural land near Leuven, Belgium. *IAHS Publication 224*, 77–86.

Quinton, J.C. 1994. Validation of physically-based models, with particular reference to Eurosem. In R.J. Rickson (ed.), *Conserving Soil Resources: European Perspectives*. CAB International, Cambridge, pp. 300–313.

Sivapalan, M. and Wood, E.F. 1987. A multiscale model of nonstationary space–time rainfall at a catchment scale. *Water Resources Research*, **23**, 1289–1299.

Smith, R.E. 1981. A kinematic model for surface mine sediment yield. *Transactions of the ASAE*, 1508–1514.

Wicks, J.M. and Bathurst, J.C. 1996. SHESED: a physically-based, distributed erosion and sediment yield component for the SHE hydrological modelling system. *Journal of Hydrology*, **175**, 213–238.

Wilson, C.B., Valdes, J.B. and Rodriguez-Iturbe, I. 1979. On the influence of the spatial distribution of rainfall on storm runoff. *Water Resources Research*, **15**(2), 321–328.

Wischmeier, W.H. and Smith, D.D. 1978. Predicting rainfall erosion losses – a guide to conservation planning. U.S. Department of Agriculture, *Agricultural Handbook*, **537**, Science and Education Administration USDA, Washington D.C., 58 pp.

Woolhiser, D.A., Smith, R.E. and Goodrich, D.C. 1990. KINEROS: A kinematic runoff and erosion model: documentation and user manual. USDA-ARS, ARS-77, 130 pp.